Agricultural Biomass for the Synthesis of Value-Added Materials

This book is a comprehensive guide to bioconversion approaches based on microorganisms and enzymes for the valorization of underused wastes of diverse categories to produce new products. Optimized conditions for microbial and enzymatic valorization are discussed, along with related biotechnological considerations, environmental considerations, bioprocess development, obstacles, and future outlooks. Biofuels, bioenergy, and other platform chemicals are only some of the products that can be produced through this book's explanation of the microbiological processes involved in the bioconversion and valorization of wastes.

Agricultural Biomass for the Synthesis of Value-Added Materials

Edited by

Sankha Chakrabortty, Jayato Nayak, Shirsendu Banerjee, and Maulin P. Shah

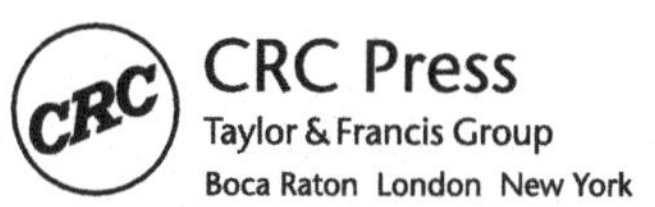

CRC Press is an imprint of the
Taylor & Francis Group, an informa business

Designed cover image: Shutterstock

First edition published 2025
by CRC Press
2385 NW Executive Center Drive, Suite 320, Boca Raton FL 33431

and by CRC Press
4 Park Square, Milton Park, Abingdon, Oxon, OX14 4RN

CRC Press is an imprint of Taylor & Francis Group, LLC

Library of Congress Cataloging-in-Publication Data
Names: Chakrabortty, Sankha, editor.
Title: Agricultural biomass for the synthesis of value-added materials /
edited by Sankha Chakrabortty, Jayato Nayak, Shirsendu Banerjee, Maulin P. Shah.
Description: First edition. | Boca Raton, FL : CRC Press, 2025. |
Includes bibliographical references and index. | Summary: "This book is a comprehensive guide to bioconversion approaches based on microorganisms and enzymes for the valorization of underused wastes of diverse categories to produce new products. Optimized conditions for microbial and enzymatic valorization are discussed, along with related biotechnological and environmental considerations, bioprocess development, obstacles, and future outlooks"—Provided by publisher.
Identifiers: LCCN 2024013217 | ISBN 9781032526539 (hardback) |
ISBN 9781032526546 (paperback) | ISBN 9781003407713 (ebook)
Subjects: LCSH: Biomass energy—Technological innovations.
Classification: LCC TP339 .A46 2025 | DDC 662/.88—dc23/eng/20240716
LC record available at https://lccn.loc.gov/2024013217

ISBN: 978-1-032-52653-9 (hbk)
ISBN: 978-1-032-52654-6 (pbk)
ISBN: 978-1-003-40771-3 (ebk)

DOI: 10.1201/9781003407713

Typeset in Times
by codeMantra

Contents

Preface viii
About the Editors x
List of Contributors xii

Chapter 1 Classification and Composition of Different Types of Agrobiomasses 1

Bhavana Gariya, Priyanka S, Devanshu Dhir, and Ranjeet Kumar Mishra

Chapter 2 Application of Different Physical and Chemical Methodologies for the Synthesis of Different Types of Value-Added Materials 24

Dr. Md. Merajul Islam and Dr. Amina Nafees

Chapter 3 Application of Microbial Fermentation Technique for the Synthesis of Fermentable Sugars from Different Agrobiomasses 44

Anjli Tanwar, Disha Mukherjee, Karan Kumar, and Vijayanand S. Moholkar

Chapter 4 Application of Enzymatic Technique for the Synthesis of Fermentable Sugars from Different Agrobiomasses 72

Meyyappan K, Sancia Morris, Vishal S K, Ashwinth K, and Kruthi Doriya

Chapter 5 Applications of Microbial Fermentation Technique for the Synthesis of Bioalcohols from Different Agrobiomasses 92

Karan Kumar, Kaustubh C. Khaire, Kuldeep Roy, and Vijayanand S. Moholkar

Chapter 6 A Review on Potential Role of Lignocellulosic Biomass in Production of Bioethanol via Microbial Fermentation 122

Soumya Banerjee

Chapter 7 Sustainable Bio-Based 2,3-Butanediol Production from Renewable Feedstocks 140

Daniel Tinôco

Chapter 8 Application of Enzymatic Technique for the Synthesis of Bioalcohols From Different Agrobiomasses 162

Shraddha M. Jadhav, Poulami Mukherjee, Karan Kumar, and Vijayanand S. Moholkar

Chapter 9 Application of Microbial Technique for the Synthesis of Organic Acid from Different Agrobiomasses 202

Karan Kumar, Shrivatsa Hegde, Ramyakrishna A.R., Anweshan, and Vijayanand S. Moholkar

Chapter 10 Application of Enzymatic Technique for the Synthesis of Organic Acids from Different Agrobiomasses 237

Mukesh Singh, Ahana Bhaduri, Ranjay Kumar Thakur, and Sudip Das

Chapter 11 Application of Microbial Technique for the Synthesis of Biodegradable Polymers from Different Agrobiomasses 261

Neha Saxena and Amit Kumar

Chapter 12 Application of Enzymatic Technique for the Synthesis of Biodegradable Polymers from Different Agrobiomasses 273

Karan Kumar, Piyal Mondal, Shraddha M. Jadhav, and Vijayanand S. Moholkar

Chapter 13 Application of Microbial Technique for Synthesis of Food-Grade Agents from Different Agrobiomasses: An Overview 300

Indrani Paul and Priyanka Sarkar

Chapter 14 Strategic Approaches for Agricultural Biomass Conversion to Bioalcohol Using Membrane-Based Microbial Technology 323

Prashant Shukla, Indrani Paul, Apurba Sarkar, Sayantan Ghosh, and Priyanka Sarkar

Chapter 15 Membrane-Based Enzymatic Technique for the Synthesis and Purification of Bioalcohols from Agrobiomass: A Review 346

Sujoy Bose, Ajay K. Shakya, and Chandan Das

Chapter 16 Application of Membrane-Based Technique for the Synthesis of Acetic Acid from Different Agrobiomasses 376

Santoshi Mohanta, Anuradha Upadhyaya, Jayato Nayak, and Sankha Chakrabortty

Chapter 17 Application of Various Thermal Treatment Methodologies for Conversion of Agrobiomass into Value-Added Products 388

Rakesh Upadhyay, Ekta Chaturvedi, and Kulbhushan Samal

Chapter 18 Biochar Revisited: A Systematic and Critical Review on Coupling of Biochar with Anaerobic Digestion and Its Role in Bioenergy Production 406

Sanjana Pal, Koustav Saha, and Ritesh Pattnaik

Chapter 19 Valorization of Agrobiomass to Ferulic Acid 427

Sourab Paul, Jayato Nayak, and Sankha Chakrabortty

Chapter 20 Agrobiomass Valorization through a Sustainability Prism 442

Neeraj Hanumante and Neeta Maitre

Index 451

Preface

In the modern world, commodity biochemical production strategies are shifting away from conventionally profit-driven and fossil-fuel stock-dependent processes towards environmentally conscious strategies with reduced profit margins. As a result, the focus of manufacturers and researchers has been directed towards the reuse and reutilization of waste biomass to transform it into value-added products, with the waste generated from agricultural fields being the primary target. Materials such as cellulose, hemicellulose, lignin, and other extractives can be found in large quantities in woody biomass, which are conventional crop wastes, and agricultural biomass that can be used as a readily available raw material for the carbon sources for fermentation or further processing. In fact, without any proper awareness, protocol, and guiding techniques, local farmers in developing countries simply burn such agro-residues, causing severe environmental pollution. Cellulose and hemicellulose are polysaccharides that can be depolymerized into simpler saccharides by breaking polysaccharide matrices. Such sugars serve as platform molecules that can be further transformed into value-added products such as bioethanol, biobutanol, and many more biomolecules and organic acids. It is possible to remove aromatic units from lignin in order to create aromatics, and lignin can also be improved through the use of specially engineered microorganisms. Large obstacles exist in utilizing lignocellulosic biomass even though lignocellulose is readily available and inexpensive. Therefore, technical methods are required to remove, separate, and degrade the components of lignocellulose biomass, and further treatments for different application purposes are required to utilize plant biomass, making the development and improvement of technical methods of great significance for improving processing procedures and utilization efficiency of plant biomass and for decreasing the accompanying adverse effects (such as environmental pollution and energy consumption). Through the same mechanism of using plant biomass, genetic engineering technologies have also been established. It is crucial to investigate and enhance the associated methods and technologies in order to make the best use of the huge lignocellulose biomass resources that will be available in the future.

This book is a comprehensive guide to bioconversion approaches based on microorganisms and enzymes for the valorization of underused wastes of diverse categories. Optimized conditions for microbial and enzymatic valorization are discussed, along with related biotechnological considerations, environmental worries, bioprocess development, obstacles, and potential. Different valuable biochemicals, biofuels, bioenergy, and other platform chemicals are only some of the products that can be produced through the book's explanation of the microbiological processes involved in the bioconversion and valorization of wastes. This book contains a wide variety of progressive bio-based methodologies through which agricultural biomass/waste materials could be valorized into high-valued commodities. In the present era of environmentally conscious legislations on the production sectors, the manufacturing industries are witnessing a paradigm shift in the production strategies, with the sector of biomolecule manufacturing completely inclining towards biochemical pathways rather than polluting routes of chemical technologies. Under such a mindset, valorization of waste biomass/agro-wastes opens up newer avenues for the conversion of polysaccharide matrices to simpler saccharides, which can further be converted into high-valued bioproducts using fermentation-based technologies. Thus, the conversion of polysaccharides to simpler sugar was extremely essential for the supply of raw materials of carbon sources, which would be critically covered in this book. The existing frameworks for such valorization of agro-wastes to simpler saccharides are not yet up to date to satisfy the global needs of the rising world. Under the dearth of the existence of any properly arranged methodological and technological interventions for the conversion of crop waste biomass for value-added products, this book covers the majority of the possible techniques in a very straightforward scientific manner satisfying the goals of circular bioeconomy in the management of bio-wastes to value-added bioproducts. Knowledge on the development of value-added goods by various means of increasing product selectivity via technology advancement will also be protected. The editors would make a strong attempt to make this book as relevant to all aspects as possible, addressing burning issues in the field of waste mitigation, management, and bioconversion to make it useful as a carbon-rich raw material for further conversion to biochemicals.

This book would be helpful in

- Presenting potential biotechnological methodologies and strategies for the transformation of agricultural waste materials into value-added products.
- Providing technical concepts for the production of various significant bioproducts under a suitable optimized condition.
- Introducing various microbial and enzymatic processes to sustainably valorize various potential agro-wastes as renewable feedstocks for the production of biofuels, biochemicals, and building block chemicals.

It is intended for researchers and graduate students in chemical engineering and technologists, bioengineering, biochemical engineering, microbial technology/microbiology, environmental engineering, and biotechnology to help them identify alternative raw materials from agrobiomass and the microbial species. Thus, they can take up new decisions, develop novel strategies, and build up newer possibilities towards the manufacturing processes of greener biochemicals manufacturing technologies from conventional waste, i.e., agrobiomass.

About the Editors

Dr. Sankha Chakrabortty is a well-known researcher with more than 10 years of experience in the fields of environmental engineering and green technology. He is currently an assistant professor at the School of Biotechnology/Chemical Technology at KIIT Deemed to be University in India. His work has been published in over 35 SCI/SCOPUS indexed high-impact factor journals, and he has also submitted patent applications for eight of his inventions. During his 10-year research career, he has been recognized by a variety of organizations and has received 14 awards or accolades. He was awarded the "Best Paper Award-2015" by the SPRINGER publisher in 2015, the "Best Researcher Award-2020" by an international organization in the previous year, and the "NESA Young Scientist of the Year 2020" title by the National Environmental Science Academy in 2020. All of these accolades were presented in 2015. In addition, Dr. Chakrabortty was presented with an additional award by KIIT University in the category of "STE-Young Scientist Award (Faculty Category)" in recognition of his outstanding research.

Dr. Jayato Nayak, Assistant Professor, Center for Life Sciences of Mahindra University, is a prominent researcher with more than 10 years of experience in the field of novel design and applications embracing the theme of process intensification. His core expertise includes water and wastewater treatment through adsorption, sludge treatment and reutilization, graphene integrated photocatalyst development and biosynthesis of value-added products. Currently, along with the national collaboration with IITs, NITs, and CSIR Labs, he has extended research collaborations with research professionals in Poland, Brazil, Malaysia, Vietnam, etc. He has more than 25 international SCI/SCOPUS indexed peer-reviewed journal papers, more than 15 book chapters, and filed 3 patents. As of now, with more than 400 Citations, he is having h-index of 10 and i10-index of 10. He is providing his expertise as a resource person for several webinars, as well as senior editor, review editor, guest editor, editorial board member, and recognized reviewer of more than 15 international journals. He has received 5 distinguished awards from various organizations.

Dr. Shirsendu Banerjee is currently working as an assistant professor at the School of Biotechnology/Chemical Technology at KIIT Deemed to be University in India. His work has been published in more than 15 SCI/SCOPUS-indexed high-impact factor journals while guiding a good number of PhD candidates under his supervision. He has also received several awards in the field of research for his outstanding research approaches.

Dr. Maulin P. Shah has been an active researcher and scientific writer in his field for over 20 years. In 1999, he received BSc in Microbiology from Gujarat University, Godhra (Gujarat), India. He also earned his PhD degree (2005) in Environmental Microbiology from Sardar Patel University, Vallabh Vidyanagar

(Gujarat), India. His research interests include biological wastewater treatment, environmental microbiology, biodegradation, bioremediation, and phytoremediation of environmental pollutants from industrial wastewater. He has published more than 350 research papers in prestigious national and international journals on various aspects of microbial biodegradation and bioremediation of environmental pollutants. He is the editor of 150 books of international repute. He has edited 25 special issues specifically in industrial wastewater research, microbial remediation, and biorefinery of wastewater treatment area. He is associated as an editorial board member in 25 highly reputed journals.

Contributors

Ahana Bhaduri
Department of Biotechnology
Haldia Institute of Technology
Haldia, West Bengal, India

Ajay K. Shakya
Department of Chemical Engineering
Indian Institute of Technology
Guwahati, Assam, India

Amina Nafees
Department of physics
IIT Delhi
New Delhi, India

Amit Kumar
Department of Petroleum Engineering and Geoengineering
Rajiv Gandhi Institute of Petroleum Technology
Jais, Uttar Pradesh, India

Anjli Tanwar
Department of Biological Sciences
Indian Institute of Science Education and Research Bhopal
Madhya Pradesh, India

Anuradha Upadhyaya
School of Chemical Technology
Kalinga Institute of Industrial Technology
Bhubaneswar, India

Anweshan
Department of Chemical Engineering
Indian Institute of Technology Guwahati
Guwahati, Assam, India

Apurba Sarkar
Department of Biotechnology
Brainware University
Barasat, Kolkata, West Bengal, India

Ashwinth K.
Institute of Chemical Technology
Mumbai ICT-IOC Bhubaneswar Campus
Odisha, India

Bhavana Gariya
Department of Chemical Engineering
Ramaiah Institute of Technology Bengaluru
Karnataka, India

Chandan Das
Department of Chemical Engineering
Indian Institute of Technology
Guwahati, Assam

M. Chithra P. Hariprasanth
Department of Chemical Engineering
Kongu Engineering College
Erode, Tamil Nadu, India

Daniel Tinôco
Biochemical Engineering Department
School of Chemistry
Federal University of Rio de Janeiro
Rio de Janeiro, RJ, Brazil

Devanshu Dhir
Department of Chemical Engineering
Ramaiah Institute of Technology Bengaluru
Karnataka, India

Disha Mukherjee
School of Biosciences and Technology
Vellore Institute of Technology
Vellore, Tamil Nadu, India

Ekta Chaturvedi
CSIR-CMERI
Durgapur, West Bengal, India

Indrani Paul
Department of Biotechnology
Brainware University
Kolkata, West Bengal, India

Jayato Nayak
Center for Life Sciences
Mahindra University
Hyderabad, India

Karan Kumar
School of Energy Science and Engineering
Indian Institute of Technology Guwahati
Guwahati, Assam, India

Kaustubh C. Khaire
School of Energy Science and Engineering
Indian Institute of Technology Guwahati
Guwahati, Assam, India

Koustav Saha
Indian Institute of Technology Guwahati
Guwahati, Assam, India

Kruthi Doriya
Institute of Chemical Technology
Mumbai ICT-IOC Bhubaneswar Campus
Odisha, India

Kulbhushan Samal
Academy of Scientific and Innovative Research (AcSIR)
Ghaziabad, Uttar Pradesh, India
CSIR- Central Mechanical Engineering Research Institute Durgapur
West Bengal, India

Kuldeep Roy
School of Chemical Engineering
Vellore Institute of Technology
Vellore Campus, Vellore, Tamil Nadu, India

Meyyappan K.
Institute of Chemical Technology
Mumbai ICT-IOC Bhubaneswar Campus
Odisha, India

Merajul Islam
School of Basic Sciences & Technology
IIMT University
Meerut, Uttar Pradesh, India

Mukesh Singh
Kabi Nazrul College
Birbhum, West Bengal, India

Neeraj Hanumante
The University of Texas at Austin
Austin, Texas, USA

Neeta Maitre
Cummins College of Engineering for Women
Pune, India

Neha Saxena
School of Basic Sciences & Technology
IIMT University
Meerut, Uttar Pradesh, India

Piyal Mondal
Department of Chemical Engineering
Indian Institute of Technology Guwahati, Guwahati, Assam, India

Prashant Shukla
Department of Biotechnology
Brainware University
Barasat, Kolkata, West Bengal, India

Poulami Mukherjee
School of Energy Science and Engineering
Indian Institute of Technology Guwahati
Guwahati, Assam, India

Priyanka S.
Department of Chemical Engineering
Ramaiah Institute of Technology Bengaluru
Karnataka, India

Priyanka Sarkar
Department of Biotechnology
Techno India University
Kolkata, West Bengal, India

Rakesh Upadhyay
Academy of Scientific and Innovative Research (AcSIR)
Ghaziabad, Uttar Pradesh, India
CSIR- Central Mechanical Engineering Research Institute Durgapur
West Bengal, India

Ramyakrishna A.R.
Department of Biotechnology and Bioinformatics
JSS Academy of Higher Education and Research
Mysore, Karnataka, India

Ranjay Kumar Thakur
Department of Food Technology
Haldia Institute of Technology
Haldia, West Bengal, India

Ranjeet Kumar Mishra
Department of Chemical Engineering
Manipal Institute of Technology
Manipal, Karnataka, India

Ritesh Pattnaik Pattnaik
Assistant Professor Kiit School of Biotechnology
Bhubaneswar, Odisha, India

S. Bharath
Department of Chemical Engineering
Kongu Engineering College
Erode, Tamil Nadu

Sancia Morris
Institute of Chemical Technology
Mumbai ICT-IOC Bhubaneswar Campus
Odisha, India

Sanjana Pal
Kiit School of Biotechnology
Bhubaneswar, Odisha, India

Sankha Chakrabortty
School of Chemical Technology
Kalinga Institute of Industrial Technology
Bhubaneswar, India

Santoshi Mohanta
School of Chemical Technology
Kalinga Institute of Industrial Technology
Bhubaneswar, India

Sayantan Ghosh
Department of Biotechnology
Brainware University
Barasat, Kolkata, West Bengal, India

K. Senthilkumar
Department of Chemical Engineering
Kongu Engineering College
Erode, Tamil Nadu, India

Shraddha M. Jadhav
School of Energy Science and Engineering
Indian Institute of Technology Guwahati
Guwahati, Assam, India

Shrivatsa Hegde
Department of Biotechnology and Bioinformatics
JSS Academy of Higher Education and Research
Mysore, Karnataka, India

Soumya Banerjee
Department of Basic Science and Humanities
Hooghly Engineering & Technology College
Chinsurah, West Bengal, India

Sourab Paul
Centre for Life Sciences
Mahindra University
Hyderabad, India

Sudip Das
Department of Biotechnology
Haldia Institute of Technology
Haldia, West Bengal, India

Sujoy Bose
Indian Institute of Chemical Engineers
Dr. H.L. Roy Building
Jadavpur University Campus
Kolkata

Vijayanand S. Moholkar
Department of Chemical Engineering
Indian Institute of Technology Guwahati
Assam, India

Vishal S. K.
Institute of Chemical Technology
Mumbai ICT-IOC Bhubaneswar Campus
Odisha, India

1 Classification and Composition of Different Types of Agrobiomasses

Bhavana Gariya, Priyanka S, Devanshu Dhir, and Ranjeet Kumar Mishra

1.1 INTRODUCTION

The Earth's atmosphere is being harmed by the ongoing rise in environmental pollution, which also significantly impacts modern society. Heavy and hazardous metals, residues, and waste from numerous sectors are among these environmental contaminants. In developed nations like the United States, China, and Russia, as well as developing nations like India and other places, one of the most important problems is environmental pollution. Additionally, the quality of the environment is continuously declining worldwide, leading to a decline in biological diversity and vegetation, as well as a rise in harmful compounds in the atmosphere, all of which have a major negative impact on every industry directly or indirectly. Additionally, environmental pollutants are typically classified as either biodegradable or non-biodegradable depending on how readily they can be broken down. Different microorganisms break down biodegradable wastes from their highly complex forms into basic chemicals and molecules, producing water and carbon dioxide as by-products. Additionally, biodegradation can be accomplished using a variety of techniques, including composting, aerobic and anaerobic digestion, or certain natural processes. Waste that degrades quickly includes a variety of biomasses, including food waste, animal trash, kitchen waste, etc. On the other hand, toxic substances and compounds that cannot be broken down naturally are included in non-biodegradable wastes. Non-biodegradable trash includes metals, plastics, chemicals, water bottles, glasses, and many synthetic polymers. Approximately 14% of energy in 2016 was supplied from biomass, according to figures from the World Bioenergy Association, and by 2030, approximately 44% of energy would be needed. Municipal solid waste (MSW), agricultural crops, crop residues, and forest residues are a few types of biomass that can be separated (Srivastava et al., 2021). Garbage products from various industries, such as agriculture and industry, as well as waste from rural and urban life, are created globally. Agricultural residues, which are cheap, easily accessible, and cost-effective, could replace fossil fuels (Yahya et al., 2015). Climate change mitigation, energy supply security, efficient energy production, and energy use efficiency are now top priorities in the energy sector. To direct these tasks more effectively, a diverse range of solid

DOI: 10.1201/9781003407713-1

biofuels have been harnessed as sustainable energy resources. Due to the significant annual production of agro-mass and its calorific values close to those of woody biomass, the use of agro-mass has recently developed a growing level of interest. Herbaceous plants are a class of plants that consist of stems and leaves that decompose to the soil level when approaching the end of the rising period. These plants do not have any wood; hence, lignin and cellulose fibers are more loosely connected in their structure. As a result, herbaceous plants have a lower lignin content than woody plants (Mishra et al., 2022). Biomass is widely accessible and an alternative to more traditional energy sources when they are in short supply. Pure oxygen is produced by biomass, which also contributes to the reduction of greenhouse gases in the environment. Biomass can be established by combining a variety of substances (Figure 1.1) (Mishra et al., 2022). Evaluating the spatial and temporal accessibility of such wastes is a vital aspect in creating biomass energy sources from residues. Decision-makers could benefit from this investigation by learning more about the possibilities for utilizing biomass remains for energy uses in many nations. Conducting extensive studies and experimental investigations is essential to utilize agro- masses effectively in

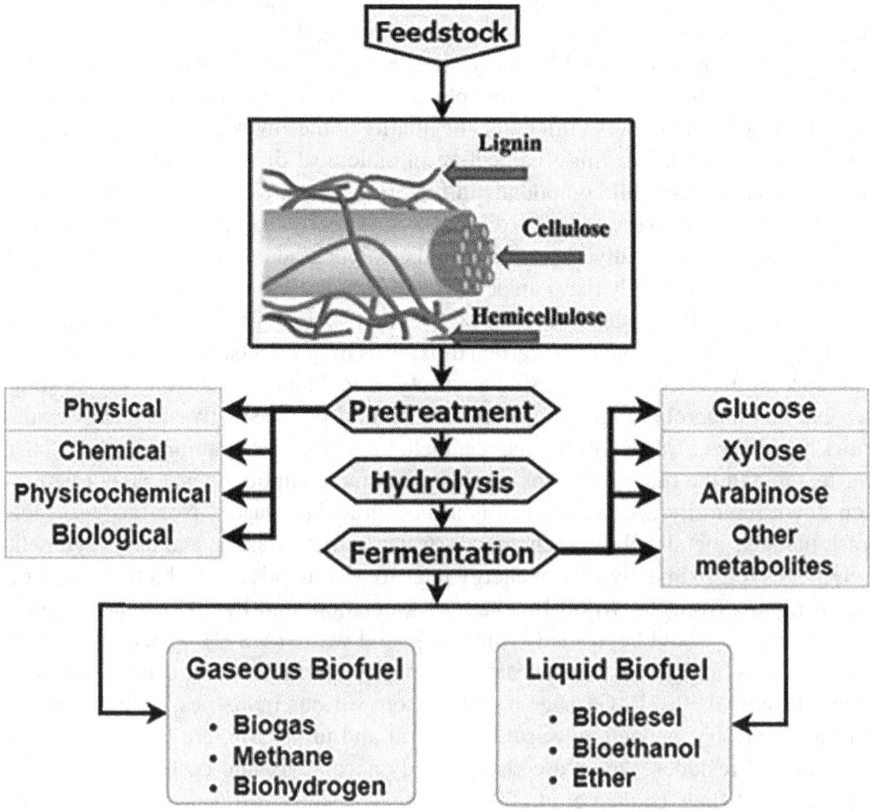

FIGURE 1.1 An overview of the biomass employed at different stages during the production of biofuel.

heating appliances. The use of agro-mass in thermochemical conversion processes (TCP) has demonstrated that, in comparison to wood fuel, a number of tasks relating to collection, storage, pretreatment, and undesirable combustion facts need to be resolved (Li et al., 2015). The establishment of gaseous and liquid substances that are engendered at greater temperatures and leave surface deposits of incineration devices, flying ash particles, and these negative phenomena are just a few of the problems that arise due to low ash fusion temperature (Niu et al., 2010; Striūgas et al., 2019). The following are three crucial barriers to effective combustion in this intricate process: (i) slag formation, sticky ash particles, and ash softening at relatively low temperatures all contribute to agglomeration; (ii) slagging, which is the deposit of flammable ash particles and particles that have melted at high temperatures on the furnace's hot surfaces; and (iii) fouling, also referred to as the build-up of deposits on equipment's colder convective surfaces. As a result of the phenomena that occur during combustion, the heat transfer efficiency declines, equipment corrodes and erodes, its lifetime is shortened, maintenance and cleaning of the equipment takes longer, and the likelihood of failure rises (Striūgas et al., 2019).

The variation in composition of wood and agricultural wastes leads to these undesired processes, resulting in unique characteristics in ash production. Woody fuel ash mostly consists of Ca, K, Mg, and Si (Vassilev et al., 2010). Other constituents, including N, Ti, Al, Fe, Na, S, and P, are present in much smaller levels, while Cl is the smallest of all the other components (Vassilev et al., 2010). Si, K, and Ca make up most of the components of agro-mass fuel, with Al, Fe, Na, P, Mg, and S following in decreasing order but not necessarily in the order listed (Vassilev et al., 2013b). Agro-mass has much higher levels of Cl and S than wood, and these two elements, along with alkali metals (Na and K) and alkaline earth metals (Mg and Ca), also lead to the majority of ash-related issues (Vassilev et al., 2013a). Due to fertilization or post-harvest material processing, a sizeable proportion of certain constituents is present in agro-mass. The quick growth age and the vast range of classes and elements that make up the innermost structure, which occasionally deviates from wood, are additional factors that affect the constitution of agro-mass (Ramage et al., 2017). The abundance of chemically active alkaline metals clearly indicates the formation of low-melting-point substances during the thermochemical conversion process. Faster equipment corrosion and the production of accumulations on the surfaces of incineration equipment are caused by volatile chemical emissions, such as potassium salts (KCl, K_2SO_4, and K_2CO_3). The type of agro-mass has a remarkable impact on the composition and volume of ash. Based on an analysis comparing the composition and average quantities of components in ash derived from cereal straw in existing literature, Si, K, and Ca make up the majority of the ash, accounting for 62%–72% of the total quantity (Nunes et al., 2016). Cl, S, P, Fe, Na, Mg, Al, etc. have other reaming parts. The ash content is linked to the composition of the feed from which the ash originated, even though there are instances where considerable variations from the suggested sequence are documented. The temperature during combustion affects ash composition as well. The findings of Misra et al. (1993) demonstrated how temperature variations from 500°C to 1,400°C affect the amount of ash in wood. Ca and K carbonates ($CaCO_3$ or $K_2Ca(CO_3)$) predominate at 600°C, whereas at 1,300°C, the ash solely contains Ca and Mg oxides (Misra et al., 1993). In the most recent instance, the total mass of the ash dropped by 23%–48%, and ash accumulation was also noticed; however, fusion was not observed. This discovery demonstrates that lower temperatures during the initial phases of agro-mass incineration produce fewer stable compounds. However, a higher temperature causes more ash to form more stably. Ash configuration is supplemented by component decay, deformation, and the establishment of stable derivatives as agro-mass is made up of different kinds of chemical components depending on the type of feed and evolution conditions. More researchers have concentrated on the following two tasks: (i) figuring out ways to raise the ash fusion temperatures of shrinkage starting temperature (SST), hemisphere temperature (HT), deformation temperature (DT), and flow temperature (FT), and (ii) recognizing indicators linked to agro-mass ash agglomeration. The first objective has been the subject of extensive browsing history, but little headway has been accomplished. Typically, the objective is to regulate the optimal elemental constituents of agro-mass due to TCP and to devise strategies for mitigating ash fusion at lower temperatures. This approach relies on comprehensive studies of carbon incineration, forming the basis for an index that quantifies the ratio of basic and acidic components in ash. Due to rising levels in P, S, Si, Cl, etc., in agro-mass, assessment of the alkalinity index in the burning of agricultural waste demonstrates that such indices cannot realistically be utilized to agro-mass (Ramage et al., 2017). As a result, scientists have tried to use altered versions of this basic-to-acid ratio or to create new applications for the alkalinity index. Specific assessments of the latter index in biomass burning and carbon mixture were the focus of a study by Blomberg (2007). The investigation highlights that creating indexes using agro-mass compositional data is more logical than using ash compositional data. Research investigations have been conducted for the second task, including studies that examine additives often made of significant amounts of CaO, SiO_2, and Al_2O_3 (Zhu et al., 2014). The later study demonstrates that a rise in ash fusion temperatures is possible because the ash produces Ca, K-Ca, and K-Al silicates.

1.2 LIGNOCELLULOSIC BIOMASS

1.2.1 Variety of Feedstock

The biological material derived from various species of living things found in the biosphere is known as biomass. Green plants use the process of photosynthesis to capture carbon from the surroundings and reintroduce it back into the atmosphere. Data show that every year, plants convert 4,500 EJ (exajoules) of solar energy and 120 GT (gigatonnes) of atmospheric carbon into oxygen (Association, 2016). The availability of various kinds of biomass resources may be seen in Figure 1.2, which shows how widespread biomass is across the world.

Global Bioenergy Statistics (WBA) will conduct a 150 EJ biomass energy potential by 2035 that has the potential to provide around 25% of the world's energy reservoir (Association, 2016). According to Global Bioenergy Supply and Demand Projections, biomass has a bright future as a resource of global renewable energy. Renewable biomass, regarded as waste or a by-product, stands out as one of the most vital energy sources, but it has not yet reached commercialization. With the additional capacity

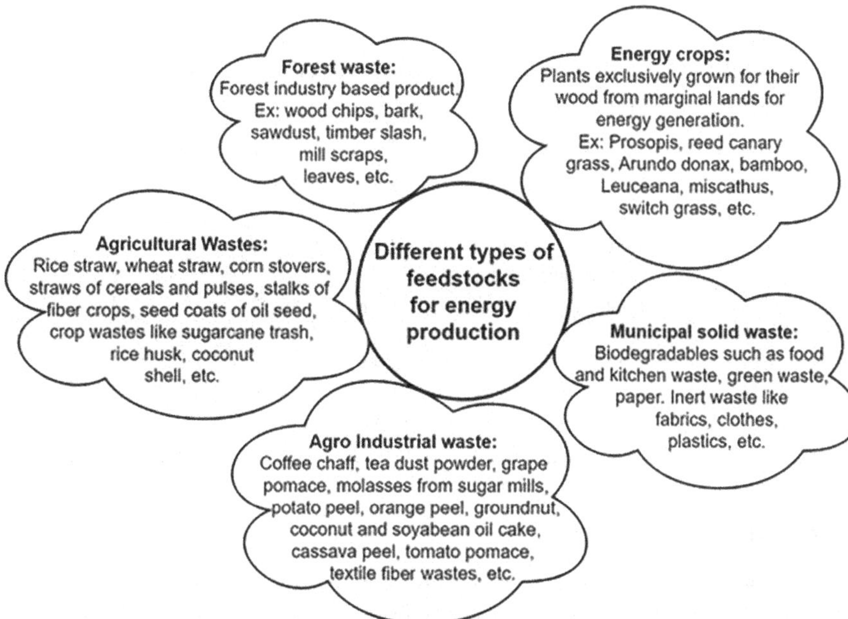

FIGURE 1.2 Diverse availability of biomass worldwide.

of committed energy crops of 30 EJ and certain organic waste ranging from 8 to 15 EJ, this quantity is expected to be greater than 34 EJ. Improved clarity is required to comprehend the preceding clause (Association, 2016). By 2035, the amount of biomass energy added to primary energy could increase from 56 EJ in 2012 to nearly 150 EJ (Association, 2016). Primary energy (PE) is defined as the energy originating in nature and not undergoing any artificially induced alteration. It is the energy contained in unprocessed fuels and other energy sources introduced as system inputs (Gholizadeh et al., 2019). By 2035, and possibly 100% a few years later, renewables might secure more than 50% of the world's energy needs in conjunction with the rapid dissemination of alternative renewable technologies (Association, 2016). If the government formulates new policies that support the exploitation of bioenergy as energy or fuel, this bioenergy capability can be utilized. These initiatives include training, education, financial considerations for integrated and sustainable growth ideologies for agriculture and sustainable forestry, the security of agricultural land, sustainable management, and reforestation of forests, among other agriculture-related initiatives that the police have endorsed (Mishra et al., 2022). The growth value of bioenergy has increased in numerous nations or regions of the world. Stopping subsidies for fossil fuels, enacting a CO_2 tax on their use, and offering encouragement or funding for the implementation of machinery to convert biomass into energy or fuel are all part of this. Installations of biogas, wood gasifiers, and biomass combustion units coupled with heat and power units for the heating of residential and commercial areas might all fall under this category. The projections in Table 1.1 show the biomass ability of the world in the years 2012 and 2035.

TABLE 1.1
The Global Capacity of Biomass in the Year 2012 and 2035 (Association, 2016)

Main Sector	Subsector	2012	2035	2035 (Average)
Agriculture	Devoted crops – Main product	3.50	26–34	30
	Waste or by-products and Manure residues	2.10	30–38	34
	Total agriculture residue	5.60	56–72	64
Forestry	–	48.90	72–84	78
Organic waste	–	1.70	6–10	8
Total	–	56.20	134–166	150

1.2.2 Types of Agro-Mass

The investigation used samples of several residual agro-mass kinds (Table 1.2). The following groups were created from the samples:

1. Group G (grains and seeds) consists of unconditioned grains and seeds that are commonly utilized as pasture; the leftovers may be used to produce energy, but this is not a typical or frequently used method.
2. Group D (non-edible seeds) mainly consists of seeds derived from non-edible sources, adding food security.
3. Group H (harvest leftovers) consists of hay and straw from a variety of cultivated plants that were produced during the harvesting process.
4. Group H* (other harvest wastes) comprises leftovers such as potato haulm and sunflower stems.
5. Group R includes leftovers from processing grains and seeds.
6. Group W contains examples of wood.
7. Group X contains mainly grass, which has a high heating value.

1.3 RESOURCE OF BIOMASS

Biomass is any substance or plant waste that can be converted into energy. Lignin, starch, proteins, lipids, and carbohydrates are the main components of biomass. Depending on the geographical setting and the source, these elements change. Energy crops, agricultural crops and their by-products, wood and wood wastes, municipal wastes, aquatic plants, and algae are only a few of the many biomass resources that are easily accessible. The biomass generated by microorganisms, plants, animals, fruits, and vegetables can be separated into four main types (Figure 1.3a). It is possible to accurately forecast the structure and composition of the corresponding carbons produced from biomass by identifying the chemical and elemental mixtures of various

TABLE 1.2
The Elemental Composition of Biomass

Group	Species Name	Proximate Study (wt.%) (On a Dry Basis)				MJ/kg	Elemental Analysis (wt.%) (on a Day Basis)							Reference
		Moisture	Volatile Matter	Ash Content	Fixed Carbon	HHV	Carbon	Hydrogen	Oxygen	Nitrogen	Sulfur (S)	Chlorine (Cl)	2S/Cl	
G														
1G	Sunflower seeds	6.45	76.68	3.44	19.88	25.4	55.96	7.54	30.30	2.60	0.16	0.79	0.41	Praspaliauskas et al. (2020)
2G	Buckwheat seeds	11.49	73.73	2.40	23.87	19.4	47.20	6.91	39.72	3.52	0.24	0.03	16.00	Praspaliauskas et al. (2020)
3G	Wheat grains	13.47	70.81	2.38	26.81	18.3	46.71	6.22	42.22	2.26	0.22	0.09	4.89	Praspaliauskas et al. (2020)
4G	Pigweed seeds	11.40	75.81	2.96	21.23	19.7	46.27	6.82	40.38	3.34	0.23	0.03	15.33	Praspaliauskas et al. (2020)
5G	Corncob	7.82	68.76	5.75	25.49	18.3	45.02	5.88	40.90	2.31	0.13	0.08	3.25	Praspaliauskas et al. (2020)
6G	Peas	13.55	72.94	3.30	23.76	18.7	44.62	6.64	41.33	3.95	0.16	0.04	8.00	Praspaliauskas et al. (2020)
7G	Oat grains	15.08	70.43	3.47	26.10	17.8	43.87	6.45	43.98	2.08	0.15	0.09	3.33	Praspaliauskas et al. (2020)
8G	Barley grains	18.44	62.93	3.05	34.02	17.1	42.30	6.36	46.28	1.90	0.12	0.14	1.71	Praspaliauskas et al. (2020)
9G	Rape seeds	–	80.55	3.80	15.65	29.1	61.89	8.99	21.55	3.44	0.33	–	–	Praspaliauskas et al. (2020)
10G	Flax seeds	–	77.41	4.88	17.71	27.7	53.50	7.10	29.66	4.84	0.22	–	–	Praspaliauskas et al. (2020)

(Continued)

TABLE 1.2 (*Continued*)
The Elemental Composition of Biomass

Group	Species Name	Proximate Study (wt.%) (On a Dry Basis)				MJ/kg	Elemental Analysis (wt.%) (on a Day Basis)							Reference
		Moisture	Volatile Matter	Ash Content	Fixed Carbon	HHV	Carbon	Hydrogen	Oxygen	Nitrogen	Sulfur (S)	Chlorine (Cl)	2S/Cl	
D														
1D	*Cascabela thevetia* seeds	4.97	78.05	2.19	14.78	19.20	54.93	9.99	31.07	3.33	0.66	0.85	1.55	Mishra and Mohanty (2018a)
2D	*Delonix regia*	6.85	76.27	2.78	14.10	19.70	53.5	6.93	32.55	6.99	–	1.26	–	Mishra and Mohanty (2018a)
3D	*Samanea saman*	6.19	76.00	3.06	14.74	18.70	48.46	6.75	37.47	7.30	–	1.15	–	Mishra and Mohanty (2018a)
4D	*Phyllanthus emblica*	6.56	75.20	2.69	15.55	19.20	48.76	5.91	43.31	2.01	–	1.05	–	Mishra and Mohanty (2018a)
5D	*Sapodilla* seeds	8.07	77.02	1.19	13.73	18.70	52.67	6.74	38.78	1.46	0.34	0.95	0.72	Mishra and Mohanty (2018a)
17D	*Azadirachta indica* seeds	13.4	76.87	5.78	16.7	16.5	48.98	6.76	36.4	1.6	0.07	0.1	1.4	Mishra and Mohanty (2018b)

(*Continued*)

TABLE 1.2 (*Continued*)
The Elemental Composition of Biomass

		Proximate Study (wt.%) (On a Dry Basis)				MJ/kg	Elemental Analysis (wt.%) (on a Day Basis)							
Group	**Species Name**	**Moisture**	**Volatile Matter**	**Ash Content**	**Fixed Carbon**	**HHV**	**Carbon**	**Hydrogen**	**Oxygen**	**Nitrogen**	**Sulfur (S)**	**Chlorine (Cl)**	**2S/Cl**	**Reference**
H														
1H	Oat straw	28.83	72.97	3.47	23.56	19.5	47.78	6.05	41.08	1.47	0.14	0.26	1.08	Praspaliauskas et al. (2020)
2H	Rye straw	8.97	67.89	4.32	27.79	17.8	45.48	6.01	43.51	0.63	0.05	0.09	1.11	Praspaliauskas et al. (2020)
3H	Triticale straw	11.32	67.37	3.90	28.73	17.8	45.21	6,17	44.04	0.63	0.05	0.11	0.91	Praspaliauskas et al. (2020)
4H	Hay	11.39	62.84	6.20	30.96	17.4	44.22	5.88	42.44	1.19	0.08	0.21	0.76	Praspaliauskas et al. (2020)
5H	Barley straw	11.71	65.46	6.57	27.97	18.1	44.12	5.73	42.49	0.99	0.10	0.54	0.37	Praspaliauskas et al. (2020)
6H	Rapeseed straw	15.78	61.27	6.97	31.76	17.0	43.23	5.97	42.51	1.08	0.23	0.29	1.59	Praspaliauskas et al. (2020)
7H	Wheat straw	10.34	64.75	3.89	31.36	18.0	42.68	5.41	46.90	0.99	0.13	0.08	3.25	Praspaliauskas et al. (2020)
8H	Buckwheat straw	14.68	61.11	6.72	32.17	17.3	40.92	5.24	45.89	0.98	0.24	0.50	0.96	Praspaliauskas et al. (2020)
9H	Mustard stalk	9.70	70	7.90	12.30	17.60	43.80	5.90	43.50	0.30	0.30	–	–	Raj et al. (2015)
10H	Cotton stalk	8.90	71	3.50	16.60	19.20	46.80	6.40	46.80	0.30	0.20	–	–	Raj et al. (2015)

(*Continued*)

TABLE 1.2 (*Continued*)
The Elemental Composition of Biomass

		Proximate Study (wt.%) (On a Dry Basis)				MJ/kg	Elemental Analysis (wt.%) (on a Day Basis)							
Group	Species Name	Moisture	Volatile Matter	Ash Content	Fixed Carbon	HHV	Carbon	Hydrogen	Oxygen	Nitrogen	Sulfur (S)	Chlorine (Cl)	2S/Cl	Reference
H*														
1H*	Sunflower stem*	82.15	64.41	11.20	24.39	17.2	41.95	5.19	40.05	1.43	0.18	0.06	6.00	Praspaliauskas et al. (2020)
2H*	Potato haulm*	19.59	62.22	12.42	25.36	16.9	39.48	5.08	41.68	1.25	0.10	0.19	1.05	Praspaliauskas et al. (2020)
R														
1R	Wheat press cake	10.27	66.06	6.16	27.78	18.4	46.99	5.60	39.21	1.92	0.13	0.03	8.67	Praspaliauskas et al. (2020)
2R	Grain oilcake	11.30	63.22	8.02	28.76	17.0	43.10	5.90	41.36	1.53	0.09	0.31	0.58	Praspaliauskas et al. (2020)
3R	Flax oilcake (c.p.)	–	70.50	5.07	24.43	17.1	52.09	7.00	30.23	5.34	0.27	–	–	Praspaliauskas et al. (2020)
4R	Flax oilcake (h.p.)	–	73.38	4.88	21.87	16.8	53.3	7.10	29.66	4.84	0.22	–	–	Praspaliauskas et al. (2020)
5R	Rapeseed oilcake	–	51.14	6.89	41.97	22.7	49.11	6.07	31.45	5.96	0.53	–	–	Praspaliauskas et al. (2020)

(*Continued*)

TABLE 1.2 (*Continued*)
The Elemental Composition of Biomass

		Proximate Study (wt.%) (On a Dry Basis)				MJ/kg	Elemental Analysis (wt.%) (on a Day Basis)							
Group	Species Name	Moisture	Volatile Matter	Ash Content	Fixed Carbon	HHV	Carbon	Hydrogen	Oxygen	Nitrogen	Sulfur (S)	Chlorine (Cl)	2S/Cl	Reference
W														
1W	Alder (wood+bark)	–	–	0.84	–	19.6	49.08	5.94	43.20	0.34	–	–	–	Praspaliauskas et al. (2020)
2W	Ashwood (wood+bark)	–	–	1.35	–	19.4	49.00	5.84	42.81	0.27	–	–	–	Praspaliauskas et al. (2020)
3W	Oak (wood+bark)	–	–	0.56	nd	19.5	48.28	5.93	44.28	0.21	–	–	–	Praspaliauskas et al. (2020)
4W	Birch (wood+bark)	–	–	0.86	–	19.9	49.55	6.06	42.78	0.30	–	–	–	Praspaliauskas et al. (2020)
5W	Oak briquettes	–	–	0.32	–	19.9	49.72	6.04	44.12	0.29	–	–	–	Praspaliauskas et al. (2020)
6W	Black Alder briquettes	–	–	1.49	–	20.2	50.08	6.29	43.85	0.35	0.02	–	–	Praspaliauskas et al. (2020)
X														
X1	Dube grass	3.20	70.89	11.34	14.57	17.96	44.86	5.57	47.64	1.23	0.70	–	–	Kumar Mishra and Mohanty (2020)
X2	Waste Dalia flower	5.79	69.50	11.66	13.05	16.52	44.86	5.57	45.07	3.66	0.84	–	–	Mishra et al. (2020)
X3	*Miscanthus sinensis* grass	5.79	78.21	3.89	12.14	17.52	41.86	5.59	52.07	0.20	0.35	–	–	Kumar and Mishra (2022)
X4	Switch grass	6.25	71.21	3.40	19.14	18.21	43.26	5.89	50.23	0.52	0.16	–	–	Mishra (2022)
X5	Elephant grass	10	77	6.90	6.10	15.60	41.60	6	34.40	1.0	–	–	–	Braga et al. (2014)

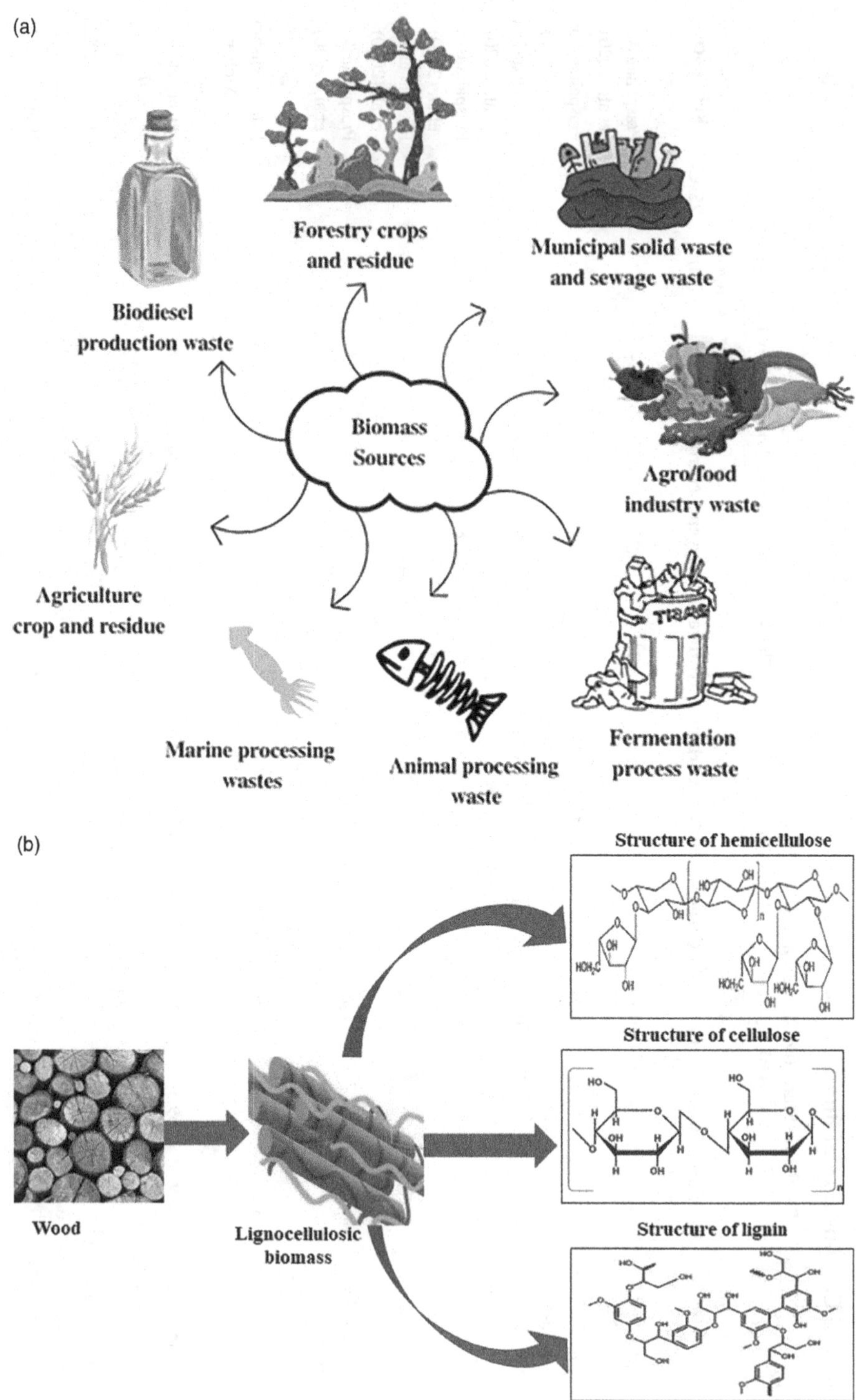

FIGURE 1.3 (a) Schematic presentation of major sources of biomass. (b) Structure of lignocellulosic biomass (hemicellulose, cellulose, and lignin).

forms of biomass. Through the process of photosynthesis, plants can store chemical energy that has been converted from solar energy. This has led to the recognition of biomass energy as source of renewable energy, alongside the sun, wind, hydroelectricity, and geothermal energy. Utilizing biomass energy also stops the release of CO_2 into the atmosphere. Currently, energy from biomass is extensively used to meet the world's rising energy needs, including transportation fuel, electricity production, and cooking and heating. Early and thorough research has revealed that biomass is a rich source of carbon, hydrogen, oxygen, and nitrogen with just minor levels of chlorine and sulfur (Mishra et al., 2022). There are various ways to use biomass to create energy without burning it. Methane gas or automobile fuels like ethanol and biodiesel can be produced from biomass and then converted into other useful energy sources. The main element of natural gas is CH_4. Methane gas is released by unpleasant items such as rotting rubbish, agricultural waste, and human waste. The most complex type of biofuel is referred to as a pellet, which is created by compressing biofuels with binders into tiny cylinders 6–12 mm in diameter and 10–30 mm in length. The fuels utilized in pellets, in particular, have a low energy to volume ratio (Williams et al., 2018). Despite being utilized less frequently than other standardized fuels such as fuelwood, wood chips, and pellets, fruit husks, seeds, and stones are also a rising source of solid biofuel. Mango stones, peanut shells, and sunflower seed husks have been shown to have a higher heating value (HHV) (12–22 MJ/kg) than other commercially available biofuels (Perea-Moreno et al., 2018). This, combined with the fact that their production is increasing globally, makes them especially attractive for the production of thermal energy and the reduction of CO_2 emissions.

Biomass consists of wood, sawdust, straw, seed waste, manure, paper trash, household waste, wastewater, and more. In contrast to other electrode materials, biomass-derived carbon has several advantages for energy and environmental applications, including more accessibility, greater availability, ecological friendliness, in situ nanoporous structure formation, and processing adaptability. The composition of the biochar, the formation of the microstructure, and the heteroatom doping of the porous carbons (PCs) are all influenced by the chemical and elemental makeup of the feedstock. The production of various necessary chemicals from biomass, in addition to being burned or indirectly converted into fuels like ethanol or biodiesel, presents a difficult biological and chemical problem for the environmentally responsible and sustainable use of biomass waste (Rodionova et al., 2017). Two of the most popular products made from biomass are biochar and ACs. To make biochar, biomass can be burned or pyrolyzed in an oxygen-depleted atmosphere. This process typically results in a structure that is extremely porous and has a large specific surface area. The ability of this specific porosity structure to absorb nutritional cations and anions can promote the development of soil-beneficial microbes, boosting soil fertility and providing important agricultural benefits (Odelade & Babalola, 2019). Some materials can be used as fuels right away because of their characteristics, whereas others require several pretreatment procedures involving a variety of technologies. Biomass has benefits in addition to disadvantages. Only two of its many benefits are its capacity to provide jobs and its ability to serve as a forest. The forest would also be cleaned, which would aid to put out forest fires. Biomass generates steady employment by taking resources out of the forest and the bush. The utilization of biomass as a biofuel has already piqued the curiosity of the scientific community. The parameters for choosing carbon from biomass were made public by Paul Thomas and colleagues (Thomas et al., 2019). They contend that raising the nitrogen content promotes the in situ production of nitrogen-doped carbon with improved conductivity and improves cycle stability. To avoid hindering the production of aromatic carbon, less oxygen precursors must be used. Chitin, keratin, and lignin are all high molecular weight, firmly crosslinked, and thermally stable bio-macromolecules that contribute to the production of aromatic carbon and high-quality biochar during carbonization. Low concentrations of high molecular weight, non-crosslinked aliphatic compounds must be avoided since they stifle the synthesis of aromatic carbon (Thomas et al., 2019). It is crucial to find a biomass precursor since the source influences the amount of carbon produced and the material's structure, both of which are crucial for energy storage and environmental protection applications.

1.4 ELEMENTAL COMPOSITION OF BIOMASS

Carbon, oxygen, sulfur, nitrogen, hydrogen, and oxygen comprise biomass. Organic matter and biomass are primarily made up of carbon, which was initially absorbed during the photosynthetic activity and is crucial in determining the fuel's ability to provide heat. Carbon is also converted into CO_2 and discharged into the atmosphere after combustion. Additionally, incomplete combustion causes the conversion of carbon in a number of combustion processes into CO or poly-aromatic hydrocarbons (PAHs) (Gollakota et al., 2018). Typically, the biomass components such as cellulose, hemicellulose, and lignin are utilized to calculate the fuel's carbon content. Depending on the region and climate, biomass has varying levels of carbon. Typically, herbaceous biomass has a lower carbon content than wood biomass. Another crucial component that modifies the fuel's ability to heat is hydrogen. Hydrogen is transformed into water during burning, considerably increasing the total gross heating value (Gollakota et al., 2018). The hydrogen content of the herbaceous material is lower (5.5%–6.0%) than that of the wood biomass (6%–8%) due to the differing chemical components (Gollakota et al., 2018). Nitrogen is a fundamental component of biomass in addition to carbon and hydrogen, and it is applied to soil as fertilizer to enhance plant growth and output by providing essential nutrients. Additionally, the breakdown rate in biochemical processes like fermentation and digestion is significantly accelerated by N_2 digestion. Due to the usage of N_2 in agricultural practices over the 20th and 21st centuries, herbaceous biomass currently comprises 0.4–1.0 wt.% nitrogen on a dry basis, as opposed to woody biomass, which has 1%–3% nitrogen (Kumar et al., 2015). Nitrogen is not oxidized during burning, which contributes to the heating properties. For plants to flourish, they require the essential component sulfur, which can be found in the structures of proteins, amino acids, and enzymes. Typically, the sulfur content of herbaceous crops or residues is higher (0%–2% or more) than that of woody biomass (0%–1%) (Gollakota et al., 2018; Mishra et al., 2022). Even though sulfur is present in biomass, it has negative effects on combustion due to corrosion, gaseous emissions (SO_x), and syngas cleansing in gasification (Gollakota et al., 2018; Mishra et al., 2022). Another essential component of biomass that comes from the process of photosynthesis is oxygen. The fuel's gross heating value (GHV) is changed when oxygen is present in biomass, which also reduces the amount of bio-sourced substances that may be used. The latter is due to the oxygen concentration of bio-oil or bio-increased crude, which is harder to break down, especially when paired with the phenolic chemicals present in the biomass. The fact that the percentage of oxygen content and the amount of dry ash content are calculated by subtracting 100 from all the constituent parts is also notable (Praspaliauskas et al., 2020). In addition, a CHNS/O analyzer can directly measure it. Overall, the yields, properties, and heating value of bio-crude derived from various kinds of biomass were significantly different from one another. Understanding the fundamentals of biomass is crucial before beginning any test or procedure to get the most out of it. Table 1.2 shows the elemental composition of a variety of biomass in the energy sector.

1.4.1 Types of Biomass

Biomass is used because of its inherent chemical characteristics, which produce a wide range of chemical reactions. Hemicellulose, cellulose, lignin, extractives, and a few other minor compounds are the main elements of lignocellulose biomass (Mishra et al., 2022). In general, hemicellulose, cellulose, and lignin elements are used to build woody and lignocellulose biomass, and as a result, the structure's strong crystallinity renders it unyielding and challenging to handle. The arrangement of the biomass disrupts the solvents' solubility criterion, which makes sustaining a continuous flow more difficult (Mishra et al., 2022). The processing of lignocellulosic materials on an industrial scale is constrained by excess oxygen content, despite the complex benefit system of carbon and hydrogen. The main components of agriculture waste and lignocellulosic biomass are listed below.

1.4.1.1 Cellulose

The chemical formula for cellulose is $(C_6H_{10}O_5)_n$. It is a lengthy chain of polysaccharides with a large molecular mass and a high degree of polymerization. These cellulosic molecules, which exclusively exist in plants and are thought to be the most prevalent polysaccharide molecules in the environment, contain 5,000 and 10,000 monomer units (Srivastava et al., 2021). Furthermore, it is anticipated that cellulose molecules make up 40%–50% of the carbon on earth. Cellulose, hemicelluloses, and lignin are the most common lignocellulosic components in biomass. It typically comprises 40–50 wt% of dry biomass. It is a linear homopolysaccharide made of d-glucose that is bound together by 1,4-glycoside linkages. Cellobiose, the basic repeating unit in cellulose chains, is produced as a result of this interaction and is essentially two units of anhydrous glucose. The cellulose chains are connected by intramolecular and intermolecular hydrogen bonds between the –OH groups, forming a crystalline super-molecular construction. Small amounts of cellulose are amorphous, but most of its chains are very crystalline. Since cellulose molecules are linear in form and have hydrogen bonds between each unit due to their ribbon-like structure, the rotation gets reserved in the molecules (Srivastava et al., 2021). While the hydrophilic group is placed laterally, the hydrophobic group is located at the surface of this ribbon-like arrangement. As a result of this specific configuration within the ribbon-like chain, a grouping of the polymer and a fractal-like characteristic develop. Additionally, the induced pressures within the molecule prevent translational motion, making the molecule more elastic and flexible (Srivastava et al., 2021). The plant cell's structure has alternately amorphous and crystalline sections and exhibits high resistance to cellulose enzymes. The cellulase enzyme is constantly capable of attacking an amorphous portion, leading to the breakdown of a glucose unit. The primary use of this cellulosic component can have substantial relevance for the production of biofuel because employing these materials has various advantages over using other sources (Srivastava et al., 2021). Additionally, this method uses agricultural residues instead of food and grain components, which are by-products of the cultivation process. The chemical production of cellulose is shown in Figure 1.3b. Table 1.3 shows the biochemical composition of a variety of biomass.

TABLE 1.3
Biochemical Composition of a Variety of Biomass

Feedstock	Hemicellulose (%)	Cellulose (%)	Lignin (%)	Protein (%)	Reference
Rice straw	19.27	23.47	9.90	2.20	Sravan et al. (2018)
Wheat straw	23.68	34.20	13.88	2.33	Miranda Neto (2018)
Barley straw	20.36	33.25	17.13	3.62	Kaushik and Singh (2011)
Corn straw	21.30	42.60	15.10	4.00	García-Aparicio et al. (2007)
Oat straw	20.0–26.0	31.0–35.0	10.0–15.0	–	Zhang et al. (2015)
Corn stalk	16.8–35.0	35.0–39.6	7.0–18.4	–	Zhang et al. (2015)
Hazelnut shell	30.4	28.8	42.9	–	Uyan et al. (2020)
Tea waste	19.9	30.20	40	–	Kim et al. (2020)
Grasses	35–50	25–40	10–30	–	van der Weijde et al. (2013)
Olive husk	23.6	24	48.4	–	Messineo et al. (2012)
Sunflower	34.6	48.8	17	–	Yigezu and Muthukumar (2015)
Swine grass	28	6.0	–	–	Xie et al. (2011)
Leaves	80–85	15–20	0	–	Martins et al. (2016)
Hardwood	24–40	44–55	18–25	–	Santos et al. (2012)
Corncobs	31.9–36.0	33.7–41.2	6.1–15.9	–	Zhang et al. (2015)
Sorghum	24.0–27.0	32.0–35.0	15.0–21.0	–	Passoth and Sandgren (2019)
Switch grass	10–40	30–50	5–20	–	David and Ragauskas (2010)
Miscanthus grass	28.22	34.22	10.33		Kumar and Mishra (2022)
Switchgrass	23.22	43.56	18.46		Mishra (2022)

(Continued)

TABLE 1.3 (*Continued*)
Biochemical Composition of a Variety of Biomass

Feedstock	Hemicellulose (%	Cellulose (%)	Lignin (%)	Protein (%)	Reference
Elephant grass	18.70	39.50	22.50		Braga et al. (2014)
Cotton stalk	19.20	39.40	23.20		Raj et al. (2015)
Cascabela Thevetia seeds	21.01	36.80	15.23		Mishra and Mohanty (2018a)
Delonix regia seeds	27.22	48.16	14.06		Mishra and Mohanty (2018a)
Samanea Saman seeds	26.55	30.81	10.59		Mishra and Mohanty (2018a)
Phyllanthus emblica seeds	21.98	48.11	5.48		Mishra and Mohanty (2018a)
Sapodilla seeds	26.55	34.03	7.61		Mishra and Mohanty (2018a)
Azadirachta indica seeds	21.94	38.04	13.58		Mishra and Mohanty (2018b)

1.4.1.2 Hemicellulose

Hemicellulose is the second most major lignocellulosic constituent of biomass. Heteropolysaccharide hemicellulose, in contrast to cellulose, comprises several monosaccharide units. For dry biomass, hemicellulose typically ranges from 25 to 35 wt.% (Srivastava et al., 2021). The monomeric components of hemicellulose (methyl glucuronic and galacturonic acids) are pentose sugars (xylose and arabinose), hexose sugars (glucose, mannose, and galactose), and sugar acids. Amorphous and having just transient branching are the hemicellulose polymer chains (Mishra et al., 2022). The schematic representation of hemicellulose's sugar units is shown in Figure 1.3b. Further, Table 1.3 shows the biochemical composition of a variety of biomass.

1.4.1.2.1 Mannans

The primary components of hemicelluloses are mannans, which are often found in bigger plants' cell walls. Hemicellulose exhibits a strong attraction for cellulose, which mannans help to bind within the wood. There are four distinct types of mannans: galactoglucomannans, linear mannans, galactomannans, and glucomannans (Srivastava et al., 2021). The quantity of mannose, glucose, or mannans units joined by a (1–4) glycoside bond determines the type of chemical composition that mannans will adopt. The 1,4-linked-D-mannopyranosyl units and significant amounts of sugar, notably in the form of galactose, are both present in linear mannans (Naidjonoka et al., 2020). These mannans are highly sought-after and used in a variety of procedures in the dairy and food sectors. Mannans are frequently used for stabilizing, coating, producing gels, producing wet absorbents, increasing textures, and altering the viscosity of liquids. Mannanases aid in the breakdown of mannans units, making them useful for a variety of processes, including the creation of fruit juice, fuel, and a decrease in the viscosity of coffee extract (Singh et al., 2018).

1.4.1.2.2 Xylans

Xylans are a type of polysaccharide polymer made up of many monomeric xylem molecules. The main chains of xylans comprise D-xylopyranose, to which the monomeric molecule is joined by a 1,4 link (Srivastava et al., 2021). Numerous small chains including mannose, rhamnose, xylose, arabinose, or 4-o-methyl glucuronic acid can be found throughout the linear polymeric chain. To increase the solubility of xylans in water, the polymerization level of the unit might be decreased (Mishra et al., 2022). According to the structure, xylans are a common type of hemicellulose found in the hardwood of the plant and as a large component of the remaining harvests. Every living creature relies on xylans for daily survival, and their presence has a direct impact on both the quality of grain flour and the firmness of the dough. It also contributes significantly to the medical field in areas like preventing dental caries prevention and as a sweetener for diabetes patients. Application of the Xylanase enzyme results in the formation of Xylooligosaccharides, which boosts biofuel generation (Srivastava et al., 2021).

1.4.1.2.3 Galactans

The form of galactans known as arabinogalactans is more frequently found in larch trees than in other types of plants, in contrast to the polysaccharide molecules in the groups. Galactans are composed of a long polymeric chain of galactose connected by 1, 3, and 16 links (Srivastava et al., 2021). This polymeric chain has a long, straight chain to which 3-D-galactopyranosyl and 4-galactopyranosyl have been alternately linked. In algae, seeds, and some varieties of buds and flowers, these molecules contain a polysaccharide structure. Some of the most popular galactan varieties include those that have been isolated from seeds such as yellow lupin, larch, algae, and other kinds of seeds (Srivastava et al., 2021). Galactans are used in a broad variety of sectors because they aid in the formation of cheese's texture and the stabilization of dairy products' viscosity. Additionally, toothpaste contains them. As a stabilizing, thickening, and gelling agent in addition to these applications, it makes a substantial contribution to the pharmaceutical business (Li & Wu, 2020; Srivastava et al., 2021).

1.4.1.2.4 Chitin and Peptidoglycan

A type of hemicellulose known as chitin has a lengthy polysaccharide chain made up of many monomeric N-acetylglucosamine units joined by –1, 4 links (Srivastava et al., 2021). This bond is similar to the linkages that bind many glucose units together in the cellulose molecule. In contrast to chitin, which has an acetyl amine group at the C2 position, cellulose molecules have hydroxyl groups at the C2 position (Thakur & Thakur, 2014). Peptidoglycan polymer, which is readily available and also creates a thin layer on bacteria's cell walls, is what gives cells their stiffness. This lengthy polymeric link comprises many N-acetylglucosamine and N-acetylmuramic subunits that are bonded together to create peptidoglycan polymer, also known as glycan (Srivastava et al., 2021).

1.4.1.3 Lignin

Lignin is the third significant lignocellulosic component of biomass. According to Figure 1.3b, it consists of multiple phenylpropane units, notably guaiacyl, syringyl, and p-hydroxy-phenyls, which are connected by C-O and C-C bonds to form a branching chain, conjugated, three-dimensional aromatic polymer. Lignin, which exists between the outer layers of the fibers, binds the fibrous cellulosic elements together and provides structural rigidity. It has a high molecular weight, constitutes about 20–35 wt.% of biomass, and is amorphous (Motta et al., 2013). Lignin helps to bind, consolidate, and organize with the fibers to increase the wood's resilience and compactness as a protective layer for lignocellulosic biomass. It stops cellulose and hemicellulose from being harmed by foreign microbes or activities (He et al., 2020). To extract the cellulose and hemicelluloses from lignocellulosic biomass, it is crucial to remove the lignin. Three different aromatic hydroxycinnamyl alcohol structures, p-sinapyl, coumaryl, and coniferyl alcohol, are present in the lignin molecules, depending on the degree of methoxylation. Lignocellulosic biomass must undergo several pretreatment procedures, including boiling, heating, pressurizing, and biological degradation, to remove the lignin (Tan et al., 2020). Pretreatment can often be accomplished through biological, physical, or chemical methods. Each method has advantages and disadvantages. Pretreatment in physical methods can be carried out through operations like grinding, milling, or pressurizing. The primary drawbacks of physical pretreatment methods are their high operating costs and energy requirements. Enhanced lignin solubilization and cellulose degradation occur in the chemical pretreatment process. The pretreatment process for this method includes using alkalis, a steam explosion, oxidation, ozonolysis, and acids as well as an ionic liquid. These actions boost the effectiveness of pretreatment procedures, but they negatively influence the environment, lowering their usefulness (Mishra et al., 2022). In the same way, various fungi, such as white rot or brown rot, that alter lignin's structure, are crucial for the biological pretreatment procedure. These procedures increase energy output, cut back on chemical consumption, and cut costs. The major disadvantage of this strategy is that it requires more time and involves slower hydrolysis processes than other pretreatment techniques. As a result, physical and chemical methods are usually preferred. Under mild weather conditions, Yan et al. (2020) executed a pretreatment experiment on grass waste using diluted NaOH and H_2O_2 (Yan et al., 2020). The holocellulose recovery was found to be around 73.80%, and the elimination of about 73.20% of lignin by this approach was also discovered (Yan et al., 2020). A change was made to the standard alkaline H_2O_2 pretreatment method by Huang et al. (2020). This method enhances the elimination of lignin from 74.90% to 80.0% at 100°C by introducing ethanol into the solution. Some carbohydrates were hydrolyzed throughout the process and later recovered, with hemicelluloses accounting for 67.60% of the total and glucan for 83.30% (Huang et al., 2020). In an experiment conducted by Sheng et al. (2021), the impact of ascorbic acid was noted on wheat straw, corn stover, and a corncob. It was found that these weak, diluted acids could increase hydrolysis by 12.47%, 18.78%, and 13.57% (Sheng et al., 2021). Table 1.3 shows the biochemical composition of a variety of biomass.

1.5 CONCLUSIONS

This study comprehensively investigated the composition, categorization, and application of lignocellulosic biomass in the production of bioenergy. It is possible to successfully convert various chemical and biological components contained in biomass into products with added value. Among these biological components are lignin, cellulose, hemicelluloses, fat, starch, water-soluble sugar, amino acids, and a few other intricate compounds. Additionally, the combination of mannans, xylans, arabinogalactans, and galactans results in the hemicellulose structure. According to their uses and potential futures, biomass can be divided into various sections. However, there is no set standard or method for classifying biomasses; therefore, we can divide them into many categories according to their quality, composition, use, and presence. Bioprocessing used in producing liquid and gaseous biofuels generated from lignocellulosic biomass in the energy sector may show possible environmental influence.

ACKNOWLEDGMENTS

The author would like to thank the Analytic Laboratory, Department of Chemical Engineering, M.S. Ramaiah Institute of Technology Bangalore, for providing all the facilities.

FUNDING

The authors have no relevant financial or non-financial interests to disclose.

AVAILABILITY OF DATA AND MATERIALS

The datasets generated during and/or analyzed during the current study are available from the corresponding author upon reasonable request.

AUTHOR CONTRIBUTIONS

Bhavana Gariya, Priyanka S., and **Devanshu Dhir** worked on visualization, review, and data collection; **Ranjeet Kumar Mishra** worked on conceptualization, data curation, investigation, experimentation, visualization, original draft writing.

DECLARATION OF COMPETING INTEREST

The authors announce that they have no known competing financial interests or personal relationships that could have appeared to influence the work testified in this paper.

REFERENCES

Association, W.B. 2016. *Global Biomass Potential towards 2035*. World Bioenergy Association: Stockholm, Sweden.

Blomberg, T. 2007. Free alkali-index for optimizing the fuel mixture in biomass co-firing. *Heat Exchanger Fouling and Cleaning*, **7**, 14.

Braga, R.M., Costa, T.R., Freitas, J.C., Barros, J.M., Melo, D.M., Melo, M.A. 2014. Pyrolysis kinetics of elephant grass pretreated biomasses. *Journal of Thermal Analysis and Calorimetry*, **117**, 1341–1348.

David, K., Ragauskas, A.J. 2010. Switchgrass as an energy crop for biofuel production: a review of its ligno-cellulosic chemical properties. *Energy & Environmental Science*, **3**(9), 1182–1190.

García-Aparicio, M.P., Ballesteros, M., Manzanares, P., Ballesteros, I., González, A., José Negro, M. 2007. Xylanase contribution to the efficiency of cellulose enzymatic hydrolysis of barley straw. *Applied Biochemistry and Biotechnology*, **137**, 353–365.

Gholizadeh, M., Hu, X., Liu, Q. 2019. Progress of using biochar as a catalyst in thermal conversion of biomass. *Reviews in Chemical Engineering*, **1**, 37.

Gollakota, A., Kishore, N., Gu, S. 2018. A review on hydrothermal liquefaction of biomass. *Renewable and Sustainable Energy Reviews*, **81**, 1378–1392.

He, J., Huang, C., Lai, C., Huang, C., Li, M., Pu, Y., Ragauskas, A.J., Yong, Q. 2020. The effect of lignin degradation products on the generation of pseudo-lignin during dilute acid pretreatment. *Industrial Crops and Products*, **146**, 112205.

Huang, C., Fang, G., Yu, L., Zhou, Y., Meng, X., Deng, Y., Shen, K., Ragauskas, A.J. 2020. Maximizing enzymatic hydrolysis efficiency of bamboo with a mild ethanol-assistant alkaline peroxide pretreatment. *Bioresource Technology*, **299**, 122568.

Kaushik, A., Singh, M. 2011. Isolation and characterization of cellulose nanofibrils from wheat straw using steam explosion coupled with high shear homogenization. *Carbohydrate Research*, **346**(1), 76–85.

Kim, J.-H., Jung, S., Park, Y.-K., Kwon, E.E. 2020. CO_2-cofed catalytic pyrolysis of tea waste over Ni/SiO2 for the enhanced formation of syngas. *Journal of Hazardous Materials*, **396**, 122637.

Kumar, A., Kumar, N., Baredar, P., Shukla, A. 2015. A review on biomass energy resources, potential, conversion and policy in India. *Renewable and Sustainable Energy Reviews*, **45**, 530–539.

Kumar Mishra, R., Mohanty, K. 2020. Effect of low-cost catalysts on yield and properties of fuel from waste biomass for hydrocarbon-rich oil production. *Materials Science for Energy Technologies*, **3**, 526–535.

Kumar, A., Mishra, R.K. 2022. Pyrolysis of low-value waste miscanthus grass: physicochemical characterization, pyrolysis kinetics, and characterization of pyrolytic end products. *Process Safety and Environmental Protection*, **163**, 68–81.

Li, R., Wu, G. 2020. Preparation of polysaccharide-based hydrogels via radiation technique. *Hydrogels Based on Natural Polymers*, **2020**, 119–148.

Li, F., Xu, M., Wang, T., Fang, Y., Ma, M. 2015. An investigation on the fusibility characteristics of low-rank coals and biomass mixtures. *Fuel*, **158**, 884–890.

Martins, M.T.B., de Souza, W.R., da Cunha, B.A.D.B., Basso, M.F., de Oliveira, N.G., Vinecky, F., Martins, P.K., de Oliveira, P.A., Arenque-Musa, B.C., de Souza, A.P. 2016. Characterization of sugarcane (Saccharum spp.) leaf senescence: implications for biofuel production. *Biotechnology for Biofuels*, **9**(1), 1–17.

Messineo, A., Volpe, R., Asdrubali, F. 2012. Evaluation of net energy obtainable from combustion of stabilised olive mill by-products. *Energies*, **5**(5), 1384–1397.

Miranda Neto, M. 2018. Desenvolvimento de processo hidrotérmico e enzimático para a obtenção de açúcares redutores a partir da palha de arroz-BRS AG.

Mishra, R.K. 2022. Pyrolysis of low-value waste switchgrass: physicochemical characterization, kinetic investigation, and online characterization of hot pyrolysis vapours. *Bioresource Technology*, **347**, 126720.

Mishra, R.K., Mohanty, K. 2018a. Characterization of non-edible lignocellulosic biomass in terms of their candidacy towards alternative renewable fuels. *Biomass Conversion and Biorefinery*, **8**, 799–812.

Mishra, R.K., Mohanty, K. 2018b. Thermocatalytic conversion of non-edible Neem seeds towards clean fuel and chemicals. *Journal of Analytical and Applied Pyrolysis*, **134**, 83–92.

Misra, M.K., Ragland, K.W., Baker, A.J. 1993. Wood ash composition as a function of furnace temperature. *Biomass and Bioenergy*, **4**(2), 103–116.

Mishra, R.K., Mohanty, K., Wang, X. 2020. Pyrolysis kinetic behavior and Py-GC-MS analysis of waste dahlia flowers into renewable fuel and value-added chemicals. *Fuel*, **260**, 116338.

Mishra, R.K., Kumar, P., Mohanty, K. 2022. Hydrothermal liquefaction of biomass for bio-crude production: a review on feedstocks, chemical compositions, operating parameters, reaction kinetics, techno-economic study, and life cycle assessment. *Fuel*, **316**, 123377.

Motta, F., Andrade, C., Santana, M. 2013. A review of xylanase production by the fermentation of xylan: classification, characterization and applications. *Sustainable Degradation of Lignocellulosic Biomass-Techniques, Applications and Commercialization*, **1**, 251–276.

Naidjonoka, P., Hernandez, M.A., Pálsson, G.K., Heinrich, F., Stålbrand, H., Nylander, T. 2020. On the interaction of softwood hemicellulose with cellulose surfaces in relation to molecular structure and physicochemical properties of hemicellulose. *Soft Matter*, **16**(30), 7063–7076.

Niu, Y., Tan, H., Wang, X., Liu, Z., Liu, H., Liu, Y., Xu, T. 2010. Study on fusion characteristics of biomass ash. *Bioresource Technology*, **101**(23), 9373–9381.

Nunes, L., Matias, J., Catalão, J. 2016. Biomass combustion systems: a review on the physical and chemical properties of the ashes. *Renewable and Sustainable Energy Reviews*, **53**, 235–242.

Odelade, K.A., Babalola, O.O. 2019. Bacteria, fungi and archaea domains in rhizospheric soil and their effects in enhancing agricultural productivity. *International Journal of Environmental Research and Public Health*, **16**(20), 3873.

Passoth, V., Sandgren, M. 2019. Biofuel production from straw hydrolysates: current achievements and perspectives. *Applied Microbiology and Biotechnology*, **103**, 5105–5116.

Perea-Moreno, A.-J., Perea-Moreno, M.-Á., Dorado, M.P., Manzano-Agugliaro, F. 2018. Mango stone properties as biofuel and its potential for reducing CO_2 emissions. *Journal of Cleaner Production*, **190**, 53–62.

Praspaliauskas, M., Pedišius, N., Čepauskienė, D., Valantinavičius, M. 2020. Study of chemical composition of agricultural residues from various agro-mass types. *Biomass Conversion and Biorefinery*, **10**, 937–948.

Raj, T., Kapoor, M., Gaur, R., Christopher, J., Lamba, B., Tuli, D.K., Kumar, R. 2015. Physical and chemical characterization of various Indian agriculture residues for biofuels production. *Energy & Fuels*, **29**(5), 3111–3118.

Ramage, M.H., Burridge, H., Busse-Wicher, M., Fereday, G., Reynolds, T., Shah, D.U., Wu, G., Yu, L., Fleming, P., Densley-Tingley, D. 2017. The wood from the trees: the use of timber in construction. *Renewable and Sustainable Energy Reviews*, **68**, 333–359.

Rodionova, M.V., Poudyal, R.S., Tiwari, I., Voloshin, R.A., Zharmukhamedov, S.K., Nam, H.G., Zayadan, B.K., Bruce, B.D., Hou, H.J., Allakhverdiev, S.I. 2017. Biofuel production: challenges and opportunities. *International Journal of Hydrogen Energy*, **42**(12), 8450–8461.

Santos, R.B., Lee, J.M., Jameel, H., Chang, H.-M., Lucia, L.A. 2012. Effects of hardwood structural and chemical characteristics on enzymatic hydrolysis for biofuel production. *Bioresource Technology*, **110**, 232–238.

Sheng, Y., Tan, X., Gu, Y., Zhou, X., Tu, M., Xu, Y. 2021. Effect of ascorbic acid assisted dilute acid pretreatment on lignin removal and enzyme digestibility of agricultural residues. *Renewable Energy*, **163**, 732–739.

Singh, S., Singh, G., Arya, S.K. 2018. Mannans: an overview of properties and application in food products. *International Journal of Biological Macromolecules*, **119**, 79–95.

Sravan, J.S., Butti, S.K., Sarkar, O., Krishna, K.V., Mohan, S.V. 2018. Electrofermentation of food waste-regulating acidogenesis towards enhanced volatile fatty acids production. *Chemical Engineering Journal*, **334**, 1709–1718.

Srivastava, N., Shrivastav, A., Singh, R., Abohashrh, M., Srivastava, K., Irfan, S., Srivastava, M., Mishra, P., Gupta, V.K., Thakur, V.K. 2021. Advances in the structural composition of biomass: fundamental and bioenergy applications. *Journal of Renewable Materials*, **9**(4), 615–636.

Striūgas, N., Sadeckas, M., Paulauskas, R. 2019. Investigation of K*, Na* and Ca* flame emission during single biomass particle combustion. *Combustion Science and Technology*, **191**(1), 151–162.

Tan, B., Yin, R., Yang, W., Zhang, J., Xu, Z., Liu, Y., He, S., Zhou, W., Zhang, L., Li, H. 2020. Soil fauna show different degradation patterns of lignin and cellulose along an elevational gradient. *Applied Soil Ecology*, **155**, 103673.

Thakur, V.K., Thakur, M.K. 2014. Recent advances in graft copolymerization and applications of chitosan: a review. *ACS Sustainable Chemistry & Engineering*, **2**(12), 2637–2652.

Thomas, P., Lai, C.W., Johan, M.R.B. 2019. Recent developments in biomass-derived carbon as a potential sustainable material for super-capacitor-based energy storage and environmental applications. *Journal of Analytical and Applied Pyrolysis*, **140**, 54–85.

Uyan, M., Alptekin, F.M., Cebi, D., Celiktas, M.S. 2020. Bioconversion of hazelnut shell using near critical water pretreatment for second generation biofuel production. *Fuel*, **273**, 117641.

van der Weijde, T., Alvim Kamei, C.L., Torres, A.F., Vermerris, W., Dolstra, O., Visser, R.G., Trindade, L.M. 2013. The potential of C4 grasses for cellulosic biofuel production. *Frontiers in Plant Science*, **4**, 107.

Vassilev, S.V., Baxter, D., Andersen, L.K., Vassileva, C.G. 2010. An overview of the chemical composition of biomass. *Fuel*, **89**(5), 913–933.

Vassilev, S.V., Baxter, D., Andersen, L.K., Vassileva, C.G. 2013a. An overview of the composition and application of biomass ash. Part 1. Phase-mineral and chemical composition and classification. *Fuel*, **105**, 40–76.

Vassilev, S.V., Baxter, D., Vassileva, C.G. 2013b. An overview of the behaviour of biomass during combustion: Part I. Phase-mineral transformations of organic and inorganic matter. *Fuel*, **112**, 391–449.

Williams, O., Taylor, S., Lester, E., Kingman, S., Giddings, D., Eastwick, C. 2018. Applicability of mechanical tests for biomass pellet characterisation for bioenergy applications. *Materials*, **11**(8), 1329.

Xie, S., Lawlor, P.G., Frost, J., Hu, Z., Zhan, X. 2011. Effect of pig manure to grass silage ratio on methane production in batch anaerobic co-digestion of concentrated pig manure and grass silage. *Bioresource Technology*, **102**(10), 5728–5733.

Yahya, M.A., Al-Qodah, Z., Ngah, C.Z. 2015. Agricultural bio-waste materials as potential sustainable precursors used for activated carbon production: a review. *Renewable and Sustainable Energy Reviews*, **46**, 218–235.

Yan, X., Cheng, J.-R., Wang, Y.-T., Zhu, M.-J. 2020. Enhanced lignin removal and enzymolysis efficiency of grass waste by hydrogen peroxide synergized dilute alkali pretreatment. *Bioresource Technology*, **301**, 122756.

Yigezu, Z.D., Muthukumar, K. 2015. Biofuel production by catalytic cracking of sunflower oil using vanadium pentoxide. *Journal of Analytical and Applied Pyrolysis*, **112**, 341–347.

Zhang, P., Dong, S.-J., M*a, H.-H., Zhang, B.-X., Wang, Y.-F., Hu, X.-M. 2015. Fractionation of corn stover into cellulose, hemicellulose and lignin using a series of ionic liquids. *Industrial Crops and Products*, **76**, 688–696.

Zhu, Y., Niu, Y., Tan, H., Wang, X. 2014. Short review on the origin and countermeasure of biomass slagging in grate furnace. *Frontiers in Energy Research*, **2**, 7.

2 Application of Different Physical and Chemical Methodologies for the Synthesis of Different Types of Value-Added Materials

Dr. Md. Merajul Islam and Dr. Amina Nafees

2.1 INTRODUCTION

The over-reliance on fossil fuels has resulted in an increase in worldwide concerns regarding energy and the environment, which in turn has led to an increased passion in the usage of green and sustainable energy sources. The transformation of biomass into a compound with additional value, liquid biofuels, and carbon-based biomaterials has garnered an increasing amount of interest in recent years. There is only biomass that is a sustainable and renewable resource of organic carbon that may be encountered in nature (Corma et al., 2007; Huber et al., 2006; Zhang et al., 2017; Zhou et al., 2011). With the help of technological and scientific advances, we will learn more about the complex and varied characteristics of biomass, which will increase our interest in converting natural assets into fuels and additional value items such as fuels, materials, chemicals, and by-products. This is because technological and scientific advances are enhancing our comprehension of how biomass can be broken down into different components. In addition, advancements are being made not only in the catalysts themselves but also in the methods that make use of these catalysts. As a consequence of this, the capacity to produce a diverse assortment of goods is being opened up. Some instances include the manufacturing of cellulosic nanomaterials via the astute implementation of chosen enzymes (de Aguiar et al., 2020; Squinca et al., 2020), the creation of sugars that are fermentable, and the creation of biofuels with a small carbon impact via either bioconversion or thermochemical handling.

Biomass sources possess an opportunity to be utilised in the manufacturing of a vast assortment of products with additional value, including biofuels (such as ethanol and hydrogen), the bioproducts (such as sugar and sugar alcohols), and vital industrial chemicals (such as solvents) (Clark & Deswarte, 2015). Transformation can be

 DOI: 10.1201/9781003407713-2

accomplished through the use of a wide range of processes, including chemical, biochemical, and thermochemical operations. When it comes to producing a high quantity of a particular good, each approach has a unique set of benefits and drawbacks.

Biomass can indeed be made from discarded wood materials such as by-products of trees, shrubs, and other types of energy sources such as kenaf, sorghum, switchgrass, miscanthus, sugarcane, and corn, as well as any leftovers from agricultural production such as wheat straw, corn stover, and other similar materials. It is possible to convert biomass into usable energy. It is possible for the various ingredients of biomass, including lignin, hemicellulose, and cellulose, to influence the transformation processes that are used in the manufacturing of valuable goods.

2.2 TOP BIOMASS-BASED VALUE-ADDED MATERIALS

2.2.1 Biofuels

Ethanol is an environmentally friendly propellant that has the potential to yield a wide diversification of plant and animal resources containing carbohydrates, such as maize, sugarcane, and sweet sorghum. Corn is extensively used throughout the ethanol fuel industry in the United States. The fermentation of carbohydrates leads to the production of a diluted alcoholic beverage fluid, which is then proceeded by distillation and extraction of water (or dehydration) to generate ethanol of propellant grade. The production process begins with the sugars. To facilitate its use in automobiles, gasoline can contain varying percentages of ethanol blends (E10, E15, and E85). When compared to fuels made from petroleum, it was found that using mixtures containing 10% ethanol (E10) results in a reduction of greenhouse gas emissions of between 12% and 19% (Saini et al., 2015).

Butanol is another type of sustainable fuel that can be produced from biomass by fermenting biomass feedstocks with alcohol (Hansen et al., 2009). Most of the time, fossil fuels are used to produce butanol. However, it is also possible to produce butanol from plants. Because of its longer carbon chain, butanol is a superior gasoline additive than ethanol. As a result, the hydrocarbon chain is consequently less polar. The amount of energy contained in butanol is comparable to that of petroleum (Jin et al., 2011). However, in contrast to ethanol, it is not considered to be a commercially feasible biofuel due to the constraints and challenges associated with its production. The ABE reaction (acetone, butanol, and ethanol) is what gives rise to biobutanol. It is a procedure of fermentation utilising a variety of substrates. The ABE fermentation process has a number of drawbacks and challenges, such as reduced butanol levels, reduced butanol output, extremely expensive substrates (such as molasses and grain), end-product blockage, and expensive product recovery by distillation (Sabra et al., 2014).

Methanol is yet another type of alcohol that can be utilised as a fuel due to its high-octane rating, ease of distribution, and high stability. Ethanol and methanol are both used as fuels and have chemical and physical properties that are comparable. Methanol is generated in an array of ways, the most common of which is through the transformation of syngas, which is obtained from fossil fuels, via catalysis. However, lignocellulosic biomass substances can also be utilised to produce methanol (Yin et al., 2005). Methanol can be used instantly as propellant, as propellant using fuel cells, and in the role of raw material for the manufacture of methyl tertiary butyl ether, which is a supplement for gasoline.

Moreover, the lignocellulosic fraction of biomass is not needed for the manufacture of biodiesel, which is another most extensively used liquid biofuel. Through a process known as transesterification, biodiesel can be manufactured using either vegetable oil or fat from animals, as well as in the capacity of both an alcohol and a catalyst.

The process of gasification of biomass substances leads to the creation of a gaseous mixture known as producer gas. It is made up of CO, N_2, CO_2, and compounds with a small molecular weight like methyl chloride. Producer gas has multiple applications, including its use as a gaseous fuel for the generation of heat or electricity (Wang et al., 2008). Syngas, alternatively referred to as synthesis gas, is a composite mixture of carbon monoxide and hydrogen. Synthesis gas has the potential to take the place of organic gas as a fuel that is superior in terms of its thermal efficiency. The power that is obtained from the combustion of syngas can be converted into electrical energy. It is also possible to employ it as a source of propellant or as an intermediary during the manufacturing of other types of chemicals (De María et al., 2013; Khandan et al., 2012).

Many people have high hopes for hydrogen as a potential gas fuel. In comparison to other types of traditional fuels, hydrogen has the highest specific calorific value. The primary application for its use as fuel is in fuel cells. Due to their increased fuel economy and lack of carbon emissions, fuel cells are currently being regarded as a potentially feasible choice for the generation of electricity (Dresselhaus et al., 2001). Pure water is the only waste or by-product produced, in contrast to the tremendous quantities of carbon dioxide that hydrocarbon fuels generate, which is a greenhouse gas. The concern of developing methods for the manufacture of hydrocarbons that are less costly and more efficient is a major challenge that restricts their use. There is reason to be optimistic about the aqueous-phase reforming (APR) technique as a means of producing large yields of H_2 gas (Meryemoglu et al., 2010; Cortright et al., 2002). This method is one of the many different conversion processes that are currently in use.

2.2.2 Industrial Importance of Value-Added Materials

Inedible lignocellulosic biomass products are garnering a growing amount of interest as a renewable resource that is also relatively cheap and readily available. This is done in an effort to lessen reliance on petroleum resources and reduce the expenses of energy and raw material. In addition to being used to generate energy and propellant, biomass can also be converted into valuable compounds and components that are based on carbon. These are referred to as bioproducts. These goods include furfurals, cellulose fibre, sugars and sugar alcohols, as well as their substitutes, bioplastics, glycerine, resins, carbonaceous materials, and a variety of other substances.

One of the organic precursors to furan-based compounds is called furfural. It is regarded an essential component in the production of chemicals that are not extracted from fossil fuel, the innovative creation of bioplastics, and possible biofuels or fuel additives. Studies on furfural and its analogues have led to the creation of a novel source of biofuels and bioplastics, as well as the synthesis of range alkanes for aircraft and gasoline (Huber et al., 2006; Xing et al., 2010). The biomass-based production of furfural can be visualised in Figure 2.1 (Cañada-Barcala et al., 2021). Both 5-hydroxymethylfurfural (or HMF) and furfural, which are products of the removal of water (dehydration) from C_5 and C_6 sugars (carbohydrates), are considered to be pivotal system molecules for the fabrication of value-added pharmaceutical and industrial substances, as well as highly customisable intermediate products in the process of converting biomass (Hu et al., 2017). HMF is able to generate high biofuels like 2,5-dimethylfuran (DMF), 2,5-dimethyltetrahydrofuran (DMTHF), and 5-ethoxymethylfurfural (EMF), as well as the development of high chemical products like 2,5-diformylfuran (DFF), levulinic acid (LA), and 2,5-dihydroxymethylfuran (DHMF), thanks to its marvellous and reactive configuration (Hu et al., 2018) Figure 2.2.

Activated carbons, prepared from lignocellulosic biomass, have been utilised in a variety of applications, including wastewater treatment (as an adsorbent and filter) (Baccar et al., 2009), catalyst (Zhou et al., 2015), catalyst support (Tsyntsarski et al., 2015), storage material (Ramesh et al., 2015), and others. Mesoporous carbons are utilised in a wide variety of applications, such as, but not restricted to: adsorbents, membranes, supercapacitors, chemical sensors, and catalyst supports (Saha et al., 2010; Oh et al., 1999). Lignin can be used to make mesoporous carbons, which can then have their mesoporosity increased using either physical or chemical activation techniques.

Sugar alcohols are materials that are extremely valuable for the food sector. As an example, xylitol is a pentose sugar alcohol that is utilised in the food industry as a sugar substitute because it contains fewer calories than sugar and is known to inhibit

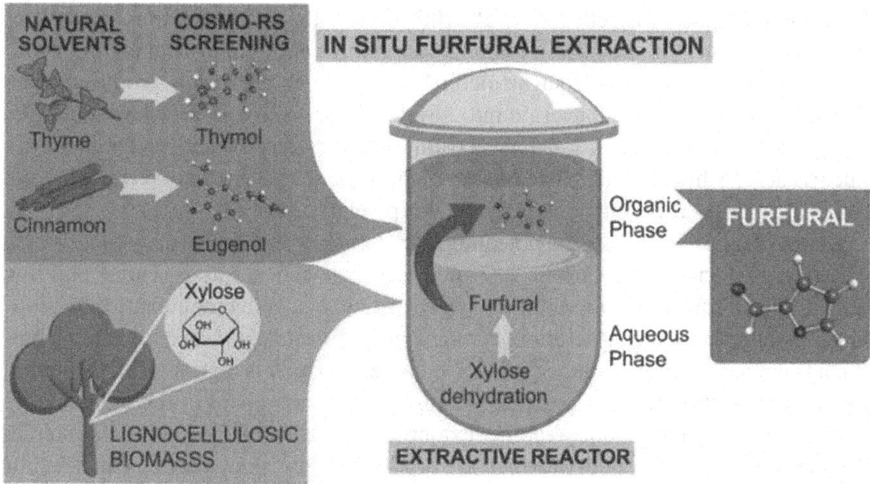

FIGURE 2.1 Schematic of production of furfural from biomass (Cañada-Barcala et al., 2021).

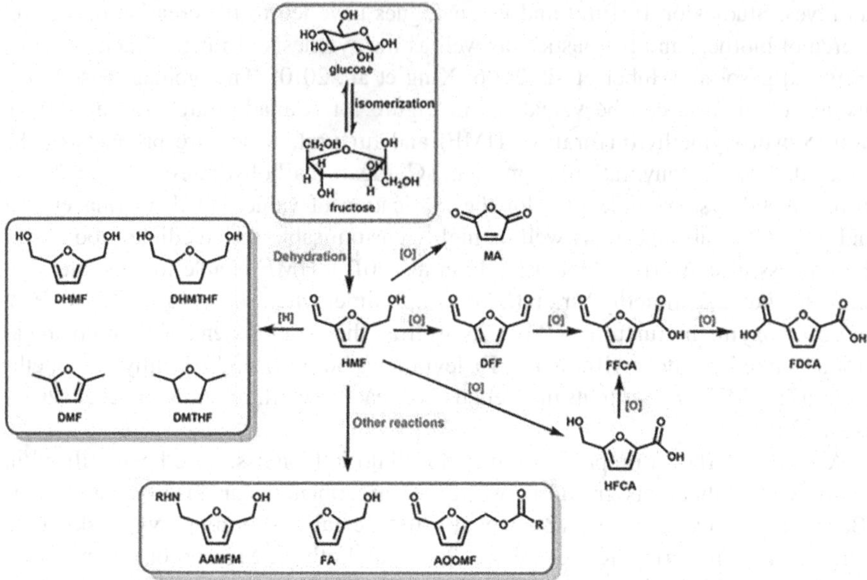

FIGURE 2.2 The catalytic conversion of hydroxymethylfurfural (HMF) into many useful compounds. In the study conducted by Hu et al. (2018), various compounds were investigated, including DFF (2,5-diformylfuran), FFCA (5-formyl-2-furancarboxylic acid), HFCA (5-hydroxymethyl-2-furancarboxylic acid), MA (maleic anhydride), FDCA (2,5-furandicarboxylic acid), DHMF (2,5-dihydroxymethylfuran), DHMTHF (2,5-dihydroxymeth yltetrahydrofuran), DMTHF (2,5-dimethyltetrahydrofuran), DMF (2,5-dimethylfuran), AAMFM (5-arylaminomethyl-2-furanmethanol), FA (furfuryl alcohol), and AOOMF (5-alkanoyloxymethylfurfural).

the growth of cancerous cells (Ko et al., 2006). More than that, xylitol is suitable for use as a component in the generation of a vast assortment of industrial compounds.

Following the addition of sugar alcohols, framework chemicals produced from C_5 and C_6 sugar could be converted into novel categories of valuable substances like 2,5-furan dicarboxylic acid, 1,4-diacids (fumaric acid, succinic acid, and malic acid), glucaric acid, 3-hydroxybutyrolactone, LA, glutamic acid, 3-hydroxy propionic acid, glycerol, aspartic acid, and itaconic acid (Werpy et al., 2004).

Biomass feedstocks have the potential to undergo a metamorphosis into a diverse spectrum of forms with various intermediates that can act as essential building blocks. These find applications in a variety of fields, such as manufacturing, shipping, fabric production, agriculture, ecology, and real estate. One example of a component that serves in this capacity is 1,3-butadiene, which is a structural element in the manufacture of rubbers. Rubbers, in turn, are a component in the manufacture of tyres for small automobile. Another example is the alcohol and lactic acid interaction that produces ester, both of which are produced using plant matter, which results in the production of the biodegradable solvent known as ethyl lactate. It is utilised in industrial applications to serve as a replacement for volatile organic compounds obtained from petroleum. Lactic acid, which would be generated almost exclusively through the fermentation of sugars and starches by microorganisms, features a broad variety of utilisation, including those in the food production industry, the polymer industry, the pharmaceutical industry, and other areas. Succinic acid, a dicarboxylic acid, is used as a predecessor in the generation of valuable items like industrial polymers, chemicals, solvents, and surfactants.

Polymers made from lignocellulose have recently come under scrutiny for their potential to contribute to both environmental and sustainability issues. Using a variety of distinct reaction pathways, one can produce a broad variety of polymers by starting with biomass derivatives (Werpy et al., 2004). For example, C_5 and C_6 carbohydrates, as well as their analogues, can either be integrated into the polymeric matrix or utilised as terminal groups in order to create glycopolymers that imitate both in terms of the structure and function of glycoproteins (Godula et al., 2010).

2.3 YIELD-AFFECTING PARAMETERS

2.3.1 Pretreatments

The intricate and rigid frameworks of biomass materials restrict their utilisation to cover a wide variety of uses, regardless of their potential. To transform biomass substances into a variety of products in an effective way, the biomass must first be cracked into smaller pieces with lower molecular weights (e.g., oligosaccharides and monosaccharides). The reason for performing pretreatment is to prepare the cellulose to undergo the hydrolysis process, which is necessary for its subsequent transformation into fuels or products with added value. In terms of physical and chemical makeup, lignocellulosic biomass is altered through the application of many approaches to pretreatment, and ultimately, the consequence is a rise in the rates of hydrolysis. By modifying the essential characteristics of the biomass, such as removing lignin and lowering the cellulose's crystalline structure, pretreatment allows for the ability to dismantle the biomass. This is accomplished by increasing the porosity of the biomass. To successfully generate biofuels and other bioproducts from lignocellulosic biomass, certain pretreatment and deconstruction methods must be utilised, and the properties of the biomass, both physically and chemically, must also be taken into consideration. For the production of carbohydrates with a lower molecular weight from feedstock in order to generate a variety of biofuels and bioproducts as well as other valuable goods, it is obligatory to utilise an efficient pretreatment method, followed by solubilisation in aqueous media. This must be done without the use of toxic and hazardous chemicals. Some pretreatment procedures are outlined in the subsequent sub-sections.

2.3.1.1 Alkali Pretreatments

For the purpose of preparing biomass for further processing, alkali pretreatment entails the utilisation of bases like hydroxides of Na, K, Ca, and NH_4^+ ions. In this particular context, sodium hydroxide is the base that is most frequently utilised. Although alkali retreatment operates best at ambient temperatures, the time required for the process can range from hours to days instead of seconds or minutes (Kumar et al., 2009). Lignin's composition is altered by alkali pretreatment, which also results in the cellulose going through a process of incomplete recrystallisation, the removal of acetyl and different modifications of uronic acid found on hemicellulose, and an increase in the ease with which enzymes can handle both cellulose and hemicellulose (Chang et al., 2000). Before enzymatic hydrolysis can begin, a neutralising step must take place, during which lignin and hindrances (such as phenolic acids, salts, aldehydes, and furfural) are eliminated. Alkaline reactions lead to a decrease in sugar breakdown when compared to acidic reactions, and a greater proportion of the caustic salts can be reclaimed and/or generated new ones (Kumar et al., 2009).

2.3.1.2 Acid Pretreatments

The basic concept behind acid pretreatment is to incorporate feedstock that has been concentrated in the hydrolysis section; however, the application of an acid with a significant concentration can be considered a treatment rather than a pretreatment. The acid that is most commonly utilised in acid pretreatment is diluted sulphuric acid (H_2SO_4). It is also possible to make use of various additional acids like HCl, H_3PO_4, and HNO_3. For the pretreatment of organic acids like fumaric acid and maleic acid, dilute acids can also be used as a substitute for these inorganic acids (Kootstra et al., 2009). The amount of time required for the application could range from a matter of minutes to several hours, possibly longer based on the acid, temperature, and the concentration utilised during the process. This pretreatment method not only dissolves the hemicellulose fraction, but it also liberates cell wall components such as monomeric sugars and soluble oligomers, which are then carried over into the hydrolysate. The removal of hemicellulose results in an enhancement in porosity, which subsequently improves the digestibility of the material by enzymes. On the other hand, the sugars released when hemicellulose is broken down can even more deteriorate into furfural and hydroxymethyl or diluted acids at temperatures ranging from 130°C to 210°C. According to the information provided in furfural, both of these compounds have an inhibitory effect on the fermentation pathway.

Oxidative delignification is another important biomass pretreatment process. This includes (i) peroxidase-catalysed lignin breakdown with hydrogen peroxide; (ii) ozone-initiated breakdown of lignin (ozonolysis) that targets and fractures aromatic ring frameworks; hemicellulose and cellulose are rarely decomposed in comparison (Sun & Cheng, 2002); and (iii) the main effects of wet oxidation are breaking down the organic bonds between lignin and hemicellulose with methanol, which dissolves hemicellulose, and taking lignin out of the structure of the biomass. Decomposition of lignin results in the release of water, carboxylic acids, and carbon dioxide. In this process, potent antioxidants like furfural and 5-hydroxymethylfurfural are produced at negligible levels.

2.3.1.3 Ammonia Fibre Explosion

To prepare biomass for use in the ammonia fibre explosion (AFEX) pretreatment technique, liquid ammonia is combined with the material at temperatures of 70°C–200°C and pressures of 0.7–2.8 MPa, after which the pressure is suddenly released. There are a few main variables that affect the breakdown, including the residence time, reaction temperature, and ammonia concentration (Mosier et al., 2005). Explosive fibre fragmentation caused by pressure release causes hemicellulose, cellulose, and lignin polymers to fragment and micropore dimension and density to rise in biomass.

2.3.1.4 Steam Explosion

The steam explosion pretreatment is a commonly employed physicochemical pretreatment for lignocellulosic biomass. Saturated steam at high pressure (0.69–4.83 MPa) is applied to biomass for a certain amount of time, ranging from a few seconds to several minutes, at an elevated temperature (160°C–260°C). A violent explosion occurs in the substances due to the destruction of the fibril structure (Sun & Cheng, 2002). Afterwards, hemicellulose is broken down, some lignin is rendered soluble, and cellulose binding is lessened. The major drawbacks of the steam explosion method are the incomplete breakdown of hemicellulose and the formation of substances that are poisonous. The efficiency of fermentation processes can be negatively impacted by the presence of toxic and/or inhibitory substances (Olivia et al., 2003). Phenolic substances and other aliphatic acids, aromatics, inorganic ions, furan aldehydes, and bioalcohols or products of additional fermentations have all been suggested as potential microorganism inhibitors (Jönsson et al., 2013).

2.3.1.5 Other Methods

Moreover, there exist certain pretreatments that involve the utilisation of microwave, ultrasound, and radiofrequency radiations. These pretreatments are employed together with other methodologies. Ultrasound generates sonochemical and mechano-acoustic phenomena that exert an influence on the chemical processes and the physical structure of organic matter. According to Bussemaker et al. (2013), the application of mechano-acoustic forces induces changes in the composition of the surface of feedstock. Additionally, the sonochemical process generates oxidising radicals that exhibit effectiveness on various components of biomass. According to the research, the application of sonication to a mixture of biomass and water at a frequency of 20 kHz ± 50 Hz for a duration of 8 minutes did not result in any significant alteration to the hydrolysis process conducted in water below the critical point subsequent to sonication. The percentage of solid biomass hydrolysed exhibited little change, but the molar mass of the polysaccharide portions included within the hydrolysates experienced a significant drop due to breakdown. Sonicated hydrolysates have a high concentration of monosaccharides, which was found to be significant, demonstrating the beneficial impact of sonication on the breakdown of polysaccharides. As an example, the hydrolysis of kenaf biomass at a temperature of 250°C using subcritical water treatment led to a xylose release of at least 10% higher in the sonicated hydrolysate compared to the non-sonicated hydrolysate (Öztürk et al., 2010).

On the contrary, microwave pretreatment has been shown to accelerate enzymatic saccharification by 53% greater in comparison to traditional heating (Hu & Wen, 2008). This is because microwaves have been shown to accelerate disturbances on the surface and the breakdown of lignin frameworks in switchgrass. The pretreatment of switchgrass with microwaves produces a beneficial effect on the switchgrass dissolution in near-critical water. The proportion of biomass that hydrolyses and the amount of total organic carbons that are liberated into solution both increase when biomass is microwave processed. The amount of solubilisation produced is proportionally increased when the microwave pretreatment is carried out at a higher temperature.

2.3.2 Hydrolysis

Hydrolysis is the chemical reaction most commonly responsible for the breakdown of cellulose or any such molecules. Either the use of potentially harmful chemicals like acids or alkalis or a longer retention period due to the utilisation of a variety of enzymes is required. In general, pretreatments with an alkaline or creating the conditions environment are used to break down lignin, whereas pretreatments with mineral acids environments are used to break down hemicellulose molecules. Recovering catalyst in chemistry is an important step in ensuring the success of these operations (Alvira et al., 2010; Godula et al., 2010).

2.3.3 Biomass Chemical Composition

The manufacturing of bio-based materials is in some way influenced by the chemical makeup of the biomass that is being used. There is a large amount of variation between species when it comes to the proportions of cellulose, hemicellulose, and lignin found in agricultural leftover. Straws and leaves, for example, have a greater proportion of hemicellulosic portions, whereas hardwoods have an adequate quantity of cellulose distributed throughout their frameworks (Jørgensen et al., 2007). Biodiversity in biomass materials influences the manufacturing of bio-based goods as well as the effectiveness with which they can be solubilised (Irmak et al., 2010).

2.4 SYNTHESIS METHODS FOR VALUE-ADDED MATERIALS

The synthesis of biomass can be accomplished through a number of different methods, the most common of which are chemical, thermochemical, biochemical, and photocatalytic conversion. The stages of combustion, pyrolysis, gasification, and liquefaction are involved in the process of thermochemical conversion. Using this method, a diverse selection of items with additional value could be manufactured from biomass. For instance, sugars or carbohydrates can be reduced thermochemically to produce its alcohols, which opens the door to new applications for sugar. In contrast, the biochemical conversion utilises a variety of enzymes in order to break down the material and lead to the creation of a plethora of products (Wainaina et al., 2018). Digestion in the absence of oxygen and fermentation are two of the most promising biochemical techniques (McKendry, 2002). Hybrid conversion, as an alternative approach, presents a superior amalgamation of benefits as compared to the cumulative advantages of conventional independent conversion processes. Some of these benefits include durability, improved product specificity, rapid pyrolysis, and a decrease in the inherent resistance exhibited by biomass (Jarboe et al., 2011; Shen et al., 2015). In recent times, the photocatalytic transformation of biomass into a variety of materials with value addition has evolved as the most novel strategy (Granone et al., 2018; Liu et al., 2019). The preferential conversion of biomass through the utilisation of renewable solar energy by means of photocatalysis has garnered significant attention from an ever-expanding scholarly community. The "dream reaction" nature of photocatalysis can be attributed to its one-of-a-kind photogenerated reactive intermediates (such as h^+, e^-, $\bullet OH$, $\bullet O_2$, and O_2), its innovative response processes, and the fairly benign reaction requirements (Chen et al., 2021).

2.4.1 Chemical

With the help of a reducing agent like sodium borohydride ($NaBH_4$), sugars extracted from biomass have the potential to be hydrogenated into C5-to-6 polyols, also known as sugar alcohols. Another potential use for biomass hydrolysates is the immediate generation of derivative products like hydroxymethyl furfural, furfural, and/or LA. This is just one of many possible applications. Hydrogenolysis is another method that can be utilised to bring about the production of C2–3 glycols from cellulose hydrolysates. Using either oxidation or halogenation reactions, sugar-containing hydrolysates can be improved to a higher quality (De Wit et al., 1993). Sugars generated from the hydrolysis of biomass can be transformed into N-heterocyclic constituents, aromatics, and pyrones, all of which can then become an assortment of different chemical intermediates after being transformed.

In addition to its use in a wide variety of different chemical applications, furfural is also an important precursor in the production of a substantial number of different fine compounds and biofuels. Its industrial production necessitates the hydrolysis and dehydration of pentoses, specifically xylose, in lignocellulosic raw materials (sunflower stalk, sugarcane bagasse, maize cobs, etc.) at temperatures ranging from 153°C to 240°C (Agirrezabal-Telleria et al., 2013). Hydrolysis of hemicellulose produces xylans, which produce pentose carbohydrates, which are then converted further into furfural in the second step. The production of furfural on a commercial scale makes use of H_2SO_4 as a homogeneous catalyst. To remove the furfural and prevent the chemical from degrading any further, the procedure makes heavy use of steam in significant quantities.

At this time, xylose, found in sources like birch wood scraps and the hemicellulose hydrolysate of sugarcane bagasse, is chemically reduced to produce xylitol. The chemical process that has been tailored for the manufacturing of xylitol from xylan-rich biomass requires greater manufacturing expenses as a result of the temperature and pressure contribution, as well as the creation of substances that need costly segregation and purification stages (Rafiqul et al., 2013). Additionally, the manufacturing of xylitol from xylan-rich biomass requires a high level of contribution.

2.4.2 Thermochemical

The process of thermochemical conversion includes combustion, pyrolysis, gasification, and liquefaction in its progression.

2.4.2.1 Combustion

The transformation of plant matter into energy that involves the least amount of complexity is combustion of biomass. The temperatures of 800°C–1,000°C are ideal for the combustion process. Under conditions of complete combustion, heat is produced as a by-product of the transformation of carbonaceous and hydrogen-rich material into carbon dioxide and water via the oxidation process. It is not possible to use biomass with a high percentage of moisture in this process; in some instances, pre-drying will be required. The high-pressure steam that is generated as a by-product of the method has multiple potential applications, including the generation of hot air, hot water, steam, and warm gases from electricity. The combustion systems utilised to generate electricity and heat are very similar to those found in most power plants that are fired by fossil fuels. High levels of NO_x emissions can be produced by combustion technologies.

2.4.2.2 Pyrolysis

The term "pyrolysis" refers to the chemical reaction of thermally degrading lignocellulosic biomass around 500°C in anaerobic conditions. The final outcome is a mixture of gaseous (or biogas) and solid (or biochar) fractions, as well as liquid bio-oil (or bio-crude). Fuel can be made from these products either directly or after being processed in some way. Both biochar and bio-oil possess the ability to be utilised in the context of manufacturing of compounds and products with additional value (Yin et al., 2013). The processes of gasification and combustion are both included in pyrolysis which involves the thermal breakdown of raw biomass into volatile compounds in the absence of an oxidising substance. This process is known as thermal degradation. When subjected to conditions in which there is neither air nor oxygen present, the temperature at which the organic constituents of a substance begin to decompose in response to heat ranges from 350°C to 550°C and then rises to 700°C–800°C (Fisher et al., 2002). The amount of time needed to convert the feedstock into pyrolysis products determines how quickly or slowly the pyrolysis process can be carried out. The production of primarily biochar takes place at lower temperatures during slow pyrolysis. The method of fast pyrolysis results in the production of bio-oils at higher temperatures. During the process of slow pyrolysis, the plant matter usually gets heated using low flame temperatures (up to 10°C–20°C/min), and adequate time is provided for repolymerisation processes in order to maximise the amount of substantial yields. On the other hand, fast pyrolysis makes use of significantly greater heating rates (ranging from 10°C to 200°C/s), an elevated working temperature, and reduced vapour stay times (fewer than 2 seconds), which results in the yield of bio-oil that is between 50 and 70 wt% (or dry biomass basis). The feedstock for fast pyrolysis must typically be ground to a finer consistency (less than 1 mm), whereas slow hydrolysis can tolerate a much wider range of particle sizes (5–50 mm) (Chhiti et al., 2013).

2.4.2.3 Gasification

At temperatures between 700°C and 1,600°C, the gasification process transforms biomass into a mixture of combustion products. The generated gas has a high concentration of CO, as well as H_2, CH_4, and CO_2. After being cleaned, this combination of combustible gases can be used for a variety of uses. The pristine gas may be utilised immediately as fuel for a motor, upgraded to fluid fuels, or transformed into chemical raw materials using a variety of different techniques such as organic decomposition or catalytic enhancing using the Fischer–Tropsch process (Huber et al., 2006). The Fischer–Tropsch reaction, commonly known as the FT reaction, is a method that turns syngas into hydrocarbons, primarily for use in the production of fuel but also for use in the synthesis of other useful compounds (Jahangiri et al., 2014).

2.4.2.4 Hydrothermal Gasification

Hydrothermal gasification processes allow for the transformation of lignocellulosic biomass towards various propellants, including liquid and gaseous forms. The hydrothermal gasification processes can be carried out in the gaseous component (or steam reforming), supercritical water phase, or liquid phase (known as aqueous phase reforming). Before lignocellulosic material can be gasified through the steam reforming process, it must first be dried. This conversion operates at high temperatures, reaching at least 800°C, which leads to the generation of a remarkable amount of tar and char. Water has thermodynamic critical points at a temperature of 374.3°C and pressure of 221.2 bars. For lignocellulosic materials to be completely gasified at temperatures above 600°C, as an illustration, the temperature of the water that is supercritical must be raised to quite a bit higher than the critical point (66). Because transformation processes can happen at temperature ranges (225°C–265°C) and pressures (27–54 bars), the APR gasification method is an appealing approach. This is because a transforming catalyst is needed for the chemical reactions to take place. This makes the APR gasification method very attractive (such as precious metal, Pt). At temperatures relevant to the processing of materials, the water-gas shift reaction ($CO + H_2O = H_2 + CO_2$) is thermodynamically favourable. It is possible to successfully convert biomass hydrolysis products, which consist of blended materials produced from biomass oxygenated hydrocarbons, into gaseous substances that are abundant in hydrogen (Cortright et al., 2002; Irmak et al., 2013).

2.4.2.5 Hydrothermal Liquification

The term "hydrothermal liquefaction of feedstock" refers to the process of thermochemically converting feedstock into liquid fuels at temperatures that vary between 280°C and 370°C and pressures varying from 10 to 25 MPa (Behrendt et al., 2008). Water plays a vital role as both a reactant and a catalyst throughout this procedure. It continues to be in its liquid state and possesses a variety of interesting characteristics when subjected to the circumstances necessary for liquefaction (Behrendt et al., 2008). During this procedure, lignocellulose is converted into products with a low molecular weight that is soluble. These products then go through additional condensation, cyclisation, and polymerisation reactions, which ultimately result in the formation of oil from plants that can be dissolved in organic solutions. The most common application for these bio-oils is as a substitute for fuels made from petroleum, but they are able to be used as a biomass for the generation of an extensive variety of products that are valuable. For instance, phenolics that come from lignin can be used to hydrogenate to produce aromatic compounds, and carbohydrates fermenting them result in the production of hydrogen or are used in the production of hydrogen via a catalytic process (Wei et al., 2015).

Using thermochemical processes, it is viable to produce or synthesise lots of different kinds of value-added products from feedstock or compounds derived from biomass. By the way of illustration, the thermochemical reduction of sugars can result in the formation of sugar alcohols. In an industrial setting, the manufacturing of sorbitol by reducing carbohydrates through heat is typically executed using reactors consisting of agitated tanks at temperatures ranging from 100°C to 180°C, with 5–15 MPa of hydrogen pressure, and the presence of a catalyst, which is typically

TABLE 2.1
The Reduction of Glucose in a Thermochemical Process Leading to the Generation of Sugar Alcohols at Varying Temperatures (Irmak et al., 2015)

	Sugar Alcohols (% Concentration)				
Temperature (°C)	Glucose	Mannitol	Sorbitol	Xylitol	Other Products[b]
80	8.8	0.9	90.3	n. d.	n. d.
100	n. d.[a]	2.2	97.3	n. d.	n. d.
120	n. d.	5.9	88.6	n. d.	5.5
140	n. d.	16.5	59.7	2.7	21.1
160	n. d.	21.8	41.0	3.6	33.6
180	n. d.	23.4	20.7	4.1	51.8
200	n. d.	19.7	12.1	2.1	66.1

[a] n. d.: Not detected
[b] Glycerol, erythritol, and any other reduced products added together.

of the Raney-type nickel or ruthenium variety. In Table 2.1, we see how glucose, a prototypical component of biomass, can be converted thermochemically into sugar alcohol. This production takes place at a range of operating temperatures, under a pressure of 2.0 MPa H_2, and in the involvement of a ruthenium catalyst backed by carbon. It took 60 minutes for the reaction to take place (Irmak et al., 2015).

2.4.3 Biochemical

In addition, biochemical transformation is able to disintegrate cellulose in carbohydrates, which will be able to transform into biofuels (gaseous or liquid fuels) and the production of biological products with the help of microbes and the enzymes. Biochemical transformation additionally has the potential to degrade biomass into fuels. This method is typically utilised for treating organic wastes that contain significant amounts of moisture. The conversion of biomass into useful compounds requires a longer amount of time and is accomplished through biochemical processes, which are relatively slow (Balan et al., 2008). Digestion in the absence of oxygen and fermentation are two of the most common biochemical processes (McKendry, 2002). Biomass derived from a variety of sources is subjected to biochemical degradation through the process of anaerobic digestion in conditions that are tightly controlled and devoid of oxygen. This results in the production of biogas, which is primarily composed of methane and carbon dioxide. Burning biogas directly for the production of heat or steam and converting it into electricity are all viable options. It is also possible to convert it into a fuel based on hydrogen or upgrade it to biomethane. Microorganisms are utilised throughout the course of the anaerobic digestion procedure, which degrade carbohydrates into sugars that can then be metabolised by other bacteria. This process is called anaerobic digestion. The breakdown of these digestible components results in the production of CH_4 and CO_2.

The generation of ethanol on a large-scale commercial basis from materials rich in carbohydrates requires the utilisation of fermentation. Enzymes are utilised in the process of converting the structural carbohydrates found in biomass into sugars. Afterwards, during the fermentation process, microorganisms change the sugars that have been released into alcohols, organic acids, or hydrocarbons. In order to acquire additional value-added chemicals, the intermediate sugars can also be utilised in the production process. In this process, the conversions take place at temperatures between 25°C and 70°C and under atmospheric pressure conditions.

2.4.4 Photocatalytic

The novel integration of photocatalysis and the utilisation of biomass exemplifies a fresh and potentially fruitful strategy for achieving a higher level of viability in the science field. The utilisation of photocatalytic devices has enabled significant advancements to be made in spite of the difficult nature of a number of biological material streams, which slows down their technological treatment. Because of its intricate configuration, biomass conversion is an energy-consuming procedure that also makes it difficult to make greater yields of the substance or fuel of interest (Liu et al., 2019). For the purpose of converting biomass, there has been investigation into biological mechanisms and chemocatalytic techniques. By using light rather than heat or electricity to power chemical reactions, photocatalysis has a number of advantages. The reaction could be performed at room temperature because photons are used as the energy source rather than heat. Hazardous chemicals, which typically have a limited biodegradability under ambient settings, can also be broken down. The advantages of photocatalysis are not restricted to the elimination of pollutants. There is a great fit between the notion of green chemistry and the three primary usages for photocatalysis: the treatment of air and water, the manufacture of fuel from wastewater or biomass that is renewable, and the manufacturing of fine chemicals (Granone et al., 2018). Figure 2.3 shows the schematic of plausible process of conversion of biomass into value-added materials using photocatalysis (Granone et al., 2018).

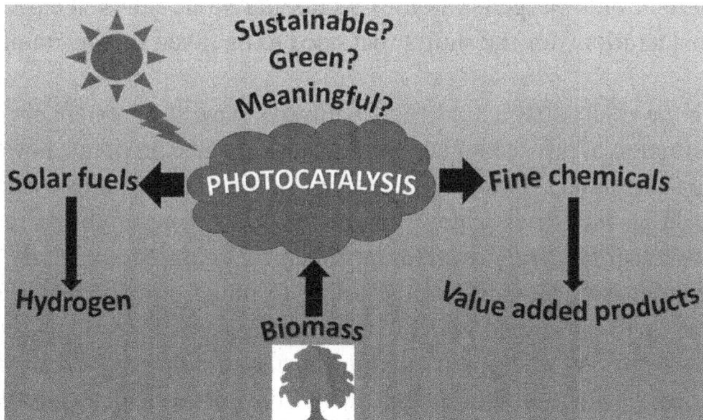

FIGURE 2.3 Schematic showing the transformation of biomass via photocatalysis to useful chemicals and fuels (Granone et al., 2018).

In order to break down cellulose or starch, biological methods such as fermentation and enzymatic reactions in the production of alcohol, for example, are utilised (Liu et al., 2019). These processes make use of specific enzymes, bacteria, and other microbes. When compared to chemocatalytic techniques, the physical conditions of biochemical functions are typically less stringent. The complexity of lignocellulose, on the other hand, makes the cellulose in the material less accessible to the enzymatic hydrolysis process. This contributes to the surprisingly low conversion efficiency and the plausibly increased cost of pretreatment (Liu et al., 2019). These obstacles prevented the biological processes from being scaled up to a greater extent. Certain chemocatalytic processes, including gasification, hydrogenation, and pyrolysis, usually need elevated pressure or/and temperature which results in a rigorous utilisation of thermal energy. Numerous different chemocatalytic processes, including the process of acid catalysis, are utilised for cellulose breakdown in lignocellulosic material by hydrolysis. This process, however, calls for extremely high temperatures and pressures. In today's world, the application of solid acid catalysts—materials like acidic resins, acid-modified amorphous carbon, and stacked transition metal oxides are examples—has garnered a significant amount of focus and interest. In comparison, solar light, which is an endless source of energy, can be used to take the position of thermal energy if harnessed properly (Liu et al., 2019). As a direct consequence of this, the evolution of innovative approaches that one might employ to utilise solar energy to transform environmentally friendly assets, including biomass, into unprocessed products with higher value and sources of energy has become an important focus for the manufacturing industry (Liu et al., 2019).

2.5 FUTURE PERSPECTIVES

The possibility exists for constant enhancement of biomass analysis and the achievement of ecologically sound and financially viable transformation of biomass to chemicals, fuels, and products with additional value. This opportunity must be accompanied by the ongoing advancement of more efficiently produced and utilised biomass feedstock and the generation of additional value items. The following are some key considerations for the future that need to be taken into account:

i. There is no doubt that a variety of pathways have been established for the transformation of refuse biomass into additional value products; however, as a result of substantial running expenses, only a few of these pathways are viable for use on an industrial scale. Consequently, the issue might be resolved if innovative energy strategies were implemented alongside tax relief.
ii. Nanotechnologies have made it possible to make a lot of different nanomaterials with additional benefit from biomass waste products, which are suitable for things like green energy, sensing, diagnosing, catalysis, smart devices, etc., which can't be done with conventional materials. But such technologies are still hard to use in daily life as they are expensive and could be bad for people's health. These problems require to be fixed.

iii. Most vegetation is converted into carbon nanomaterials by subjecting it to elevated temperatures (between 450°C and 1,200°C) and pressure for extended periods of time. At the same time, the majority of the suggested methods for creating carbon nanomaterials from biomass also call for the application of potentially harmful chemicals like H_2SO_4, H_3PO_4, $ZnCl_2$, heavy metal catalysts, NaOH, $KMnO_4$, etc., which are not benign to the atmosphere in any way. This means that these issues will need to be fixed soon.
iv. Anomalies in the way a biorefinery works can be avoided (and even predicted) if their causes and effects are fully understood and studied. In conjunction with the continued advancement of further efficiently produced and analysed lignocellulosic substrates and the obtaining of additional value items, the prospect exists for constant advancement of such biomass handling and the accomplishing of viable and commercially appealing transformation of such biomass into fuels, chemicals, and additional value items.
v. Because of their superb oxidative and reductive reactions, TiO_2- and g-C_3N_4-based photocatalysts are among the most commonly employed in biomass valorisation. Therefore, it becomes crucial to develop novel, more efficient, and noble-metal-free photocatalysts that transform carbohydrates into advantageous things that exhibit excellent transformation and selectivity by matching semiconductor band edge positions with the intended oxidation and reduction half reactions. A one-pot reaction for avoiding intermediary isolation costs is additionally important.

2.6 CONCLUSION

Biomass resources have the ability to be employed in the generation of a vast variety of additional value substances like various bioproducts products (for example, glycerine, furfurals, sugars and sugar alcohols, cellulose fibre and its substitutes, resins, carbonaceous materials, bioplastics, and a variety of other substances), and essential industrial chemicals. Among the products that could be produced using biomass sources is a vast assortment of additional value products. The process of transformation can be achieved through the utilisation of a broad variety of processes, including chemical, biochemical, thermochemical, and photocatalytic operations. Photocatalysis, which uses renewable solar energy to preferentially convert biomass, has attracted the interest of a rapidly growing academic community. Each strategy has its own set of benefits and drawbacks when it comes to the mass production of a specific item in question.

REFERENCES

Agirrezabal-Telleria, I., Gandarias, I., Arias, P. L. (2013). Production of furfural from pentosan-rich biomass: analysis of process parameters during simultaneous furfural stripping. *Bioresour. Technol.*, 143; 258–264. https://doi.org/10.1016/j.biortech.2013.05.082

Alvira, P., Tomás-Pejó, E., Ballesteros, M., Negro, M. J. (2010). Pretreatment technologies for an efficient bioethanol production process based on enzymatic hydrolysis: a review. *Bioresour. Technol.*, 101; 4851–4861. https://doi.org/10.1016/j.biortech.2009.11.093

Baccar, R., Bouzid, J., Feki, M., Montiel, A. (2009). Preparation of activated carbon from Tunisian olive-waste cakes and its application for adsorption of heavy metal ions. *J. Hazard. Mater.*, 162; 1522–1529. https://doi.org/10.1016/j.jhazmat.2008.06.041

Balan, V., da Costa Sousa, L., Chundawat, S. P., Vismeh, R., Jones, A. D., Dale, B. E. (2008). Mushroom spent straw: a potential substrate for an ethanol-based biorefinery. *J. Ind. Microbiol. Biotechnol.*, 35; 293–301. https://doi.org/10.1007/s10295-007-0294-5

Behrendt, F., Neubauer, Y., Oevermann, M., Wilmes, B., Zobel, N. (2008). Direct liquefaction of biomass. *Chem. Eng. Technol.*, 31; 667–677. https://doi.org/10.1002/ceat.200800077

Bussemaker, M. J., Zhang, D. (2013). Effect of ultrasound on lignocellulosic biomass as a pretreatment for biorefinery and biofuel applications. *Ind. Eng. Chem. Res.*, 52; 3563–3580. https://doi.org/10.1021/ie3022785

Cañada-Barcala, A., Rodriguez-Llorente, D., Lopez, L., Navarro, P., Hernandez, E., Águeda, V. I., Larriba, M. (2021). Sustainable production of furfural in biphasic reactors using terpenoids and hydrophobic eutectic solvents. *ACS Sustain. Chem. Eng.*, 9; 10266–10275. https://doi.org/10.1021/acssuschemeng.1c02798

Chang, V. S., Holtzapple, M. T. (2000). Fundamental factors affecting biomass enzymatic reactivity. *Appl. Biochem. Biotechnol.*, 84; 5–37. https://doi.org/10.1385/ABAB:84-86:1-9:5

Chen, H., Wan, K., Zheng, F., Zhang, Z., Zhang, H., Zhang, Y., Long, D. (2021). Recent advances in photocatalytic transformation of carbohydrates into valuable platform chemicals. *Front. Chem. Sci. Eng.*, 3; 615309. https://doi.org/10.3389/fceng.2021.615309

Chhiti, Y., Kemiha, M. (2013). Thermal conversion of biomass, pyrolysis and gasification. *Int. J. Eng. Sci. (IJES)*, 2; 75–85.

Clark, J., Deswarte, F. (2015). *Introduction to Chemicals from Biomass*. 2nd ed. Chichester: Wiley; pp. 114–281.

Corma, A., Iborra, S., Velty, A. (2007). Chemical routes for the transformation of biomass into chemicals. *Chem. Rev.*, 107; 2411–2502. https://doi.org/10.1021/cr050989d

Cortright, R. D., Davda, R. R., Dumesic, J. A. (2002). Hydrogen from catalytic reforming of biomass-derived hydrocarbons in liquid water. *Nature*, 418; 964–967. https://doi.org/10.1038/nature01009

de Aguiar, J., Bondancia, T. J., Claro, P. I. C., Mattoso, L. H. C., Farinas, C. S., Marconcini, J. M. (2020). Enzymatic deconstruction of sugarcane bagasse and straw to obtain cellulose nanomaterials. *ACS Sustainable Chem. Eng.*, 8; 2287–2299. https://doi.org/10.1021/acssuschemeng.9b06806

De María, R., Díaz, I., Rodríguez, M., Sáiz, A. (2013). Industrial methanol from syngas: kinetic study and process simulation. *Int. J. Chem. Reac. Eng.*, 11; 469–477. https://doi.org/10.1515/ijcre-2013-0061

De Wit, D., Maat, L., Kieboom, A. P. G. (1993). Carbohydrates as industrial raw materials. *Ind. Crops Products*, 2; 1–12. https://doi.org/10.1016/0926-6690(93)90004-S

Dresselhaus, M. S., Thomas, I. L. (2001). Alternative energy technologies. *Nature*, 414; 332–337. https://doi.org/10.1038/35104599

Fisher, T., Hajaligol, M., Waymack, B., Kellogg, D. (2002). Pyrolysis behavior and kinetics of biomass derived materials. *J. Anal. Appl. Pyrol.*, 62; 331–349. https://doi.org/10.1016/S0165-2370(01)00129-2

Godula, K., Bertozzi, C. R. (2010). Synthesis of glycopolymers for microarray applications via ligation of reducing sugars to a poly (acryloyl hydrazide) scaffold. *J. Am. Chem. Soc.*, 132; 9963–9965. https://doi.org/10.1021/ja103009d

Granone, L. I., Sieland, F., Zheng, N., Dillert, R., Bahnemann, D. W. (2018). Photocatalytic onversion of biomass into valuable products: a meaningful approach? *Green Chem.*, 20; 1169–1192. https://doi.org/10.1039/C7GC03522E

Hansen, A. C., Kyritsis, D. C., Lee, C. F. (2009). Characteristics of biofuels and renewable fuel standards. In: Vertes, A. A., Blaschek, H. P., Yukawa, H., Qureshi, N., editors. *Biomass to Biofuels-Strategies for Global Industries*. New York: John Wiley; pp. 1–26. https://doi.org/10.1002/9780470750025.ch1

Hu, Z., Wen, Z. (2008). Enhancing enzymatic digestibility of switchgrass by microwave-assisted alkali pretreatment. *Biochem. Eng. J.*, 38; 369–378. https://doi.org/10.1016/j.bej.2007.08.001

Hu, L., Lin, L., Wu, Z., Zhou, S., Liu, S. (2017). Recent advances in catalytic transformation of biomass-derived 5-hydroxymethylfurfural into the innovative fuels and chemicals. *Renew. Sustain. Energy Rev.*, 74; 230–257. https://doi.org/10.1016/j.rser.2017.02.042

Hu, L., Xu, J., Zhou, S., He, A., Tang, X., Lin, L., & Zhao, Y. (2018). Catalytic advances in the production and application of biomass-derived 2, 5-dihydroxymethylfuran. *ACS Catalysis*, 8; 2959–2980. https://doi.org/10.1021/acscatal.7b03530

Huber, G. W., Iborra, S., Corma, A. (2006). Synthesis of transportation fuels from biomass: chemistry, catalysts, and engineering. *Chem. Rev.*, 106; 4044–4098. https://doi.org/10.1021/cr068360d

Irmak, S., Öztürk, İ. (2010). Hydrogen rich gas production by thermocatalytic decomposition of kenaf biomass. *Int. J. Hydrogen Energy*, 35; 5312–5317. https://doi.org/10.1016/j.ijhydene.2010.03.081

Irmak, S., Kurtuluş, M., Hasanoğlu, A., Erbatur, O. (2013). Gasification efficiencies of cellulose, hemicellulose and lignin fractions of biomass in aqueous media by using Pt on activated carbon catalyst. *Biomass Bioenerg.*, 49; 102–108. https://doi.org/10.1016/j.biombioe.2012.12.016

Irmak, S., Meryemoglu, B., Hasanoglu, A., Erbatur, O. (2015). Does reduced or non-reduced biomass feed produce more gas in aqueous-phase reforming process? *Fuel*, 139; 160–163. https://doi.org/10.1016/j.fuel.2014.08.028

Jahangiri, H., Bennett, J., Mahjoubi, P., Wilson, K., Gu, S. (2014). A review of advanced catalyst development for Fischer-Tropsch synthesis of hydrocarbons from biomass derived syn-gas. *Catal Sci Technol.*, 4(8); 2210–2229. https://doi.org/10.1039/C4CY00327F

Jarboe, L. R., Wen, Z., Choi, D., & Brown, R. C. (2011). Hybrid thermochemical processing: fermentation of pyrolysis-derived bio-oil. *Appl. Microbiol. Biotechnol.*, 91; 1519–1523. https://doi.org/10.1007/s00253-011-3495-9

Jin, C., Yao, M., Liu, H., Chia-fon, F. L., Ji, J. (2011). Progress in the production and application of n-butanol as a biofuel. *Renew. Sustain. Energy Rev.*, 15; 4080–4106. https://doi.org/10.1016/j.rser.2011.06.001

Jönsson, L. J., Alriksson, B., Nilvebrant, N. O. (2013). Bioconversion of lignocellulose: inhibitors and detoxification. *Biotechnol. Biofuels*, 6; 16. https://doi.org/10.1186/1754-6834-6-16

Jørgensen, H., Kristensen, J. B., Felby, C. (2007). Enzymatic conversion of lignocellulose into fermentable sugars: challenges and opportunities. *Biofuels, Bioprod. Biorefin.*, 1; 119–134. https://doi.org/10.1002/bbb.4

Khandan, N., Kazemeini, M., Aghaziarati, M. (2012). Direct production of dimethyl ether from synthesis gas utilizing bifunctional catalysts. *Appl. Petrochem. Res.*, 1; 21–27. https://doi.org/10.1007/s13203-011-0002-2

Ko, B. S., Kim, J., Kim, J. H. (2006). Production of xylitol from D-xylose by a xylitol dehydrogenase gene-disrupted mutant of Candida tropicalis. *Appl. Environ. Microbiol.*, 72; 4207–4213. https://doi.org/10.1128/AEM.02699-05

Kootstra, A. M. J., Beeftink, H. H., Scott, E. L., Sanders, J. P. (2009). Comparison of dilute mineral and organic acid pretreatment for enzymatic hydrolysis of wheat straw. *Biochem. Eng. J.*, 46; 126–131. https://doi.org/10.1016/j.bej.2009.04.020

Kumar, P., Barrett, D. M., Delwiche, M. J., Stroeve, P. (2009). Methods for pretreatment of lignocellulosic biomass for efficient hydrolysis and biofuel production. *Ind. Eng. Chem. Res.*, 48; 3713–3729. https://doi.org/10.1021/ie801542g

Liu, X., Duan, X., Wei, W., Wang, S., Ni, B. J. (2019). Photocatalytic conversion of lignocellulosic biomass to valuable products. *Green Chem.*, 21; 4266–4289. https://doi.org/10.1039/C9GC01728C

McKendry, P. (2002). Energy production from biomass (part 2): conversion technologies. *Bioresour. Technol.*, 83; 47–54. https://doi.org/10.1016/S0960-8524(01)00119-5

Meryemoglu, B., Hesenov, A., Irmak, S., Atanur, O. M., Erbatur, O. (2010). Aqueous-phase reforming of biomass using various types of supported precious metal and raney-nickel catalysts for hydrogen production. *Int. J. Hydrogen Energy*, 35; 12580–12587. https://doi.org/10.1016/j.ijhydene.2010.08.046

Mosier, N., Wyman, C., Dale, B.E., Elander, R., Lee, Y.Y., Holtzapple, M., Ladisch, M. (2005). Features of promising technologies for pretreatment of lignocellulosic biomass. *Bioresour. Technol.*, 96; 673–686. https://doi.org/10.1016/j.biortech.2004.06.025

Oh, S., Kim, K. (1999). Synthesis of a new mesoporous carbon and its application to electrochemical double-layer capacitors. *Chem. Commun.*, 21; 2177–2178. https://doi.org/10.1039/A906872D

Olivia, J.M., Sáez, F., Ballesteros, I., Gónzalez, A., Negro, M.J., Manzanares, P., Ballesteros, M. (2003). Effect of lignocellulosic degradation compounds from steam explosion pretreatment on ethanol fermentation by thermotolerant yeast Kluyverinyces narxianus. *Appl. Microbiol. Biotechnol.*, 105; 141–153. https://doi.org/10.1385/ABAB:105:1-3:141

Öztürk, İ., Irmak, S., Hesenov, A., Erbatur, O. (2010). Hydrolysis of kenaf (Hibiscus cannabinus L.) stems by catalytical thermal treatment in subcritical water. *Biomass Bioenerg.*, 34; 1578–1585. https://doi.org/10.1016/j.biombioe.2010.06.005

Rafiqul, I. S. M., Sakinah, A. M. (2013). Processes for the production of xylitol-a review. *Food Rev. Int.*, 29; 127–156. https://doi.org/10.1080/87559129.2012.714434

Ramesh, T., Rajalakshmi, N., Dhathathreyan, K. S. (2015). Activated carbons derived from tamarind seeds for hydrogen storage. *J. Energ. Storage*, 4; 89–95. https://doi.org/10.1016/j.est.2015.09.005

Sabra, W., Groeger, C., Sharma, P. N., Zeng, A. P. (2014). Improved n-butanol production by a non-acetone producing Clostridium pasteurianum DSMZ 525 in mixed substrate fermentation. *Appl. Microbiol. Biotechnol.*, 98; 4267–4276. https://doi.org/10.1007/s00253-014-5588-8

Saha, D., Deng, S. (2010). Adsorption equilibrium and kinetics of CO_2, CH_4, N_2O, and NH_3 on ordered mesoporous carbon. *J. Colloid Interface Sci.*, 345; 402–409. https://doi.org/10.1016/j.jcis.2010.01.076

Saini, J. K., Saini, R., Tewari, L. (2015). Lignocellulosic agriculture wastes as biomass feedstocks for second-generation bioethanol production: concepts and recent developments. *3 Biotech.*, 5; 337–353. https://doi.org/10.1007/s13205-014-0246-5

Shen, Y., Jarboe, L., Brown, R., Wen, Z. (2015). A thermochemical-biochemical hybrid processing of lignocellulosic biomass for producing fuels and chemicals. *Biotechnol. Adv.*, 33; 1799–1813. https://doi.org/10.1016/j.biotechadv.2015.10.006

Squinca, P., Bilatto, S., Badino, A. C., Farinas, C. S. (2020). Nanocellulose production in future biorefineries: an integrated approach using tailor-made enzymes. *ACS Sustainable Chem. Eng.*, 8; 2277–2286. https://doi.org/10.1021/acssuschemeng.9b06790

Sun, Y., Cheng, J. (2002). Hydrolysis of lignocellulosic materials for ethanol production: a review. *Bioresour Technol.*, 83; 1–11. https://doi.org/10.1016/S0960-8524(01)00212-7

Tsyntsarski, B., Stoycheva, I., Tsoncheva, T., Genova, I., Dimitrov, M., Petrova, B.,... Petrov, N. (2015). Activated carbons from waste biomass and low rank coals as catalyst supports for hydrogen production by methanol decomposition. *Fuel Process. Technol.*, 137; 139–147. https://doi.org/10.1016/j.fuproc.2015.04.016

Wainaina, S., Horváth, I. S., Taherzadeh, M. J. (2018). Biochemicals from food waste and recalcitrant biomass via syngas fermentation: a review. *Bioresour. Technol.*, 248; 113–121. https://doi.org/10.1016/j.biortech.2017.06.07

Wang, L., Weller, C. L., Jones, D. D., Hanna, M. A. (2008). Contemporary issues in thermal gasification of biomass and its application to electricity and fuel production. *Biomass Bioenerg.*, 32; 573–581. https://doi.org/10.1016/j.biombioe.2007.12.007

Wei, Z., Zeng, G., Huang, F., Kosa, M., Sun, Q., Meng, X.,... Ragauskas, A. J. (2015). Microbial lipid production by oleaginous Rhodococci cultured in lignocellulosic autohydrolysates. *Appl. Microbiol. Biotechnol.*, 99; 7369–7377. https://doi.org/10.1007/s00253-015-6752-5

Werpy, T., Petersen, G. (2004). Top value-added chemicals from biomass: volume I--results of screening for potential candidates from sugars and synthesis gas (No. DOE/GO-102004-1992). National Renewable Energy Lab., Golden, CO (US). https://doi.org/10.2172/15008859

Xing, R., Subrahmanyam, A. V., Olcay, H., Qi, W., van Walsum, G. P., Pendse, H., Huber, G. W. (2010). Production of jet and diesel fuel range alkanes from waste hemicellulose-derived aqueous solutions. *Green Chem.*, 12; 1933–1946. https://doi.org/10.1039/C0GC00263A

Yin, X., Leung, D. Y., Chang, J., Wang, J., Fu, Y., Wu, C. (2005). Characteristics of the synthesis of methanol using biomass-derived syngas. *Energ. & Fuels.*, 19; 305–310. https://doi.org/10.1021/ef0498622

Yin, R., Liu, R., Mei, Y., Fei, W., Sun, X. (2013). Characterization of bio-oil and bio-char obtained from sweet sorghum bagasse fast pyrolysis with fractional condensers. *Fuel*, 112; 96–104. https://doi.org/10.1016/j.fuel.2013.04.090m

Zhang, Z., Song, J., Han, B. (2017). Catalytic transformation of lignocellulose into chemicals and fuel products in ionic liquids. *Chem. Rev.*, 117; 6834–6880. https://doi.org/10.1021/acs.chemrev.6b00457

Zhou, C.H., Xia, X., Lin, C.X., Tong, D.S., Beltramini, J. (2011). Catalytic conversion of lignocellulosic biomass to fine chemicals and fuels. *Chem. Soc. Rev.*, 40; 5588–5617. https://doi.org/10.1039/C1CS15124J

Zhou, F., Lu, C., Yao, Y., Sun, L., Gong, F., Li, D., Chen, W. (2015). Activated carbon fibers as an effective metal-free catalyst for peracetic acid activation: implications for the removal of organic pollutants. *Chem. Eng. J.*, 281; 953–960. https://doi.org/10.1016/j.cej.2015.07.034

3 Application of Microbial Fermentation Technique for the Synthesis of Fermentable Sugars from Different Agrobiomasses

Anjli Tanwar, Disha Mukherjee, Karan Kumar, and Vijayanand S. Moholkar*

3.1 INTRODUCTION

Agrobiomass is a promising source of sustainable platform chemicals and alternative fuels. However, to fully realize its potential, finding an economically viable way to convert lignocellulosic biomass to desirable fermentable sugars is crucial. Several lignocellulosic biomasses, including rice straw, wheat straw, risk husk, sugarcane bagasse, corn straw, cotton straw, and other plant wastes, are available in abundance and can be used for this purpose (Tan et al., 2021). Cellulose, hemicellulose, and lignin are components of lignocellulosic biomass, with cellulose being the key component. Cellulose can be used as a renewable energy and chemical source (Tan et al., 2021). However, lignocellulosic biomass must first be converted into fermentable sugars to produce biofuels and biochemicals. The subsequent purification and downstream processing steps are critical to achieve high yield, cost-effectiveness, and desired product quality (Kumar et al. 2022, 2024a).

Purification involves removing impurities and contaminants from the crude product obtained during fermentation, often through filtration, centrifugation, chromatography, and crystallization (Mahato et al., 2019; Chen et al., 2020). Downstream processing involves separation and purification techniques such as centrifugation, filtration, chromatography, drying, precipitation, adsorption, extraction, membrane filtration, and crystallization to obtain high-purity fermentable sugars. These processes are very quick and efficient to remove impurities from the agrobiomass. However, they come up with certain limitations: (i) the presence of small contaminants during filtration, precipitation, and membrane filtration; (ii) filter clogging in both filtration and membrane filtration; (iii) the high energy requirement associated with centrifugation and adsorption; (iv) time-consuming aspects within crystallization and chromatography; and (v) the requirement of careful optimization across all processes.

 DOI: 10.1201/9781003407713-3

Recent studies have shown that agrobiomass can produce fermentable sugars using microbial fermentation techniques (Yamada et al., 2013). However, the lack of uniformity in the composition of the materials used, the possibility of contamination of the fermentation process, the presence of inhibitory compounds, and the high cost of production are a few of the difficulties and limitations that this process faces despite its promising potential (Baral & Shah, 2014; Kim et al., 2011).

Genetic and metabolic engineering can improve microbial strains that produce fermentable sugars from agrobiomass, which is more effective and economical. These sugars can produce organic acids, biogas, artificial sweeteners, biofuels, and other products with added value. Optimization and strain engineering are essential to build efficient and sustainable microbial fermentation processes and lessen our reliance on fossil fuels (Kumar et al., 2022; Valles et al., 2021).

However, many challenges are faced during optimization and strain engineering, which include selecting and optimizing the appropriate microbial strains influenced by factors such as productivity, resistance to inhibitors, and growth rate, alongside optimization process conditions, dealing with inhibitors in agrobiomass, genetic instability resulting in reduced productivity, phenotypic trade-offs decreasing strain's growth rate or stress tolerance, metabolic burden, and scale-up (Anand et al., 2024; Kumar et al., 2023a, 2023b, 2024b, 2024c; Singhvi et al., 2014). These challenges can be complex and require careful consideration to overcome.

In conclusion, agrobiomass-derived fermentable sugars present a promising avenue for producing sustainable chemicals and fuels. Continued research and development in this field can lead to more efficient and cost-effective production methods, reducing the environmental impact of traditional chemical and fuel production methods while promoting economic growth and enhancing environmental sustainability.

3.2 MANUFACTURING OF FERMENTABLE SUGARS

Most agrobiomass comprises lignocellulosic wastes (Mujtaba et al., 2023), which are primarily divided into cellulose, hemicellulose, and lignin. Cellulose and hemicellulose covalently bind to lignin, which often makes it difficult for the microorganisms to uncover the carbohydrates (Huang et al., 2023; Macedo et al., 2020). On the other hand, fermentable sugars can be easily accessed by the respective microorganisms (here, lactic acid producing microbes like *Lactobacillus paracasei* and *Clostridium sensu stricto*) (Nancib et al., 2015). This greatly simplifies the production process. Similarly, high starch content biomass can be converted into the following ways:

a. hydrolysis into glucose followed by fermentation to obtain lactic acid;
b. direct fermentation by amylolytic lactic acid bacteria (Kunasundari et al., 2017).

It is also important to remember that the microbes vary in their tolerance to the environment which can affect their ability to convert complex matter to simple forms (Dutta et al., 2023). Other more conventional pretreatment methods involve physical methods like decreasing the material size and chemical methods like acid and alkaline pretreatment (Mankar et al., 2021). Other factors involve ensuring adequate nutrients to both sustain microbial activity and increase sugar production, as well as removing toxic by-products which may accumulate during the microbial process (Bala et al., 2022). However, the products obtained need to be refined further to obtain a pure product.

3.2.1 Purification

The crude product obtained is unsuitable for immediate use, bringing us to our next point of discussion—the purification of the same. It refers to removing impurities, contaminants, and other unwanted by-products generated during fermentation. Removing these impurities early in the process reduces the load on and fouling in. Similarly, ion exchange resins and chromatography separation equipment significantly lower operating costs and improve the overall yield. Depending on its properties and desired purification level, a purified solution of the desired product is obtained with options to optimize the process. It can be suitably combined with various purification techniques to enrich the concentrate further (Mahato et al., 2019). Some standard methods of purification include filtration and centrifugation.

The filtration process involves using membranes or sieves to remove impurities and other particles from the solution. At the same time, centrifugation consists of using centrifugal force to separate particles in a solution based on their density (Van Reis & Zydney, 2001). Both are particularly useful for removing larger particles, such as cells and debris generated during fermentation.

After filtering the supernatant, it may be subjected to diafiltration (Farchaus et al. 1998), where permeable molecules like salts, small proteins, and solvents are removed or separated with the help of filters to obtain a pure solution. It is a critical step that is often introduced to facilitate buffer exchange. Additionally, chromatographic techniques can be employed to eliminate any remaining nucleic acid and peptide impurities. These techniques encompass column chromatography, high-speed countercurrent chromatography (HSCC), and high-performance liquid chromatography (HPLC).

Unlike HSCC, HPLC employs a two-phase solvent system, with the mobile phase being gaseous in nature. The stationary phase consists of a solid adsorbent, and the utilization of a high-pressure pump serves to enhance the efficiency of the separation process by tightly packing the column with smaller particles, thereby increasing output and separation power. It can be equipped with various detectors which measure ultraviolet and refractive indexes (Debebe et al. 2018). HPLC can also be fully automated. However, the output result is based on a preparative run including only three factors. Thus, a decision must be made to opt for high purity, compromising throughput and yield. HSCC proves to be better when considering sample loss due to a liquid–liquid partition technique. Any sample loss due to the adsorption of a solid matrix is eliminated. It is also a much faster method with one-step isolation and often needs no sample pretreatment.

Flash chromatography is another alternative method of purification, wherein a medium application of pressure results in the separation of compounds in bulk. This technology has further enhanced efficiency by modifying existing equipment with computerized fraction collectors and online detection units (Sandesh et al. 2021).

Cation/anion exchange chromatography is another important technique, primarily separating and purifying particles based on charge properties. In cation exchange chromatography, compounds with a negative charge interact with a positively charged resin, such as sodium polystyrene sulfonate. Conversely, in anion exchange chromatography, compounds bearing a positive charge adhere to a negatively charged resin, like chloride.

It is important to note that if a solvent extraction step is used in the above techniques, it results in the early removal of dissolved impurities, eliminating the need for a surfactant (Tehrani et al., 2023). It further improves efficiency in the crystallization processes and reduces the residual liquid volume. Crystallization involves the precipitation of a compound from a solution by cooling it or adding solvents that are useful in purifying sugars and other compounds (Shekunov & York, 2000).

Within industrial environments, particularly in the refinement of raw beet juice, the primary feed stream undergoes an initial pass through a pre-liming tank. A segment of lime juice is extracted from the central liming tank and introduced into the feed. This combined mixture is subsequently directed to the central liming tank, where additional lime is introduced to induce the precipitation of non-sugar components, consequently enhancing their concentration. The mixture is then conveyed to a carbonation tank, where carbon dioxide is injected at the tank's base, facilitating the transformation of lime into calcium carbonate. During this process, the precipitating calcium carbonate has the capacity to capture colloidal material as it progresses, effectively adsorbing potential contaminants in the process. Finally, a decanter removes and separates the precipitated solids.

Purification can also be integrated with other downstream processing steps (discussed later) to enhance the final product's quality further. It must be carefully optimized to minimize the loss of the desired product and maximize the recovery of the purified solution. It is critical for the commercial viability of microbial fermentation. Impure or contaminated products can result in high production costs, low yield, and reduced product value, reducing profitability. Therefore, efficient purification methods are necessary to achieve the desired product quality, high output, and cost-effectiveness.

3.2.2 Downstreaming Process

As discussed earlier, the desired product contains several impurities and contaminants. They can affect the quality of the final product and may also hinder downstream processing. As a result, downstream processing is an important step that involves separating and purifying the outcome of interest from the original stock. It may also be paired with other purification techniques to boost production efficiency.

It is essential to utilize the appropriate downstream process technique as it is inherently linked to the quality of the end product. Furthermore, the final product's quality depends on the downstream processing techniques. Various techniques like centrifugation, filtration, chromatography, and drying are employed in downstream processing to obtain high-purity fermentable sugars (see Table 3.1):

- **Centrifugation:** Centrifugation is one of the most straightforward and quick methods to separate impurities, which employs centrifugal force to obtain the desired product. The fermentation broth is subjected to a high

TABLE 3.1
Comparing Downstream Performance and Separation Efficiencies of Common Processes

Process	Downstream Performance	Separation Efficiency	References
Centrifugation	High throughput (HT) separation based on differing densities	High for solid impurities, low for smaller particles	Masri et al. (2017)
Filtration	HT, strict selection, possibilities of achieving high purity	Variable for large particles	Sun et al. (2023)
Chromatography	High selectivity, effective for complex mixtures, can separate similar sized particles	High based on chemical properties of molecules	Urh et al. (2009)
Precipitation	Integration with other techniques possible, scalable	Not appropriate for sensitive particles	Du et al. (2020)
Adsorption	Removing impurities from solution, excellent for functional group binding	High affinity for target molecules, dependent on chemical properties	Assfalg (2021); Andler and Goddard (2018)
Membrane filtration	High yield of non-denatured proteins, low energy consumed, flexible use, integration with other techniques desirable	Increase in yield and highly efficient, subject to membrane fouling	Noorimotlagh et al. (2021)
Crystallization	Separate and purify solids from solutions, concentrating solutes, pure end products	Dependent on growth of crystal, extremely high separation, minimum contaminants	Rajoub et al. (2023)

spin rotation in a centrifuge, triggering the denser components to move to the exterior edges of the vessel (Grzybowski et al., 2022). In comparison, the lighter components remain in the center. Thus, components are separated based on differing densities. Mostly used to separate cell debris from the broth, the speed required for separation is inversely proportional to the mass of the particles (Gregori Tena & Javierre Cazador, 2023; Kumar & Bhattacharya, 2023). Thus, high-speed centrifugation separates most of the solid impurities, whereas low-speed centrifugation separates larger fragments from the liquid. Studies by Mathur et al. (2023) and Khalili-Samani et al. (2024) have depicted its importance in removing 99% xylose from a mixed fermentation broth, with a recovery rate of 95%. Similarly, Bai et al. (2023) demonstrated the removal of an important chemical 2,3-Butanediol (2,3-BDO), stressing on the possibility of scaling up the process.

- **Filtration:** Filtration is another common downstream processing technique. It is particularly beneficial to remove smaller fragments from the fermentation broth. Usually, the liquid is passed through a filter media, while the solid particles are retained. It is helpful in the removal of smaller particles from the fermentation broth. Filtration involves the use of a filter media that allows the liquid to pass through while retaining the solid particles (Leam et al. 2020). The filter media choice largely depends on the size of the particles required to be removed. In downstream processing, microfiltration, ultrafiltration, and nanofiltration are the most prevalent techniques (Kumar & Bhattacharya, 2023; Vergauwen et al., 2023). Microfiltration eliminates larger particles, whereas ultrafiltration and nanofiltration remove smaller particles like proteins and other contaminants.

- **Chromatography:** It is yet another method to separate certain specific components from the fermentation broth. The components are separated using an inert and mobile phase based on physical and chemical characteristics (Sarkar & Ahmed, 2023). In most cases, the stationary phase is a resin or gel, while the mobile phase is a gas or liquid. Various chromatography methods are used depending on the chemical makeup of the components to be separated (Bernardo et al. 2022). For example, gel permeation chromatography is used to separate molecules based on their size, while paper chromatography is based on the retention capacity of the chromatographic paper used. Recent studies of Saari et al. (2023) have used several ionic exchange chromatographies to assess the purity of extracellular vesicles, which is essential in developing therapies.
- **Drying:** It is the final step in downstream processing that involves eliminating the fluid from the extracted product. Again, the drying technique to be used largely depends on the product's chemical makeup and the envisioned final form. Freeze-drying and spray-drying are some of the commonly used methods. The latter involves converting the fluid component into granular form via heat and a spray nozzle, while the former involves removing water via sublimation under vacuum (Smith 2012).
- **Precipitation:** It is frequently used for purifying biomolecules like proteins and enzymes. A precipitating agent like alcohol, ammonium sulfate, or acetone causes the desired product to form a solid pellet-like precipitate that can be removed from the liquid phase (Schirmeister & Mülhaupt, 2022). Precipitation can be used independently or in tandem with alternate techniques like centrifugation or filtration. For example, the polar components are separated using an opposite acetone solution (Kushida et al., 2022; Liu et al., 2021). Epigallocatechin-3-gallate which is a polyphenol derived from green tea has been used as a precipitating agent for protein valorization from soy whey wastewater and was shown to achieve a high recovery of 60.7%, although the purity achieved was just a little over 50% (Xu et al., 2019).
- **Adsorption:** It is used to remove impurities from the liquid phase using the adsorption capabilities of a solid surface. It is usually a resin or gel with a high affinity for contaminants and selectively adsorbs them. The desired product is left behind in the liquid phase. Adsorption is a successful strategy to discard economically unimportant fragments and other contaminants while separating proteins and enzymes (Obodo & Aigbe, 2023).
- **Extraction:** The product of interest is obtained by separating it from a complex mixture by selectively dissolving it in an appropriate solvent. Extraction often removes non-polar impurities like lipids and waxes from the fermentation broth. Extraction can be accomplished using a variety of solvents like ethanol, hexane, and acetone.

- **Membrane filtration:** It is a technique that involves the use of a membrane to separate components based on their size or charge. The membrane functions as a barrier, allowing small molecules to pass through while retaining larger molecules. They can also be modified to separate viruses, bacteria, and endotoxins, which improve the quality of the product. Studies conducted in the early twentieth century like those of Tuovinen and Kelly (1973) have used this method to generate higher yields of *Thiobacillus ferrooxidans*, which are important in the bioleaching process.
- **Crystallization:** Representing the purest form of the end product, it separates impurities during crystal formation. The solid needs to be super saturated to obtain crystals by preparing an excess of solid in a hot liquid (Nordstrom et al., 2023). The crystals can be separated from the liquid phase by filtration or centrifugation. When the crystal is formed in the surrounding microenvironment, the concentration of the solid decreases (Chiou & Riegelman, 1971; Hermans et al., 2022). Thus, the crystal tends to dissolve back into the solution, and they need to be removed promptly after their formation. The sugarcane industry has been extremely active in utilizing this process to generate high-purity fermentable sugars (Hua et al., 2023; Lee et al., 2023).

3.3 CHARACTERIZATION OF FERMENTABLE SUGARS

Determining the characteristics of fermentable sugars derived from agrobiomass is crucial for a number of uses, such as the manufacturing of biofuel and biorefineries. Characterization entails figuring out what kinds and how much of each sugar are present, evaluating how pure they are, and comprehending how much of them can ferment. An overview of the main procedures needed to characterize fermentable sugars from agrobiomass is provided below:

- **Determination of fermentable sugars:** Identifying the variety of sugars present, including glucose, xylose, arabinose, mannose, galactose, and sucrose, constitutes a crucial step. These commonly found fermentable sugars can be distinguished and characterized using analytical techniques such as gas chromatography (GC) or HPLC (Saraiva et al., 2022).
- **Quantification of fermentable sugars:** For quantifying the amount of sugar present, quantitative techniques like spectrophotometry or particular sugar assays are frequently employed to find out how much of each sugar present in a sample (Aguila Rodriguez et al., 2019; Magwaza & Opara, 2015).
- **Purity analysis:** Examining how pure the sugars that were extracted are and measuring the presence of contaminants or impurities, which could include organic acids, phenolics, or other undesirable substances, will reveal purity (Lips, 2021).
- **Degree of polymerization analysis:** Abbreviated as DP, it recognizes sugars and categorizes them according to their dimerization. They may exist

as monomers, dimers, or even higher order oligomers, which are usually recognized by mass spectrography (Vyas & Nidetzky, 2023).

- **Fermentability ability:** Optimizing fermentation processes is possible only if the sugar undergoes fermentation. Fermentation trials are conducted using enzymatic assays or specific microbes to assess this property (Bier et al., 2019).

3.4 VARIOUS MICROBES FOR SYNTHESIS OF FERMENTABLE SUGARS FROM AGROBIOMASS

Lignocellulosic biomass comprises cellulose, hemicellulose, lignin, and various soluble compounds as its primary components (Zoghlami & Paës, 2019). Cellulose accounts for most of the biomass, while hemicellulose and lignin comprise significant portions (Srivastava et al., 2019). These components hold immense potential as renewable sources of energy and chemicals, and enzymatic hydrolysis is a sustainable strategy for converting them into fermentable sugars. Cellulases, hemicellulases, and ligninases are the three primary enzymes in the enzymatic system, along with a few accessory enzymes (Okeke et al., 2022). Diverse microorganisms, including fungi, bacteria, and archaea, produce these enzymes (Kuhad et al., 2016; Suleiman et al., 2020).

Hydrolytic enzymes, including cellulases and hemicellulases, play a crucial role in the degradation of cellulose and hemicellulose chains. Cellulases constitute three main types of enzymes: endoglucanases, exoglucanases, and β-glucosidases (Sharma et al., 2016). Hemicellulases can be categorized into two groups: those with depolymerizing activity, which cleave the glycosidic bonds in the primary chain, and accessory enzymes, which are responsible for breaking the ester bonds and glycosidic bonds within the side chains of hemicellulose (Shallom & Shoham, 2003; Wyman, 2018). Ligninases include peroxidases and oxidases that degrade lignin (Plácido & Capareda, 2015).

Enzymatic hydrolysis is a highly efficient strategy for producing fermentable sugars by degrading the complex structure of lignocellulosic biomass. An extensive variety of both aerobic and anaerobic microorganisms engage in this process, possessing the capability to generate lignocellulolytic enzymes. The subsequent list comprises microorganisms known for their proficiency in producing lignocellulolytic enzymes:

- **Archaea and eubacteria:** Archaea, including *Pyrococcus*, *Sulfolobus*, *Thermogladius*, and *Thermofilum*, have demonstrated potential in degrading lignocellulosic biomass, despite their resilience in extreme conditions characterized by elevated temperatures, pH levels, and salinity. In tandem with archaea, they represent essential sources of these enzymes (Benatti and Polizeli 2023).

  While various genera may contain species with lignocellulolytic potential, the predominant lignocellulolytic bacteria are typically affiliated with the *Bacillus*, *Acinetobacter*, *Cellulomonas*, *Clostridium*, and *Pseudomonas* genera. *Actinobacteria*, a category of aerobic microorganisms, encompass species such as *Cellulomonas flavigena*, *Cellulomonas fimi*, *Actinomycosis bovis*, *Xylanimonas cellulosilytica*, and *Thermobifida fusca*. These microorganisms are adept at producing cellulolytic and lignocellulolytic enzymes, rendering them well-suited for cellulose hydrolysis and lignin modification (Saini et al., 2015, 2016). Additionally, several other bacteria have been explored for their capacity to degrade lignin. These bacteria encompass *Bacillus*, *Streptomyces*, *Sphingomonas*, *Pseudomonas*, *Rhodococcus*, and *Nocardia*. Within the realm of anaerobic bacteria, the *Clostridium* genus, notably *Clostridium thermocellum*, houses lignocellulolytic enzyme producers (Lee et al., 2019).

- **Fungi:** Among microorganisms, filamentous fungi have received the most extensive research attention when it comes to lignocellulose breakdown. Particularly, species within the *Aspergillus* and *Trichoderma* genera, both falling under the *Ascomycota* phylum, are highly recognized and widely utilized as prime examples of cellulolytic fungi in various industries. Detailed proteomic studies have unveiled their extensive secretome, which plays a pivotal role in the degradation of lignocellulose. Other genera such as *Penicillium*, *Fusarium*, and *Rhizopus* are also among the major industrial producers of these enzymes (Passos et al., 2018; Benatti & Polizeli, 2023).

  In addition to *Ascomycetes*, fungi belonging to the *Basidiomycota* phylum possess both enzymatic systems: a hydrolytic one for cellulose and hemicellulose degradation and an oxidative one for lignin oxidation and degradation, with species of some genera such as *Phlebia*, *Pleurotus*, *Phanerochaete*, *Trametes*, *Polyporus*, and *Lentinus* standing out (Liu et al., 2020; Okal et al., 2020). These fungi are capable of breaking down both cellulose and lignin, making them highly valuable in biotechnological applications (Benatti & Polizeli, 2023).

3.5 CHALLENGES AND LIMITATIONS ASSOCIATED WITH MICROBIAL FERMENTATION OF AGROBIOMASS

Agrobiomass fermentation by microbial techniques is promising for biofuels and fermentable sugars production. However, several challenges and limitations associated with this process exist (Nautiyal, Boesl, and Agarwal 2020):

- **Lack of uniformity/consistency** in the materials' composition can result in erroneous fermentation. Agrobiomass materials vary significantly in their chemical and physical properties, which lead to the selection of different pretreatment methods and microbial strains for optimal fermentation. To address this challenge, researchers have focused on developing more efficient and versatile pretreatment strategies which effectively break down plant cell walls, improving access to fermentable sugars. These methods include physical methods like a steam explosion, chemical methods like acid or alkali pretreatment, and biological methods like enzymatic hydrolysis. Meanwhile, pretreatment methods depend on factors like the type of agrobiomass material, the desired end product, and the cost-effectiveness of resources (Tan et al., 2021).

- **Contamination** of the fermentation process by unwanted microorganisms poses a significant risk regarding the purity of the final product. It also reduces the efficiency of the process (Brexó & Sant'Ana, 2017; Mahboubi et al., 2018; Sharma et al., 2024). Strict environmental controls and sanitation procedures may be implemented, including attempts to maintain a sterile environment, utilize aseptic techniques, and monitor the fermentation process regularly for possible signs of contamination.
- **The presence of inhibitory compounds** in agrobiomass can also impact the efficiency of the fermentation process. This is discussed later while discussing strain-specific optimization and related problems (Kumar et al., 2023). To address this issue, researchers have focused on developing microbial strains that are more tolerant to inhibitory compounds and on optimizing fermentation conditions to minimize the impact of these compounds on the fermentation process.
- **The complex structure of plant cell walls** significantly limits microbial fermentation of agrobiomass (Dixon 2013). Carbohydrates present in the cell walls, such as cellulose and hemicellulose, must be broken down into fermentable sugars to be utilized by microorganisms. However, this process can be tricky and inefficient, often leading to low conversion efficiency to fermentable sugars. Thus, researchers have focused on developing specialized and more effective enzymes to break down the complex structure of plant cell walls quickly (Amin et al., 2017). This includes several enzymes like cellulases, hemicellulases, and ligninases, which can break down the respective components of the plant cell walls. The possibility of using microbial consortia is another exciting prospect, as these consortia work in tandem to break down different parts of a cell wall, improving the efficiency of the process.
- **Inhibitory effects of the final product** on microbial growth can lead to high product concentrations, which cause feedback inhibition. Continuous fermentation and product removal techniques have been utilized to tackle this issue to maintain optimal concentrations and limit the corresponding effects. Furthermore, researchers have also explored the possibility of using microbial strains far more tolerant to high product concentrations to improve the overall fermentation efficiency (Cubas-Cano et al., 2018).
- **High production costs** are also a significant concern. The process needs expensive equipment; skilled labor is often energy-intensive, ultimately increasing expenditure. Additionally, the need for strict environmental control and susceptibility to microbial contamination can significantly limit the capacity of the fermentation process (Zabed et al., 2017). Research is ongoing to develop more efficient and cost-effective fermentation systems.

It is necessary to overcome these limitations to realize the potential of microbial fermentation of agrobiomass and make it a sustainable and economical option for biofuel production.

3.6 WAYS TO ENHANCE THE PRODUCTION OF FERMENTABLE SUGARS

In traditional methods, agrobiomass produces fermentable sugars, which are more expensive and time-consuming. Genetic and metabolic engineering can improve microbial strains involved in the more efficient and cost-effective production of fermentable sugars from agrobiomass.

3.6.1 Genetic Engineering of Microbial Strains

The production of fermentable sugars can be enhanced by genetically altering microbial strains to metabolize sugars. By rerouting sugar metabolism toward the desired outcome, this procedure can involve changing the microbe's natural pathways which can be done by introducing new genes, deleting existing genes, or tweaking existing genes.

Another strategy involves the introduction of genes encoding enzymes that can convert complicated sugars into simpler fermentable sugars. This strategy is beneficial when working with complicated substrates like lignocellulosic biomass, which are challenging to degrade using conventional techniques.

To carry out these modifications, a variety of techniques, such as gene cloning, site-directed mutagenesis, and genome editing, are used by genetic engineers. Genetic tools such as CRISPR-Cas9, zinc finger nucleases (ZFN), and transcription-activator-like effector nucleases (TALEN) are among the most commonly employed techniques in the field of microbial engineering (Shapiro et al., 2018).

The following are examples of genetically engineered microbial strains that can enhance fermentable sugar production from agrobiomass:

- ***Penicillium oxalicum*:** ClrB, XlnR, and AraR are three transcriptional activators used in the combinatorial engineering of *Penicillium oxalicum*. The genetic modification led to the development of a novel strain named MCAX, which demonstrates a noteworthy enhancement in the production of lignocellulolytic enzymes when compared to the original strain. Specifically, the lignocellulolytic enzyme production of the MCAX strain has surged to a remarkable 51.0-fold increase from an initial 3.1-fold improvement. As a result, corn fiber releases more fermentable sugars (Gao et al., 2021).
- ***Bacillus clausii*:** A previously unidentified GH43 β-xylosidase (XYLO) originating from *Bacillus clausii* TCCC 11004 underwent genetic modifications to generate a recombinant β-xylosidase, denoted as rXYLO. With the maximum activity measured at pH 8.0 and a broad pH range stability from 7.0 to 11.0, the rXYLO enzyme has remarkable features. It is an ideal option for usage in a variety of industrial applications due to its exceptional alkali tolerance. Additionally, this alkali-tolerant β-xylosidase has high xylose tolerance and endo-xylanase activity, leading to increased xylose production from agricultural waste (Wang et al., 2022).
- ***Bacillus subtilis*:** *Bacillus subtilis* (SSL26) has been genetically modified to secrete xylanase that can successfully depolymerize xylan in situ and

produce up to 7.1 g/L of xylose. This represents 66.7% of the 13.3 g/L of xylan's total xylose content. This finding significantly impacts increasing xylose production (Mhatre et al., 2022).

- ***Trichoderma reesei*:** *Trichoderma reesei* has been genetically modified to produce a strain capable of producing laccase, manganese peroxide, and lignin peroxidase. These enzymes are essential for biodelignifying lignocellulosic biomass, which involves the breakdown of lignin in agrobiomass. The enhanced production of these enzymes in the genetically engineered strain leads to the faster degradation of lignin, improving fermentable sugar production (Iyyappan et al., 2023).
- ***Caldicellulosiruptor bescii*:** Genetic engineering of *Caldicellulosiruptor bescii* has been shown to impact both growth and sugar release from carboxymethylcellulose significantly. In one study, the expression of E1 endoglucanase from *Acidothermus cellulolyticus* in *C. bescii* was found to enhance its ability to deconstruct crystalline cellulose (Chung et al., 2015). This improvement was further enhanced in another study, where GuxA from *A. cellulolyticus* was integrated into the chromosome of the *C. bescii* strain containing E1 endoglucanase from *A. cellulolyticus*, leading to increased cellulolytic activity (Kim et al., 2019). The faster degradation of cellulose in agrobiomass due to this genetic engineering has significant implications for the production of fermentable sugars.
- ***Escherichia coli:*** The bioconversion of xylan to sugar is an essential process in biomass utilization, and genetically engineered *Escherichia coli* has shown promise in this regard. In a novel approach, a xylanase B gene from *Thermotoga neapolitana* (Tne) was cloned and overexpressed in *E. coli* using a heat-shock vector *pHsh*. The resulting enzyme showed the highest endo-β-1,4-xylan hydrolase activity at 75°C and pH 6.0. This genetically engineered xylanase is highly efficient in a hyperthermic environment, making it a valuable tool in the xylan-to-sugar biotransformation process (Sha et al., 2020).
- ***Aspergillus niger*:** Pectinolytic enzymes are well-known for being produced by *Aspergillus niger* and have many industrial uses. Three chimeric GaaR-XlnR transcription factors were developed in *A. niger* using CRISPR/Cas9 genome editing. The xlnR promoter was used to express these factors, and in *A. niger*, their N-terminal regions swapped with those of the GaaR. In the presence of the inducing sugar D-xylose, the resultant chimeric GaaR-XlnR transcription factors induced pectinolytic activity. Since pectin is prevalent in plant biomass and can be a significant obstacle to the formation of fermentable sugars, this effective pectin breakdown is especially beneficial (Kun et al., 2021).
- ***Trichoderma reesei*:** The overexpression of XYR1 in *Trichoderma reesei* effectively enhances the production of (hemi)cellulases. In a specific case, a strain of *T. reesei* lacking four major cellulases (CBHI, CBHII, EGI, and EGII) was engineered to overexpress XYR1. As a result, this strain showed a significant increase in glucose release. It demonstrated remarkable performance in the hydrolysis of unpretreated corn fiber, releasing high amounts of xylose, arabinose, and mannose (Zhang et al., 2022).

3.6.2 Metabolic Engineering of Microbial Strains

Metabolic engineering is a powerful tool that can engineer microbial strains capable of efficiently transforming agrobiomass into fermentable sugars. The process involves designing and optimizing metabolic pathways within microorganisms to improve their ability to break down complex carbohydrates found in agrobiomass and convert them into fermentable sugars. The selection of the specific agrobiomass that will be used as a feedstock is the first stage in metabolic engineering, and various resources such as crop residues, forestry by-products, and waste streams from food processing and other industries can be included. Microorganisms employed for fermentation are chosen or engineered after the agrobiomass has been identified. Usually, bacteria or yeast are used, and these microorganisms have been altered to have the desirable metabolic pathways and characteristics. Metabolic engineering aims to maximize the conversion of agrobiomass into fermentable sugars by optimizing the metabolic pathways within the microorganisms. This is accomplished by either choosing or genetically modifying microorganisms equipped with specific enzymes capable of efficiently breaking down intricate carbohydrates, including cellulose and hemicellulose, into more readily fermentable simple sugars. A substantial challenge in the economically viable bioconversion of plant cell wall hydrolysates into fuels and chemicals has been the limited ability of microorganisms to effectively co-ferment both pentose and hexose sugars, particularly glucose and xylose. Nonetheless, metabolic engineering has been harnessed to cultivate microbial strains that possess the capability for effective xylose utilization, which is a pivotal requirement for establishing an economically viable microbial conversion pathway for terrestrial lignocellulosic biomass into biofuels and biochemicals.

Examples of metabolically engineered microbial strains that can enhance fermentable sugar production from agrobiomass are given below and in Table 3.2:

- ***Saccharomyces cerevisiae*:** The engineered *Saccharomyces cerevisiae* strain, which contains 20 cellulase genes from *T. reesei* in a specific ratio, has demonstrated impressive results. This strain has a more excellent cellulose hydrolysis activity due to the integration of cellulase genes into the transposable elements' delta sites. Cellulose genes—*egl2*, *cbh2*, and *bgl1*, respectively—are most effective at the ratio of 13:6:1. Compared to other *S. cerevisiae* strains with varying proportions of cellulose hydrolysis activity, this engineered strain can produce fermentable sugars at higher rates (Yamada et al., 2013; Kricka et al., 2014).
- ***Yarrowia lipolytica*:** An engineered *Yarrowia lipolytica* strain has been developed with the ability to express xylanases, which can break down xylan, a major component of hemicellulose, into xylose. To achieve this, the xylanase enzyme has been either secreted or displayed on the cell surfaces of *Y. lipolytica*. In the case of secretion, the enzyme is synthesized within the cell and then transported outside, where it can interact with the xylan substrate. Alternatively, when the enzyme is displayed on the cell surface, it can interact with the substrate directly without secretion. The ability to express xylanases on the cell surface or secrete them is an essential factor

TABLE 3.2
Showing the Performance of Wild Strains and the Strains After Genetic and Metabolic Engineering

		Performance		
		Wild Strain	**Engineered Strain**	**References**
Genetic Engineering	*Penicillium oxalicum*	3.1-fold increase in lignocellulolytic enzyme production	51.0-fold increase in lignocellulolytic enzyme production	Gao et al. (2021)
	Bacillus clausii	Stable under alkaline conditions, low xylose tolerance and endo-xylanase activity production, low xylose production from agrobiomass	Maximum activity at pH 8.0 and broad pH range stability from 7.0 to 11.0, high xylose tolerance and endo-xylanase activity, increased xylose production from agrobiomass	Wang et al. (2022)
	Bacillus subtilis	Low xylose production	Xylose production up to 7.1 g/L, which means 66.7% of the 13.3 g/L of xylan's total xylose content	Mhatre et al. (2022)
	Trichoderma reesei	Lack of complete ligninolytic activity	Increased ligninolytic activity, faster degradation of lignin	Iyyappan et al. (2023)
	Caldicellulosiruptor bescii	Ability to degrade cellulose present in agrobiomass	Increased cellulolytic activity, faster degradation of cellulose present in agrobiomass	Chung et al., (2015) and Kim et al. (2019)
	Escherichia coli	Unable to degrade agrobiomass at high temperature	Agrobiomass degradation at high temperature	Sha et al., (2020)
	Aspergillus niger	Unable to degrade pectin present in agricultural residues	Induced pectinolytic activity, faster degaradation of pectin	Kun et al. (2021)
Metabolic engineering	*Saccharomyces cerevisiae*	Ability to hydrolyze cellulose	Increased cellulose hydrolysis activity, produce fermentable sugars at higher rates	Yamada et al. (2013), Kricka et al. (2014)
	Yarrowia lipolytica	Unable to express xylanases	Ability to express xylanases, increased xylose production	Ledesma-Amaro and Nicaud (2016)
	Saccharomyces pastorianus	Less efficient in breaking down cellulose into fermentable sugars	Hydrolyze cellulose into glucose more efficiently, making the process more economically viable	Fitzpatrick et al. (2014)
	Aureobasidium melanogenum	Produces pullulan in low amount	Produces 71 g/L of pullulan, which is then converted into glucose and other fermentable sugars by the pullulanase enzyme	Kumar et al. (2019)

in increasing the efficiency and cost-effectiveness of the process. This is because it eliminates the need to purify the enzyme, which can be time-consuming and expensive. Furthermore, the stability and activity of the enzyme can be increased by displaying it on the cell surface, thus enhancing the process efficiency (Ledesma-Amaro & Nicaud, 2016).

- ***Aureobasidium melanogenum*:** To produce vast amounts of pullulan, a valuable polysaccharide utilized in many sectors, from the agrobiomass, an engineered strain of the bacterium *Aureobasidium melanogenum* has been developed. The inulinase gene from *Kluyveromyces marxianus* was inserted into the bacterium's genome to accomplish this. The integration of the *inulinase* gene has produced 71 g/L of pullulan by the engineered *A. melanogenum* strain (Hossain et al., 2016). Pullulan is a polymer made up of maltotriose units that are easily converted into glucose and other fermentable sugars by the pullulanase enzyme. This enhanced production of fermentable sugars (Kumar et al., 2019).
- ***Saccharomyces pastorianus***: Recombinant enzymes from *Saccharomyces pastorianus* are more efficient than a commercial cocktail of *T. reesei* in breaking down cellulose into fermentable sugars. These enzymes consist of cDNA copies of genes representing three significant classes of cellulases—*endoglucanases*, *cellobiohydrolases*, and *β-glucosidases*—from *T. reesei*, expressed in *S. pastorianus*. The resulting enzymes are secreted and act in synergy to hydrolyze cellulose. By breaking down cellulose into glucose more efficiently, the overall cost and time required for the production process can be reduced, while the yield of biofuels and other products can be increased, making the process more economically viable (Fitzpatrick et al., 2014).
- ***Saccharomyces cerevisiae*:** The addition of two *T. reesei* enzymes, CBH II (*cellobiohydrolase* III) and EGII (*endoglucanase* II), results in an engineered strain of *S. cerevisiae* that produces more glucose from cellulose. To do this, *CipA3* from *Clostridium thermocellum*, a mini scaffolding, was used to arrange the two enzymes nearby before it was displayed on the cell surface of *S. cerevisiae*. Compared to unaltered strains, this led to an increase in the synthesis of glucose from cellulose (Wen et al., 2010).
- ***Yarrowia lipolytica***: An engineered strain of *Y. lipolytica* overexpresses endogenous or heterologous β-glucosidases, which allows it to grow in cellobiose. Cellobiose is a disaccharide produced during cellulose hydrolysis, and many microorganisms cannot directly utilize it. However, by overexpressing *β-glucosidases*, *Y. lipolytica* can convert cellobiose into its glucose subunits. This enhanced production of fermentable sugars (Ledesma-Amaro & Nicaud, 2016).

The generation of fermentable sugars from agrobiomass through the application of genetic and metabolic engineering holds the promise of diminishing our reliance on fossil fuels while advocating for sustainable production methodologies. Such an approach can create a viable market for agricultural waste products, thereby presenting fresh economic prospects for farmers and rural communities. Through the conversion of agrobiomass into fermentable sugars, this technology not only aids in curtailing the volume of waste stemming from agricultural practices, but also serves as a sustainable wellspring of energy and chemical resources. Additionally, using agrobiomass as a feedstock for fermentable sugar production can reduce competition for arable land between food and fuel crops, which is a significant concern in many parts of the world. Overall, using genetic and metabolic engineering to produce fermentable sugars from agrobiomass can support sustainable development and create new economic opportunities while reducing our dependence on fossil fuels.

3.7 UTILIZATION OF FERMENTABLE SUGARS IN VARIOUS WAYS

Agrobiomass is an ecologically sound source that can produce fermentable sugars used in multiple applications. The process of microbial fermentation, employing a range of microorganisms like bacteria, fungi, and yeast, can be harnessed for the conversion of these sugars into valuable end products. Some of the potential applications of fermentable sugars derived from agrobiomass include:

Production of biofuels: Ethanol is the most common biofuel obtained from fermentable sugars and is used as an important transportation fuel additive (Farrell et al., 2006). Butanol stands as another biofuel option, offering a higher energy content in comparison to ethanol, and it can effectively serve as a substitute for gasoline. Furthermore, fermentable sugars can be transformed into biodiesel using a technique known as transesterification (Demirbas et al., 2016).

Production of high-value chemicals: Biodegradable polymers like polylactic acid (PLA) and succinic acid, which act as a precursor for producing various chemicals, can be derived from lactic acid (Doungsri et al., 2019). Citric acid is used as a food additive and in the pharmaceutical industry (Priya et al., 2022).

Exploring substrates for microbial biomass production: Microbial biomass can be generated by a diverse array of microorganisms and can be harnessed as a protein source (Fu et al., 2011) for livestock feed, potentially replacing soybean meal and other protein-rich feed ingredients.

Conversion into hydrogen gas: Fermentable sugars can serve as a fuel source for fuel cells, resulting in the production of solely water as an end product (Ferraren-De Cagalitan & Abundo, 2021), making them a sustainable and environmentally friendly alternative source of energy.

Biogas generation via anaerobic digestion: A renewable blend of methane, carbon dioxide, and additional gases, biogas finds utility in heating, electricity generation, and as a transportation fuel (Ramonet Marques et al., 2023; Tsipis et al., 2023).

Organic acid synthesis: Organic acids like acetic acid, propionic acid, and butyric acid serve as essential building blocks for the production of diverse chemicals. They hold the potential to be employed in the creation of renewable plastics, adhesives, and solvents.

Production of bio-based surfactants: Bio-based surfactants are advantageous over petroleum-based surfactants due to their lower toxicity and biodegradability.

Production of bio-based adhesives: Bio-based adhesives are more environmentally friendly than petroleum-based adhesives with significantly lower emissions and are primarily used in the manufacture of various products like plywood and particleboard (Ma et al., 2023).

Production of bio-based plastics: Bio-based plastics offer multiple benefits: Bio-based plastics exhibit a range of advantages compared to conventional petroleum-based plastics, notably including reduced greenhouse gas emissions and biodegradability.

Production of fermented beverages: Fermentable sugars derived from agrobiomass can provide a steady supply of sugars for beverages like wine and beer.

Production of bio-based fertilizers: Bio-based fertilizers have several advantages over traditional chemical fertilizers, including improved soil health and reduced environmental impact (Kurniawati et al., 2023).

Enzyme production: Enzymes serve a diverse array of industrial purposes, encompassing food processing, textile manufacturing, and pharmaceuticals, among others.

Probiotics are essential for maintaining gut health and immune function, and they are typically produced using lactic acid bacteria and other strains that can ferment different types of sugars (Kaur et al., 2023). Agrobiomass-derived fermentable sugars can provide a sustainable source for producing these beneficial microorganisms.

Bio-based chemicals, such as amino acids, vitamins, and pigments, can also be produced from fermentable sugars derived from agrobiomass (García-Peñas & Sharma, 2023). These chemicals find wide-ranging applications in diverse industries, spanning food and beverage, pharmaceuticals, and cosmetics.

In addition, agrobiomass-derived fermentable sugars can produce bioremediation agents that can help improve the environmental health of contaminated sites. Microorganisms can degrade organic contaminants, such as pesticides and petroleum products, into harmless by-products, making them a sustainable and environmentally friendly solution for cleaning up contaminated soil and water (Patel & Sharma, 2023).

In conclusion, the use of fermentable sugars from agrobiomass presents a promising avenue for sustainable product development across various industries. However, challenges and limitations still need to be addressed, such as the need for efficient and cost-effective fermentation processes and proper management of waste streams. Sustained research and development in this domain has the potential to yield more efficient and sustainable production approaches, consequently diminishing the environmental footprint of conventional chemical and fuel manufacturing techniques. This, in turn, can stimulate economic growth and bolster environmental sustainability.

3.8 CHALLENGES FACED DURING OPTIMIZATION AND STRAIN ENGINEERING

Optimization and strain engineering are powerful techniques that can be used to develop efficient and sustainable microbial fermentation processes, reducing our reliance on fossil fuels (Adegboye et al., 2021). However, these techniques are not without challenges. Some of the challenges faced during optimization and strain engineering include:

- **Inhibitors in agrobiomass:** Lignin, hemicellulose, and other chemicals found in agrobiomass can inhibit microbial growth and fermentation. Substances like organic acids, furfural, and hydroxymethylfurfural can also be toxic to microbial cells, reducing the effectiveness of fermentation (Nigam & Singh, 2011; Taherzadeh & Karimi, 2008). For example, lignin can prevent cellulose from fermenting by blocking microbial enzymes' access to cellulose.
- **Optimization of process conditions:** The process conditions, including temperature, pH, and nutrient concentration, must be optimized to convert agrobiomass into fermentable sugars efficiently. However, the optimization process can be complex and time-consuming, as it involves finding a balance between several factors affecting microbial growth and metabolism. For instance, optimizing factors such as temperature, pH, and inoculum size is crucial for producing ethanol from maize stover using *S. cerevisiae.*
- **Strain selection and optimization:** Choosing and optimizing appropriate microbial strains for agrobiomass fermentation can be challenging, as it requires balancing factors like productivity, resistance to inhibitors, and growth rate (Adegboye et al., 2021).
- **Genetic instability:** Genetic alterations made to microbial strains during strain engineering can make the strains unstable or more prone to genetic mutations, resulting in reduced productivity or even the loss of the desired phenotype (Makino et al., 2011).
- **Phenotypic trade-offs:** Improving one feature of a strain may come at the expense of another. For example, improving product yield may decrease the strain's growth rate or stress tolerance (Kim et al., 2020; Ekkers et al., 2022).
- **Metabolic burden:** The introduction of new metabolic pathways or alterations to existing pathways can create a metabolic load on the host organism, resulting in slower growth or lower overall productivity (Peng et al., 2017).
- **Scale-up:** The upscaling of fermentation procedures from laboratory settings to an industrial scale can be a complex task, encompassing the management of greater biomass volumes and the refinement of the process to ensure efficient mass and heat transfer. The scaling-up process is additionally intricate due to the necessity for diverse processing techniques, which may vary depending on the specific type of agrobiomass under consideration. For example, mass and heat transfer must be optimized to maintain effective fermentation rates when producing ethanol from sugarcane bagasse at a larger scale.

3.9 SUMMARY, CONCLUSIONS, AND FUTURE PROSPECTIVE

Using agrobiomass as a sustainable source for producing platform chemicals and alternative fuels is a promising approach that offers an attractive solution for converting recalcitrant lignocellulose into useful intermediates (Zhu et al., 2017). Apart from its most common advantages of cost-effectiveness and efficiency, microbial techniques are extremely versatile and find use in different types of agrobiomasses. They almost always result in up to 90% and above purification of end products, as evidenced by yeast strains converting high amounts of ethanol from glucose (Queiroz et al., 2023). Cellulose, one of the significant components of agrobiomass, is an ideal source for biofuels and biochemical production due to its relative abundance and sustainability. The initial stride in the production of biofuels and biochemicals from agrobiomass entails the transformation of lignocellulosic biomass into fermentable sugars.

Manufacturing fermentable sugars involves purification and downstream processing, primarily dependent on the nature of the product and the impurities present in the fermentation broth (Furlong et al., 2019). During purification and downstream processing, various techniques like filtration, centrifugation, diafiltration, and chromatography methods such as HPLC and HSCC may be employed. An additional technique involves passing the original feed through a pre-liming tank with carbonation tanks aiding in removing non-sugars and impurities.

Enzymatic hydrolysis of lignocellulosic biomass is a sustainable strategy for converting it into fermentable sugars. Lignocellulolytic enzymes are produced by a variety of microorganisms, including fungi, bacteria, and archaea. Both aerobic and anaerobic microorganisms have demonstrated potential in degrading lignocellulosic biomass, making them highly sought after in biotechnological applications. Genetic engineering and the CRISPR/Cas9 genome editing technique have also shown promise in enhancing the yield of fermentable sugars from agrobiomass. There have also been attempts to bio-delignify lignocellulosic biomass using genetically engineered microorganisms (Shapiro et al., 2018; Costa et al., 2020)

Metabolic engineering is a powerful tool for enhancing the ability of microorganisms to convert lignocellulosic biomass into fermentable sugars. Ongoing research focuses on modifying microorganisms' metabolic pathways to improve the bioconversion process's efficiency. The concept of an integrated biorefinery, which processes various streams of agrobiomasses using different techniques to obtain different products, has gained significant attention in recent years. High-throughput screening techniques have also been used to identify new microbial strains with improved capabilities for fermentable sugar production (Sarnaik et al., 2020).

Continued research in the areas mentioned above has the potential to revolutionize the approach to producing energy and other valuable products. Developing more efficient and sustainable lignocellulosic biorefineries could significantly reduce the environmental impact of traditional chemical and fuel production methods, promote economic growth, and enhance environmental sustainability. In the global shift toward a low-carbon economy, it is anticipated that lignocellulosic biomass will assume a progressively vital role in fulfilling our daily energy and chemical requirements.

ACKNOWLEDGMENTS

Dr. Karan Kumar is grateful to the Prime Minister's Research Fellowship provided by Ministry of Education, Government of India. Ms. Anjli and Ms. Disha, two authors of this chapter, acknowledge the online WINBIOCB-2022 internship/training provided by Prof. V.S. Moholkar.

NOTE

* Corresponding author

REFERENCES

Anand et al., 2024: Anand, A., Kumar, K., & Moholkar, V. S. (2024). Various Routes for Hydrogen Production and Its Utilization for Sustainable Economy. In *Biohydrogen-Advances and Processes* (pp. 503–527). Cham: Springer Nature Switzerland.

Adegboye, M. F., Ojuederie, O. B., Talia, P. M., & Babalola, O. O. (2021). Bioprospecting of microbial strains for biofuel production: Metabolic engineering, applications, and challenges. *Biotechnology for Biofuels*, *14*(1), 5. https://doi.org/10.1186/s13068-020-01853-2

Aguila Rodriguez, G., Arias Duque, N. P., Gonzalez Sanchez, B. E., Sandoval Gonzalez, O. O., Giraldo Osorio, O. H., Trujillo Romero, C. J., Wilches Torres, M. A., & Flores Cuautle, J. D. J. A. (2019). Sugar concentration measurement system using radiofrequency sensor. *Sensors*, *19*(10), 2354. https://doi.org/10.3390/s19102354

Amin, F. R., Khalid, H., Zhang, H., Rahman, S. U., Zhang, R., Liu, G., & Chen, C. (2017). Pretreatment methods of lignocellulosic biomass for anaerobic digestion. *Amb Express*, *7*, 1–12.

Andler, S. M., & Goddard, J. M. (2018). Transforming food waste: How immobilized enzymes can valorize waste streams into revenue streams. *NPJ Science of Food*, *2*, 19. https://doi.org/10.1038/s41538-018-0028-2

Assfalg, M. (2021). Protein adsorption and conformational changes. *Molecules (Basel, Switzerland)*, *26*(23), 7079. https://doi.org/10.3390/molecules26237079

Bai, Y., Feng, H., Liu, N., & Zhao, X. (2023). Biomass-derived 2,3-butanediol and its application in biofuels production. *Energies*, *16*(15), 5802.

Bala, S., Garg, D., Thirumalesh, B. V., Sharma, M., Sridhar, K., Inbaraj, B. S., & Tripathi, M. (2022). Recent strategies for bioremediation of emerging pollutants: A review for a green and sustainable environment. *Toxics*, *10*(8), 484.

Baral, N. R., & Shah, A. (2014). Microbial inhibitors: Formation and effects on acetone-butanol-ethanol fermentation of lignocellulosic biomass. *Applied Microbiology and Biotechnology*, *98*(22), 9151–9172. https://doi.org/10.1007/s00253-014-6106-8

Benatti, A. L. T., & Polizeli, M. de L. T. de M. (2023). Lignocellulolytic biocatalysts: The main players involved in multiple biotechnological processes for biomass valorization. *Microorganisms*, *11*(1), 162. https://doi.org/10.3390/microorganisms11010162

Bernardo, S. C., Carapito, R., Neves, M. C., Freire, M. G., & Sousa, F. (2022). Supported ionic liquids used as chromatographic matrices in bioseparation—An overview. *Molecules*, *27*(5), 1618.

Bier, M. C. J., Medeiros, A. B. P., De Kimpe, N., & Soccol, C. R. (2019). Evaluation of antioxidant activity of the fermented product from the biotransformation of R-(+)-limonene in solid-state fermentation of orange waste by Diaporthe sp. *Biotechnology Research and Innovation*, *3*(1), 168–176.

Brexó, R. P., & Sant'Ana, A. S. (2017). Impact and significance of microbial contamination during fermentation for bioethanol production. *Renewable and Sustainable Energy Reviews*, *73*, 423–434. https://doi.org/10.1016/j.rser.2017.01.151

Chen, X., Diao, W., Ma, Y., & Mao, Z. (2020). Extraction and purification of ε-poly- L -lysine from fermentation broth using an ethanol/ammonium sulfate aqueous two-phase system combined with ultrafiltration. *RSC Advances*, *10*(49), 29587–29593. https://doi.org/10.1039/D0RA04245E

Chiou, W. L., & Riegelman, S. (1971). Pharmaceutical applications of solid dispersion systems. *Journal of Pharmaceutical Sciences*, *60*(9), 1281–1302.

Chung, D., Young, J., Cha, M., Brunecky, R., Bomble, Y. J., Himmel, M. E., & Westpheling, J. (2015). Expression of the Acidothermus cellulolyticus E1 endoglucanase in Caldicellulosiruptor bescii enhances its ability to deconstruct crystalline cellulose. *Biotechnology for Biofuels*, *8*(1), 113. https://doi.org/10.1186/s13068-015-0296-x

Costa, F. F., Oliveira, D. T. D., Brito, Y. P., Rocha Filho, G. N. D., Alvarado, C. G., Balu, A. M., Luque, R., & Nascimento, L. A. S. D. (2020). Lignocellulosics to biofuels: An overview of recent and relevant advances. *Current Opinion in Green and Sustainable Chemistry*, *24*, 21–25. https://doi.org/10.1016/j.cogsc.2020.01.001

Cubas-Cano, E., González-Fernández, C., Ballesteros, M., & Tomás-Pejó, E. (2018). Biotechnological advances in lactic acid production by lactic acid bacteria: Lignocellulose as novel substrate. *Biofuels, Bioproducts and Biorefining*, *12*(2), 290–303.

Debebe, A., Temesgen, S., Redi-Abshiro, M., Chandravanshi, B. S., & Ele, E. (2018). Improvement in analytical methods for determination of sugars in fermented alcoholic beverages. *Journal of analytical methods in chemistry, 2018*(1), 4010298.

Demirbas, A., Bafail, A., Ahmad, W., & Sheikh, M. (2016). Biodiesel production from non-edible plant oils. *Energy Exploration & Exploitation*, *34*(2), 290–318. https://doi.org/10.1177/0144598716630166

Dixon, R. A. (2013). Microbiology: Break down the walls. *Nature, 493*(7430), 36–37.

Doungsri, S., Rattanaphanee, P., & Wongkoblap, A. (2019). Production of lactic acid from cellulose using solid catalyst. *MATEC Web of Conferences*, *268*, 07006. https://doi.org/10.1051/matecconf/201926807006

Du, L., Arauzo, P. J., Meza Zavala, M. F., Cao, Z., Olszewski, M. P., & Kruse, A. (2020). Towards the Properties of Different Biomass-Derived Proteins via Various Extraction Methods. *Molecules (Basel, Switzerland), 25*(3), 488. https://doi.org/10.3390/molecules25030488

Dutta, N., Usman, M., Ashraf, M. A., Luo, G., Gamal El-Din, M., & Zhang, S. (2023). Methods to convert lignocellulosic waste into biohydrogen, biogas, bioethanol, biodiesel and value-added chemicals: A review. *Environmental Chemistry Letters*, *21*(2), 803–820.

Ekkers, D. M., Tusso, S., Moreno-Gamez, S., Rillo, M. C., Kuipers, O. P., & Van Doorn, G. S. (2022). Trade-offs predicted by metabolic network structure give rise to evolutionary specialization and phenotypic diversification. *Molecular Biology and Evolution*, *39*(6), msac124. https://doi.org/10.1093/molbev/msac124

Farchaus, J. W., Ribot, W. J., Jendrek, S., & Little, S. F. (1998). Fermentation, purification, and characterization of protective antigen from a recombinant, avirulent strain of Bacillus anthracis. *Applied and environmental microbiology, 64*(3), 982–991.

Farrell, A. E., Plevin, R. J., Turner, B. T., Jones, A. D., O'Hare, M., & Kammen, D. M. (2006). Ethanol can contribute to energy and environmental goals. *Science*, *311*(5760), 506–508. https://doi.org/10.1126/science.1121416

Ferraren-De Cagalitan, D. D. T., & Abundo, M. L. S. (2021). A review of biohydrogen production technology for application towards hydrogen fuel cells. *Renewable and Sustainable Energy Reviews*, *151*, 111413. https://doi.org/10.1016/j.rser.2021.111413

Fitzpatrick, J., Kricka, W., James, T. C., & Bond, U. (2014). Expression of three *Trichoderma reesei* cellulase genes in *Saccharomyces pastorianus* for the development of a two-step process of hydrolysis and fermentation of cellulose. *Journal of Applied Microbiology*, *117*(1), 96–108. https://doi.org/10.1111/jam.12494

Fu, C., Mielenz, J. R., Xiao, X., Ge, Y., Hamilton, C. Y., Rodriguez Jr, M., Chen, F., Foston, M., Ragauskas, A., & Bouton, J. (2011). Genetic manipulation of lignin reduces recalcitrance and improves ethanol production from switchgrass. *Proceedings of the National Academy of Sciences*, *108*(9), 3803–3808.

Furlong, V. B., Corrêa, L. J., Giordano, R. C., & Ribeiro, M. P. A. (2019). Fuzzy-enhanced modeling of lignocellulosic biomass enzymatic saccharification. *Energies*, *12*(11), 2110. https://doi.org/10.3390/en12112110

Gao, L., He, X., Guo, Y., Wu, Z., Zhao, J., Liu, G., & Qu, Y. (2021). Combinatorial engineering of transcriptional activators in *Penicillium oxalicum* for improved production of corn-fiber-degrading enzymes. *Journal of Agricultural and Food Chemistry*, *69*(8), 2539–2548. https://doi.org/10.1021/acs.jafc.0c07659

García-Peñas, A., & Sharma, G. (2023). *New Materials for a Circular Economy*. Materials Research Forum LLC, Millersville, PA.

Gregori Tena, A., & Javierre Cazador, D. (2023). *Centrifugal method to clean waste cooking oil: A study of the feasibility of constructing a centrifuge to remove unwanted particles from waste cooking oil using reused materials.* Bachelor Thesis for Bachelor of Engineering, Vaasa, Finland.

Grzybowski, B. A., Sobolev, Y. I., Cybulski, O., & Mikulak-Klucznik, B. (2022). Materials, assemblies and reaction systems under rotation. *Nature Reviews Materials*, *7*(5), 338–354.

Hermans, A., Milsmann, J., Li, H., Jede, C., Moir, A., Hens, B., Morgado, J., Wu, T., & Cohen, M. (2022). Challenges and strategies for solubility measurements and dissolution method development for amorphous solid dispersion formulations. *The AAPS Journal*, *25*(1), 11.

Hossain, G. S., Shin, H., Li, J., Wang, M., Du, G., Chen, J., & Liu, L. (2016). Metabolic engineering for amino-, oligo-, and polysugar production in microbes. *Applied Microbiology and Biotechnology*, *100*(6), 2523–2533. https://doi.org/10.1007/s00253-015-7215-8

Hua, X., Han, J., Liu, X., & Xu, Y. (2023). Cascading and precise regulation of the selective bio-production of 2- or 5-ketogluconic acid from glucose with whole-cell catalysis technology. *Green Chemistry*, *25*(6), 2378–2386.

Huang, J., Wang, J., & Liu, S. (2023). Advanced fermentation techniques for lactic acid production from agricultural waste. *Fermentation*, *9*(8), 765.

Iyyappan, J., Pravin, R., Al-Ghanim, K. A., Govindarajan, M., Nicoletti, M., & Baskar, G. (2023). Dual strategy for bioconversion of elephant grass biomass into fermentable sugars using Trichoderma reesei towards bioethanol production. *Bioresource Technology*, *374*, 128804. https://doi.org/10.1016/j.biortech.2023.128804

Kaur, M., Mohammad Said Al-Tawaha, A. R., & Karnwal, A. (2023). The beneficial impact of microbes in food production, health, and sustainability. In A. Karnwal & A. R. Mohammad Said Al-Tawaha (Eds), *Food Microbial Sustainability: Integration of Food Production and Food Safety* (pp. 289–309). Springer Nature, New York. https://doi.org/10.1007/978-981-99-4784-3_14

Khalili-Samani, A., Rezahasani, R., Satari, B., Aghbashlo, M., Amiri, H., Tabatabaei, M., & Nizami, A.-S. (2024). Sugar fermentation: C2 (ethanolic) platform. In H. Amiri, M. Tabatabaei, & A-S. Nizami (Eds), *Higher Alcohols Production Platforms* (pp. 99–123). Elsevier, Amsterdam, The Netherlands.

Kim, Y., Ximenes, E., Mosier, N. S., & Ladisch, M. R. (2011). Soluble inhibitors/deactivators of cellulase enzymes from lignocellulosic biomass. *Enzyme and Microbial Technology*, *48*(4–5), 408–415. https://doi.org/10.1016/j.enzmictec.2011.01.007

Kim, S.-K., Chung, D., Himmel, M. E., Bomble, Y. J., & Westpheling, J. (2019). Heterologous co-expression of two β-glucanases and a cellobiose phosphorylase resulted in a significant increase in the cellulolytic activity of the *Caldicellulosiruptor bescii* exoproteome. *Journal of Industrial Microbiology and Biotechnology*, *46*(5), 687–695. https://doi.org/10.1007/s10295-019-02150-0

Kim, J., Darlington, A., Salvador, M., Utrilla, J., & Jiménez, J. I. (2020). Trade-offs between gene expression, growth and phenotypic diversity in microbial populations. *Current Opinion in Biotechnology*, *62*, 29–37. https://doi.org/10.1016/j.copbio.2019.08.004

Kricka, W., Fitzpatrick, J., & Bond, U. (2014). Metabolic engineering of yeasts by heterologous enzyme production for degradation of cellulose and hemicellulose from biomass: A perspective. *Frontiers in Microbiology*, *5*, 174. https://doi.org/10.3389/fmicb.2014.00174

Kuhad, R. C., Deswal, D., Sharma, S., Bhattacharya, A., Jain, K. K., Kaur, A., Pletschke, B. I., Singh, A., & Karp, M. (2016). Revisiting cellulase production and redefining current strategies based on major challenges. *Renewable and Sustainable Energy Reviews*, *55*, 249–272. https://doi.org/10.1016/j.rser.2015.10.132

Kumar, K., Shah, H., & Moholkar, V. S. (2022). Genetic algorithm for optimization of fermentation processes of various enzyme productions. In Eswari J. S., Suryawanshi, N. (eds). *Optimization of sustainable enzymes production* (pp. 121–144). Chapman and Hall/CRC.

Kumar K, Barbora L, Moholkar V. S. (2023a). Genomic insights into clostridia in bioenergy production: comparison of metabolic capabilities and evolutionary relationships. *Biotechnology and Bioengineering* *121*(4): 1297–1312. https://doi.org/10.1002/bit.28610

Kumar, K., and V. S. Moholkar. (2023b). Mechanistic aspects of enhanced kinetics in sonoenzymatic processes using three simultaneous approaches. In: V. S. Moholkar, K. Mohanty, and V. V. Goud (eds). *Sustainable Energy Generation and Storage*, 41–57. Singapore: Springer Nature Singapore. https://doi.org/10.1007/978-981-99-2088-4_5.

Kumar, K., Anand, A., & Moholkar, V. S. (2024a). Molecular Hydrogen (H2) Metabolism in Microbes: A Special Focus on Biohydrogen Production. In *Biohydrogen-Advances and Processes* (pp. 25–58). Cham: Springer Nature Switzerland.

Kumar, K., Kumar, S., Goswami, A., & Moholkar, V. S. (2024b). Advancing sustainable biofuel production: A computational insight into microbial systems for isopropanol synthesis and beyond. *Process Safety and Environmental Protection,* 188, 1118–1132. https://doi.org/10.1016/j.psep.2024.06.024

Kumar, K., Jadhav, S. M., & Moholkar, V. S. (2024c). Acetone-Butanol-Ethanol (ABE) fermentation with clostridial co-cultures for enhanced biobutanol production. *Process Safety and Environmental Protection,* 185, 277–285. https://doi.org/10.1016/j.psep.2024.03.027

Kumar, R. R., & Bhattacharya, C. (2023). Design and operation of new microbial product bioprocessing system. In A. Sarkar, & I. A. Ahmed (Eds), *Microbial Products for Future Industrialization* (pp. 23–54). Springer, New York.

Kumar, A., Yadav, M., & Sehrawat, N. (Eds). (2019). *Microbial Enzymes and Additives for the Food Industry*. Nova Science Pubishers, New York.

Kumar, J. A., Sathish, S., Prabu, D., Renita, A. A., Saravanan, A., Deivayanai, V. C., Anish, M., Jayaprabakar, J., Baigenzhenov, O., & Hosseini-Bandegharaei, A. (2023). Agricultural waste biomass for sustainable bioenergy production: Feedstock, characterization and pre-treatment methodologies. *Chemosphere*, *331*, 138680. https://doi.org/10.1016/j.chemosphere.2023.138680

Kun, R. S., Garrigues, S., Di Falco, M., Tsang, A., & de Vries, R. P. (2021). The chimeric GaaR-XlnR transcription factor induces pectinolytic activities in the presence of D-xylose in Aspergillus niger. *Applied Microbiology and Biotechnology*, *105*(13), 5553–5564. https://doi.org/10.1007/s00253-021-11428-2

Kunasundari, B., Zulkeple, M. F., & Teoh, Y. P. (2017). Screening for direct production of lactic acid from rice starch waste by Geobacillus stearothermophilus. *MATEC Web of Conferences*, *97*, 01049.

Kurniawati, A., Toth, G., Ylivainio, K., & Toth, Z. (2023). Opportunities and challenges of bio-based fertilizers utilization for improving soil health. *Organic Agriculture*, *13*(3), 335–350. https://doi.org/10.1007/s13165-023-00432-7

Kushida, W., Gonzales, R. R., Shintani, T., Matsuoka, A., Nakagawa, K., Yoshioka, T., & Matsuyama, H. (2022). Organic solvent mixture separation using fluorine-incorporated thin film composite reverse osmosis membrane. *Journal of Materials Chemistry A*, *10*(8), 4146–4156.

Leam, J. J., Bilad, M. R., Wibisono, Y., Wirzal, M. D. H., & Ahmed, I. (2020). Membrane technology for microalgae harvesting. In *Microalgae Cultivation for Biofuels Production* (pp. 97–110). Academic Press.

Ledesma-Amaro, R., & Nicaud, J.-M. (2016). Metabolic engineering for expanding the substrate range of yarrowia lipolytica. *Trends in Biotechnology*, *34*(10), 798–809. https://doi.org/10.1016/j.tibtech.2016.04.010

Lee, S., Kang, M., Bae, J.-H., Sohn, J.-H., & Sung, B. H. (2019). Bacterial valorization of lignin: Strains, enzymes, conversion pathways, biosensors, and perspectives. *Frontiers in Bioengineering and Biotechnology*, *7*, 209. https://doi.org/10.3389/fbioe.2019.00209

Lee, H., Sohn, Y. J., Jeon, S., Yang, H., Son, J., Kim, Y. J., & Park, S. J. (2023). Sugarcane wastes as microbial feedstocks: A review of the biorefinery framework from resource recovery to production of value-added products. *Bioresource Technology*, *376*, 128879.

Lips, D. (2022). Fuelling the future of sustainable sugar fermentation across generations. *Engineering Biology*, *6*(1), 3–16. https://doi.org/10.1049/enb2.12017

Liu, J., Yang, J., Wang, R., Liu, L., Zhang, Y., Bao, H., Jang, J. M., Wang, E., & Yuan, H. (2020). Comparative characterization of extracellular enzymes secreted by Phanerochaete chrysosporium during solid-state and submerged fermentation. *International Journal of Biological Macromolecules*, *152*, 288–294. https://doi.org/10.1016/j.ijbiomac.2020.02.256

Liu, C., Dong, G., Tsuru, T., & Matsuyama, H. (2021). Organic solvent reverse osmosis membranes for organic liquid mixture separation: A review. *Journal of Membrane Science*, *620*, 118882.

Ma, Y., Kou, Z., Hu, Y., Zhou, J., Bei, Y., Hu, L., Huang, Q., Jia, P., & Zhou, Y. (2023). Research advances in bio-based adhesives. *International Journal of Adhesion and Adhesives*, *126*, 103444. https://doi.org/10.1016/j.ijadhadh.2023.103444

Macedo, J. V. C., de Barros Ranke, F. F., Escaramboni, B., Campioni, T. S., Núñez, E. G. F., & de Oliva Neto, P. (2020). Cost-effective lactic acid production by fermentation of agro-industrial residues. *Biocatalysis and Agricultural Biotechnology*, *27*, 101706.

Magwaza, L. S., & Opara, U. L. (2015). Analytical methods for determination of sugars and sweetness of horticultural products-A review. *Scientia Horticulturae*, *184*, 179–192. https://doi.org/10.1016/j.scienta.2015.01.001

Mahato, S., Sharma, K., & Cho, M. H. (2019). Modern extraction and purification techniques for obtaining high purity food-grade bioactive compounds and value-added co-products from citrus wastes. *Foods*, *8*(11), 523. https://doi.org/10.3390/foods8110523

Mahboubi, A., Cayli, B., Bulkan, G., Doyen, W., De Wever, H., & Taherzadeh, M. J. (2018). Removal of bacterial contamination from bioethanol fermentation system using membrane bioreactor. *Fermentation*, *4*(4), 88.

Makino, T., Skretas, G., & Georgiou, G. (2011). Strain engineering for improved expression of recombinant proteins in bacteria. *Microbial Cell Factories*, *10*(1), 32. https://doi.org/10.1186/1475-2859-10-32

Mankar, A. R., Pandey, A., Modak, A., & Pant, K. (2021). Pretreatment of lignocellulosic biomass: A review on recent advances. *Bioresource Technology*, *334*, 125235.

Masri, F., Hoeve, M. A., De Sousa, P. A., & Willoughby, N. A. (2017). Challenges and advances in scale-up of label-free downstream processing for allogeneic cell therapies. *Cell and Gene Therapy Insights*, *3*(6), 447–467.

Mathur, S., Kumar, D., Kumar, V., Dantas, A., Verma, R., & Kuca, K. (2023). Xylitol: Production strategies with emphasis on biotechnological approach, scale up, and market trends. *Sustainable Chemistry and Pharmacy*, *35*, 101203.

Mhatre, A., Kalscheur, B., Mckeown, H., Bhakta, K., Sarnaik, A. P., Flores, A., Nielsen, D. R., Wang, X., Soundappan, T., & Varman, A. M. (2022). Consolidated bioprocessing of hemicellulose to fuels and chemicals through an engineered Bacillus subtilis-Escherichia coli consortium. *Renewable Energy*, *193*, 288–298. https://doi.org/10.1016/j.renene.2022.04.124

Mujtaba, M., Fraceto, L., Fazeli, M., Mukherjee, S., Savassa, S. M., de Medeiros, G. A., Santo Pereira, A. do E., Mancini, S. D., Lipponen, J., & Vilaplana, F. (2023). Lignocellulosic biomass from agricultural waste to the circular economy: A review with focus on biofuels, biocomposites and bioplastics. *Journal of Cleaner Production*, *402*, 136815.

Nancib, A., Nancib, N., Boubendir, A., & Boudrant, J. (2015). The use of date waste for lactic acid production by a fed-batch culture using Lactobacillus casei subsp. Rhamnosus. *Brazilian Journal of Microbiology*, *46*, 893–902.

Nautiyal, P., Boesl, B., Agarwal, A., Nautiyal, P., Boesl, B., & Agarwal, A. (2020). Challenges during in-situ mechanical testing: Some practical considerations and limitations. In-situ Mechanics of Materials: Principles, Tools, Techniques and Applications, 227–238.

Nigam, P. S., & Singh, A. (2011). Production of liquid biofuels from renewable resources. *Progress in Energy and Combustion Science*, *37*(1), 52–68. https://doi.org/10.1016/j.pecs.2010.01.003

Noorimotlagh, Z., Mirzaee, S. A., Jaafarzadeh, N., Maleki, M., Kalvandi, G., & Karami, C. (2021). A systematic review of emerging human coronavirus (SARS-CoV-2) outbreak: Focus on disinfection methods, environmental survival, and control and prevention strategies. *Environmental Science and Pollution Research International*, *28*(1), 1–15. https://doi.org/10.1007/s11356-020-11060-z

Nordstrom, F. L., Sirota, E., Hartmanshenn, C., Kwok, T. T., Paolello, M., Li, H., Abeyta, V., Bramante, T., Madrigal, E., & Behre, T. (2023). Prevalence of impurity retention mechanisms in pharmaceutical crystallizations. *Organic Process Research & Development*, *27*(4), 723–741.

Obodo, K., & Aigbe, U. O. (2023). The types, characteristics, and management options (reusability/recyclability/final disposal) of commonly used adsorbents in environmental sustainability. In K. E. Ukhurebor, U. O. Aigbe, & R. B. Onyancha (Eds), *Adsorption Applications for Environmental Sustainability* (pp. 2–25). IOP Publishing, Bristol, UK.

Okal, E. J., Aslam, M. M., Karanja, J. K., & Nyimbo, W. J. (2020). Mini review: Advances in understanding regulation of cellulase enzyme in white-rot basidiomycetes. *Microbial Pathogenesis*, *147*, 104410. https://doi.org/10.1016/j.micpath.2020.104410

Okeke, E. S., Ezugwu, A. L., Anaduaka, E. G., Mayel, M. H., Ezike, T. C., & Ossai, E. C. (2022). Ligninolytic and cellulolytic enzymes-biocatalysts for green agenda. *Biomass Conversion and Biorefinery*, *2022*, 3011–3055. https://doi.org/10.1007/s13399-022-02777-7

Passos, D. D. F., Pereira, N., & Castro, A. M. D. (2018). A comparative review of recent advances in cellulases production by *Aspergillus*, *Penicillium* and *Trichoderma* strains and their use for lignocellulose deconstruction. *Current Opinion in Green and Sustainable Chemistry*, *14*, 60–66. https://doi.org/10.1016/j.cogsc.2018.06.003

Patel, A. K., & Sharma, A. K. (2023). *Sustainable Production Innovations: Bioremediation and Other Biotechnologies*. John Wiley & Sons, Hoboken, NJ.

Peng, B., Plan, M. R., Carpenter, A., Nielsen, L. K., & Vickers, C. E. (2017). Coupling gene regulatory patterns to bioprocess conditions to optimize synthetic metabolic modules for improved sesquiterpene production in yeast. *Biotechnology for Biofuels*, *10*(1), 43. https://doi.org/10.1186/s13068-017-0728-x

Plácido, J., & Capareda, S. (2015). Ligninolytic enzymes: A biotechnological alternative for bioethanol production. *Bioresources and Bioprocessing*, *2*(1), 23. https://doi.org/10.1186/s40643-015-0049-5

Priya, A., Dutta, K., & Daverey, A. (2022). A comprehensive biotechnological and molecular insight into plastic degradation by microbial community. *Journal of Chemical Technology & Biotechnology*, *97*(2), 381–390.

Queiroz, S. de S., Jofre, F. M., Bianchini, I. de A., Boaes, T. da S., Bordini, F. W., Chandel, A. K., & Felipe, M. das G. de A. (2023). Current advances in Candida tropicalis: Yeast overview and biotechnological applications. *Biotechnology and Applied Biochemistry*, *70*, 2069–2087.

Rajoub, N., Gerard, C. J., Pantuso, E., Fontananova, E., Caliandro, R., Belviso, B. D., Curcio, E., Nicoletta, F. P., Pullen, J., & Chen, W. (2023). A workflow for the development of template-assisted membrane crystallization downstream processing for monoclonal antibody purification. *Nature Protocols*, *18*, 1–52.

Ramonet Marques, F., Gleeson, T., Galvin, M., Haddadi Sisakht, B., & Harasek, M. (2023). Anaerobic digestion as a tool to mitigate greenhouse gas emissions from animal slurries. *Atmosphere*, *14*, 120. https://repositum.tuwien.at/handle/20.500.12708/188836

Saari, H., Pusa, R., Marttila, H., Yliperttula, M., & Laitinen, S. (2023). Development of tandem cation exchange chromatography for high purity extracellular vesicle isolation: The effect of ligand steric availability. *Journal of Chromatography A*, *1707*, 464293. https://doi.org/10.1016/j.chroma.2023.464293

Saini, A., Aggarwal, N. K., Sharma, A., & Yadav, A. (2015). Actinomycetes: A source of lignocellulolytic enzymes. *Enzyme Research*, *2015*, 1–15. https://doi.org/10.1155/2015/279381

Saini, A., Aggarwal, N. K., & Yadav, A. (2016). Cellulolytic potential of actinomycetes isolated from different habitats. *Bioengineering and Bioscience*, *4*(5), 88–94. https://doi.org/10.13189/bb.2016.040503

Sandesh, J. S. S., Shyamala, S. K., Balaiah, S., & Sharma, J. V. C. (2021). A review on flash chromatography and its pharmaceutical applications. *J. Biomed. Pharm. Res*, 10, 120–124.

Saraiva, A., Carrascosa, C., Ramos, F., Raheem, D., & Raposo, A. (2022). Agave syrup: Chemical analysis and nutritional profile, applications in the food industry and health impacts. *International Journal of Environmental Research and Public Health*, *19*(12), 7022. https://doi.org/10.3390/ijerph19127022

Sarkar, A., & Ahmed, I. A. (2023). *Microbial Products for Future Industrialization*. Springer Nature, New York.

Sarnaik, A., Liu, A., Nielsen, D., & Varman, A. M. (2020). High-throughput screening for efficient microbial biotechnology. *Current Opinion in Biotechnology*, *64*, 141–150. https://doi.org/10.1016/j.copbio.2020.02.019

Schirmeister, C. G., & Mülhaupt, R. (2022). Closing the carbon loop in the circular plastics economy. *Macromolecular Rapid Communications*, *43*(13), 2200247.

Sha, C., Sadaqat, B., Wang, H., Guo, X., & Shao, W. (2020). Efficient xylan-to-sugar biotransformation using an engineered xylanase in hyperthermic environment. *International Journal of Biological Macromolecules*, *157*, 17–23. https://doi.org/10.1016/j.ijbiomac.2020.04.145

Shallom, D., & Shoham, Y. (2003). Microbial hemicellulases. *Current Opinion in Microbiology*, *6*(3), 219–228. https://doi.org/10.1016/S1369-5274(03)00056-0

Shapiro, R. S., Chavez, A., & Collins, J. J. (2018). CRISPR-based genomic tools for the manipulation of genetically intractable microorganisms. *Nature Reviews Microbiology*, *16*(6), 333–339. https://doi.org/10.1038/s41579-018-0002-7

Sharma, A., Tewari, R., Rana, S. S., Soni, R., & Soni, S. K. (2016). Cellulases: Classification, methods of determination and industrial applications. *Applied Biochemistry and Biotechnology*, *179*(8), 1346–1380. https://doi.org/10.1007/s12010-016-2070-3

Sharma, A., Rana, A., Sharma, N., & Sharma, S. (2024). Application of fermentation techniques in the production of genetically engineered microorganisms (GMOs). In Inamuddin, C. O. Adetunji, M. F. Ahmer, & T. A. Altalhi (Eds), *Genetically Engineered Organisms in Bioremediation* (pp. 63–73). CRC Press, Boca Raton, FL.

Shekunov, B. Y., & York, P. (2000). Crystallization processes in pharmaceutical technology and drug delivery design. *Journal of Crystal Growth*, *211*(1–4), 122–136.

Singhvi, M. S., Chaudhari, S., & Gokhale, D. V. (2014). Lignocellulose processing: A current challenge. *RSC Advances*, *4*(16), 8271. https://doi.org/10.1039/c3ra46112b

Smith, D. (2012). Culture collections. In *Advances in applied microbiology* (Vol. 79, pp. 73–118). Academic Press.

Srivastava, N., Mishra, K., Srivastava, M., Srivastava, K. R., Gupta, V. K., Ramteke, P. W., & Mishra, P. K. (2019). Role of compositional analysis of lignocellulosic biomass for efficient biofuel production. In V. K. Gupta (Eds), *New and Future Developments in Microbial Biotechnology and Bioengineering* (pp. 29–43). Elsevier, Amsterdam, The Netherlands. https://doi.org/10.1016/B978-0-444-64223-3.00003-5

Suleiman, M., Krüger, A., & Antranikian, G. (2020). Biomass-degrading glycoside hydrolases of archaeal origin. *Biotechnology for Biofuels*, *13*(1), 153. https://doi.org/10.1186/s13068-020-01792-y

Sun, B., Zhao, J., Wang, T., Li, Y., Yang, X., Tan, F., Li, Y., Chen, C., & Sun, D. (2023). Highly efficient construction of sustainable bacterial cellulose aerogels with boosting PM filter efficiency by tuning functional group. *Carbohydrate Polymers*, *309*, 120664.

Taherzadeh, M. J., & Karimi, K. (2008). Pretreatment of lignocellulosic wastes to improve ethanol and biogas production: A review. *International Journal of Molecular Sciences*, *9*(9), 1621–1651.

Tan, J., Li, Y., Tan, X., Wu, H., Li, H., & Yang, S. (2021). Advances in pretreatment of straw biomass for sugar production. *Frontiers in Chemistry*, *9*, 696030. https://doi.org/10.3389/fchem.2021.696030

Tehrani, S. F., Bharadwaj, P., Chain, J. L., & Roullin, V. G. (2023). Purification processes of polymeric nanoparticles: How to improve their clinical translation? *Journal of Controlled Release*, *360*, 591–612.

Tsipis, E. V., Agarkov, D. A., Borisov, Yu . A., Kiseleva, S. V., Tarasenko, A. B., Bredikhin, S. I., & Kharton, V. V. (2023). Waste gas utilization potential for solid oxide fuel cells: A brief review. *Renewable and Sustainable Energy Reviews*, *188*, 113880. https://doi.org/10.1016/j.rser.2023.113880

Tuovinen, O. H., & Kelly, D. P. (1973). Studies on the growth of Thiobacillus ferrooxidans: I. Use of membrane filters and ferrous iron agar to determine viable numbers, and comparison with 14CO2-fixation and iron oxidation as measures of growth. *Archiv Für Mikrobiologie*, *88*, 285–298.

Urh, M., Simpson, D., & Zhao, K. (2009). Affinity chromatography: General methods. *Methods in Enzymology*, *463*, 417–438.

Valles, A., Capilla, M., Álvarez-Hornos, F. J., García-Puchol, M., San-Valero, P., & Gabaldón, C. (2021). Optimization of alkali pretreatment to enhance rice straw conversion to butanol. *Biomass and Bioenergy*, *150*, 106131. https://doi.org/10.1016/j.biombioe.2021.106131

Van Reis, R., & Zydney, A. (2001). Membrane separations in biotechnology. *Current Opinion in Biotechnology*, *12*(2), 208–211.

Vergauwen, L., Scanlan, C., Krishnan, R., Pandey, S. K., Loong, D., & Bhushan, A. (2023). Filtration principles and techniques for bioprocessing of viral vector-based therapeutics. In S. Gautam, A. I. Chiramel, & R. Pach (Eds), *Bioprocess and Analytics Development for Virus-based Advanced Therapeutics and Medicinal Products (ATMPs)* (pp. 125–143). Springer, New York.

Vyas, A., & Nidetzky, B. (2023). Energetics of the glycosyl transfer reactions of sucrose phosphorylase. *Biochemistry*, *62*, 1953–1963.

Wang, F., Ge, X., Yuan, Z., Zhang, X., Chu, X., Lu, F., & Liu, Y. (2022). Insights into the mechanism for the high-alkaline activity of a novel GH43 β-xylosidase from Bacillus clausii with a promising application to produce xylose. *Bioorganic Chemistry*, *126*, 105887. https://doi.org/10.1016/j.bioorg.2022.105887

Wen, F., Sun, J., & Zhao, H. (2010). Yeast surface display of trifunctional minicellulosomes for simultaneous saccharification and fermentation of cellulose to ethanol. *Applied and Environmental Microbiology*, *76*(4), 1251–1260. https://doi.org/10.1128/AEM.01687-09

Wyman, C. (2018). *Handbook on Bioethanol: Production and Utilization*. In C. E. Wyman (Ed) (1st ed, pp. 179–212). Routledge, Boca Raton, FL. https://doi.org/10.1201/9780203752456

Xu, Z., Hao, N., Li, L., Zhang, Y., Yu, L., Jiang, L., & Sui, X. (2019). Valorization of soy whey wastewater: How epigallocatechin-3-gallate regulates protein precipitation. *ACS Sustainable Chemistry & Engineering*, *7*(18), 15504–15513. https://doi.org/10.1021/acssuschemeng.9b03208

Yamada, R., Hasunuma, T., & Kondo, A. (2013). Endowing non-cellulolytic microorganisms with cellulolytic activity aiming for consolidated bioprocessing. *Biotechnology Advances*, *31*(6), 754–763. https://doi.org/10.1016/j.biotechadv.2013.02.007

Zabed, H., Sahu, J. N., Suely, A., Boyce, A. N., & Faruq, G. (2017). Bioethanol production from renewable sources: Current perspectives and technological progress. *Renewable and Sustainable Energy Reviews*, *71*, 475–501. https://doi.org/10.1016/j.rser.2016.12.076

Zhang, W., Guo, J., Wu, X., Ren, Y., Li, C., Meng, X., & Liu, W. (2022). Reformulating the hydrolytic enzyme cocktail of *Trichoderma reesei* by combining XYR1 overexpression and elimination of four major cellulases to improve saccharification of corn fiber. *Journal of Agricultural and Food Chemistry*, *70*(1), 211–222. https://doi.org/10.1021/acs.jafc.1c05946

Zhu, L., Gao, N., & Cong, R.-G. (2017). Application of biotechnology for the production of biomass-based fuels. *BioMed Research International*, *2017*, 1–2. https://doi.org/10.1155/2017/3896505

Zoghlami, A., & Paës, G. (2019). Lignocellulosic biomass: Understanding recalcitrance and predicting hydrolysis. *Frontiers in Chemistry*, *7*, 874. https://doi.org/10.3389/fchem.2019.00874

4 Application of Enzymatic Technique for the Synthesis of Fermentable Sugars from Different Agrobiomasses

Meyyappan K, Sancia Morris, Vishal S K, Ashwinth K, and Kruthi Doriya

4.1 INTRODUCTION

Renewable energy systems, such as bioethanol, offer many advantages, achieving a stable balance between energy supply and demand, safeguarding food security and economic equilibrium, and mitigating pollution. Nonetheless, for them to be a sustainable long-term option, the production costs must be lowered to effectively compete with petroleum (Periyasamy et al., 2022). Biofuels can utilize a diverse range of raw materials, including sugars, starchy substances, agricultural products, forestry resources, municipal waste, industrial by-products, and urban residues, rendering them well-suited for biofuel production (Bušić et al., 2018).

Lignocellulosic agricultural waste, which contains cellulose, hemicellulose, lignin, and other components, is a promising biofuel source (Akhtar et al., 2016). Lignocellulosic biomasses are renewable resources that do not emit CO_2 into the atmosphere, unlike starch-based feedstocks that may compete with food production. Hence, agrobiomass is increasingly favored as a feedstock to generate fermentable sugars, enabling the production of diverse value-added products like biofuels, chemicals, and materials (Tse et al., 2021; Blasi et al., 2023).

Furthermore, the use of agricultural waste derived from lignocellulosic sources in biorefineries promotes a circular bioeconomy. This not only improves the quality of life, but also reduces dependence on petroleum-based resources and expands waste management options (Blasi et al., 2023). Agro-food residues and agrobiomass primarily consist of lignocellulosic biomass, which includes cellulose, hemicellulose, and lignin. Cellulose is a polymer made up of D-glucose molecules connected by β-1,4-glycosidic bonds, while hemicellulose is a heteropolymer composed of various sugar monomers. Lignin, on the other hand, acts as a protective barrier, making enzymatic hydrolysis and the degradation of biomass more challenging (Diaz et al., 2018).

DOI: 10.1201/9781003407713-4

Methods for pretreatment can be broadly categorized into four groups: physical, chemical, physicochemical, and biological. Biological pretreatment methods use microorganisms, such as fungi or bacteria, or enzymes, such as cellulase, xylanase, β-glucosidase, and endoxylanase; cellulases (cellobiohydrolase, endoglucanase, and β-glucosidase) are used to break down cellulose and xylanases (endoxylanase and β-xylosidase) are used to break down hemicellulose. Advancement and innovations in the field of biofuels and bioproducts are currently focused on utilizing enzymatic techniques to synthesize fermentable sugars from various types of agrobiomasses (Bušić et al., 2018).

The conversion of biomass into biofuels and other valuable products primarily involves two fermentation approaches: submerged fermentation (SmF) and solid-state fermentation (SSF). In SmF, microorganisms are cultivated in a liquid medium containing biomass and necessary nutrients. In contrast, SSF occurs on solid biomass with adequate moisture content, which supports microbial growth and the production of desired products. Both bioethanol and the generation of postbiotics can be accomplished using either SmF or SSF (Ruslan & Ahmad, 2023). Notably, SSF characterized by restricted water availability and higher substrate concentrations, demonstrates remarkable ethanol yield (up to 30%) and accelerated production compared to SmF (Passadis et al., 2022).

Figure 4.1 depicts the process of transforming agrobiomass residues into bioethanol. This transformation involves the initial depolymerization of lignin and cellulose through pretreatment, followed by hydrolysis to extract individual components. Subsequently, these components can undergo fermentation or chemical treatment for

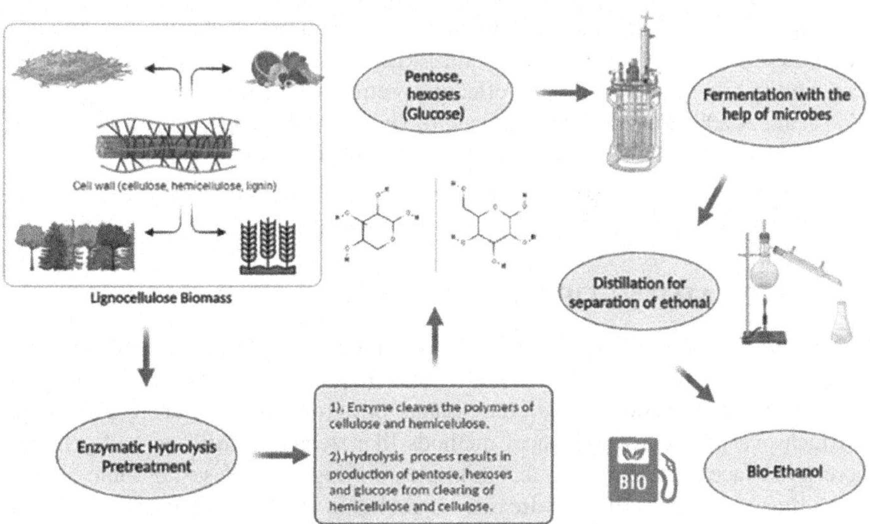

FIGURE 4.1 Schematic representation of lignocellulosic biomass biotransformation into biofuel.

the production of bioethanol. This chapter offers a thorough examination of agrobiomass and its pretreatment techniques, with a specific emphasis on the involvement of enzymes and the subsequent conversion processes. The chapter also explores the production techniques of hydrolytic enzymes and discusses the future prospects in this field.

4.2 COMPOSITION OF AGROBIOMASS

Renewable biomass sourced from agriculture is a cost-effective and readily available energy option. It comprises of three main components: cellulose, hemicelluloses, and lignin. Among these, lignin stands out as an abundant aromatic biopolymer with the potential to replace petroleum-based resources (Isikgor & Becer, 2015). Maj's research in 2018 has laid a strong foundation for considering agrobiomass, including materials like wheat straw, oat grains, larch needles, and rapeseed pods, as a sustainable and environmentally friendly fuel source. This work also demonstrated that the biofuels under investigation align with the principles of sustainable development (Maj, 2018).

Lignocellulosic biomass, which includes energy crops, agricultural residues, forest remnants, and more, represents a sustainable and environmentally friendly alternative to petroleum. Its key biopolymers consist of cellulose, hemicelluloses, and lignin. Cellulose is a sturdy, insoluble polysaccharide with a high degree of polymerization, breaking down at approximately 180°C. Hemicelluloses are sugars with five or six carbon atoms, and are easily hydrolyzed compared to cellulose (Periyasamy et al., 2023). Lignin, constituting 15%–40% of the non-carbohydrate components in lignocellulosic biomass, is a complex aromatic biopolymer. It acts as a protective and strengthening element for polysaccharides, shielding them from microbial degradation (Yoo et al., 2020a). The research by Yoo et al. underscores how lignin characteristics, including content, composition, and molecular weight, influence the recalcitrance and conversion of biomass. Lignin's presence can hinder cellulose accessibility and sugar release, but effective pretreatment can mitigate these effects (Yoo et al., 2020b). Furthermore, lignin carbohydrate complex (LCC) forms through covalent bonds between lignin and various carbohydrates like glucose, galactose, mannose, arabinose, and xylose. These bonds include linkages (Periyasamy et al., 2023) (Table 4.1).

4.3 PRETREATMENT OF AGROBIOMASS

Pretreatment of agrobiomass for sugar production involves several steps to improve the enzymatic access to cellulose and hemicellulose. Amin et al. (2017) provide an extensive examination of the pretreatment of agrobiomass, offering a thorough and inclusive perspective. Physical methods like steaming, milling, and grinding increase surface area. Chemical techniques employing acids, bases, or ionic liquids, e.g., sulfuric acid and sodium hydroxide, target lignin and hemicellulose breakdown. Biological approaches, often combined with other methods, employ microorganisms, while thermal methods like steam explosion and microwave heating degrade lignin and hemicellulose through high temperatures (Amin et al., 2017).

TABLE 4.1
Cellulose, Hemicellulose, and Lignin Content in Different Types of Agrobiomasses

Agrobiomass	Cellulose (%)	Hemicellulose (%)	Lignin (%)	Reference
Sugarcane bagasse	32–45	20–32	17–32	Alokika et al. (2021)
Cornstalk	38	29	23	Zhang et al. (2022)
Sweet sorghum bagasse	17	17	30	Mafa et al. (2020)
Wheat straw	35–45	20–30	8–15	Tufail et al. (2021)
Rapeseed straw	49	13	17	Ji et al. (2014)
Rice straw	35–50	20–35	10–25	Chen et al. (2020)
Corn cob	24	33	22	Mafa et al. (2020)
Milled barley husk	41	22	25	El Halal et al. (2015)
Soybean hull	29	20	13	Qing et al. (2017)
Soybean straw	42	17	22	Qing et al. (2017)

4.3.1 Physical Pretreatment Methods

Generally, physical pretreatment methods are favored over chemical pretreatment methods because they are less expensive, non-toxic, non-corrosive, and have a minimal environmental footprint.

Mechanical pretreatment methods such as ball-milling, rod-milling, ultra-fine grinding, and extrusion are employed for the pretreatment of the various substrates such as rice straw, wheat straw, *Phragmites australis*, and corn stover (Jędrzejczyk et al., 2019). For example, rod-milling pretreatment of wheat straw enhances pyrolysis efficiency by reducing particle size and cellulose crystallinity, increasing specific surface area and lowering thermal degradation temperature. This results in lower activation energy values (Bai et al., 2018). Extrusion is a physical method used for pretreating biomass in bioethanol production, offering several advantages such as versatile control options, absence of sugar degradation products, compatibility with other processes, and the ability to handle high solid loads (Zheng & Rehmann, 2014). The extrusion pretreatment process employs tightly fitted extruder screws to create high shear forces, generating pressure and temperature to modify the biomass structure. Parameters like screw type, rotation speed, temperature, and moisture content significantly influence the outcome. Increasing screw speed enhances glucose yield by reducing fiber length and increasing enzyme accessibility (Han et al., 2013). Ai et al. utilized a twin-screw extruder and a neutral-pH deep eutectic solvents to pretreat sorghum bagasse biomass at solid loadings of up to 50%. Experimental investigation suggests that the continuous extrusion process achieved over 85% glucose and xylose yields through enzymatic saccharification, while maintaining the biomass's chemical composition (Ai et al., 2020).

Integrating ultrasonic energy into pretreatment and conversion processes provides advantages by facilitating lignocellulosic structure degradation, biomass component fractionation, and increases yields of sugars, bioethanol, and gas products. High-frequency ultrasound enables biomass harvesting and process monitoring, while high-intensity ultrasound activates heterogeneous and enzymatic catalysis, offering positive process benefits for biomass-to-biofuels conversion (Luo et al., 2014). Zhang et al. conducted a study demonstrating that ultrasound-assisted hydrochloric acid-catalyzed pretreatment with ionic liquid solutions significantly enhanced the production of reducing sugar, cellulose conversion, and delignification in rice straw. Temperature, acid concentration, and time were found to have a notable impact on enzymatic hydrolysis, while ultrasound played a crucial role in disrupting the morphology, chemical structure, and crystallinity of the rice straw (Zhang et al., 2021). Further, optimizing pretreatment parameters is crucial for maximizing sugar release, minimizing inhibitors, and achieving higher ethanol concentrations without the need for detoxification or added costs (Mikulski & Kłosowski, 2020).

4.3.2 Chemical Methods

Several chemical techniques, such as alkali treatments, dilute acids, oxidizing agents, and organic solvents, are utilized in the chemical pretreatment of biomass. These methods enhance the biomass's suitability for subsequent processing or conversion. Effective pretreatment processes take into account factors such as cost-effective chemical usage, minimal chemical consumption, the preservation of hemicellulose and cellulose integrity, low energy requirements, economical size reduction, and the production of reactive cellulosic fibers (Nauman Aftab et al., 2019).

Alkaline pretreatment utilizes bases like NaOH, KOH, $Ca(OH)_2$, Na_2CO_3, and liquid ammonia to remove lignin, hemicellulose, and cellulose, improving porosity and structure (Saratale & Oh, 2015). Although NaOH is commonly used for biogas enhancement, it can lead to inhibitory phenolic compounds and additional pH control costs. Alkali pretreatment is effective for cellulosic materials at lower temperatures and pressures, with sodium hydroxide extensively studied and cost-effective calcium hydroxide that is capable of being regenerated to calcium carbonate using lime kiln technology (Kim et al., 2016).

Acid pretreatment offers advantages in lignocellulosic matrix disruption and amorphous cellulose conversion compared to other methods. Although it is considered a promising technology nearing commercialization, it produces inhibitory compounds like aldehydes, ketones, and phenolic acids from sugar degradation and lignin decomposition (Jung & Kim, 2015). This acid pretreatment can be categorized into wet and dry methods based on solids loading. Wet acid pretreatment, commonly used for hemicellulose disruption in second-generation raw materials, utilizes mineral and organic acids as catalysts. Concentrated acid pretreatment offers lower operation costs with lower temperatures, but equipment must be corrosion-resistant. Dilute acid-based pretreatment has drawbacks like high-pressure and high-temperature requirements, the need for small biomass particle sizes, neutralization and salt removal, and the formation of inhibitors (Hoang et al., 2021). For instance, in the helically agitated reactor, dry dilute acid pretreatment of corn stover and dilute sulfuric acid but without impregnation achieved a cellulose conversion of 84.77% and ethanol yield of 79.80%, similar to well-impregnated corn stover (He et al., 2014). In general, sulfuric acid (H_2SO_4) is the preferred catalyst for biomass acid pretreatment because it effectively targets the breakdown of carbon-oxygen and alkyl-aryl ether bonds in biomass and lignin polymers (Hoang et al., 2021).

Ozonolysis is increasingly popular for its efficient and gentle nature, reducing lignin content in lignocellulosic wastes, enhancing digestibility, and leaving no toxic residues (Travaini et al., 2016). Over recent decades, extensive research has documented the ozonolysis of diverse biomasses, such as cereal straws. Current investigations are emphasizing the importance of exploring gentler, eco-friendly methods like microwave and ultrasound activation, along with advanced oxidative processes, to improve the effectiveness of lignocellulosic biomass pretreatment (Den et al., 2018). As an example, employing a two-stage ozonolysis process with an intermediate washing step resulted in a substantial decrease in the acid-insoluble lignin content of wheat straw, lowering it from 13.04% to 9.34%. Simultaneously, it led to a notable increase in the yield of fermentable sugars, elevating it from 60% to 80% (Al jibouri et al., 2015).

Organic solvent pretreatment efficiently separates high-purity cellulose with minimal degradation, allowing for increased cellulose accessibility and bioethanol yield. It also yields high-quality lignin suitable for various industrial applications. Additionally, the process offers effective hemicellulose fractionation, enabling the production of valuable chemicals like furfural and xylitol. The method's low input requirements make it feasible for small-scale plants in flexible locations (Zhang et al., 2016). A recent study compared six organosolv methods for extracting lignin from wood sawdust at atmospheric pressure. The isolated lignins showed varying yields, thermal properties, molecular weights, and pyrolysis properties, making them suitable for diverse value-added applications in the bio-based economy (Başakçılardan Kabakcı & Tanış, 2021).

To evaluate the efficiency and suitability of pretreatments for glucose recovery, various factors such as energy use, chemical/enzyme expenses, plant setup costs, waste generation, and environmental impact should be considered. The application of mechanical pretreatment is a crucial step in augmenting the enzymatic hydrolysis of lignocellulosic biomass. Several methods, including wet disk milling, hammer milling, and ball-milling, have been proven effective in this regard. Some studies have also suggested that combining different pretreatment techniques can improve the glucose yield from biomass (Ahmad Rizal et al., 2018).

4.3.3 Biological Pretreatment

Biological pretreatment of lignocellulosic biomass uses microorganisms like fungi and bacteria to degrade lignin and increase digestibility with hydrolytic enzymes. White-rot fungi, specializing in efficient lignin degradation via high redox potential oxidoreductases, have been extensively studied (Vasco-Correa et al., 2016). Taha et al. examined 30 bacteria and 18 fungi isolates across various straw substrates, uncovering notable synergies. For instance, *Neosartorya fischeri–Myceliophthora thermophila* dual isolates achieved a 3-fold increase in saccharification, underscoring co-culturing's viability for commercial bioethanol-focused microbial consortia (Taha et al., 2015). Fungal pretreatment, particularly with white-rot fungi, is efficient but time-consuming, while bacterial and enzymatic approaches offer faster results. Despite significant yield improvements (up to 485% with enzymatic pre-treatment), large-scale implementation faces techno-economic challenges requiring further research (Zabed et al., 2019). Benedetti et al. performed an investigation where *Chlamydomonas reinhardtii* was genetically modified to synthesize thermostable cell wall-degrading enzymes and phosphite dehydrogenase. This enhancement facilitated effective lignocellulose conversion and growth in non-axenic conditions. The transgenic microalgal hydrolysates boosted biogas production and growth of *C.vulgaris,* overcoming alkaline-treated biomass challenges and enabling efficient sugar utilization (Benedetti et al., 2021).

4.4 ENZYMATIC HYDROLYSIS

Bioethanol production can utilize three main types of biomasses: sugar-containing materials (e.g., sugarcane and sweet sorghum), starch-containing feedstocks (e.g., corn and cassava), and lignocellulosic biomass (e.g., straw and agricultural waste), with the latter being a promising alternative due to its widespread availability, low cost, and non-competition with food and feed crops (Bušić et al., 2018). Ayodele et al. recently reviewed multiple experimental studies investigating 1G bioethanol production from sugar-based feedstocks (sugarcane, sweet sorghum, watermelon, sugar beet, and cashew apple). In lab-scale production, the commonly used batch mode fermenter relies on critical parameters, including pH (3.7–5.5), temperature (30°C–40°C), and agitation speed (150–300 rpm), to optimize the conversion of sugar to bioethanol. These controlled conditions ensure efficient sugar-to-bioethanol conversion during the process. Careful feedstock selection, including non-food sources like agricultural waste, is crucial to address the logistical challenge of biofuel production. Starch-based agricultural wastes, such as cassava and potato peels, show potential for bioethanol production, and integrating them with lignocellulosic substrates can enhance multiple biofuel streams. Additionally, utilizing waste streams from agricultural waste-biofuel processes can boost biorefinery profitability while minimizing environmental impacts. To achieve efficient bioethanol production, pretreatment is essential to release sugars from starch-based agricultural wastes and lignocellulosic substrates, enabling effective enzymatic hydrolysis and microbial fermentation. In recent years, enzymatic hydrolysis has become a key technology for converting biomass into renewable fuels and chemicals (Sweeney & Xu, 2012).

During the fractionation of biomass materials, such as sugarcane residues, corn cobs, and rice straw, monosaccharides and water-soluble oligomers, including xylo-oligomers and cello-oligomers rich in xylan and glucan, are released (Bhatia et al., 2019). Enzymes such as cellulases and hemicellulases are used to cleave cellulose and hemicellulose polymers into various pentoses and hexoses. The complex structure of lignocellulose, which consists of cellulose, hemicellulose, and lignin, makes it difficult for enzymes to break it down. Moreover, some enzymes get stuck to the condensed lignin, which reduces the hydrolysis yield by forming non-specific linkages (Zoghlami & Paës, 2019). In addition, high levels of lignin can inhibit enzyme accessibility, reducing the rate and yield of hydrolysis (Yuan et al., 2021). Factors affecting the formation of monomeric sugars include the liquid-to-solid ratio, type of acid used, temperature, reaction time, particle size, polymerization degree of cellulose, and connection with other plant cell wall structures (Das et al., 2023) (Table 4.2).

TABLE 4.2
Various Enzymes that are Used in the Conversion of Agrobiomasses to Reducing Sugars

Enzymes	Key Points	Process Condition	Reference
Cellulases	Cellulases refer to a group of enzymes that catalyze the breakdown of cellulose into glucose. They are frequently employed in the hydrolysis of agricultural by-products like corn stover, rice straw, wheat straw, and sugarcane bagasse	The optimal conditions for cellulase activity are pH 4.5–5.0 and temperature of 50°C–60°C	Lynd et al. (2002)
Hemicellulase	Hemicellulases are enzymes that break down hemicellulose into a mixture of pentoses and hexoses, commonly used to hydrolyze agricultural residues to produce biofuels like ethanol. The resulting xylose, arabinose, and mannose can be fermented for this purpose	These enzymes are active at a pH of 5.0–6.5 and a temperature range of 50°C–60°C	Kumar et al. (2009)
Ligninases	Ligninases are enzymes that break down lignin, a polymer that binds cellulose and hemicellulose, making them inaccessible to hydrolysis. They are used in the pretreatment of lignocellulosic biomass to enhance the accessibility of cellulases and hemicellulases	Ligninase activity is optimal at a pH of 4.5–5.5 and a temperature range of 30°C–40°C	Couto et al. (2006)
Amylases	Amylases can break down starch into glucose in starchy biomass like corn and cassava	The optimal conditions for amylase activity are pH 7.0–8.0 and temperature of 50°C–60°C	Chethana et al. (2011)
Proteases	Proteases break down proteins into amino acids and are commonly applied to protein-rich biomass like soybean meal and distiller grains	The optimal conditions for protease activity are pH 7.0–9.0 and temperature of 40°C–60°C	Champasri et al. (2021)
Xylanases	Xylanases break down xylan in plant cell walls to xylose and are frequently used for hydrolyzing agricultural residues such as corn stover, wheat straw, and sugarcane bagasse	The optimal conditions for xylanase activity are pH 5.0–6.0 and temperature of 50°C–60°C	Kumar et al. (2009)
Invertases	Invertases hydrolyze sucrose into glucose and fructose commonly used for sugarcane and sugar beet pulp hydrolysis	The optimal conditions for invertase activity are pH 4.5–5.5 and temperature of 50°C–60°C	Sarkar et al. (2012)
α-Amylases	α-Amylases are enzymes that break down starch into maltose and glucose by cleaving α-1,4-glycosidic bonds, and they are commonly utilized for corn and cassava starch hydrolysis	The optimal conditions for α-amylase activity are pH 5.5–7.0 and temperature of 70°C–90°C	Pandit et al. (2022)
Glucoamylase	Glucoamylases convert starch into glucose and are utilized to produce fermentable sugars from starch-rich biomass	Glucoamylases work best at a pH of 4.0–5.0 and a temperature range of 35°C–65°C	Benassi et al. (2014)

(*Continued*)

TABLE 4.2 (*Continued*)
Various Enzymes that are Used in the Conversion of Agrobiomasses to Reducing Sugars

Enzymes	Key Points	Process Condition	Reference
β-Glucosidase	β-Glucosidase is an enzyme that breaks down cellobiose and other β-glucosides into glucose, frequently utilized to hydrolyze cellulose- and hemicellulose-rich biomass such as sugarcane bagasse and switchgrass	The optimal conditions for β-glucosidase activity are pH 3.5–4.0 and temperature of 50°C–60°C	da Cunha et al. (2023)
Mannanase	Mannanases are hemicellulase enzymes that degrade mannans, a component of plant cell walls, and are employed to convert hemicellulosic biomass into fermentable sugars	The optimal conditions for mannanase activity are typically pH 5.0–6.5 and a temperature range of 50°C–60°C	Chandra et al. (2011)
Laccases	Laccases are enzymes that oxidize phenolic compounds, typically found in lignocellulosic biomass and hindering enzymatic hydrolysis, and are frequently utilized for pretreatment to enhance enzymatic hydrolysis efficacy	The optimal conditions for laccase activity are pH 7.5–9.5 and temperature of 80°C–100°C	Al-kahem Al-balawi et al. (2017)
Galactosidase	Galactosidases hydrolyze galactooligosaccharides in lactose-rich biomass to galactose and glucose, facilitating the production of galactose-rich syrups and biofuels	Galactosidases are active at a pH of 3.0–5.0 and a temperature range of 55°C–65°C	Ramos and Malcata (2011)
β-Xylosidase	Xylosidases, including β-xylosidase (EC 3.2.1.37), play a vital role in breaking down xylan into xylose and oligosaccharides, contributing to applications in industries and biofuel production. These enzymes are categorized into various glycosyl hydrolase (GH) families, such as GH3, GH39, GH43, GH52, and GH54, based on their sequence similarities	β-xylosidase displayed peak activity at pH 6.0 and 95°C, retaining over 50% activity for 2 hours at 75°C–85°C (pH 6.0), with a half-life of around 2 hours at 85°C	Shi et al. (2013)
Cryovial/ADY Cellerity® 1.0	Cellerity® 1.0 is an advanced yeast strain that allows to co-ferment both C5 and C6 sugars. Cellerity® 1.0 also delivers fast xylose use in fermentation, resulting in higher ethanol yields	Cellerity® 1.0 typically has a pH of 4.5–5.0 and a temperature range of 45°C–50°C	NOVOZYMES
Cellic® Ctec2	The Cellic® Ctec2 combines cellulases, β-glucosidases, and hemicellulases that exhibit strong hydrolytic activity and can effectively operate on various pretreated lignocellulosic substrates	The optimal process conditions for Cellic® Ctec2 are typically a pH of 4.5–5.0 and a temperature range of 50°C–60°C	NOVOZYMES
Cellic® Ctec3 HS	Cellic® Ctec3 HS is a highly efficient cellulase and hemicellulase complex with proprietary enzyme activities. It unlocks new opportunities for process optimization in pretreatment, hydrolysis, and fermentation. That helps the industrial plant to achieve its lowest possible cost of production	The optimal process conditions for Cellic® Ctec3 HS are typically a pH of 4.5–5.0 and a temperature range of 50°C–60°C	NOVOZYMES

4.4.1 Cellulase

Numerous bacteria and aerobic fungi are known to produce cellulase enzymes, which find extensive applications in various industrial processes. These cellulase enzymes are categorized into three main types: endocellulases, exocellulases, and β-glucosidases (Sharma et al., 2016).

i. **Endocellulases (Endoglucanases):** These enzymes target the amorphous regions of cellulose, which are the less structured and more easily accessible areas. Endocellulases cleave internal bonds within the cellulose chains, creating shorter cellulose fragments with free chain ends. This process increases the accessibility of the substrate for further enzymatic action.
ii. **Exocellulases (Cellobiohydrolases):** Exocellulases work on the reducing and non-reducing ends of the cellulose chains. Their primary function is to detach cellobiose units, which are composed of two glucose molecules linked together, from the exposed chain ends generated by endocellulases. Exocellulases can be categorized into two distinct types based on their mode of action: cellobiohydrolases I (CBH I), which releases cellobiose from the reducing end, and cellobiohydrolases II (CBH II), which releases cellobiose from the non-reducing end.
iii. **β-Glucosidases:** These enzymes play a critical role in the final stage of cellulose hydrolysis. They break down cellobiose into individual glucose molecules. β-Glucosidases are essential for efficient cellulose degradation because they prevent the accumulation of cellobiose, which can hinder the activity of other cellulase enzymes.

The cooperative action of these three distinct cellulase enzyme types is of utmost importance for the thorough hydrolysis of cellulose into glucose (Sharma et al., 2016).

Cellobiohydrolases play a pivotal role in cleaving cellulose bonds and are frequently the predominant components in commercial cellulase blends utilized for cellulose breakdown (Harrison et al., 2014). Notably, extremophile-derived thermostable endoglucanases have emerged as promising candidates for biofuel production due to their remarkable attributes, including high-temperature operability and resilience to extreme pH variations, which are essential in biofuel production processes. Their capacity to perform optimally under elevated temperatures allows them to retain their effectiveness even under challenging reaction conditions, resulting in enhanced biofuel yields (Yennamalli et al., 2013).

Microorganisms such as *Trichoderma reesei* and *Aspergillus niger* are widely investigated for cellulase enzyme production (Legodi et al., 2023). However, the cellulase system of *T. reesei* suffers from a deficiency in β-glucosidase (BGL) activity, resulting in incomplete hydrolysis of cellobiose and subsequent enzyme inhibition. To mitigate this issue, adding extra BGL, such as from *A. niger*, can be employed to improve cellobiose hydrolysis and prevent enzyme inhibition. Therefore, efforts are consistently made to identify an effective cellulase producer for biomass hydrolysis (Abdella et al., 2016). Recent discoveries have highlighted the significance of lytic polysaccharide monooxygenases (LPMOs) in cellulose degradation. Nonetheless, the reaction conditions and production costs associated with these enzymes pose challenges, leading researchers to explore novel microorganisms to produce more efficient cellulases for biomass utilization (Singh et al., 2021).

The hydrolytic efficiency of a multi-enzyme complex is contingent on the individual properties of the enzymes and their relative proportions within the cocktail. Despite advancements in comprehending cellulose degradation and cellulase production, there remains a need to identify highly active cellulases to create optimal multi-enzyme mixtures that can effectively break down lignocellulosic materials, a goal achievable through genetic techniques.

4.4.2 Xylanase

Xylan, a hemicellulose component in lignocellulosic biomass, serves as a barrier that limits the access of cellulases to cellulose. Several researchers have attempted to remove xylan through pretreatment and enzymatic hydrolysis to enhance cellulose accessibility. However, severe treatment to eliminate remaining xylan leads to the degradation of significant amounts of cellulose (Lee et al., 2010).

Xylanase (3.2.1.8) is an enzyme responsible for breaking down xylan, a complex carbohydrate present in plant cell walls, into simpler sugars like xylose. Xylanase is produced by many microorganisms, including bacteria, fungi, and yeasts, as well as some higher plants and animals. It has found utility in various industries, including pentose and fuel production, fruit-juice clarification, enhancement of rumen digestion, the paper industry, as well as xylan removal from kraft pulp, and the brewing industry (Bhardwaj et al., 2019).

Xylanases derived from diverse sources, such as *Thermoactinomyces thalophilus, Bacillus* sp., *Humicola insolens*, and *Bispora* sp. The production of xylooligosaccharides (XOS) has seen an upswing through the utilization of endoxylanase, exoxylanase, and β-xylosidase enzyme activities in the hydrolysis of lignocellulosic materials with high xylan content. These materials include rice straw, wheat straw, corn cobs, sugarcane bagasse, among others (Dong et al., 2023). Notably, xylanases isolated from *Bacillus halodurans* have been used to produce XOS from wheat bran, and commercial xylanase from *Trichoderma viride* has been employed to generate XOS from alkali-treated sugarcane bagasse (Manisha & Yadav, 2017).

4.4.3 Amylase

Most of the agrobiomass sources contain starch, which is a complex molecule that needs to be broken down into simple sugars before it can be fermented. This process requires the use of different hydrolytic enzymes. One commonly used enzyme for this purpose is microbial amylase (EC 3.2.1.1; 1,4-α-D-glucan glucanohydrolase), which is an extracellular hydrolase (Vasic et al., 2021).

In the same vein, Pervez et al. screened and isolated *Aspergillus fumigatus* KIBGE-IB33 as amylase-producing filamentous fungi. They concentrated the mixture of amylolytic enzymes and introduced it into a cassava starch slurry, resulting in the highest yield of glucose formation. Subsequently, the generated glucose was subjected to fermentation using *Saccharomyces cerevisiae* to produce bioethanol, with a remarkable yield of 84.0% (Pervez et al., 2014). In another study conducted by Kumar et al., bacterial-based amylase pretreatment produced a better yield of ethanol from starch-based solid waste than acid pretreatment (Kumar et al., 2016).

4.4.4 Amyloglucosidase

Amyloglucosidases, also known as glucoamylases (E.C. 3.1.2.3), are enzymes responsible for breaking down starch by cleaving its α-1,4 linkages, liberating glucose molecules. Operating as an exoamylase, it releases β-D glucose from the non-reducing ends of substances like amylose, amylopectin, and glycogen. It's worth noting that it can also break α-1,6 glycosidic bonds, albeit at a slower pace (Ravindran et al., 2018). In the examination of various enzymatic techniques for extracting sugar from microalgal biomass, it was discovered that the hydrolysis rate of carbohydrates improved significantly with the simultaneous addition of enzymes, specifically in the sequence of cellulases first, followed by α-amylase, and then amyloglucosidase (Shokrkar & Ebrahimi, 2018). A recent study demonstrated enhanced hydrolysis of microalgal carbohydrates for bioethanol production by synergistically combining cellulases and amylolytic enzymes. β-Glucosidase played a role in improving cellulose hydrolysis, while a combination of α-amylase and amyloglucosidase facilitated starch breakdown. This approach led to a reduction in hydrolysis time and yielded 0.13 g ethanol per gram of dried algae in subsequent *S. cerevisiae* fermentation (Shokrkar & Ebrahimi, 2014).

4.5 HYDROLYTIC ENZYMES PRODUCTION FOR CONVERSION OF AGROBIOMASS TO FERMENTABLE SUGARS

The saccharification of lignocellulosic materials with the aid of microbial enzymes represents a notably demanding and rate-restricting phase in the biofuel sector. Additionally, the utilization of commercial lignocellulolytic enzymes from various sources can pose a considerable cost barrier. Many microorganisms have been identified to produce one or two lignocellulolytic enzymes, and these enzymes can be generated through SmF or SSF methods (Østby et al., 2020).

SmF stands as the preferred approach for the industrial-scale production of hydrolytic enzymes such as cellulase, hemicellulase, xylanase, and amylase due to its convenience and extensive adoption within the industry (Mrudula & Murugammal, 2011). In a recent investigation, researchers explored the influence of various inducer sources, encompassing nitrogen sources (such as yeast extract, potassium nitrate, sodium nitrate, and ammonium sulfate), carbon sources (including malt extract, glucose, fructose, carboxymethyl cellulose, starch, and xylose), and various agricultural biomass materials (like stevia straw, wheat straw, etc.), on cellulase production by

Pleurotus ostreatus and *Phanerochaete chrysosporium* through SmF conditions. The findings revealed that wheat straw and corn stover as agrobiomass sources exhibited improved cellulase activity (Datsomor et al., 2022). Another study aimed to explore the feasibility of producing lignocellulosic enzymes utilizing economical agricultural residues. Findings from this study suggest that *Sphingobacterium* sp. ksn-11 holds promise as the prime candidate for large-scale manufacturing of a variety of lignocellulosic enzymes via submerged fermentation (Datsomor et al., 2022). Overall, the SmF technique has enabled the production of large quantities of enzymes to meet demand.

SSF holds significant potential in the efficient utilization of agricultural waste materials for enzyme production. This is primarily attributed to the favorable physical and chemical properties of lignocellulosic substrates, which render them well-suited for cultivation in a solid-phase environment (Verduzco-Oliva & Gutierrez-Uribe, 2020). Moreover, SSF-paired filamentous fungi is a practical, cost-effective, and sustainable method for biorefinery processes and manufacturing biofuels or other industrial products using lignocellulosic residues (Salomão et al., 2019). This approach involves microorganisms fermenting solid substrates like lignocellulosic residues, producing a variety of beneficial enzymes and metabolites without a free-flowing liquid. Therefore, various value-added products, including ethanol, butanol, methane, industrial enzymes, organic acids, and pigments, can be produced by this technique (Ravindran et al., 2018).

For instance, Moran-Aguilar and colleagues examined various strains of *A. niger* to assess their capability in producing xylanase and cellulase enzymes via SSF, employing agro-residues, sugarcane bagasse, and brewery spent grain as substrates. The aforementioned lignocellulosic residues underwent three distinct pretreatments, namely alkaline treatment, boiling water treatment, and autoclave treatment, and have shown maximum cellulase and xylanase activity (Moran-Aguilar et al., 2021). Analogously, Balakrishnan et al. conducted an optimization study on the synthesis of α-amylase using SSF with *Aspergillus oryzae* and edible cakes, including groundnut, coconut, and sesame oil cake. Their emphasis was on fine-tuning process variables, including temperature, pH, and incubation duration (Balakrishnan et al., 2021). A recent development involves the assessment of floral waste from vegetal sources as a medium for the concurrent generation of cellulases, hemicellulases, and reducing sugars through a single-step SSF process (Zamora et al., 2021). In the field of SSF, several advancements were made, such as the usage of mixed cultures, and advances in bioreactor design have improved the ability to regulate operational variables like temperature, moisture levels, and aeration, resulting in greater consistency and replicable outcomes (Webb, 2017).

4.6 CONCLUSION AND FUTURE PERSPECTIVE

Utilizing agrobiomass as a source material for bioenergy production shows immense potential for the future. Enzymatic biomass conversion to sugars offers advantages over traditional chemical processes, such as lower energy requirements and higher yield. However, challenges remain in optimizing enzyme production and addressing biomass variability. Future research may focus on improving enzyme systems, increasing scalability, and exploring new feedstock sourcing and biorefinery processes. For instance, several researchers have reported that enzyme cocktails containing multiple types of cellulases, hemicellulases, and xylanases from diverse origins can enhance the saccharification process's efficacy.

Further, the enzymes' stability and functionality can be enhanced via protein engineering and various other strategies. Additionally, microorganisms can produce enzymes and be engineered for higher efficiency and specificity. To improve sustainability and efficiency, research should also focus on optimizing enzyme production from low-cost agro-wastes, identifying optimal fermentation conditions, developing more efficient enzyme cocktails, and integrating lignocellulosic waste streams into biorefinery processes to generate valuable commodities such as biofuels, bioplastics, and chemicals. Overall, enzymatic conversion holds promise for a sustainable and eco-friendly bioeconomy.

REFERENCES

Abdella, A., Mazeed, T. E. S., El-Baz, A. F., & Yang, S. T. (2016). Production of β-glucosidase from wheat bran and glycerol by *Aspergillus niger* in stirred tank and rotating fibrous bed bioreactors. *Process Biochemistry*, *51*(10), 1331–1337. https://doi.org/10.1016/j.procbio.2016.07.004

Ahmad Rizal, N. F. A., Ibrahim, M. F., Zakaria, M. R., Abd-Aziz, S., Yee, P. L., & Hassan, M. A. (2018). Pre-treatment of oil palm biomass for fermentable sugars production. *Molecules*, *23*(6), 1–14. https://doi.org/10.3390/molecules23061381

Ai, B., Li, W., Woomer, J., Li, M., Pu, Y., Sheng, Z., et al., (2020). Natural deep eutectic solvent mediated extrusion for continuous high-solid pretreatment of lignocellulosic biomass. *Green Chemistry*, *22*(19), 6372–6383. https://doi.org/10.1039/d0gc01560a

Akhtar, N., Gupta, K., Goyal, D., & Goyal, A. (2016). Recent advances in pretreatment technologies for efficient hydrolysis of lignocellulosic biomass. *Environmental Progress and Sustainable Energy*, *35*(2), 489–511. https://doi.org/10.1002/ep

Al Jibouri, A. K. H., Turcotte, G., Wu, J., & Cheng, C. H. (2015). Ozone pretreatment of humid wheat straw for biofuel production. *Energy Science and Engineering*, *3*(6), 541–548. https://doi.org/10.1002/ese3.93

Al-kahem Al-balawi, T. H., Wood, A. L., Solis, A., Cooper, T., & Barabote, R. D. (2017). Anoxybacillus sp. strain UARK-01, a new thermophilic soil bacterium with hyperthermostable alkaline laccase activity. *Current Microbiology*, *74*(6), 762–771. https://doi.org/10.1007/s00284-017-1239-5

Alokika, Anu, Kumar, A., Kumar, V., & Singh, B. (2021). Cellulosic and hemicellulosic fractions of sugarcane bagasse: Potential, challenges and future perspective. *International Journal of Biological Macromolecules*, *169*, 564–582. https://doi.org/10.1016/j.ijbiomac.2020.12.175

Amin, F. R., Khalid, H., Zhang, H., Rahman, S., Zhang, R., Liu, G., & Chen, C. (2017). Pretreatment methods of lignocellulosic biomass for anaerobic digestion. *AMB Express*, *7*(1), 72. https://doi.org/10.1186/s13568-017-0375-4

Bai, X., Wang, G., Yu, Y., Wang, D., & Wang, Z. (2018). Changes in the physicochemical structure and pyrolysis characteristics of wheat straw after rod-milling pretreatment. *Bioresource Technology*, *250*, 770–776. https://doi.org/10.1016/j.biortech.2017.11.085

Balakrishnan, M., Jeevarathinam, G., Kumar, S. K. S., Muniraj, I., & Uthandi, S. (2021). Optimization and scale-up of α-amylase production by Aspergillus oryzae using solid-state fermentation of edible oil cakes. *BMC Biotechnology*, *21*(1), 1–11. https://doi.org/10.1186/s12896-021-00686-7

Başakçılardan Kabakcı, S., & Tanış, M. H. (2021). Pretreatment of lignocellulosic biomass at atmospheric conditions by using different organosolv liquors: A comparison of lignins. *Biomass Conversion and Biorefinery*, *11*(6), 2869–2880. https://doi.org/10.1007/s13399-020-00677-2

Benassi, V. M., Pasin, T. M., Facchini, F. D. A., Jorge, J. A., & De Moraes Polizeli, M. de L. T. (2014). A novel glucoamylase activated by manganese and calcium produced in submerged fermentation by Aspergillus phoenicis. *Journal of Basic Microbiology*, *54*(5), 333–339. https://doi.org/10.1002/jobm.201200515

Benedetti, M., Barera, S., Longoni, P., Guardini, Z., Herrero Garcia, N., Bolzonella, D., et al., (2021). A microalgal-based preparation with synergistic cellulolytic and detoxifying action towards chemical-treated lignocellulose. *Plant Biotechnology Journal*, 19, 124–137. https://doi.org/10.1111/pbi.13447

Bhardwaj, N., Kumar, B., & Verma, P. (2019). A detailed overview of xylanases: An emerging biomolecule for current and future prospective. *Bioresources and Bioprocessing*, *6*(1), 40. https://doi.org/10.1186/s40643-019-0276-2

Bhatia, L., Sharma, A., Bachheti, R. K., & Chandel, A. K. (2019). Lignocellulose derived functional oligosaccharides: Production, properties, and health benefits. *Preparative Biochemistry and Biotechnology*, *49*(8), 744–758. https://doi.org/10.1080/10826068.2019.1608446

Blasi, A., Verardi, A., Lopresto, C. G., Siciliano, S., & Sangiorgio, P. (2023). Lignocellulosic agricultural waste valorization to obtain valuable products : An overview. *Recycling*, *8*(4), 1–46.

Bušić, A., Mardetko, N., Kundas, S., Morzak, G., Belskaya, H., Šantek, M. I., et al., (2018). Bioethanol production from renewable raw materials and its separation and purification: A review. *Food Technology and Biotechnology*, *56*(3), 289–311. https://doi.org/10.17113/ftb.56.03.18.5546

Champasri, C., Phetlum, S., & Pornchoo, C. (2021). Diverse activities and biochemical properties of amylase and proteases from six freshwater fish species. *Scientific Reports*, *11*(1), 1–11. https://doi.org/10.1038/s41598-021-85258-7

Chandra, M. R. S., Lee, Y. S., Park, I. H., Zhou, Y., Kim, K. K., & Choi, Y. L. (2011). Isolation, purification and characterization of a thermostable β-mannanase from Paenibacillus sp. DZ3. *Journal of Applied Biological Chemistry*, *54*(3), 325–331. https://doi.org/10.3839/jksabc.2011.052

Chen, C., Chen, Z., Chen, J., Huang, J., Li, H., Sun, S., et al., (2020). Profiling of chemical and structural composition of lignocellulosic biomasses in tetraploid rice straw. *Polymers*, *12*(2), 340. https://doi.org/10.3390/polym12020340

Chethana, S., Pratap, B., Roy, S., Jaiswal, A., SD, S., & AB, V. (2011). Bioethanol production from rice water waste: A low cost motor fuel. *Pharmacologyonline*, *3*(11), 125–134.

Couto, S. R., Moldes, D., & Sanromán, M. A. (2006). Optimum stability conditions of pH and temperature for ligninase and manganese-dependent peroxidase from Phanerochaete chrysosporium. Application to in vitro decolorization of Poly R-478 by MnP. *World Journal of Microbiology and Biotechnology*, *22*(6), 607–612. https://doi.org/10.1007/s11274-005-9078-0

Das, N., Jena, P. K., Padhi, D., Kumar Mohanty, M., & Sahoo, G. (2023). A comprehensive review of characterization, pretreatment and its applications on different lignocellulosic biomass for bioethanol production. *Biomass Conversion and Biorefinery*, *13*(2), 1503–1527. https://doi.org/10.1007/s13399-021-01294-3

Datsomor, O., Yan, Q., Opoku-Mensah, L., Zhao, G., & Miao, L. (2022). Effect of different inducer sources on cellulase enzyme production by white-rot basidiomycetes *pleurotus ostreatus* and phanerochaete chrysosporium under submerged fermentation. *Fermentation*, *8*(10), 561. https://doi.org/10.3390/fermentation8100561

Den, W., Sharma, V. K., Lee, M., Nadadur, G., & Varma, R. S. (2018). Lignocellulosic biomass transformations via greener oxidative pretreatment processes: Access to energy and value added chemicals. *Frontiers in Chemistry*, *6*, 1–23. https://doi.org/10.3389/fchem.2018.00141

Diaz, A. B., Blandino, A., & Caro, I. (2018). Value added products from fermentation of sugars derived from agro-food residues. *Trends in Food Science and Technology*, *71*, 52–64. https://doi.org/10.1016/j.tifs.2017.10.016

Dong, C. Di, Tsai, M. L., Nargotra, P., Kour, B., Chen, C. W., Sun, P. P., & Sharma, V. (2023). Bioprocess development for the production of xylooligosaccharide prebiotics from agro-industrial lignocellulosic waste. *Heliyon*, *9*(7), e18316. https://doi.org/10.1016/j.heliyon.2023.e18316

El Halal, S. L. M., Colussi, R., Deon, V. G., Pinto, V. Z., Villanova, F. A., Carreño, N. L. V., et al., (2015). Films based on oxidized starch and cellulose from barley. *Carbohydrate Polymers*, *133*, 644–653. https://doi.org/10.1016/j.carbpol.2015.07.024

Han, M., Kang, K. E., Kim, Y., & Choi, G. W. (2013). High efficiency bioethanol production from barley straw using a continuous pretreatment reactor. *Process Biochemistry*, *48*(3), 488–495. https://doi.org/10.1016/j.procbio.2013.01.007

Harrison, M. D., Zhang, Z., Shand, K., Chong, B. F., Nichols, J., Oeller, P., et al., (2014). The combination of plant-expressed cellobiohydrolase and low dosages of cellulases for the hydrolysis of sugar cane bagasse. *Biotechnology for Biofuels*, *7*(1), 1–14. https://doi.org/10.1186/s13068-014-0131-9

He, Y., Zhang, J., & Bao, J. (2014). Dry dilute acid pretreatment by co-currently feeding of corn stover feedstock and dilute acid solution without impregnation. *Bioresource Technology*, *158*, 360–364. https://doi.org/10.1016/j.biortech.2014.02.074

Hoang, A. T., Nizetic, S., Ong, H. C., Chong, C. T., Atabani, A. E., & Pham, V. V. (2021). Acid-based lignocellulosic biomass biorefinery for bioenergy production: Advantages, application constraints, and perspectives. *Journal of Environmental Management*, *296*(July), 113194. https://doi.org/10.1016/j.jenvman.2021.113194

Isikgor, F. H., & Becer, C. R. (2015). Lignocellulosic biomass: A sustainable platform for the production of bio-based chemicals and polymers. *Polymer Chemistry*, *6*(25), 4497–4559. https://doi.org/10.1039/c5py00263j

Jędrzejczyk, M., Soszka, E., Czapnik, M., Ruppert, A. M., & Grams, J. (2019). Physical and chemical pretreatment of lignocellulosic biomass. In A. Basile, & F. Dalena (Eds) *Second and Third Generation of Feedstocks: The Evolution of Biofuels* (pp. 143–196). Elsevier, Amsterdam, The Netherlands. https://doi.org/10.1016/B978-0-12-815162-4.00006-9

Ji, W., Shen, Z., & Wen, Y. (2014). A continuous hydrothermal saccharification approach of rape straw using dilute sulfuric acid. *Bioenergy Research*, *7*(4), 1392–1401. https://doi.org/10.1007/s12155-014-9468-y

Jung, Y. H., & Kim, K. H. (2015). *Acidic Pretreatment. Pretreatment of Biomass: Processes and Technologies*. Elsevier B.V., Amsterdam, The Netherlands. https://doi.org/10.1016/B978-0-12-800080-9.00003-7

Kim, J. S., Lee, Y. Y., & Kim, T. H. (2016). A review on alkaline pretreatment technology for bioconversion of lignocellulosic biomass. *Bioresource Technology*, *199*, 42–48. https://doi.org/10.1016/j.biortech.2015.08.085

Kohli, U., Nigam, P., Singh, D., & Chaudhary, K. (2001). Thermostable, alkalophilic and cellulase free xylanase production by Thermoactinomyces thalophilus subgroup C. *Enzyme and Microbial Technology*, *28*(7–8), 606–610. https://doi.org/10.1016/S0141-0229(01)00320-9

Kumar, P., Barrett, D. M., Delwiche, M. J., & Stroeve, P. (2009). Methods for pretreatment of lignocellulosic biomass for efficient hydrolysis and biofuel production. *Industrial and Engineering Chemistry Research*, *48*(8), 3713–3729. https://doi.org/10.1021/ie801542g

Kumar, V., Nanda, M., & Singh, A. (2016). Effect of bacterial amylase pretreatment on bioethanol production from starch-based solid waste (SBSW). *Energy Sources, Part A: Recovery, Utilization and Environmental Effects*, *38*(17), 2604–2609. https://doi.org/10.1080/15567036.2015.1098745

Lee, J. W., Rodrigues, R. C. L. B., Kim, H. J., Choi, I. G., & Jeffries, T. W. (2010). The roles of xylan and lignin in oxalic acid pretreated corncob during separate enzymatic hydrolysis and ethanol fermentation. *Bioresource Technology*, *101*(12), 4379–4385. https://doi.org/10.1016/j.biortech.2009.12.112

Legodi, L. M., La Grange, D. C., & van Rensburg, E. L. J. (2023). Production of the cellulase enzyme system by locally isolated trichoderma and aspergillus species cultivated on banana pseudostem during solid-state fermentation. *Fermentation*, *9*(5), 412. https://doi.org/10.3390/fermentation9050412

Luo, H., Li, J., Yang, J., Wang, H., Yang, Y., Huang, H., et al., (2009). A thermophilic and acid stable family-10 xylanase from the acidophilic fungus Bispora sp. MEY-1. *Extremophiles : Life Under Extreme Conditions*, *13*(5), 849–857. https://doi.org/10.1007/s00792-009-0272-0

Luo, J., Fang, Z., & Smith, R. L. (2014). Ultrasound-enhanced conversion of biomass to biofuels. *Progress in Energy and Combustion Science*, *41*(1), 56–93. https://doi.org/10.1016/j.pecs.2013.11.001

Lynd, L. R., Weimer, P. J., Zyl, W. H. Van, & Isak, S. (2002). Microbial cellulose utilization : Fundamentals and biotechnology. *Microbiology and Molecular Biology Reviews*, *66*(3), 506–577. https://doi.org/10.1128/MMBR.66.3.506

Mafa, M. S., Malgas, S., Bhattacharya, A., & Rashamuse, K. (2020). The effects of alkaline pretreatment on agricultural biomasses (corn cob and sweet sorghum bagasse). *Agronomy*, *10*(1211), 1–13.

Maj, G. (2018). Emission factors and energy properties of agro and forest biomass in aspect of sustainability of energy sector. *Energies*, 11(6), 1516. https://doi.org/10.3390/en11061516

Manisha, & Yadav, S. K. (2017). Technological advances and applications of hydrolytic enzymes for valorization of lignocellulosic biomass. *Bioresource Technology*, *245*, 1727–1739. https://doi.org/10.1016/j.biortech.2017.05.066

Mikulski, D., & Kłosowski, G. (2020). Microwave-assisted dilute acid pretreatment in bioethanol production from wheat and rye stillages. *Biomass and Bioenergy*, *136*, 105528. https://doi.org/10.1016/j.biombioe.2020.105528

Moran-Aguilar, M. G., Costa-Trigo, I., Calderón-Santoyo, M., Domínguez, J. M., & Aguilar-Uscanga, M. G. (2021). Production of cellulases and xylanases in solid-state fermentation by different strains of *Aspergillus niger* using sugarcane bagasse and brewery spent grain. *Biochemical Engineering Journal*, *172*, 108060. https://doi.org/10.1016/j.bej.2021.108060

Mrudula, S., & Murugammal, R. (2011). Production of cellulase by *Aspergillus niger* under submerged and solid state fermentation using coir waste as a substrate. *Brazilian Journal of Microbiology*, *42*(3), 1119–1127. https://doi.org/10.1590/S1517-83822011000300033

Nauman Aftab, M., Iqbal, I., Riaz, F., Karadag, A., & Tabatabaei, M. (2019). Different pretreatment methods of lignocellulosic biomass for use in biofuel production. *Biomass for Bioenergy - Recent Trends and Future Challenges*, 2(4), 48–50. https://doi.org/10.5772/intechopen.84995

Østby, H., Hansen, L. D., Horn, S. J., Eijsink, V. G. H., & Várnai, A. (2020). Enzymatic processing of lignocellulosic biomass: Principles, recent advances and perspectives. *Journal of Industrial Microbiology and Biotechnology*, *47*, 623–657. https://doi.org/10.1007/s10295-020-02301-8

Passadis, K., Christianides, D., Malamis, D., Barampouti, E. M., & Mai, S. (2022). Valorisation of source-separated food waste to bioethanol: Pilot-scale demonstration. *Biomass Conversion and Biorefinery*, *12*(10), 4599–4609. https://doi.org/10.1007/s13399-022-02732-6

Periyasamy, S., Senthil, V. K. P., Isabel, J. B., & Temesgen, T. (2022). Chemical, physical and biological methods to convert lignocellulosic waste into value - added products. A review. *Environmental Chemistry Letters*, *20*(2), 1129–1152. https://doi.org/10.1007/s10311-021-01374-w

Periyasamy, S., Isabel, J. B., Kavitha, S., Karthik, V., Mohamed, B. A., Getachew, D., et al., (2023). Recent advances in consolidated bioprocessing for conversion of lignocellulosic biomass into bioethanol - A review. *Chemical Engineering Journal*, *453*(P1), 139783. https://doi.org/10.1016/j.cej.2022.139783

Pervez, S., Aman, A., Iqbal, S., Siddiqui, N. N., & Ul Qader, S. A. (2014). Saccharification and liquefaction of cassava starch: An alternative source for the production of bioethanol using amylolytic enzymes by double fermentation process. *BMC Biotechnology*, *14*, 1–10. https://doi.org/10.1186/1472-6750-14-49

Qing, Q., Guo, Q., Zhou, L., Gao, X., Lu, X., & Zhang, Y. (2017). Comparison of alkaline and acid pretreatments for enzymatic hydrolysis of soybean hull and soybean straw to produce fermentable sugars. *Industrial Crops and Products*, *109*(May), 391–397. https://doi.org/10.1016/j.indcrop.2017.08.051

Ramos, O. S., & Malcata, F. X. (2011). Food-grade enzymes. In A. Moreira (Ed) *Comprehensive Biotechnology* (2nd ed., Vol. 3, pp. 587–603). Elsevier B.V., Amsterdam, The Netherlands. https://doi.org/10.1016/B978-0-08-088504-9.00213-0

Ravindran, R., Hassan, S. S., Williams, G. A., & Jaiswal, A. K. (2018). A review on bioconversion of agro-industrial wastes to industrially important enzymes. *Bioengineering*, *5*(4), 1–20. https://doi.org/10.3390/bioengineering5040093

Ruslan, N. F., & Ahmad, N. (2023). Sustainable bioethanol production by solid state fermentation : A systematic review. *Research Square Platform LLC*, 1–21.

Salomão, G. S. B., Agnezi, J. C., Paulino, L. B., Hencker, L. B., de Lira, T. S., Tardioli, P. W., & Pinotti, L. M. (2019). Production of cellulases by solid state fermentation using natural and pretreated sugarcane bagasse with different fungi. *Biocatalysis and Agricultural Biotechnology*, *17*, 1–6. https://doi.org/10.1016/j.bcab.2018.10.019

Sarkar, N., Ghosh, S. K., Bannerjee, S., & Aikat, K. (2012). Bioethanol production from agricultural wastes: An overview. *Renewable Energy*, *37*(1), 19–27. https://doi.org/10.1016/j.renene.2011.06.045

Saratale, G. D., & Oh, M. K. (2015). Improving alkaline pretreatment method for preparation of whole rice waste biomass feedstock and bioethanol production. *RSC Advances*, *5*(118), 97171–97179. https://doi.org/10.1039/c5ra17797a

Sharma, A., Tewari, R., Rana, S. S., Soni, R., & Soni, S. K. (2016). Cellulases: Classification, methods of determination and industrial applications. *Applied Biochemistry and Biotechnology*, *179*(8), 1346–1380. https://doi.org/10.1007/s12010-016-2070-3

Shi, H., Li, X., Gu, H., Zhang, Y., Huang, Y., Wang, L., & Wang, F. (2013). Biochemical properties of a novel thermostable and highly xylose-tolerant β-xylosidase/α-arabinosidase from Thermotoga thermarum. *Biotechnology for Biofuels*, *6*(1), 1–10. https://doi.org/10.1186/1754-6834-6-27

Shokrkar, H., & Ebrahimi, S. (2014). Synergism of cellulases and amylolytic enzymes in the hydrolysis of microalgal carbohydrates. *Biofuels, Bioproducts and Biorefining*, *8*(6), 743. https://doi.org/10.1002/BBB

Shokrkar, H., & Ebrahimi, S. (2018). Evaluation of different enzymatic treatment procedures on sugar extraction from microalgal biomass, experimental and kinetic study. *Energy*, *148*, 258–268. https://doi.org/10.1016/j.energy.2018.01.124

Singh, A., Bajar, S., Devi, A., & Pant, D. (2021). An overview on the recent developments in fungal cellulase production and their industrial applications. *Bioresource Technology Reports*, *14*(February), 100652. https://doi.org/10.1016/j.biteb.2021.100652

Sweeney, M. D., & Xu, F. (2012). Biomass converting enzymes as industrial biocatalysts for fuels and chemicals: Recent developments. *Catalysts*, *2*(2), 244–263. https://doi.org/10.3390/catal2020244

Taha, M., Shahsavari, E., Al-Hothaly, K., Mouradov, A., Smith, A. T., Ball, A. S., & Adetutu, E. M. (2015). Enhanced biological straw saccharification through coculturing of lignocellulose-degrading microorganisms. *Applied Biochemistry and Biotechnology*, *175*(8), 3709–3728. https://doi.org/10.1007/s12010-015-1539-9

Travaini, R., Martín-Juárez, J., Lorenzo-Hernando, A., & Bolado-Rodríguez, S. (2016). Ozonolysis: An advantageous pretreatment for lignocellulosic biomass revisited. *Bioresource Technology*, *199*, 2–12. https://doi.org/10.1016/j.biortech.2015.08.143

Tse, T. J., Wiens, D. J., & Reaney, M. J. T. (2021). Production of bioethanol-a review of factors affecting ethanol yield. *Fermentation*, *7*(4), 1–18. https://doi.org/10.3390/fermentation7040268

Tufail, T., Saeed, F., Afzaal, M., Ain, H. B. U., Gilani, S. A., Hussain, M., & Anjum, F. M. (2021). Wheat straw: A natural remedy against different maladies. *Food Science and Nutrition*, *9*(4), 2335–2344. https://doi.org/10.1002/fsn3.2030

Vasco-Correa, J., Ge, X., & Li, Y. (2016). Biological pretreatment of lignocellulosic biomass. *Biomass Fractionation Technologies for a Lignocellulosic Feedstock Based Biorefinery*, *2016*, 561–585. https://doi.org/10.1016/B978-0-12-802323-5.00024-4

Vasic, K., Knez, Z., & Leitgeb, M. (2021). Bioethanol production by enzymatic hydrolysis from different. *Molecules*, *26*(753), 1–23.

Verduzco-Oliva, R., & Gutierrez-Uribe, J. A. (2020). Beyond enzyme production: Solid state fermentation (SSF) as an alternative approach to produce antioxidant polysaccharides. *Sustainability (Switzerland)*, *12*(2), 495. https://doi.org/10.3390/su12020495

Webb, C. (2017). Design aspects of solid state fermentation as applied to microbial bioprocessing. *Journal of Applied Biotechnology & Bioengineering*, *4*(1), 511–532. https://doi.org/10.15406/jabb.2017.04.00094

Yennamalli, R. M., Rader, A. J., Kenny, A. J., Wolt, J. D., & Sen, T. Z. (2013). Endoglucanases: Insights into thermostability for biofuel applications. *Biotechnology for Biofuels*, *6*(1), 1. https://doi.org/10.1186/1754-6834-6-136

Yoo, C. G., Meng, X., Pu, Y., & Ragauskas, A. J. (2020b). The critical role of lignin in lignocellulosic biomass conversion and recent pretreatment strategies: A comprehensive review. *Bioresource Technology*, *301*, 122784. https://doi.org/10.1016/j.biortech.2020.122784

Yuan, Y., Jiang, B., Chen, H., Wu, W., Wu, S., Jin, Y., & Xiao, H. (2021). Recent advances in understanding the effects of lignin structural characteristics on enzymatic hydrolysis. *Biotechnology for Biofuels*, *14*(1), 1–20. https://doi.org/10.1186/s13068-021-02054-1

Zabed, H. M., Akter, S., Yun, J., Zhang, G., Awad, F. N., Qi, X., & Sahu, J. N. (2019). Recent advances in biological pretreatment of microalgae and lignocellulosic biomass for biofuel production. *Renewable and Sustainable Energy Reviews*, *105*, 105–128. https://doi.org/10.1016/j.rser.2019.01.048

Zamora, H. D., Silva, T. A. L., Varão, L. H. R., Baffi, M. A., & Pasquini, D. (2021). Simultaneous production of cellulases, hemicellulases, and reducing sugars by *Pleurotus ostreatus* growth in one-pot solid state fermentation using Alstroemeria sp. waste. *Biomass Conversion and Biorefinery*, *13*, 4879–4892. https://doi.org/10.1007/s13399-021-01723-3

Zhang, K., Pei, Z., & Wang, D. (2016). Organic solvent pretreatment of lignocellulosic biomass for biofuels and biochemicals: A review. *Bioresource Technology*, *199*, 21–33. https://doi.org/10.1016/j.biortech.2015.08.102

Zhang, W., Liu, J., Wang, Y., Sun, J., Huang, P., & Chang, K. (2021). Effect of ultrasound on ionic liquid-hydrochloric acid pretreatment with rice straw. *Biomass Conversion and Biorefinery*, *11*(5), 1749–1757. https://doi.org/10.1007/s13399-019-00595-y

Zhang, K., Wei, L., Sun, Q., Sun, J., Li, K., Zhai, S., et al., (2022). Effects of formaldehyde on fermentable sugars production in the low-cost pretreatment of corn stalk based on ionic liquids. *Chinese Journal of Chemical Engineering*, *42*, 406–414. https://doi.org/10.1016/j.cjche.2021.01.001

Zheng, J., & Rehmann, L. (2014). Extrusion pretreatment of lignocellulosic biomass: A review. *International Journal of Molecular Sciences*, *15*(10), 18967–18984. https://doi.org/10.3390/ijms151018967

Zoghlami, A., & Paës, G. (2019). Lignocellulosic biomass: Understanding recalcitrance and predicting hydrolysis. *Frontiers in Chemistry*, *7*(December), 874. https://doi.org/10.3389/fchem.2019.00874

5 Applications of Microbial Fermentation Technique for the Synthesis of Bioalcohols from Different Agrobiomasses

Karan Kumar, Kaustubh C. Khaire, Kuldeep Roy, and Vijayanand S. Moholkar**

5.1 INTRODUCTION

The depletion of non-renewable energy sources, combined with the increasing global demand for energy, has led to exploring alternative and sustainable energy options (Kumar & Moholkar, 2023; Kumar et al., 2021; Malani et al., 2019). Bioalcohols, such as ethanol and butanol, have emerged as promising candidates due to their renewable nature, high energy content, and reduced greenhouse gas emissions. Microbial fermentation techniques for bioalcohol synthesis from various agrobiomasses have gained significant attention in recent years, offering the potential to convert waste materials into valuable and environmentally friendly products (Kumar et al., 2022, 2023a, 2023b, 2024a, 2024b, 2024c; Dhabhai et al., 2018). However, the recalcitrance of agrarian-based lignocellulosic biomass (LB) poses challenges to its effective utilization by fermentative microorganisms (Hernawan et al., 2017). Various techniques have been developed to convert cellulosic waste materials into glucose to address these challenges, providing an alternative and sustainable approach to fuel production (Adeboye et al., 2014; Amiri & Karimi, 2018). Enzymatic pretreatment of LB has shown promise in enhancing biofuel production. Additionally, advanced high-throughput techniques, including metagenomics, next-generation sequencing, metatranscriptomics, and synthetic biology, have facilitated the discovery of novel microorganisms and powerful enzymes with superior activity, thermostability, and pH stability (Guilherme et al., 2017; Bussamra et al., 2015; Ostadjoo et al., 2019).

Agrarian-based LB has emerged as an attractive, cost-effective feedstock for microbial bioprocesses producing biofuels and valuable chemicals. LB primarily consists of polysaccharides, such as cellulose, hemicelluloses, and lignin, interconnected through covalent and non-covalent bonds, providing structural rigidity (Periyasamy et al., 2023). However, the inherent recalcitrance of LB poses challenges to its effective utilization by fermentative microorganisms. Nevertheless, significant

 DOI: 10.1201/9781003407713-5

progress has been made in microbial fermentation technologies, incorporating genome-based systems biology and metabolic engineering approaches. These have opened new avenues for innovative bio-based refineries utilizing engineered cells. Using LB for sustainable bioalcohol synthesis offers numerous benefits, including reduced environmental pollution and establishing a sustainable biorefinery (Saini & Sharma, 2021). Microbial fermentation techniques play a vital role in bioalcohol synthesis from LB, and recent studies have focused on understanding the unique metabolic pathways of microbes capable of metabolizing LB. Co-culture and consolidated bioprocessing techniques have also been investigated to enhance LB conversion (Du et al., 2020). By leveraging these advancements, using LB for sustainable bioalcohol synthesis provides a promising solution to address the challenges of fossil fuels and environmental pollution. The urgent need for alternative and sustainable energy sources arises from the increasing population and energy demand worldwide. Proper management of agricultural waste is essential to mitigate environmental impacts.

5.1.1 Aim and Scope of This Chapter

This chapter focuses on the optimization of bioalcohol production, with particular emphasis on recent developments in microbial fermentation techniques involving agrarian-based LB. It not only highlights the diversity of microorganisms capable of metabolizing LB, but also explores their specific metabolic pathways, all geared toward enhancing LB utilization and maximizing bioalcohol yields. Furthermore, innovative strategies, including co-culture and consolidated bioprocessing techniques, are thoroughly examined to boost LB conversion efficiency and elevate bioalcohol production (Dhabhai et al., 2018; Ahuja et al., 2023). These pioneering methods harness the synergistic potential of diverse microorganisms within consortia, unlocking the full potential of LB and paving the way for increased bioalcohol output. The exploration of consolidated bioprocessing, where a single microorganism performs multiple conversion steps, promises significant streamlining of bioalcohol production from agrarian-based lignocellulosic biomass (Periyasamy et al., 2023). In essence, this chapter offers a comprehensive perspective on the cutting-edge approaches dedicated to advancing bioalcohol production.

5.2 ACCESSING THE POTENTIAL UTILITY OF AGROBIOMASS AS FEEDSTOCK FOR FERMENTATION

Agrobiomass is the organic matter produced by agricultural activities having a high potential to be utilized for various applications (S & P, 2019; Arhin et al., 2023), such as animal feed, fertilizer, soil amendment, source for bioenergy production, chemical production, textile production, paper production, and environmental remediation. These applications depend on their origin and type of agrobiomass. Some of its types are as follows: residual crops (straw, husks, and leaves), energy crops (switchgrass, willow, and miscanthus), animal wastes (manure, poultry litter, etc.), wood, food waste (food scraps, spoiled food, etc.), oil crops (soybeans, rapeseed, and sunflowers), algae, and aquatic plants (water hyacinth and duckweed). However, while all of these biomasses can be utilized for various purposes, using certain agrobiomass, such as rice straw, risk husk, wheat straw, cotton straw, corn straw, sugarcane bagasse, and other plant residues, as feedstocks for microbial fermentation is very attractive. These agrobiomasses are abundant and readily available, which makes them a cost-effective and sustainable feedstock for energy generation, including bioalcohol production (Hernawan et al., 2017; Amiri & Karimi, 2018; Jiang et al., 2018).

The process of manufacturing bioalcohols with the help of microorganisms is commonly known as bioalcoholic microbial fermentation (BMF). In BMF, microbes such as yeast or bacteria convert sugars into bioalcohols such as ethanol, butanol, and propanol. These alcohols produced from BMF are utilized as biofuel, a promising alternative to traditional fossil fuels. The process of BMF typically involves four major steps: feedstock preparation, inoculation, fermentation, and distillation of the desired products, respectively (Arhin et al., 2023). However, using agrobiomass as feedstock for the BMF seems appealing owing to its availability and cost-effective choice, which has several benefits. It is full of challenges (Arhin et al., 2023). Table 5.1 summarizes the benefits and challenges associated with using agrobiomass as feedstocks for BMF. Later on, in this section, we will also provide an overview of various microbial fermentation techniques for utilizing agrobiomass that exists till now.

5.2.1 Potential of LB

Research indicates that the integration of various sources, including crops, agricultural remnants, trees, forest waste, as well as the utilization of conservation reserve land for production, has the potential to yield approximately 1.3 billion dry tons of biomass each year. The viability of biomass as a feedstock for biofuel is underscored

TABLE 5.1
Pros and Cons of Utilizing Agrobiomass as Feedstocks for Bioalcoholic Microbial Fermentation

Pros	Cons
Abundant and renewable resource	High production costs
Reduces reliance on fossil fuels	Technological and infrastructure challenges
Carbon-neutral or low carbon footprint	Variability in feedstock composition
Potential for waste and by-product valorization	Land and water resource requirements
Supports rural economies and job creation	Competition with food and feed production
Diversifies energy sources	Potential environmental impacts
Can contribute to circular economy	Regulatory and policy uncertainties
Enhances energy security	Need for efficient logistics and transportation
Mitigates greenhouse gas emissions	Requires appropriate harvesting and storage methods
Offers potential for decentralized production	Integration with existing energy infrastructure
Enables utilization of marginal lands	Challenges in scaling up production
Provides opportunities for biorefineries	Sensitivity to climate and weather conditions
Supports sustainable agricultural practices	Requires continuous research and development
Enhances rural development	Challenges in securing investment and funding
Promotes regional self-sufficiency	Market and economic uncertainties

by its recognition within the Renewable Fuel Standard (RFS), as stipulated in the Energy Independence and Security Act of 2007. LB, which forms the primary component of biofuel feedstock, holds significant promise for advancing biofuel production and the effort to reduce carbon dioxide emissions (Elfasakhany, 2016). Aside from fuel (e.g., ethanol or butanol), LB has the potential to yield a diverse array of chemicals and value-added products. Once biorefinery technologies are developed and brought into commercial use, LB holds the promise of being a versatile source for an extensive range of chemicals and biofuels, encompassing olefins, plastics, solvents, chemical intermediates, biogasoline, bioalcohols, and biodiesel (Saini & Sharma, 2021; Liu et al., 2021).

5.2.2 Challenges Linked to the Utilization of Lignocellulosic Biomass in the Fuel and Chemical Industry

The usage of LB in the fuel and chemical industry presents several challenges. LB's complex structure and chemical composition, including lignin, cellulose, and hemicellulose, make saccharification and conversion into biofuels and chemicals enzymatically challenging. The recalcitrance of LB is also affected by factors such as crystallinity, surface area, acetyl group content, and the distribution of lignin and hemicelluloses (Zoghlami & Paës, 2019). Pretreatment is essential to increase the availability of LB for cellulase enzymes, improve its digestibility, and increase product yield. The composition and structural arrangement of cellulose, hemicelluloses, and lignin differ significantly among plant species, posing a formidable challenge in fine-tuning the bioconversion process for lignocellulosic biomass (LB) (Zoghlami & Paës, 2019). Nonetheless, LB represents an abundant and renewable reservoir of cellulosic glucose, signifying its critical role as an alternative source for the generation of biofuels and a spectrum of biochemicals that serve as foundational elements for the creation of novel materials. In the forthcoming era, biorefineries are poised to harmonize diverse biomass conversion techniques, aiming to yield not only biofuels but also power, heat, and high-value chemicals, marking a significant stride toward a sustainable and multifaceted energy and chemical industry (Liu et al., 2021; Adegboye et al., 2021). The challenges associated with using LB in the fuel and chemical industry can be overcome through continued research and development of innovative technologies and processes.

5.2.3 Other Challenges Associated

Several other challenges are associated with using lignocellulose in the fuel and chemical industry. A notable challenge arises when dealing with biomass materials containing elevated moisture levels, rendering them unsuitable as feedstock for traditional thermochemical conversion methods like gasification and pyrolysis (Karunarathna et al., 2019). Elevated moisture content in biomass can lead to issues such as biological degradation, the formation of mold, and losses in organic content during storage. These problems can significantly diminish the efficiency and effectiveness of conversion processes. The utilization of concentrated acid for biomass hydrolysis presents a separate challenge, with various drawbacks including increased energy consumption, equipment corrosion, handling of hazardous chemicals, the need for an additional step of acid neutralization, and the formation of by-products that can hinder the fermentation process (Chang et al., 2021; Kumar et al., 2022). The thermochemical conversion of LB has the advantage of accommodating heterogeneous biomass mixtures. Yet, for single-step processes like pyrolysis to efficiently deconstruct LB at a lower cost and produce bio-oil, high temperatures are often necessary (Chen et al., 2019). Bio-oil, derived from LB, is a complex mixture comprising hundreds of compounds, which poses challenges in terms of convenience and difficulty in separating specific chemicals and fuels through single-step methods. Overall, these challenges highlight the need for continued research and development of innovative technologies and processes to overcome these obstacles and optimize the bioconversion process of LB.

5.3 MICROORGANISMS AND THEIR SACCHAROLYTIC PATHWAYS

Microorganisms play a crucial role in the conversion of agrobiomass into valuable products, such as bioalcohols. Understanding the diverse saccharolytic pathways employed by different microorganisms is essential for optimizing bioalcohol production. This section aims to provide an overview of the microorganisms involved in saccharification and their metabolic pathways, highlighting their potential for efficient utilization of agrobiomass.

5.3.1 Microbes Capable of Utilizing Agricultural Waste and Synthesizing Bioalcohols

Microbes play a crucial role in the synthesis of bioalcohols from agricultural waste biomass, offering a sustainable and renewable alternative to fossil fuels. These microorganisms possess the metabolic pathways and enzymes necessary for the conversion of fermentable sugars derived from agricultural waste into bioalcohols (Anand et al., 2024; Gheshlaghi et al., 2009; Kumar et al., 2022). In this subsection, we have discussed few examples of microbes that have demonstrated the ability to synthesize bioalcohols from agricultural waste biomass.

5.3.1.1 Yeasts

***Saccharomyces cerevisiae*:** *Saccharomyces cerevisiae* is a well-known yeast species widely used in bioalcohol production. It can efficiently ferment glucose and other sugars derived from agricultural waste biomass into bioalcohols, primarily ethanol (Lee et al., 2022). *S. cerevisiae* has been extensively engineered to enhance its bioethanol production capabilities, including the development of strains capable of fermenting pentose sugars like xylose.

***Pichia stipitis*:** *Pichia stipitis* is another yeast species known for its ability to ferment pentose sugars, such as xylose, which are abundant in agricultural waste biomass (Selim et al., 2020). It can convert xylose into bioethanol, making it a promising candidate for lignocellulosic bioethanol production from agricultural residues.

5.3.1.2 Bacteria

Zymomonas mobilis: *Zymomonas mobilis* is a bacterium capable of fermenting glucose and fructose into ethanol with high efficiency (Shabbir et al., 2023). It has been widely studied for its potential in bioethanol production from agricultural waste biomass. *Z. mobilis* exhibits high ethanol yields and is known for its tolerance to inhibitory compounds present in agricultural waste, such as lignin-derived compounds and organic acids.

***Clostridium* species:** Certain species of the *Clostridium* genus, such as *Clostridium ljungdahlii* and *Clostridium autoethanogenum*, are capable of synthesizing bioalcohols, including ethanol and higher alcohols, through the gas fermentation process (Arhin et al., 2023). These bacteria can utilize carbon monoxide and carbon dioxide, which can be derived from agricultural waste biomass, as carbon sources for bioalcohol production. Solventogenic species from this genera such as *Clostridium acetobutylicum*, *Clostridium beijerinckii*, and *Clostridium pasteurianum* can utilize the carbon sources to produce alcohol through acetone-butanol-ethanol (ABE) fermentation pathway (Du et al., 2020; Ahlawat et al., 2019).

5.3.1.3 Fungi

***Trichoderma reesei*:** *Trichoderma reesei*, a filamentous fungus, not only possesses cellulolytic capabilities, but also produces cellulase enzymes that can convert agricultural waste biomass into fermentable sugars (Jun et al., 2013). The released sugars can be subsequently fermented by other microbes to synthesize bioalcohols.

***Neurospora crassa*:** *Neurospora crassa* is a filamentous fungus known for its metabolic versatility. It can utilize a wide range of carbon sources, including sugars derived from agricultural waste biomass, to produce bioalcohols like ethanol (Periyasamy et al., 2023). *N. crassa* has been explored for its potential in consolidated bioprocessing (CBP), where it can saccharify LB and simultaneously ferment the released sugars into bioalcohols.

These examples illustrate the diversity of microbes capable of synthesizing bioalcohols from agricultural waste biomass. Each microorganism possesses unique metabolic pathways and enzymatic capabilities that enable efficient bioalcohol production. Ongoing research in genetic engineering, metabolic engineering, and fermentation optimization continues to enhance the capabilities of these microbes and expand the range of agricultural waste biomass feedstocks that can be utilized for bioalcohol synthesis (Selim et al., 2020; Kumar et al., 2022; Adegboye et al., 2021).

5.3.2 Microbial Saccharolytic Pathways for Utilization of Agrobiomass

Microbial saccharolytic pathways play a crucial role in the efficient utilization of agrobiomass for the production of valuable products such as biofuels and biochemicals. Agrobiomass, which includes various plant-based materials like LB, agricultural residues, and energy crops, is rich in complex polysaccharides such as cellulose and hemicellulose (Zoghlami & Paës, 2019). Microorganisms possess specific enzymatic machinery and metabolic pathways that allow them to break down these complex polysaccharides into fermentable sugars and further metabolize them for energy and product synthesis (Hernawan et al., 2017; Hon et al., 2017). Understanding these saccharolytic pathways is essential for optimizing the utilization of agrobiomass. In this section, we discuss the microbial saccharolytic pathways employed by different microorganisms for the efficient utilization of agrobiomass.

5.3.2.1 Cellulolytic Pathways

Cellulose is a major component of agrobiomass and represents a highly abundant and renewable feedstock for bioconversion. Microorganisms utilize cellulolytic pathways to hydrolyze cellulose into glucose monomers, which can be further metabolized (Saini & Sharma, 2021). The saccharolytic pathways involved in cellulose degradation typically include the action of cellulases, which are enzymes that break down cellulose into glucose units. Examples of microorganisms with efficient cellulolytic pathways include cellulolytic bacteria like *Clostridium Thermocellum* and *Clostridium cellulolyticum*, cellulolytic yeasts like *S. cerevisiae*, and cellulolytic fungi like *T. reesei* (Periyasamy et al., 2023).

5.3.2.2 Hemicellulolytic Pathways

Hemicellulose is another major component of agrobiomass, consisting of a complex mixture of sugars such as xylose, arabinose, mannose, and galactose (Khaire et al., 2021a). Microorganisms employ hemicellulolytic pathways to break down hemicellulose into fermentable sugars. The saccharolytic pathways involved in hemicellulose degradation typically include the action of hemicellulases, which are enzymes that hydrolyze hemicellulose into its constituent sugars (Khambhaty & Reena, 2023). Various microorganisms, including bacteria, yeasts, and fungi, possess specific hemicellulolytic enzymes and pathways for the efficient utilization of hemicellulose-derived sugars.

5.3.2.3 Lignocellulolytic Pathways

LB represents a complex substrate for microbial utilization. Microorganisms employ lignocellulolytic pathways to break down the complex structure of lignocellulose and extract fermentable sugars. These pathways involve the concerted action of cellulases, hemicellulases, and lignin-degrading enzymes (Saini & Sharma, 2021). Filamentous fungi, such as *T. reesei* and *Phanerochaete chrysosporium*, are known for their efficient lignocellulolytic capabilities and have been extensively studied for their potential in lignocellulosic bioconversion (Shimizu & Toya, 2021).

5.3.2.4 Co-utilization of Sugars

Some microorganisms possess versatile metabolic pathways that allow them to co-utilize multiple sugars derived from agrobiomass. For example, certain yeasts, including *S. cerevisiae* and *P. stipitis*, are capable of fermenting both glucose and xylose, which are the major sugars found in agrobiomass (Nazar et al., 2022). These microorganisms utilize specific metabolic pathways to efficiently metabolize and convert different sugars, enabling efficient utilization of agrobiomass-derived sugars. Understanding and optimizing microbial saccharolytic pathways is essential for the efficient utilization of agrobiomass in bioconversion processes (Fox & Prather, 2020). Genetic and metabolic engineering approaches can be employed to enhance the capabilities of microorganisms by improving the expression of key enzymes (Adegboye et al., 2021).

5.4 OPTIMIZATION OF MICROBIAL FERMENTATION TECHNIQUES FOR ENHANCED BIOALCOHOL PRODUCTION

BMF is a key process for the production of bioalcohols, such as ethanol and butanol, from various feedstocks including agrobiomass. The optimization of fermentation techniques plays a crucial role in improving bioalcohol yields and overall process efficiency. This section aims to discuss different aspects of microbial fermentation optimization, including media optimization, strain optimization using genetic and metabolic engineering, and the challenges associated with strain optimization techniques.

5.4.1 Optimization of Media Components

Media optimization is a crucial aspect of BMF process. The composition of the fermentation medium significantly influences the growth, metabolic activity, and ultimately, the product yields (Kumar et al., 2022). By adjusting various factors/components such as amount of carbon and nitrogen sources, pH, and the presence of essential vitamins and minerals, it is possible to create an optimized environment that promotes microbial growth and enhances the efficiency of bioalcohol production (Mu et al., 2021). This subsection aims to discuss the importance of media optimization and highlight various strategies employed to enhance bioalcohol yields through medium optimization. Techniques like Response Surface Methodology (RSM) and statistical experimental designs are commonly employed to identify the optimal media conditions (Kumar & Moholkar, 2023).

5.4.1.1 Importance of Media Optimization

The choice of fermentation medium and its composition have a significant impact on the overall performance of microbial fermentation for bioalcohol production. The medium provides essential nutrients, energy sources, and growth factors necessary for microbial growth and metabolism (Kumar et al., 2022). Through media optimization, it is possible to enhance the utilization of substrates, improve microbial growth rates, and maximize bioalcohol yields (Tiwari et al., 2022). Additionally, optimizing the medium can help mitigate the formation of undesirable by-products and increase the tolerance of microbial strains to inhibitory compounds, thereby improving the efficiency of the fermentation process (Kumar et al., 2022).

5.4.1.2 Factors to Consider during Media Optimization

Carbon source concentration: The selection and concentration of carbon sources in the fermentation medium are critical for bioalcohol production. Different carbon sources, such as glucose, xylose, and mannose, can be utilized by various microbial strains. Optimization involves determining the optimal concentration of the carbon source to balance microbial growth and bioalcohol production rates (Jun et al., 2013).

Nitrogen sources: Nitrogen is an essential component for microbial growth and metabolism. The choice of nitrogen sources, such as ammonium salts, amino acids, or complex nitrogenous compounds, can significantly impact bioalcohol yields.

Optimization involves identifying the most suitable nitrogen sources and their concentrations to achieve optimal microbial growth and bioalcohol production (Peng et al., 2022).

pH and buffering system: The pH of the fermentation medium affects microbial growth, enzyme activity, and product formation. Optimization of pH involves selecting an appropriate pH range and implementing a buffering system to maintain the desired pH throughout the fermentation process. This ensures optimal microbial performance and bioalcohol yields (Nazar et al., 2022).

Essential nutrients: The presence of essential nutrients, including vitamins, minerals, and trace elements, is crucial for microbial growth and bioalcohol production. Optimization involves determining the optimal concentrations and sources of these nutrients to support robust microbial metabolism and maximize bioalcohol yields (Kumar et al., 2022).

5.4.1.3 Strategies for Media Optimization

One-factor-at-a-time (OFAT) approach: This approach involves varying one-factor-at-a-time while keeping all other factors constant. It allows for the evaluation of the individual effects of each factor on bioalcohol production. Although simple, this method may not capture potential interactions between different factors (Singh et al., 2022).

Statistical experimental designs: Techniques such as RSM, Box–Behnken Design (BBD), and Central Composite Design (CCD) enable the simultaneous evaluation of multiple factors and their interactions (Khaire et al., 2021b; Kumar et al., 2022). These designs help identify the optimal levels of different variables and provide a more comprehensive understanding of their effects on bioalcohol yields.

Metabolic engineering: Metabolic engineering approaches can be employed to modify microbial strains to enhance their substrate utilization efficiency and bioalcohol production capabilities (Adegboye et al., 2021). Genetic modifications can be made to redirect metabolic fluxes toward desired pathways and improve the overall performance of the microbial strains in specific media conditions.

The production yield and productivity in bioalcohol production vary significantly across different feedstocks and microbial species. Ethanol yields range from 13.6 g/L for coffee husk with *S. cerevisiae* to 68.7% for corn stover during enzymatic hydrolysis. This variation can be attributed to factors such as the type of feedstock, pretreatment methods, and the choice of microorganisms. For instance, wheat straw, pretreated with alkaline hydrogen peroxide, yields 31.1 g/L of ethanol with the *S. cerevisiae* SR8u strain. In contrast, coconut husks, subjected to alkaline pretreatment, achieve an ethanol yield of 59.6% during enzymatic hydrolysis. Additionally, the choice of microorganisms, such as *S. cerevisiae* or *Enterobacter aerogenes*, can influence the final ethanol yield. It is evident that these factors interact to determine the overall productivity and efficiency of bioalcohol production processes (Table 5.3).

Furthermore, these findings highlighted in Table 5.3 the importance of tailored approaches for bioalcohol production, emphasizing the significance of feedstock selection and pretreatment methods. These considerations, along with the choice of microorganisms and enzymatic hydrolysis, have a profound impact on the ultimate success of bioalcohol production. Understanding these variations is essential for optimizing processes, reducing environmental impact, and achieving sustainable bioalcohol production in the field of renewable energy.

5.4.1.4 Challenges and Future Perspectives in Media Optimization

Media optimization for enhanced bioalcohol production is a complex task that requires careful consideration of multiple factors. Several challenges persist, such as the need for extensive screening of different carbon and nitrogen sources, the identification of optimal concentrations, and the potential trade-offs between microbial growth and bioalcohol production (Amiri, 2020). Furthermore, the scalability of optimized fermentation conditions from lab-scale to industrial-scale processes poses additional challenges. Table 5.2 shows the optimization techniques for enhanced bioalcohol production in microbial fermentation. The combination and customization of these techniques can greatly enhance bioalcohol production in microbial fermentation processes.

5.4.2 Strain Optimization Using Genetic and Metabolic Engineering

Strain optimization through genetic and metabolic engineering has emerged as a powerful approach to enhance bioalcohol production (Shimizu & Toya, 2021). By manipulating the genetic makeup and metabolic pathways of microbial strains, it is possible to improve their ability to convert substrates into bioalcohols efficiently. Metabolic engineering approaches aim to redirect metabolic fluxes toward desired bioalcohol production pathways. These strategies have proven successful in improving bioalcohol yields, substrate utilization efficiency, and tolerance to inhibitory compounds (Adegboye et al., 2021). This section focuses on the significance of strain

TABLE 5.2
Techniques Involved in the Enhanced Bioethanol Production

Optimization Technique	Description
Strain Engineering	Genetic modification of microorganisms to improve their bioalcohol production capabilities, such as enhancing metabolic pathways or introducing new enzymes
Medium Optimization	Optimization of nutrient composition and concentrations in the fermentation medium to provide ideal conditions for microbial growth and bioalcohol production
pH Control	Maintaining an optimal pH level throughout the fermentation process to maximize bioalcohol production and microbial activity
Temperature Control	Controlling the temperature during fermentation to create an environment that promotes the growth and bioalcohol production of the selected microorganism
Oxygen Management	Manipulating the oxygen levels in the fermentation process, such as using anaerobic conditions, to enhance bioalcohol production and prevent unwanted by-products
Fed-Batch Fermentation	A strategy where additional nutrients or substrates are added at specific intervals during fermentation to optimize bioalcohol production and microbial growth
Enzyme Optimization	Using specific enzymes or enzyme cocktails to improve the efficiency of saccharification and bioalcohol production from complex biomass substrates

optimization and highlights various strategies employed to enhance bioalcohol yields through genetic and metabolic engineering.

5.4.2.1 Importance of Strain Optimization

Microbial strains used for bioalcohol production often exhibit limitations in terms of inefficient substrate utilization, low bioalcohol yields, and lower tolerance to inhibitory compounds. Strain optimization aims to overcome these limitations by engineering microbial genomes to enhance key metabolic pathways involved in bioalcohol synthesis (Adegboye et al., 2021). This approach enables the development of superior microbial strains with improved fermentation characteristics, leading to higher bioalcohol yields, enhanced substrate utilization efficiency, and increased tolerance to inhibitory compounds (Kumar et al., 2022; Burgard et al., 2003). Table 5.3 summarizes the literature on agrobiomass utilized as feedstock, pretreatment methods, microorganism utilized, fermentation process, and the respective ethanol yield.

5.4.2.2 Strategies for Strain Optimization

Overexpression of key enzymes: One strategy in strain optimization techniques involves the overexpression of genes encoding key enzymes involved in bioalcohol synthesis. By increasing the expression levels of enzymes such as alcohol/aldehyde dehydrogenases (*AADs*) or pyruvate decarboxylases, it is possible to enhance the flux of carbon toward bioalcohol production pathways (Islam et al., 2023; Kumar et al., 2022). This approach can improve bioalcohol yields and substrate utilization efficiency.

Deletion or downregulation of competing pathways: Competing metabolic pathways within microbial strains can divert carbon away from bioalcohol production. By deleting or downregulating genes involved in competing pathways, it is possible to redirect metabolic fluxes toward desired bioalcohol synthesis pathways (Das et al., 2020). This strategy can enhance bioalcohol yields and reduce the formation of unwanted by-products.

Introduction of heterologous genes: Genetic engineering allows for the introduction of genes from other organisms to confer novel capabilities to microbial strains. By introducing new genes encoding enzymes with superior catalytic properties involved in alternative bioalcohol synthesis pathways, it is possible to enhance bioalcohol production (Adegboye et al., 2021). This approach provides opportunities to expand the range of bioalcohols that can be synthesized by microbial strains.

Metabolic engineering for flux redirection: Metabolic engineering strategies aim to redirect metabolic fluxes within microbial strains to optimize bioalcohol production. This can be achieved through the manipulation of enzymes, transporters, or regulatory elements involved in metabolic pathways (Adegboye et al., 2021). By fine-tuning metabolic fluxes, it is possible to improve bioalcohol yields and substrate utilization efficiency.

Despite significant advancements in strain optimization, several challenges remain unsolved. The complexity of microbial metabolism and the interplay between different metabolic pathways make it challenging to predict the outcomes of genetic modifications accurately. Balancing metabolic fluxes, minimizing by-product formation, and maintaining strain stability pose additional challenges (Kaushal et al., 2018;

TABLE 5.3
Summary of Agrobiomass Utilized as Feedstocks, Pretreatment Method, Microorganism Utilized, Fermentation Method, and Ethanol Yield

Biomass	Pretreatment	Microorganism	Fermentation	Ethanol Yield	References
Coffee husk	–	*Saccharomyces cerevisiae*	–	13.6 g/L	Gouvea et al. (2009)
Soybean hulls	–	*S. cerevisiae* D_5A	–	25.3 g/L	Mielenz et al. (2009)
Wheat straw	Alkaline hydrogen peroxide (AHP) pretreatment	*S. cerevisiae* SR8u strain	–	31.1 g/L	Gamage et al. (2010)
Corn stover	Hydrothermal pretreatment	–	Enzymatic hydrolysis (EH)	68.7 %	Saha et al. (2013)
Lemon peel	Steam explosion	*S. cerevisiae*	EH	60 L/1,000 kg	Boluda-Aguilar and López-Gómez (2013)
Pineapple peel	Acid pretreated	*S. cerevisiae* TISTR 5048 and *Enterobacter aerogenes* TISTR 1468	EH	9.7 g/L	Choonut et al. (2014)
Rapeseed straw	Acid pretreated	*S. cerevisiae* (Fermentis ethanol red, France)	EH	39.9 g/L	López-Linares et al. (2014)
Sugarcane bagasse	H_3PO_4 + Steam explosion	*S. cerevisiae*	Acid hydrolysis	19 g/L	Gupta et al. (2014)

(Continued)

TABLE 5.3 (*Continued*)
Summary of Agrobiomass Utilized as Feedstocks, Pretreatment Method, Microorganism Utilized, Fermentation Method, and Ethanol Yield

Biomass	Pretreatment	Microorganism	Fermentation	Ethanol Yield	References
Coconut husks	Alkaline pretreatment	*S. cerevisiae*	Enzymatic hydrolysis	59.6%	Cabral et al. (2016)
Cotton stalk	Microwave assisted $FeCl_3$	*S. cerevisiae* MTCC 174	EH	0.4 g_p/g_s	Singh et al. (2017)
Sugarcane tops	Acid hydrolysis	*S. cerevisiae*	Cellulysin®	13.7 g/L	Dodo et al. (2017)
Sugarcane bagasse	Alkaline Pretreatment	*S. cerevisiae*	–	6.8%	Hernawan et al. (2017)
Banana peals	Acid pretreated	–	Enzymatic reaction	21 g/L	Palacios et al. (2017)
Bagasse	Dilute acid-base pretreated	–	Dilute acid-base hydrolysis conditions	55.4%	Philippini et al. (2019)
Sesame plant residue	Microbial assisted pretreatment	*S. cerevisiae*	–	1.9 g/L	Kumar et al. (2020)
Barley straw	Explosive decompression pretreatment	–	Enzymatic Hydrolysis	4.3–9 g/100 g	Raud et al. (2021)
Sugarcane tops	Alkali pretreatment	*S. cerevisiae*	Enzymatic hydrolysis	1.9 g/g	Khaire et al. (2021c)
Cassava peels	–	*S. cerevisiae*	–	3.8 g/L	Mardina et al. (2021)
Carrot pulps	–	*S. cerevisiae*	–	40.6 g/L	Khoshkho et al. (2022)

Shimizu & Toya, 2021). The selection of suitable promoters, expression systems, and genetic tools for specific microbial hosts can also impact strain optimization outcomes. Furthermore, the scale-up of optimized strains from laboratory to industrial fermentation processes may require further adjustments and optimization to maintain productivity and stability (Hollinshead et al., 2014).

Future research should focus on addressing the complexities of microbial metabolism and improving our understanding of the underlying cellular mechanisms. The development of novel genetic tools, advanced modeling techniques, and high-throughput screening methods will enable more efficient and targeted strain optimization (Ko et al., 2020). Additionally, exploring synthetic biology approaches and multi-omics analysis can provide valuable insights into the design and optimization of microbial strains for enhanced bioalcohol production (Cao et al., 2020).

Strain optimization through genetic and metabolic engineering offers significant potential for enhancing bioalcohol production. By manipulating microbial genomes and metabolic pathways, it is possible to develop superior strains with improved fermentation characteristics, leading to higher bioalcohol yields, enhanced substrate utilization efficiency, and increased tolerance to inhibitory compounds (Chang et al., 2021; Chae et al., 2017). Continued advancements in genetic tools, metabolic engineering strategies, and systems biology approaches will drive further progress in strain optimization for bioalcohol production (Chen et al., 2020; Choi et al., 2019; Liu et al., 2019).

5.4.3 Problems and Bottlenecks in Strain Optimization Techniques

Strain optimization techniques, including genetic and metabolic engineering, have shown promise in enhancing bioalcohol production. However, several challenges and bottlenecks hinder the successful implementation of these techniques (Chen et al., 2020; Caspeta & Nielsen, 2013). The complexity of microbial metabolism and the interplay between different metabolic pathways make it difficult to predict the outcomes of genetic modifications accurately. Balancing metabolic fluxes, minimizing by-product formation, and maintaining strain stability pose additional challenges (Ko et al., 2020). The selection of suitable promoters, expression systems, and genetic tools for specific microbial hosts can also impact strain optimization outcomes (Lozano Terol et al., 2021). Moreover, the scale-up of optimized strains from laboratory to industrial fermentation processes may require further adjustments and optimization to maintain productivity and stability (Tiso et al., 2020). This subsection discusses the problems and bottlenecks encountered during strain optimization for bioalcohol production and explores potential solutions to overcome these obstacles (Wang, 2021; Whitford, 2022).

5.4.3.1 Complexity of Microbial Metabolism

One of the primary challenges in strain optimization is the complexity of microbial metabolism. Microorganisms have intricated metabolic networks comprising interconnected pathways that interact and influence each other (Kwon et al., 2020; Sarma et al., 2017). Modifying a single gene or pathway may have unexpected consequences on the overall metabolic balance, leading to undesired outcomes (Choi et al., 2019;

Chae et al., 2017). Predicting the impact of genetic modifications accurately becomes increasingly difficult as the complexity of the system increases.

5.4.3.2 Balancing Metabolic Fluxes

Redirecting metabolic fluxes toward desired bioalcohol synthesis pathways while minimizing by-product formation is a key objective in strain optimization. Achieving a proper balance of metabolic fluxes requires a comprehensive understanding of the underlying metabolic pathways and regulatory mechanisms. However, fine-tuning the metabolic network to optimize bioalcohol production without disrupting cellular homeostasis remains a significant challenge. Researchers can mitigate these challenges by performing comparative genomic analysis and reconstructing the strain-specific genome-scale metabolic models (Kumar et al., 2022).

5.4.3.3 Strain Stability

Maintaining strain stability is crucial for long-term bioalcohol production. Genetic modifications can introduce genetic instability, leading to decreased productivity or loss of desired traits over time. Genetic instability can result from factors such as genetic rearrangements, mutations, or plasmid instability (Das et al., 2020). Ensuring strain stability throughout fermentation processes, especially in industrial-scale applications, is essential for consistent and sustainable bioalcohol production.

5.4.3.4 Selection of Suitable Promoters and Expression Systems

The choice of promoters and expression systems for gene expression plays a vital role in strain optimization. Promoters and expression systems must be compatible with the host organism and provide appropriate regulation of gene expression. However, not all promoters and expression systems function optimally in every microbial host. Selecting suitable regulatory elements that ensure high expression levels, proper regulation, and minimal metabolic burden is a challenging task.

5.4.3.5 Genetic Tools and Engineering Techniques

Effective strain optimization relies on the availability of robust genetic tools and engineering techniques. However, not all microbial hosts have well-developed genetic tools and efficient transformation methods. Limited genetic manipulation tools hinder the ability to introduce and manipulate genes, limiting the scope of strain optimization. Advancements in genetic tools and engineering techniques are necessary to overcome this bottleneck.

5.4.3.6 Scale-up from Laboratory to Industrial Settings

Successful strain optimization in the laboratory does not always guarantee similar performance at an industrial scale. Various factors, such as differences in fermentation conditions, substrate composition, and process scale, can impact the performance of optimized strains. Optimized strains need to be evaluated and adjusted accordingly during the scale-up process to ensure consistent bioalcohol production under industrial conditions.

Overcoming the problems and bottlenecks associated with strain optimization requires interdisciplinary research efforts. Integration of computational modeling, synthetic biology, and systems biology approaches can provide valuable insights into the complexity of microbial metabolism and aid in the rational design of optimized strains. Advances in high-throughput screening techniques, such as omics-based analysis and directed evolution, can expedite strain optimization processes. Furthermore, the continuous development of robust genetic tools and expression systems will facilitate more efficient and targeted genetic modifications.

Strain optimization techniques for enhanced bioalcohol production face several challenges and bottlenecks, including the complexity of microbial metabolism, balancing metabolic fluxes, strain stability, selection of suitable promoters and expression systems, limited genetic tools, and scale-up issues. Addressing these challenges requires a combination of experimental and computational approaches, as well as advancements in genetic tools and engineering techniques. By overcoming these obstacles, strain optimization techniques can be harnessed to improve bioalcohol production and contribute to the development of sustainable and renewable energy sources.

5.5 CO-CULTURE AND CONSOLIDATED BIOPROCESSING TECHNIQUES FOR BIOALCOHOL SYNTHESIS

Co-culture and Consolidated Bioprocessing (CBP) techniques are promising strategies for producing bulk chemicals and biofuels directly from LB. CBP, characterized by the integration of hydrolysis and fermentation into a single step without the need for additional cellulases, is widely acknowledged as a highly promising approach for the cost-effective production of biofuels and chemicals derived from biomass. Cellulolytic organisms such as *Trichoderma* and *Myceliophthora* are used to directly produce biofuels and biochemicals like ethanol, itaconic acid, and butanol (Anand et al., 2024; Kumar et al., 2024). Ethanol stands as the predominant renewable transportation biofuel in the United States, with production reaching 13.3 billion gallons in 2012. A significant proportion of environmental challenges arises from the utilization of conventional energy sources, and the adoption of biomass resources holds substantial potential to curtail the nation's reliance on imported fossil fuels. LB is distinctly well-suited as a sustainable feedstock for the biotechnological manufacturing of alternative fuels and chemicals. Thermophilic anaerobic non-cellulolytic *Thermoanaerobacter* species play a pivotal role in the biotechnological production of cellulosic ethanol. Their unique ability to concurrently ferment hexose and pentose compounds, leading to high ethanol yields, underscores their crucial significance in this context.

A range of economic and environmental sustainability considerations, coupled with consumer demand for bio-based products derived from natural sources, has set the stage for the growth and advancement of biorefining technologies. Microalgae, with their superior areal productivity compared to traditional crops and their reduced land and water requirements, offer a promising and innovative source of biomass for biofuel production. Overall, continued research and development of innovative technologies and processes are necessary to optimize the bioconversion process of LB and increase efficiency while reducing costs and environmental impact.

In 1981, Thomas K. Ng and his collaborators at the University of Wisconsin-Madison co-cultured *C. thermocellum* LQRI and *Clostridium thermohydrosulfuricum* 39E strains to double the ethanol yield from various cellulose and hemicellulose substrates. Since then, co-culturing techniques have evolved into several industries, including food and drink, supplements, bioremediation, and the drug industry. Successful co-culture techniques can be instrumental in designing efficient, robust, and economic biosynthetic pathways by tapping into the unexplored allelopathic interaction territory. However, monocultured genetically engineered microbes remain in the center stage of the bioprocessing industry despite their proven inadequacies.

Fine-tuning of many parameters, such as preprocessing of the biomass, compositions of lignocellulosic substrates, operating temperature, solvent pH, and understanding of interaction, as well as in-depth knowledge of the growth and interaction dynamics of the co-cultured microbes, are needed to transfer a shake bottle experiment to a successful industrial CBP scheme. To understand the slowly shifting focus of the current industry toward co-culture techniques, recent academic and industrial developments can be categorized according to their advantages, challenges, and scopes.

5.5.1 Monoculture versus Co-culture Pathways: Division of Labor

In principle, engineering a pure microbial culture by incorporating all necessary genes is sufficient in producing biosynthetic products. However, a pure microbial culture often fails to achieve the expected level of product yield due to the metabolic burden, cell stress, and other technical challenges. Mixed communities have been shown to produce more power in microbial fuel cells than pure cultures. Progress in molecular biology has supplemented traditional microbiological methods in the field of microbial ecology (Peng et al., 2022). The impact of herbicides on soil microbial activities has been the subject of investigation by numerous researchers, both in pure culture and mixed microbial populations. Synthetic microbial consortia facilitate the swift assembly of pure translation machinery, yet the pure culture of prokaryotes remains indispensable in uncovering the functions and roles of these organisms (Cao et al., 2020). Microbial resources sourced from the human gut have potential applications in various fields, including empirical research on the microbiome, the creation of probiotic products, and bacteriotherapy.

The microbial community structure within traditional fermented foods is notably intricate, leading to a lack of clarity regarding the relationships between different strains. Given the substantial metabolic demands and limited efficiency of pure microbial cultures in biotechnological processes for biochemical production, scientists have turned their attention to microbial co-cultivation systems (Peng et al., 2022; Benito-Vaquerizo et al., 2020). These systems enable the division of labor and facilitate the accomplishment of more complex tasks. Online platforms enable the remote and real-time monitoring of disruptions. This metabolic burden results in the depravity of resources in the host cell and alters its physiological and biochemical properties. This, in turn, can cause unwanted accumulation of intermediate products, chemical toxicity, and cell stress (Shabbir et al., 2023; Jiang et al., 2018). Division of metabolic pathway load among different strains in co-cultured microbes can often address the above issues to achieve higher yield.

Co-culture with multiple bacteria often shows complicated dynamics, such as neutralism, mutualism, and commensalism. Neutralism occurs when two species consume different substrates and do not produce inhibitory compounds for co-evolving species (Wang, 2021; Yao & Nokes, 2014). Mutualism occurs when two species exchange metabolic resources or mutually remove inhibitors or toxins to help each other grow. Commensalism occurs when one species produces metabolic resources or removes inhibitors to facilitate the growth of other species. These co-evolution dynamics can increase the efficiency and robustness of the biosynthetic pathway. Other co-evolution dynamics, such as amensalism, competition, and predation, also exist and can impact the success of co-culture techniques. Understanding the full spectrum of these dynamics is necessary to tap into the full potential of co-culture techniques (Periyasamy et al., 2023).

5.5.2 Challenges and Bottlenecks in CBP and Potential Solutions

Microbial consortia in CBP techniques face several challenges that can hinder their metabolic activity and efficiency. Some of the bottlenecks and challenges associated with CBP and co-culture are discussed in this subsection.

5.5.2.1 Metabolic Competition

In microbial consortia, different microorganisms possess distinct metabolic pathways that might compete for substrates and essential nutrients, limiting their productivity. Metabolic competition refers to the phenomenon where different microbial strains within a consortium competing for the same carbon source and nutrients, which lead to reduced efficiency or failure of the CBP (Du et al., 2020; Bao et al., 2016). Some of the major metabolic challenges posed by metabolic competition in microbial consortia include:

Substrate preferences: Different microbes may preferentially use different components of the agrobiomass, such as hemicellulose or lignin, which can lead to incomplete substrate utilization and reduced overall efficiency of the CBP process.

Toxic metabolites: Some microbes might produce toxic metabolites that can potentially inhibit the growth of other microorganism in the consortium, leading to reduced overall efficiency or even failure of the CBP process.

Metabolic imbalance: The production rates of different metabolites within a consortium may be imbalanced, leading to the accumulation of certain metabolites that may be toxic to the microbial community or inhibit the activity of other enzymes.

Maintenance energy: Each microbial strain in the consortium requires a minimum amount of energy for their maintenance, which can compete with the overall efficiency of the CBP process, especially when the media has limited substrate.

These challenges can be mitigated via various strategies such as selecting microbial strains with complementary metabolic profiles, optimizing growth conditions to reduce competition, and developing metabolically engineered strains. This will help to improve the overall process efficiency and limit the production of toxic metabolites.

5.5.2.2 Microbial Stability

Maintaining microbial stability and diversity over long periods is essential for sustained bioconversion. However, microbial interactions can lead to instability and loss of consortia integrity, which can limit process productivity. Microbial stability is also a significant challenge in microbial consortia-based CBP techniques. The composition of microbial consortia is often complex, with different microorganisms having varying growth rates and requirements. This can lead to imbalances in the population dynamics of the consortia, resulting in the dominance of certain microorganisms and the decline of others. In some cases, this can lead to the loss of critical microorganisms necessary for efficient CBP processes, resulting in reduced productivity or even process failure.

Additionally, environmental stresses such as fluctuations in temperature, pH, and nutrient availability can also impact the stability of microbial consortia. Developing strategies to maintain microbial stability in CBP processes, such as optimizing growth conditions and controlling population dynamics, is crucial to ensuring the long-term viability and success of these techniques.

5.5.2.3 Oxygen Transfer

Oxygen transfer is one of the challenges in microbial consortia in CBP techniques. As CBP processes are typically aerobic, adequate oxygen transfer is crucial to maintain microbial activity and growth. However, in microbial consortia, different microorganisms have different oxygen requirements and tolerance limits, and ensuring adequate oxygen supply to all members of the consortium can be a challenge (Wang et al., 2021). Oxygen transfer limitations can lead to reduced microbial activity and growth, as well as changes in the microbial community composition, affecting the overall performance of the CBP process. Strategies such as optimizing the reactor design, improving agitation and aeration, and controlling dissolved oxygen levels can be used to overcome this challenge (Kumar et al., 2022).

5.5.2.4 Genetic Instability

Genetic instability can also be a challenge in microbial consortia used in CBP techniques. In some cases, mutations or genetic changes can occur in the microorganisms over time, which can affect their metabolic capabilities and stability within the consortium. Additionally, horizontal gene transfer (HGT) can occur between different microorganisms within the consortium, leading to the acquisition of new genetic traits that may or may not be beneficial for bioconversion. HGT can also introduce genetic instability, as it can lead to the loss of plasmids or other genetic elements that are necessary for certain metabolic pathways. Therefore, it is important to monitor the genetic stability of microbial consortia used in CBP techniques and to implement strategies to maintain their stability over time. This can include using selective pressure to maintain the desired traits, employing genetic engineering techniques to modify the microorganisms for better stability and performance, or developing strategies to prevent or minimize HGT.

5.5.2.5 Process Monitoring and Control

The dynamic and complex nature of microbial consortia requires sophisticated monitoring and control strategies to maintain the desired performance, and implementing these can be challenging. Process monitoring and control is another challenge in microbial consortia in CBP techniques. It is difficult to monitor and control the activities of multiple microorganisms in a consortium as they interact with each other and with the environment.

Process monitoring involves measuring and analyzing various parameters such as pH, temperature, dissolved oxygen, substrate concentration, product formation, and microbial growth. However, in a microbial consortium, these parameters are not only influenced by the metabolic activities of individual microorganisms, but also by their interactions. Therefore, it is crucial to develop advanced analytical tools and techniques to accurately monitor and control the process. Furthermore, microbial consortia are dynamic and can undergo changes in their composition and function over time, which can affect the process performance. Therefore, it is essential to have a reliable and efficient system for detecting and responding to changes in the microbial consortium to maintain process stability and performance. Overall, process monitoring and control in microbial consortia present significant challenges, and the development of advanced tools and techniques for accurate monitoring and control is necessary to ensure the efficient and stable operation of CBP processes.

5.5.2.6 Scale-up Challenges and Economic Viability

Scale-up challenges and economic viability are important considerations when transitioning microbial consortia-based processes from the laboratory to the industrial scale. The differences in reactor design, substrate availability, and other factors can pose challenges in scaling up these processes (Berlanga & Guerrero, 2016). Additionally, the high cost of enzyme production and the complexities associated with scaling up microbial consortia-based processes can limit their economic viability (Che & Men, 2019).

To overcome these challenges, continued research and development of microbial consortia-based processes is necessary. Optimization of process conditions, bioreactor design, and monitoring and control strategies are also crucial (Berlanga & Guerrero, 2016). Understanding the underlying molecular mechanisms of microbial interactions is essential for the rational design and optimization of defined consortia (Che & Men, 2019). Synthetic biology approaches can be employed to construct synthetic microbial consortia with specific functionalities (Song et al., 2014).

Recent studies have explored the potential of microbial consortia in lignocellulose degradation, which is relevant to the economic viability of biofuel production. Soil-derived microbial consortia enriched with different plant biomasses have revealed distinct players involved in lignocellulose degradation (De Lima Brossi et al., 2016). This knowledge can inform the development of strategies to enhance lignocellulose degradation efficiency in industrial-scale processes. Furthermore, the design, analysis, and application of synthetic microbial consortia have been investigated. Synthetic microbial consortia can be programmed using quorum sensing-based cell–cell communication, enabling the construction of isogenic microbial communities with specific functionalities (Song et al., 2014). These synthetic consortia have potential applications in various fields, including drug production, human health, and environmental management (Jia et al., 2016).

In summary, scaling up microbial consortia-based processes from the laboratory to the industrial scale presents challenges related to reactor design, substrate availability, and cost. However, ongoing research and development efforts, optimization of process conditions and bioreactor design, and the use of synthetic biology approaches offer potential solutions to overcome these challenges and improve the economic viability of microbial consortia-based processes.

5.5.3 Downstream Processing and Purification Techniques for Bioalcohols

Downstream processing and purification techniques are crucial steps in the production of bioalcohols. Recent publications have explored various strategies to optimize these processes and improve the efficiency of bioalcohol purification.

One study by Basen et al. (2014) focused on the construction of a fundamentally different synthetic pathway for bioalcohol production using a thermophilic archaeon, *Pyrococcus furiosus*. The researchers inserted the gene for bacterial alcohol dehydrogenase (AdhA) into *P. furiosus*, resulting in bioalcohol production at 70°C. This innovative approach highlights the potential of genetic engineering to enhance bioalcohol production and offers insights into novel downstream processing techniques. Scalability is a critical factor in downstream processing, especially for large-scale production of bioalcohols. Qu et al. (2015) discussed scalable downstream strategies for the purification of bioalcohols. Although the focus of the study was on genetic vectors, the principles and techniques discussed are applicable to bioalcohol purification. The authors emphasized the importance of both upstream and downstream processing steps in achieving efficient purification.

Continuous purification techniques have also been explored for the downstream processing of bioalcohols. Rosa et al. (2013) investigated the continuous purification of bioalcohols using aqueous two-phase systems (ATPS). This approach involves clarification, concentration, selective purification steps, and virus clearance. The study demonstrated the feasibility of continuous purification techniques and their potential application in the bioalcohol production process. Furthermore, Soares et al. (2015) conducted a comprehensive analysis of partitioning in ATPS. This review article discussed the strengths, weaknesses, opportunities, and threats associated with ATPS in various bioprocesses, including bioalcohol production. The authors highlighted the potential of ATPS for liquid–liquid extraction and its relevance to biomanufacturing.

In summary, recent publications have explored various downstream processing and purification techniques for bioalcohols. These studies have focused on genetic engineering approaches, scalability, continuous purification techniques, and the use of ATPS. These advancements offer valuable insights into optimizing the production and purification of bioalcohols, contributing to the development of sustainable and efficient bioalcohol production processes.

5.6 OVERVIEW AND CONCLUSIONS

The synthesis of bioalcohols from a variety of agrobiomass sources using BMF emerges as a promising and sustainable solution to meet the escalating need for renewable energy resources. The versatility of microbial fermentation processes enables the conversion of diverse agrobiomass types, encompassing agricultural residues, lignocellulosic materials, and energy crops, into valuable bioalcohol products, including ethanol, butanol, isobutanol, and higher alcohols.

Critical to this process is the pivotal role played by microorganisms, spanning bacteria, yeasts, and fungi, in efficiently transforming sugars derived from agrobiomass into bioalcohols. Leveraging genetic and metabolic engineering approaches has proven instrumental in enhancing microbial strains' fermentation capabilities, bolstering their proficiency in substrate utilization, and augmenting their resistance to inhibitory compounds inherent to agrobiomass.

The optimization of key process parameters within BMF, including pH, temperature, substrate concentration, and the presence or absence of oxygen, stands as a pivotal determinant in maximizing bioalcohol yields and overall process efficiency. Complementary to this is the implementation of pretreatment techniques, strategically devised to render agrobiomass carbohydrates more accessible to microbial enzymes, thereby elevating the effectiveness of bioalcohol production. Continuous research and development endeavors focusing on microbial fermentation techniques, genetic engineering, metabolic engineering, and process refinement are anticipated to propel agrobiomass utilization for bioalcohol synthesis. These advancements hold the promise of fostering sustainable and eco-friendly biofuel production, consequently reducing dependence on fossil fuels and ameliorating the consequences of climate change.

In the context of CBP involving lignocellulosic feedstocks, a distinctive metabolic feat emerges, largely orchestrated by genetically modified microorganisms working singularly. A compelling strategy has emerged, wherein cellulase and xylanase act synergistically to facilitate the efficient hydrolysis of LB. This collaborative approach achieves optimal LB deconstruction, an instrumental milestone in the broader bioalcohol production process.

The pressing global need for alternative energy sources has been precipitated by the overwhelming burden on fossil fuels, giving rise to acute energy crises and associated climate change challenges. In response, harnessing agrobiomass resources for sustainable bioalcohol synthesis has emerged as a promising approach to redress these pressing issues and pave the way for a sustainable energy future.

ACRONYMS AND ABBREVIATIONS

Acronyms	Full form
AADs	Alcohol/aldehyde dehydrogenases
ATPS	Aqueous two-phase systems
BBD	Box–Behnken Design
BMF	Bioalcoholic microbial fermentation
CBP	Consolidate bioprocessing
CCD	Central Composite Design
HGT	Horizontal gene transfer
LB	Lignocellulosic biomass
OFAT	One-Factor-at-a-Time
RFS	Renewable Fuel Standard
RSM	Response Surface Methodology

ACKNOWLEDGMENT

Dr. Karan Kumar is grateful to Prime Minister's Research Fellowship provided by Ministry of Education, Government of India.

NOTE

* Corresponding author

REFERENCES

Adeboye, P. T., M. Bettiga, and L. Olsson. 2014. The chemical nature of phenolic compounds determines their toxicity and induces distinct physiological responses in saccharomyces cerevisiae in lignocellulose hydrolysates. *AMB Express* 4 (1): 46. https://doi.org/10.1186/s13568-014-0046-7.

Adegboye, M. F., O. B. Ojuederie, P. M. Talia, and O. O. Babalola. 2021. Bioprospecting of microbial strains for biofuel production: metabolic engineering, applications, and challenges. *Biotechnology for Biofuels* 14 (1): 5. https://doi.org/10.1186/s13068-020-01853-2.

Ahlawat, S., M. Kaushal, B. Palabhanvi, M. Muthuraj, G. Goswami, and D. Das. 2019. Nutrient modulation based process engineering strategy for improved butanol production from clostridium acetobutylicum. *Biotechnology Progress* 35 (2): e2771. https://doi.org/10.1002/btpr.2771.

Ahuja, V., A. K. Bhatt, B. Ravindran, Y. -H. Yang, and S. K. Bhatia. 2023. A mini-review on syngas fermentation to bio-alcohols: current status and challenges. *Sustainability (Switzerland)* 15 (4): 3765. https://doi.org/10.3390/su15043765.

Amiri, H. 2020. Recent innovations for reviving the abe fermentation for production of butanol as a drop-in liquid biofuel. *Biofuel Research Journal* 7 (4): 1256–1266. https://doi.org/10.18331/BRJ2020.7.4.4.

Amiri, H., and K. Karimi. 2018. Pretreatment and hydrolysis of lignocellulosic wastes for butanol production: challenges and perspectives. *Bioresource Technology* 270 (December): 702–721. https://doi.org/10.1016/j.biortech.2018.08.117.

Anand, A., Kumar, K., & Moholkar, V. S. (2024). Various Routes for Hydrogen Production and Its Utilization for Sustainable Economy. In *Biohydrogen-Advances and Processes* (pp. 503–527). Cham: Springer Nature Switzerland.

Arhin, S. G., A. Cesaro, F. Di Capua, and G. Esposito. 2023. Recent progress and challenges in biotechnological valorization of lignocellulosic materials: towards sustainable biofuels and platform chemicals synthesis. *The Science of the Total Environment* 857 (Pt 1): 159333. https://doi.org/10.1016/j.scitotenv.2022.159333.

Bao, H., C. Chen, L. Jiang, Y. Liu, M. Shen, W. Liu, and A. Wang. 2016. Optimization of key factors affecting biohydrogen production from microcrystalline cellulose by the co-culture of clostridium acetobutylicum X9 + ethanoigenens harbinense B2. *RSC Advances* 6 (5): 3421–3427. https://doi.org/10.1039/C5RA14192C.

Basen, M., G. J. Schut, D. Nguyen, G. L. Lipscomb, R. A. Benn, C. J. Prybol, B. J. Vaccaro, F. L. Poole, R. M. Kelly, and M. W. W. Adams. 2014. Single gene insertion drives bioalcohol production by a thermophilic archaeon. *Proceedings of the National Academy of Sciences* 111: 49. https://doi.org/10.1073/pnas.1413789111.

Benito-Vaquerizo, S., M. Diender, I. P. Olm, V. A. P. Martins dos Santos, P. J. Schaap, D. Z. Sousa, and M. Suarez-Diez. 2020. Modeling a co-culture of clostridium autoethanogenum and clostridium kluyveri to increase syngas conversion to medium-chain fatty-acids. *Computational and Structural Biotechnology Journal* 18: 3255–3266. https://doi.org/10.1016/j.csbj.2020.10.003.

Berlanga, M., and R. Guerrero. 2016. Living together in biofilms: the microbial cell factory and its biotechnological implications. *Microbial Cell Factories* 15 (1): 165. https://doi.org/10.1186/s12934-016-0569-5.

Boluda-Aguilar, M., and A. López-Gómez. 2013. Production of bioethanol by fermentation of lemon (Citrus Limon L.) peel wastes pretreated with steam explosion. *Industrial Crops and Products* 41 (January): 188–197. https://doi.org/10.1016/j.indcrop.2012.04.031.

Burgard, A. P., P. Pharkya, and C. D. Maranas. 2003. Optknock: a bilevel programming framework for identifying gene knockout strategies for microbial strain optimization. *Biotechnology and Bioengineering* 84 (6): 647–657. https://doi.org/10.1002/bit.10803.

Bussamra, B. C., S. Freitas, and A. C. da Costa. 2015. Improvement on sugar cane bagasse hydrolysis using enzymatic mixture designed cocktail. *Bioresource Technology* 187 (July): 173–181. https://doi.org/10.1016/j.biortech.2015.03.117.

Cabral, M. M. S., A. K. de Souza Abud, C. E. de Farias Silva, and R. M. R. G. Almeida. 2016. A produção de bioetanol a partir de fibra de casca de coco. *Ciência Rural* 46 (October): 1872–1877. https://doi.org/10.1590/0103-8478cr20151331.

Cao, M., V. G Tran, and H. Zhao. 2020. Unlocking nature's biosynthetic potential by directed genome evolution. *Current Opinion in Biotechnology* 66 (December): 95–104. https://doi.org/10.1016/j.copbio.2020.06.012.

Caspeta, L., and J. Nielsen. 2013. Toward systems metabolic engineering of *aspergillus* and *pichia* species for the production of chemicals and biofuels. *Biotechnology Journal* 8 (5): 534–544. https://doi.org/10.1002/biot.201200345.

Chae, T. U., S. Y. Choi, J. W. Kim, Y. -S. Ko, and S. Y. Lee. 2017. Recent advances in systems metabolic engineering tools and strategies. *Current Opinion in Biotechnology* 47 (October): 67–82. https://doi.org/10.1016/j.copbio.2017.06.007.

Chang, D., Z. U. Islam, J. Zheng, J. Zhao, X. Cui, and Z. Yu. 2021. Inhibitor tolerance and bioethanol fermentability of levoglucosan-utilizing escherichia coli were enhanced by overexpression of stress-responsive gene ycfR: the proteomics-guided metabolic engineering. *Synthetic and Systems Biotechnology* 6 (4): 384–395. https://doi.org/10.1016/j.synbio.2021.11.003.

Che, S., and Y. Men. 2019. Synthetic microbial consortia for biosynthesis and biodegradation: promises and challenges. *Journal of Industrial Microbiology and Biotechnology* 46 (9–10): 1343–1358. https://doi.org/10.1007/s10295-019-02211-4.

Chen, W.-H., Y.-Y. Lin, H.-C. Liu, T.-C. Chen, C.-H. Hung, C.-H. Chen, and H. C. Ong. 2019. A comprehensive analysis of food waste derived liquefaction bio-oil properties for industrial application. *Applied Energy* 237 (March): 283–291. https://doi.org/10.1016/j.apenergy.2018.12.084.

Chen, Y., D. Banerjee, A. Mukhopadhyay, and C. J. Petzold. 2020. Systems and synthetic biology tools for advanced bioproduction hosts. *Current Opinion in Biotechnology* 64 (August): 101–109. https://doi.org/10.1016/j.copbio.2019.12.007.

Choi, K. R., W. D. Jang, D. Yang, J. S. Cho, D. Park, and S. Y. Lee. 2019. Systems metabolic engineering strategies: integrating systems and synthetic biology with metabolic engineering. *Trends in Biotechnology* 37 (8): 817–837. https://doi.org/10.1016/j.tibtech.2019.01.003.

Choonut, A., M. Saejong, and K. Sangkharak. 2014. The production of ethanol and hydrogen from pineapple peel by saccharomyces cerevisiae and enterobacter aerogenes. *Energy Procedia, 2013 International Conference on Alternative Energy in Developing Countries and Emerging Economies (2013 AEDCEE)* 52 (January): 242–249. https://doi.org/10.1016/j.egypro.2014.07.075.

Das, M., P. Patra, and A. Ghosh. 2020. Metabolic engineering for enhancing microbial biosynthesis of advanced biofuels. *Renewable and Sustainable Energy Reviews* 119 (March): 109562. https://doi.org/10.1016/j.rser.2019.109562.

De Lima Brossi, M. J., D. J. Jiménez, L. Cortes-Tolalpa, and J. D. Van Elsas. 2016. Soil-derived microbial consortia enriched with different plant biomass reveal distinct players acting in lignocellulose degradation. *Microbial Ecology* 71 (3): 616–627. https://doi.org/10.1007/s00248-015-0683-7.

Dhabhai, R., C. H. Niu, and A. K. Dalai. 2018. Agricultural byproducts-based biosorbents for purification of bioalcohols: a review. *Bioresources and Bioprocessing* 5 (1): 37. https://doi.org/10.1186/s40643-018-0223-7.

Dodo, C. M., S. Mamphweli, and O. Okoh. 2017. Bioethanol production from lignocellulosic sugarcane leaves and tops. *Journal of Energy in Southern Africa* 28 (3): 1–11. https://doi.org/10.17159/2413-3051/2017/v28i3a2354.

Du, Y., W. Zou, K. Zhang, G. Ye, and J. Yang. 2020. Advances and applications of clostridium co-culture systems in biotechnology. *Frontiers in Microbiology* 11: 560223. https://www.frontiersin.org/articles/10.3389/fmicb.2020.560223.

Elfasakhany, A. 2016. Experimental study of dual n-butanol and iso-butanol additives on spark-ignition engine performance and emissions. *Fuel* 163 (January): 166–174. https://doi.org/10.1016/j.fuel.2015.09.059.

Fox, K. J, and K. L. J. Prather. 2020. Carbon catabolite repression relaxation in escherichia coli: global and sugar-specific methods for glucose and secondary sugar co-utilization. *Current Opinion in Chemical Engineering* 30 (December): 9–16. https://doi.org/10.1016/j.coche.2020.05.005.

Gheshlaghi, R., J. M. Scharer, M. Moo-Young, and C. P. Chou. 2009. Metabolic pathways of clostridia for producing butanol. *Biotechnology Advances* 27 (6): 764–781. https://doi.org/10.1016/j.biotechadv.2009.06.002.

Gouvea, B. M., C. Torres, A. S. Franca, L. S. Oliveira, and E. S. Oliveira. 2009. Feasibility of ethanol production from coffee husks. *Biotechnology Letters* 31 (9): 1315–1319. https://doi.org/10.1007/s10529-009-0023-4.

Guilherme, A. de A., P. V. F. Dantas, J. C. J. Soares, E. S. dos Santos, F. A. N. Fernandes, and G. R. de Macedo. 2017. Pretreatments and enzymatic hydrolysis of sugarcane bagasse aiming at the enhancement of the yield of glucose and xylose. *Brazilian Journal of Chemical Engineering* 34 (4): 937–947. https://doi.org/10.1590/0104-6632.20170344s20160225.

Gupta, A., S. P. Das, A. Ghosh, R. Choudhary, D. Das, and A. Goyal. 2014. Bioethanol production from hemicellulose rich populus nigra involving recombinant hemicellulases from clostridium thermocellum. *Bioresource Technology* 165 (August): 205–213. https://doi.org/10.1016/j.biortech.2014.03.132.

Hernawan, R. M., D. Pratiwi, S. K. Wahono, C. Darsih, S. N. Hayati, C. D. Poeloengasih, et al., 2017. Bioethanol production from sugarcane bagasse by simultaneous sacarification and fermentation using saccharomyces cerevisiae. *AIP Conference Proceedings* 1823 (1): 020026. https://doi.org/10.1063/1.4978099.

Hollinshead, W., L. He, and Y. J. Tang. 2014. Biofuel production: an odyssey from metabolic engineering to fermentation scale-up. *Frontiers in Microbiology* 5 (July): 00344. https://doi.org/10.3389/fmicb.2014.00344.

Hon, S., D. G. Olson, E. K. Holwerda, A. A. Lanahan, S. J. L. Murphy, M. I. Maloney, T. Zheng, B. Papanek, A. M. Guss, and L. R. Lynd. 2017. The ethanol pathway from thermoanaerobacterium saccharolyticum improves ethanol production in clostridium thermocellum. *Metabolic Engineering* 42 (July): 175–184. https://doi.org/10.1016/j.ymben.2017.06.011.

Islam, T., T. P. Nguyen-Vo, V. K. Gaur, J. Lee, and S. Park. 2023. Metabolic engineering of escherichia coli for biological production of 1, 3-butanediol. *Bioresource Technology* 376 (May): 128911. https://doi.org/10.1016/j.biortech.2023.128911.

Jia, X., C. Liu, H. Song, M. Ding, J. Du, Q. Ma, and Y. Yuan. 2016. Design, analysis and application of synthetic C. *Synthetic and Systems Biotechnology* 1 (2): 109–117. https://doi.org/10.1016/j.synbio.2016.02.001.

Jiang, Y., D. Guo, J. Lu, P. Dürre, W. Dong, W. Yan, W. Zhang, J. Ma, M. Jiang, and F. Xin. 2018. Consolidated bioprocessing of butanol production from xylan by a thermophilic and butanologenic thermoanaerobacterium Sp. M5. *Biotechnology for Biofuels* 11 (1): 89. https://doi.org/10.1186/s13068-018-1092-1.

Jun, H., H. Guangye, and C. Daiwen. 2013. Insights into enzyme secretion by filamentous fungi: comparative proteome analysis of trichoderma reesei grown on different carbon sources. *Journal of Proteomics* 89 (August): 191–201. https://doi.org/10.1016/j.jprot.2013.06.014.

Karunarathna, M. S., M. K. Lauer, T. Thiounn, R. C. Smith, and A. G. Tennyson. 2019. Valorisation of waste to yield recyclable composites of elemental sulfur and lignin. *Journal of Materials Chemistry A* 7 (26): 15683–15690. https://doi.org/10.1039/C9TA03222C.

Kaushal, M., K. V. N. Chary, S. Ahlawat, B. Palabhanvi, G. Goswami, and D. Das. 2018. Understanding regulation in substrate dependent modulation of growth and production of alcohols in clostridium sporogenes NCIM 2918 through metabolic network reconstruction and flux balance analysis. *Bioresource Technology* 249 (February): 767–776. https://doi.org/10.1016/j.biortech.2017.10.080.

Khaire, K. C., V. S. Moholkar, and A. Goyal. 2021a. Bioconversion of sugarcane tops to bioethanol and other value added products: an overview. *Materials Science for Energy Technologies* 4 (January): 54–68. https://doi.org/10.1016/j.mset.2020.12.004.

Khaire, K. C., V. S. Moholkar, and A. Goyal. 2021b. Separation and characterization of cellulose from sugarcane tops and its saccharification by recombinant cellulolytic enzymes. *Preparative Biochemistry & Biotechnology* 51 (8): 811–820. https://doi.org/10.1080/10826068.2020.1861011.

Khaire, K. C., Suryakant Moholkar, V. S., & Goyal, A. (2021c). Alkaline pretreatment and response surface methodology based recombinant enzymatic saccharification and fermentation of sugarcane tops. *Bioresource technology*, 341, 125837. https://doi.org/10.1016/j.biortech.2021.125837

Khambhaty, Y., and R. Reena. 2023. Chapter 10- macroalgal polysaccharides: biocatalysts in biofuel/bioenergy production. In R. Goldbeck and P. Poletto (Eds) *Polysaccharide-Degrading Biocatalysts, Foundations and Frontiers in Enzymology*, pp. 227–273. Academic Press: New York. https://doi.org/10.1016/B978-0-323-99986-1.00009-0.

Khoshkho, S. M., M. Mahdavian, F. Karimi, H. Karimi-Maleh, and P. Razaghi. 2022. Production of bioethanol from carrot pulp in the presence of saccharomyces cerevisiae and beet molasses inoculum; a biomass based investigation. *Chemosphere* 286 (January): 131688. https://doi.org/10.1016/j.chemosphere.2021.131688.

Ko, Y.-S., J. W. Kim, J. A. Lee, T. Han, G. B. Kim, J. E. Park, and S. Y. Lee. 2020. Tools and strategies of systems metabolic engineering for the development of microbial cell factories for chemical production. *Chemical Society Reviews* 49 (14): 4615–4636. https://doi.org/10.1039/D0CS00155D.

Kumar, K., and V. S. Moholkar. 2023. Mechanistic aspects of enhanced kinetics in sonoenzymatic processes using three simultaneous approaches. In V. S. Moholkar, K. Mohanty, and V. V. Goud (Eds) *Sustainable Energy Generation and Storage* (pp. 41–57). Springer Nature: Singapore. https://doi.org/10.1007/978-981-99-2088-4_5.

Kumar, K., L. Barbora, and V. S. Moholkar. (2023a). Genomic insights into clostridia in bioenergy production: comparison of metabolic capabilities and evolutionary relationships. *Biotechnology and Bioengineering* 121 (4): 1298–1313. https://doi.org/10.1002/bit.28610.

Kumar, K., P. Patro, U Raut, V Yadav, L Barbora, and VS. Moholkar. (2023b). Elucidating the molecular mechanism of ultrasound-enhanced lipase-catalyzed biodiesel synthesis: a computational study. *Biomass Conversion and Biorefinery*. https://doi.org/10.1007/s13399-023-04742-4.

Kumar, K., Jadhav, S. M., & Moholkar, V. S. (2024). Acetone-Butanol-Ethanol (ABE) fermentation with clostridial co-cultures for enhanced biobutanol production. *Process Safety and Environmental Protection*, 185, 277-285. https://doi.org/10.1016/j.psep.2024.03.027

Kumar, K., Anand, A., & Moholkar, V. S. (2024a). Molecular Hydrogen (H2) Metabolism in Microbes: A Special Focus on Biohydrogen Production. In *Biohydrogen-Advances and Processes* (pp. 25–58). Cham: Springer Nature Switzerland.

Kumar, K., Jadhav, S. M., & Moholkar, V. S. (2024b). Acetone-Butanol-Ethanol (ABE) fermentation with clostridial co-cultures for enhanced biobutanol production. *Process Safety and Environmental Protection*, 185, 277–285. https://doi.org/10.1016/j.psep.2024.03.027

Kumar, K., Kumar, S., Goswami, A., & Moholkar, V. S. (2024c). Advancing sustainable biofuel production: A computational insight into microbial systems for isopropanol synthesis and beyond. *Process Safety and Environmental Protection*, 188, 1118–1132. https://doi.org/10.1016/j.psep.2024.06.024

Kumar, P., V. Kumar, S. Kumar, J. Singh, and P. Kumar. 2020. Bioethanol production from sesame (Sesamum Indicum L.) plant residue by combined physical, microbial and chemical pretreatments. *Bioresource Technology* 297 (February): 122484. https://doi.org/10.1016/j.biortech.2019.122484.

Kumar, K., K. Roy, and V. S. Moholkar. 2021. Mechanistic investigations in sonoenzymatic synthesis of N-butyl levulinate. *Process Biochemistry* 111 (December): 147–158. https://doi.org/10.1016/j.procbio.2021.09.005.

Kumar, K., L. Barbora, and V. S. Moholkar. 2022. Comparative genomic analysis and constraint-based analysis of genome-scale metabolic models of the genus clostridia. Preprint. https://doi.org/10.22541/au.166858569.91782041/v1.

Kumar, K., A. S. Pragati, P. C. Saanya Yadav, M. Kori, A. Shiv Ram, L. Barbora, and V. S. Moholkar. 2022. Computational investigations in inhibition of alcohol/aldehyde dehydrogenase in lignocellulosic hydrolysates. *Preprint. Bioinformatics*. https://doi.org/10.1101/2022.11.19.517192.

Kumar, K., H. Shah, and V. S. Moholkar. 2022. Genetic algorithm for optimization of fermentation processes of various enzyme productions. In J. S. Eswari, & N. Suryawanshi (Eds), *Optimization of Sustainable Enzymes Production*. Chapman and Hall/CRC: Boca Raton, FL.

Kwon, M. S., B. T. Lee, S. Y. Lee, and H. U. Kim. 2020. Modeling regulatory networks using machine learning for systems metabolic engineering. *Current Opinion in Biotechnology* 65 (October): 163–170. https://doi.org/10.1016/j.copbio.2020.02.014.

Lee, J.-Y., S.-E. Lee, and D.-W. Lee. 2022. current status and future prospects of biological routes to bio-based products using raw materials, wastes, and residues as renewable resources. *Critical Reviews in Environmental Science and Technology* 52 (14): 2453–2509. https://doi.org/10.1080/10643389.2021.1880259.

Liu, J., S. H. J. Chan, J. Chen, C. Solem, and P. R. Jensen. 2019. Systems biology - a guide for understanding and developing improved strains of lactic acid bacteria. *Frontiers in Microbiology* 10 (April): 876. https://doi.org/10.3389/fmicb.2019.00876.

Liu, Y., Y. Tang, H. Gao, W. Zhang, Y. Jiang, F. Xin, and M. Jiang. 2021. Challenges and future perspectives of promising biotechnologies for lignocellulosic biorefinery. *Molecules* 26 (17): 5411. https://doi.org/10.3390/molecules26175411.

López-Linares, J. C., I. Romero, C. Cara, E. Ruiz, M. Moya, and E. Castro. 2014. Bioethanol production from rapeseed straw at high solids loading with different process configurations. *Fuel* 122 (April): 112–118. https://doi.org/10.1016/j.fuel.2014.01.024.

Lozano, T., G. Julia Gallego-Jara, R. A. S. Martínez, A. M. Vivancos, M. C. Díaz, and T. de Diego Puente. 2021. Impact of the expression system on recombinant protein production in escherichia Coli BL21. *Frontiers in Microbiology* 12: 682001. https://www.frontiersin.org/articles/10.3389/fmicb.2021.682001.

Malani, R. S., S. B. Umriwad, K. Kumar, A. Goyal, and V. S. Moholkar. 2019. Ultrasound-assisted enzymatic biodiesel production using blended feedstock of non-edible oils: kinetic analysis. *Energy Conversion and Management* 188 (May): 142–150. https://doi.org/10.1016/j.enconman.2019.03.052.

Mardina, P., C. Irawan, M. D. Putra, S. B. Priscilla, M. Misnawati, and I. F. Nata. 2021. Bioethanol production from cassava peel treated with sulfonated carbon catalyzed hydrolysis. *Jurnal Kimia Sains Dan Aplikasi* 24 (1): 1–8. https://doi.org/10.14710/jksa.24.1.1-8.

Mielenz, J. R., J. S. Bardsley, and C. E. Wyman. 2009. Fermentation of soybean hulls to ethanol while preserving protein value. *Bioresource Technology* 100 (14): 3532–3539. https://doi.org/10.1016/j.biortech.2009.02.044.

Mu, R., Y. Jia, G. Ma, L. Liu, K. Hao, F. Qi, and Y. Shao. 2021. Advances in the use of microalgal-bacterial consortia for wastewater treatment: community structures, interactions, economic resource reclamation, and study techniques. *Water Environment Research: A Research Publication of the Water Environment Federation* 93 (8): 1217–1230. https://doi.org/10.1002/wer.1496.

Nazar, M., L. Xu, M. W.Ullah, J. M. Moradian, Y. Wang, S. Sethupathy, B.Iqbal, M. Z. Nawaz, and Da. Zhu. 2022. biological delignification of rice straw using laccase from bacillus ligniniphilus L1 for bioethanol production: a clean approach for agro-biomass utilization. *Journal of Cleaner Production* 360 (August): 132171. https://doi.org/10.1016/j.jclepro.2022.132171.

Ostadjoo, S., F. Hammerer, K. Dietrich, M.-J. Dumont, T. Friscic, and K.Auclair. 2019. Efficient enzymatic hydrolysis of biomass hemicellulose in the absence of bulk water. *Molecules* 24 (23): 4206. https://doi.org/10.3390/molecules24234206.

Palacios, S., H. A. Ruiz, R. Ramos-Gonzalez, J. Martínez, E. Segura, M. Aguilar, A. Aguilera, G. Michelena, C. Aguilar, and A. Ilyina. 2017. Comparison of physicochemical pretreatments of banana peels for bioethanol production. *Food Science and Biotechnology* 26 (4): 993–1001. https://doi.org/10.1007/s10068-017-0128-9.

Peng, X, Y. Liao, K. Ren, Y. Liu, M. Wang, A. Yu, T. Tian, et al., 2022. Fermentation performance, nutrient composition, and flavor volatiles in soy milk after mixed culture fermentation. *Process Biochemistry* 121 (October): 286–297. https://doi.org/10.1016/j.procbio.2022.07.018.

Periyasamy, S., J. B. Isabel, S. Kavitha, V. Karthik, B. A. Mohamed, D. G. Gizaw, P. Sivashanmugam, and T. M. Aminabhavi. 2023. Recent advances in consolidated bioprocessing for conversion of lignocellulosic biomass into bioethanol - a review. *Chemical Engineering Journal* 453 (February): 139783. https://doi.org/10.1016/j.cej.2022.139783.

Philippini, R. R., S. E. Martiniano, A. K. Chandel, W.de Carvalho, and S. S. da Silva. 2019. Pretreatment of sugarcane bagasse from cane hybrids: effects on chemical composition and 2G sugars recovery. *Waste and Biomass Valorization* 10 (6): 1561–1570. https://doi.org/10.1007/s12649-017-0162-0.

Qu, W., M. Wang, Y. Wu, and R.-h. Xu. 2015. Scalable downstream strategies for purification of recombinant adeno-associated virus vectors in light of the properties. *Current Pharmaceutical Biotechnology* 16 (8): 684–695. https://doi.org/10.2174/1389201016666150505122228.

Raud, M., L. Rocha-Meneses, D. J. Lane, O. Sippula, N. J. Shurpali, and Timo Kikas. 2021. Utilization of barley straw as feedstock for the production of different energy vectors. *Processes* 9 (4): 726. https://doi.org/10.3390/pr9040726.

Rosa, P. A . J., A. M. Azevedo, S. D. Sommerfeld, M. Mutter, W. Bäcker, and M. R. Aires-Barros. 2013. Continuous purification of antibodies from cell culture supernatant with aqueous two-phase systems: from concept to process. *Biotechnology Journal* 8 (3): 352–362. https://doi.org/10.1002/biot.201200031.

Saha, B. C., T. Yoshida, M. A. Cotta, and K. Sonomoto. 2013. Hydrothermal pretreatment and enzymatic saccharification of corn stover for efficient ethanol production. *Industrial Crops and Products* 44 (January): 367–372. https://doi.org/10.1016/j.indcrop.2012.11.025.

Saini, S., and K. K.. Sharma. 2021. Fungal lignocellulolytic enzymes and lignocellulose: a critical review on their contribution to multiproduct biorefinery and global biofuel research. *International Journal of Biological Macromolecules* 193 (December): 2304–2319. https://doi.org/10.1016/j.ijbiomac.2021.11.063.

Sarma, S., A. Anand, V. K. Dubey, and V. S. Moholkar. 2017. Metabolic flux network analysis of hydrogen production from crude glycerol by clostridium pasteurianum. *Bioresource Technology* 242 (October): 169–177. https://doi.org/10.1016/j.biortech.2017.03.168.

Selim, K. A., S. M. Easa, and A. I. El-Diwany. 2020. The xylose metabolizing yeast spathaspora passalidarum is a promising genetic treasure for improving bioethanol production. *Fermentation* 6 (1): 33. https://doi.org/10.3390/fermentation6010033.

Shabbir, S., W. Wang, M. Nawaz, P. Boruah, M. F.-E.-A. Kulyar, M. Chen, B. Wu, et al., 2023. Molecular mechanism of engineered zymomonas mobilis to furfural and acetic acid stress. *Microbial Cell Factories* 22 (1): 88. https://doi.org/10.1186/s12934-023-02095-1.

Shimizu, H.I, and Y. Toya. 2021. Recent advances in metabolic engineering-integration of in silico design and experimental analysis of metabolic pathways. *Journal of Bioscience and Bioengineering* 132 (5): 429–436. https://doi.org/10.1016/j.jbiosc.2021.08.002.

Singh, A., S. Bajar, and N. R. Bishnoi. 2017. Physico-chemical pretreatment and enzymatic hydrolysis of cotton stalk for ethanol production by saccharomyces cerevisiae. *Bioresource Technology* 244 (November): 71–77. https://doi.org/10.1016/j.biortech.2017.07.123.

Singh, N., K. Kumar, A. Goyal, and V. S. Moholkar. 2022. Ultrasound-assisted biodiesel synthesis by in-situ transesterification of microalgal biomass: optimization and kinetic analysis. *Algal Research* 61 (January): 102582. https://doi.org/10.1016/j.algal.2021.102582.

Soares, Ruben R. G., A. M. Azevedo, J. M. Van Alstine, and M. R. Aires-Barros. 2015. Partitioning in aqueous two-phase systems: analysis of strengths, weaknesses, opportunities and threats. *Biotechnology Journal* 10 (8): 1158–1169. https://doi.org/10.1002/biot.201400532.

Song, H., M.-Z. Ding, X.-Q.g Jia, Q. Ma, and Y.-J. Yuan. 2014. Synthetic microbial consortia: from systematic analysis to construction and applications. *Chemocal Society Review* 43 (20): 6954–6981. https://doi.org/10.1039/C4CS00114A.

Tiso, T., N. Ihling, S. Kubicki, A. Biselli, A. Schonhoff, I. Bator, S. Thies, et al., 2020. Integration of genetic and process engineering for optimized rhamnolipid production using pseudomonas putida. *Frontiers in Bioengineering and Biotechnology* 8: 976. https://doi.org/10.3389/fbioe.2020.00976.

Tiwari, S., E. Beliya, M. Vaswani, K. Khawase, D. Verma, N. Gupta, J. S. Paul, and S. K. Jadhav. 2022. Rice husk: A potent lignocellulosic biomass for second generation bioethanol production from klebsiella oxytoca ATCC 13182. *Waste and Biomass Valorization* 13 (5): 2749–2767. https://doi.org/10.1007/s12649-022-01681-5.

Wang, S., R. Tian, B. Liu, H. Wang, J. Liu, C. Li, M. Li, S.E. Evivie, and B. Li. 2021. Effects of carbon concentration, oxygen, and controlled ph on the engineering strain lactiplantibacillus casei E1 in the production of bioethanol from sugarcane molasses. *AMB Express* 11 (1): 95. https://doi.org/10.1186/s13568-021-01257-x.

Whitford, W. G. 2022. Bioprocess intensification: technologies and goals. In G. Subramanian (Ed), *Process Control, Intensification, and Digitalisation in Continuous Biomanufacturing*, 93–136. John Wiley & Sons, Ltd, New York. https://doi.org/10.1002/9783527827343.ch4.

Yao, W., and S. E. Nokes. 2014. First proof of concept of sustainable metabolite production from high solids fermentation of lignocellulosic biomass using a bacterial co-culture and cycling flush system. *Bioresource Technology* 173 (December): 216–223. https://doi.org/10.1016/j.biortech.2014.08.113.

Zoghlami, A., and G.l Paës. 2019. Lignocellulosic biomass: understanding recalcitrance and predicting hydrolysis. *Frontiers in Chemistry* 7 (December): 874. https://doi.org/10.3389/fchem.2019.00874.

6 A Review on Potential Role of Lignocellulosic Biomass in Production of Bioethanol via Microbial Fermentation

Soumya Banerjee

6.1 INTRODUCTION

Since prehistoric times, humans have been found to be highly dependable on natural resources for survival. Thus, it could be stated that the modern form of anthropogenic civilization could not be achieved if there was no constant supply of such resources. Since inception, humans have been using dried plant parts or dried dungs as fuel, but their dependence on fossils fuel was more prominent as it were used in major areas. This includes petroleum, coal and natural gas, which are exploited for chemicals, fuels, supplies and energy. Such a dependency on fossil fuel or dried plants although have fulfilled the requirement in continuation of an ever growing civilization but at the expense of environment. These damages resulted in catastrophic climatic outcome which in turn have affected our livelihood. This has been evident from melting of glaciers to alteration in seasonal cycles and annual increase in temperature by 2°C. For the past few decades, there have been reports on increased greenhouse gases (GHGs), viz., carbon monoxide (CO), carbon dioxide (CO_2), methane (CH_4) and nitrous oxide (N_2O), and side by side, constant fall in the supply of fossil fuel due to overexploitation of crude resource. Unfortunately, the fatality of such consequences has been reported by various international agencies which stated that emission of GHGs might increase up to 35% globally by 2035 (Priyanka et al., 2019).

Constant decrease in fossil fuel reservoir has urged for a substitute and this has made biofuels a sustainable alternative. Various countries have been considered as potential biofuel producers, which could decrease the overall GHGs by 80%. So far, 2% of fuel-driven transport has been substituted with biofuels and this trend might increase with innovation in technology. In contrast to fossil fuel, biofuels could be far more sustainable which might limit both organic and inorganic pollution. The oil crisis in 1970s led many countries opting for an alternative sustainable source of fuel which could be both eco-friendly and cost-effective. So far, traditional fuels like coal, natural gas and petroleum are being used for fuel, chemical and power generation.

 DOI: 10.1201/9781003407713-6

Unfortunately, very little expenses were made on investigating the possibilities of bioalcohols, biogases, biodiesels, biochar, vegetable oils, and biosynthetic gases as alternative energy sources. Production of bioethanol concentrates on the usage of plants or edible products, which might create havoc scarcity on food and economic instability. Thus, this might raise serious concerns regarding feasibility of bioethanol as a better alternative. As a result, non-edible plants and their residue or agricultural wastes have been considered for generation of bioethanol. The commonly found laboratory reagent ethanol could be the answer for the depleting layer of fossil fuel. Traditionally, any raw materials consisting of fermentable sugars ($C_6H_{12}O_6$) could be used by microbes under anaerobic condition for the production of ethanol (C_2H_5OH) and carbon dioxide (CO_2) (Eq. 6.1). This age-old practice could be a solution for both energy crisis and bio-waste (Robak & Balcerek, 2020).

$$C_6H_{12}O_6 \xrightarrow{\text{yeast}} 2C_2H_5OH + 2CO_2 \tag{6.1}$$

6.1.1 History of Bioethanol

The fermentation technique used for the production of ethanol could be traced back before microorganisms were invented. On the other hand, the science behind fermentation and production of ethanol via distillation did not exist before 12th–14th centuries. Production of pure ethanol started with a starch-containing feedstock in Ireland. In the Middle Ages, ethanol was used as an effective solvent for preparation of medicines, perfumes and colours. Nowadays, ethanol could be found in the making of polishes and explosives. Production of ethanol at industrial scale started somewhere around the 19th century due to its low cost and versatile usage.

The first instance of using ethanol as fuel in an internal combustion engine was demonstrated by a German inventor named Nicholas Otto in 1869. Later, Henry Ford constructed an engine that ran solely on ethanol, which inspired him to construct a T engine that could run on either ethanol or gasoline or in a mixture of both. The result obtained from the experiment made by Henry Ford realized that ethanol is the "future fuel". He also documented that ethanol could be produced easily from sugar-containing feedstock and it could be used as an alternative to fossil fuel. Upon mixing with gasoline, anhydrous ethanol was found to increase the octane number, resulting in more stability of the fuel (Roran et al., 2022). Hence, the utility of ethanol as a fuel in automobiles gained its popularity by the 20th century, encouraging the installation of multiple production units.

6.1.2 Characterization of Ethanol

Ethanol can be described as a transparent and colourless organic solvent with pleasant aroma. Ethanol varies in taste based on its concentration, where it could be pungent if concentrated and sweet if diluted. It has a melting and boiling point of −117.1°C and 78.5°C, respectively, and other physiochemical properties have been enlisted in Table 6.1. Ethanol as a solvent has been found to be effective in making both polar and non-polar solutions. There have been various reports on ethanol that has been used for preparation of tincture solutions (Kwon et al., 2012).

TABLE 6.1
Physiological Characters of Ethanol (Priyanka et al., 2019)

Particulars	Features
Molecular Formula	C_2H_5OH
Molecular Mass	46.07 g/mol
Density	0.789 Kg/L
Flash point	12.8°C
Ignition temperature	365°C
Octane number	99
Vapour pressure at 38 °C	50 mm Hg
Specific heat	2.46 J/g°C
Viscosity	1.200 mPa s (20°C)
PKa	15.9

6.1.3 Gasoline versus Ethanol

Popularity of ethanol as fuel increased due to various advantages among which it has got 66% more energy than the same amount of gasoline and a higher octane number. Thus, gasoline performs better upon mixing with ethanol. Along with octane value, evaporation enthalpy, laminar flame speed and heat of vaporization are also higher than gasoline. As a result, these criteria confirmed volumetric efficacy of ethanol-blended gasoline over pure gasoline with the maximum energy output. Ethanol and gasoline could be compared based on their oxygen percentage. Ethanol has been found to be 34.7% more oxygenated than gasoline with zero oxygen. As a result, ethanol has 15% more combustion efficacy over gasoline, while also helping to reduce the emission of nitrogen oxides and particulate matters. Again, ethanol contains negligible amount of sulphur, thus mixing it with gasoline would reduce overall sulphur emissions upon combustion. Methyl tertiary butyl ether or MTBE helps in the maintenance of octane number in gasoline. In various reports, MTBE has been found to emit carbon dioxide and carbon monoxide upon combustion. As a result, this MTBE affects food chain and causes harmful effect on both human health and nature. Therefore, in some countries, government authorities under the Toxic Substance Control Act have proposed limiting the usage of MTBE in gasoline (Zabed et al., 2017).

6.1.4 Current Global Scenario of Bioethanol

Production of bioethanol depends solely on indigenous production and the availability of feedstock. Raw materials like lignocellulosic, cellulosic, starchy and other biomass have been found to be suitable for bioethanol production (Broda et al., 2022). In today's time, countries like United States and Brazil have been listed as leading producers of bioethanol. This has been attributed to mass cultivation of sugarcane and corn in Brazil and the United States, respectively. Several other countries like China and India have been busy in establishing infrastructural facilities for bioethanol production.

6.1.5 Global Policy on Bioethanol Production

Over the years, various countries have undertaken different measures to lessen the excessive burden on fossil fuel demand. Therefore, different nations have been coming up with newer ways of preserving fossil fuel for future generation. Thus, the regulation of ethanol as biofuel has been taking a gradual but steady momentum. In this aspect, Brazil should be praised for their policies on fuel which they have undertaken post-1970s oil crisis. The National Ethanol Fuel Program launched by the Brazilian government helped in improving technical aspects of bioethanol production in order to reduce import oil cost and boost the sugar industry laterally (Tan et al., 2015). By 2000, the country observed steady decrease in oil price due to mixing of 20% anhydrous ethanol with gasoline. Later, the proportion was increased gradually to 22% and 25% in 2005. Interestingly, in today's time, the automobiles are being run with 98% ethanol. As per a report published in 2016, around 80% of active vehicles in Brazil were run by ethanol-blended gasoline.

In another Latin America country like Columbia, blending of ethanol with gasoline by 10% in cities has been made mandatory. This law was realized in 2005, and astonishingly by 2009, around 75% of urban vehicles were running in ethanol-blended gasoline. As a result, this reinvigorated the government to set the target up to 25% by 2020.

Extensive oil demand in highly populated countries is not common. Also, such demand creates a huge import burden if not produced indigenously. Hence, in over-populated countries like India, an alternative to conventional fuel should be prioritized considerably. In India, molasses has been used as feedstock for the production of bioethanol, and its blending with gasoline was made operative in 2008. As of 2017, there was another report which proposed 20% bioethanol to be mixed with gasoline along with diesel. Unfortunately, the Oil Marketing Companies have accepted with 5% of bioethanol to be mixed with gasoline and used in 20 Indian states and 4 Union Territories. Fortunately, the Indian government has confirmed the use of 10% of bioethanol with gasoline by 2022, and they have been keen on increasing the proportion up to 20% by 2025. The Ministry of New and Renewable Energy has granted the sugar industries to produce bioethanol from sugarcane juices and to increase their annual outcome gradually so that it could be used as biofuel.

6.2 LIGNOCELLULOSIC BIOMASS

Global agricultural activities produce around 1.3 billion tonnes of waste annually. This waste consists of lignocellulosic biomass (LCB), which could be used as an alternative source for production of chemicals and biomass (Kumar & Sharma, 2017; Baruah et al., 2018). The first-generation feedstock such as wheat, sugarcane, corn, etc. or other food crops were used for biofuel production. As a result, such feedstock later was found to be illogical in the long run since it might cause global food crisis, extensive exploitation of agricultural lands, manpower and destruction of ecosystem. Interestingly, the second-generation feedstock like LCB was found to be a rational choice due to its abundance and non-conflict with food supply. In comparison to the first-generation feedstock, retrieval of LCB has not been found to restrict to a single source. This could also be retrieved from forest residue and from paper and pulp industries. Also, the usage of LCB would help reducing both environmental pollution and unwanted dumping of biomass and would create opportunities for revenue generation.

LCB is a three-dimensional complex composed of cellulose (35%–50%), lignin (10%–25%) and hemicellulose (20%–35%), as well as small fraction of oils, proteins, pectin and ashes. Therefore, the proportion of cellulose, hemicellulose and lignin varies depending on the nature of its source (Wei et al., 2017). Fermentable sugars, viz., xylan and glucan and glucose needed for bioalcohol production, are supplied from polysaccharides like hemicellulose and cellulose, respectively (Kim et al., 2015).

6.2.1 Cellulose

It comprises the hard, water insoluble and linear polymer which resides in the woody region of plant. It is composed of 3,000–4,000 glucose monomers arranged in linear fashion bound together in 1,4-glycosidic bond (Kumar & Sharma, 2017). Cellulose exhibits fibrous texture in plants obtained from clustering of 60–70 polymers via multiple hydrogen bonds. Interestingly, these hydrogen bonds have been found to be the reason behind restricted hydrolysis of LCB. Cellulose contains higher amount of reactive hydroxyl groups with hydrophilic features yet insoluble in water, has elevated tensile strength and is biocompatible, making it a suitable candidate for biofuel production. Thus, cellulose provides enlarged surface area capable of retaining water on its surface and prevents itself from any kind of hydrolytic enzymes. As a result, the usage of cellulose for biofuel production could only be possible with prior degradation of its complex structure.

6.2.2 Hemicellulose

Hemicellulose is a heterogeneous polysaccharide and the second most abundant carbohydrate found in plant after cellulose. This contains diverse group of carbohydrates like pentose, xylan, hexose, glucomannan and galactomannan linked together in 1,4-glycoscidic bond (Zhou et al., 2017). In plant cell wall, cellulose, hemicellulose and lignin are bound together via hydrogen bonds, which provide them the structural flexibility. In general, monosacharaides could be obtained from hemicellulose via enzymatic actions. This degradation depends on various factors like degree of polymerization and crystallinity.

6.2.3 Lignin

Lignin is an aromatic heteropolymeric component of plant cell wall that binds cellulose and hemicellulose covalently in a three-dimensional structure, preventing the cell from hydrolysis and degradation (Jung et al., 2015). Lignin provides a protective layer on cellulose and hemicellulose, which makes it a major hurdle in inaccessibility of LCB during biofuel production. Pretreatment of the polymer results in synthesis of free radicals out of phenylpropionic alcohols via peroxide-mediated dehydrogenation. As a result, retrieving of cellulose and hemicellulose from plant cell wall could only be achieved by degradation of lignin via co-metabolism. Lignin is composed of phenolic compounds and also consists of guaiacyl, p-hydroxyphenyl and syringyl. Thus, production of biofuel from LCB could be made possible with proper pretreatment of stubborn biochemical components of plant cell, viz., cellulose, hemicellulose and lignin.

6.3 PRETREATMENT OF LIGNOCELLULOSIC BIOMASS

Being a storehouse of carbohydrates, waste LCB could be converted into bioethanol than combusting it for energy and space. Structural complexity of LCB (Figure 6.1) could be attributed to the covalent and non-covalent linkages. Structural complexity of LCB amplifies with the presence of lignin, which makes its molecular fragmentation difficult. As listed in Table 6.2, compositional percentage of cellulose and hemicellulose changes due to pretreatment. Several studies have substantiated that around 40% of total production cost of ethanol are incurred in pretreatment of LCB. The economic feasibility of bioethanol production depends greatly on structural composition of LCB and its pretreatment, which affects consumption of energy, chemical and manpower as well as post-production stage. Pretreatment includes physical, chemical, biological, mechanical and physiochemical methods that are effective for conversion of polymeric sugars into monomers (Banu et al., 2019; Saha et al., 2016; Divyalakshmi et al., 2018; Kavitha et al., 2018; Eswari et al., 2016). In accordance to biomass type, pretreatment could be further classified as single, amalgamated and phase-separated processes.

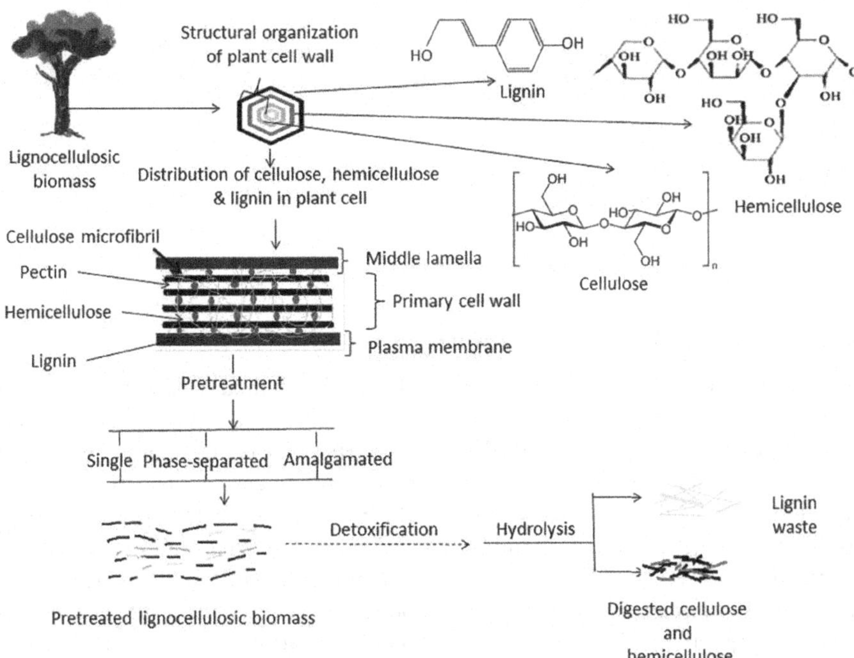

FIGURE 6.1 Structure of lignocellulosic biomass and its pretreatment.

TABLE 6.2
Compositional Variation in Raw and Pretreated LCB (Robak & Balcerek, 2020)

Raw Material	Pretreatment Type	Composition of Raw LCB		Composition of Pretreated LCB	
		Cellulose %	Hemicellulose %	Cellulose %	Hemicellulose %
Grass waste	Diluted Sodium hydroxide (1% w/v)	31.2	19.9	43.3	21.1
Grass waste	Hydrogen Peroxide (2% w/v)	31.2	19.9	35.1	22.9
Corn stalk	Diluted Sodium hydroxide (2% w/v)	35.6	28.3	46.7	36.7
Wheat straw	Formic acid (78% w/v)	35.1	23.4	72.2	10.5
Poplar wood	Choline Chloride/lactic acid mixture (1.2 M)	43.2	14.7	79.8	7.1
Spruce wood chips	Ethanol (52% w/w)	37.6	17.6	69.9	4.0
Olive stone	Steam explosion (205°C, 10 minutes)	20.9	26.0	35.4	2.6
Corncob	Sulphuric acid (1% v/v)	40.7	31.1	72.1	8.2
Rapeseed straw	Sulphuric acid (1% w/w)	30.0	12.7	53.5	2.3
Lignin reduced sugarcane bagasse	Sulphuric acid (0.2 % w/v)	38.3	23.1	47.4	16.4
Wheat straw	Sulphuric acid (0.5 % w/v)	35.1	23.4	53.6	7.7
Industrial hemp stem	Sulphuric acid (1% w/w)	40.1	12.5	62.1	0.8
Sugarcane bagasse	Sulphuric acid (0.2% w/v)	37.3	21.8	46.5	16.1
Sugarcane bagasse	Sodium hydroxide (2% w/v)	34.8	25.0	50.1	24.8

6.3.1 Single Pretreatment Methods

This mode of pretreatment employs any single kind of method chosen based on biomass and infrastructural facilities. Physical pretreatment, a kind of single pretreatment process, concentrates on particle size reduction for better removal of lignin. This could be achieved via thermal or microwave treatment. In the case of microwave treatment, collision of molecules could be made by dielectric polarization, resulting in the generation of thermal energy. On the other hand, LCB is heated at a temperature range of 150°C–240°C and a constant pressure of 35 bar during thermal pretreatment. In many cases, it has been reported that heating could enhance biofuel production by multiple folds (Rajput & Visvanathan, 2018).

Similarly, modification of the structural chemistry of LCB could easily alter the system pH. With alteration of pH to highly acidic or alkaline condition, defragmentation of cellulose, hemicellulose and lignin could be achieved easily. The change in pH could be made with hydrochloric acid, sulphuric acid, calcium hydroxide and sodium hydroxide (Dahunsi, 2019). Chemical treatment of biomasses has been favoured in many cases due to its minimal usage of resources.

In the case of stubborn biomasses mechanical pretreatment has been found to be effective. This could be achieved via techniques like ultra-sonication. The modification includes the formation of monomeric sugar units from hemicellulose and cellulose, which would be used for biofuel production (Kumar & Sharma, 2017). On the other hand, disperser could be used to reduce particle size of biomass. Also, this method destroys structure and crystallinity of LCB.

Biological techniques, on the other hand, have been gaining popularity as a pretreatment method due to their positive attributes. This could be substantiated with least consumption of energy, chemicals and larger apparatus. Effective microbial tools like bacteria and fungi are used that feeds on LCB and solubilizes it via enzymatic actions. Fungal species has been found to be effective in delignification of LCB, producing cellulose and hemicellulose for bioethanol production. Fungal species could solubilize lignin via secretion of oxidative enzymes like lignin peroxidase, manganese peroxidase and laccase (Shirkavand et al., 2016). Fungal enzymes have been found to be equally effective in solubilizing cellulose and hemicellulose via secretion of enzymes like endocellulase and β-glucosidases, α-arabinofuranosidase and esterases, respectively.

6.3.2 Amalgamated Pretreatment

Lignin being a stubborn organic polymer should be removed from LCB for ethanol production. Hence, in some cases, the amalgamation of techniques has proved to be highly resourceful for the complete removal of lignin. Apart from lignin solubilization, such technique provides substrate with better surface area suitable for the enzymatic actions needed for ethanol production (Galbe & Wallberg, 2019). As a result, combination of techniques like physiochemical-mechanical, biochemical, chemo-mechanical and physicochemical has been tested and documented for the processing of persistent LCB for the production of biofuel.

6.3.3 Phase-Separated Pretreatment

In this kind of technique, pretreatment occurs via lignin separation and saccharification of cellulose and hemicellulose separately but in a single unit. Enzymatic separation helps in the execution of this technique. In this method, lignin separation from the cell wall occurs first followed by the extraction of polysaccharides (Qing et al., 2017). This technique could be conducted by either physicochemical or chemo-biological method.

6.3.4 Factors Regulating Pretreatment of Lignocellulosic Biomass

Pretreatment of biomass could be considered as a difficult process that affects the feedstock both physically and chemically. Such treatment of biomass results in the formation of strong bonded unwanted organic compounds, which interfere with the sugar content of LCB and final ethanol production. Factors like carbon sources, phenolic recalcitrant compounds and sugar affect pretreatment of LCB. Apart from biochemical aspects, the size of the components also interferes in the process. Pretreatment processes could be delayed due to improper sizing of LCB; thus, systematic size reduction requires prime concern. Again, such physical parameters or chemical processes require high energy for proper function. Pretreatment of biomass could alter surface area by changing its porosity; hence, hydrolytic activities have to be optimized properly. The efficacy of bioethanol production depends greatly on the availability of fermentable sugars retrieved from cellulose and hemicellulose. Thus, such pretreatment methods should be opted which would proliferate the activity of β-glucosidase for better ethanol production (Karimi & Taherzadeh, 2016).

6.4 BIOETHANOL PRODUCTION VIA MICROBIAL FERMENTATION

Energy produced within a plant cell via photosynthesis serves as the raw material for ethanol production via fermentation. Ethanol synthesis consists of multiple stages (Figure 6.2), viz., detoxification, hydrolysis, fermentation, condensation and dehydration (Robak & Balcerek, 2020; Das et al., 2021). After processing, pretreated biomass

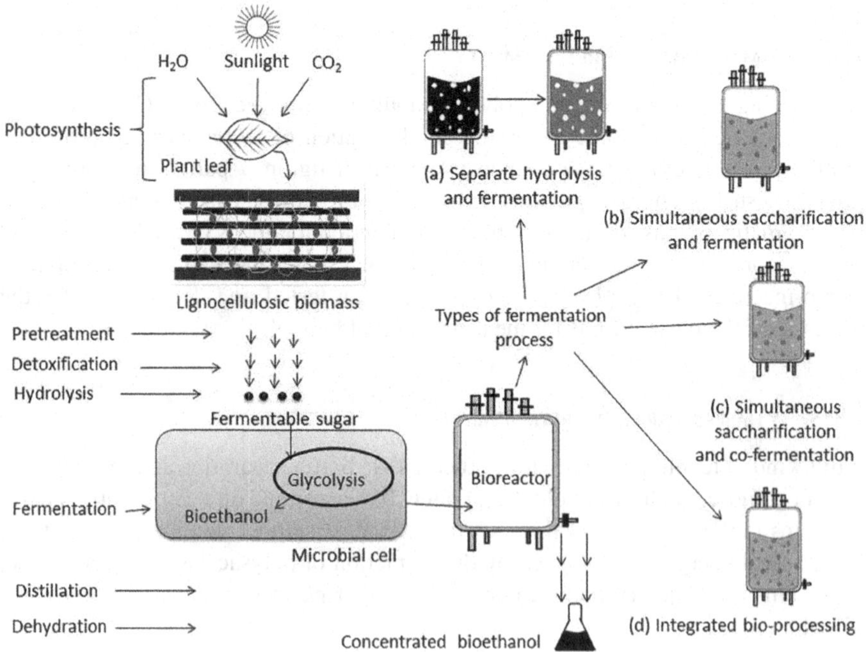

FIGURE 6.2 Schematic representation of bioethanol production from lignocellulosic biomass under different experimental setups in a bioreactor.

has been found to be covered with unwanted toxic compounds such as furan aldehydes, aliphatic acids and phenolic complexes. Such components should be removed prior to fermentation to avoid inhibition during microbial activities. Several methods have been found to be effective in the elimination of such unwanted compounds from pretreated biomass, thus facilitating bioethanol production by increasing microbial activities. Techniques like membrane abstraction, solvent abstraction, ion exchange, membrane bioreactors, biosorption and microbial consortium or by designed microorganisms, etc. could be used for the elimination of unwanted contaminants from pretreated LCB (Robak & Balcerek, 2020).

It is considered as one of the important processes where fermentable sugars used by microorganisms during fermentation are retrieved via enzymatic activities. The polysaccharides of LCB are composed of pentose and hexose, which are extracted via enzymatic hydrolysis, and in some cases, acids or alkalis are also used (Robak & Balcerek, 2020; Tsai & Meyer, 2014).

Pretreated biomass has to be processed further into hydrolysis which further simplifies LCB. Decomposition of polysaccharides like cellulose converts into its monomeric unit glucose via an enzyme complex known as cellulase. This enzyme comprises enzymes like endo-1,4-β-D-glucanase, exo-1,4-β-D-glucanase/exo-cellobiohydrolase and β-glycosidase. The first set of cellulase hydrolyzes glycosidic bonds, whereas the second set of enzyme subtracts monomers and dimers from glucose chains. Finally, the third variety of enzyme participates in breaking of glycosidic bonds of dimers and other short cellulose complexes into their monomers. Thus, the overall conversion of polysaccharides requires combined action of three sets of enzymes. By nature, cellulase can only be obtained from certain bacteria and fungi of aerobic, anaerobic, mesophilic and thermophilic types. Bacterial genera capable of producing cellulase include *Erwinia, Streptomyces, Ruminococcus, Clostridium, Cellvibrio, Acetivibrio, Cellulomonas, Bacillus* and *Bacteroides.* Likewise, fungal genera include *Aspergillus, Caecomyces, Humicola, Neocallimastix, Oprinomyces, Penicillium, Schizophyllum* and *Trichoderma*, and Actinobacterial genera include *Microbispora* and *Thermomonospora.* Similarly, fungal species including *Sclerotium rolfsii* and *Phanerochaete chrysosporium* could be employed for production of cellulase (Srivastava et al., 2018; Singhania et al., 2017; Mohanram et al., 2013). Hydrolysis of cellulose could become difficult due to its microfibrils which stabilized the structure by both internal and external hydrogen bonds. Cellulose network gets a further covering of hemicellulose through covalent and hydrogen bonds (Figure 6.1). Hence, enzymatic hydrolysis enhances bioethanol production (Tsai & Meyer, 2014).

30% of LCB contains hemicellulose and its conversion into fermentable sugars is equally important for higher bioethanol yield. Conversion of hemicellulose has been found to be comparatively easier than cellulose. A core depolymerizing enzyme helps in breaking the central backbone of hemicellulose and allows the maximum availability of fermentable sugars. The core enzyme consists of 1,4-mannanases, endo-1,4-xylanases, xylan 1,4-xylosidases, acetylxylan esterase, α-L-arabinofuranosidase, glucuronidase, ferulic acid esterase and p-coumaric acid esterase. Like cellulose, various kinds of microorganisms help in enzymatic hydrolysis of hemicellulose. This includes fungal species like *Cochliobolus carbonum, Trichoderma reesei, Aspergillus niger, Penicillium wortmanii, Aspergillus awamori* and *Agaricus bisporus.* Again, fungal species from genera like *Sclerotium, Agaricus, Trichoderma* and *Aspergillus* have been found to produce such enzymes. Similarly, bacterial species like *Clostridium thermocellum, Thermobacillus xylanilyticus, Thermotoga maritima, Clostridium cellulovorans, Paenibacillus polymyxa* cel44C-man26A, *Cellvibrio japonicus, Caldibacillus cellulovorans, Caldicellulosiruptor* Rt8b, *Bacillus* spp., *Caldocellum saccharolyticum* and *Streptomyces* spp. have been found to produce hemicellulose hydrolyzing enzymes. Thus, it can be stated that the overall yield of bioethanol depends largely on combined actions of various microorganisms facilitating depolymerization of cellulose and hemicellulose (Mohanram et al., 2013; Bhattacharya et al., 2015). Breaking of LCB includes series of parameters like solid loading, sugar concentrations, enzymes packing, shaking speed, contact time, presence of inhibitors and involvement of additives. Enzymatic saccharification has been found to be a challenging process which consumes around 30% of the total production cost making it a prime criterion for techno-chemical analysis.

6.4.1 Fermentation of LCB

Fermentation of LCB for ethanol production includes a wide range of sugars, viz., glucose, fructose, sucrose, xylose, mannose, galactose and arabinose (Robak & Balcerek, 2020). Fermentation of glucose on an industrial scale has been reportedly achieved with *Zymomonas mobilis* and *Saccharomyces cerevisiae* because of their capability to produce higher units of ethanol and resistance towards higher concentration of ethanol (120 g/L). Unfortunately, these microbial species could not ferment pentose, making them inappropriate for ethanol production from LCB. Again, microbial species like *Pachysolen tannophilus, Candida shehatae* and *Scheffersomyces stipitis* could ferment pentose to a moderate level. Again, pentose fermenting yeasts could produce ethanol, but they cannot resist alcohol beyond 40 g/L. Again, such microorganisms could not ferment xylose at lower pH. These microbes show catabolite repression in the presence of glucose and also need microaerophilic condition to grow. Again, under mixed sugar condition, these microbes could only ferment xylose under limited glucose conditions (Robak & Balcerek, 2020).

Thus, the above discussion suggests a crisis in terms of naturally occurring microbes that could digest both pentose and hexose sugars simultaneously. As a result, in most of the cases, engineered microorganisms are being used which could ferment both pentose and hexose sugars easily. Metabolically designed microbes are being made by genetic alteration which helps them in fermentation of sugars present in LCB, survival under unfavourable conditions, resistance against inhibitors and the ability to produce a combination of enzymes that could hydrolyze the stubborn biomass. These engineered microbes are so designed that it could propagate easily and abundantly under metabolically challenged conditions. Presently, engineered microbial species like *Fusarium oxysporum, Corynebacterium glutamicum, S. cerevisiae, Thermoanaerobacter mathranii* and *Z. mobilis* have been used in production of bioethanol from LCB. Thus, it could be said that a perfectly engineered microbial species would be able to ferment every available sugars in LCB leading to maximum bioethanol yield (Robak & Balcerek, 2020; Zaldivar et al., 2001; Kim et al., 2022; Anasontzis et al., 2016; Yao & Mikkelsen, 2010).

Production of ethanol from LCB could be performed with immobilized recombinant microbial cells. In this technique, live cells are entrapped in a porous matrix which is fixed by covalent bonds or cross-linking in such a manner that it does interfere normal enzymatic activities of the microbial cells. Carrier material thus used for this technique should be non-toxic, biodegradable and cost-efficient which includes pre-polymers, carrageenan, chitosan, calcium alginate, cellulose and silica-hydrogel. Fermentation of LCB could be conducted in both fed-batch and continuous bioreactors. Commercially, fed-batch stirred tank bioreactors are preferred for bioethanol production since they offer maximum functionality of the microbes during fermentation. Likewise, at industrial level, several techniques have been found to be effective in bioethanol synthesis from LCB (Figure 6.2). This includes:

Separate hydrolysis and fermentation (SHF): In this process, the breaking down of substrate and fermentation are performed in two separate chambers. At first, the pretreated substrates are solubilized in the hydrolyzing unit followed by its fermentation in the second unit. This technique has been found to be time-consuming due to longer standing time needed for complete hydrolysis and enzyme loading. Similarly, the process questions its economic feasibility due to doubled fabrication cost and the elimination of unwanted end-product inhibition required during fermentation (Su et al., 2020).

Simultaneous saccharification and fermentation (SSF): This process facilitates substrate breaking and fermentation in the same vessel. As compared to SHF, this technique exhibits superiority by continuously removing sugar monomers by yeast and breaking down enzyme-product complexes simultaneously during fermentation. Unfortunately, this SSF encounters a shortcoming in terms of its temperature regulation. This could be attributed to the temperature requirement in substrate breaking and fermentation, which interferes with enzymatic activities of microbes during bioethanol production (Su et al., 2020).

Simultaneous saccharification and co-fermentation (SSCF): Here, in this process, both substrate breaking and fermentation of LCB are performed in the same chamber where co-fermentation of 5-carbon and 6-carbon sugars is performed by pentose fermenting microbial strains. Thus, the overall ethanol production increases. This method has been found to be effective for xylose-rich biomasses like hardwoods and agricultural wastes, but upon comparison with SSF, the overall ethanol production could be comparatively low (Su et al., 2020).

Integrated bioprocessing (IBP): This technique supports substrate breaking, fermentation and enzyme synthesis within the same chamber. In this case, microbes are used either in consortium, such as strains of yeast and *C. thermocellum* or as genetically modified single strain. The microbes used are designed to hydrolyze biomass enzymatically followed by the fermentation of sugar molecules to ethanol. This method has been found to have the potential to curb expenses made on infrastructure and resources. This might change the course of bioethanol production in terms of both production cost and environmental friendliness. Unfortunately, this technology is still under research and developmental stage since the overall ethanol produced does not meet industrial standards (Liu et al., 2020).

6.4.2 Distillation and Dehydration

Alcohol produced out of fermentation could not be used as fuel without passing it through distillation and dehydration. Distillation helps in separation of ethanol from fermentation broth via series of evaporation and condensation based on their individual volatilities. In post-fermented mixture, the water content is generally high up to 80%, which could affect the functioning of ethanol in every aspect. Thus, concentrating ethanol via dehydration has to be performed effectively. Concentration of fermented ethanol requires a huge amount of energy, making the final product costlier. Initially, the ethanol produced from fermentation would be around 37% by volume, and after dehydration, the final product of 99.5% by volume will be obtained (Robak & Balcerek 2020; Zaldivar et al., 2001).

So far, several methods have been employed in separation and production of commercial grade bioethanol, including adsorption condensation, membrane procedures, azeotropic condensation, diffusion condensation, extractive condensation, pervaporation, vacuum condensation and chemical desiccation. Each of these processes is employed depending on the operation cost and effectiveness of the procedure (Aditiya et al., 2016). Among all these processes, membrane distillation and pervaporation are commonly used for their economic feasibility.

Membrane distillation is frequently used for bioethanol separation from fermentation broth because it provides a comparatively lesser energy consumption. Flat or tubed and porous membranes with 70%–80% absorbency, hydrophobic and high thermal resistance are used. The process becomes feasible due to molecular differences in components found in gaseous phases. Also, the membrane distillation process has been developed based on specific process demand which includes contact, air-gap, vacuum and sweeping gas membrane condensation. Unlike any other distillation process, this technique could be highly effective even at lower temperature. As a result, cost incurred in boiling water to the boiling point of ethanol is not needed, making it a highly cost-efficient technique. Apart from being a cheaper method, this technique could remove 100% of non-volatile components from ethanol, making the final product more superior than ethanol distilled from other conventional methods. Also, membrane distillation allows fermentation and distillation to a continuous process with simultaneous ethanol separation (Aditiya et al., 2016; Li et al., 2018).

Pervaporation, in contrast, is a membrane technique used for obtaining dehydrated ethanol at the industrial level. This method employs differences in alcohol concentration on both sides of an asymmetric profuse polymer membrane. Thus, the parting will be grounded on the attraction of water and ethanol on both sides of the membrane, i.e., their diffusion and dissolving capability. Finally, the ethanol obtained from fermenting LCB will be 99.8% (Aditiya et al., 2016; Gaykawad et al 2013).

6.5 TECHNO-ECONOMIC ANALYSIS

In today's time, the availability of sustainable fuel source has been a greater challenge. Also, the major energy consumers like transportation continue with fossil fuel. Although various researches have presented reports of promising alternatives, the actual problem is more with its marketing than its production. Thus, the prime concern should be more into generating awareness among the common mass regarding the availability of fossil fuel and its significance on environment and future generations. Ideally, any green product could be marketed as low-carbon emitting or sustainable resource, but in the case of liquid, fuel regulation needs wider reach for its approval from the market. At the same time, flexible infrastructure also has to be provided for creating a convenient ecosystem of suppliers, traders and consumers. Initially, the focus was primarily on the product and its production, which failed to create the market needed for biofuel. Unlikely, in recent times, more actions are taken in terms of its commercialization and acceptance by preaching the benefits of bioethanol as a form of sustainable fuel for a better and greener future. Thus, techno-economic analysis or TEA could provide a solution in bridging up laboratory- or pilot-scale research and commercial production of bioethanol (Phillips, 2007; Rentizelas et al., 2009). As a result, techno-economic analysis could be summarized as a process which could help in process fabrication, product design, simulation and cash exchange. Thus, this would help in maintaining mass and energy balance, as well as the variation of economic metrics, which could support scaling technical visibility even before it has been conceived. As a result, TEA could help in preparing complete system blueprint, simulated scale-up and estimation formulated with empirical data if possible. Again, mass and energy balances could impart necessary inputs for life-cycle assessments. Hence, in recent reports, techno-economic analysis of biofuel and bioproducts is being formulized including every possible complexities that could be used for system upgradation necessary for future process modelling.

Traditionally, TEA on biofuel production and its by-products concentrated on process design and simulations. In such cases, software like SuperPro Designer® and Aspen Plus® has been used (Yang & Rosentrater, 2019; Ruddy et al., 2019; Bbosa et al., 2018; Alonso et al., 2017; Granjo et al., 2017; Koutinas et al., 2014; Kumar & Murthy 2011; Davis et al., 2011; Klein-Maruschamer et al., 2010). In another report, Aspen Plus® were used for simulating bioethanol synthesis from corn stover and explained the merits of the software needed for process designing (Humbird et al., 2011). Such commercially available software could be used for establishing a base for an efficient model that could manage biorefinery complexities along with multiple recycling loops. In other words, such model facilitates proper handling of process shortcomings like integration of temperature controlling utilities, effluent treatment and on-site co-generation. Most of the available commercial software deals with serial modular modes, which are convenient for simplifying calculations and equation-oriented models providing simpler results of a process at a faster rate. In some cases, these models could be effective in determining the cost driving parameters or factors of bubble or stirred tank reactors.

Like any other process that includes the involvement of feedstock and their fermentation towards the production of particular products, there are several technical and financial constraints. These shortcomings may not be controlled or fully understood with the help of existing modules, but several simulation studies and optimization of parameters have to be performed to overcome the lacunas hindering in establishing a proper connection among the parameters. This could be attributed to proper understanding of this software on microbial activities under varied conditions and their role in fermentation.

6.6 FUTURE SCOPE

The production and use of second-generation biofuel has proven to have various positive effects on current crisis on fossil fuel. It has been reported in various forums and by various organizations that dependency on fossil fuel and increasing level of carbon emission would affect not only fuel reservoir but also global climatic conditions. It has been an established fact that, from past several decades, quality of life started degrading at the community level, which ultimately contributes to global environmental catastrophe. Cities and metros have got maximum dependency on fossil fuel for transportation, which has caused several health hazards among many urban dwellers. Thus, an alternative form of fuel has always been in the question, which would help in reducing dependency on both fossil fuel and carbon emission. Thus, feedstock like LCB for alternative and sustainable fuel should be encouraged for more exploitation. There have been various forms of carbon neutral fuels that have been optimized and are in use, but the fate of biofuels is still under consideration. This could be attributed to a lack of tax on carbon fuels and absence of stringent law on the use of fossil fuel and general awareness. Again, there are very fewer facilities and infrastructure available for biofuel production, as well as the usage of agricultural crops or its residue as feedstock for biofuel production. It has to be remembered that production cost of bioethanol is comparatively higher than fossil fuel. Again, the usage of fossil fuel for transportation has been costing us our lives and future. Thus, preference of fuel in accordance to its purchase cost should be made rationally and scientifically. Again, it has been observed that the technologies used for biofuel production could be used for production of other products, which are commonly obtained as petroleum by-products. As reported, LCB could be a promising feedstock for bioethanol production, but the problem remains in its procurement, processing, production and distribution. A greater amount of fermentable sugar gets wasted in the process due to lack of infrastructural facilities. In terms of biological treatment of LCB and its fermentation, the availability of naturally occurring microbial species is still absent from the picture. Thus, the survival of bioethanol depends greatly on finding such microbial species. Although genetically engineered microbial species have been made for fermentation of LCB, this increases synthesis and their metabolic capability also gets challenged. Thus, the future of ethanol as biofuel depends greatly on legal, infrastructural, and social awareness and encouragement.

REFERENCES

Aditiya, H.B., Mahlia, T.M.I., Chong, W.T., Nur, H., Sebayang, A.H. (2016) Second generation bioethanol production: a critical review. *Renew Sustain Energy Rev*. 66, 631–653.

Alonso, D.M., Hakim, S.H., Zhou, S., Won, W., Hosseinaei, O., Tao, J., Garcia-Negron, V., Motagamwala, A.H., Mellmer, M.A., Huang, K. et al. (2017) Increasing the revenue from lignocellulosic biomass: maximizing feedstock utilization. *Sci Adv*. 3, 3160–3301.

Anasontzis, G.E., Kourtoglou, E., Villas-Boâs, S.G., Hatzinikolaou, D.G., Christakopoulos, P. (2016) Metabolic engineering of *Fusarium Oxysporum* to improve its ethanol-producing capability. *Front Microbiol*. 7, 632.

Banu, J., Parvathy Eswari, A., Kavitha, S., Kannah, R.Y., Kumar, G., Jamal, M.T., Saratale, G.D., Nguyen, D.D., Lee, D.G., Chang, S.W. (2019) Energetically efficient microwave disintegration of waste activated sludge for biofuel production by zeolite: quantification of energy and biodegradability modeling. *Int J Hydrogen Energ.* 44, 2274–2288.

Baruah, J., Nath, B., Sharma, R., Kumar, S., Deka, R., Baruah, D., Kalita, E. (2018) Recent trends in the pretreatment of lignocellulosic biomass for value-added products. *Front Energy Res.* 6, 141.

Bbosa, D., Mba-Wright, M., Brown, R.C. (2018) Moe than ethanol: a techno-economic analysis of a corn stover-ethanol biorefinery integrated with a hydrothermal liquefaction process to convert lignin into biochemical. *Biofuels Bioprod Bioref.* 12, 497–509.

Bhattacharya, A.S., Bhattacharya, A., Pletschke, B.I. (2015) Synergism of fungal and bacterial cellulases and hemicellulases: a novel perspective for enhanced bio-ethanol production. *Biotechnol Lett.* 37, 1117–1129.

Broda, M., Yelle, D.J., Serwanska, K. (2022) Bioethanol production from lignocellulosic biomass-challenges and solutions. *Molecule.* 27, 8717. https://doi.org/ 10.3390/molecules27248717

Dahunsi, S.O. (2019) Liquefaction of pineapple peel: pretreatment and process optimization. *Energy.* 185, 1017–1031.

Das, N., Jena, P.K., Padhi, D., Mohanty, K.M., Sahoo, G. (2021) A comprehensive review of characterization, pretreatment and its applications on different lignocellulosic biomass for bioethanol production. *Biomass Conv Bioref.* 82, 1–25.

Davis, R., Aden, A., Pienkos, P.T. (2011) Techno-economic analysis of autotrophic microalgae for fuel production. *Appl Energy.* 88, 3524–3531.

Divyalakshmi, P., Murugan, D., Sivarajan, M., Sivasamy, A., Saravanan, P., Rai, C.L. (2018) Effect of ultrasonic pretreatment on secondary sludge and anaerobic biomass to enhance biogas production. *J Mater Cycles Waste Manage.* 20, 481–488.

Eswari, A.P., Kavitha, S., Kaliappan, S., Yeom, I.T., Banu, J.R. (2016) Enhancement of sludge anaerobic biodegradability by combined microwave-H pretreatment in acidic conditions. *Environ Sci Pollut Res.* 23, 13467–13479.

Galbe, M., Wallberg, O. (2019) Pretreatment for biorefineries: a review of common methods for efficient utilisation of lignocellulosic materials. *Biotechnol Biofuels.* 12, 294.

Gaykawad, S.S., Zha, Y., Punt, P.J., van Groenestijn, J.W., van der Wielen, L.A.M., Straathof, A.J.J. (2013) Pervaporation of ethanol from lignocellulosic fermentation broth. *Bioresour Technol.* 129, 469–476.

Granjo, J.F.O., Duarte, B.P.M., Oliveira, N.M.C. (2017) Integrated production of biodiesel in a soybean biorefinery: modelling, simulation and economical assessment. *Energy.* 129, 273–291.

Humbird, D., Davis, R., Tao, L., Kinchin, C., Hsu, D., Aden, A., Schoen, P., Lukas, J., Olthof, B., Worley, M., Sexton, D., Dudgeon, D. (2011) Process design and economics and biochemical conversion of lignocellulosic biomass to ethanol: dilute-acid pretreatment and enzymatic hydrolysis of corn stover, National Renew Energy Laboratory. Technical Report, NREL/TP-5100-47764.

Jung, S.J., Kim, S.H., Chung, I.M. (2015) Comparison of lignin, cellulose, and hemicellulose contents for biofuels utilization among 4 types of lignocellulosic crops. *Biomass Bioenerg.* 83, 322–327.

Karimi, K., Taherzadeh, M.J. (2016) A critical review on analysis in pretreatment of lignocelluloses: degree of polymerization, adsorption/desorption, and accessibility. *Bioresour Technol.* 203, 348–356.

Kavitha, S., Banu, J.R., Kumar, G., Kaliappan, S., Yeom, I.T. (2018) Profitable ultrasonic assisted microwave disintegration of sludge biomass: modelling of biomethanation and energy parameter analysis. *Bioresour Technol.* 254, 203–213.

Kim, J., Hwang, S., Lee, S.M. (2022) Metabolic engineering for the utilization of carbohydrate portions of lignocellulosic biomass. *Metab Eng*. 71, 2–12.

Kim, J.S., Lee, Y.Y., Kim, T.H. (2015) A review on alkaline pretreatment technology for bioconversion of lignocellulosic biomass. *Biores Technol*. 199, 42–48.

Klein-Marcuschamer, D., Oleskowicz-Popiel, P., Simmons, B.A., Blanch, H.W. (2010) Technoeconomic analysis of biofuels: a wiki-based platform for lignocellulosic biorefineries. *Biomass Bioenerg*. 34, 1914–1921.

Koutinas, A.A., Chatzifragkou, A., Kopsahelis, N., Papanikolaou, S., Kookos, I.K. (2014) Design and techno-economic evaluation of microbial oil production as renewable resources for biodiseal and oleochemical production. *Fuel*. 116, 566–577.

Kumar, D., Murthy, G.S. (2011) Impact of pretreatment and downstream processing technologies on economic and energy in cellulosic ethanol production. *Biotechnol Biofuels*. 4, 27.

Kumar, K.A., Sharma, S. (2017) Recent updates on different methods of pretreatment of lignocellulosic feedstocks: a review. *Bioresour Bioprocess*. 4, 1–19.

Kwon, E. E., Jean, Y. J., Yi, H. (2012) New candidate for biofuel feedstock beyond terrestrial biomass for thermo-chemical process (pyrolysis/gasification) enhanced by carbon dioxide (CO). *Bioresour. Technol*. 123, 673–677.

Li, J., Zhou, W., Fan, S., Xiao, Z., Liu, Y., Liu, J., Qiu, B., Wang, Y. (2018) Bioethanol production in vacuum membrane distillation bioreactor by permeate fractional condensation and mechanical vapor compression with Polytetrafluoroethylene (PTFE) membrane. *Bioresour Technol*. 268, 708–714.

Liu, Y., Xie, X., Liu, W., Xu, H., Cao, Y. (2020) Consolidated bioprocess for bioethanol production from lignocellulosic biomass using *Clostridium Thermocellum* DSM 1237. *BioResource*. 15, 8355–8368.

Mohanram, S., Amat, D., Choudhary, J., Arora, A., Nain, L. (2013) novel perspectives for evolving enzyme cocktails for lignocellulose hydrolysis in biorefineries. *Sustain Chem Processes*. 1, 15.

Philips, S.D. (2007) Technoeconomic analysis of a lignocellulosic biomass indirect gasification process to make ethanol via mixed alcohols synthesis. *Ind Eng Chem Res*. 46, 8887–8897.

Priyanka, M., Kumar, D., Shankar, U., Yadav, A., Yadav, K. (2019) Agricultural waste management for bioethanol production. In Information Resources Management Association (IRMA) (Ed), *Biotechnology: Concepts, Methodologies, Tools, and Applications*, IGI Global Publisher, Hershey, PE, pp. 492–524.

Qing, Q., Zhou, L., Guo, Q., Gao, X., Zhang, Y., He, Y., Zhang, Y. (2017) Mild alkaline presoaking and organosolv pretreatment of corn stover and their impacts on corn stover composition, structure, and digestibility. *Bioresour Technol*. 233, 284–290.

Rajput, A.A., Visvanathan, C. (2018) Effect of thermal pretreatment on chemical composition, physical structure and biogas production kinetics of wheat straw. *J Environ Manag*. 221, 45–52.

Rentizelas, A., Karellas, S., Kakaras, E., Tatsiopoulos, I. (2009) Comparative techno-economic analysis of ORC and gasification for bioenergy applications. *Energy Convers Manage*. 50, 674–681.

Robak, K., Balcerek, M. (2020) Current state-of-the-art in ethanol production from lignocellulosic feedstocks. *Microbiol Res*. 240, 126534.

Roran, P., Pierre, B., Christine, M.R., Guillaume, D., Fabien, H. (2022) Laminar flame speed of ethanol/ammonia blends-An experimental and kinetic study. *Fuel Com*. 12, 100052.

Ruddy, D.A., Hensley, J.E., Nash, C.P., Tan, E.C.D., Christensen, E., Farberow, C.A., Baddour, F.G., VanAllsburg, K.M., Schaidle, J.A. (2019) Methanol to high-octane gasoline within market-responsive biorefinery concept enabled by catalysis. *Nat Catal*. 2, 632–640.

Saha, S., Kurade, M.B., El-Dalatony, M.M., Chatterjee, P.K., Lee, D.S., Jeon, H.H. (2016) Improving bioavailability of fruit wastes using organic acids: an exploratory study of biomass pretreatment for fermentation. *Energy Convers Manage*. 127, 256–264.

Shirkavand, E., Baroutian, S., Gapes, D.J., Young, B.R. (2016) Combination of fungal and physicochemical processes for lignocellulosic biomass pretreatment - a review. *Renew Sustain Energy Rev*. 54, 217–234.

Singhania, R.R., Adsul, M., Pandey, A., Patel, A.K. (2017) 4–Cellulases. In *Current Developments in Biotechnology and Bioengineering*, Edited by: Pandey, A., Negi, S., Soccol, C.R. Elsevier Publisher, Amsterdam, Netherlands. pp. 73–101.

Srivastava, N., Srivastava, M., Mishra, P.K., Gupta, V.K., Molina, G., Rodriguez-Couto, S., Manikanta, A., Ramteke, P.W. (2018) Applications of fungal cellulases in biofuel production: advances and limitations. *Renew Sustain Energy Rev*. 82, 2379–2386.

Su, T., Zhao, D., Khodadadi, M., Len, C. (2020) Lignocellulosic biomass for bioethanol: recent advances, technology trends, and barriers to industrial development. *Curr Opin Green Sustain Chem*. 24, 56–60.

Tan, L., Sun, Z. Y., Okamoto, S., Takaki, M., Tang, Y. Q., Morimura, S., Kida, K. (2015) Production of ethanol from raw juice and thick juice of sugar beet by continuous ethanol fermentation with flocculating yeast strain KF-7. *Biomass Bioenerg*. 81, 265–272.

Tsai, C.-T., Meyer, A.S. (2014) Enzymatic cellulose hydrolysis: enzyme reusability and visualization of -glucosidase immobilized in calcium alginate. *Moleclule*. 19, 19390–19406.

Wei, H., Yingting, Y., Jingjing, G., Wenshi, Y., Junhong, T. (2017) Lignocellulosic biomass valorization: production of ethanol. In *Encyclopaedia of Sustainable Technologies*, Edited by: Martin, A. Abraham. Elsevier Publishers, Amsterdam, Netherlands. pp. 601–604.

Yang, M., Rosentrater, K.A. (2019) Techno-economic analysis of the production process of structural bio-adhesive derived from glycerol. *J Clean Prod*. 228, 388–398.

Yao, S., Mikkelsen, M.J. (2010) Metabolic Engineering to Improve Ethanol Production in *Thermoanaerobacter Mathranii*. *Appl Microbiol Biotechnol*. 88, 199–208.

Zabed, H., Sahu, J.N., Suely, A., Boyce, A.N., Faruq, G. (2017) Bioethanol production from renewable sources: current prospectives and technological progress. *Renew Sus Energy Rev*. 71, 475–501.

Zaldivar, J., Nielsen, J., Olsson, L. (2001) Fuel ethanol production from lignocellulose: a challenge for metabolic engineering and process integration. *Appl Microbiol Biotechnol*. 56, 17–34.

Zhou, S., Raouche, S., Grisel, S., Sigoillot, J., Herpoël-Gimbert, I. (2017) Efficient biomass pretreatment using the white-rot fungus *Polyporus brumalis*. *Fungal Genom Biol*. 7, 2.

7 Sustainable Bio-Based 2,3-Butanediol Production from Renewable Feedstocks

Daniel Tinôco

7.1 INTRODUCTION

Dependence on oil, increased environmental concerns, and current conflicts in countries with non-renewable resources responsible for successive fluctuations in crude oil market prices have accelerated the transition from the petrochemical to the bio-based industry (Hazeena et al., 2020; Priya & Lal, 2019). The production of energy, fuels, and value-added products from renewable feedstocks has become an eco-friendly and economically attractive alternative, contributing to the emergence and consolidation of integrated biorefinery processes with a zero carbon circular economy (Narisetty et al., 2022a; Priya & Lal, 2019). The agro-industrial waste use has therefore helped to achieve the sustainable development goals proposed by the United Nations, mainly the 7th (Affordable and clean energy) and the 12th (Responsible consumption and production) (UNO, 2022).

Many platform chemicals, commonly synthesized by the chemical route, can be produced by fermentation processes. 2,3-butanediol (2,3-BDO) is an example of a multifunctional chemical and a valuable commodity used in different industrial applications and the manufacture of fuel, synthetic rubber, antifreeze, flavoring agent, cosmetics, and pharmaceuticals (Lee et al., 2023; Rehman et al., 2021), which can be produced by the biological route. The interest in this bioproduct is due to its rising market, which is predicted to grow at a CAGR (compound annual growth rate) of 3.5% in 2020–2030 and reach around US$300 million by 2030 (Transparency Market Research, 2022). Furthermore, 2,3-BDO can be converted into other high-value chemicals, whose average annual production of 32 million tons is responsible for approximately US$ 43 billion in sales (Köpke et al., 2011). Thus, bio-based 2,3-BDO production can become a sustainable route for different industrial segments (Parate et al., 2021).

A wide range of renewable wastes can be used for bio-based 2,3-BDO production, including lignocellulosic hydrolysates, molasses, distillery residues, dairy products, and glycerol (Sathesh-Prabu et al., 2020; Yang et al., 2022). These raw materials do not compete with the food and feed industry, have low cost and high abundance, and can provide different fermentable substrates to microbial producers (Kuenz et al., 2020).

 DOI: 10.1201/9781003407713-7

The substrate availability depends on the feedstock processing, which is carried out through physical–chemical and enzymatic conversion processes, preceded or not by a treatment step to increase biomass digestibility (An et al., 2022). Together, the saccharification and fermentation of renewable sources require less energy, release smaller toxic waste amounts into the environment, and are more cost-effective than the 2,3-BDO chemical synthesis, thus making its industrial application promising (Ge et al., 2016; Parate et al., 2021).

An article prospection in the *Scopus* database (Elsevier) from 2019 to March 2023 was carried out to collect information on the current state-of-the-art bio-based 2,3-BDO production from residual biomass. A preliminary analysis of the selected documents identified the main trends in natural resource utilization, highlighting the renewable feedstock types and the most used treatment and conversion technologies. The titer, yield, and productivity of 2,3-BDO from each carbon source analyzed were also presented to demonstrate the productive capacity of the sustainable, green, and inexpensive biological route. This chapter presents an initial overview of the 2,3-BDO biorefinery and its potential large-scale application.

7.2 2,3-BUTANEDIOL (2,3-BDO)

7.2.1 Properties and Applications

2,3-BDO is a high-value chemical with different physicochemical properties that characterize it as a platform chemical, fuel, agricultural biostimulant, and antifreeze agent (Mailaram et al., 2022). It is also a green chemical, odorless, colorless, and transparent at ambient temperature, which can be degraded by different microorganisms without causing significant risks to the environment and its users (Priya & Lal, 2019; Yu et al., 2022). Different market segments employ 2,3-BDO and its derivatives in their production processes, mainly the pharmaceutical, cosmetics, polymers, food, fuel, and agricultural industries (Rehman et al., 2021).

As a bulk chemical, 2,3-BDO can be converted to 1,3-butadiene and methyl ethyl ketone (MEK) by dehydration, diacetyl and acetoin by dehydrogenation, 2,3-BDO diester by esterification, and polyurethane by polymerization. These compounds are used as synthetic rubber precursor, solvent and aviation fuel additive, flavoring and bacteriostatic agents, drug and cosmetic precursors, and chain initiator and extender of polymer intermediates, respectively (Parate et al., 2021; Tinôco et al., 2020). The ability to form high-value derivatives allows the manufacture and marketing of various chemicals such as moistening foods, antifreeze, printing inks, plasticizers, fumigants, perfumes, and pharmaceuticals (Cha et al., 2020).

The high octane number and high heating value of around 27.2 kJ/g make 2,3-BDO a good fuel, which is comparable to traditional compounds such as methanol (22.1 kJ/g), ethanol (29.1 kJ/g) (Alvarez-Guzmán et al., 2020), and *n*-butanol (27.2 kJ/g) (Amraoui et al., 2022). Furthermore, 2,3-BDO has a low vapor pressure (0.23 hPa), contributing to its combustible properties (Narisetty et al., 2022b). As a result, 2,3-BDO can be used as a fuel additive in a gasoline engine (Mailaram et al., 2022).

The 2,3-BDO biostimulant capacity results from the metabolic modulation of hormones linked to the protection of plants and other crops (Lee et al., 2023). Through systemic resistance mechanisms, 2,3-BDO can act against infections by pathogens and insect pests, inducing phytohormones production and antioxidant responses (Duraisamy et al., 2022). Furthermore, 2,3-BDO can control the salicylic acid accumulation, which is responsible for inducing systemic tolerance to drought in several plant species (Lee et al., 2023).

The 2,3-BDO antifreeze properties are related to its optically active isomer (2R,3R)-BDO, called *levo*-isomer, with a low freezing point of −60°C (Wang et al., 2021). Besides this form, 2,3-BDO can be found as a dextro-isomer (2S,3S-BDO) and meso-isomer (2R,3S-BDO) (Figure 7.1) (Maina et al., 2019). Each 2,3-BDO isomer is used in a specific application according to its characteristic properties. *Levo*- and *dextro*-2,3-BDO have chiral properties that allow them to be used in the asymmetric synthesis of pharmaceuticals and fine chemicals (Tinôco et al., 2021b). In turn, *meso*-2,3-BDO is used in cosmetic applications since it has antibacterial properties capable of treating skin diseases such as eczema infected by bacteria and fungi (Lee et al., 2022).

7.2.2 Industrial 2,3-BDO Production

The 2,3-BDO production can be carried out by the chemical and biological routes, as illustrated in Figure 7.1 (Celińska & Grajek, 2009; Gräfje et al., 2019). Although the interest in the platform chemicals production from renewable raw materials has been growing in recent years, the 2,3-BDO industrial production is still based on fossil energy sources such as oil due to the limitations observed in the large-scale bio-based production process (Amraoui et al., 2021; Ge et al., 2016). Different strategies have been investigated to overcome these limitations, such as using low-cost and low-environmental impact wastes (Hazeena et al., 2022).

7.2.2.1 Chemical Route

The 2,3-BDO production by chemical route starts with the naphtha cracking process at 700°C–800°C, obtained from crude oil (Gräfje et al., 2019). A crack gas product is generated, and the removal of butadiene and isobutene results in a C4-hydrocarbons fraction known as C4-raffinate II, which is composed of butane (77% on average), butanes and isobutanes (23% on average) (Xie et al., 2022). The high energy demand and the greenhouse gas emissions make the chemical-based 2,3-BDO process economically costly and environmentally unfriendly (Priya & Lal, 2019).

Then, the C4 raffinate II fraction is subjected to the chlorohydrination with a chlorine solution in water, followed by the cyclization of the chlorohydrins formed by the reaction with sodium hydroxide. A mixture of 55% trans-2,3-butene oxide, 30% cis-2,3-butene oxide, and 15% 1,2-butene oxide is obtained. The hydrolysis of these butene oxides at 160°C–220°C and 50 bar ($\Delta H = -42$ kJ/mol) under excess water to avoid the polyethers formation leads to the formation of *meso*-2,3-BDO from trans-2-butene via trans-2,3-butene oxide, *levo*- and *dextro*-BDO from cis-2-butene via cis-2,3-butene oxide, and 1,2-BDO from 1,2-butene oxide (Gräfje et al., 2019) (Figure 7.1a).

The final butanediol mixture has low optical purity (Ge et al., 2016), contributing to increased process costs since a complex downstream step is required for efficient

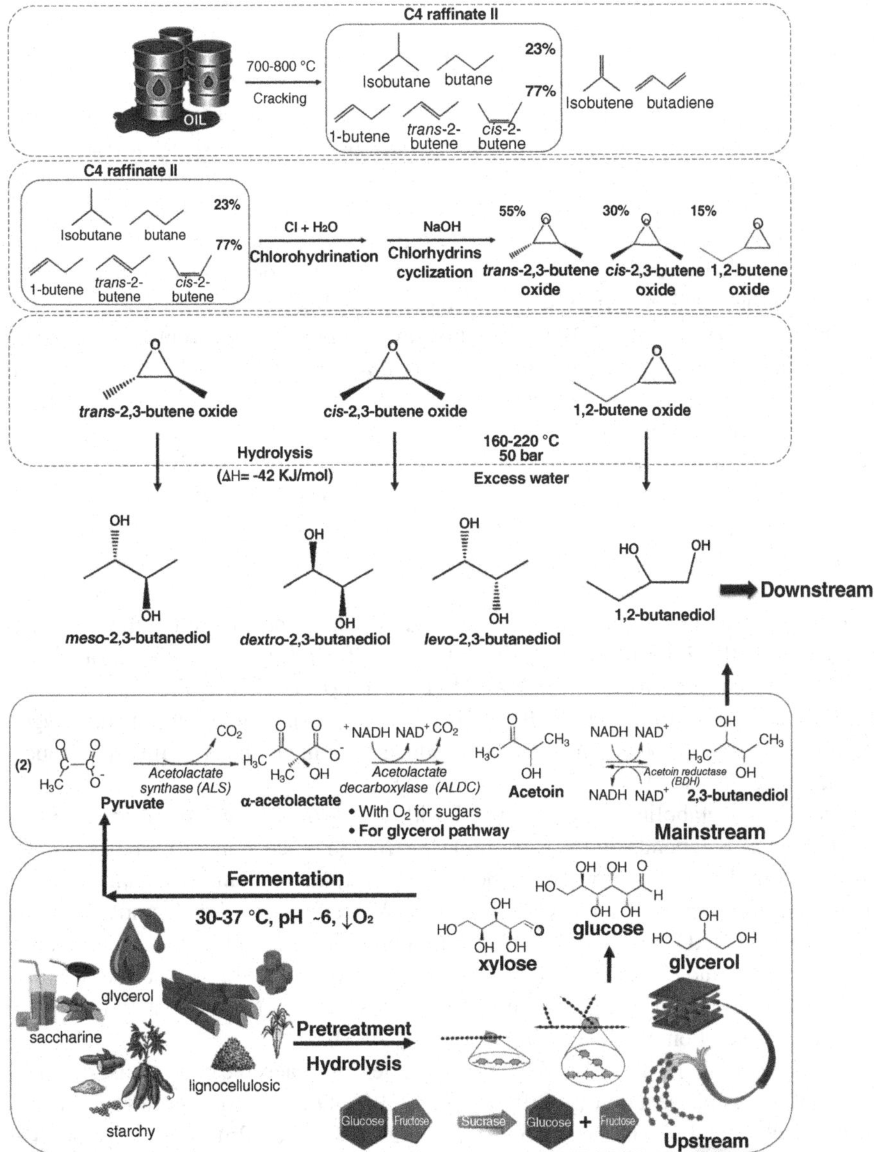

FIGURE 7.1 2,3-BDO industrial production by (a) chemical and (b) biological routes. Based on Ge et al. (2016), Priya and Lal (2019), Tinôco et al. (2021b), and Xie et al. (2022).

recovery and purification of 2,3-BDO isomers (Xie et al., 2022). Therefore, the high energy requirements with multiple unit operations and heat integration schemes, besides the sustainable and environmental aspects, have pressured the 2,3-BDO sector to transition from petroleum to bio-based approaches with better cost-effectiveness (Priya & Lal, 2019).

7.2.2.2 Biological Route

The biological route for 2,3-BDO production consists of upstream, mainstream, and downstream steps (Figure 7.1b). The mainstream step corresponds to a mixed acid–butanediol fermentation in which the pyruvate from the catabolism of hexoses, pentoses, and glycerol under glycolysis and pentose phosphate pathways (Ourique et al., 2020) is converted into 2,3-BDO. This conversion depends on the three sequential enzymes: α-acetolactate synthase (ALS), α-acetolactate decarboxylase (ALDC), and 2,3-butanediol dehydrogenase (BDH) (Ji et al., 2011). By-products such as acetoin, ethanol, acetic, lactic, succinic, and formic acid can be coproduced from reactions parallel to the 2,3-BDO synthesis (Tinôco et al., 2021b).

The key enzymes of 2,3-BDO metabolism are found in different bacterial species, mainly from the genera *Bacillus* and *Paenibacillus* (generally recognized as safe – GRAS microorganisms) and *Enterobacter*, *Klebsiella*, and *Serratia* (non-GRAS microorganisms) (Ji et al., 2011). The maximum enzymatic activity of ALS, ALDC, and BDH depends on the culture conditions, mainly the pH, which should be slightly acidic, around 5–6 (Celińska & Grajek, 2009). The temperature must be maintained within optimal ranges according to each microorganism used. Generally, bio-based 2,3-BDO production occurs at 30°C–37°C (Celińska & Grajek, 2009; Tinôco et al., 2021b).

Oxygen availability is another culture condition that is considered as the most critical factor in the 2,3-BDO metabolism (Ji et al., 2011). It regulates the intracellular redox balance represented by the $NADH/NAD^+$ ratio, controlling the α-acetolactate conversion into acetoin and 2,3-BDO (Dai et al., 2014). Depending on the oxygen supply, 2,3-BDO may or may not be produced, along with its main by-products (Celińska & Grajek, 2009).

As it is a metabolite semi-associated with cell growth, the bio-based 2,3-BDO production also depends on the respiratory process (Tinôco et al., 2021a). Its synthesis occurs in the deceleration phase, between the exponential and stationary steps of microbial growth (Celińska & Grajek, 2009). While fermentation promotes NADH oxidation and NAD^+ regeneration, respiration provides microbial biomass necessary for conversion and directing carbon flux for 2,3-BDO synthesis. Therefore, the balance between respiration and fermentation processes must be established for a high 2,3-BDO production efficiency (Ourique et al., 2020).

The maximum 2,3-BDO yield is reached under microaerobic conditions, in which the final reversible reaction is shifted to the 2,3-BDO formation instead of acetoin, its direct chemical precursor (Tinôco et al., 2021b) (Figure 7.2b). In the oxygen presence, α-acetolactate undergoes spontaneous decarboxylation leading to the diacetyl formation. Diacetyl reductase (DAR) is then activated, converting diacetyl to acetoin, which is converted to 2,3-BDO by the BDH action (Celińska & Grajek, 2009).

Depending on the BDH stereospecificity (Ge et al., 2016), the activity of the genes present in microbial producers related to the 2,3-BDO synthesis (Chen et al., 2014), and the culture conditions (Dias et al., 2018), the three isomers can be formed by the acetoin conversion. Meanwhile, a high optical purity of at least 99.2% of one of the 2,3-BDO isomers is typically verified through biological synthesis (Priya & Lal, 2019).

Compared to the chemical route, the biological route is an eco-friendly process with no greenhouse gas emissions, low energy requirement, and capable of producing 2,3-BDO with high isomeric optical purity (Ge et al., 2016; Priya & Lal, 2019). Despite these competitive advantages, industrial bio-based 2,3-BDO production is still scarce. According to Tinôco et al. (2020), only six companies produce bio-based 2,3-BDO on a large scale: Global Bio-Chem, LanzaTech, Orochem, Biokemik, GS Caltex, and Praj.

Among the main limiting factors for industrial bio-based 2,3-BDO production, the culture medium (upstream stage) and downstream processing stand out (Narisetty et al., 2021; Yang et al., 2022), as they account for at least 50% of production costs, which may compromise the bioprocess techno-economic viability (Hazeena et al., 2020; Tinôco et al., 2020). The amount and nature of the unit operations used to recover and purify 2,3-BDO depend on the fermentation broth composition. Therefore, its costs are linked to the raw materials and other nutrients used in the bioprocess (Xie et al., 2022).

The final cost of bio-based 2,3-BDO can be significantly reduced using waste products and residual biomass like glycerol, whey, lignocellulosic hydrolysates, and other agro-industrial residues as carbon and energy sources (Ji et al., 2011). Therefore, the choice and preparation of the culture medium carried out in the upstream step are critical for the overall economics of bio-based 2,3-BDO production and should be considered when designing large-scale fermentations (Tinôco et al., 2020).

7.3 SUSTAINABLE 2,3-BDO PRODUCTION

7.3.1 Renewable Resources

The primary low-cost and renewable resources can be divided into two broad groups according to nature and chemical composition: glycidic and oleaginous feedstocks. Glycidic feedstocks are represented by saccharine, starch, and lignocellulosic carbohydrates, while oleaginous feedstocks are formed by triglyceride molecules and free fatty acids (Melero et al., 2012).

About 53 studies on bio-based 2,3-BDO production from residual biomass were identified in 2019–2023, of which 82% used glycidic feedstocks and 18% oleaginous feedstocks (Figure 7.2a). Carbohydrates correspond to approximately 75% (w/w) of the residual biomass, while oils and fats make up 5% (w/w) of other remaining compounds (Chatterjee et al., 2015). Several microorganisms can quickly assimilate mono- and disaccharides and polysaccharides after chemical and enzymatic pretreatment. In contrast, only fungi and bacteria of the genus *Pseudomonas* with a β-oxidation pathway have been reported as microorganisms capable of using triglycerides as the sole carbon source (Li et al., 2019). Global glycidic material production is much higher than waste lipids (Melero et al., 2012). While about 2.4 billion tons of sugar (7.5%), wheat (32.6%), rice (21.3%), and milk and derivatives (38.7%) are forecast for 2022/23, an average production of 255.4 million tons of oils and fats is expected for the same period, according to the Biannual Report on Global Food Markets by the Food and Agriculture Organization of the United Nations (FAO, 2022). Therefore, the glycidic feedstock predominance in bio-based 2,3-BDO

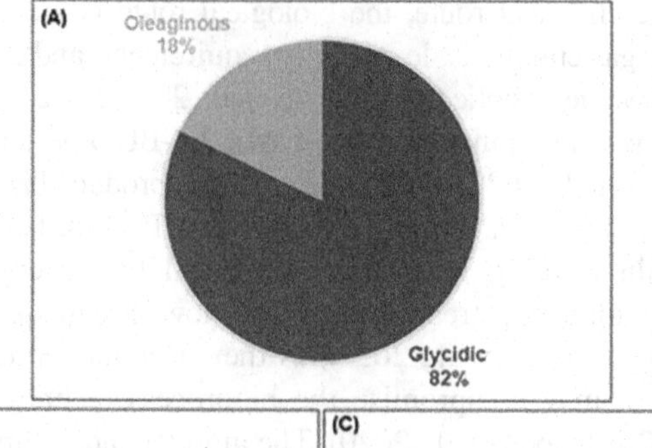

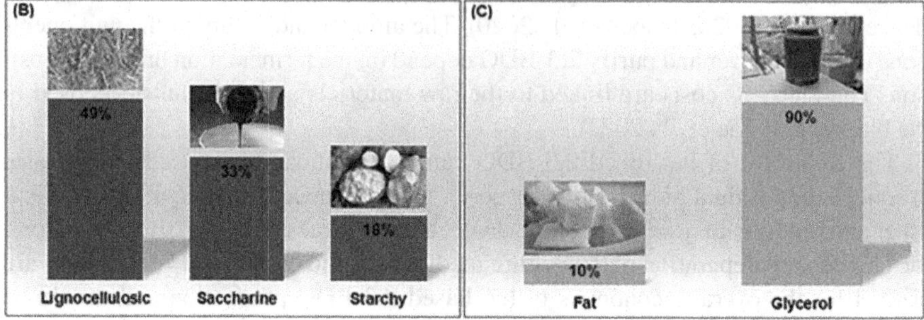

FIGURE 7.2 Main feedstocks used for sustainable bio-based 2,3-BDO production in 2019–2023: (a) chemical classification, (b) carbohydrate compounds, and (c) lipid compounds.

production can be justified by its high fraction in the total biomass composition, greater metabolization, and sizeable global production.

Lignocellulosic residues were the primary type of glycidic biomass reported in 49% of the studies. Then, sugar and starch compounds were found in 33% and 18% of the documents, respectively (Figure 7.2b). The interest in lignocellulosic biomass can be justified by its high availability in nature, with an expected volume of more than 500 million dry tons/year by 2040 (Affandy et al., 2023), a wide variety of low-cost feedstocks such as agricultural and forestry hydrolysates capable of generating fermentable sugars without competing with the food and feed industry (Kuenz et al., 2020), and the circular economy and climate change mitigation by integrating a management system for less CO_2 emission with a value-added chemical production from biorefinery concepts (Hazeena et al., 2022; Kim et al., 2020).

Regarding lipid-based compounds, glycerol has been widely investigated in 90% of the studies, while fats were reported in only 10% (Figure 7.2c). Several microorganisms can catabolize glycerol, although it is an energy-poor carbon source, mainly compared to glucose (Ripoll et al., 2020). Despite this limitation, glycerol use is profitable due to its relationship with biodiesel production. The integration of glycerol biorefineries for the coproduction of energy, biofuels, and chemicals such as 2,3-BDO indicates a global trend capable of uniting economic interests with increasingly latent environmental issues (Ripoll et al., 2021). As a result, disposal problems can be overcome, and its low market value of 0 $/kg improved, making glycerol an essential substrate for bio-based 2,3-BDO industrial production (Priya & Lal, 2019).

7.3.2 Biomass Pretreatment Processes

Each carbon source requires specific prior treatment to make its constituent substrates available to microbial producers (Chatterjee et al., 2015). In general, pretreatments should be able to improve biomass characteristics, leading to greater energy utilization efficiency (Anukam & Berghel, 2021). Furthermore, they must have low capital and operating costs, low energy demand, and generate as little waste and inhibitory compounds as possible (Agbor et al., 2011). The pretreatment step is a bottleneck in biomass-based processes. Therefore, it must be controlled and optimized for the biorefinery to be successful, mainly on a large scale (Tinôco, 2022).

The main pretreatment types reported in the selected documents were physical (35%), chemical (28%), physical–chemical (21%), and biological (3%) methods (Figure 7.3a). Most of them were related to lignocellulosic biomass due to its structural complexity requiring more robust processing for efficient metabolization (An et al., 2022). Meanwhile, the other agro-industrial waste, such as molasses and glycerol, did not require any pretreatment type (13%) except for a step to remove contaminants capable of affecting the fermentation process (Tinôco et al., 2021a).

Among the most used physical methods (Figure 7.3b), drying, grinding/milling, sieving, and washing were reported in 22%, 30%, 20%, and 8% of the documents, respectively. These methods correspond to upstream processes of preparation, manipulation, and biomass pre-conditioning, mainly lignocellulosic materials (Agbor et al., 2011). The grinding/milling step aims to reduce the particle size, increasing the surface area that will be in contact with the physical–chemical and enzymatic agents of the following steps. These particles can be classified into different size ranges through the sieving process, which ensures biomass uniformity and hydrolysis efficiency (Tinôco, 2022). Other physical methods reported in 20% of the studies were maceration (Maina et al., 2021b), homogenization, and dilution (OHair et al., 2021), which were used mainly with saccharine and starchy materials.

The pH-based methods were the primary chemical pretreatment method reported in 89% of the documents (Figure 7.3c). The lignocellulosic biomass chemical pretreatment aims to hydrolyze plant fibers, solubilizing the hemicellulosic fraction and partially removing lignin (Narisetty et al., 2022a). The most used acid and alkali to treat the biomass used for the bio-based 2,3-BDO production in the last 5 years were sulfuric acid and sodium hydroxide, respectively. Among the studies employing acid pretreatment, 84.6% used H_2SO_4 in the average concentration range of 0.5-2% (v/v). Only two documents reported other inorganic acids, such as HCl (Narisetty et al., 2022b) and a mixture of H_3PO_4 and HNO_3 (Ra et al., 2020). Other chemical pretreatments, including ionic liquids, were reported in 10.5% of the documents. Ionic liquids are special saline solvents formed by a small inorganic anion and a large organic cation (Tinôco, 2022). In general, ionic liquids have green characteristics such as low corrosiveness, low volatility, and excellent recyclability. Meanwhile, its application is still limited to the lab scale due to its high cost and toxicity to enzymes and microorganisms in large-scale fermentations. To overcome these limitations, low cost and non-toxic ionic liquids such as ethanolammonium acetate have been investigated (An et al., 2022).

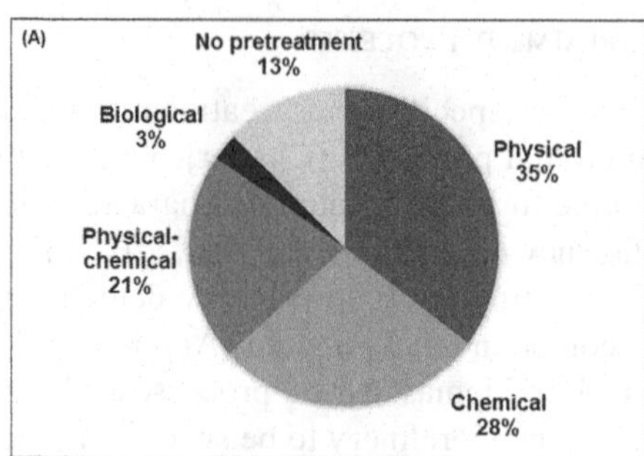

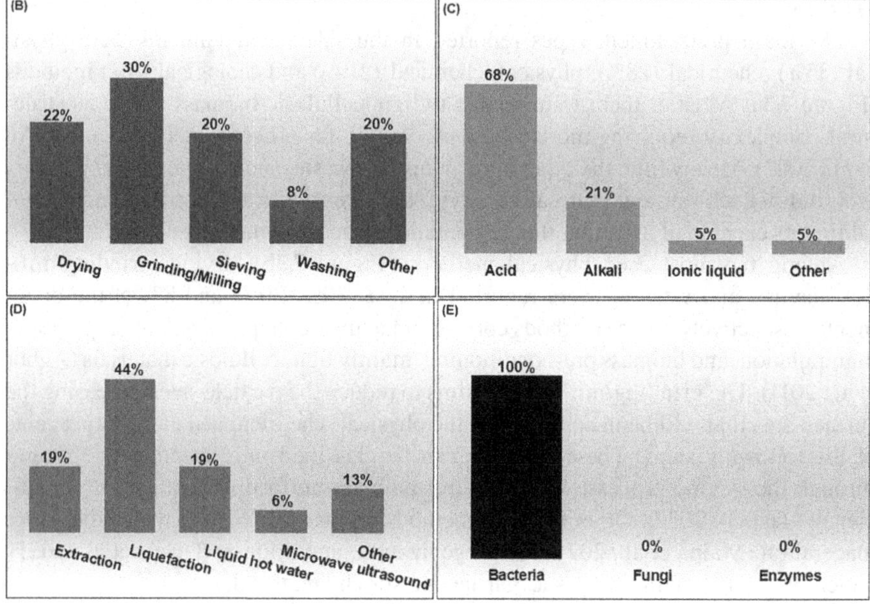

FIGURE 7.3 Main pretreatment methods used for sustainable bio-based 2,3-BDO production in 2019–2023: (a) pretreatment classification, (b) physical pretreatment methods, (c) chemical pretreatment methods, (d) physical–chemical pretreatment methods, and (e) biological pretreatment methods.

Physical–chemical pretreatments combine temperature and pressure with chemicals for a specific time to improve biomass digestibility (Agbor et al., 2011). The main methods used with sugary, starchy, and lignocellulosic materials were extraction (19%), liquefaction (44%), and liquid hot water treatment (19%), respectively (Figure 7.3d). Extraction is used to recover sugars such as glucose, fructose, and sucrose through water or another chemical, such as ethanol, at high temperatures (OHair et al., 2021). Liquefaction allows amylolytic enzymes, mainly α-amylase, to partially hydrolyze the starch at high temperatures. This process precedes the saccharification step, in which glucoamylases are used to release glucose (Lee & Seo, 2019). Liquid hot water is a solvent capable of penetrating the biomass and hydrolyzing the lignocellulosic complex under high temperatures and pressures

(Tinôco, 2022). This method is considered simple and low cost, as it is based only on the water at high temperatures, not requiring addition of any chemical product and the washing of biomass to recover these compounds. Usually, cellulose crystallinity tends to increase slightly while the polymerization degree decreases, favoring its enzymatic hydrolysis. Furthermore, hemicellulose is degraded to oligo- and monosaccharides, and lignin is partially dissolved through this hydrothermal method (Kuenz et al., 2020; Lunze et al., 2021). Other physical–chemical methods reported in 19% of the selected documents included microwave-assisted alkali pretreatment (Amraoui et al., 2022), deacetylation, and mechanical refining pretreatment process (Affandy et al., 2023).

Finally, the biological pretreatment was characterized by the lignocellulolytic bacteria application, although fungi and isolated enzymes can also be used (Figure 7.3e). Biological pretreatment is considered a greener and cheaper strategy capable of depolymerizing and degrading lignin while saccharifying cellulose and hemicellulose under mild conditions, low energy consumption, and minimal environmental impact. This method is based on the lignocellulolytic microbes with high enzymatic activity, mainly cellulases, hemicellulases, and ligninases (Li et al., 2023; Tinôco, 2022). Two lignocellulolytic bacteria have been reported to produce bio-based 2,3-BDO from renewable feedstocks in the last 5 years. *Klebsiella pneumoniae* PX14 isolated from insect gut was used to hydrolyze unpretreated bamboo biomass since it can secrete proteins with high enzymatic activity (Li et al., 2023). *Pseudomonas alcaligenes* PA-3 was co-cultured with *K. pneumoniae* AA405 for waste lard metabolization. *P. alcaligenes* PA-3 secreted lipases to hydrolyze long-chain triglycerides into glycerol and fatty acids, while *K. pneumoniae* AA405 coproduced 1,3-propanediol (1,3-PDO) and 2,3-BDO from released glycerol (Li et al., 2019).

7.3.3 Recent Researches

7.3.3.1 Glycidic Feedstocks

7.3.3.1.1 Saccharine Biomass

Saccharine biomass (sugars) consists of mono- and disaccharides, mainly hexoses (glucose, fructose, and galactose), pentoses (xylose and arabinose), and compounds resulting from hexoses linked by glycosidic bonds (maltose, sucrose, and lactose) (Chatterjee et al., 2015). Saccharine materials generally do not require severe chemical or enzymatic pretreatment to make their sugars available. Only an extraction step using acid solutions such as acetic acid and hydrochloric acid, followed or not by dilution, depending on the feedstock used, is enough to break the glycosidic bonds and release the sugars (Chatterjee et al., 2015; Mathewson, 1980).

The most commonly used saccharine biomasses for bio-based 2,3-BDO production in the last 5 years have been whey, molasses, and food waste. Fernández-Gutiérrez et al. (2022) investigated the whey and its permeate fermentation after an initial autoclaving. Two batch cultures using whey and its permeate with about 31 and 34 g/L of lactose (first batch) and 51 and 47 g/L of lactose (second batch) were conducted in a 2 L bioreactor, respectively. An aerated batch assay using diluted whey with 31 g/L lactose was also carried out. No pretreatment process was reported for whey and its permeate. Palaiogeorgou et al. (2019) investigated the fermentation of sugarcane molasses composed of 50%–52% (w/v) sucrose, 6% (w/v) protein, 5% (w/v) fructose, and 14% (w/v) ashes. After autoclaving, the initial sugarcane molasses concentration in fed-batch cultures was adjusted to obtain approximately 30 g/L of total sugars. A concentrated 600 g/L sugarcane molasses solution containing 5% (w/v) yeast extract was fed into a 2 L bioreactor whenever the carbon source was low. No pretreatment process was reported for sugarcane molasses. Sindhu et al. (2020) studied the food and kitchen waste valorization for the integrated production of 2,3-BDO (fine chemical), poly-3-hydroxybutyrate (biopolymer), bioethanol (biofuel), and pectinase (enzyme). Dried fruit and vegetable waste were used as carbon sources. No sterilization or pretreatment steps have been reported.

7.3.3.1.2 Starchy Biomass

Starchy biomass is formed by polysaccharides composed of glucose molecules linked through α-1,4 and α-1,6 bondings, which give rise to amylose and amylopectin structures, respectively (Melero et al., 2012). Amylose corresponds to 10%–20% of starch, characterized by a long linear unbranched chain and soluble in water. In contrast, amylopectin corresponds to 80%–90% of starch and is characterized by a long-branched chain. Amylopectin is responsible for the starch granules formation due to its helical shape, which confers greater crystallinity to the polysaccharide and high viscosity in water (Chatterjee et al., 2015).

Starchy materials require liquefaction followed by an acid or enzymatic treatment step before the fermentation process to gelatinize and hydrolyze the starch, respectively. Many microorganisms cannot use starch directly as a substrate (Khunnonkwao et al., 2021; Moon et al., 2021). The use of raw materials in the grains form also requires initial milling, followed or not by dilution, for better starch availability (Mathewson, 1980).

The primary starchy resources used for bio-based 2,3-BDO production in the last 5 years have been cassava, Jerusalem artichoke tuber (JAT), and bakery waste. These biomasses were subjected to liquefaction and enzymatic hydrolysis to make their fermentable sugars available.

Lee and Seo (2019) subjected ground and sieved cassava chips to 1 mm containing approximately 72% dry weight starch to liquefaction at 100°C for 90 min using a commercial α-amylase at 0.7 g/kg dry matter. Then, saccharification was performed aseptically at 50°C for 24 h using a commercial glucoamylase at 0.5 g/kg dry matter. About 100 g glucose could be obtained from 126 g treated cassava chips. Dai et al. (2020) used JAT containing 64.5% sugar, initially sliced, dried, and ground to 30~200 mesh. JAT extract was pretreated with H_2SO_4 at 115°C, pH 3 for 20 minutes. Under these conditions, around 93% inulin at an initial concentration of 202.6 g/L from JAT extract with added commercial inulin was consumed by a newly isolated *K. pneumoniae*-producing inulinase in batch fermentation. Narisetty et al. (2022b) investigated acid and enzymatic hydrolysis to pretreat bread waste containing 46% (w/w) starch. Acid hydrolysis was performed using 2% (v/v) H_2SO_4 at 121°C for 15 minutes and a solid loading of 20% (w/v). A glucose yield of 0.37 g/g was achieved. In turn, enzymatic hydrolysis was performed in two parts. In the first step, 20% (w/v) solid loading was gelatinized at 121°C, pH 4.3 for 15 minutes in an autoclave.

In the second step, a commercial glucoamylase at 0.6 mg/g bread waste was used to hydrolyze the liquefied bread waste at 45°C, 250 rpm for 48 hours. A glucose yield of 0.41 g/g was reached.

7.3.3.1.3 Lignocellulosic Biomass

Lignocellulosic biomass is a polymeric organic structure formed by cellulose, hemicellulose, and lignin (Melero et al., 2012). Cellulose and hemicellulose are polysaccharides composed chiefly of glucose and xylose, respectively. In turn, lignin is formed by methoxylated phenylpropane units (aromatic compounds), commonly used to generate energy by burning it (Hazeena et al., 2020).

The lignocellulosic complex is highly recalcitrant, resulting from cellulose and hemicellulose microfibrils crosslinking with lignin through hydrogen bonds in its crystalline structure (Lunze et al., 2021). A pretreatment step is required to disrupt the lignocellulosic matrix, increasing the accessibility of hydrolytic enzymes to cellulose and hemicellulose molecules through improved surface area and solubility (An et al., 2022).

Pretreatment processes must be able to make available the maximum amount of fermentable sugars possible with a minimum of microbial inhibitory compounds (Hazeena et al., 2020) since these inhibitors can affect the intracellular redox balance by consuming reducing agents such as NADH and NAD^+, reducing the yield of glucose and xylose released (Okonkwo et al., 2021). According to nature, pretreatments can be classified as physical (washing, drying, grinding, and sieving), chemical (acid, alkali, organic solvent, and ionic liquid), physical–chemical (steam explosion, liquid hot water, ammonia fiber expansion, and microwave-/ultrasound-assisted technologies), and biological (Fungi/bacteria and enzymes) (Tinôco, 2022).

The most recent studies on the bio-based 2,3-BDO production have investigated the use of bagasse, straw, and hull hydrolysates as lignocellulosic sources of fermentable sugars, which are released mainly after chemical pretreatment with inorganic acid at high temperatures followed by enzymatic hydrolysis.

Pasaye-Anaya et al. (2019) investigated the use of mezcal-waste agave. Initially, agave bagasse was washed with distilled water at 80°C to remove free sugars, dried at room temperature, and ground to approximately 1 cm. Then, it was treated with 5% (v/v) H_2SO_4 at 121°C for 1 hour, with a solid:liquid ratio of 1:10 (w/v). Enzymatic hydrolysis was performed at 50°C using commercial cellulases and hemicellulases at 5 IU/g, pH=5, 150 rpm for 48 hour. The liquor obtained contained 25.8 g/L of glucose, xylose, and arabinose, with a sugar conversion efficiency of approximately 52.5%. Okonkwo et al. (2021) used wheat straw bagasse (WSH) ground into 1 mm particles and treated with 1% (v/v) H_2SO_4 at 121°C for 1 hour, with a solids loading of 15% (w/v). The pH was adjusted to 5 using ammonium hydroxide. Acid-pretreated WSH was hydrolyzed by commercial cellulase and xylanase at 50°C and 80 rpm for 120 hours. About 98.9 g/L of glucose, xylose, and arabinose in the ratio 8.5:12.5:1 were released, with a yield of 0.66 g_{sugar}/g_{WSH}. No detoxification process was carried out. Cortivo et al. (2019) subjected soybean hull to pretreatment with 1% (v/v) H_2SO_4 at 121°C for 40 minutes, with a solid:liquid ratio of 1:10 (w/v), after initial drying. Enzymatic hydrolysis was performed using a commercial enzyme preparation at 15 FPU/g_{solids}, which resulted in the solubilization of 62% hemicellulose and 78% cellulose. Glucose, xylose, and arabinose were released with furfural and hydroxymethylfurfural (<0.3 g/L). No detoxification process was carried out. Both biomass fractions were used for 2,3-BDO production.

Other lignocellulosic feedstocks were also investigated. Amraoui et al. (2022) used Brewers' spent grain, a major by-product of breweries. The authors applied microwave-assisted alkali to pretreat the biomass, which was suspended in a 0.5% (w/v) NaOH solution and subjected to microwave radiation at 400 W for 60 seconds. The treated biomass was washed with distilled water until the pH was adjusted to around 6 and then dried. Enzymatic hydrolysis was evaluated under different conditions, where a maximum yield of 0.25 $g_{glucose}/g_{biomass}$ was achieved using 100 U/mL of a commercial cellulase at 50°C, pH 6, 10% (w/v) hydrolysate, and 2% (v/v) enzyme loading. Han et al. (2022) evaluated tobacco waste after initial drying, comminution, and sieving. Uncrushed tobacco residues were boiled for 30 minutes with a solid:liquid ratio of 1:20 (w/v), centrifuged, and cooled. About 30.69 mg/g of soluble sugar was present in the tobacco extract. Li et al. (2023) investigated a biological treatment by *Klebsiella pneumoniae* PX14 to solubilize bamboo lignocellulosic biomass, which was initially sliced and crushed into powder. *K. pneumoniae* PX14 was cultured at 37°C, pH 7, 200 rpm for 7 days in a potassium morpholinopropane sulfonate (MOPS) medium supplied with 10% (w/v) bamboo powder, which resulted in a degradation rate of 47.2%. The lignocellulolytic enzymes secreted by this bacterium were mixed with a commercial cellulase forming an enzyme cocktail in which 42.9 g/L of reducing sugars were produced from 10% (w/v) solid loading in 48 hours. The fermentation process was conducted in fed-batch cultures based on the results found with bamboo powder saccharification.

Hazeena et al. (2019) used an initially washed oil palm front, which was cut into 3–4 cm long pieces, dried at 60°C, ground, and sieved to 0.5 mm. The biomass was pretreated with 3% (w/v) NaOH at 121°C for 20 minutes, with a solid loading of 15% (w/v). Enzymatic hydrolysis was performed using commercial cellulase at 20 FPU/g, 37°C, and 200 rpm for 24 hours. About 49.9 g/L of total reducing sugars were released. Ra et al. (2020) evaluated red seaweed *Gelidium amansii* hydrolysate use. The authors submitted about 12% (w/v) seaweed slurry to a hyperthermal acid hydrolysis pretreatment using a 180 mM acid mixture of H_3PO_4 and HNO_3 (5:5 ratio) at 150°C for 10 minutes. Blended enzymes containing cellulase, ß-glucanase (endo-1,3 or 1,4), and hemicellulase at 16 U/mL were used for biomass saccharification at 50°C and 150 rpm for 24 hours. Approximately 49.8 g/L of reducing sugars with an efficiency of 54.8% were obtained from 90.96 g/L of *G. amanssi* slurry.

7.3.3.2 Oleaginous Feedstocks

7.3.3.2.1 Triglycerides and Free Fatty Acids

Triglycerides are compounds formed by glycerol and long-chain alkyl fatty acids linked by ester bonds, commonly called oils and fats, and obtained from vegetable and animal raw materials (Melero et al., 2012). Glycerol and fatty acids are available by triglycerides hydrolysis catalyzed by acid or lipases capable of breaking ester bonds (Li et al., 2019). Processed and low-grade oleaginous materials have more free fatty acids from hydrolyzed triglycerides (Melero et al., 2012). Glycerol can also be obtained by transesterification of oils and fats, as in biodiesel production, where about 10% (v/v) of glycerol is released as a by-product (Kongjan et al., 2021). Glycerol has several industrial applications, including bio-based 2,3-BDO production (Parate et al., 2021).

Crude glycerol is the primary oleaginous feedstock used as a carbon source for bio-based 2,3-BDO production in the last 5 years. Maina et al. (2021) studied different crude glycerols obtained from two industrial biodiesel production facilities through chemical transesterification. The crude glycerols contained 100% soybean oil or 10%–35% tallow fat. They were subjected to overnight decantation to remove impurities. The best bio-based 2,3-BDO production was achieved using crude glycerol 100% soybean oil containing 79.4% (w/w) glycerol after decanting in fed-batch fermentation. Priya and Lal (2019) used waste glycerol from Jatropha-based biodiesel containing about 22%–30% glycerol. The Jatropha plant is a non-food crop whose oleaginous fraction can be used for value-added chemical production in a circular economy and biorefinery concept. Meanwhile, glycerol shows greater difficulty in cell assimilation than glucose due to its reducing nature, which compromises its use, especially on a large scale. To overcome this limitation, the authors investigated different control strategies, mainly pH and agitation, which resulted in a higher 2,3-BDO titer and reduced by-products in a shorter fermentation time. No pretreatment was performed. Ripoll et al. (2020) obtained a high 2,3-BDO production using crude glycerol containing 55%–85% (w/w) glycerol in fed-batch fermentation. No pretreatment was performed. It was found that a 2,3-BDO titer was about 11% lower using crude glycerol than pure glycerol, while the yield of 2,3-BDO did not change significantly.

Fat remains and crude glycerol supplementation with glycidic feedstocks have also been investigated for enhanced bio-based 2,3-BDO production. Li et al. (2019) studied the coproduction of 1,3-PDO and 2,3-BDO from waste lard as the sole carbon source for the first time. Waste lard is widely used for biodiesel production through methylation reactions due to its high content of long-chain saturated triglycerides. Their hydrolysis produces glycerol and free fatty acids oxidized through glycerol and ß-oxidation pathways, respectively. The authors pretreated waste lard to extract proteins and free amino acids using petroleum ether. Then, *P. alcaligenes* PA-3 was used to promote the long-chain triglycerides hydrolysis and the intracellular oxidation of fatty acids. This strain was co-cultivated with *K. pneumoniae* AA405, which is responsible for synthesizing 1,3-PDO and 2,3-BDO from released glycerol. Tinôco et al. (2021a) supplemented a crude glycerol-based medium with sugarcane molasses to improve the bio-based 2,3-BDO production. Crude glycerol was obtained through chemical transesterification of 10% (v/v) tallow fat and 90% (v/v) soybean oil, containing about 79.4% (v/v) glycerol. Sugarcane molasses were composed of approximately 34% (w/w) sucrose, 5% (w/w) glucose, and 6% (w/w) fructose. Crude glycerol and sugarcane molasses were previously treated to remove impurities. The 2,3-BDO titer was improved by approximately 31% with supplementation of 10 g/L sugarcane molasses in a medium containing 50 g/L crude glycerol compared to the glycerol-based medium alone.

7.3.3.3 Titer, Yield, and Productivity of Sustainable 2,3-BDO

Table 7.1 summarizes the title, yield, and productivity of bio-based 2,3-BDO from different residual biomasses reported in the previously presented studies to demonstrate the productive capacity of renewable feedstocks after pretreatment and saccharification, where applicable. The best-combined results of the three

TABLE 7.1
Fermentation Parameters of Sustainable Bio-Based 2,3-BDO Production

Microorganism	Carbon Source	Fermentation Mode	2,3-BDO (g/L)	Yield (g/g)	Productivity (g/L/h)	Reference
Enterobacter ludwigii (engineered-type)	Brewers' spent grain hydrolysate (glucose-rich)	Fed-batch	118.5	0.43	1.65	Amraoui et al. (2022)
Co-cultivation of *Klebsiella pneumoniae* BLh-1 and *Pantoea agglomerans* BL1 (wild-type)	Soybean hull hydrolysate (non-detoxified)	Shake flask	21.9 (hemicellulose) and 20.1 (cellulose)	0.40 (hemicellulose) and 0.50 (cellulose)	0.30 (hemicellulose) and 0.28 (cellulose)	Cortivo et al, (2019)
K. pneumoniae H3 (wild-type)	Jerusalem Artichoke Tubers extract + inulin	Batch	80.4	0.43	2.23	Dai et al. (2020)
Escherichia coli K12 MG1655 (engineered-type)	Diluted whey	Batch (aerated)	12.5	0.41	0.52	Fernández-Gutiérrez et al. (2022)
Bacillus amyloliquefaciens strain T4 (wild-type)	Tobacco waste	Shake flask	1.54	0.031 (tobacco waste)	0.032	Han et al. (2022)
Enterobacter cloacae sp. SG1 (wild-type)	Oil palm front hydrolysate	Shake flask	30.7	ND	0.32	Hazeena et al. (2019)
Saccharomyces cerevisiae YG01_SDBN (engineered-type)	Cassava hydrolysate	Fed-batch	132	0.32	1.92	Lee and Seo (2019)
Co-cultivation of *Pseudomonas alcaligenes* PA-3 and *K. pneumoniae* AA405 (wild-type)	Waste lard	Batch	4.29	NA	NA	Li et al. (2019)
K. pneumoniae PX14 (wild-type)	Bamboo powder (unpretreated)	Fed-batch	43.5	NA	0.36	Li et al. (2023)

(Continued)

TABLE 7.1 (*Continued*)
Fermentation Parameters of Sustainable Bio-Based 2,3-BDO Production

Microorganism	Carbon Source	Fermentation Mode	2,3-BDO (g/L)	Yield (g/g)	Productivity (g/L/h)	Reference
Klebsiella michiganensis ID18024 (wild-type)	Crude glycerol (others)	Fed-batch	76.1	0.41	1.38	Maina et al. (2021a)
Enterobacter ludwigii (engineered-type)	Bread waste hydrolysates	Fed-batch	135.4 (acid hydrolysis) and 138.8 (enzymatic hydrolysis)	0.42 (acid hydrolysis) and 0.48 (enzymatic hydrolysis)	1.41 (acid hydrolysis) and 1.45 (enzymatic hydrolysis)	Narisetty et al. (2022b)
Paenibacillus polymyxa DSM 365 (wild-type)	Wheat straw hydrolysate (non-detoxified at 60%)	Shake flask	32.5	0.33	0.45	Okonkwo et al. (2021)
Enterobacter sp. FMCC-208 (wild-type)	Sugarcane molasses	Fed-batch	52.0	0.23	0.81	Palaiogeorgou et al. (2019)
Klebsiella oxytoca UM2-17 (wild-type)	Agave bagasse	Shake flask	10.3	0.40	0.06	Pasaye-Anaya et al. (2019)
E. cloacae TERI BD 18 (wild-type)	Crude glycerol	Fed-batch	41.4 (acetoin + 2,3-BDO)	0.48 (acetoin + 2,3-BDO)	0.27 (acetoin + 2,3-BDO)	Priya and Lal (2019)
S. cerevisiae BD4 (engineered-type)	Red seaweed hydrolysate	Shake flask	14.8	0.30	0.21	Ra et al. (2020)
Raoultella terrigena CECT 4519 (wild-type)	Crude glycerol	Fed-batch	80.5	0.44	0.40	Ripoll et al. (2020)
E. cloacae SG1 (wild-type)	Fruit and vegetable waste	Shake flask	3.0	NA	0.03	Sindhu et al. (2020)
P. polymyxa PM 3605 (wild-type)	Crude glycerol supplemented with sugarcane molasses	Shake flask	19.0	0.35	0.13	Tinôco et al. (2021a)

fermentation parameters were achieved using starchy cassava hydrolysates (Lee & Seo, 2019), waste bread (Narisetty et al., 2022b), and JAT extract (Dai et al., 2020). Next, lignocellulosic hydrolysates from Brewers' spent grain (Amraoui et al., 2022) and crude glycerol (Maina et al., 2021a; Ripoll et al., 2020) were the glycidic and oleaginous feedstocks that led to the best bio-based 2,3-BDO production, respectively.

7.4 CONCLUSION

The production of bulk chemicals such as 2,3-BDO by biological route is a global trend that should consolidate in the coming years due to the increasing progress of the transition process from the fossil resource-based to renewable resource-based industry. The residual biomass use contributes to the 2,3-BDO production economy due to its low cost, natural abundance, and low energy requirement while assisting in waste management, solving environmental disposal problems. Reusing and recycling natural resources such as saccharine, starch, lignocellulosic, and glycerol compounds, whether or not submitted to pretreatment and saccharification processes for the bio-based 2,3-BDO production, contribute to the development of a more sustainable and green circular economy. As a result, responsible consumption and production, clean and affordable energy generation, and a zero waste policy can be achieved, strengthening 2,3-BDO biorefineries as potential and competitive large-scale production systems.

REFERENCES

Affandy, M., Zhu, C., Swita, M., Hofstad, B., Cronin, D., Elander, R., Lebarbier Dagle, V. 2023. Production and catalytic upgrading of 2,3-butanediol fermentation broth into sustainable aviation fuel blendstock and fuel properties measurement. *Fuel*. 333, 126328. https://doi.org/10.1016/j.fuel.2022.126328

Agbor, V.B., Cicek, N., Sparling, R., Berlin, A., Levin, D.B. 2011. Biomass pretreatment: Fundamentals toward application. *Biotechnol. Adv*. 29, 675–685. https://doi.org/10.1016/j.biotechadv.2011.05.005

Alvarez-Guzmán, C.L., Balderas-Hernández, V.E., De Leon-Rodriguez, A. 2020. Coproduction of hydrogen, ethanol and 2,3-butanediol from agro-industrial residues by the Antarctic psychrophilic GA0F bacterium. *Int. J. Hydrogen Energ*. 45, 26179–26187. https://doi.org/10.1016/j.ijhydene.2020.02.105

Amraoui, Y., Narisetty, V., Coulon, F., Agrawal, D., Chandel, A.K., Maina, S., Koutinas, A., Kumar, V. 2021. Integrated fermentative production and downstream processing of 2,3-butanediol from sugarcane bagasse-derived xylose by mutant strain of enterobacter ludwigii. *ACS Sustain. Chem. Eng*. 9, 10381–10391. https://doi.org/10.1021/acssuschemeng.1c03951

Amraoui, Y., Prabhu, A.A., Narisetty, V., Coulon, F., Kumar Chandel, A., Willoughby, N., Jacob, S., Koutinas, A., Kumar, V. 2022. Enhanced 2,3-butanediol production by mutant enterobacter ludwigii using brewers' spent grain hydrolysate: process optimization for a pragmatic biorefinery loom. *Chem. Eng. J*. 427, 130851. https://doi.org/10.1016/j.cej.2021.130851

An, Y.-M., Zhuang, J., Li, Y., Dai, J.-Y., Xiu, Z.-L. 2022. Pretreatment of Jerusalem artichoke stalk using hydroxylammonium ionic liquids and their influences on 2,3-butanediol fermentation by Bacillus subtilis. *Bioresour. Technol*. 354, 127219. https://doi.org/10.1016/j.biortech.2022.127219

Anukam, A., Berghel, J. 2021. Biomass pretreatment and characterization: a review. In: T. P. Basso, T. O. Basso, & L. C. Basso (Eds), *Biotechnological Applications of Biomass.* IntechOpen, London, UK. https://doi.org/10.5772/intechopen.93607

Celińska, E., Grajek, W. 2009. Biotechnological production of 2,3-butanediol-current state and prospects. *Biotechnol. Adv.* 27, 715–725. https://doi.org/10.1016/j.biotechadv.2009.05.002

Cha, J.W., Jang, S.H., Kim, Y.J., Chang, Y.K., Jeong, K.J. 2020. Engineering of Klebsiella oxytoca for production of 2,3-butanediol using mixed sugars derived from lignocellulosic hydrolysates. *GCB Bioenerg.* 12, 275–286. https://doi.org/10.1111/gcbb.12674

Chatterjee, C., Pong, F., Sen, A. 2015. Chemical conversion pathways for carbohydrates. *Green Chem.* 17, 40–71. https://doi.org/10.1039/C4GC01062K

Chen, C., Wei, D., Shi, J., Wang, M., Hao, J. 2014. Mechanism of 2,3-butanediol stereoisomer formation in Klebsiella pneumoniae. *Appl. Microbiol. Biotechnol.* 98, 4603–4613. https://doi.org/10.1007/s00253-014-5526-9

Cortivo, P.R.D., Machado, J., Hickert, L.R., Rossi, D.M., Ayub, M.A.Z. 2019. Production of 2,3-butanediol by Klebsiella pneumoniae BLh-1 and Pantoea agglomerans BL1 cultivated in acid and enzymatic hydrolysates of soybean hull. *Biotechnol. Prog.* 35, 1–8. https://doi.org/10.1002/btpr.2793

Dai, J.-J., Cheng, J.-S., Liang, Y.-Q., Jiang, T., Yuan, Y.-J. 2014. Regulation of extracellular oxidoreduction potential enhanced (R,R)-2,3-butanediol production by Paenibacillus polymyxa CJX518. *Bioresour. Technol.* 167, 433–440. https://doi.org/10.1016/j.biortech.2014.06.044

Dai, J.Y., Guan, W.T., Xiu, Z.L. 2020. Bioconversion of inulin to 2,3-butanediol by a newly isolated Klebsiella pneumoniae producing inulinase. *Process Biochem.* 98, 247–253. https://doi.org/10.1016/j.procbio.2020.08.018

Dias, B. do C., Lima, M.E. do N.V., Vollú, R.E., da Mota, F.F., da Silva, A.J.R., de Castro, A.M., Freire, D.M.G., Seldin, L. 2018. 2,3–Butanediol production by the non-pathogenic bacterium Paenibacillus brasilensis. *Appl. Microbiol. Biotechnol.* 102, 8773–8782. https://doi.org/10.1007/s00253-018-9312-y

Duraisamy, K., Ha, A., Kim, J., Park, A.R., Kim, B., Song, C.W., Song, H., Kim, J.C. 2022. Enhancement of disease control efficacy of chemical fungicides combined with plant resistance inducer 2,3-butanediol against turfgrass fungal diseases. *Plant Pathol. J.* 38, 182–193. https://doi.org/10.5423/PPJ.OA.02.2022.0022

FAO. 2022. *Food Outlook - Biannual Report on Global Food Markets.* FAO, Rome. https://doi.org/10.4060/cc2864en

Fernández-Gutiérrez, D., Veillette, M., Ávalos Ramirez, A., Giroir-Fendler, A., Faucheux, N., Heitz, M. 2022. Fermentation of whey and its permeate using a genetically modified strain of Escherichia coli K12 MG1655 to produce 2,3-butanediol. *Environ. Qual. Manag.* 31, 329–345. https://doi.org/10.1002/tqem.21788

Ge, Y., Li, K., Li, L., Gao, C., Zhang, L., Ma, C., Xu, P. 2016. Contracted but effective: production of enantiopure 2,3-butanediol by thermophilic and GRAS: Bacillus licheniformis. *Green Chem.* 18, 4693–4703. https://doi.org/10.1039/c6gc01023g

Gräfje, H., Körnig, W., Weitz, H.-M., Reiß, W., Steffan, G., Diehl, H., Bosche, H., Schneider, K., Kieczka, H., Pinkos, R. 2019. Butanediols, butenediol, and butynediol. In: C. Ley (Ed), *Ullmann's Encyclopedia of Industrial Chemistry.* Wiley-VCH Verlag GmbH & Co. KGaA, Weinheim, Germany, pp. 1–12. https://doi.org/10.1002/14356007.a04_455.pub2

Han, J., Wang, F., Li, Z., Liu, L., Zhang, G., Chen, G., Liu, J., Zhang, H. 2022. Isolation and identification of an osmotolerant Bacillus amyloliquefaciens strain T4 for 2, 3-butanediol production with tobacco waste. *Prep. Biochem. Biotechnol.* 52, 210–217. https://doi.org/10.1080/10826068.2021.1925912

Hazeena, S.H., Nair Salini, C., Sindhu, R., Pandey, A., Binod, P. 2019. Simultaneous saccharification and fermentation of oil palm front for the production of 2,3-butanediol. *Bioresour. Technol.* 278, 145–149. https://doi.org/10.1016/j.biortech.2019.01.042

Hazeena, S.H., Shurpali, N.J., Siljanen, H., Lappalainen, R., Anoop, P., Adarsh, V.P., Sindhu, R., Pandey, A., Binod, P. 2022. Bioprocess development of 2, 3-butanediol production using agro-industrial residues. *Bioprocess Biosyst. Eng.* 45, 1527–1537. https://doi.org/10.1007/s00449-022-02761-5

Hazeena, S.H., Sindhu, R., Pandey, A., Binod, P. 2020. Lignocellulosic bio-refinery approach for microbial 2, 3-butanediol production. *Bioresour. Technol.* 122873. https://doi.org/10.1016/j.biortech.2020.122873

Ji, X., Huang, H., Ouyang, P. 2011. Microbial 2, 3-butanediol production : a state-of-the-art review. *Biotechnol. Adv.* 29, 351–364. https://doi.org/10.1016/j.biotechadv.2011.01.007

Khunnonkwao, P., Jantama, S.S., Jantama, K., Joannis-Cassan, C., Taillandier, P. 2021. Sequential coupling of enzymatic hydrolysis and fermentation platform for high yield and economical production of 2, 3-butanediol from cassava by metabolically engineered *Klebsiella oxytoca. J. Chem. Technol. Biotechnol.* 96, 1292–1301. https://doi.org/10.1002/jctb.6643

Kim, D.G., Yoo, S.W., Kim, M., Ko, J.K., Um, Y., Oh, M.K. 2020. Improved 2,3-butanediol yield and productivity from lignocellulose biomass hydrolysate in metabolically engineered Enterobacter aerogenes. *Bioresour. Technol.* 309, 123386. https://doi.org/10.1016/j.biortech.2020.123386

Kongjan, P., Jariyaboon, R., Reungsang, A., Sittijunda, S. 2021. Co-fermentation of 1,3-propanediol and 2,3-butanediol from crude glycerol derived from the biodiesel production process by newly isolated Enterobacter sp.: Optimization factors affecting. *Bioresour. Technol. Rep.* 13, 100616. https://doi.org/10.1016/j.biteb.2020.100616

Köpke, M., Mihalcea, C., Liew, F.M., Tizard, J.H., Ali, M.S., Conolly, J.J., Al-Sinawi, B., Simpson, S.D. 2011. 2,3–Butanediol production by acetogenic bacteria, an alternative route to chemical synthesis, using industrial waste gas. *Appl. Environ. Microbiol.* 77, 5467–5475. https://doi.org/10.1128/AEM.00355-11

Kuenz, A., Jäger, M., Niemi, H., Kallioinen, M., Mänttäri, M., Prüße, U. 2020. Conversion of xylose from birch hemicellulose hydrolysate to 2,3-butanediol with bacillus vallismortis. *Fermentation* 6, 86. https://doi.org/10.3390/fermentation6030086

Lee, J.W., Bhagwat, S.S., Kuanyshev, N., Cho, Y.B., Sun, L., Lee, Y.-G., Cortés-Peña, Y.R., Li, Y., Rao, C.V, Guest, J.S., Jin, Y. 2023. Rewiring yeast metabolism for producing 2,3-butanediol and two downstream applications: techno-economic analysis and life cycle assessment of methyl ethyl ketone (MEK) and agricultural biostimulant production. *Chem. Eng. J.* 451, 138886. https://doi.org/10.1016/j.cej.2022.138886

Lee, Y.-G., Bae, J.-M., Kim, S.-J. 2022. Enantiopure meso-2,3-butanediol production by metabolically engineered Saccharomyces cerevisiae expressing 2,3-butanediol dehydrogenase from Klebsiella oxytoca. *J. Biotechnol.* 354, 1–9. https://doi.org/10.1016/j.jbiotec.2022.05.001

Lee, Y.-G., Seo, J.-H., 2019. Production of 2,3-butanediol from glucose and cassava hydrolysates by metabolically engineered industrial polyploid Saccharomyces cerevisiae. *Biotechnol. Biofuels.* 12, 204. https://doi.org/10.1186/s13068-019-1545-1

Li, Y.-Q., Wang, M.-J., Gan, X.-F., Luo, C.-B. 2023. Cleaner 2,3-butanediol production from unpretreated lignocellulosic biomass by a newly isolated Klebsiella pneumoniae PX14. *Chem. Eng. J.* 455, 140479. https://doi.org/10.1016/j.cej.2022.140479

Li, Y., Zhu, S., Ge, X. 2019. Co-production of 1,3-propanediol and 2,3-butanediol from waste lard by co-cultivation of pseudomonas alcaligenes and Klebsiella pneumoniae. *Curr. Microbiol.* 76, 415–424. https://doi.org/10.1007/s00284-019-01628-5

Lunze, A., Heyman, B., Chammakhi, Y., Eichhorn, M., Büchs, J., Anders, N., Spiess, A.C. 2021. Investigation of silphium perfoliatum as feedstock for a liquid hot water-based biorefinery process towards 2,3-butanediol. *BioEnergy Res.* 14, 799–814. https://doi.org/10.1007/s12155-020-10194-9

Mailaram, S., Narisetty, V., Ranade, V. V., Kumar, V., Maity, S.K. 2022. Techno-economic analysis for the production of 2,3-butanediol from brewers' spent grain using pinch technology. Ind. Eng. Chem. Res. 61, 2195–2205. https://doi.org/10.1021/acs.iecr.1c04410

Maina, S., Dheskali, E., Papapostolou, H., Castro, A.M. de, Guimaraes Freire, D.M., Nychas, G.J.E., Papanikolaou, S., Kookos, I.K., Koutinas, A. 2021a. Bioprocess development for 2,3-butanediol production from crude glycerol and conceptual process design for aqueous conversion into methyl ethyl ketone. *ACS Sustain. Chem. Eng.* 9, 8692–8705. https://doi.org/10.1021/acssuschemeng.1c00253

Maina, S., Schneider, R., Alexandri, M., Papapostolou, H., Nychas, G.-J., Koutinas, A., Venus, J. 2021b. Volumetric oxygen transfer coefficient as fermentation control parameter to manipulate the production of either acetoin or D-2,3-butanediol using bakery waste. *Bioresour. Technol.* 335, 125155. https://doi.org/10.1016/j.biortech.2021.125155

Maina, S., Stylianou, E., Vogiatzi, E., Vlysidis, A., Mallouchos, A., Nychas, G.J.E., de Castro, A.M., Dheskali, E., Kookos, I.K., Koutinas, A. 2019. Improvement on bioprocess economics for 2,3-butanediol production from very high polarity cane sugar via optimisation of bioreactor operation. *Bioresour. Technol.* 274, 343–352. https://doi.org/10.1016/j.biortech.2018.11.001

Mathewson, S.W. 1980. *The Manual for the Home and Farm Production of Alcohol Fuel.* Ten Speed Press, Emeryville, CA.

Melero, J.A., Iglesias, J., Garcia, A. 2012. Biomass as renewable feedstock in standard refinery units. Feasibility, opportunities and challenges. *Energy Environ. Sci.* 5, 7393. https://doi.org/10.1039/c2ee21231e

Moon, S.K., Chung, J.H., Min, J. 2021. Integrated process development for the efficient and industrial production of 2,3-butanediol and bioethanol using dried cassava chips as commercial biomass. J. *Chem. Technol. Biotechnol.* 96, 3342–3348. https://doi.org/10.1002/jctb.6886

Narisetty, V., Amraoui, Y., Abdullah, A., Ahmad, E., Agrawal, D., Parameswaran, B., Pandey, A., Goel, S., Kumar, V. 2021. High yield recovery of 2,3-butanediol from fermented broth accumulated on xylose rich sugarcane bagasse hydrolysate using aqueous two-phase extraction system. *Bioresour. Technol.* 337, 125463. https://doi.org/10.1016/j.biortech.2021.125463

Narisetty, V., Narisetty, S., Jacob, S., Kumar, D., Leeke, G.A., Chandel, A.K., Singh, V., Srivastava, V.C., Kumar, V. 2022a. Biological production and recovery of 2,3-butanediol using arabinose from sugar beet pulp by Enterobacter ludwigii. *Renew. Energ.* 191, 394–404. https://doi.org/10.1016/j.renene.2022.04.024

Narisetty, V., Zhang, L., Zhang, J., Sze Ki Lin, C., Wah Tong, Y., Loke Show, P., Kant Bhatia, S., Misra, A., Kumar, V. 2022b. Fermentative production of 2,3-Butanediol using bread waste - A green approach for sustainable management of food waste. *Bioresour. Technol.* 358, 127381. https://doi.org/10.1016/j.biortech.2022.127381

OHair, J., Jin, Q., Yu, D., Wu, J., Wang, H., Zhou, S., Huang, H. 2021. Non-sterile fermentation of food waste using thermophilic and alkaliphilic Bacillus licheniformis YNP5-TSU for 2,3-butanediol production. *Waste Manag.* 120, 248–256. https://doi.org/10.1016/j.wasman.2020.11.029

Okonkwo, C.C., Ujor, V., Ezeji, T.C. 2021. Production of 2,3-Butanediol from non-detoxified wheat straw hydrolysate: impact of microbial inhibitors on Paenibacillus polymyxa DSM 365. *Ind. Crops Prod.* 159, 113047. https://doi.org/10.1016/j.indcrop.2020.113047

Ourique, L.J., Rocha, C.C., Gomes, R.C.D., Rossi, D.M., Ayub, M.A.Z., 2020. Bioreactor production of 2,3-butanediol by Pantoea agglomerans using soybean hull acid hydrolysate as substrate. *Bioprocess Biosyst. Eng.* 43, 1689–1701. https://doi.org/10.1007/s00449-020-02362-0

Palaiogeorgou, A.M., Papanikolaou, S., de Castro, A.M., Freire, D.M.G., Kookos, I.K., Koutinas, A.A. 2019. A newly isolated Enterobacter sp. strain produces 2,3-butanediol during its cultivation on low-cost carbohydrate-based substrates. *FEMS Microbiol. Lett.* 366, 1–15. https://doi.org/10.1093/femsle/fny280

Parate, R., Borgave, M., Dharne, M., Rode, C. 2021. Bioglycerol (C3) upgrading to 2,3-butanediol (C4) by cell-free extracts of *Enterobacter aerogenes* NCIM 2695. *J. Chem. Technol. Biotechnol.* 96, 1316–1325. https://doi.org/10.1002/jctb.6650

Pasaye-Anaya, L., Vargas-Tah, A., Martínez-Cámara, C., Castro-Montoya, A.J., Campos-García, J. 2019. Production of 2,3-butanediol by fermentation of enzymatic hydrolysed bagasse from agave mezcal-waste using the native Klebsiella oxytoca UM2-17 strain. *J. Chem. Technol. Biotechnol.* 94, 3915–3923. https://doi.org/10.1002/jctb.6190

Priya, A., Lal, B. 2019. Efficient valorization of waste glycerol to 2,3-butanediol using Enterobacter cloacae TERI BD 18 as a biocatalyst. *Fuel.* 250, 292–305. https://doi.org/10.1016/j.fuel.2019.03.146

Ra, C.H., Seo, J.-H., Jeong, G.-T., Kim, S.-K. 2020. Evaluation of 2,3-Butanediol Production from Red Seaweed Gelidium amansii Hydrolysates Using Engineered Saccharomyces cerevisiae. *J. Microbiol. Biotechnol.* 30, 1912–1918. https://doi.org/10.4014/jmb.2007.07037

Rehman, S., Khairul Islam, M., Khalid Khanzada, N., Kyoungjin An, A., Chaiprapat, S., Leu, S.Y. 2021. Whole sugar 2,3-butanediol fermentation for oil palm empty fruit bunches biorefinery by a newly isolated Klebsiella pneumoniae PM2. *Bioresour. Technol.* 333, 125206. https://doi.org/10.1016/j.biortech.2021.125206

Ripoll, V., Ladero, M., Santos, V.E. 2021. Kinetic modelling of 2,3-butanediol production by Raoultella terrigena CECT 4519 resting cells: Effect of fluid dynamics conditions and initial glycerol concentration. *Biochem. Eng. J.* 176, 108185. https://doi.org/10.1016/j.bej.2021.108185

Ripoll, V., Rodríguez, A., Ladero, M., Santos, V.E. 2020. High 2,3-butanediol production from glycerol by Raoultella terrigena CECT 4519. *Bioprocess Biosyst. Eng.* 43, 685–692. https://doi.org/10.1007/s00449-019-02266-8

Sathesh-Prabu, C., Kim, D., Lee, S.K. 2020. Metabolic engineering of Escherichia coli for 2,3-butanediol production from cellulosic biomass by using glucose-inducible gene expression system. *Bioresour. Technol.* 309, 123361. https://doi.org/10.1016/j.biortech.2020.123361

Sindhu, R., Manju, A., Mohan, P., Rajesh, R.O., Madhavan, A., Arun, K.B., Hazeena, S.H., Mohandas, A., Rajamani, S.P., Puthiyamadam, A., Binod, P., Reshmy, R. 2020. Valorization of food and kitchen waste: an integrated strategy adopted for the production of poly-3-hydroxybutyrate, bioethanol, pectinase and 2, 3–butanediol. *Bioresour. Technol.* 310, 123515. https://doi.org/10.1016/j.biortech.2020.123515

Tinôco, D. 2022. Biotechnology development of bioethanol from sweet sorghum bagasse. In: A. Y. H. Almanza, N. Balagurusamy, H. R. Leza, & C. N. Aguilar (Eds), *Bioethanol.* Apple Academic Press, Boca Raton, FL, pp. 339–370. https://doi.org/10.1201/9781003277132-12

Tinôco, D., Borschiver, S., Coutinho, P.L., Freire, D.M.G. 2020. Technological development of the bio-based 2,3-butanediol process. *Biofuels, Bioprod. Bioref.* 2, 1–20. https://doi.org/10.1002/bbb.2173

Tinôco, D., de Castro, A.M., Seldin, L., Freire, D.M.G. 2021a. Production of (2R,3R)-butanediol by Paenibacillus polymyxa PM 3605 from crude glycerol supplemented with sugarcane molasses. *Process Biochem.* 106, 88–95. https://doi.org/10.1016/j.procbio.2021.03.030

Tinôco, D., Pateraki, C., Koutinas, A.A., Freire, D.M.G. 2021b. Bioprocess development for 2,3-butanediol production by paenibacillus strains. *ChemBioEng Rev.* 8, 1–20. https://doi.org/10.1002/cben.202000022

Transparency Market Research. 2022. Global 2,3-Butanediol Market 2019-2030 [WWW Document]. URL https://www.transparencymarketresearch.com/2-3-butanediol-market.html (accessed 8.5.22).

UNO. 2022. The sustainable development goals report 2022. United Nations Publ. issued by Dep. Econ. Soc. Aff. 64.

Wang, D., Oh, B.R., Lee, S., Kim, D.H., Joe, M.H. 2021. Process optimization for mass production of 2,3-butanediol by Bacillus subtilis CS13. *Biotechnol. Biofuels*. 14, 1–11. https://doi.org/10.1186/s13068-020-01859-w

Xie, S., Li, Z., Zhu, G., Song, W., Yi, C. 2022. Cleaner production and downstream processing of bio-based 2,3-butanediol: a review. *J. Clean. Prod.* 343, 131033. https://doi.org/10.1016/j.jclepro.2022.131033

Yang, Y., Deng, T., Cao, W., Shen, F., Liu, S., Zhang, J., Liang, X., Wan, Y. 2022. Effectively converting cane molasses into 2,3-butanediol using clostridium ljungdahlii by an integrated fermentation and membrane separation process. *Molecules*. 27, 954. https://doi.org/10.3390/molecules27030954

Yu, D., O'Hair, J., Poe, N., Jin, Q., Pinton, S., He, Y., Huang, H. 2022. Conversion of food waste into 2,3-butanediol via thermophilic fermentation: effects of carbohydrate content and nutrient supplementation. *Foods*. 11, 169. https://doi.org/10.3390/foods11020169

8 Application of Enzymatic Technique for the Synthesis of Bioalcohols From Different Agrobiomasses

Shraddha M. Jadhav, Poulami Mukherjee, Karan Kumar, and Vijayanand S. Moholkar*

8.1 INTRODUCTION

Enzymatic hydrolysis (EH) is one of the long-standing buzz-word in the happening field of the bioalcohol industry. A quick look into the role of bioalcohol in the fuel industry is enough to understand the hype behind the EH. Samuel Morey used ethanol, the most popular bioalcohol to date, in a combustion engine prototype for the first time in 1827 (Bowyer et al., 2017). The usage of ethanol-blended petroleum-based fuels, such as gasoline and kerosene, was popular, even during World War II. Large automobile companies used to produce engines on large scales to be fueled by such ethanol-blended petrol (EBP) till the 1950s. Post-1950, oil-producing companies started producing enough fossil fuels to maintain a cheap and stable supply to the fuel industry. During 1950–1960, fossil fuels became economically viable and showed better performance compared to EBP. Automobile companies started opting for petrol-based engines over EBP-based engines, and the usage of ethanol in the fuel industry has faded away. During the geopolitical and economic instability in the 1970s, ethanol-based engines saw a briefly renewed interest, particularly in Brazil, but they phased out quickly with the discovery of large petroleum reserves in Brazil (Bowyer et al., 2017; Cho et al., 2019). The moral of the story is that ethanol (bioalcohol in general) always stood second to petroleum-based fuels in the history of the fuel industry for two main reasons:

a. Lower cost of production and easy availability of fossil fuels.
b. Higher energy conversion efficiency of fossil fuels (Said et al., 2021).

The use of ethanol as an alternate energy source became popular only when the supply of fossil fuels was disrupted due to economic or geopolitical crises like

DOI: 10.1201/9781003407713-8

war (Lamichhane et al., 2021). The ever-depleting sources of non-renewable fossil fuels, however, cannot match the ever-increasing demands of energy in our society, and Governments across the countries have started making policy shifts toward alternative sources like bioalcohols to complement fossil fuels. The usage of bioalcohols has started to become popular with the Government policy shifts (see Figure 8.1 for year-wise bioalcohol usage in the United States). To make bioalcohol popular and economically viable in the worldwide fuel industry, even without Government intervention, an efficient, low-cost production scheme should be advocated. Developing such a scheme would require understanding five key steps (see Figure 8.2) in bioalcohol production. The food and beverage industries have long developed steps associated with fermentation and distillation. In these industries, food crops are primarily used to produce alcohol (Kumar et al., 2022, 2024a, 2024b, 2024c). Using food grains to make bioalcohol threatens food security (Acharya et al., 2021). Moreover, the competition between the food market and the fuel industry increases costs of biomass resources and incurs costs in bioalcohol production (Malani et al., 2019; Kumar et al., 2021, 2022). To avoid such competition, researchers started focusing on alternate sources, such as agro/municipality wastes, algae, and engineered crops, to reduce the direct use of food grains. These alternative sources are heterogeneous and contain a large percentage of hemicellulose. The hydrolysis of these alternative biomass sources usually needs high use of chemicals. These chemicals often go down the line in the production pathway and affect the performances of fermenting microbes. Moreover, enzymatic hydrolysis (EH) of hemicellulose generates many inhibitors and decreases production yields. This bleak picture of the bioalcohol industry started changing in 1980 with the discovery of β-glucosidase in *Trichoderma reesei* Rut C30 (Zhang et al., 2022). Furthermore, the usage of enzymes to remove inhibitors does not leave toxic material in the production pathway. Post this discovery, the bioalcohol product has seen exponential growth (See Figure 8.2 for illustration). A major workforce since then has been directed toward finding and inventing new enzymes with higher activity, enhanced selectivity, and broader scopes in terms of substrates.

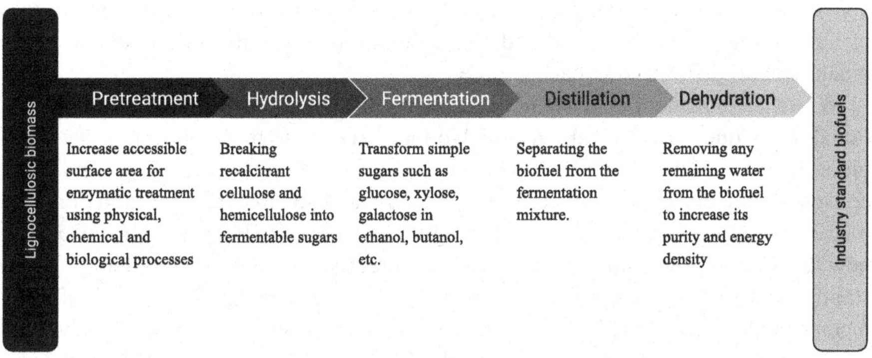

FIGURE 8.1 Steps to produce bioalcohols from lignocellulosic biomass.

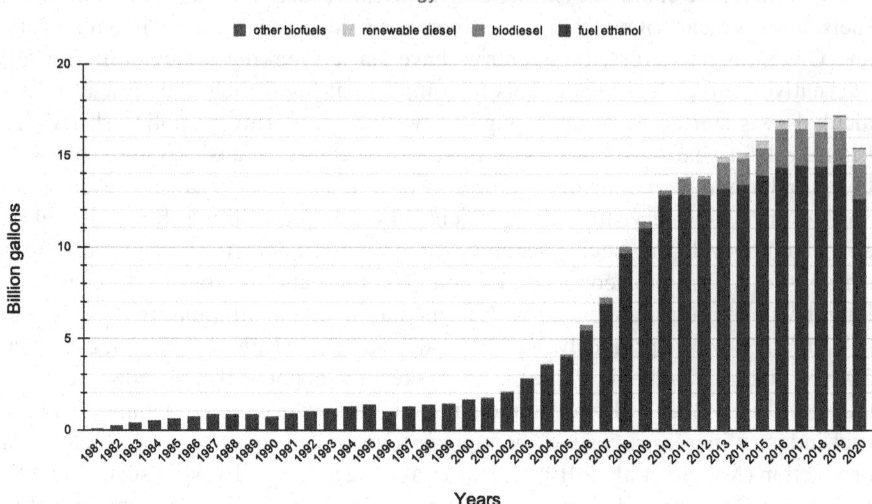

FIGURE 8.2 Biofuel production in the United States (data acquired from Total Energy Monthly Data – U.S. Energy Information Administration (EIA), April 2022, Tables 10.3 and 10.4a–c). We see fast exponential growth in bioalcohol production till 2011. Bioalcohol (mainly ethanol) production in the United States primarily uses maize. After the food crisis in 2009 (Shane et al., 2009), researchers started looking into alternate biomass sources.

8.1.1 Aim and Scope of this Chapter

The current chapter focuses on applications of different enzymatic techniques in bioalcohol production, and it is structured in the following way to convey the development and the current status of enzymatic hydrolysis in the industry. The next section will first discuss three main components of lignocellulosic biomasses: cellulose, hemicellulose, and lignin. Understanding their chemical properties and roles in bioalcohol production would be essential in understanding the current advancements and frontiers in enzymatic bioalcohol research. We will discuss different sources of lignocelluloses, both natural and engineered, and the challenges associated with bioalcohol production from each source.

While advancement in enzyme engineering minimizes the roles of other conventional techniques, all current, sustainable methods in the bioalcohol industry combine other techniques with enzymatic treatments. In Section 8.3, we discuss other conventional techniques, such as acid hydrolysis and steam explosion, and compare them with EH. In Section 8.4, we go through recent trends in EH research and explain how different types of enzymes work on different substrates. To completely understand the current research directions in this field, we go through the major inhibiting factors and bottlenecks in EH methods. Section 8.5 focuses on genetic and enzymatic engineering methods to address those inhibiting factors. Finally, in Section 8.6, we wrap this chapter with a brief discussion on the current status and future perspectives of enzyme hydrolysis in the bioalcohol industry.

8.2 TYPES OF AGROBIOMASSES AND THEIR COMPOSITIONS

Agrobiomass and lignocellulosic sources are plant materials that can be used to produce bioalcohols (Kumar et al., 2022). These materials typically contain a high amount of lignocellulose. Lignocellulosic biomass is mainly composed of three things: cellulose (30%–50%), hemicellulose (20%–35%), and lignin (10%–30%) (Wei et al., 2017). The chemical properties of these three components are described in this section.

a. **Cellulose** constitutes 30%–50% of lignocellulose. It is a linear polymer of glucose units connected by β-1,4 glycosidic bonds. Cellulose provides structural support to the plant and gives biomass its mechanical strength. For bioalcohol production, however, the recalcitrant structure of cellulose can make it difficult to break down into its constituent sugars. This step stands as a major challenge in bioalcohol production. Various physical, chemical, and biological pretreatment methods (Meenakshisundaram et al., 2022) are adopted to break down the complex structure of cellulose and make it more accessible to enzymes. In recent years, EH has gained more popularity in the bioalcohol industry, but to date, the pretreatment of cellulose still plays a significant role in improving the efficiency of EH techniques.
b. **Hemicelluloses** belong to the heterogeneous polysaccharides family and generally consist of a matrix, made of a complex mixture of hexoses (glucose, mannose, and galactose) and pentoses (xylose and arabinose) (Wei et al., 2017). The exact chemical composition of hemicellulose varies from source to source. In wood biomasses, hemicelluloses are the second most abundant components (20%–35%) after cellulose. The hemicellulosic matrix binds the cellulose and lignin together, providing additional structural and mechanical support to the plant (Wei et al., 2017). The usage of hemicellulose in bioalcohol production has a potential role in cost reduction in the fuel industry. Due to its less compact structure, hemicellulose is easier to hydrolyze than cellulose, and it can be used as a complementary feedstock to increase bioalcohol yield. In the current biosynthesis industry, bioalcohol production is mainly geared toward food-based feedstocks, such as sugarcane and corn (Khoshkho et al., 2022). Producing bioalcohols from hemicellulose can shift focus toward other wood-based resources and reduce the competition for food-based feedstocks. The usage of hemicellulose in bioalcohol production, however, is a relatively new field of research. Chemical heterogeneity of hemicellulose poses several challenges in the pretreatment part of bioalcohol production, and degradation of hemicellulose can produce inhibitors affecting fermentation pathways (Zhai et al., 2018).
c. **Lignin** is a complex heterogeneous amorphous polymer that primarily consists of coniferyl, p-coumaryl, sinapyl alcohols, and other phenylpropanoid units. Lignins are generally highly branched and are usually characterized by their crosslinked structures (Zhang & Naebe, 2021a). Lignins, along with the hemicellulosic matrix, provide additional mechanical strength to

the plant. Lignin constitutes 10%–30% (of dry weight) of lignocellulose (Wei et al., 2017), but it is generally considered a side product of bioalcohol production rather than a feedstock due to its low energy content and chemical complexity. Its hydrophobic structural features and hydrogen bonding in lignin affect the accessibility of the enzymes to cellulose. Lignin also reduces enzyme activities by adsorbing cellulases irreversibly during EH (Janusz et al., 2017). Despite being traditionally viewed as a side product and inhibitor of bioalcohol production, lignin has different potential usage in developing renewable biomaterials. It can be broken down into smaller chemical compounds to be used as feedstocks for the production of adhesives (resins) (Li et al., 2018), bisphenol (Koelewijn et al., 2017) carbon fibers, bioplastics, antioxidants, etc. Lignin can be used as a soil amendment to improve soil quality, water retention, and nutrient uptake.

In addition to cellulose, hemicellulose, and lignin, lignocellulosic biomass may comprise many different constituents like extractives such as fats, waxes, terpenes, minerals, and ash. The exact composition of lignocellulose can fluctuate depending on various factors like the plant's age, the kind of biomass, the growth location, and environmental conditions.

There are several sources of agrobiomass, but below are some important classifications of agrobiomass:

a. **Energy crops** such as corn, beans, and sugarcane were used to produce bioalcohols. These bioalcohols are termed first-generation bioalcohols and compete with food sources and push up the price of grain expenses. Grains, which are used to make bioalcohols to fill a regular family car twice, are sufficient to nourish a youngster for an entire year (Malode et al., 2021). Energy crops, such as switchgrass, miscanthus, and sorghum, are specifically grown for energy production. These crops have a high biomass yield and can be grown on marginal land, making them a sustainable alternative to traditional food crops (Carrillo-Nieves et al., 2019).
b. **Agricultural residues** are the by-products of crops such as straw, corn stover, rice husks, and sugarcane bagasse. Agricultural residues are typically abundant and low cost, making them an attractive feedstock for bioconversion. Production and storage of these agricultural residues often utilize a large land mass (Verma et al., 2023). The usage of agricultural residues also invokes socioeconomic and environmental problems. In bioalcohol-delivering nations, such as African or Latin American countries, lots of forests are eliminated for bioalcohol production (Chisti, 2013).
c. **Forestry residues** include materials such as bark, sawdust, and wood chips that are generated during the processing of timber. These materials are typically abundant in areas with high levels of forest activity and can be used as a feedstock for bioconversion (Goswami et al., 2023).
d. **Municipal solid waste**, especially, the organic portion of municipal solid waste, such as food and yard waste, can be processed through anaerobic decomposition to generate biogas or compost (Sipra et al., 2018).

e. **Algae**, Marine algae, seaweed, and other aquatic plants are very promising biomass sources. While second-generation bioalcohols from agricultural, forestry, and municipal wastes are the most cost-effective biosynthetic pathways, Marine algae are considered to be primary resources for third-generation bioalcohols (Singh et al., 2022). The fast and sustainable growth of algae solves problems regarding the storage and transport of biomass. Bioprocessing of algae is, however, comparatively costlier than fossil fuels (Chisti, 2013).

8.3 METHODS TO HYDROLYZE AGROBIOMASS

Agrobiomass, abundant in the form of crop residues, wood chips, and grasses, plays a significant role in this transition. Lignin, present in lignocellulosic materials, acts as a protective barrier against the degradation of plant cell structures by fungi and bacteria, hindering their conversion into fuel. To transform biomass into bioalcohols, it is essential to break down cellulose and hemicellulose into their respective monomers (sugars) (Gandla et al., 2019). Various physicochemical, structural, and compositional factors impede the digestibility of cellulose within lignocellulosic biomass. To expose the cellulose within plant fibers for efficient conversion to fuel, biomass requires pretreatment. This pretreatment involves a range of techniques (Figure 8.3), including ammonia fiber explosion, chemical treatment, biological treatment, and steam explosion, which modify the structure of cellulosic biomass to enhance cellulose accessibility (Zhai et al., 2018; Liu & Chen, 2015; Meenakshisundaram et al., 2022). While Figure 8.1 provides a schematic representation of the biomass-to-fuel conversion process, it encompasses the hydrolysis of various components within lignocellulosic materials to yield fermentable reducing sugars and their subsequent fermentation into fuels like ethanol and butanol.

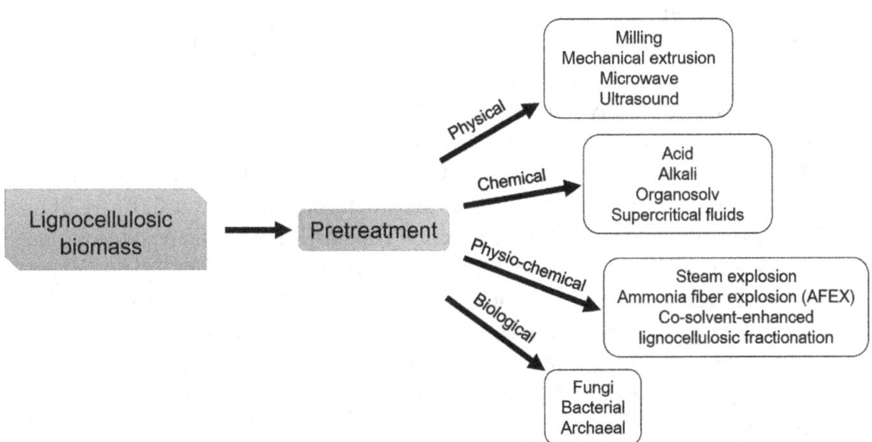

FIGURE 8.3 Types of pretreatment methods.

8.3.1 Conventional Techniques to Hydrolyze Agrobiomass

Hydrolysis is a key process in the conversion of agrobiomass into usable products. After partial removal and hydrolysis of lignin and hemicellulose, the remaining cellulose is further hydrolyzed to release a significant fraction of sugars. This process, known as saccharification, generates many high-quality saccharides that can be readily used as the substrate for the H_2-producing bacteria (Amândio et al., 2023). The hydrolysis process can be conducted by various methods, which rely mainly on physical, chemical, or biological methods. Some of the conventional techniques are listed below:

a. **Acid hydrolysis** stands as one of the most frequently employed techniques for breaking down agrobiomass. This method involves subjecting biomass to various acids, such as H_2SO_4 and HCl, resulting in the production of xylose, arabinose, glucose, and acetic acid. This reaction is achieved through the cleavage of the β-1,4 linkages found within the glucose or xylose monomers, liberating acetyl groups and other products residing in the cellulose and hemicellulose components of biomass (Lee & Jeffries, 2011).

   Typically, acid hydrolysis is conducted at elevated temperatures and pressures to facilitate the disintegration of cellulose and hemicellulose into simpler sugars. The dilute acid process, for instance, necessitates temperatures between 120°C and 200°C and high pressures ranging from 15 to 75 psi, with the reaction extending from 30 minutes to 2 hours, thus allowing for continuous processing. In dilute acid hydrolysis, the feedstock is usually pressed before undergoing pretreatment, either through dewatering presses or compression screw feeders available in various commercially accessible continuous hydrolysis reactors (Roy et al., 2020). Conversely, the concentrated acid process has proven to be more fruitful, yielding higher quantities of sugar. This method frequently involves the use of 60%–90% sulfuric acid, employing milder temperatures and moderate pressures by conveying materials from one vessel to another for efficient hydrolysis. Additionally, other acids like nitric acid, hydrochloric acid, and phosphoric acid have been subject to testing (Lamichhane et al., 2021).

   In a distinct study conducted by Pu et al. (2013), various pretreatment experiments were executed using different sulfuric acid concentrations (2%, 4%, and 6%) and temperatures (80°C, 100°C, and 120°C). The optimal conditions for pretreating corn stover were determined to be a 2.0% H_2SO_4 concentration and a 43-minute reaction time at 120°C. This resulted in an impressive 77% xylose yield, albeit with a lower 8.4% glucose yield (Zhao et al., 2012). Notably, the treated solid phase displayed a high susceptibility to EH, generating solutions containing up to 42.1 g of glucose per 100 g of substrate, representing a 70% conversion yield under these optimized conditions. However, acidic hydrolysis alone has several challenges, such as acid recovery in practice to make the process economically feasible (Oriez et al., 2020).

b. **Steam explosion** occurs when the agrobiomass is heated under pressure to high temperatures and then rapidly depressurized. The parameters affecting this technique are particle size, temperature, resident time, etc. (Liu & Chen, 2015). Recent studies have investigated hydrothermal pretreatment via steam explosion (Ruiz et al., 2013; Liu et al., 2013). They found that varying reaction times and temperatures (160°C–230°C, 1–30 minutes) influenced hemicellulose solubilization and de-acetylation, while slight lignin removal and modification with a mild increase in crystallinity were observed, making this method valuable for biomass conversion in industrial contexts. The use of acid is also explored during this process which increases cellulose digestion, even though this invites some disadvantages like hemicellulose degradation and the generation of some toxic compounds like furfural and 5-hydroxymethyl furfural (Ostadjoo et al., 2019; Adeboye et al., 2014). Generation of these compounds, therefore, induces additional costs for removing such compounds.
c. **Alkaline hydrolysis** involves using strong bases (e.g., sodium hydroxide) to break down the biomass into simple sugars. Generally, alkaline hydrolysis is slow and necessitates neutralization and retrieval of the added alkali. The OH^- ion attacks the anomeric carbon atom during this process and cleaves the ether bridge. After that, the uptake of water molecules and the liberation of hydroxyl radicals lead to glucose formation. Treatment of a dilute NaOH on lignocellulosic biomass results in swelling, which expands the internal surface area. This process also reduces crystallinity, separates structural connections between lignin and carbohydrates, and breaks down the lignin structure (Oriez et al., 2020; Fuertez-Córdoba et al., 2021). Acquiring a substantial amount of sugar through this technique is arduous since mono- and dimeric carbohydrates, such as glucose, fructose, or cellobiose, are profoundly affected by alkalis at temperatures lower than 100°C. Also, it generates organic acids, which also lead to alkali consumption. Alkaline hydrolysis is better suited for herbaceous crops and agricultural waste rather than woody biomass due to the latter's higher lignin content (Oriez et al., 2020; Khaire et al., 2021).
d. **Organosolv** is a technique that involves using organic or aqueous solvents such as ethanol, acetone, ethylene glycol, and tetrahydrofurfuryl alcohol, methanol, to dissolve the lignin and break down cellulose and hemicellulose into simple sugars (Amiri & Karimi, 2016; González-Peñas et al., 2018). The extent of hemicellulose solubilization can vary depending on the operational conditions employed. Key parameters like temperature, ranging from 90°C to 250°C, and reaction time, spanning from 30 to 60 minutes, may be adjusted according to the specific process being utilized. This method primarily influences the degree of polymerization and crystallinity, two critical factors to be taken into consideration in the realm of biomass conversion (Xin et al., 2019; Kaschuk & Frollini, 2018). The resulting sugar solution is then fermented to produce bioalcohols. Solvents must be separated using appropriate techniques like evaporation and condensation because they may

hinder EH and fermentative microorganisms. Additionally, the expensive cost of solvents is a crucial factor to consider for industrial use.

8.3.2 Major Challenges in Conventional Methods

Conventional methods for hydrolyzing agrobiomass have been long developed. However, several challenges need to be addressed to make these methods more efficient and economically viable. Some of the major challenges are as follows:

a. **High energy and resource consumption:** Conventional methods of hydrolysis often require high temperatures, high pressure, and the use of strong acids or bases. These conditions are generally energy-intensive and require large amounts of water and other resources, making the process expensive and environmentally unsustainable. High reaction temperature, low recyclability, waste treatment, and vessel corrosion are examples of these conditions (Goswami et al., 2023). Also, some inhibitory compounds like furfural and phenolic compounds, methylfurfural, acetic acid, formic acid, and levulinic acid are released, which hamper fermentation (Tiwari & Baghela, 2020).
b. **Side reaction:** Sometimes harsh reactions can give rise to a variety of side reactions, such as overhydrolysis. During this, the target molecule is completely destroyed. Also, racemization can be observed, where the stereochemistry of the product is altered. Concentrated acid serves to disrupt the hydrogen bonding between cellulose chains, causing it to transition into a fully amorphous state. Once the decrystallization of cellulose occurs, it undergoes a transformation, forming a uniform gel-like substance with the acid (Zhang et al., 2022; Galbe & Wallberg, 2019). Conventional techniques can also cause desired product degradation or decrease the quality of bioalcohols.
c. **Incomplete hydrolysis:** Incomplete hydrolysis can be a challenge, particularly when dealing with certain agrobiomass types like lignocellulosic materials, which resist complete breakdown using conventional methods. Such incomplete hydrolysis can lead to increased energy and resource demands within the process, as additional steps may be necessary for the recovery and utilization of unconverted biomass (Oriez et al., 2020). Moreover, it can pose obstacles to downstream fermentation or conversion processes, as the presence of unhydrolyzed components may impede the activity of microorganisms or catalysts (Baral & Shah, 2014; Rasmussen et al., 2017). Ultimately, this results in a reduced yield of simple sugars, negatively impacting the economic feasibility of the entire process.
d. **Corrosion and safety concerns:** The use of strong acids and bases in conventional methods can cause corrosion of equipment and pose safety hazards for operators. This can increase the process's cost and pose risks to workers (Chukwuma et al., 2021a; Galbe & Wallberg, 2019).

8.3.3 EH Technologies and their Advantages Over Conventional Methods

EH technologies refer to using enzymes to break down complex biopolymers, such as cellulose and hemicellulose, into oligomers. These oligomers are further hydrolyzed into mono-/disaccharides. EH is advantageous over conventional methods on several fronts, but it is most effective when preceded by pretreatment procedures to make the substrate more accessible for enzymatic activities (Ostadjoo et al., 2019; Gandla et al., 2018). The following pointers often drive the immense potential and increasing popularity of EH over conventional methods.

Specificity: Enzyme activities are particular, and they can break down complex biopolymers into their constituent sugars without damaging the sugars or other valuable compounds. The specificity of enzymes can be increased by modifying both the enzyme and the substrate (Anand et al., 2024; Weiss et al., 2019). We will discuss this in detail while discussing different classes of cellulolytic enzymes.

High efficiency: Enzymes can work under mild conditions (temp 25°C–60°C, pH 4–5) and achieve high conversion rates in a relatively short time (Saha et al., 2017). This reduces the amount of energy and resources required for biomass conversion.

Reduced environmental impact: The environmental impact of the conversion process can be reduced, and its sustainability can be improved using EH techniques (Amândio et al., 2023).

Reduced cost: EH can be carried out using lower-cost, non-corrosive equipment and can reduce the overall cost of the process. Cost of the enzymes can further be reduced by using enzyme consortia instead of developing single enzyme (Brondi et al., 2019).

Improved product quality: EH can produce high-quality sugars and other valuable compounds with fewer impurities (Ostadjoo et al., 2019).

Recent literature has emphasized the development of novel enzymes and enzyme systems with enhanced properties, such as increased activity, stability, and specificity, as a critical area of research in EH. These advancements hold the potential to improve the efficiency of EH technologies and reduce their cost, thereby offering distinct advantages over conventional methods (Pu et al., 2013; Lopes et al., 2018; Yuan et al., 2021; Battısta & Bolzonella, 2018).

Studies have highlighted the molecular structure basis for biomass recalcitrance during dilute acid and hydrothermal pretreatments, shedding light on the structural transformation of major biomass biopolymers related to the reduction of recalcitrance (Pu et al., 2013). Additionally, the development of enzymatic cocktails for lignocellulose breakdown has been a subject of review, reflecting the increasing attention given to alternative energy sources (Lopes et al., 2018). Furthermore, recent research has focused on understanding the effects of lignin structural characteristics on EH, providing new insights into improving EH efficiency (Yuan et al., 2021). Moreover, recent literature has explored the kinetics of enzymatic high-solid hydrolysis of LB, motivated by industrial relevance and growing interest in high-solid hydrolysis. Additionally, recent studies have addressed critical aspects of EH at high dry-matter content, contributing to a comprehensive understanding of this area (Battısta & Bolzonella, 2018).

8.4 RECENT TRENDS IN VARIOUS EH PATHWAYS FOR BIOALCOHOL SYNTHESIS

EH pathways are a key research area in bioalcohol synthesis. Recent trends in various EH pathways of bioalcohol synthesis include the study of multi-enzyme systems (to increase substrate diversity) (Brondi et al., 2019), immobilization of enzymes (to increase enzyme recyclability) (Califano et al., 2021), integration of enzyme hydrolysis with fermentation (to streamline bioalcohol production pathways) (Martínez-Avila et al., 2021), and development of novel, industrial grade enzymes (Ejaz et al., 2021). We will first discuss the scope of research in each of these fields and later discuss different types of enzymes used in the bioalcohol industry.

a. **Multi-enzyme systems:** The compositions and structures of celluloses and hemicelluloses, in particular, vary significantly with sources. Traditional EH of these materials usually removes hemicelluloses by physical and chemical pretreatment processes before EH. In this procedure, we typically lose one-third of the bioalcohol resources. Researchers are exploring with multi-enzyme systems that can work synergistically to efficiently break down complex biopolymers into their constituent sugars (Abu et al., 2015; Brondi et al., 2019). These multi-enzyme systems can include a combination of cellulases, hemicellulases, and other debranching enzymes such as acetyl xylan esterase, β-mannosidases, α-galactosidases, α-arabinofuranosidase, and α-glucuronidase.
b. **Immobilization of enzymes:** Immobilization of enzymes on solid supports, such as scratch, pectin, gelatin, zeolites, magnetic beads, and nanoparticles, is gaining popularity in bioalcohol research (Califano et al., 2021). Compared to free enzymes, immobilized enzymes show lower activity due to less access to substrate surfaces. The attachment of enzymes to the solid surface can increase their stability and reusability in the hydrolysis process. The heterogeneity of the immobilized enzymes allows easy alteration of the bioreactor design and modular study of the enzyme action on the substrate. Several methods are used for enzyme immobilization, including physical adsorption, covalent binding, entrapment to support matrices, such as gel or membrane, and *crosslinking* to enzyme–support complexes (Palma et al., 2020; Homaei et al., 2013).
c. **Integration with fermentation:** Researchers are exploring the integration of EH with fermentation to produce bioalcohols in a single step. Enzymes break down the complex polysaccharides in the biomass into simple sugars, which microorganisms can ferment to produce bioalcohol. The enzymes used in this step typically include a mixture of cellulases, hemicellulases, and accessory enzymes, which efficiently convert the biomass into fermentable sugars (Martínez-Avila et al., 2021; Jin et al., 2020). The fermentation step usually uses yeasts or other microorganisms capable of converting the sugars into ethanol. It can reduce the overall production time and is associated with separate steps related to hydrolysis and fermentation (Martínez-Avila et al., 2021). Additionally, simultaneous hydrolysis and

fermentation can increase ethanol production yield from the same amount of biomass, boosting the economic viability of bioalcohol production. However, there are also challenges associated, including the need to control the pH and temperature of the process carefully, as well as the potential for microbial contamination.

d. **Development of novel enzymes:** Developing novel enzymes with higher activity, specificity, and stability is key to a more efficient and cost-reduced hydrolysis process. The goal is to identify and engineer more efficient, stable, and specific enzymes toward complex biomolecules, such as cellulose, hemicellulose, and lignin (as side products). One approach to developing novel enzymes is to screen microbial and other natural sources for enzymes that have properties that are desirable for hydrolysis (Zhao et al., 2021; Yuan et al., 2021). For example, microbes (bacteria, fungi, etc.) that can grow on plants may secrete enzymes adapted to break down the complex polysaccharides found in biomass. Another approach is to use directed evolution techniques to modify existing enzymes to enhance their activity, specificity, or stability (Packer & Liu, 2015a; Ejaz et al., 2021). This involves generating random mutations in the enzyme genes and then selecting variants with improved properties. Over time, this can lead to the development of enzymes with optimized properties for a given hydrolysis application. Recent advances in protein engineering have enabled the design and synthesis of engineered enzymes with tailored properties. Enzymes can be designed using computational tools to optimize their catalytic activity, specificity, and stability. Synthetic biology approaches, such as DNA synthesis and genome editing, can also be used to engineer enzymes with specific properties (Vasina et al., 2022; Ranganathan et al., 2022).

e. **Use of non-conventional substrates:** Researchers are exploring non-conventional substrates (Weiss et al., 2019), such as lignocellulosic biomass and industrial by-products, as feedstocks for EH to produce fourth-generation bioalcohols. This can increase the availability of bioalcohol sources (Abdullah et al., 2019) and reduce the environmental impact of the conversion pathways. Fourth-generation bioalcohols, however, also carry certain risks and challenges. Risks and apprehensions are associated with this technology on the environmental and economic fronts. The large-scale production of algae or other feedstocks used for fourth-generation bioalcohols can potentially have negative environmental impacts, such as eutrophication, algal blooms, and water pollution (Cheng et al., 2015; Kim et al., 2012; Ajala et al., 2022). The use of land for feedstock production can lead to land use change, which can have negative impacts on biodiversity, ecosystem services, and food security. The production of fourth-generation bioalcohols may require significant investment, and economic risks may be associated with developing and scaling up these technologies. It is important to adopt strategies that can mitigate these risks and ensure the sustainable development and deployment of these technologies (Abdullah et al., 2019).

8.4.1 Different Cellulolytic Enzymes for Bioalcohol Synthesis

Lignocellulosic biomass primarily contains three components: cellulose, hemicellulose, and lignin. Because enzymes are particular toward their substrates, we need enzymes that are broadly classified into three blocks: cellulases, hemicellulases, and ligninolytic enzymes.

a. **Cellulases** break down cellulose into smaller, soluble sugars, such as glucose, cellobiose, and cellotriose, which can then be utilized as a substrate for bioalcohols. A variety of microbes, such as bacteria, fungi, and some protozoa, produce cellulases (Ranganathan et al., 2022; Barbosa et al., 2020; Kumar et al., 2023). There are three main types of cellulases that cleave 1,4-β-glycosidic bonds: endoglucanases, exoglucanases, and β-glucosidases. Moreover, oxidative cellulases hydrolyze cellulose via oxidation, whereas phosphorylases reduce cellobiose using phosphates.

   Endoglucanases are 1,4-β-D-glucan-4-glucan-hydrolases that can break down the internal bonds within the cellulose chains by attacking the amorphous region, creating smaller cellulose fragments and fibrillation. By striking the crystalline areas of the cellulose structure, endoglucanases produce new chain ends accessible for oxidative cellulase activity (Tong et al., 2023).

   Exoglucanases attack the cellulose chains at their ends, cleaving 1,4-β-D-glycosidic linkages in cello-oligosaccharides to release glucose and other simple sugars. There are two main types of exoglucanases: cellobiohydrolases (1,4-β-D-glucan cellobiohydrolases) and cellodextrinases (1,4-β-D-glucan glucanohydrolases). Cellodextrinases produce D-glucose from cellodextrins and cellulose (Fan et al., 2021). Cellobiohydrolases release D-cellobiose from cellulose chain ends made by endoglucanases (Fan et al., 2021; Wang et al., 2023). Cellobiose, however, strongly inhibits endoglucanases and cellobiohydrolases (Barbosa et al., 2020)

b. **Cellulose oxidases**, such as cellobiose dehydrogenase, polysaccharide monooxygenases, and lactonase, are essential in aerobic microbes to initiate cellulose degradation (Forsberg et al., 2011). Cellobiose dehydrogenase and lactonase together remove cellobioses to increase end product yield. Phosphorylases also remove cellobiose to facilitate EH.

c. **Hemicellulases** are a group of enzymes that hydrolyze hemicellulose (Bhatia et al., 2021), a complex polysaccharide found in plant cell walls along with cellulose and lignin. Hemicellulose is composed of different monosaccharide units, including xylose, arabinose, mannose, and galactose. Different types of hemicellulases exist, including xylanases, mannanases, arabinofuranosidases, and galactanases (Kaschuk & Frollini, 2018).

   Xylanases are enzymes that break down the 1,4-β-xylosidic linkages in hemicellulose xylan, which is composed of xylose units (Bajpai, 2014; Zhang et al., 2022). They can be endo-xylanases, exo-xylanases, or β-xylosidases. Endo-xylanases break the internal links in the xylan chain, exo-xylanases remove the xylose units from the xylan chain ends, and β-xylosidases produce xylose by cleaving the non-reducing ends of xylobioses produced by endo-xylanases.

Mannanases hydrolyze the 1,4-β-glycosidic linkages in mannan, a type of hemicellulose. Mannanases have different industrial applications in producing bioalcohol, paper, and food and beverage items (Dawood & Ma, 2020). The most common types of mannanases are endo-β-mannanases and exo-β-mannanases.

Arabinofuranosidase cleaves the α-1,2, α-1,3, and α-1,5 linkages between L-arabinofuranose residues and other sugars in plant cell wall polysaccharides (Xin et al., 2019). They can be endo-arabinofuranosidases or exo-arabinofuranosidases.

Galactanase catalyzes the hydrolysis of 1,4-β-glycosidic bonds in galactans (Zhou et al., 2022), which are complex polysaccharides composed of galactose molecules. They can also be endo-galactanases or exo-galactanases.

Hemicellulases are used in conjunction with cellulases to hydrolyze the hemicellulose fraction of the biomass, releasing different sugar units to be used as substrates for the bioalcohols, such as ethanol and butanol, and other value-added products (Zhang et al., 2022).

d. **Ligninolytic enzymes** can effectively be produced by a certain class of fungi (i.e., white-rot fungi) and bacteria, and they are capable of breaking down lignin into smaller, more easily degradable fragments (Kumar & Chandra, 2020). Three different ligninolytic enzymes are extensively used in the bioalcohol industry: lignin peroxidase (LiP), manganese peroxidase (MnP), and laccase. These enzymes work in concert to degrade lignin through a series of oxidation reactions.

LiPs are a member of the heme-peroxidase family that are commonly found in white-rot fungi (Kumar & Chandra, 2020). The chemical structure of lignin peroxidase is composed of a heme group, a calcium ion, and a protein backbone. The heme group is a complex organic molecule that contains an iron atom and is responsible for the enzyme's catalytic activity. The calcium ion plays a structural role and helps to stabilize the protein structure. LiPs cannot penetrate a plant cell as it is too large; thus, they act only on exposed regions of the lumen (Goodell, 2020). They are the most effective peroxidases capable of oxidizing both phenolic and non-phenolic compounds. LiPs possess higher redox potentials (>1 V) and act as strong oxidants due to their porphyrin ring iron component's increased electron deficiency. They are relatively less specific to substrates and act on a broad range of substrates.

MnPs are also heme enzymes that use manganese as a cofactor and are similar to LiPs at the molecular level (Kumar & Chandra, 2020). Their primary function is to oxidize Mn^{2+} to Mn^{3+} using H_2O_2. Mn^{3+} then interacts with oxalates in lignin to form Mn^{3+}-oxalate complexes. Unlike LiPs, MnPs are not able to attack non-phenolic units of lignin, but they can oxidize phenolic compounds and non-phenolic lignin model compounds in the presence of Mn^{+2} via peroxidation of unsaturated lipids. White-rot fungi that produce MnP, but not LiP, may produce mediators to enable MnP to cleave non-phenolic lignin substrates (Dhagat & Jujjavarapu, 2022).

Laccase is a copper-containing enzyme and oxidizes a wide range of phenolic and non-phenolic lignin-related compounds (Malhotra & Suman, 2021). Unlike peroxidases, laccases do not require H_2O_2 as a co-substrate and can work under aerobic conditions.

Lignins, as discussed above, are not considered bioalcohol sources due to their low energy contents. Ligninolytic enzymes are, however, important in the bioconversion of lignocellulosic biomass into bioalcohols and bio-products. They are used in conjunction with cellulases and hemicellulases to depolymerize the lignocellulose matrix. Lignin fragments, produced by ligninolytic enzymes, can be used as substrates for different value-added products, such as resins. In Table 8.1, we list some of the popular enzymes used in the bioalcohol industry.

e. **Cellulosomes** are large, multi-enzyme complexes produced by certain microorganisms that are capable of breaking down plant cell walls, particularly the cellulose component. These complexes consist of a non-catalytic scaffoldin protein that anchors multiple catalytic enzymes, which work together to hydrolyze the cellulose into simple sugars that can be used as a carbon source in bioalcohol production (Kovács et al., 2013; Thompson et al., 2016; Kumar et al., 2023). The enzymes attached to the scaffoldin can include cellulases, hemicellulases, and other polysaccharide-degrading enzymes. The scaffoldin protein of the cellulosome contains multiple cohesin domains that bind to complementary dockerin domains present on the catalytic enzymes. This allows the enzymes to be held together in close proximity, increasing their efficiency and allowing for more effective degradation of the cellulose.

Cellulosomes are found primarily in anaerobic bacteria, particularly in the gut of ruminants such as cows and sheep, which rely on cellulose as a major component of their diet (Froidurot & Julliand, 2022). However, they have also been found in other microbial communities, such as soil and aquatic environments. Cellulosomes are usually absent in aerobic bacteria. The main reason behind the absence of cellulosomes in aerobic bacteria is that the aerobic degradation of cellulose generates reactive oxygen species (ROS) that can damage enzymes, and so the enzymes involved in this process need to be constantly replenished (Barbosa et al., 2020). Some rare examples of aerobic bacteria that produce cellulosomes include species of Streptomyces, a genus of filamentous bacteria that are found in soil and are known for their ability to degrade a wide range of complex organic compounds (Verma et al., 2023). These bacteria produce modified forms of the scaffoldin protein that are adapted to aerobic conditions and can efficiently degrade plant material even in the presence of oxygen.

Aerobic bacteria can enzymatically hydrolyze lignocellulose using a variety of enzymes, including cellulases, hemicellulases, and lignin-degrading enzymes. The breakdown of lignocellulose by aerobic bacteria typically involves the secretion of a large number of different enzymes rather than the use of a multi-enzyme complex like the cellulosome (Fontes & Gilbert, 2010). The process of lignocellulose

TABLE 8.1
Natural Enzymes Used for Enzymatic Hydrolysis in the Bioalcohol Industry

Enzyme Type	Enzyme Name	Source
Endoglucanase (Cellulase)	*Trichoderma reesei* endoglucanase I	Fungal
	Clostridium thermocellum endoglucanase	Bacterial
	Cellulomonas fimi endoglucanase	Bacterial
	Bacillus subtilis endoglucanase	Bacterial
	Humicola insolens endoglucanase	Fungal
Exoglucanases (Cellulase)	*Trichoderma reesei* exoglucanase I	Fungal
	Thermobifida fusca exoglucanase	Bacterial
	Cellulomonas fimi exoglucanase	Bacterial
	Humicola insolens exoglucanase	Fungal
	Aspergillus niger exoglucanase	Fungal
β-Glucanases (Cellulase)	*Trichoderma reesei* β-glucanase	Bacterial
	Bacillus subtilis β-glucanase	Fungal
	Aspergillus niger β-glucanase	Funga
	Streptomyces sp. β-glucanase	Bacterial
	Cellulomonas fimi β-glucanase	Bacterial
Endo-xylanases (Hemicellulase)	*Aspergillus niger* endo-xylosidase	Fungal
	Paenibacillus barcinonensis endo-xylosidase	Bacterial
	Trichoderma reesei endo-xylosidase	Fungal
	Streptomyces olivaceoviridis endo-xylosidase	Bacterial
	Thermomyces lanuginosus endo-xylosidase	Fungal
Exo-xylanases (Hemicellulase)	*Aspergillus niger* exo-xylosidase	Fungal
	Bifidobacterium bifidum exo-xylosidase	Bacterial
	Paenibacillus sp. SYBC-H1 exo-xylosidase	Bacterial
	Talaromyces cellulolyticus exo-xylosidase	Fungal
	Streptomyces thermovulgaris exo-xylosidase	Bacterial
β-xylosidases (Hemicellulase)	*Aspergillus niger* β-xylosidase	Fungal
	Thermotoga maritima β-xylosidase	Bacterial
	Trichoderma reesei β-xylosidase	Fungal
	Bacillus subtilis β-xylosidase	Bacterial
	Cellulomonas uda β-xylosidase	Bacterial
Endo-β-mannanases (Hemicellulase)	*Bacillus licheniformis* endo-β-mannanase	Bacterial
	Aspergillus niger endo-β-mannanase	Fungal
	Trichoderma reesei endo-β-mannanase	Fungal
	Clostridium thermocellum endo-β-mannanase	Bacterial
	Paenibacillus polymyxa endo-β-mannanase	Bacterial
Exo-β-mannanases (Hemicellulase)	*Trichoderma reesei* exo-β-mannanase	Fungal
	Streptomyces olivaceoviridis exo-β-mannanase	Bacterial
	Bacillus sp. MITS 2017 exo-β-mannanase	Bacterial
	Aspergillus aculeatus exo-β-mannanase	Fungal
	Penicillium funiculosum exo-β-mannanase	Fungal

(*Continued*)

TABLE 8.1 (*Continued*)
Natural Enzymes Used for Enzymatic Hydrolysis in the Bioalcohol Industry

Enzyme Type	Enzyme Name	Source
Endo-arabinofuranosidase (Hemicellulase)	*Trichoderma reesei* endo-arabinofuranosidase	Fungal
	Bacillus subtilis endo-arabinofuranosidase	Bacterial
	Aspergillus niger endo-arabinofuranosidase	Fungal
	Paenibacillus sp. TN-16 endo-arabinofuranosidase	Bacterial
	Neosartorya fischeri endo-arabinofuranosidase	Fungal
Exo-arabinofuranosidase (Hemicellulase)	*Aspergillus niger* exo-arabinofuranosidase	Fungal
	Penicillium chrysogenum exo-arabinofuranosidase	Fungal
	Bifidobacterium adolescentis exo-arabinofuranosidase	Bacterial
	Bacillus subtilis exo-arabinofuranosidase	Bacterial
	Neosartorya fischeri exo-arabinofuranosidase	Fungal
Endo-galactanases (Hemicellulase)	*Aspergillus niger* endo-galactanase	Fungal
	Penicillium chrysogenum endo-galactanase	Fungal
	Bacillus licheniformis endo-galactanase	Bacterial
	Paenibacillus polymyxa endo-galactanase	Bacterial
	Clostridium thermocellum endo-galactanase	Bacterial
Exo-galactanases (Hemicellulase)	*Talaromyces cellulolyticus* exo-galactanase	Fungal
	Streptomyces sp. S27 exo-galactanase	Bacterial
	Rhodothermus marinus exo-galactanase	Bacterial
	Aspergillus oryzae exo-galactanase	Fungal
	Bacillus sp. HJ14 exo-galactanase	Bacterial
Lignin peroxidase	*Phanerochaete chrysosporium* lignin peroxidase	Fungal
	Bjerkandera adusta lignin peroxidase	Fungal
	Pleurotus ostreatus lignin peroxidase	Fungal
	Ganoderma lucidum lignin peroxidase	Fungal
	Phlebia radiata lignin peroxidase	Fungal
Manganese peroxidase	*Phanerochaete chrysosporium* manganese peroxidase	Fungal
	Pleurotus ostreatus manganese peroxidase	Fungal
	Trametes versicolor manganese peroxidase	Fungal
	Bjerkandera adusta manganese peroxidase	Fungal
	Irpex lacteus manganese peroxidase	Fungal
Laccase	*Trametes versicolor* laccase	Fungal
	Pleurotus ostreatus laccase	Fungal
	Rhus vernicifera laccase	Fungal
	Bacillus subtilis laccase	Fungal
	Myceliophthora thermophila laccase	Fungal

breakdown by aerobic bacteria begins with the secretion of cellulases, which break down the cellulose component of the lignocellulose into smaller sugar molecules, such as glucose. These sugar molecules are then transported into the cell, where they can be used as a source of energy (Chukwuma et al., 2021b). In addition to cellulases, aerobic bacteria also secrete a variety of hemicellulases, which break down the hemicellulose component of the lignocellulose into smaller sugar molecules, such as xylose, arabinose, and mannose. These sugar molecules can also be used as a source of energy by the bacteria. Finally, lignin-degrading enzymes are produced by some aerobic bacteria to break down the lignin component of the lignocellulose (Vélez-Mercado et al., 2021). As discussed above, these enzymes, such as lignin peroxidase and manganese peroxidase, can oxidize and degrade the lignin molecules, making it easier for the cellulases and hemicellulases to access the cellulose and hemicellulose components of the lignocellulose.

8.4.2 Factors and Bottlenecks in EH Methods for Bioalcohol Synthesis

The efficiency of EH process is highly influenced by a number of different factors such as:

8.4.2.1 Factors Related to Substrate

The presence of a considerable quantity of lignin in biomass is a crucial aspect that restricts the breakdown of lignocellulosic biomass by cellulolytic and hemicellulolytic enzymes. The enzymatic conversion of lignocellulosic biomass is substantially influenced by the structural characteristics of cellulose.

a. **Crystallinity of cellulose:** The crystallinity of cellulose in agrobiomass can have a significant impact on the efficiency of EH (Kaschuk & Frollini, 2018). Crystallinity refers to the degree to which cellulose molecules are tightly packed together in a rigid, crystalline structure (Park et al., 2010). In highly crystalline cellulose, the individual cellulose molecules are closely packed, making it difficult for enzymes to access and break down the cellulose into smaller sugar molecules. EH relies on enzymes, such as cellulases, to break down the cellulose in agrobiomass into simple sugars, such as glucose, which can be used for further processing. Suppose the cellulose in the biomass is highly crystalline. In that case, the enzymes will have a harder time accessing and breaking down the cellulose, leading to a slower rate of hydrolysis and lower overall yield of sugars. Therefore, reducing the crystallinity of cellulose in agrobiomass is an important factor in increasing the efficiency of EH. This can be achieved through various pretreatment methods, such as steam explosion (Marcos et al., 2013), acid or alkali pretreatment, or biological pretreatment, which can break down the rigid structure of cellulose and make it more accessible to enzymes. Also, cellulases complex responsible for hydrolysis of complex cellulose is discovered (Thompson et al., 2016). Overall, the crystallinity of cellulose is an important factor to consider in EH of agrobiomass, and appropriate pretreatment methods should be chosen to optimize the hydrolysis process and increase the yield of sugars.

b. **Accessible surface area (ASA):** It refers to the amount of surface area on a molecule or substrate available for enzymatic attack. Therefore, increasing the ASA improves the efficiency of enzymatic attack. The accessible area varies during EH. It is generally observed to be high in early and low in the later stages (Rasmussen et al., 2017).
c. **Substrate concentration:** In general, a higher substrate concentration in the hydrolysis reaction can lead to a higher yield of simple sugars, as there are more cellulose and hemicellulose molecules available to be broken down by the enzymes. However, high substrate concentrations can also lead to several challenges that need to be considered. One challenge is that a high substrate concentration can result in a high viscosity of the reaction mixture, making it more difficult for enzymes to access the cellulose and hemicellulose (Zhai et al., 2018; Kumar et al., 2022). This can lead to a decrease in the rate of hydrolysis and lower overall yield of simple sugars. Another challenge is that a high substrate concentration can result in the build-up of inhibitory compounds, such as acetic acid, which can reduce the efficiency of EH. Therefore, finding an optimal substrate concentration is important to maximize the efficiency of EH. The optimal concentration will depend on several factors, including the type of biomass, the enzymes used, and the hydrolysis conditions (Kumar et al., 2022). Generally, it is recommended to use a concentration of substrate that is high enough to maximize yield, but not so high that it leads to significant viscosity or the build-up of inhibitory compounds. Overall, substrate concentration is an important factor to consider in EH of agrobiomass, and it is important to find an optimal concentration that balances high yield with efficient EH (Muthukumar & Rajendran, 2013).
d. **Lignin content and distribution:** High amount of hemicellulose and lignin retards the access of cellulase enzymes further decreasing the efficiency of hydrolysis. It generally acts as a physical barrier while hydrolysis by non-productively associating cellulolytic enzymes (Lee & Ra, 2021; Yuan et al., 2021). Many new techniques have been explored to prevent ineffective adsorption, such as adding polyethylene glycol, tween, and bovine serum albumin. However, the addition of these reagents introduces additional cost (Luo et al., 2021; Zhang & Naebe, 2021b).
e. **Porosity:** Recent studies have indicated that the primary hindrance is the size of pores on the substrate compared to the enzymes. Removal of the excess hemicellulose content facilitates an increase in mean pore size which further elevates the accessibility of cellulose for hydrolysis (Van Dyk & Pletschke, 2012). Cellulases generally get trap in the pores if the internal area is high as compared to the external area.

8.4.2.2 Factors Related to Enzyme

a. **Enzyme selection:** The choice of enzyme can significantly impact the efficiency of EH. Different enzymes have varying substrate specificities, and some enzymes work better than others for breaking down specific types of polysaccharides. The cost of the enzyme is also a factor to consider.

The optimization of specific enzyme ratios is considered a crucial step in achieving efficient degradation while managing enzyme costs (Vasić et al., 2021). Given the wide variety of lignocellulose substrates, researchers generally believe that pretreatments, enzyme combinations, and ratios should be customized for each substrate type.

Studies by various groups of researchers highlighted that enzyme ratios are influenced not only by concentrations of lignin, cellulose, and hemicellulose but also by cell wall anatomy and microstructure (Gao et al., 2011). In many cases, there is no straightforward relationship between the importance of a particular enzyme and the abundance of the specific bonds it forms (Van Dyk & Pletschke, 2012). Therefore, the required enzyme loading for a particular enzyme may not necessarily correlate with the abundance of the bonds it acts upon.

b. **Enzyme stability:** Enzymes are often sensitive to changes in temperature, pH, and other environmental factors. Therefore, maintaining enzyme stability during hydrolysis is essential to ensure the efficiency and consistency of the process.
c. **Substrate pretreatment:** LB is typically very resistant to hydrolysis and requires pretreatment to make it more amenable to EH. Pretreatment methods such as acid or alkaline treatment, steam explosion, or milling can increase the accessibility of enzymes to the biomass.
d. **Enzyme loading:** The amount of enzyme added to the hydrolysis reaction can impact the efficiency of the process. Factors affecting enzyme loadings include the presence of lignin, which may require higher enzyme quantities due to non-productive enzyme adsorption. Additionally, the type of enzymes used, such as cellulase or xylanase, can impact the required enzyme loadings. It is noted that while increased enzyme loadings may enhance hydrolysis up to a certain point, further increases might slow the process due to factors like competition for binding sites (Wyman et al., 2011). High substrate loadings can similarly slow down hydrolysis due to factors like reduced enzyme binding and difficulties in enzyme diffusion in viscous mediums. The literature shows variations in enzyme and substrate loadings based on different substrates and pretreatments (Van Dyk & Pletschke, 2012).
e. **Inhibitors:** The pretreatment of substrates can lead to the formation of inhibitory compounds, which may affect the performance of saccharolytic enzymes and the microorganisms involved in fermenting sugars into bioethanol. Furthermore, enzyme activity can also be influenced by the by-products of their own reactions and those formed by other enzymes (Kumar et al., 2022). During pretreatment, a range of inhibitors may be generated, including phenolics, furfural, 5-hydroxymethyl furfural, ρ-coumaric acid, gallic acid, acetic acid, formic acid, levulinic acid, vanillic acid, acetovanillone, guaiacol, protocatechuic acid, hydroxy-cinnamic, and 4-hydroxybenzoic acids (Malani et al., 2019). Subsequently, during EH, the formation of oligosaccharides, disaccharides, and monomers can potentially inhibit the enzymes involved. When these compounds accumulate to high

concentrations, they can have a detrimental impact on hydrolysis efficiency (García-Aparicio et al., 2006).

f. **Fermentation:** The efficiency of the fermentation process can also impact the overall efficiency and cost-effectiveness of the bioalcohol production process. Factors such as the choice of microorganism, fermentation conditions, and the presence of inhibitors can all affect the yield and quality of the bioalcohol product.

In summary, selecting the right enzyme, optimizing the hydrolysis conditions, and carefully managing inhibitors are all critical factors in achieving efficient and cost-effective EH for bioalcohol synthesis.

8.4.2.3 Physical Parameters

Physical parameters are important factors that can impact the efficiency of EH of agrobiomass. These parameters include temperature, pH, agitation, and substrate size and shape (Kumar et al., 2022).

a. **Temperature:** In general, enzymes work best within an optimum temperature range, and different enzymes have different optimal temperatures. Higher temperatures can increase the rate of hydrolysis, but if the temperature is too high, it can lead to enzyme denaturation, which can decrease the efficiency of the process.
b. **pH:** Enzymes have an optimal pH range in which they are most active. If the pH is too high or too low, it can lead to decreased enzymatic activity and decreased efficiency of the hydrolysis process.
c. **Agitation:** Agitation helps to mix the substrate and enzymes, which can improve contact and increase the efficiency of the process. However, excessive agitation can also lead to enzyme denaturation or degradation, and therefore, the level of agitation needs to be optimized for the specific process (Roy et al., 2020).
d. **Substrate size and shape:** The size and shape of the substrate can impact the accessibility of the enzymes to the cellulose and hemicellulose molecules. Generally, smaller particle size can increase the efficiency of EH, as it increases the surface area for enzyme action.

8.5 ADVANCED TECHNOLOGIES FOR ENHANCING EFFICIENCY OF EH

Researchers have found many naturally occurring enzymes for hydrolyzing lignocellulosic biomasses. With the constant demand from the fuel industry for a more economically viable pathway to produce industrial standard bioalcohol, a search for enzymes with higher activity, specificity, and substrate range is always in order.

a. **Enzyme recycling** to improve efficiency and reduce the cost of EH by reusing the enzymes released during hydrolysis. This can be achieved through membrane filtration, immobilization, and recycling reactors (Su et al., 2020).

b. **In situ product removal** to improve the efficiency and yield of bioalcohol production by removing the products as they are formed, which can alleviate product inhibition and improve the driving force of the reaction. This can be achieved through pervaporation, membrane distillation, and adsorption (Santos et al., 2021).
c. **Advanced pretreatment methods** to improve the accessibility of cellulose and hemicellulose to the enzymes, increase their surface area, and reduce their crystallinity (Roy et al., 2020). Pretreatment can be done using physical, chemical, or biological methods, such as steam explosion, acid hydrolysis, alkaline treatment, or microwave irradiation.
d. **Multi-enzyme systems** to achieve higher yields from a broad spectrum of substrates.
e. **Enzyme engineering** with protein and genetic engineering methods. With the computation-aided modeling of the individual and multi-enzyme systems, researchers use rational modeling and directed evolution to propose engineering modifications toward enzymes. These proposed enzymes are experimentally verified using high-throughput screening (HTS).

HTS is a powerful approach used in enzyme engineering to rapidly evaluate large numbers of variants for desired properties such as improved activity, specificity, stability, and selectivity. HTS methods typically involve the use of automated systems that can process and analyze large numbers of samples in a short time. HTS starts with generating an extensive library of enzyme variants using methods such as directed evolution, rational design, or a combination of both (Vasina et al., 2022). The library should contain a diverse set of enzyme variants with varying properties. The next step is the development of a high-throughput assay that should be able to rapidly screen a large number of enzyme variants for the desired properties. The library of enzyme variants is then screened using the high-throughput assay. The assay can be carried out in microplates or other automated systems that can process a large number of samples at once. The data from the high-throughput assay is finally analyzed to identify enzyme variants with improved properties. The identified enzyme variants can be further optimized using additional rounds of directed evolution or rational design (we will discuss these processes later in this section) to improve their properties (Sarnaik et al., 2020; Marcellin & Nielsen, 2018). HTS methods have been used successfully in enzyme engineering to identify variants with improved properties. However, HTS methods have limitations, including the need for high-quality assays, the cost of reagents and equipment, and the complexity of data analysis.

In addition, computational modeling can be used to study the behavior of enzymes under different conditions, such as temperature, pH, and substrate concentration. This allows for the prediction of the optimal conditions for enzyme function, which can be used to optimize the performance of enzymes in industrial processes (Vasina et al., 2022). Overall, these advanced technologies can improve efficiency and reduce the cost of EH for bioalcohol production, as well as expand the range of feedstocks that can be used for bioalcohol synthesis (Wang et al., 2017). Below, we will discuss two main aspects of these engineering methods, namely, genome and enzyme engineering.

8.5.1 Genome Engineering

In the past, researchers have disrupted or deleted genes encoding for undesired products in order to elevate the production of desired products. Additionally, hydrolytic enzyme genes from one organism have been expressed in different organisms to increase expression. Recently, a variety of techniques and mechanisms have been developed to explore, harbor, and elevate hydrolytic enzymes and their applications at both the experimental and industrial levels (Kwon et al., 2020; Hsu et al., 2014). Regarding the breakdown of lignocellulosic biomass, cellulose-degrading enzymes that are crucial for the process are usually not naturally occurring in forms that can withstand extreme temperatures or pH levels, nor are they able to function properly under conditions with high levels of salt, metal ions, or solvents (Kim et al., 2019; Choi et al., 2019). Therefore, genome engineering plays a pivotal role in creating and evolving microorganisms that can survive in extreme environments, as well as for developing enzymes that are highly stable.

Genome engineering involves the manipulation of an organism's genetic material to create desired traits or characteristics. This can be done through various techniques, such as CRISPR-Cas9, which allows for precise editing of specific genes (Bruder et al., 2016). By using genome engineering, researchers can create microorganisms that are better suited for industrial processes, such as LB hydrolysis. For example, they can create microorganisms that produce more efficient and stable enzymes, or that can survive in extreme environments (Barbosa et al., 2020). Overall, the development of new techniques and mechanisms for exploring, harboring, and elevating hydrolytic enzymes, as well as the use of genome engineering, are important tools for advancing the production of bioalcohols and other value-added products from lignocellulosic feedstock (Hille & Charpentier, 2016; Loureiro & da Silva, 2019). These advancements have the potential to make the process more efficient, cost-effective, and environmentally friendly. Different genome engineering techniques for exploring, harboring, and elevating hydrolytic enzymes include the following steps (Kwon et al., 2020):

a. **Genome sequencing:** This technique involves sequencing the entire genome of an organism to identify genes encoding hydrolytic enzymes. This information can be used to identify new enzymes or engineer existing enzymes for specific applications.
b. **Codon optimization:** Codon optimization is a powerful tool in genome engineering that involves modifying the DNA sequence of a gene to replace infrequently used codons with more frequently used ones. This can improve the efficiency of protein synthesis and reduce the likelihood of translation errors (Wang et al., 2019). In the case of hydrolytic enzymes, codon optimization can also be used to enhance the activity of the enzyme. This involves introducing codons that encode amino acids that are preferred at specific positions within the active site of enzyme. By optimizing the codons that are used to encode the amino acid residues within the active site, the enzyme can be designed to have improved activity, stability, and specificity (Tani et al., 2014; Wang et al., 2019). For example, in the case of cellulases, codon

optimization can be used to introduce codons that encode for amino acids that are known to be important for cellulose binding and hydrolysis. This can improve the efficiency of the enzyme in breaking down cellulose and increase its activity under specific conditions, such as high temperature or low pH (Abreu et al., 2019).

c. **Directed evolution:** Directed evolution for genome engineering is a powerful strategy that involves enhancing the efficacy, selectivity, and resilience of enzymes through a process of multiple cycles of random mutation and selection. During the process of directed evolution, a library of enzyme variants is created by introducing random mutations into the gene encoding the enzyme (Yu et al., 2022). These mutations can be generated through techniques such as error-prone PCR, DNA shuffling, or saturation mutagenesis. The resulting library of enzyme variants is then screened for improved activity, specificity, or stability under specific conditions. For example, in the case of improving the activity of a hydrolytic enzyme, a library of enzyme variants may be screened for increased activity toward a particular substrate as well as to improve stability under high temperature and pH (Packer & Liu, 2015a). Enzymes that show improved activity are then selected for further rounds of mutation and selection. The process of directed evolution can be repeated multiple times, with each round of mutation and selection leading to the identification of enzyme variants with improved properties. Over time, this process can lead to the creation of enzymes that are highly efficient, specific, and stable under a wide range of conditions (Goedegebuur et al., 2017; Østby et al., 2020). By introducing random mutations and selecting for improved activity, specificity, and stability, researchers can create enzymes that are better suited for industrial processes, such as lignocellulosic biomass hydrolysis (Cao et al., 2020; Packer & Liu, 2015b).

d. **Fusion with binding domains:** Enzymes often have two distinct domains: a catalytic domain and a binding domain. The catalytic domain contains the active site where the substrate is transformed into the product. The binding domain is responsible for recognizing and binding the substrate, and it also helps to position the substrate in the correct orientation for catalysis to occur (Vasina et al., 2022). Fusion with binding domains in the case of hydrolytic enzymes can increase their substrate specificity and catalytic efficiency. This involves fusing the catalytic domain of one enzyme with the binding domain of another enzyme that recognizes a different substrate. The resulting enzyme is able to catalyze the hydrolysis of the substrate recognized by the catalytic domain while also recognizing the substrate recognized by the binding domain. This approach allows the enzyme to recognize and bind the substrate more efficiently, leading to a higher rate of catalysis. For example, in the case of cellulases, the catalytic domain may be fused with a binding domain that recognizes specific types of cellulose, such as crystalline cellulose or amorphous cellulose. This can increase the efficiency of the enzyme in breaking down the targeted cellulose substrate.

e. **Cellulosome designing:** Cellulosomes are multi-enzyme complexes that are capable of efficiently breaking down cellulose. The cellulosome is composed of a scaffoldin protein that anchors multiple enzymatic subunits, enabling synergistic activity to break down the cellulose fibers (Rosa et al., 2010; Sharma et al., 2022). Designing cellulosomes with hydrolytic enzymes involves the identification of the desired enzyme components and their appropriate assembly onto the scaffoldin protein. Therefore, the enzymes must be selected based on their ability to break down the specific type of cellulose that needs to be targeted. For example, different enzymes may be required for the hydrolysis of crystalline cellulose as compared to amorphous cellulose (Janusz et al., 2017; Fontes & Gilbert, 2010).

There are several approaches that can be used to design cellulosomes with hydrolytic enzymes. One approach is to use genetic engineering techniques to introduce the desired genes for the enzymes and scaffoldin protein into a suitable host organism, such as a bacterium. This can allow for the production of the cellulosome complex in large quantities. The genes can be introduced using plasmids or other vectors, and the expression of the genes can be controlled using inducible promoters or other regulatory elements (Thompson et al., 2016; Liang et al., 2014). Another approach is to use synthetic biology techniques to design and construct the cellulosome in vitro. This involves the assembly of the scaffoldin protein and enzymatic subunits in a test tube, followed by purification and characterization of the resulting complex. Synthetic biology techniques can be used to optimize the assembly process and introduce modifications to the scaffoldin protein or enzymatic subunits to improve the stability or activity of the resulting complex (Kwon et al., 2020; Choi et al., 2019).

8.5.2 Enzyme Engineering

Enzyme engineering is an age-old process that focuses on modifying and improving enzymes for specific applications. Enzymes are proteins that catalyze chemical reactions. Enzyme engineering requires a deep understanding of the chemical properties of different functional groups present in the enzyme and requires chemically controlled mechanisms to alter those groups to study the activity of the modified enzymes (Sharma et al., 2022; Cho et al., 2019). Enzyme engineering is widely used in various industries, including biotechnology, pharmaceuticals, and food processing. Enzyme engineering involves several techniques. We will discuss a few of those techniques below:

a. **Rational design** makes targeted modifications to the amino acid sequence of an enzyme based on an understanding of its structure and function. Computational tools for molecular modeling are used to predict the effects of specific amino acid substitutions on the activity, stability, and specificity of an enzyme (Pongsupasa et al., 2022). The rational method typically involves three main steps. The first step is to identify the specific amino acid residues that are likely to affect the activity, stability, or specificity of enzyme. This is typically done by analyzing the enzyme's crystal structure and comparing it to related enzymes. Based on the information gathered in the first step, specific amino acid substitutions are designed to improve the enzyme's performance (Kumar et al., 2021, 2023). These substitutions can include changes to the active site, allosteric sites, or other regions of the enzyme. The last step of testing the newly designed enzyme typically involves site-directed mutagenesis to introduce the desired amino acid substitutions, followed by enzyme assays and other biochemical tests (Beg et al., 2018; Ostadjoo et al., 2019).

b. **Semi-rational design** mixes rigor or rational design with random experimental screening to reduce the search parameters for novel enzyme design. The advantage of rational methods is that they can be highly targeted and efficient in introducing mutations that are more likely to improve the enzyme's properties. However, the approach is limited to the knowledge available of the enzyme's structure and function. Semi-rational methods, on the other hand, can be used to identify mutations that improve enzyme properties even with limited knowledge of the enzyme (Sarnaik et al., 2020; Choi et al., 2019). This approach is, however, less efficient in introducing targeted mutations, and it relies on experimental screening to identify effective mutations.

   The semi-rational method generally begins with the identification of the target enzyme, its structural and functional features, and the analysis of the enzyme's active site and its interaction with substrates (Biundo et al., 2021). A library of mutations that can modify the enzyme's properties is generated. Next, screening of the mutations is done using high-throughput assays to identify variants with improved activity and stability. Finally, structural and biochemical analysis of the selected variants to understand the molecular basis of their improved properties. Mutation, screening, and analysis are iterated to optimize the enzyme's properties (Kumar et al., 2022; Kumar & Moholkar, 2023). The semi-rational method allows for the rapid exploration of sequence space while still leveraging the knowledge of the enzyme's structure and function. It can be a powerful tool in the development of enzymes with improved performance for a variety of industrial and biotechnological applications (Pongsupasa et al., 2022).

c. **Directed evolution** is primarily used in genetic engineering, as we have already discussed above. Directed evolution is also used to improve or modify the enzymes through iterative rounds of mutagenesis and screening (Yu et al., 2022). The directed evolution process typically involves several steps. The starting molecule is subjected to random or targeted mutagenesis, using methods such as error-prone PCR, site-directed mutagenesis, or gene shuffling. This generates a library of variant molecules, each with a slightly different sequence or structure. The library is screened or selected for variants with improved properties, using methods such as activity assays, binding assays, or functional screens. The selected variants are then used as the starting point for the next round of mutagenesis and screening. The process is repeated until a desired level of improvement is achieved.

Directed evolution can generate variants with properties that may not have been predictable or even possible to achieve through rational design. Additionally, directed evolution can be used to optimize properties that are difficult to predict or model, such as stability or activity under specific conditions (Yu et al., 2022; Cao et al., 2020).

d. **Combinatorial method** in enzyme engineering is a technique to engineer enzymes for specific applications (Victorino da Silva Amatto et al., 2022). Rational methods and directed evolution are two distinct approaches to enzyme engineering, but they can be combined to create a combinatorial method that leverages the strengths of both approaches.

The combinatorial method involves the generation of a large library of mutant enzymes by randomly introducing mutations into the enzyme's DNA. This is typically done using error-prone PCR or DNA shuffling. The resulting library of mutant enzymes is then screened for variants with desired properties. The rational and directed evolution approaches are used in the next step to guide the design of the mutations introduced into the library of mutant enzymes (Pongsupasa et al., 2022). The rational approach can be used to design mutations in specific areas of the enzyme, such as the active site, based on knowledge of the enzyme's structure and function. Directed evolution, on the other hand, can be used to introduce random mutations throughout the enzyme's sequence, allowing for the exploration of a broader range of mutational space (Goedegebuur et al., 2017). The resulting library of mutant enzymes can be screened using a high-throughput assay to identify variants with desired properties. The variants can then be further optimized using rational design or directed evolution to improve their properties. The combinatorial approach allows the generation of a large library of diverse enzymes, providing a greater chance of identifying variants with improved properties (Santos et al., 2012).

8.6 CHALLENGES IN SCALE-UP AND DOWNSTREAM PROCESSING FOR PURIFICATION

8.6.1 Challenges in Scale-up of Bioalcohol Production

Scaling up the hydrolysis of agrobiomass to produce bioalcohols presents a multitude of complex challenges that need to be carefully addressed to ensure the efficiency and economic viability of this promising process. One of the primary challenges is the variability in agrobiomass feedstocks, which can differ in composition, moisture content, and other factors (Hollinshead et al., 2014; Goswami et al., 2023). Ensuring consistent feedstock quality at a larger scale is crucial to maintain the reliability and reproducibility of the hydrolysis process. Another critical factor is the production and cost of enzymes used in EH technique. As production scales up, there is an increased demand for enzymes, which can be both expensive and logistically challenging to procure in sufficient quantities (Kumar et al., 2022). Also, the design of hydrolysis reactors for large-scale operations is also a significant challenge. Issues related to heat and mass transfer, mixing, and efficient enzyme–substrate interactions become increasingly complex at larger scales. The choice of reactor type and configuration significantly influences the process like efficient heat and mass transfer (Su et al., 2020; Calinescu et al., 2019). These are vital for large-scale hydrolysis. Maintaining optimal temperature and pH throughout the reactor while ensuring uniform enzyme distribution is crucial. Poor heat and mass transfer can lead to reduced hydrolysis efficiency. Managing inhibition and maintaining high conversion rates in large-scale reactors is a challenging task. Safety and environmental concerns become more pronounced as the process scales up. Energy consumption is another critical consideration. Scaling up can lead to higher energy demands for mixing, temperature control, and other process requirements. Minimizing energy consumption while maximizing productivity is essential for economic viability. Fermentation, separation, and recovery of bioalcohols in downstream processes pose their own set of scale-up challenges. Efficient fermentation and cost-effective recovery processes are essential for the overall success of the bioalcohol production chain. Addressing these multifaceted challenges is essential for the successful implementation of large-scale bioalcohol production processes. Researchers and engineers are working to overcome these challenges and develop more efficient and sustainable methods for hydrolyzing agro-biomass. Some of the strategies being explored include developing new catalysts, optimizing process conditions, improving enzyme stability and efficiency, and developing new separation technologies (Kumar et al., 2023).

8.6.2 Downstream Processing Methods and their Challenges

Downstream processing methods play a crucial role in the purification of alcohols, ensuring the production of high-quality products. Several techniques are employed, each with its own set of challenges. Here are common downstream processing methods for the purification of alcohols and their respective challenges:

a. **Distillation**

   Method: Distillation separates alcohols based on their boiling points.

   Challenges: Energy-intensive, especially for high boiling point alcohols. Azeotropic mixtures can complicate separation.

b. **Adsorption**

   Method: Adsorption processes use solid materials (adsorbents) to selectively capture alcohol molecules.

   Challenges: Limited capacity for adsorption, regeneration of adsorbents, and potential contamination.

c. **Membrane separation**

   Method: Membrane processes exploit differences in molecular size to separate alcohols.

   Challenges: Fouling of membranes, low selectivity for some alcohols, and membrane degradation.

d. **Crystallization**

   Method: Crystallization involves cooling a solution to precipitate purified alcohol crystals.

   Challenges: Crystal purity, solvent removal, and control of crystal size.

e. **Extraction**

Method: Extraction uses solvents to selectively remove alcohols from a mixture.

Challenges: Choice of solvent, solvent recovery, and potential co-extraction of impurities.

f. **Chromatography**

Method: Chromatographic techniques separate alcohols based on their interactions with a stationary phase.

Challenges: High cost, scalability, and development of suitable stationary phases.

g. **Hybrid processes**

Method: Combination of multiple downstream methods for enhanced purification.

Challenges: Process complexity, optimization, and potential high capital costs.

Each method presents a trade-off between purity, yield, and cost-effectiveness. The choice of the downstream processing method depends on the specific requirements of the alcohol production process and the desired end product.

8.7 COMMERCIAL STATUS OF BIOALCOHOL SYNTHESIS FROM AGROBIOMASS

The use of agrobiomass as a feedstock for bioalcohol production has gained significant interest in recent years as an alternative to fossil fuels. Bioethanol is the most widely produced bioalcohol from agrobiomass, with several commercial-scale facilities around the world, particularly in countries with policies and incentives supporting renewable energy (Lamichhane et al., 2021). However, the production of other bioalcohols, such as butanol and propanol, is still in the research and development stage and has yet to reach commercialization (Goswami et al., 2023).

The production of bioalcohols from agrobiomass faces several challenges, including feedstock availability, cost, and competition with fossil fuels. The availability and cost of agrobiomass vary depending on factors such as climate, soil quality, and land use. Additionally, the cost of bioalcohol production from agrobiomass is often higher than that of fossil fuels due to the higher cost of feedstock and processing technologies (Malode et al., 2021; Kushwaha et al., 2019; Amiri, 2020; Ferreira Dos Santos Vieira et al., 2021). However, continued research and development in this field, as well as supportive government policies, may help to reduce the cost and increase the commercial viability of bioalcohol synthesis from agrobiomass.

The commercial status of bioalcohol synthesis from agrobiomass is mixed, with some technologies in the early stages of commercialization and others still in the research and development stage (O' et al., 2019). At present, the biochemical transformation of lignocellulosic biomass into ethanol has advanced to a Technology Readiness Level (TRL) of 7–8, marked by the emergence of a limited number of plants that have entered the initial stages of commercialization (Vasilakou et al., 2023).

Continued investment and development of technologies for bioalcohol synthesis from agrobiomass could lead to increased commercialization and greater use of sustainable, renewable energy resources (Vasilakou et al., 2023). The use of agrobiomass as a feedstock for bioalcohol production has several advantages, including the potential to reduce greenhouse gas emissions and to provide a sustainable source of energy.

8.8 OVERVIEW, CONCLUSIONS, AND FUTURE PERSPECTIVES

EH of agrobiomass is a promising method for producing bioalcohols, offering several advantages over traditional chemical methods. However, there are still challenges that need to be addressed to make this process more efficient and cost-effective. Some of the future perspectives of EH of agrobiomass include:

a. **Development of more efficient enzymes:** The cost and efficiency of enzymes used in EH are major challenges. Ongoing research is focused on developing more efficient enzymes that can break down the complex structure of cellulose and hemicellulose, leading to increased efficiency and reduced costs.
b. **Use of mixed enzymes:** Combining multiple enzymes in a single hydrolysis process has the potential to improve efficiency. Researchers are developing enzyme cocktails that can work together to break down the complex structure of agrobiomass more efficiently.
c. **Development of HTS techniques:** HTS techniques allow researchers to quickly test and evaluate large numbers of enzymes for their ability to break down agrobiomass. These techniques will help identify more efficient enzymes and enzyme cocktails, speeding up the development of commercial-scale processes.
d. **Optimization of hydrolysis conditions:** Physical parameters such as temperature, pH, and agitation are crucial for EH. Optimization of these conditions can improve the efficiency of the process and reduce costs.

EH of agrobiomass has numerous benefits, including concentrated sugars, lower expenses, and less energy input, making it a promising method for producing lignocellulosic ethanol on an industrial scale. With ongoing research and development, EH has the potential to become a more efficient and cost-effective method for producing bioalcohols from agrobiomass.

ACKNOWLEDGMENTS

Dr. Karan Kumar is grateful to the Prime Minister's Research Fellowship provided by Ministry of Education, Government of India. Ms. Poulami and Ms. Shraddha, two authors of this chapter, acknowledge the WINBIOCB-2022 internship/training provided by Prof. V.S. Moholkar at IIT Guwahati campus.

NOTE

* Corresponding author

REFERENCES

Abdullah, B., S. A. F. S. Muhammad, Z. Shokravi, S. Ismail, K. A. Kassim, A. N. Mahmood, and M. M. A. Aziz. 2019. Fourth generation biofuel: a review on risks and mitigation strategies. *Renewable and Sustainable Energy Reviews* 107 (June): 37–50. https://doi.org/10.1016/j.rser.2019.02.018.

Abreu, F. P. D., R. P. Menin, N. S. de Oliveira, A. R. Lenz, S. Ávila e Silva, A. J. P. D., and M. Camassola. 2019. Papel da regulação gênica na ação de enzimas lignocelulolíticas. *Interdisciplinary Journal of Applied Science* 4 (7): 43–46.

Abu, R., M. T. Gundersen, and J. M. Woodley. 2015. Thermodynamic calculations for systems biocatalysis. *Computer Aided Chemical Engineering* 37: 233–238. https://doi.org/10.1016/B978-0-444-63578-5.50034-7.

Acharya, S., S. Liyanage, P. Parajuli, S. S. Rumi, J. L. Shamshina, and N. Abidi. 2021. Utilization of cellulose to its full potential: a review on cellulose dissolution, regeneration, and applications. *Polymers* 13 (24): 4344. https://doi.org/10.3390/polym13244344.

Adeboye, P. T., M. Bettiga, and L. Olsson. 2014. The chemical nature of phenolic compounds determines their toxicity and induces distinct physiological responses in saccharomyces cerevisiae in lignocellulose hydrolysates. *AMB Express* 4 (1): 46. https://doi.org/10.1186/s13568-014-0046-7.

Ajala, E. O., M. A. Ajala, G. S. Akinpelu, and V. C. Akubude. 2022. Cultivation and processing of microalgae for its sustainability as a feedstock for biodiesel production. *Nigerian Journal of Technological Development* 18 (4): 322–343. https://doi.org/10.4314/njtd.v18i4.8.

Amândio, M. S. T., J. M. S. Rocha, and A. M. R. B. Xavier. 2023. Enzymatic hydrolysis strategies for cellulosic sugars production to obtain bioethanol from eucalyptus globulus bark. *Fermentation* 9 (3): 241. https://doi.org/10.3390/fermentation9030241.

Amiri, H.. 2020. Recent innovations for reviving the abe fermentation for production of butanol as a drop-in liquid biofuel. *Biofuel Research Journal* 7 (4): 1256–1266. https://doi.org/10.18331/BRJ2020.7.4.4.

Amiri, H., and K. Karimi. 2016. Integration of autohydrolysis and organosolv delignification for efficient acetone, butanol, and ethanol production and lignin recovery. *Industrial & Engineering Chemistry Research* 55 (17): 4836–4845. https://doi.org/10.1021/acs.iecr.6b00110.

Anand, A., Kumar, K., & Moholkar, V. S. (2024). Various Routes for Hydrogen Production and Its Utilization for Sustainable Economy. In *Biohydrogen-Advances and Processes* (pp. 503–527). Cham: Springer Nature Switzerland.

Bajpai, P. 2014. *Xylanolytic Enzymes*. Amsterdam: Elsevier/AP, Academic Press is an imprint of Elsevier.

Baral, N. R., and A. Shah. 2014. Microbial inhibitors: formation and effects on acetone-butanol-ethanol fermentation of lignocellulosic biomass. *Applied Microbiology and Biotechnology* 98 (22): 9151–9172. https://doi.org/10.1007/s00253-014-6106-8.

Barbosa, F. C., M. A. Silvello, and R. Goldbeck. 2020. Cellulase and oxidative enzymes: new approaches, challenges and perspectives on cellulose degradation for bioethanol production. *Biotechnology Letters* 42 (6): 875–884. https://doi.org/10.1007/s10529-020-02875-4.

Battısta, F., and D. Bolzonella. 2018. Some critical aspects of the enzymatic hydrolysis at high dry-matter content: a review. *Biofuels, Bioproducts and Biorefining* 12 (4): 711–723. https://doi.org/10.1002/bbb.1883.

Beg, M. A., S. Shivangi, S. C. Thakur, and L. S. Meena. 2018. Structural prediction and mutational analysis of Rv3906c gene of *Mycobacterium Tuberculosis* H37 Rv to determine its essentiality in survival. *Advances in Bioinformatics* 2018: 1–12. https://doi.org/10.1155/2018/6152014.

Bhatia, S. K., S. S. Jagtap, A. A. Bedekar, R. K. Bhatia, K. Rajendran, A. Pugazhendhi, C. V. Rao, A. E. Atabani, G. Kumar, and Y.-H.Yang. 2021. Renewable biohydrogen production from lignocellulosic biomass using fermentation and integration of systems with other energy generation technologies. *Science of The Total Environment* 765 (April): 144429. https://doi.org/10.1016/j.scitotenv.2020.144429.

Biundo, A., P. Saénz-Méndez, and T. Görbe. 2021. Enzyme modification. In *Biocatalysis for Practitioners*, edited by G. de Gonzalo and I. Lavandera, 33–62. New York: John Wiley & Sons, Ltd. https://doi.org/10.1002/9783527824465.ch2.

Bowyer, J., J. Howe, R. A. Levins, H. Groot, K. Fernholz, and E. Pepke. 2017. *The Once and Future Bioeconomy and the Role of Forests*. Minneapolis, MN: Dovetail Partners.

Brondi, M. G., V. M. Vasconcellos, R. C. Giordano, and C. S. Farinas. 2019. Alternative low-cost additives to improve the saccharification of lignocellulosic biomass. *Applied Biochemistry and Biotechnology* 187 (2): 461–473. https://doi.org/10.1007/s12010-018-2834-z.

Bruder, M. R., M. E. Pyne, M. Moo-Young, D. A. Chung, and C. P. Chou. 2016. Extending CRISPR-Cas9 technology from genome editing to transcriptional engineering in the genus clostridium. edited by M. Kivisaar. *Applied and Environmental Microbiology* 82 (20): 6109–6119. https://doi.org/10.1128/AEM.02128-16.

Califano, D., B. L. Patenall, M.A .S. Kadowaki, D. Mattia, J. L. Scott, and K. J. Edler. 2021. Enzyme-functionalized cellulose beads as a promising antimicrobial material. *Biomacromolecules* 22 (2): 754–762. https://doi.org/10.1021/acs.biomac.0c01536.

Calinescu, I., A. Vartolomei, I. -A. Gavrila, M. Vinatoru, and T .J. Mason. 2019. A reactor designed for the ultrasonic stimulation of enzymatic esterification. *Ultrasonics Sonochemistry* 54 (June): 32–38. https://doi.org/10.1016/j.ultsonch.2019.02.018.

Cao, M., V. G Tran, and H. Zhao. 2020. Unlocking nature's biosynthetic potential by directed genome evolution. *Current Opinion in Biotechnology* 66 (December): 95–104. https://doi.org/10.1016/j.copbio.2020.06.012.

Carrillo-Nieves, D., M.J. R. Alanís, R. de la Cruz Quiroz, H. A. Ruiz, H. M..N. Iqbal, and R. Parra-Saldívar. 2019. Current status and future trends of bioethanol production from agro-industrial wastes in Mexico. *Renewable and Sustainable Energy Reviews* 102 (March): 63–74. https://doi.org/10.1016/j.rser.2018.11.031.

Chandrakant K. K., V. S.Moholkar, and A. Goyal. 2021. Alkaline pretreatment and response surface methodology based recombinant enzymatic saccharification and fermentation of sugarcane tops. *Bioresource Technology* 341 (December): 125837. https://doi.org/10.1016/j.biortech.2021.125837.

Cheng, H.-H., L.-M. Whang, K.-C. Chan, M.-C. Chung, S.-H. Wu, C.-P. Liu, S.-Y.Tien, S.-Y. Chen, J.-S.Chang, and W.-J. Lee. 2015. Biological butanol production from microalgae-based biodiesel residues by clostridium acetobutylicum. *Bioresource Technology* 184 (May): 379–385. https://doi.org/10.1016/j.biortech.2014.11.017.

Chisti, Y. 2013. Constraints to commercialization of algal fuels. *Journal of Biotechnology* 167 (3): 201–214. https://doi.org/10.1016/j.jbiotec.2013.07.020.

Cho, C., S. Hong, H. Gi Moon, Y.-S. Jang, D. Kim, and S. Y. Lee. 2019. Engineering clostridial aldehyde/alcohol dehydrogenase for selective butanol production. edited by Derek R. Lovley. *mBio* 10 (1): e02683–18. https://doi.org/10.1128/mBio.02683-18.

Choi, K. R., W. D. Jang, D. Yang, J. S. Cho, D. Park, and S. Y. Lee. 2019. Systems metabolic engineering strategies: integrating systems and synthetic biology with metabolic engineering. *Trends in Biotechnology* 37 (8): 817–837. https://doi.org/10.1016/j.tibtech.2019.01.003.

Chukwuma, O., M. Rafatullah, H. A. Tajarudin, and N. Ismail. 2021a. A review on bacterial contribution to lignocellulose breakdown into useful bio-products. *International Journal of Environmental Research and Public Health* 18 (11): 6001. https://doi.org/10.3390/ijerph18116001.

Chukwuma, O. B., M. Rafatullah, H. A. Tajarudin, and N. Ismail. 2021b. A review on bacterial contribution to lignocellulose breakdown into useful bio-products. *International Journal of Environmental Research and Public Health* 18 (11): 6001. https://doi.org/10.3390/ijerph18116001.

Dawood, A., and K. Ma. 2020. Applications of microbial β-mannanases. *Frontiers in Bioengineering and Biotechnology* 8 (December): 598630. https://doi.org/10.3389/fbioe.2020.598630.

Dhagat, S., and S. E., Jujjavarapu. 2022. Utility of lignin-modifying enzymes: a green technology for organic compound mycodegradation. *Journal of Chemical Technology & Biotechnology* 97 (2): 343–358. https://doi.org/10.1002/jctb.6807.

Ejaz, U., M. Sohail, and A. Ghanemi. 2021. Cellulases: From bioactivity to a variety of industrial applications. *Biomimetics* 6 (3): 44. https://doi.org/10.3390/biomimetics6030044.

Fan, M. Z., W. Wang, L. Cheng, J. Chen, W. Fan, and M.Wang. 2021. Metagenomic discovery and characterization of multi-functional and monomodular processive endoglucanases as biocatalysts. *Applied Sciences* 11 (11): 5150. https://doi.org/10.3390/app11115150.

Ferreira Dos Santos, V., C., Augusto Duzi Sia, F. M. Filho, R. Ma. Filho, and A. P. Mariano. 2021. Isopropanol-butanol-ethanol production by cell-immobilized vacuum fermentation. *Bioresource Technology* 344 (Pt B): 126313. https://doi.org/10.1016/j.biortech.2021.126313.

Fontes, C. M.G.A., and H. J. Gilbert. 2010. Cellulosomes: highly efficient nanomachines designed to deconstruct plant cell wall complex carbohydrates. *Annual Review of Biochemistry* 79 (1): 655–681. https://doi.org/10.1146/annurev-biochem-091208-085603.

Forsberg, Z., G. Vaaje-Kolstad, B. Westereng, A. C. Bunæs, Y. Stenstrøm, A. MacKenzie, M.Sørlie, S. J. Horn, and V. G .H. Eijsink. 2011. Cleavage of cellulose by a CBM33 protein. *Protein Science* 20 (9): 1479–1483. https://doi.org/10.1002/pro.689.

Froidurot, A., and V. Julliand. 2022. Cellulolytic bacteria in the large intestine of mammals. *Gut Microbes* 14 (1): 2031694. https://doi.org/10.1080/19490976.2022.2031694.

Fuertez-Córdoba, J. M., J. C. Acosta-Pavas, and Á. A. Ruiz-Colorado. 2021. Alkaline delignification of lignocellulosic biomass for the production of fermentable sugar syrups. *DYNA* 88 (218): 168–177.

Galbe, M., and O. Wallberg. 2019. Pretreatment for biorefineries: a review of common methods for efficient utilisation of lignocellulosic materials. *Biotechnology for Biofuels* 12 (1): 294. https://doi.org/10.1186/s13068-019-1634-1.

Gandla, M. L., C. Martín, and L. J. Jönsson. 2018. Analytical enzymatic saccharification of lignocellulosic biomass for conversion to biofuels and bio-based chemicals. *Energies* 11 (11): 2936. https://doi.org/10.3390/en11112936.

Gao, D., N. Uppugundla, S. P. S. Chundawat, X. Yu, S. Hermanson, K. Gowda, P. Brumm, D. Mead, V. Balan, and B. E. Dale. 2011. Hemicellulases and auxiliary enzymes for improved conversion of lignocellulosic biomass to monosaccharides. *Biotechnology for Biofuels* 4 (1): 5. https://doi.org/10.1186/1754-6834-4-5.

Goedegebuur, F., L. Dankmeyer, P. Gualfetti, S. Karkehabadi, H. Hansson, S. Jana, V. Huynh, et al., 2017. Improving the thermal stability of cellobiohydrolase Cel7A from hypocrea jecorina by directed evolution. *The Journal of Biological Chemistry* 292 (42): 17418. https://doi.org/10.1074/jbc.M117.803270.

González-Peñas, H., T. A. Lú-Chau, N. Botana, M. T. Moreira, J. M. Lema, and G. Eibes. 2018. Organosolv pretreated beech wood as a substrate for acetone butanol ethanol extractive fermentation. *Holzforschung* 73 (1): 55–64. https://doi.org/10.1515/hf-2018-0098.

Goodell, B. 2020. Fungi involved in the biodeterioration and bioconversion of lignocellulose substrates. In *Genetics and Biotechnology*, edited by J. P. Benz and K. Schipper, 369–397. Cham: Springer International Publishing. https://doi.org/10.1007/978-3-030-49924-2_15.

Goswami, G., M. Mukherjee, J. K. Katari, S. Datta, and D. Das. 2023. Lignocellulosic bio-butanol production: challenges and solution. In *Advances and Developments in Biobutanol Production*, edited by J. G. Segovia-Hernandez, S. Behera, and E. Sanchez-Ramirez, 261–277. Amsterdam: Elsevier. https://doi.org/10.1016/B978-0-323-91178-8.00009-6.

Hille, F., and E. Charpentier. 2016. CRISPR-cas: biology, mechanisms and relevance. *Philosophical Transactions of the Royal Society B: Biological Sciences* 371 (1707): 20150496. https://doi.org/10.1098/rstb.2015.0496.

Hollinshead, W., L. He, and Y. J. Tang. 2014. Biofuel production: an odyssey from metabolic engineering to fermentation scale-up. *Frontiers in Microbiology* 5 (July): 344. https://doi.org/10.3389/fmicb.2014.00344.

Homaei, A. A., R. Sariri, F. Vianello, and R. Stevanato. 2013. Enzyme immobilization: an update. *Journal of Chemical Biology* 6 (4): 185–205. https://doi.org/10.1007/s12154-013-0102-9.

Hsu, P. D., E. S. Lander, and F. Zhang. 2014. Development and applications of CRISPR-Cas9 for genome engineering. *Cell* 157 (6): 1262–1278. https://doi.org/10.1016/j.cell.2014.05.010.

Janusz, G., A. Pawlik, J. Sulej, U. Świderska-Burek, A. Jarosz-Wilkołazka, and A. Paszczyński. 2017. Lignin degradation: microorganisms, enzymes involved, genomes analysis and evolution. *FEMS Microbiology Reviews* 41 (6): 941–962. https://doi.org/10.1093/femsre/fux049.

Jin, Q., Z. An, A. Damle, N. Poe, J. Wu, H. Wang, Z. Wang, and H. Huang. 2020. High acetone-butanol-ethanol production from food waste by recombinant clostridium saccharoperbutylacetonicum in batch and continuous immobilized-cell fermentation. *ACS Sustainable Chemistry & Engineering* 8 (26): 9822–9832. https://doi.org/10.1021/acssuschemeng.0c02529.

Kaschuk, J. J., and E. Frollini. 2018. Effects of average molar weight, crystallinity, and hemicelluloses content on the enzymatic hydrolysis of sisal pulp, filter paper, and microcrystalline cellulose. *Industrial Crops and Products* 115 (May): 280–289. https://doi.org/10.1016/j.indcrop.2018.02.011.

Khoshkho, S. M., M. Mahdavian, F. Karimi, H. Karimi-Maleh, and P. Razaghi. 2022. Production of bioethanol from carrot pulp in the presence of saccharomyces cerevisiae and beet molasses inoculum; a biomass based investigation. *Chemosphere* 286 (January): 131688. https://doi.org/10.1016/j.chemosphere.2021.131688.

Kim, J. D., G. W. Chae, H. J. Seo, N. Chaudhary, Y. H. Yoon, T. S. Shin, and M. Y. Kim. 2012. Bioalcohol production with microalgae, microcystis aeruginosa. *KSBB Journal* 27 (6): 335–340. https://doi.org/10.7841/ksbbj.2012.27.6.335.

Kim, J., M. Tremaine, J. A. Grass, H. M. Purdy, R. Landick, P. J. Kiley, and J. L. Reed. 2019. Systems metabolic engineering of *Escherichia Coli* improves coconversion of lignocellulose-derived sugars. *Biotechnology Journal* 14 (9): 1800441. https://doi.org/10.1002/biot.201800441.

Koelewijn, S.-F., S. Van Den Bosch, T. Renders, W. Schutyser, B. Lagrain, M. Smet, J. Thomas, et al., 2017. Sustainable bisphenols from renewable softwood lignin feedstock for polycarbonates and cyanate ester resins. *Green Chemistry* 19 (11): 2561–2570. https://doi.org/10.1039/C7GC00776K.

Kovács, K., B. J. Willson, K. Schwarz, J. T. Heap, A. Jackson, D. N. Bolam, K. Winzer, and N. P. Minton. 2013. Secretion and assembly of functional mini-cellulosomes from synthetic chromosomal operons in clostridium acetobutylicum ATCC 824. *Biotechnology for Biofuels* 6 (1): 117. https://doi.org/10.1186/1754-6834-6-117.

Kumar, A., and R. Chandra. 2020. Ligninolytic enzymes and its mechanisms for degradation of lignocellulosic waste in environment. *Heliyon* 6 (2): e03170. https://doi.org/10.1016/j.heliyon.2020.e03170.

Kumar, K., and V. S. Moholkar. 2023. Mechanistic aspects of enhanced kinetics in sonoenzymatic processes using three simultaneous approaches. In *Sustainable Energy Generation and Storage*, edited by V. S. Moholkar, K. Mohanty, and V. V. Goud, 41–57. Singapore: Springer Nature Singapore. https://doi.org/10.1007/978-981-99-2088-4_5.

Kumar, K., Anand, A., & Moholkar, V. S. (2024a). Molecular Hydrogen (H2) Metabolism in Microbes: A Special Focus on Biohydrogen Production. In *Biohydrogen-Advances and Processes* (pp. 25–58). Cham: Springer Nature Switzerland.

Kumar, K., Jadhav, S. M., & Moholkar, V. S. (2024b). Acetone-Butanol-Ethanol (ABE) fermentation with clostridial co-cultures for enhanced biobutanol production. *Process Safety and Environmental Protection*, 185, 277–285. https://doi.org/10.1016/j.psep.2024.03.027

Kumar, K., Kumar, S., Goswami, A., & Moholkar, V. S. (2024c). Advancing sustainable biofuel production: A computational insight into microbial systems for isopropanol synthesis and beyond. *Process Safety and Environmental Protection*, 188, 1118-1132. https://doi.org/10.1016/j.psep.2024.06.024

Kumar, K., K. Roy, and V. S. Moholkar. 2021. Mechanistic investigations in sonoenzymatic synthesis of N-butyl levulinate. *Process Biochemistry* 111 (December): 147–158. https://doi.org/10.1016/j.procbio.2021.09.005.

Kumar, K., Shah, H., & Moholkar, V. S. (2022). Genetic algorithm for optimization of fermentation processes of various enzyme productions. In Eswari J. S., Suryawanshi, N. (eds). *Optimization of sustainable enzymes production* (pp. 121–144). Chapman and Hall/CRC.

Kumar, K., A. Siddiqa, P. Chandane, S. Yadav, M. Kori, A. Shivram, L. Barbora, and V. S. Moholkar. 2022. Computational investigations in inhibition of alcohol/aldehyde dehydrogenase in lignocellulosic hydrolysates. *Bioinformatics*. https://doi.org/10.1101/2022.11.19.517192.

Kumar, K., L. Barbora, and V. S. Moholkar. 2023. Genomic insights into clostridia in bioenergy production: comparison of metabolic capabilities and evolutionary relationships. *Biotechnology and Bioengineering* 121 (4): 1298–1313. https://doi.org/10.1002/bit.28610.

Kumar, K., P. Patro, U. Raut, V. Yadav, L. Barbora, and V. S. Moholkar. 2023. Elucidating the molecular mechanism of ultrasound-enhanced lipase-catalyzed biodiesel synthesis: a computational study. *Biomass Conversion and Biorefinery* 2023: 1–12. https://doi.org/10.1007/s13399-023-04742-4.

Kushwaha, D., N. Srivastava, I. Mishra, S. Nath Upadhyay, and P. K. Mishra. 2019. Recent trends in biobutanol production. *Reviews in Chemical Engineering* 35 (4): 475–504. https://doi.org/10.1515/revce-2017-0041.

Kwon, S. W., K. A. Paari, A. Malaviya, and Y.-S. Jang. 2020. Synthetic biology tools for genome and transcriptome engineering of solventogenic clostridium. *Frontiers in Bioengineering and Biotechnology* 8 (April): 282. https://doi.org/10.3389/fbioe.2020.00282.

Lamichhane, G., A. Acharya, D. K. Poudel, B. Aryal, N. Gyawali, P. Niraula, S. R. Phuyal, P. Budhathoki, B. K. Ganesh, and N. Parajuli. 2021. Recent advances in bioethanol production from lignocellulosic biomass. *International Journal of Green Energy* 18 (7): 731–744. https://doi.org/10.1080/15435075.2021.1880910.

Lee, J.-W., and T. W. Jeffries. 2011. Efficiencies of acid catalysts in the hydrolysis of lignocellulosic biomass over a range of combined severity factors. *Bioresource Technology* 102 (10): 5884–5890. https://doi.org/10.1016/j.biortech.2011.02.048.

Lee, S. Y., and C. H. Ra. 2021. Comparison of liquid and solid-state fermentation processes for the production of enzymes and beta-glucan from hulled barley. *Journal of Microbiology and Biotechnology* 32 (3): 317–323. https://doi.org/10.4014/jmb.2111.11002.

Liang, Y., T. Si, E. L. Ang, and H. Zhao. 2014. Engineered pentafunctional minicellulosome for simultaneous saccharification and ethanol fermentation in saccharomyces cerevisiae. *Applied and Environmental Microbiology* 80 (21): 6677–6684. https://doi.org/10.1128/aem.02070-14.

Liu, Z.-H., and H.-Z. Chen. 2015. Xylose production from corn stover biomass by steam explosion combined with enzymatic digestibility. *Bioresource Technology* 193 (October): 345–356. https://doi.org/10.1016/j.biortech.2015.06.114.

Liu, Z.-H., L. Qin, F. Pang, M.-J. Jin, B.-Z. Li, Y. Kang, B. E. Dale, and Y.-J. Yuan. 2013. Effects of biomass particle size on steam explosion pretreatment performance for improving the enzyme digestibility of corn stover. *Industrial Crops and Products* 44 (January): 176–184. https://doi.org/10.1016/j.indcrop.2012.11.009.

Lopes, A. M., E. X. Ferreira Filho, and L. R. S. Moreira. 2018. An update on enzymatic cocktails for lignocellulose breakdown. *Journal of Applied Microbiology* 125 (3): 632–645. https://doi.org/10.1111/jam.13923.

Loureiro, A., and G. J. da Silva. 2019. CRISPR-Cas: converting a bacterial defence mechanism into a state-of-the-art genetic manipulation tool. *Antibiotics* 8 (1): 18. https://doi.org/10.3390/antibiotics8010018.

Luo, H., Z. Liu, F. Xie, M. Bilal, and F. Peng. 2021. Lignocellulosic biomass to biobutanol: toxic effects and response mechanism of the combined stress of lignin-derived phenolic acids and phenolic aldehydes to clostridium acetobutylicum. *Industrial Crops and Products* 170 (October): 113722. https://doi.org/10.1016/j.indcrop.2021.113722.

Malani, R. S., S. B. Umriwad, K. Kumar, A. Goyal, and V. S. Moholkar. 2019. Ultrasound-assisted enzymatic biodiesel production using blended feedstock of non-edible oils: kinetic analysis. *Energy Conversion and Management* 188 (May): 142–150. https://doi.org/10.1016/j.enconman.2019.03.052.

Malhotra, M., and S. K. Suman. 2021. Laccase-mediated delignification and detoxification of lignocellulosic biomass: removing obstacles in energy generation. *Environmental Science and Pollution Research* 28 (42): 58929–58944. https://doi.org/10.1007/s11356-021-13283-0.

Malode, S. J., K. K. Prabhu, R. J. Mascarenhas, N. P. Shetti, and T. M. Aminabhavi. 2021. Recent advances and viability in biofuel production. *Energy Conversion and Management: X* 10 (June): 100070. https://doi.org/10.1016/j.ecmx.2020.100070.

Marcellin, E., and L. K. Nielsen. 2018. Advances in analytical tools for high throughput strain engineering. *Current Opinion in Biotechnology* 54 (December): 33–40. https://doi.org/10.1016/j.copbio.2018.01.027.

Marcos, M., M. T. García-Cubero, G. González-Benito, M. Coca, S. Bolado, and S. Lucas. 2013. Optimization of the enzymatic hydrolysis conditions of steam-exploded wheat straw for maximum glucose and xylose recovery: enzymatic hydrolysis of steam-exploded wheat straw. *Journal of Chemical Technology & Biotechnology* 88 (2): 237–246. https://doi.org/10.1002/jctb.3820.

Martínez-Avila, O., J. Llimós, and S. Ponsá. 2021. Integrated solid-state enzymatic hydrolysis and solid-state fermentation for producing sustainable polyhydroxyalkanoates from low-cost agro-industrial residues. *Food and Bioproducts Processing* 126 (March): 334–344. https://doi.org/10.1016/j.fbp.2021.01.015.

Meenakshisundaram, S., A. Fayeulle, E. Léonard, C. Ceballos, X. Liu, and A. Pauss. 2022. combined biological and chemical/physicochemical pretreatment methods of lignocellulosic biomass for bioethanol and biomethane energy production-a review. *Applied Microbiology* 2 (4): 716–734. https://doi.org/10.3390/applmicrobiol2040055.

Muthukumar, S., and L. Rajendran. 2013. Analytical expressions of the concentrations of substrate and product in enzyme inhibition process. *Natural Science* 05 (09): 1047–1055. https://doi.org/10.4236/ns.2013.59129.

O'connell, A. P., C. Prussi, M. Padella, A.Konti, and L. Lonza. 2019. Sustainable advanced biofuels: technology market report. *JRC Publications Repository*: 1–200. https://doi.org/10.2760/487802.

Oriez, V., J. Peydecastaing, and P.-Y. Pontalier. 2020. Lignocellulosic biomass mild alkaline fractionation and resulting extract purification processes: conditions, yields, and purities. *Clean Technologies* 2 (1): 91–115. https://doi.org/10.3390/cleantechnol2010007.

Ostadjoo, H., K. Dietrich, M. J. Dumont, T. Friščić, and K. Auclair. 2019. Efficient enzymatic hydrolysis of biomass hemicellulose in the absence of bulk water. *Molecules* 24 (23): 4206. https://doi.org/10.3390/molecules24234206.

Ostadjoo, S., F. Hammerer, K. Dietrich, M.-J. Dumont, T. Friscic, and K. Auclair. 2019. Efficient enzymatic hydrolysis of biomass hemicellulose in the absence of bulk water. *Molecules* 24 (23): 4206. https://doi.org/10.3390/molecules24234206.

Østby, H., L. D. Hansen, S. J. Horn, V. G. H. Eijsink, and A. Várnai. 2020. Enzymatic processing of lignocellulosic biomass: principles, recent advances and perspectives. *Journal of Industrial Microbiology & Biotechnology* 47 (9): 623–657. https://doi.org/10.1007/s10295-020-02301-8.

Packer, M. S., and D. R. Liu. 2015a. Methods for the directed evolution of proteins. *Nature Reviews Genetics* 16 (7): 379–394. https://doi.org/10.1038/nrg3927.

Packer, M. S., and D. R. Liu. 2015b. Methods for the directed evolution of proteins. *Nature Reviews Genetics* 16 (7): 379–394. https://doi.org/10.1038/nrg3927.

Park, S., J. O. Baker, M. E. Himmel, P. A. Parilla, and D. K. Johnson. 2010. Cellulose crystallinity index: measurement techniques and their impact on interpreting cellulase performance. *Biotechnology for Biofuels* 3 (1): 10. https://doi.org/10.1186/1754-6834-3-10.

Palma, G. B., R. A. C. Leão, R. O. M.A. de Souza, and O. G. Pandoli. 2020. Immobilization of lipases on lignocellulosic bamboo powder for biocatalytic transformations in batch and continuous flow. *Catalysis Today*, 381: 280–287. https://doi.org/10.1016/j.cattod.2020.04.041.

Pongsupasa, V., P. Anuwan, S. Maenpuen, and T. Wongnate. 2022. Rational-design engineering to improve enzyme thermostability. *Methods in Molecular Biology (Clifton, N.J.)* 2397: 159–178. https://doi.org/10.1007/978-1-0716-1826-4_9.

Pu, Y., F. Hu, F. Huang, B. H. Davison, and A. J. Ragauskas. 2013. Assessing the molecular structure basis for biomass recalcitrance during dilute acid and hydrothermal pretreatments. *Biotechnology for Biofuels* 6 (1): 15. https://doi.org/10.1186/1754-6834-6-15.

Ranganathan, S., S. Mahesh, S. Suresh, A. Nagarajan, T. Z. Sen, and R. M.Yennamalli. 2022. Experimental and computational studies of cellulases as bioethanol enzymes. *Bioengineered* 13 (5): 14028–14046. https://doi.org/10.1080/21655979.2022.2085541.

Rasmussen, H., D. Tanner, H. R. Sørensen, and A. S. Meyer. 2017. New degradation compounds from lignocellulosic biomass pretreatment: routes for formation of potent oligophenolic enzyme inhibitors. *Green Chemistry* 19 (2): 464–473. https://doi.org/10.1039/C6GC01809B.

Rosa, M. F., E. S. Medeiros, J. A. Malmonge, K. S. Gregorski, D. F. Wood, L. H. C. Mattoso, G. Glenn, W. J. Orts, and S. H. Imam. 2010. Cellulose nanowhiskers from coconut husk fibers: effect of preparation conditions on their thermal and morphological behavior. *Carbohydrate Polymers* 81 (1): 83–92. https://doi.org/10.1016/j.carbpol.2010.01.059.

Roy, R., M. S. Rahman, and D. E. Raynie. 2020. Recent advances of greener pretreatment technologies of lignocellulose. *Current Research in Green and Sustainable Chemistry* 3 (June): 100035. https://doi.org/10.1016/j.crgsc.2020.100035.

Ruiz, H. A., R. M. Rodríguez-Jasso, B. D. Fernandes, A. A. Vicente, and J. A. Teixeira. 2013. Hydrothermal processing, as an alternative for upgrading agriculture residues and marine biomass according to the biorefinery concept: a review. *Renewable and Sustainable Energy Reviews* 21 (May): 35–51. https://doi.org/10.1016/j.rser.2012.11.069.

Saha, K., U. M. R, J. Sikder, S. Chakraborty, S. S. Da Silva, and J. C. D. Santos. 2017. Membranes as a tool to support biorefineries: applications in enzymatic hydrolysis, fermentation and dehydration for bioethanol production. *Renewable and Sustainable Energy Reviews* 74 (July): 873–890. https://doi.org/10.1016/j.rser.2017.03.015.

Said, M. S. M., W. A.W. A. K. Ghani, H. B. Tan, and D. K. S. Ng. 2021. Prediction and optimisation of syngas production from air gasification of napier grass via stoichiometric equilibrium model. *Energy Conversion and Management: X* 10 (June): 100057. https://doi.org/10.1016/j.ecmx.2020.100057.

Santos, C. N. S., W. Xiao, and G. Stephanopoulos. 2012. Rational, combinatorial, and genomic approaches for engineering L-tyrosine production in *Escherichia Coli. Proceedings of the National Academy of Sciences* 109 (34): 13538–13543. https://doi.org/10.1073/pnas.1206346109.

Santos, A. G., T. L. de Albuquerque, B. D. Ribeiro, and M. A. Z. Coelho. 2021. In situ product recovery techniques aiming to obtain biotechnological products: a glance to current knowledge. *Biotechnology and Applied Biochemistry* 68 (5): 1044–1057. https://doi.org/10.1002/bab.2024.

Sarnaik, A., A. Liu, D. Nielsen, and A. M Varman. 2020. High-throughput screening for efficient microbial biotechnology. *Current Opinion in Biotechnology* 64 (August): 141–150. https://doi.org/10.1016/j.copbio.2020.02.019.

Sharma, A., S. Balda, N. Capalash, and P. Sharma. 2022. Engineering multifunctional enzymes for agro-biomass utilization. *Bioresource Technology* 347 (March): 126706. https://doi.org/10.1016/j.biortech.2022.126706.

Singh, N., K. Kumar, A. Goyal, and V. S. Moholkar. 2022. Ultrasound-assisted biodiesel synthesis by in-situ transesterification of microalgal biomass: optimization and kinetic analysis. *Algal Research* 61 (January): 102582. https://doi.org/10.1016/j.algal.2021.102582.

Sipra, A.T., N. Gao, and H. Sarwar. 2018. Municipal solid waste (MSW) pyrolysis for bio-fuel production: a review of effects of MSW components and catalysts. *Fuel Processing Technology* 175 (June): 131–147. https://doi.org/10.1016/j.fuproc.2018.02.012.

Su, Z., J. Luo, X. Li, and M. Pinelo. 2020. Enzyme membrane reactors for production of oligosaccharides: a review on the interdependence between enzyme reaction and membrane separation. *Separation and Purification Technology* 243 (July): 116840. https://doi.org/10.1016/j.seppur.2020.116840.

Tani, S., T. Kawaguchi, and T. Kobayashi. 2014. Complex regulation of hydrolytic enzyme genes for cellulosic biomass degradation in filamentous fungi. *Applied Microbiology and Biotechnology* 98 (11): 4829–4837. https://doi.org/10.1007/s00253-014-5707-6.

Thompson, R. A., S. Dahal, S. Garcia, In. Nookaew, and C. T. Trinh. 2016. Exploring complex cellular phenotypes and model-guided strain design with a novel genome-scale metabolic model of clostridium thermocellum DSM 1313 implementing an adjustable cellulosome. *Biotechnology for Biofuels* 9 (1): 194. https://doi.org/10.1186/s13068-016-0607-x.

Tiwari, S., and A. Baghela. 2020. Challenges and prospects of xylitol production by conventional and non-conventional yeasts. In *New and Future Developments in Microbial Biotechnology and Bioengineering: Recent Advances in Application of Fungi and Fungal Metabolites: Environmental and Industrial Aspects*, edited by J. Singh and P. Gehlot, 211–222. Amsterdam: Elsevier. https://doi.org/10.1016/B978-0-12-821007-9.00016-4.

Tong, X., Z. He, L. Zheng, H. Pande, and Y. Ni. 2023. Enzymatic treatment processes for the production of cellulose nanomaterials: a review. *Carbohydrate Polymers* 299 (January): 120199. https://doi.org/10.1016/j.carbpol.2022.120199.

Van Dyk, J. S., and B. I. Pletschke. 2012. A review of lignocellulose bioconversion using enzymatic hydrolysis and synergistic cooperation between enzymes-factors affecting enzymes, conversion and synergy. *Biotechnology Advances* 30 (6): 1458–1480. https://doi.org/10.1016/j.biotechadv.2012.03.002.

Vasić, K., Ž. Knez, and M. Leitgeb. 2021. Bioethanol production by enzymatic hydrolysis from different lignocellulosic sources. *Molecules* 26 (3): 753. https://doi.org/10.3390/molecules26030753.

Vasilakou, K., P. Nimmegeers, G. Thomassen, P. Billen, and S. Van Passel. 2023. Assessing the future of second-generation bioethanol by 2030 - a techno-economic assessment integrating technology learning curves. *Applied Energy* 344 (August): 121263. https://doi.org/10.1016/j.apenergy.2023.121263.

Vasina, M., J. Velecký, J. Planas-Iglesias, S. M. Marques, J. Skarupova, J. Damborsky, D. Bednar, S. Mazurenko, and Z. Prokop. 2022. Tools for computational design and high-throughput screening of therapeutic enzymes. *Advanced Drug Delivery Reviews* 183 (April): 114143. https://doi.org/10.1016/j.addr.2022.114143.

Vélez-Mercado, M. I., A. G. Talavera-Caro, K. M. Escobedo-Uribe, S. Sánchez-Muñoz, M. P. Luévanos-Escareño, F. Hernández-Terán, A. Alvarado, and N. Balagurusamy. 2021. Bioconversion of lignocellulosic biomass into value added products under anaerobic conditions: insight into proteomic studies. *International Journal of Molecular Sciences* 22 (22): 12249. https://doi.org/10.3390/ijms222212249.

Verma, S., A. Kumar, S. Joshi, S. Gangola, and A. Rani. 2023. Role of microorganisms in agricultural waste management. In *Advanced Microbial Technology for Sustainable Agriculture and Environment*, edited by S. Gangola, S. Kumar, S. Joshi, and P. Bhatt, 137–153. Amsterdam: Elsevier. https://doi.org/10.1016/B978-0-323-95090-9.00007-8.

Victorino da Silva Amatto, I., N. Gonsales da Rosa-Garzon, F. A. de Oliveira Simões, F. Santiago, N. P. da Silva Leite, J. R. Martins, and H. Cabral. 2022. Enzyme engineering and its industrial applications. *Biotechnology and Applied Biochemistry* 69 (2): 389–409. https://doi.org/10.1002/bab.2117.

Wang, L., S. Dash, C. Y. Ng, and C. D. Maranas. 2017. A review of computational tools for design and reconstruction of metabolic pathways. *Synthetic and Systems Biotechnology* 2 (4): 243–252. https://doi.org/10.1016/j.synbio.2017.11.002.

Wang, J., Y. Liu, X. Guo, B. Dong, and Y. Cao. 2019. High-level expression of lipase from galactomyces geotrichum Mafic-0601 by codon optimization in pichia pastoris and its application in hydrolysis of various oils. *3 Biotech* 9 (10): 354. https://doi.org/10.1007/s13205-019-1891-5.

Wang, B., M. Qi, Y. Ma, B. Zhang, and Y. Hu. 2023. microbiome diversity and cellulose decomposition processes by microorganisms on the ancient wooden seawall of Qiantang River of Hangzhou, China. *Microbial Ecology* 86 (3): 2109–2119. https://doi.org/10.1007/s00248-023-02221-x.

Wei, H., Y. Yingting, G. Jingjing, Y.Wenshi, and T. Junhong. 2017. Lignocellulosic biomass valorization: production of ethanol. In *Encyclopedia of Sustainable Technologies*, edited by M. A. Abraham, 601–604. Amsterdam: Elsevier. https://doi.org/10.1016/B978-0-12-409548-9.10239-8.

Weiss, N. D., C. Felby, and L. G. Thygesen. 2019. Enzymatic hydrolysis is limited by biomass-water interactions at high-solids: improved performance through substrate modifications. *Biotechnology for Biofuels* 12 (1): 3. https://doi.org/10.1186/s13068-018-1339-x.

Xin, D., X. Chen, P. Wen, and J. Zhang. 2019. Insight into the role of α-arabinofuranosidase in biomass hydrolysis: cellulose digestibility and inhibition by xylooligomers. *Biotechnology for Biofuels* 12 (1): 64. https://doi.org/10.1186/s13068-019-1412-0.

Xin, F., W. Dong, W. Zhang, J. Ma, and M. Jiang. 2019. Biobutanol production from crystalline cellulose through consolidated bioprocessing. *Trends in Biotechnology* 37 (2): 167–180. https://doi.org/10.1016/j.tibtech.2018.08.007.

Yu, H., S. Ma, Y. Li, and P. A. Dalby. 2022. Hot spots-making directed evolution easier. *Biotechnology Advances* 56 (May): 107926. https://doi.org/10.1016/j.biotechadv.2022.107926.

Yuan, Y., B. Jiang, H. Chen, W. Wu, S. Wu, Y. Jin, and H. Xiao. 2021. Recent advances in understanding the effects of lignin structural characteristics on enzymatic hydrolysis. *Biotechnology for Biofuels* 14 (1): 205. https://doi.org/10.1186/s13068-021-02054-1.

Zhai, R., J. Hu, and J. N. Saddler. 2018. The inhibition of hemicellulosic sugars on cellulose hydrolysis are highly dependant on the cellulase productive binding, processivity, and substrate surface charges. *Bioresource Technology* 258 (June): 79–87. https://doi.org/10.1016/j.biortech.2017.12.006.

Zhang, Y., and M. Naebe. 2021. Lignin: a review on structure, properties, and applications as a light-colored UV absorber. *ACS Sustainable Chemistry & Engineering* 9 (4): 1427–1442. https://doi.org/10.1021/acssuschemeng.0c06998.

Zhang, P., Q. Li, Y. Chen, N. Peng, W. Liu, X.Wang, and Y. Li. 2022. Induction of cellulase production in trichoderma reesei by a glucose-sophorose mixture as an inducer prepared using stevioside. *RSC Advances* 12 (27): 17392–17400. https://doi.org/10.1039/D2RA01192A.

Zhang, R., D. Lin, L. Zhang, R. Zhan, S. Wang, and K. Wang. 2022. Molecular and biochemical analyses of a novel trifunctional endoxylanase/endoglucanase/feruloyl esterase from the human colonic bacterium *Bacteroides Intestinalis* DSM 17393. *Journal of Agricultural and Food Chemistry* 70 (13): 4044–4056. https://doi.org/10.1021/acs.jafc.2c01019.

Zhao, X., L. Zhang, and D. Liu. 2012. Biomass Recalcitrance. Part II: fundamentals of different pre-treatments to increase the enzymatic digestibility of lignocellulose. *Biofuels Bioproducts and Biorefining* 6 (September): 561–579. https://doi.org/10.1002/bbb.1350.

Zhao, J., M. Ma, Z. Zeng, P. Yu, D. Gong, and S. Deng. 2021. Production, purification and biochemical characterisation of a novel lipase from a newly identified lipolytic bacterium *Staphylococcus Caprae* NCU S6. *Journal of Enzyme Inhibition and Medicinal Chemistry* 36 (1): 249–257. https://doi.org/10.1080/14756366.2020.1861607.

Zhou, T., Y. Hu, X. Yan, J. Cui, Y. Wang, F. Luo, Y. Yuan, Z. Yu, and Y. Zhou. 2022. Molecular cloning and characterization of a novel exo-β-1,3-galactanase from penicillium oxalicum Sp. *Korean Society for Microbiology and Biotechnology* 68 (8) 1064–1071. https://doi.org/10.4014/jmb.2204.04012.

9 Application of Microbial Technique for the Synthesis of Organic Acid from Different Agrobiomasses

Karan Kumar, Shrivatsa Hegde, Ramyakrishna A.R., Anweshan, and Vijayanand S. Moholkar*

9.1 INTRODUCTION

Organic acids (OAs) are a class of organic compounds that contain carbon–hydrogen and carbon–oxygen bonds and exhibit lower acidity than inorganic acids (Coban, 2020). They do not completely dissociate in water, releasing H^+ ions. OAs can be classified based on their functional groups, such as carboxylic, alcohols, sulfonic, and amino acids (Ghai et al., 2023). These acids naturally occur in plants and animals and can be produced through various industrial processes (Duan et al., 2020). Some common examples include citric acid, acetic acid, lactic acid, tartaric acid, oxalic acid, formic acid, succinic acid, malic acid, and fumaric acid (Anand et. al., 2024, Kumar et. al., 2024a, 2024b, 2024c). OAs are also produced in the metabolic pathways of organisms, including tartaric acid, acetic acid, lactic acid, formic acid, oxalic acid, malic acid, and uric acid. OAs have various applications in various industries, including food, cosmetics, textiles, and agriculture (Liu & Nielsen, 2019). In the food industry, OAs are used for food preservation, as they prevent the growth of certain bacteria by varying the internal pH and disturbing their metabolic pathway, leading to increased osmotic pressure, ultimately causing permanent incapacitation. (Abbas et al., 2022). OAs are used as exfoliants and skin brighteners in the cosmetics industry. In the textile industry, they enhance crosslinkers and polymeric material dimensions (Ambaye et al., 2021).

OAs such as lactic acid, azelaic acid, glycolic acid, and almond acid are widely used in cosmetics to remove scars and wrinkles and brighten skin. The production of OAs has a diverse impact on many industry sectors (Abbas et al., 2022). OAs have also been studied for their nematicidal action in the context of biological control. According to the market study on OA, the entire annual trade is expected to grow

 DOI: 10.1201/9781003407713-9

to $12.54 billion by 2026 from $6.94 million in 2016, with a compounding annual growth rate of 79.7% (Anwar et al., 2014).

OAs can be produced through two methods: chemical synthesis and fermentation. The fermentation process is a microbial process that is environmentally friendly and highly productive, making it the preferred method over chemical synthesis, which results in environmental pollution (Zikmanis et al., 2020). Microorganisms play a crucial role in producing OAs, and microbial synthesis of OA is one of the best and most feasible methods for obtaining chemical building blocks from carbon sources (Jadaun et al., 2021). The continuous ever increasing demand for OAs have driven the boost in production and productivity by preferring new and genetically modified strains, along with optimization of fermentation procedures and improvement in recovery and purification methods (Ning et al., 2021).

OA synthesis plays a vital role in the industry. It is an essential building block for producing other chemicals, such as vinyl acetate and polyvinyl acetate, commonly used in woodworking and paper industries and produced by acetic acid (Yankov, 2022). OAs such as tartaric acid, malic acid, and citric acid are used as preservatives in the food and beverage industry, and acetylsalicylic acid (aspirin) is used as an analgesic and anti-inflammatory agent in pharmaceuticals (Abbas et al., 2022).

OA synthesis from agrobiomass, which includes agricultural waste, forestry residues, and energy crops, represents a sustainable source of raw materials for the synthesis, reducing the dependence on non-renewable resources and minimizing environmental impacts (Cubas-Cano et al., 2018). It is also cost-effective and more sustainable than traditional raw materials such as petroleum and natural gas (Passoth & Sandgren, 2019). OA synthesis from agrobiomass can help reduce greenhouse gas emissions by utilizing waste materials that could release methane, a potent greenhouse gas released into the atmosphere (Mazzoli, 2021). The use of different microbial diversity in the production can improve the stability and resilience of the process, reducing the risk of contamination or failure and increasing the overall yield (Rastogi et al., 2020). Microbes have different metabolic pathways and can produce different OAs at varying rates and products, so microbes can optimize the production of specific OAs, which makes use of the trash and enhances the process's profitability and efficiency (Jadaun et al., 2021). Some common examples of OA synthesized from agrobiomass are lactic acid, which can be produced from corn stover, wheat straw, and sugarcane bagasse by *Lactobacillus* and *Lactococcus* bacteria (Kumar et al., 2021). *Acetobacter* species produce acetic acid from a lignocellulosic biomass, sugarcane bagasse. Citric acid is produced from steep corn liquor and sugarcane molasses by *Aspergillus niger* and *Candida oleophila* (Tan et al., 2017). *Actinobacillus succinogenes* and *Escherichia coli* produce succinic acid from wheat bran and lignocellulosic biomass. Propionic acid is produced from corn stover and wheat straw using *Propionibacterium* species (Tang et al., 2016).

In conclusion, OA synthesis from agrobiomass is a sustainable and cost-effective approach that can reduce greenhouse gas emissions and minimize environmental impacts (Roell et al., 2019). Using different microbial diversity in the production can improve the stability and resilience of the process and increase the overall yield. The ongoing advancements in strain selection, fermentation optimization, and recovery and purification techniques aim to meet the increasing demand for OAs (Carrascosa et al., 2021).

9.2 OAS AND THEIR SIGNIFICANCE

Various naturally occurring foods, such as fruits, vegetables, and fermented foods, include OAs, which may be produced from several organic substances, such as carbohydrates, lipids, and amino acids (Abbas et al., 2022). Numerous biological activities, such as cellular metabolism, food storage, and gastrointestinal health, depend heavily on OAs, which play a role in cellular pH control and energy generation (Coban, 2020). OAs are employed in food preservation to prevent the development of dangerous bacteria, yeast, and molds, increasing food goods' shelf life (Mazzoli, 2021). They play a crucial role in managing the gut flora, which is critical for general health and well-being; OAs are also vital to maintaining gut health (Ebeid & Al-Homidan, 2022). Acetic acid is a typical food preservative that aids in preventing the growth of hazardous germs. In contrast, citric acid is a key component of the citric acid cycle, a crucial metabolic route in cells (Rastogi et al., 2020).

There are several commercial and technical uses for OAs. In addition to serving as intermediates in creating numerous compounds and polymers (Zikmanis et al., 2020), they are employed as ingredients in food preservatives, solvents, and detergents. In the food and beverage industry, using OAs as preservatives to lengthen the shelf life of items is prevalent (Pylak et al., 2019). They also promote the flavor and acidity of many other foods and drinks, including vinegar, citrus fruits, and wine. In the pharmaceutical business, OAs can be used as intermediates and active ingredients in various medicinal products. They can also act as catalysts, solvents, and starting ingredients in multiple processes (Duan et al., 2020; Sun et al., 2020). In agriculture, OAs are used as fungicides, insecticides, and herbicides. They can also be used to increase the acidity of soil and water, which benefits plant health and growth. Polyvinyl acetate, a typical polymer used in adhesives and paints, is produced by the polymerization of acetic acid (Mazzoli, 2020). Some OAs, such as citric acid, have also been investigated for their potential medicinal benefits. Citric acid effectively treats kidney stones by boosting urinary citrate excretion, which lowers the concentration of calcium ions in the urine and prevents the development of calcium oxalate and calcium phosphate stones (Luo et al., 2023). Research has also been done on the possible cancer-treating effects of citric acid. It has been demonstrated to stop cancer cells from metabolizing glycolysis and to create oxidative stress, which releases reactive oxygen species (ROS) that break DNA and kill cells. It is known as "programmed cell death" or apoptosis (Bicas et al., 2016).

Here are a few more instances of OAs that have beneficial impacts on health:

Lactic acid: Lactic acid is a by-product of cellular respiration that is present in a variety of bodily tissues. Lactic acid's anti-inflammatory qualities have been demonstrated to promote wound healing and lessen pain and swelling (Castillo Martinez et al., 2013; Abedi & Hashemi, 2020).

Acetic acid: Acetic acid has been demonstrated to provide several health advantages. Studies have been conducted on acetic acid's anti-inflammatory, anti-microbial, and anti-cancer effects (Pandey et al., 2021).

Fumaric acid: This dicarboxylic acid may be discovered in several fruits and vegetables. Psoriasis is a persistent autoimmune condition that affects the skin, and fumaric acid has been used to treat it (Guo et al., 2020).

Succinic acid: This is a dicarboxylic acid found in various foods and has been shown to have anti-inflammatory and antioxidant properties. It has also been studied for its potential to treat various metabolic disorders, such as diabetes (Raj et al., 2023).

9.3 MICROBES INVOLVED IN THE PRODUCTION OF ORGANIC ACID

Microbes transform organic matter through various metabolic processes into OAs, which are produced as a result (Alonso et al., 2015). As the world places greater importance on sustainable and renewable energy sources, the biotech industry has redirected its attention to novel microorganisms with superior functional and fermentation properties (Cappelletti et al., 2020).

Utilizing microorganisms in OAs synthesis offers numerous benefits over chemical synthesis techniques, such as higher efficiency, genetic modifiability, and cost-effectiveness (Rastogi et al., 2020). The selection of microorganisms is essential for the optimal synthesis of OAs. Bacteria and fungi are two significant categories of microbes involved in synthesizing OAs (Alotaibi et al., 2022).

Bacteria: Bacteria transform organic materials into OAs through metabolic processes, including acetic acid and alcohol fermentation (Lu et al., 2020). New synthetic metabolic pathways capable of manufacturing various biochemicals have been developed for non-natural producer organisms owing to recent advances in synthetic biology and metabolic engineering (Zaaba & Jaafar, 2020). Focusing on substrate specificity and tolerance robustness to produce overproduction phenotypes, these pathways have significantly increased the chemical repertoire of microbes. Different bio-based feedstocks can be used by extending the substrate range, and increasing the tolerance robustness can result in lower costs for downstream processing. Some common bacteria used in the production of OAs include *Acetobacter, E. coli, and Klebsiella* (Alotaibi et al., 2022; Amin et al., 2021; Domingues et al., 2021).

Genetically modified bacteria, including *E. coli* and *Corynebacterium glutamicum*, can withstand harmful substances and break down carbon sources like xylose and arabinose (Hamann & Noronha, 2022). The creation of microbial platforms with increased capabilities has attracted a lot of attention as a way to overcome challenges associated with the commercial production of OAs and to introduce novel activities in production hosts such as *E. coli* (Zhang et al., 2019). *E. coli* has been chosen as a bacterial host for manufacturing due to its adaptability, plasticity, scalability, and vast understanding of genetics, leading to improved succinic, glucaric, glycolic, and malic acid yields through metabolic engineering (Roell et al., 2019). When using *E. coli* as a host for a biorefining platform, it is necessary to consider the toxicity and tolerance mechanisms of OAs, and product toxicity continues to be a significant concern (Yu et al., 2020).

Fungi: Some widespread fungi, such as *Aspergillus*, *Rhizopus*, and *Penicillium*, use metabolic processes, such as citric acid fermentation, to transform organic materials into OAs (Cappelletti et al., 2020). Since xylose is the most prevalent pentose sugar in hemicellulosic hydrolysate, *Saccharomyces. cerevisiae* is frequently utilized in low-pH commercial OA synthesis (Parapouli et al., 2020). However, it is incompatible with this process. Hemicellulase and xylose isomerases, on the other hand, can be produced by *S. cerevisiae* to convert them into OAs. *S. cerevisiae* strains have been engineered to create a range of OAs, including succinic and malic acid, by suppressing their native capacity to produce ethanol, enhancing carboxylate reactions, and expressing heterologous product exporters (Amin et al., 2021). Industries have used yeast-based manufacturing techniques at low pH levels to produce OAs like lactic acid and succinic acid. Galactonate, xylanase, and pectinases have also been produced using synthetic fungi such as *A. niger* and *Pichia kudriavzevii*. Genetically modified *Rhizopus oryzae* has shown the potential to generate fumarate from renewable carbohydrates (Panda et al., 2019).

9.3.1 *Lactobacillus*

Lactobacillus is a Gram-positive bacterium with a thick cell wall that retains a crystal violet stain and a rod-shaped structure about 0.5–2 micrometers long (Bukhari et al., 2020; N. R. et al., 2022). *Lactobacillus* is a facultative anaerobe that can survive in both oxygenated and non-oxygenated environments. It produces lactic acid through sugar fermentation (Mazzoli et al., 2020; Abedi & Hashemi, 2020). It plays a vital role in developing probiotics and medicines and producing enzymes and biochemicals. The bacteria can be genetically engineered to produce specific proteins or compounds useful in various food processing applications and pharmaceuticals (Sun et al., 2020). It plays a prominent role in medical applications, including treating bacterial vaginosis, preventing urinary tract infections, and treating inflammatory bowel disease. In the food and beverage sector, *Lactobacillus* bacteria are frequently employed to produce fermented goods like yogurt, kefir, cheese, sauerkraut, and pickles (Guimarães et al., 2018). This raises these items' nutritional content while enhancing their flavor and texture. Using *Lactobacillus*, probiotic supplements that enhance the immune system and improve digestive health are created in the pharmaceutical sector (Okoye et al., 2022). Because some strains of *Lactobacilli* can form compounds with antibacterial qualities, it is also utilized to make antibiotics. *Lactobacillus* bacteria are employed in agriculture and animal husbandry to improve the health of plants and animals (Di Biase et al., 2022). It is frequently used as a feed additive to boost animal health and production and as a soil inoculum to stimulate plant growth and yield. In the biotechnology sector, *Lactobacillus* bacteria are utilized to make enzymes and other biochemicals (Miksusanti et al., 2016). Enzymes produced by some *Lactobacilli* strains can be used to make biofuels and other industrial goods. Cosmetic items like skin creams and emulsions for personal hygiene are made using *Lactobacillus*. It contains hydrating and anti-inflammatory effects to help the skin look and feel better (Xie et al., 2022). *Lactobacillus* bacteria are widely utilized to generate OAs from biomass. OAs are essential industrial chemicals used in different industries, such as food and drinks, pharmaceuticals, and chemicals. For instance, *Lactobacilli* create the ubiquitous organic acid lactic acid from biomass (Miksusanti et al., 2016).

During fermentation, *Lactobacillus* can generate OAs from biomass. *Lactobacillus* converts the carbohydrates in the biomass during fermentation into OAs such as lactic acid, acetic acid, and propionic acid (Chen et al., 2019; Yankov, 2022). Many sources, including agricultural waste, food waste, and forest residues, can provide the biomass needed for this process (Abedi & Hashemi, 2020). There are many benefits to using *Lactobacilli* to create OAs from biomass. First, because *Lactobacillus* is a natural creature and no hazardous waste is produced during the process, it is environmentally benign. A cheap and plentiful natural resource like biomass makes the process cost-effective. Third, the method is appropriate for industrial production since it is simple to scale up and create vast amounts of OAs. A potential area of study that may offer sustainable solutions for creating vital industrial chemicals is using *Lactobacilli* in synthesizing OAs from biomass (N. R. et al., 2022; Roberto Mazzoli, 2020; Bukhari et al., 2020).

9.3.2 Streptococcus

Gram-positive, non-motile, non-sporogenous cocci or short rods in chains or pairs are known as *Streptococci*. With a fermentative metabolism that yields L-(+) lactic acid as the main by-product of glucose fermentation, they are facultatively anaerobic chemo-organotrophs with intricate dietary needs (Barak et al., 2022). Their long-chain fatty acids are mostly straight-chain or mono-unsaturated, and they have many peptidoglycan types. They are also catalase-negative (Cappelletti et al., 2020). Most species of *Streptococcus* do not produce pyrrolidonyl arylamidase. Although reported levels typically lie between 35% and 43%, the range of the guanine plus cytosine concentration is between 33% and 46%. *Streptococcus* is a varied genus of Gram-positive bacteria that is significant in both business and medicine (Solieri et al., 2022; Barak et al., 2022). While some *Streptococci* are part of the normal microbial flora, others can cause diseases ranging from mild to severe. The nomenclature for *Streptococci* is primarily based on serogroup identification (Zhang et al., 2023). Currently, 40 species in the *Streptococcus* genus have been shown to form six species groups: *pyogenic*, *mitis*, *salivarius*, *bovis*, *anginosus*, and *mutants*. *Streptococcus suis* and *Streptococcus acidominimus* are not clustered with other Streptococcal species based on rRNA sequence analysis (Jadaun et al., 2021). *Streptococcus adjacens* and *Streptococcus defectiva* have been transferred to a new genus, *Abiotrophia*. Several newly recognized species, including *Streptococcus capricious*, *Streptococcus gallactolyticus*, *Streptococcus difficile*, and *Streptococcus phocae*, have also been identified. *Streptococci* are used in various applications, such as producing OAs, antibiotics, and probiotics (Garavand et al., 2023). The fermentative metabolism of *Streptococci* produces OAs, such as lactic acid, which are used in the food industry for the fermentation and preservation of various foods. *Streptococcus thermophilus* is widely used in the dairy industry to produce yogurt and cheese due to its ability to produce lactic acid and contribute to the flavor and texture of the final product (Singh et al., 2022). *Streptococci* are also produced in the products, such as streptomycin and erythromycin. *Streptococcus pneumoniae* has a pneumococcal vaccine, which protects against pneumococcal diseases such as pneumonia and meningitis (Solieri et al., 2022). Furthermore, *Streptococci* are used as probiotics, live bacteria that can benefit human health. *Streptococcus salivarius*, for example, is found in the human oral cavity and has been shown to have probiotic effects, including the prevention of dental caries and treatment of Streptococcal pharyngitis (Hu et al., 2022; Solieri et al., 2022). Currently, there is yet to be a comprehensive identification scheme for *Streptococci*. Particularly with the viridans species, chromogenic or fluorogenic substrate assays have been beneficial when paired with conventional biochemical tests such as acetoin synthesis, arginine hydrolysis, aesculin, and carbohydrate fermentation. Standardization has been aided by the creation of commercial identification test kits, but more work has to be done before a fully functional system is accomplished (Singh et al., 2022; Zhang et al., 2023; Barak et al., 2022).

9.3.3 *SACCHAROMYCES CEREVISIAE*

The unicellular fungus *S. cerevisiae* has 16 chromosomes, a genome of approximately 12,068 kb, and about 6,000 genes, of which 5,570 are thought to encode proteins (Chen et al., 2022; Parapouli et al., 2020). Through lateral gene transfer, the organism has absorbed a few foreign genes that code for prokaryotic or eukaryotic origin proteins (Chen et al., 2022). Extra-chromosomal components, such as retroviruses, single- and double-stranded RNA molecules, mitochondrial DNA, and the 2 m circle, are also included in the *S. cerevisiae* genome. *S. cerevisiae* is frequently employed as a model organism in fundamental research and several commercial applications because of its capacity to produce and accumulate ethanol even under aerobic conditions (Yu et al., 2022; Parapouli et al., 2020). The organism's extraordinary tolerance to high sugar concentrations and capacity to synthesize aromatic and volatile compounds are particularly significant in some industrial applications (Tran & Zhao, 2022).

S. cerevisiae, commonly known as baker's or brewer's yeast, is a valuable model organism for basic research and industrial applications (Parapouli et al., 2020). This is due to its unique "make-accumulate- consumption" feature, which is based on the Crabtree effect. This allows the yeast to produce and accumulate ethanol, eliminating competition from other microbial species (Yu et al., 2022). This strategy evolved gradually before the whole-genome duplication of *S. cerevisiae* and other yeast species occurred approximately 100 million years ago (Tofalo et al., 2022).

S. cerevisiae is a versatile organism that can survive and occupy various natural niches, such as overwintering in soil, colonizing the leaves and trunks of trees, and being insect-borne (Yu et al., 2022). It is found rarely in intact grapes but more frequently in damaged grapes because of its presence in insects that feed on them. Environmental strains of *S. cerevisiae* have additional survival strategies and can bear genotypes with potentially exciting properties for biotechnological applications (Fu et al., 2023). However, using "wild" strains for industrial applications may not be straightforward because genetic diversity does not always correspond to phenotypic diversity (Zhao et al., 2022). Recent studies have found that *S. cerevisiae* strains originating from sugar-rich environments are more efficient in fermentation compared to laboratory or environmental strains (Solieri et al., 2022). The molecular basis of this improved adaptation is unknown and may involve epigenetic phenomena. *S. cerevisiae* is widely used in the fermentation of foods and beverages. The European yeast industry produces one million tons annually, with 30% of that quantity going outside (Tofalo et al., 2022). Across the world, *S. cerevisiae* is used to prepare various distilled and fermented drinks and other alcoholic beverages (Parapouli et al.,

2020). It is possible for fermentation to start either naturally or after adding a pure yeast culture. The role of *S. cerevisiae* in wine, bread, and cocoa fermentations is examined, along with the biochemical processes that define the end products, the qualities required for effective starters, and the possibility of using native strains in the industry (Solieri et al., 2022; Tofalo et al., 2022; Zhang et al., 2019). The most studied and often utilized eukaryotic organism is *S. cerevisiae,* employed in manufacturing ethanol, food, and wine, among other industrial products (Dharmalingam et al., 2023). There is still room for refining the current strains or utilizing the huge natural reservoir of environmental isolates, even though the various *S. cerevisiae* strains used in these processes have shown effective adaptability (Zhao et al., 2022).

9.3.4 *Acetobacter pasteurianus*

The gram-negative bacterium *Acetobacter pasteurianus,* a member of the *Acetobacteraceae* family, is widespread in nature and can be found in various habitats (Wu et al., 2018). This rod-shaped bacterium converts ethanol to acetic acid, an essential step in vinegar production. As a facultative anaerobe with a respiratory chain, *A. pasteurianus* may effectively employ oxygen during ethanol oxidation (Yang et al., 2019). This bacterium is also renowned for its exceptional capacity to ferment acetic acid to generate premium vinegar with a distinctive flavor and aroma (Wu et al., 2018; Di Biase et al., 2022). Additionally, *A. pasteurianus* has pH homeostasis mechanisms and membrane transporters that enable it to survive in high acetic acid concentrations. These bacteria have numerous biotechnological and commercial uses (Alonso et al., 2015). *A. pasteurianus* has multiple applications in various fields. One of its most significant applications is the production of bacterial cellulose, a valuable biomaterial with a wide range of applications in medicine, cosmetics, and food (Ebeid & Al-Homidan, 2022). *A. pasteurianus* produces large quantities of bacterial cellulose from various sugars (Yang et al., 2019). Additionally, *A. pasteurianus* can produce biosensors that can detect and measure various analytes in environmental and biological samples. These biosensors use *A. pasteurianus* as a biorecognition element to identify and bind specific analytes (Ingle et al., 2020). Studies have also revealed that *A. pasteurianus* has potential therapeutic uses because of its ability to produce biologically active compounds, such as acetic acid (Wu et al., 2018). *A. pasteurianus* can survive in the gut and may promote gut health by producing beneficial metabolites, thus making it a potential probiotic (Zheng et al., 2017). *A. pasteurianus* is being studied for its potential to remediate contaminated environments by degrading organic pollutants, such as PAHs, chlorinated solvents, and phenolic compounds, and oxidizing them to fewer toxic metabolites (Yassunaka Hata et al., 2023). Overall, *A. pasteurianus* is a versatile bacterium with a broad range of industrial and biotechnological applications, making it a promising candidate for various environmental and industrial uses (Ghai et al., 2023).

9.3.5 *Clostridium thermocellum*

Clostridium thermocellum can break down complex plant components, especially cellulose, into simpler sugars (Zamani et al., 2023). It produces various extracellular enzymes involved in cellulose degradation, including cellulases and hemicellulases.

Because *C. thermocellum* is a thermophilic bacterium, it can grow and flourish in hot environments (Zamani et al., 2023; Chen et al., 2022). It can be used in high-temperature industrial processes because of its optimal temperature range of 55°C–60°C. As an obligate anaerobe, *C. thermocellum* can only grow without oxygen (Dash et al., 2017). Its primary metabolic mechanism is fermentation. Endospores, which are highly resilient structures that enable *C. thermocellum* to survive in hostile settings, are a property of the bacteria. Around 3,000 genes make up the extensive and intricate genome of *C. thermocellum* (Schroeder et al., 2023). It is believed that this complexity is what allows it to break down complicated plant components efficiently. By its metabolic pathways, *C. thermocellum* can also produce hydrogen gas, making it helpful for the creation of bioenergy (Pylak et al., 2019; Duan et al., 2020). Due to its distinct characteristics, *C. thermocellum* is a strong candidate for use in various biotechnological processes, particularly in generating bioenergy and bioremediation (Ingle et al., 2020).

The capacity of *C. thermocellum* to effectively break down cellulose and other complex plant materials has been proven to substantially impact its potential for synthesizing OAs from biomass (Zamani et al., 2023). Cellulases and hemicellulases, two extracellular enzymes produced by *C. thermocellum*, are effective at breaking down cellulose and other intricate plant components (Cappelletti et al., 2020). For the effective conversion of biomass into OAs, this ability is crucial. Because *C. thermocellum* is a thermophilic bacterium, it can grow and flourish in hot environments (Zamani et al., 2023). Because high temperatures can increase the effectiveness of enzymatic activities, this is crucial for efficiently converting biomass into OAs (Hamann & Noronha, 2022). It is well known that *C. thermocellum* can withstand potent end-product inhibition, which is essential for the practical synthesis of organic matter. Acids with a high concentration of the final product, such as OAs, can inhibit the formation of the product by stopping the activity of the enzymes needed to produce it. Because *C. thermocellum* can withstand significant concentrations of OAs, manufacturing is effective (Sharma et al., 2023). Moreover, *C. thermocellum* can endure an acidic environment, which is crucial for forming OAs since many of the conditions created during the fermentation of OAs are acidic (Gupta et al., 2016; Chen et al., 2019).

9.3.6 *Megasphaera hexanoica*

The cocci-shaped bacterium *Megasphaera hexanoica* typically appears in groups. *M. hexanoica* is anaerobic, meaning it cannot thrive with oxygen (Sun et al., 2022). It also consumes organic molecules to produce energy, making it a chemo-organotroph. Together with other members of the oral microbiota, *M. hexanoica* is frequently found in the oral cavity (Hackmann, 2023). *M. hexanoica* is typically regarded as commensal, which means it does not harm its host, yet it is connected to some illnesses, such as endodontic infections and periodontitis (Kang, 2020). The genome of *M. hexanoica* was sequenced, and it was discovered to have about 2.5 million base pairs. It was also found to have genes related to lipids, amino acids, and carbohydrate metabolism (Kang et al., 2022). *M. hexanoica*, a component of the oral microbiota, can affect oral health. *M. hexanoica* is being studied for its potential significance in diseases, including endodontic infections and periodontitis, and there is a chance that new treatments will be created that specifically target this bacterium (Yang et al., 2019). We looked at *M. hexanoica*'s capacity to manufacture short-chain fatty acids (SCFA). It could be used in the creation of bioplastics and other industrial chemicals (Sun et al., 2020). *M. hexanoica* is particularly successful at creating hexanoic acid, a six-carbon SCFA that can be used as a precursor for nylon manufacture (Diaz-Ruano et al., 2023). SCFAs can be created by fermenting organic material (Dharmalingam et al., 2023). For use as a probiotic in feed, *M. hexanoica* has been researched. Live bacteria known as probiotics, such as *M. hexanoica*, can improve the intestinal health of livestock. Probiotics can improve the health of the host animal (Fu et al., 2023).

M. hexanoica's capacity to create SCFA through fermentation led to its identification as a promising candidate for generating OAs from agrobiomass (Roy et al., 2023). *M. hexanoica*, in particular, was discovered to be successful at producing hexanoic acid, a six-carbon SCFA that can be employed as a precursor for the synthesis of nylon (Mazzoli et al., 2020). Agrobiomasses, such as agricultural waste, have been discovered as potential renewable carbon sources for manufacturing bioplastics and other industrial chemicals (Roell et al., 2019). However, producing these goods from agrobiomass necessitates the utilization of microorganisms that can successfully ferment the raw material to produce the desired good (Liu & Nielsen, 2019). *M. hexanoica*'s capacity to ferment a wide range of substrates, including sugars and OAs, made it an excellent option for this procedure. Moreover, it has been discovered that *M. hexanoica* can withstand high levels of OAs, which can inhibit other fermentation-related microbes (Kang, 2020).

A sustainable bio-based economy can benefit from using *M. hexanoica* to produce OAs from agrobiomass, reducing reliance on fossil fuels and providing a renewable carbon source for manufacturing industrial chemicals (Dharmalingam et al., 2023; Okoye et al., 2022).

9.3.7 *Aspergillus*

Aspergillus is the filamentous fungi belonging to the genus (Aberathna et al., 2023). *Aspergillus* is widely distributed in the environment and can be found in soil, plant waste, and indoor settings. This genus has more than 300 species, is highly diverse, and has many uses in the biotechnology, food, and pharmaceutical industries (Lee et al., 2018). But *Aspergillus* is also known to cause various infections, from minor allergic reactions to fatal invasive infections in people with compromised immune systems (Lee et al., 2018; Liu & Nielsen, 2019). The hyphae, which form an intertwined network of septate filaments and suggest that they are divided by walls or septa, distinguish the fungus and allow for the transfer of nutrients and communication between the compartments of the fungus (Cubas-Cano et al., 2018). *Aspergillus hyphae* are typically unbranched, but some species can produce branches (Sani et al., 2020). One of *Aspergillus*' most notable characteristics is the presence of specialized hyphae called *conidiophores*, which produce asexual spores called conidia (Sharma et al., 2022). The conidiophores are often highly branched, and the conidia are produced on the tips of the branches (Cubas-Cano et al., 2018). The fungus *Aspergillus* is highly adaptable and can grow in various environmental settings. An aerobic organism can thrive when oxygen is present and withstand low oxygen levels (Chen et al., 2022). Additionally, *Aspergillus* species can thrive in multiple conditions, including acidic to alkaline pH levels and temperatures between 5 and 60 (Zhu et al., 2020). They can use a variety of nitrogen sources as well as a wide range of carbon sources, such as simple sugars, complex carbohydrates, and lipids. *Aspergillus* can be found in various habitats, such as soil, dead vegetation, and enclosed spaces (Singh et al., 2022). Some *Aspergillus* species are found in plants, where they may produce healthy compounds or cause disease. Others are frequently discovered in grains or other food products that have been stored, where they can lead to spoilage (Carrascosa et al., 2021). Air samples can also contain *Aspergillus* species, which can cause issues with indoor air quality (Chen et al., 2022). In biotechnology, *Aspergillus* is a significant genus because it has been used to produce a variety of enzymes, OAs, and other compounds (Ebeid & Al-Homidan, 2022). The production of citric acid by *A. niger*, used in the food and beverage industry as a flavor enhancer and preservative, is one of the most well-known examples (Yassunaka Hata et al., 2023). In addition, *Aspergillus* makes enzymes with numerous industrial uses, including amylases, proteases, and cellulases (Coban, 2020). Additionally, *Aspergillus* species are used in manufacturing antibiotics that are frequently employed in treating bacterial infections, including penicillin and cephalosporins (Fu et al., 2023). While many *Aspergillus* species are unharmful, some have been linked to human and animal diseases. Particularly in people with compromised immune systems, *Aspergillus* infections can range from minor allergic reactions to severe and life-threatening invasive conditions (Zheng et al., 2017). Inhaling *Aspergillus* spores can cause respiratory infections like Aspergillosis. *Aspergillus* is a highly diverse genus of filamentous fungi with many applications in the biotechnology, food, and pharmaceutical industries (Cornélio Favarin et al., 2013). Still, it can also cause infections in humans and animals (Aberathna et al., 2023). Its ability to adapt to a wide range of environmental conditions and utilize various carbon and nitrogen sources makes it a highly versatile organism that has been the focus of numerous studies in various fields (Cubas-Cano et al., 2018).

9.4 VARIOUS TYPES OF OAS FROM AGROBIOMASS

These acids manifest in different forms in nature, such as the biomass of crops and residues (Ning et al., 2021). OAs sourced from agrobiomass possess diverse practical applications across several industries, including food, pharmaceuticals, and energy. This information is primarily of academic significance (Vet et al., 2014).

When extracting OAs from agrobiomass, various types are available, each with unique characteristics and potential applications (Ebeid & Al-Homidan, 2022; Guimarães et al., 2018; Abbas et al., 2022). This discussion will focus on the most commonly extracted OAs from agrobiomass.

Citric acid: Citric acid, derived from citrus fruits like lemons, limes, and oranges, as well as from sugar and corn processing waste, is a prevalent OA used in the food and beverage industry (Singh & Kumar, 2021). It is a flavor enhancer, pH adjuster, and preservative in various food products, including soft drinks, jams, and jellies (Roy et al., 2023).

Lactic acid: It is a natural acid created by certain microorganisms like bacteria and fungi extracted from corn, beets, and potatoes (Chen et al., 2019; Di Biase et al., 2022). It's commonly used in the food and beverage industry to enhance flavor and as a preservative, while also utilized as an exfoliant in cosmetics and drug delivery agents in pharmaceuticals (Castillo Martinez et al., 2013).

Succinic acid: It is a dicarboxylic acid produced through agrobiomass fermentation, including crops like corn, wheat, and sugar beet (Raj et al., 2023). It's applied in producing various chemicals like resins, plastics, and solvents, while also serving as a food preservative and flavor enhancer (Raj et al., 2023; Luo et al., 2023).

Acetic acid: It is a weak OA produced through ethanol oxidation derived from agrobiomass such as sugarcane bagasse and corn cobs (Wu et al., 2018). It has chemicals like vinyl acetate and acetic anhydride while functioning as a pH adjuster, food preservative, and flavor enhancer (Yassunaka Hata et al., 2023).

Fumaric acid: It is another dicarboxylic acid created through agrobiomass fermentation of crops such as corn and sugar beet (Díaz et al., 2020). It's utilized as an acidulant, flavor enhancer in the food industry, and an ingredient in drugs for skin disorders such as psoriasis (Coban, 2020).

Malic acid: It is a dicarboxylic acid occurring naturally in fruits and vegetables like apples and grapes and can also be obtained from agrobiomass such as corn and sugar beet (Yu et al., 2022). The food and beverage industry utilizes malic acid as a flavor enhancer, acidulant, and pH adjuster. In contrast, the pharmaceutical industry employs it as an ingredient in drugs for treating dry mouth (Barak et al., 2022).

Tartaric acid: It is a naturally occurring acid found in many fruits and vegetables and commonly produced by grape fermentation (Dharmalingam et al., 2023). It has multiple applications in producing food and beverages, as a food preservative, and as a chemical intermediate for producing other chemicals (Carrascosa et al., 2021).

Levulinic acid: It is a relatively new OA produced from biomass-derived carbohydrates (Ning et al., 2021). Its applications include the production of biofuels, biodegradable plastics, and other chemicals (Sharma et al., 2023).

Itaconic acid: It is found naturally in many fruits and vegetables, and it can be produced by fermenting sugars using microorganisms such as fungi (Sun et al., 2020). It has uses in producing biodegradable plastics, as a chemical intermediate in producing other chemicals, and as a feed additive (Chen et al., 2022).

Propionic acid: It is a naturally occurring acid found in many dairy products, and it can be produced by fermenting sugars using bacteria like *Propionibacterium* (Jiang et al., 2022). It has applications in producing food and feed additives, as a chemical intermediate in making other chemicals, and as a preservative (Fu et al., 2023).

OAs derived from agrobiomass provide sustainable and eco-friendly alternatives to traditional chemical compounds (Ebeid & Al-Homidan, 2022). Extracting OAs from agrobiomass can also offer an additional source of income for farmers and reduce waste produced by the agricultural industry. Numerous other OAs can be made from different biomass sources, including lignocellulosic materials, starches, and sugars (Ning et al., 2021; Domingues et al., 2021).

9.4.1 Lactic Acid Production from Sugar Beet Pulp

Sugar beet pulp can be fermented to create lactic acid (Díaz et al., 2020). The by-product of processing sugar beets, sugar beet pulp, provides carbohydrates that lactic acid bacteria can use to make lactic acid (Chen et al., 2019). There are several methods used for the synthesis of lactic acid (Abedi & Hashemi, 2020; Castillo Martinez et al., 2013; Mazzoli et al., 2020). They are as follows:

Direct fermentation: This process involves directly fermenting sugar beet pulp with lactic acid bacteria to make lactic acid. Enzymes first convert complex carbohydrates in the pulp into simple sugars (Zaaba & Jaafar, 2020). The pulp is then inoculated with lactic acid bacteria and left to ferment for a few days. The fermented pulp is removed from the lactic acid and then purified (Pandey et al., 2021).

Enzyme hydrolysis and fermentation: By using enzymes, the beet pulp is hydrolyzed into simple sugars, which lactic acid bacteria then ferment (Chen et al., 2019). First, cellulose enzymes are used to break down the cellulose fibers in the cellulose, and then amylase enzymes are used to break down starch into simple sugars (Díaz et al., 2020). After that, lactic acid bacteria ferment the resulting mixture of simple carbohydrates to create lactic acid (Castillo Martinez et al., 2013).

Two-stage fermentation: This process has two fermentation stages. Cellulase enzymes are produced during the fermentation of the sugar beet pulp in the first phase (Zhang et al., 2019). Enzymes break down the complex carbohydrates in the pulp into simple sugars, which are utilized in the second fermentation stage. The second stage involves adding lactic acid bacteria to the fermented pulp, which produces lactic acid (Mazzoli et al., 2020).

The steps in the lactic acid production process include the following (Tan et al., 2017; Pandey et al., 2021; Abedi & Hashemi, 2020):

- **Pretreatment:** The pulp from sugar beets is cleaned and given an enzyme treatment to convert complex carbs into simple sugars.
- **Fermentation:** A bioreactor ferments the processed pulp after being combined with lactic acid bacteria. Bacteria eat sugars during fermentation and create lactic acid as a by-product.
- **Separation:** Using techniques like filtering, centrifugation, or precipitation, the lactic acid is isolated from the fermented matter after fermentation.
- **Purification:** To produce ultrapure lactic acid, the separated lactic acid is purified using several processes, including distillation, ion exchange, and crystallization.

Factors that affect the yield of lactic acid from this fermentation process are:

- **pH:** The production of lactic acid can also be impacted by the pH of the fermentation medium. pH values between 5.0 and 6.5 provide the ideal habitat for lactic acid bacteria to flourish. pH values outside this range can prevent the lactic acid generation and bacterial development.

- **Temperature:** The fermentation process's temperature influences the rate and yield of lactic acid production. At 30°C, lactic acid bacteria often thrive. Temperatures above or below this range can inhibit bacterial growth and the formation of lactic acid.
- **Substrate concentration:** The pace and quantity of lactic acid production can be influenced by the sugar beet pulp concentration. Low substrate concentrations might not provide enough carbon supply for bacterial growth and the formation of lactic acid, while high substrate concentrations might cause substrate inhibition.
- **Oxygen concentration:** As lactic acid bacteria are anaerobic, they can grow and create lactic acid without the help of oxygen. As a result, oxygen can prevent the formation of lactic acid.
- **Nutrient availability:** For growth and lactic acid production, lactic acid bacteria require nutrients like nitrogen, phosphorus, and trace elements. The availability of specific nutrients can impact the pace and quantity of lactic acid generation.
- **Type of strain:** The rate and output of lactic acid generation can also be impacted by the bacteria used. Various bacterial strains have different metabolic traits that can influence the formation of lactic acid.

Different types of bacteria are used in the synthesis of lactic acid (Abedi & Hashemi, 2020). This genus of lactic acid bacteria is typically employed to produce lactic acid in industrial settings (Tan et al., 2017). Figure 9.1 represents the diagrammatic representation of lactic acid production in industries (Sun et al., 2020). Several substrates, including sugar beet pulp, can support the growth of *Lactobacillus* species,

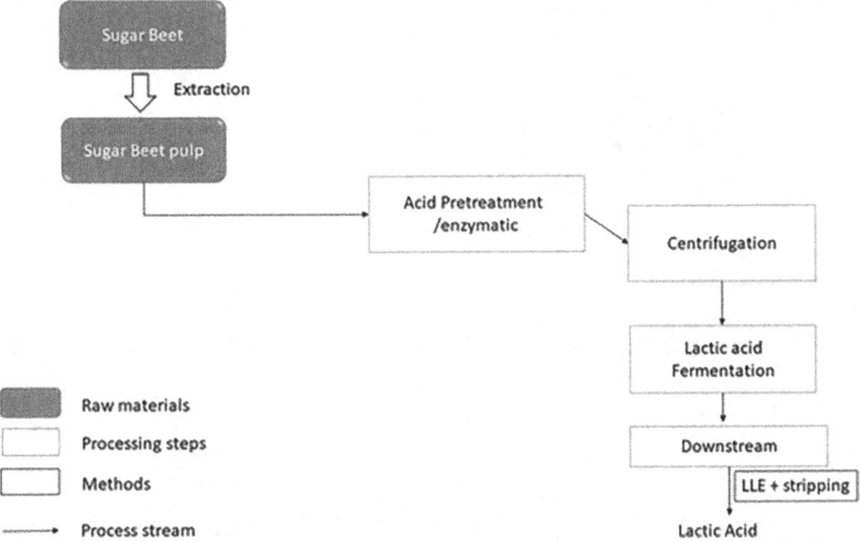

FIGURE 9.1 Diagrammatic representation of steps involved in lactic acid production from sugar beet pulp.

which can generate significant volumes of lactic acid when anaerobic circumstances prevail. A different genus of lactic acid bacteria is frequently employed to make lactic acid (Cubas-Cano et al., 2018). *Streptococcus* species may thrive on various substrates and generate large amounts of lactic acid in anaerobic environments. This type of lactic acid bacteria is frequently utilized in the dairy sector to produce cheese and yogurt (Zhu et al., 2020). Furthermore, lactic acid can be created from sugar beet pulp using *Lactococcus lactis*. It has been demonstrated that this strain of lactic acid bacteria is efficient at generating lactic acid from sugar beet pulp (Marzo et al., 2021; Díaz et al., 2020). The choice of bacteria depends on several variables, including the intended yield, the circumstances surrounding the fermentation, and the bacterium's availability. Each bacterial strain has unique metabolic properties that can impact the quantity and quality of the generated lactic acid (Castillo Martinez et al., 2013). Consequently, selecting the right bacterial strain for a particular production process is crucial (Lu et al., 2020).

9.4.2 Xylose Acid Production from Corn Stover

Five-carbon sugar called xylose is a crucial component in creating chemicals and biofuels (Domingues et al., 2021). After harvest, corn's leftover stalks, leaves, and cobs corn stover can serve as a source of xylose. From corn stover, xylose can be extracted and transformed into xylonic acid, which has numerous industrial uses (Hamann & Noronha, 2022).

Pretreatment, hydrolysis, fermentation, and purification are all steps in converting corn stover into xylonic acid (Zhu et al., 2020; Yu et al., 2020). Figure 9.2 represents the pictorial view of the industrial setup for producing xylonic acid (Liu & Chen, 2015).

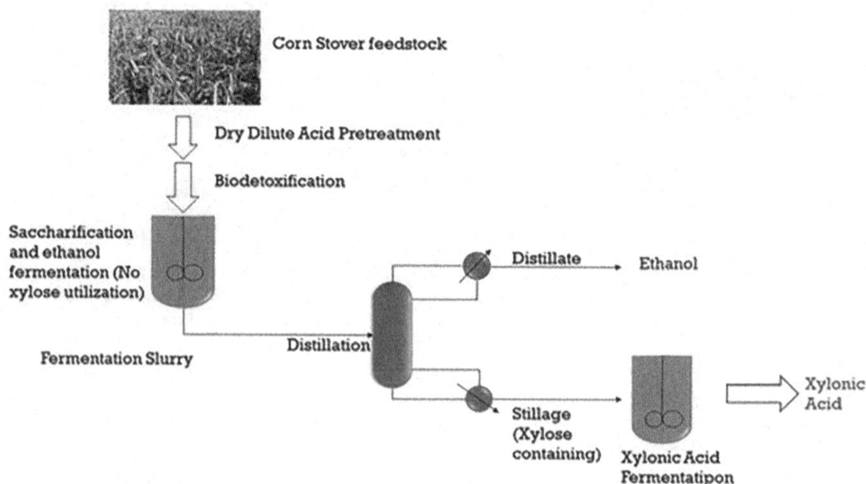

FIGURE 9.2 A graphical representation of steps involved in xylose acid production from corn stover.

- **Pretreatment:** Corn stover contains a complex matrix of cellulose, hemicellulose, and lignin and it is a lignocellulosic biomass. These components must be broken down. The pretreatment of the corn stover removes the lignin and breaks down the cellulose and hemicellulose as the first step in the procedure. Alkaline treatment, acid hydrolysis, and steam explosion are some pretreatment techniques (Zhu et al., 2020).
- **Hydrolysis:** Enzymatic hydrolysis converts the cellulose and hemicellulose in corn stover to their sugars. To catalyze the hydrolysis reaction, enzymes like cellulases and hemicellulases are added. The resulting glucose and xylose sugar mixture is then filtered to separate it from the biomass (Lekshmi Sundar & Madhavan Nampoothiri, 2022).
- **Fermentation:** The next step is to ferment the xylose sugar into xylonic acid. This can be done using a variety of microorganisms, including bacteria and yeast. One example of a bacterium used in xylonic acid production is *Gluconobacter oxydans*. This bacterium can convert xylose into xylonic acid through enzymatic reactions. The fermentation process typically occurs in a bioreactor under controlled temperature, pH, and oxygenation conditions (Liu & Chen, 2015).
- **Purification:** The final step is to purify the xylonic acid from the fermentation broth. This can be done using techniques like filtration, precipitation, and chromatography. The purified xylonic acid can be used in various industrial applications, such as producing biodegradable plastics, food additives, and pharmaceuticals (Yu et al., 2020).

The process is still in the research and development phase; it holds promise as a sustainable and renewable source of xylose for producing biofuels and chemicals (Liu & Chen, 2015; Zhang et al., 2017).

9.4.3 Acetic Acid Production

Currently, the research is prioritizing the exploration of alternate and more affordable carbon sources in order to decrease the expenses associated with raw materials. Multiple patents have been created for beverages that possess high nutritional value and distinctive flavors. These beverages are produced through the process of acetic fermentation using fruits, juices extracted from cereals, and vegetables. The invention EP1801195 describes the production of a fermented cloudberry vinegar containing an acetic acid concentration ranging from 50 to 150 g/L. The cloudberries were combined with a solution of alcohol in water, and the resulting mixture, together with sugars, underwent fermentation in the presence of *Acetobacter aceti* (Merli et al., 2021). The fermentation of strawberries has been carried out using a similar method as described in patent CN104116107 (Lin, 2014), while persimmons have been fermented according to patents CN109136055 and CN112375650A (Liu, 2019; Ren et al., 2021). Vegetables can serve as a substrate for vinegar manufacturing. The invention KR20190007329 pertains to the production of vinegar using solely 100% onion as the raw material, without the incorporation of any additional substances. This process achieves a final concentration of acetic acid of 53 g/L (Joung, 2019).

The objective of Patent CN112322444A is to create a nourishing sweet potato vinegar beverage that is suitable for regular use. This is achieved by blending sweet potatoes, which are primarily utilized as animal feed due to their unappealing taste and expensive processing for human consumption, with tea leaves (Huang et al., 2021). The invention CN103865975 presents a highly effective two-step fermentation technique that utilizes high-activity dry yeasts and AAB to process kitchen garbage. The benefits include proper waste disposal and decreased production costs for acetic acid. By employing this approach, a concentration of acetic acid amounting to 25.6 g/L was achieved, surpassing the concentration obtained without the addition of AAB and highly active dry yeast (Chai et al., 2016).

Molasses is another viable substrate to procure acetic acid. The technique for making vinegar from molasses requires concurrently conducting fermentation of both alcoholic and acetic types. The temperature for the fermentation process has been precisely adjusted to 30°C, while the yeast:vinegar ratio has been established at 2–3:1 in order to initiate yeast fermentation from the outset. Acetic acid fermentation is not carried out for molasses exceeding 15% sugar content or alcohol content exceeding 8%. Therefore, it is imperative to maintain the concentrations at a constant level. Molasses can lead to excessive oxidation because of its high sugar content (Merli et al., 2021).

9.4.4 Lactic Acid Production

Lactic acid, a versatile chiral compound crucial for various applications such as food supplements, chiral drugs, and the synthesis of biodegradable thermoplastics like polylactic acid (PLA), is synthesized from pyruvate by lactate dehydrogenase (LDH) (Abedi & Hashemi, 2020; Panagopoulos et al., 2022; Rajendran & Han, 2022).

Lactic acid bacteria (LAB), Gram-positive microbes, perform glycolysis to produce lactic acid under anaerobic conditions. Heterofermentative LAB, including *Leuconostoc* sp., *Oenococcus* sp., *Lactobacillus brevis*, *Lactobacillus fermentum*, and *Lactobacillus reuteri*, convert glucose into lactic acid, generating by-products such as acetic acid, ethanol, formic acid, acetoin, and carbon dioxide (Moestedt et al., 2020; Nagarajan et al., 2022). Heterofermentative LAB utilizes two pathways for lactic acid production: (i) phosphogluconate and (ii) phosphoketolase pathways. However, despite the utility of LAB in lactic acid production, achieving commercial success necessitates considerations such as increased productivity, yield, and titers of lactic acid, selective stereospecific production of L/D-lactic acid, and tolerance against high acid concentrations during fermentation.

For instance, *Lactobacillus* strains, recognized for their high lactic acid production, have been genetically engineered to introduce heterologous enzymes, enhancing lactic acid production (Tian et al., 2021). The incorporation of *A. niger* phosphofructokinase (PfkA) into *L. lactis* increased phosphofructokinase (PFK) activity, resulting in elevated glycolytic flux and heightened lactic acid production (up to 22.8 g lactic acid/(g CDW)h) with optimized glucose uptake (up to 1.7 μM/s(gCDW)) (Papagianni & Avramidis, 2011). Genome shuffling of *Lactobacillus rhamnosus* ATCC 11443 led to phenotypes with rapid glucose consumption (~200 g/L), achieving enhanced L-lactic acid production (up to 180 g/L) (Yu et al., 2008). Substituting native D-lactate dehydrogenase (LdhD) with L-lactate dehydrogenase (LdhL) in *Lactobacillus helveticus*

CNRZ32 improved optical purity, resulting in higher production of pure L-lactic acid (approximately over 70 g/L) from lactose.

To establish a viable commercial lactic acid production system, enhancing acid tolerance in lactic acid-producing microorganisms is imperative. Genome shuffling improved acid tolerance (up to pH 3.6) and volumetric productivity (26.5% higher than the wild type) in *L. rhamnosus* ATCC 11443 (Wang et al., 2007). Long-term adaptive evolution with lactic acid supplementation yielded the *Leuconostoc mesenteroides* LMS70 mutant, which is capable of producing 76.8 g/L of lactic acid, even in the presence of more than 60.0 g/L of lactic acid in the culture medium, double that of the wild type. Genomic and proteomic analyses identified a mutation in the ATPase ε subunit atpC gene and high intracellular ammonia as contributors to the enhanced acid tolerance (Ju et al., 2016).

9.4.5 Succinic Acid Production

Succinic acid, a four-carbon dicarboxylic acid, is a valuable platform chemical with diverse applications in various industries. It is produced as one of the fermentation products of anaerobic metabolism (Kumar et al., 2022). The production of succinic acid has garnered significant attention due to its potential as a sustainable and environmentally friendly alternative to petrochemical-based production methods. Recent research has focused on various aspects of succinic acid production, including strain development, substrate utilization, downstream purification, and metabolic engineering (Kumar et al., 2022; Kumar et al., 2024).

Genome-based metabolic engineering of *Mannheimia succiniciproducens* has been explored for succinic acid production, demonstrating the potential for genetic modifications to enhance succinic acid yield and productivity (Kumar et al., 2023). Additionally, the evaluation of organic fractions of municipal solid waste as a renewable feedstock for succinic acid production has been investigated, highlighting the importance of sustainable feedstock for succinic acid production (Kumar et al., 2022).

Furthermore, bio-based succinic acid has been identified as a promising platform chemical that could be used as a precursor for the establishment of a sustainable chemical industry. The fermentative production of succinic acid has received significant research attention, with several bacteria identified as major producers of succinic acid (Raj et al., 2023). Additionally, consolidated bioprocessing of hemicellulose-enriched lignocellulose to succinic acid through a microbial co-cultivation system has been explored, highlighting the potential for sustainable bioprocessing methods (Mazzoli, 2021).

Recent studies have also focused on the metabolic engineering of *C. glutamicum* for enhanced succinic acid production from renewable resources, emphasizing the importance of utilizing sustainable feedstocks for succinic acid production (Stylianou et al., 2020). Moreover, the design and techno-economic analysis of a succinic acid production facility has been investigated, providing insights into the economic viability of succinic acid production (Dai et al., 2019).

In summary, recent research articles have significantly advanced the understanding of succinic acid production, highlighting the potential for sustainable and environmentally friendly production methods. These studies have provided valuable insights into strain development, substrate utilization, downstream purification, and metabolic engineering, paving the way for more sustainable and efficient manufacturing processes for succinic acid production.

9.4.6 Various Challenges in the Synthesis of Organic Acid from Agrobiomass

The production of OAs from agrobiomass is a promising approach for developing a sustainable and environmentally friendly source of chemicals (Ebeid & Al-Homidan, 2022). The synthesis of OAs from agrobiomass is often associated with several challenges that need to be addressed for an efficient process some of the significant difficulties encountered during the production of OAs from agrobiomass (Lekshmi Sundar & Madhavan Nampoothiri, 2022).

The complex composition of agrobiomass: Agrobiomass is a complex mixture of various compounds, including carbohydrates, lignin, cellulose, hemicellulose, and other compounds (Singh & Kumar, 2021). These compounds' different chemical and physical properties make their conversion into OAs challenging. The composition of agrobiomass also varies depending on the plant species, growth conditions, and harvesting methods. Thus, it is essential to identify the most suitable biomass source for OA production (Dharmalingam et al., 2023; Duan et al., 2020).

Low concentration of OAs: OAs in agrobiomass are relatively low, making their extraction and purification difficult. The low concentration of OAs can also affect the yield and efficiency of the fermentation process. Developing efficient extraction and purification techniques is necessary to obtain high yields of pure OAs (Marzo et al., 2021).

Inhibitory compounds: Agrobiomass may contain inhibitory compounds such as phenols, furans, and OAs that can negatively affect the growth and metabolism of microorganisms during fermentation. These inhibitory compounds can also decrease the yield and quality of OAs produced (Zhang et al., 2023). It is essential to minimize the concentration of these compounds before fermentation.

Microbial contamination: Microbial contamination is a common problem during fermentation, which can decrease the yield and purity of the OAs produced (Sani et al., 2020). Microbial contamination can occur due to airborne microorganisms or inadequate sterilization of equipment and materials used in the process. It is essential to maintain strict measures and sterilization protocols during the fermentation process (Lekshmi Sundar & Madhavan Nampoothiri, 2022).

Strain selection and optimization: Selecting the appropriate microbial strain for OA production is critical. Different microbial strains have different metabolic pathways, which may vary in their ability to produce OAs. Therefore, it is necessary to identify the most suitable microbial strain and optimize the fermentation conditions to obtain high yields of OAs (Singh et al., 2022).

Energy and cost efficiency: OAs from agrobiomass should be economically feasible and energy-efficient. The high cost of feedstock, equipment, and labor can limit the commercialization of OA production. It is crucial to create production methods that are both economical and energy-efficient (Duan et al., 2020).

The production of OAs from agrobiomass is a promising approach for developing sustainable and environmentally friendly chemicals. Several challenges need to be addressed to achieve efficient and cost-effective production processes (Singh et al., 2022). These challenges include the complex composition of agrobiomass, low concentration of OAs, inhibitory compounds, microbial contamination, strain selection and optimization, and energy and cost efficiency. Addressing these challenges will lead to the development of efficient and sustainable processes for producing OAs from agrobiomass (Ebeid & Al-Homidan, 2022).

9.4.7 State-of-the-Art Microbial Techniques of Organic Acid from Agrobiomass

Several state-of-the-art microbial methods are used to enhance OA production from agrobiomass (Singh & Kumar, 2021). Some of these techniques include:

Strain improvement: Although microorganisms can naturally create OAs, they might need to do so in sufficient numbers for industrial use. By choosing and genetically altering microbe strains that produce more OAs, scientists can enhance them (Cubas-Cano et al., 2018). In Table 9.1, we can observe the OA production difference between different strains (Table 9.2).

TABLE 9.1
A Representation of Percentage Conversion Rate of Different Species of Microbes

Microbe	Raw Material	Product	Operational Parameters	% Conversion
Aspergillus species	Renewable feedstocks	Organic acids	Low pH (<2.0), nutrient-limited growth conditions	80%–90%
Lactic acid bacteria (*Lactobacillus* spp.)	Lactose, starch, or agrobiomass	Lactic acid	Controlled temperature and pH conditions	85%–95%
Wild *Acetobacter* species	Ethanol or acetic acid	Acetic acid	Aerobic conditions, low pH, high ethanol concentration	80%–95%
Clostridium acetobutylicum	Lignocellulosic biomass	Acetone and butanol	Anaerobic fermentation, controlled pH, and temperature	70%–90%
Engineered *Escherichia coli*	Glucose or glycerol	Succinic acid	Controlled pH, aeration, and temperature	70%–85%
Engineered *Saccharomyces cerevisiae*	Sugar or lignocellulosic biomass	Xylonic acid	Fermentation at neutral pH and aerobic conditions	75%–90%
Wild *Zymomonas mobilis*	Glucose	Acetic acid	Anaerobic fermentation, controlled pH, and temperature	80%–95%
Engineered *Corynebacterium glutamicum*	Glucose or glycerol	Glutamic acid	Controlled temperature and pH conditions	80%–95%
Wild *Penicillium* species	Sugar, starch, or agrobiomass	Citric acid	Aerobic fermentation, temperature control	70%–85%

Notes: FBB: fibrous-bed bioreactor; PFB: plant fibrous-bed bioreactor.

TABLE 9.2
Microbial Synthesis of Lactic Acid, Propionic Acid, and 3-Hydroxypropionic Acid from Agro-industrial Wastes: A Comparative Analysis

Host Strain	Carbon Source	Key Features	Titer (g/L)	Productivity (g/L/h)	Yield (g/g)	Cultivation Method	Source References
Lactic Acid Bacteria							
Lactobacillus. bulgaricus ATCC 8001, PTCC 1332	Cheese whey	Natural strain growth for lactic acid production	19.3	1.6	–	Immobilized cell reactor	Ghasemi et al. (2017)
Recombinant *Escherichia coli* B0013-070	Crude glycerol	Enhanced lactic acid production through ldhA overexpression, eliminating by-product-related genes	100.3	2.78	0.754	Fed-batch	Chen et al. (2014)
Recombinant E. coli BL21	Crude glycerol	Overexpression of *Lactobacillus helveticus* D-ldh, enhanced glpK and glpD, and by-product-related gene elimination	105	2.625	0.87	Fed-batch	Wang et al. (2015)
Recombinant *E. coli* K12 *MG1655*	Crude glycerol	Increased glpK and glpD expression, by-product-related gene elimination	45	0.536	0.83	Flask	Mazumdar et al. (2010)
Recombinant *Klebsiella pneumonia* ATCC25955	Crude glycerol	ldhA overexpression, by-product-related gene elimination	7.41	0.15	0.78	Fed-batch	Feng et al. (2014)
Sporolactobacillus inulinus NBRC 13,595	Palmyra palm jaggery	Natural strain cultivation with whey protein hydrolysate as a nitrogen source	189 ± 8.53	5.25 ±0.24	0.94	Batch	Reddy & Babu (2017)
Recombinant *Lactobacillus plantarum* NCIMB 8826	Raw corn starch	Expression of *Streptococcus bovis* 148 amyA, by-product-related gene elimination	73.2	1.53	0.85	Batch	Okano et al. (2009)
Bacillus sp. strain XZL9	Corncob molasses	Natural strain growth for lactic acid production	74.7	0.38	0.5	Fed-batch	Wang et al. (2010)
Bacillus coagulans LA204	Corn stover	Natural strain growth with NaOH-pretreated corn stover using simultaneous saccharification and fermentation	97.59	1.63	0.68	Fed-batch	Hu et al. (2015)

(*Continued*)

TABLE 9.2 (*Continued*)
Microbial Synthesis of Lactic Acid, Propionic Acid, and 3-Hydroxypropionic Acid from Agro-industrial Wastes: A Comparative Analysis

Host Strain	Carbon Source	Key Features	Titer (g/L)	Productivity (g/L/h)	Yield (g/g)	Cultivation Method	Source References
Recombinant *S. cerevisiae* OC2T T165R	Cane juice	PDC1 elimination, integration of six copies of L-LDH genes, EMS mutation, cultivation in cane juice-based medium	122	2.54	–	Batch	Ishida et al. (2006)
Recombinant *Enterococcus faecalis* W11	Crude glycerol	Expression of *Lactobacillus pentosus* ldhL1, by-product-related gene elimination	27.02	1.6	–	Shake-flask	Doi (2015)
B. coagulans CC17	Bagasse sulfite pulp	Natural strain growth with $CaCO_3$ as the buffering reagent in solid-state fermentation	20.68	0.43	0.77	Shake-flask	Zhou et al. (2016)
B. coagulans CC17	Bagasse sulfite pulp	Natural strain growth with xylanase in solid-state fermentation	110	0.573	0.72	Fed-batch	Zhou et al. (2014)
Lactobacillus helveticus	Whey	Natural strain growth with hydrolyzed whey as a nitrogen source	26.9	–	0.76	Batch	Lund et al. (1992)
Propionic Acid							
Propionibacterium shermanii	Milk whey	Natural strain growth with pasteurized milk whey	10	0.1389	–	Shake-flask	Anderson et al. (1986)
Propionibacterium freudenreichii ssp. *shermanii*	Whey lactose and crude glycerol	Natural strain growth with simultaneous usage of glycerol and whey lactose under relatively anaerobic conditions	24.80	0.05	0.60	Batch	Kośmider et al. (2010)
Recombinant *Propionibacterium acidipropionici* ATCC 4875	Whey lactose	Overexpression of native otsA; cultivation under fed-batch fermentation	135 ± 6.5	0.61 ±0.10	0.67 ± 0.06	Fed-batch	Jiang et al. (2014)
Mixed culture of *P. freundenreichii* ATCC 6207 and *Lactobacillus paracasei*	Whey cheese	Natural strain growth after preparation of mixed culture, cultivation without shaking	23	0.19	0.55	Shake-flask	Xie et al. (2022)

(*Continued*)

TABLE 9.2 (*Continued*)
Microbial Synthesis of Lactic Acid, Propionic Acid, and 3-Hydroxypropionic Acid from Agro-industrial Wastes: A Comparative Analysis

Host Strain	Carbon Source	Key Features	Titer (g/L)	Productivity (g/L/h)	Yield (g/g)	Cultivation Method	Source References
Recombinant *P. acidipropionici* ATCC 4875	Whey lactose	Overexpression of native otsA; cultivation under fed-batch fermentation	135 ± 6.5	0.61 ± 0.10	0.67 ± 0.06	Fed-batch	Jiang et al. (2014)
P. acidipropionici ATCC 4965	Sugarcane molasses	Natural strain growth under low dissolved oxygen concentration and without pH control in static incubation	6.916	0.052	0.56	Static Batch	Coral et al. (2008)
P. freudenreichii CCTCC M207015	Sugarcane molasses	Natural strain growth under plant FBB (PFB)	91.89	0.36	0.46	Fed-batch	Feng et al. (2011)
P. freudenreichii ssp. *shermanii* DSM 20,270	Sugarcane molasses and lactose	Natural strain growth with basal medium with sugarcane molasses (BMSM)	5.25	0.036	–	Fed-batch	Feng et al. (2011)
P. acidipropionici ATCC 4875	Hemicellulose hydrolysate	Natural strain growth with mineral salts medium	18.3	0.280	0.75	Batch	Ramsay et al. (1998)
P. acidipropionici ATCC 4875	Corncob molasses	Natural strain growth with slow nitrogen flux	71.8	0.28	–	Fed-batch	Liu et al. (2012)
P. acidipropionici NRRL B-3569	Wheat flour hydrolysate	Natural strain growth with soy peptone as a nitrogen source	–	–	0.54	Batch	Kagliwal et al. (2013)
P. acidipropionici ATCC 4875	Soy molasses	–	0.81	0.42	–	Sequential batch	Yang et al. (2018)
Adaptive strain of ATCC 4875; cultivation under sequential batch with corn steep liquor as the nitrogen source	Sweet sorghum bagasse	Natural strain growth under sequential batch fermentation with immobilized cells	35.3	1.17	0.617	Sequential batch	Castro et al. (2021)
Lactobacillus reuteri DSM 20,016	Crude glycerol	Overexpression of pdu pathway pduP, pduL, pduW	14	0.254	0.48	Fed-batch	Dishisha et al. (2015)

Co-cultivation: Using two or more microorganisms in a single fermentation process is known as co-cultivation. By fostering a synergistic interaction between several microbes, this method can improve the generation of OAs (Marzo et al., 2021).

Process optimization: The complex process of microbial fermentation can be improved to produce more OAs. Several fermentation parameters might be used to this end, including pH, temperature, agitation, aeration, and substrate concentration (Dharmalingam et al., 2023).

Metabolic engineering: Metabolic engineering involves the manipulation of the metabolic pathways of microorganisms to increase the production of desired metabolites, such as OAs (Lekshmi Sundar & Madhavan Nampoothiri, 2022). This can be achieved by using genetic engineering techniques to change the genetic makeup of microorganisms (Sun et al., 2022).

Substrate pretreatment: Complex carbohydrates in agrobiomasses are challenging for microorganisms to metabolize into OAs. Substrates that have undergone pretreatment are more readily available for microbial fermentation because the pretreatment simplifies complicated polysaccharides (Marzo et al., 2021).

Submerged fermentation: Submerged fermentation makes OAs from agrobiomass in free water. In this method, agrobiomass is fermented in a liquid state by microorganisms, and the resulting OAs are then extracted with the help of appropriate solvents (Hackmann, 2023; Yu et al., 2022).

Solid-state fermentation: Solid-state fermentation is a method for making OAs from agrobiomass without free water. By using microorganisms to ferment agrobiomass in a solid state, this technique produces OAs, which are then extracted using suitable solvents (Hackmann, 2023).

Enzymatic analysis: Another technique for generating OAs from agrobiomass is enzyme hydrolysis. By using enzymes, this approach transforms agrobiomass's complex polysaccharides into simple sugars that can then be fermented into OAs (Singh et al., 2019).

Fermentation: The most popular method for generating OAs from agrobiomass is fermentation. It involves converting the sugars in agrobiomass into OAs using microorganisms like bacteria, yeast, and fungi. Lactic acid, acetic acid, and citric acid are a few of the OAs frequently made using this method (Singh et al., 2022).

9.5 GENETIC ENGINEERING

Genetic engineering is a branch of biotechnology that alters an organism's genetic makeup to change its traits or abilities (Lekshmi Sundar & Madhavan Nampoothiri, 2022). Genetic engineering is modifying an organism's genetic composition by manipulating its DNA or RNA sequences or introducing new genetic material. This technique aims to enhance the organism's traits, such as increasing its capacity to produce beneficial substances or withstand illnesses (Bicas et al., 2016; Cubas-Cano et al., 2018). This methodology improves and optimizes the organism's genetic features. The utilization of this technology extends to various sectors, including agriculture, medicine, and environmental science, as per its wide range of applications (Schroeder et al., 2023). This is particularly noteworthy in academic circles. Recombinant DNA technology is a crucial tool in genetic engineering that involves the manipulation of DNA sequences to create novel gene combinations (Kumar et al., 2021). This technique enables scientists to insert, delete, or alter specific genes within an organism's DNA sequence. Another essential method is gene editing, which uses techniques like CRISPR-Cas9 to precisely modify some areas of DNA (Hsu et al., 2014). Accurate and efficient methods for manipulating genetic material have brought a significant revolution in genetic engineering (Lekshmi Sundar & Madhavan Nampoothiri, 2022). In this process, vectors introduce new genetic material into an organism. Plasmids are widely used among these vectors because they can be easily replicated and transferred between organisms (Cubas-Cano et al., 2018). However, integrating new genetic material into an organism's DNA stably and predictably remains a significant challenge in genetic engineering. Introducing new genetic material may cause unintended effects on the organism's traits or properties (Sun et al., 2020). Additionally, ensuring that genetically engineered organisms do not risk the environment or human health is another challenge that requires rigorous testing, monitoring, and regulation. To enhance the production of OAs, one can utilize genetic manipulation techniques such as introducing or overexpressing genes encoding enzymes involved in the desired metabolic pathway (Sun et al., 2022). Alternatively, one can also delete genes that encode for enzymes competing with the desired pathway. For instance, in the production of lactic acid from agrobiomass, the productivity of LAB can be improved by engineering their metabolic pathway. This can be done by overexpressing genes responsible for converting glucose to pyruvate, which is later converted to lactic acid (Luo et al., 2023). Another approach is to delete genes accountable for producing competing metabolites like ethanol or acetic acid. Moreover, genetic engineering can also improve the tolerance of microorganisms to harsh fermentation process conditions such as high temperatures, low pH, or high concentrations of OAs (Lu et al., 2020).

9.6 METABOLIC ENGINEERING

The creation and optimization of microbial cell factories for synthesizing essential substances like OAs are done using the potent instrument known as metabolic engineering (Sundar & Nampoothiri, 2022). Microorganisms like bacteria or yeast are frequently used to produce OAs from biomass. These microbes can create vast amounts of the desired OA by changing their metabolic pathways (Sun et al., 2022). For instance, metabolic engineering can boost the production of molecules like pyruvate or acetyl-CoA, which are precursors in creating OAs (Luo et al., 2023). Engineering biosynthetic pathways have emerged as a practical method to build effective microbial cell factories, thanks to the growth of metabolic engineering and synthetic biology over the past 10 years (Bicas et al., 2016).

The process of metabolic engineering includes the following:

Strain selection: Selecting a microorganism that can effectively transform biomass into the appropriate OA is the first step in metabolic engineering (Sundar &

Nampoothiri, 2022; Zhao et al., 2022). This can entail screening a variety of microorganisms and choosing the most productive ones.

Genetic modification: Genetic modification can enhance the metabolic pathways involved in manufacturing OAs after a suitable bacterium has been found. This may entail changing or removing genes that compete with the intended pathway and adding or increasing genes involved in synthesizing OAs (Fayyaz et al., 2020).

Fermentation optimization: Fermentation conditions can also be improved to boost the generation of OAs. To produce an environment that encourages the synthesis of OAs, it may entail controlling pH, temperature, oxygenation, or the availability of nutrients (Zhu et al., 2020).

Substrate engineering: The synthesis of OAs can be significantly influenced by the type and quality of biomass employed as a substrate. To increase the yields of OAs, substrate engineering entails choosing and perfecting biomass feedstock (Abedi & Hashemi, 2020).

Process integration: Last but not least, process integration entails optimizing every step of the manufacturing cycle, from biomass production to post-treatment of OAs (Lee et al., 2018). Enzymatic hydrolysis, fermentation, and separation are a few examples of unit processes that may need to be designed and optimized to do this.

9.6.1 Production of Lactate from *Corynebacterium glutamicum* Through Metabolic Engineering

A Gram-positive bacterium called *C. glutamicum* is frequently employed in biotechnology to manufacture amino acids like glutamate and lysine (Cappelletti et al., 2020). It is a promising candidate for metabolic engineering to manufacture lactic acid because it can produce lactate under specific circumstances (Abedi & Hashemi, 2020). By adding lactate dehydrogenase (LDH) genes, the enzyme that transforms pyruvate into lactate, *C. glutamicum* can be metabolically engineered to create lactate (Yankov, 2022). To boost lactate generation, endogenous genes involved in pyruvate metabolism can alter their expression levels (Chen et al., 2019).

Inhibiting rival pathways, including the formation of acetate, which can lower the efficacy of the lactate production pathway, is one technique to boost lactate production in *C. glutamicum*. This can be accomplished by overexpressing lactate-producing genes or eliminating the genes implicated in these pathways (Okoye et al., 2022). Another tactic is optimizing fermentation parameters like pH and substrate concentration to encourage lactate generation (Wu et al., 2018). For instance, *C. glutamicum* can produce more lactate when the pH and glucose levels are low.

The potential of *C. glutamicum* for the metabolic engineering of lactate generation has been shown in recent investigations (Sundar & Nampoothiri, 2022). For instance, by overexpressing the LDH gene and sabotaging the acetate synthesis route, researchers successfully created a strain of *C. glutamicum* that generates high amounts of glucose lactate. The metabolic engineering of *C. glutamicum* may develop a sustainable and profitable lactic acid production process to produce lactate (Martinez et al., 2013).

9.6.2 Engineering the Non-Oxidative Pathway for Malic Acid Production from Metabolic Engineering

Malic acid, a dicarboxylic acid, has a variety of uses in the food, drug, and chemical industries (Ghai et al., 2023). Nowadays, microorganisms, including *Aspergillus oryzae*, *R. oryzae*, and *S. cerevisiae*, are used in microbial fermentation to make it (Parapouli et al., 2020). Using metabolic engineering, wherein the microorganism is genetically altered to boost the yield and productivity of the intended product, it is possible to increase the production of malacic acid (Sundar & Nampoothiri, 2022). The non-oxidative route is a viable candidate for such alterations. A sequence of biochemical processes known as the non-oxidative pathway converts glucose-6-phosphate to pyruvate without producing NADPH, a crucial component in oxidative metabolism (Sun et al., 2022). The transaldolase and transketolase enzymes, part of the route, convert intermediates between the pentose phosphate cycle and glycolysis. Malic acid production can be boosted by accelerating the transit of the non-oxidative route by overexpressing or deleting the genes that control it (Kövilein et al., 2020). For instance, the overexpression of *A. oryzae* transketolase gene increased malic acid synthesis by 2.2 times. Engineering the microorganism to utilize different carbon sources, such as xylose or arabinose, which can be broken down non-oxidatively to create malic acid, is an additional strategy (Sun et al., 2022). This can be accomplished by including heterologous genes that encode the enzymes needed to convert these sugars into glucose-6-phosphate (Luo et al., 2023). Increasing the generation of malic acid through metabolic engineering of a non-oxidative pathway is a promising approach. This strategy can boost malic acid productivity and yield, producing more affordable and environmentally friendly production methods (Bicas et al., 2016).

9.6.3 Production of OAs in the Tricarboxylic Acid Cycle of Microorganisms through Metabolic Engineering:

Many OAs, such as citric, malic, and succinic, are produced mainly thanks to the tricarboxylic acid (TCA) cycle (Favarin et al., 2013). To maximize the production of these OAs in microbes, scientists can manipulate the genes and metabolic pathways involved in the TCA cycle (Raj et al., 2023). The overexpression of essential TCA cycle enzymes is one of the most popular methods of metabolism used to produce OAs (Dharmalingam et al., 2023). For instance, the overexpression of fumarase and aconitase can increase succinic acid production, while the overexpression of citrate synthetase and aconitase can increase citric acid production (Raj et al., 2023). Another strategy is introducing heterologous genes from other bacteria to enhance the metabolic pathways that generate OAs. For instance, *S. cerevisiae* can produce more malic acid when adding the pyruvate carboxylase gene from *A. niger* (Luo et al., 2023; Schroeder et al., 2023). In addition to these methods, scientists can target particular genes and either delete or change them using the CRISPR-Cas9 technology. For instance, *E. coli* can produce more succinic acid when deleting the gene for the pyruvate dehydrogenase kinase (Singh & Kumar, 2021). Many techniques are available in metabolic technology to help bacteria produce OAs as efficiently as possible. Researchers can increase yields and develop more effective bioprocesses to produce these valuable molecules by modifying the TCA cycle and other metabolic pathways (Hsu et al., 2014).

9.7 SYSTEMS BIOLOGICAL STUDIES

Systems biology is an interdisciplinary approach that prioritizes understanding biological systems as a whole rather than just their parts (Alon, 2019). By shifting our focus away from the study of genes and proteins and toward an understanding of a system's structure and dynamics, this strategy calls for a shift in the way we think about biology (Di Biase et al., 2022). A system cannot be fully understood by merely depicting its connections on a diagram. Instead, we need to know why and how traffic patterns form so that we can manage them. Four fundamental characteristics, such as system structures, system dynamics, control strategies, and design strategies, are the foundation for understanding biological systems (Su et al., 2017). The network of gene interactions, biochemical pathways, and the modification of the physical characteristics of intracellular and multicellular structures all fall under the heading of system structures. The study of a system's behavior over time and under various circumstances is called system dynamics, and it can be studied using methods like metabolic analysis, sensitivity analysis, phase portrait, and bifurcation analysis (Re & Mazzoli, 2023). Control methods entail systematically controlling the cell state to reduce malfunctions and provide potential therapeutic targets for treating diseases. Instead of relying on trial-and-error methods, design methods concentrate on strategies to modify and build biological systems with desired properties based on clear design principles and simulations (Chae et al., 2017). Any of the areas above would benefit significantly from significant advancements in our knowledge of computational sciences, genomics, and measurement technologies. Combining these discoveries with what is already known is essential to comprehend biological systems fully. Mathematical and computational models that describe the system's behavior based on its components' characteristics and interactions are created and tested against experimental data to hone and boost their accuracy (Subbalakshmi et al., 2022).

In systems biological studies, high-throughput data generation is a crucial tool that enables researchers to gather massive amounts of data on the molecular constituents of biological systems (Sun et al., 2020). Then, this information creates computational models that can forecast how the system will behave in various scenarios. Numerous biological systems can be studied using a systems biology approach, including cells, tissues, organs, and organisms (Alon, 2019). It can look into several biological processes, including gene regulation, signal transduction, metabolism, etc. In addition, figuring out the molecular mechanisms underlying disease states can be used to research illnesses and create new treatments (Subbalakshmi et al., 2022).

9.8 CONCLUSION AND FUTURE PERSPECTIVES

The escalating demand for OA production has spurred intensive research efforts aimed at enhancing efficiency and productivity. Strategies such as strain selection, genetic modification, fermentation optimization, and improved recovery techniques have been explored to meet the rising demands. Innovative methodologies, including genome-based engineering, transcriptional engineering, adaptive laboratory evolution (ALE), and transcriptional analysis-guided design, have showcased promising outcomes for various microorganisms. Notably, systems and synthetic biology advancements have begun unraveling the tolerance mechanisms of acid-resistant bacteria, shedding light on their resilience in low pH environments.

Despite significant progress, the commercialization of most fermentation technologies remains constrained by cost-effectiveness challenges. While numerous studies have assessed technological viability, interdisciplinary research is now imperative to overcome existing production hurdles. Emerging disciplines like systems metabolic technology, leveraging synthetic biology, and evolutionary approaches hold promise for strain improvement and omics-guided advancements. Biomass emerges as a sustainable and economically viable raw material for OA synthesis, particularly from lignocellulosic, agricultural, and food waste sources. Future studies are poised to focus on enhancing process productivity, maximizing product yield, and exploring novel biomass resources. The application of biotechnological tools, particularly synthetic biology and metabolic engineering, will play a pivotal role in achieving these objectives. The growing interest in integrated biorefinery systems underscores the potential for creating circular and sustainable economies by maximizing biomass resources.

In summary, the current trajectory of OA research emphasizes the need for interdisciplinary approaches, biotechnological innovations, and a sustainable biomass-centric paradigm to meet the challenges and opportunities in the evolving landscape of OA production.

ACKNOWLEDGMENTS

Dr. Karan Kumar is grateful to the Prime Minister's Research Fellowship provided by the Ministry of Education, Government of India. Mr. Shrivatsa and Ms. Ramyakrishna, two authors of this chapter, acknowledge the online WINBIOCB-2022 internship/training provided by Prof. V.S. Moholkar.

NOTE

* Corresponding author

REFERENCES

Abbas G, Arshad M, Saeed M, Imran S, Ali Kamboh A, Ka Al-Taey D, Asad Aslam M, Saeed Imran M, Ashraf M, Asif M, et al. 2022. An update on the promising role of organic acids in broiler and layer production. *J Anim Health Prod.* 10(3), 286. https://doi.org/10.17582/journal.jahp/2022/10.3.273.286

Abedi E, Hashemi SMB. 2020. Lactic acid production - producing microorganisms and substrates sources-state of art. *Heliyon.* 6(10):e04974. https://doi.org/10.1016/j.heliyon.2020.e04974

Aberathna A, Satharasinghe D, Jayaweera A, Weerathilake D, Prathapasinghe G, Premarathne K. 2023. Insights into the Aspergillus bio fertilizers Potential and Prospects. In: Proceedings of the 11th YSF Symposium – 2023, 7 May 2023, Singapore.

Alon U. 2019. *An Introduction to Systems Biology: Design Principles of Biological Circuits.* Second edition. Boca Raton, FL: CRC Press, Taylor & Francis Group.

Alonso S, Rendueles M, Díaz M. 2015. Microbial production of specialty organic acids from renewable and waste materials. *Crit Rev Biotechnol.* 35(4):497–513. https://doi.org/10.3109/07388551.2014.904269

Ambaye TG, Vaccari M, Bonilla-Petriciolet A, Prasad S, Van Hullebusch ED, Rtimi S. 2021. Emerging technologies for biofuel production: a critical review on recent progress, challenges and perspectives. *J Environ Manage.* 290:112627. https://doi.org/10.1016/j.jenvman.2021.112627

Anand, A., Kumar, K., & Moholkar, V. S. (2024). Various Routes for Hydrogen Production and Its Utilization for Sustainable Economy. In *Biohydrogen-Advances and Processes* (pp. 503–527). Cham: Springer Nature Switzerland.

Anderson TM, Bodie EA, Goodman N, Schwartz RD. 1986. Inhibitory effect of autoclaving whey-based medium on propionic acid production by propionibacterium shermanii. *Appl Environ Microbiol.* 51(2):427–428. https://doi.org/10.1128/aem.51.2.427-428.1986

Anwar Z, Gulfraz M, Irshad M. 2014. Agro-industrial lignocellulosic biomass a key to unlock the future bio-energy: a brief review. *J Radiat Res Appl Sci.* 7(2):163–173. https://doi.org/10.1016/j.jrras.2014.02.003

Bicas J, Maróstica Jr M, Pastore G, editors. 2016. Biotechnological production of organic acids. *Biotechnol Prod Nat Ingred Food Ind.* 2016:164–206. https://doi.org/10.2174/9781681082653116010007

Cappelletti M, Presentato A, Piacenza E, Firrincieli A, Turner RJ, Zannoni D. 2020. Biotechnology of Rhodococcus for the production of valuable compounds. *Appl Microbiol Biotechnol.* 104(20):8567–8594. https://doi.org/10.1007/s00253-020-10861-z

Carrascosa C, Raheem D, Ramos F, Saraiva A, Raposo A. 2021. Microbial biofilms in the food industry-a comprehensive review. *Int J Environ Res Public Health.* 18(4):2014. https://doi.org/10.3390/ijerph18042014

Castillo Martinez FA, Balciunas EM, Salgado JM, Domínguez González JM, Converti A, Oliveira RP de S. 2013. Lactic acid properties, applications and production: a review. *Trends Food Sci Technol.* 30(1):70–83. https://doi.org/10.1016/j.tifs.2012.11.007

Castro PGM, Maeda RN, Rocha VAL, Fernandes RP, Pereira Jr N. 2021. Improving propionic acid production from a hemicellulosic hydrolysate of sorghum bagasse by means of cell immobilization and sequential batch operation. *Biotechnol Appl Biochem.* 68(6):1120–1127. https://doi.org/10.1002/bab.2031

Chae TU, Choi SY, Kim JW, Ko Y-S, Lee SY. 2017. Recent advances in systems metabolic engineering tools and strategies. *Curr Opin Biotechnol.* 47:67–82. https://doi.org/10.1016/j.copbio.2017.06.007

Chai X, He D, Li Y, Niu D, Xie T, Zhao Y. 2016. A Kind of Method Utilizing Changing Food Waste Fermentative Production Acetic Acid Thereof. CN103865975B.

Chen C, Lu Y, Yu H, Chen Z, Tian H. 2019. Influence of 4 lactic acid bacteria on the flavor profile of fermented apple juice. *Food Biosci.* 27:30–36. https://doi.org/10.1016/j.fbio.2018.11.006

Chen X, Tian K, Niu D, Shen W, Algasan G, Singh S, Wang Z. 2014. Efficient bioconversion of crude glycerol from biodiesel to optically pure d -lactate by metabolically engineered Escherichia coli. *Green Chem.* 16(1):342–350. https://doi.org/10.1039/C3GC41769G

Coban HB. 2020. Organic acids as antimicrobial food agents: applications and microbial productions. *Bioprocess Biosyst Eng.* 43(4):569–591. https://doi.org/10.1007/s00449-019-02256-w

Coral J, Karp SG, Porto de Souza Vandenberghe L, Parada JL, Pandey A, Soccol CR. 2008. Batch fermentation model of propionic acid production by propionibacterium acidipropionici in different carbon sources. *Appl Biochem Biotechnol.* 151(2):333–341. https://doi.org/10.1007/s12010-008-8196-1

Cornélio Favarin D, Martins Teixeira M, Lemos de Andrade E, de Freitas Alves C, Lazo Chica JE, Artério Sorgi C, Faccioli LH, Paula Rogerio A. 2013. Anti-inflammatory effects of ellagic acid on acute lung injury induced by acid in mice. *Mediators Inflamm.* 2013:1–13. https://doi.org/10.1155/2013/164202

Cubas-Cano E, González-Fernández C, Ballesteros M, Tomás-Pejó E. 2018. Biotechnological advances in lactic acid production by lactic acid bacteria: lignocellulose as novel substrate. *Biofuels Bioprod Biorefining.* 12(2):290–303. https://doi.org/10.1002/bbb.1852

Dash S, Khodayari A, Zhou J, Holwerda EK, Olson DG, Lynd LR, Maranas CD. 2017. Development of a core Clostridium thermocellum kinetic metabolic model consistent with multiple genetic perturbations. *Biotechnol Biofuels.* 10(1):108. https://doi.org/10.1186/s13068-017-0792-2

Dharmalingam B, Tantayotai P, Panakkal EJ, Cheenkachorn K, Kirdponpattara S, Gundupalli MP, Cheng Y-S, Sriariyanun M. 2023. Organic acid pretreatments and optimization techniques for mixed vegetable waste biomass conversion into biofuel production. *BioEnergy Res.* 16(3):1667–1682. https://doi.org/10.1007/s12155-022-10517-y

Di Biase M, Le Marc Y, Bavaro AR, Lonigro SL, Verni M, Postollec F, Valerio F. 2022. Modeling of growth and organic acid kinetics and evolution of the protein profile and amino acid content during lactiplantibacillus plantarum ITM21B fermentation in liquid sourdough. *Foods.* 11(23):3942. https://doi.org/10.3390/foods11233942

Dishisha T, Pyo S-H, Hatti-Kaul R. 2015. Bio-based 3-hydroxypropionic- and acrylicacid production from biodiesel glycerol via integrated microbial and chemicalcatalysis. *Microb Cell Factories.* 14(1):200. https://doi.org/10.1186/s12934-015-0388-0

Doi Y. 2015. L-lactate production from biodiesel-derived crude glycerol by metabolically engineered Enterococcus faecalis: Cytotoxic evaluation of biodiesel waste and development of a glycerol-inducible gene expression system. *Appl Environ Microbiol.* 81(6):2082–2089. https://doi.org/10.1128/AEM.03418-14

Duan Y, Pandey A, Zhang Z, Awasthi MK, Bhatia SK, Taherzadeh MJ. 2020. Organic solid waste biorefinery: sustainable strategy for emerging circular bioeconomy in China. *Ind Crops Prod.* 153:112568. https://doi.org/10.1016/j.indcrop.2020.112568

Ebeid TA, Al-Homidan IH. 2022. Organic acids and their potential role for modulating the gastrointestinal tract, antioxidative status, immune response, and performance in poultry. *Worlds Poult Sci J.* 78(1):83–101. https://doi.org/10.1080/00439339.2022.1988803

Fayyaz M, Chew KW, Show PL, Ling TC, Ng I-S, Chang J-S. 2020. Genetic engineering of microalgae for enhanced biorefinery capabilities. *Biotechnol Adv.* 43:107554. https://doi.org/10.1016/j.biotechadv.2020.107554

Feng X, Chen F, Xu H, Wu B, Li H, Li S, Ouyang P. 2011. Green and economical production of propionic acid by Propionibacterium freudenreichii CCTCC M207015 in plant fibrous-bed bioreactor. *Bioresour Technol.* 102(10):6141–6146. https://doi.org/10.1016/j.biortech.2011.02.087

Feng X, Ding Y, Xian M, Xu X, Zhang R, Zhao G. 2014. Production of optically pure d-lactate from glycerol by engineered Klebsiella pneumoniae strain. *Bioresour Technol.* 172:269–275. https://doi.org/10.1016/j.biortech.2014.09.074

Ghai M, Agnihotri N, Kumar V, Agnihotri R, Kumar A, Sahu K. 2023. Global organic acids production and their industrial applications. *Phys Sci Rev.* 2023:1–20. https://doi.org/10.1515/psr-2022-0157

Ghasemi M, Ahmad A, Jafary T, Azad AK, Kakooei S, Wan Daud WR, Sedighi M. 2017. Assessment of immobilized cell reactor and microbial fuel cell for simultaneous cheese whey treatment and lactic acid/electricity production. *Int J Hydrog Energy.* 42(14):9107–9115. https://doi.org/10.1016/j.ijhydene.2016.04.136

Guo F, Wu M, Dai Z, Zhang S, Zhang W, Dong W, Zhou J, Jiang M, Xin F. 2020. Current advances on biological production of fumaric acid. *Biochem Eng J.* 153:107397. https://doi.org/10.1016/j.bej.2019.107397

Gupta N, Pal M, Sachdeva M, Yadav M, Tiwari A. 2016. Thermophilic biohydrogen production for commercial application: the whole picture: hydrogen energy. *Int J Energy Res.* 40(2):127–145. https://doi.org/10.1002/er.3438

Hsu PD, Lander ES, Zhang F. 2014. Development and applications of CRISPR-Cas9 for genome engineering. *Cell.* 157(6):1262–1278. https://doi.org/10.1016/j.cell.2014.05.010

Hu J, Zhang Z, Lin Y, Zhao S, Mei Y, Liang Y, Peng N. 2015. High-titer lactic acid production from NaOH-pretreated corn stover by Bacillus coagulans LA204 using fed-batch simultaneous saccharification and fermentation under non-sterile condition. *Bioresour Technol.* 182:251–257. https://doi.org/10.1016/j.biortech.2015.02.008

Huang J, Huang Z, Long S. 2021. Preparation Method of Sweet Potato Vinegar Beverage. There of. CN112322444 A.

Ingle AP, Philippini RR, Martiniano S, Marcelino PRF, Gupta I, Prasad S, Da Silva SS. 2020. Bioresources and their significance. *Curr Dev Biotechnol Bioeng.* 2020:3–40. https://doi.org/10.1016/B978-0-444-64309-4.00001-5

Ishida N, Saitoh S, Ohnishi T, Tokuhiro K, Nagamori E, Kitamoto K, Takahashi H. 2006. Metabolic engineering of Saccharomyces cerevisiae for efficient production of pure L-(+)-lactic acid. *Appl Biochem Biotechnol.* 131(1–3):795–807. https://doi.org/10.1385/ABAB:131:1:795

Jadaun JS, Rai AK, Singh SP. 2021. Microbial Applications in Organic Acid Production. In: Singh SP, Upadhyay SK, editors. *Bioprospecting of Microorganism-Based Industrial Molecules.* 1st ed. New York: Wiley; pp. 104–124. https://doi.org/10.1002/9781119717317.ch6

Jiang L, Cui H, Zhu L, Hu Y, Xu X, Li S, Huang H. 2014. Enhanced propionic acid production from whey lactose with immobilized Propionibacterium acidipropionici and the role of trehalose synthesis in acid tolerance. *Green Chem.* 17(1):250–259. https://doi.org/10.1039/C4GC01256A

Joung KC. 2019. Onion Vinegar and Its Preparation Method. Thereof. KR20190007329A.

Kagliwal LD, Survase SA, Singhal RS, Granström T. 2013. Wheat flour based propionic acid fermentation: an economic approach. *Bioresour Technol.* 129:694–699. https://doi.org/10.1016/j.biortech.2012.12.154

Kang S. 2020. Caproic acid production from lactate using Megasphaera hexanoica. In *International Chain Elongation Conference 2020.* Wageningen University & Research, Wageningen, Netherlands. https://doi.org/10.18174/icec2020.18024

Kang S, Kim H, Jeon BS, Choi O, Sang B-I. 2022. Chain elongation process for caproate production using lactate as electron donor in Megasphaera hexanoica. *Bioresour Technol.* 346:126660. https://doi.org/10.1016/j.biortech.2021.126660

Kośmider A, Drozdzyńska A, Blaszka K, Leja K, Czaczyk K. 2010. Propionic acid production by propionibacterium freudenreichii ssp. shermanii using crude glycerol and whey lactose industrial wastes. *Pol J Environ Stud.* 19(6):1249–1253.

Kövilein A, Kubisch C, Cai L, Ochsenreither K. 2020. Malic acid production from renewables: a review. *J Chem Technol Biotechnol.* 95(3):513–526. https://doi.org/10.1002/jctb.6269

Kumar, K., Roy, K., & Moholkar, V. S. (2021). Mechanistic investigations in sonoenzymatic synthesis of n-butyl levulinate. *Process Biochemistry,* 111, 147–158.

Kumar, K., Shah, H., & Moholkar, V. S. (2022). Genetic algorithm for optimization of fermentation processes of various enzyme productions. In Eswari J. S., Suryawanshi, N. (eds). *Optimization of sustainable enzymes production* (pp. 121–144). Chapman and Hall/CRC.

Kumar K, Barbora L, Moholkar VS. (2023). Genomic insights into clostridia in bioenergy production: comparison of metabolic capabilities and evolutionary relationships. *Biotechnology and Bioengineering* 121(4): 1297–1312. https://doi.org/10.1002/bit.28610

Kumar, K., Anand, A., & Moholkar, V. S. (2024a). Molecular Hydrogen (H2) Metabolism in Microbes: A Special Focus on Biohydrogen Production. In *Biohydrogen-Advances and Processes* (pp. 25–58). Cham: Springer Nature Switzerland.

Kumar, K., Kumar, S., Goswami, A., & Moholkar, V. S. (2024b). Advancing sustainable biofuel production: A computational insight into microbial systems for isopropanol synthesis and beyond. *Process Safety and Environmental Protection*, 188, 1118–1132. https://doi.org/10.1016/j.psep.2024.06.024

Kumar, K., Jadhav, S. M., & Moholkar, V. S. (2024c). Acetone-Butanol-Ethanol (ABE) fermentation with clostridial co-cultures for enhanced biobutanol production. Process Safety and Environmental Protection, 185, 277-285. https://doi.org/10.1016/j.psep.2024.03.027

Kumar V, Singh K, Shah MP, Singh AK, Kumar A, Kumar Y. 2021. Application of omics technologies for microbial community structure and function analysis in contaminated environment. In: Shah MP, Sarkar A, Mandal S, editors. *Wastewater Treat.* Amsterdam, The Netherlands: Elsevier; pp. 1–40. https://doi.org/10.1016/B978-0-12-821881-5.00001-5

Lee SI, Lee KJ, Chun HH, Ha S, Gwak HJ, Kim HM, Lee J-H, Choi H-J, Kim HH, Shin TS, et al. 2018. Process development of oxalic acid production in submerged culture of Aspergillus niger F22 and its biocontrol efficacy against the root-knot nematode Meloidogyne incognita. *Bioprocess Biosyst Eng.* 41(3):345–352. https://doi.org/10.1007/s00449-017-1867-y

Lekshmi Sundar MS, Madhavan Nampoothiri K. 2022. An overview of the metabolically engineered strains and innovative processes used for the value addition of biomass derived xylose to xylitol and xylonic acid. *Bioresour Technol.* 345:126548. https://doi.org/10.1016/j.biortech.2021.126548

Lin C. 2014. Natural Strawberry Vinegar and Manufacturing Process. Thereof. CN104116107.

Liu J. 2019. Persimmon Vinegar Making Method. CN109136055.

Liu Y, Nielsen J. 2019. Recent trends in metabolic engineering of microbial chemical factories. *Curr Opin Biotechnol.* 60:188–197. https://doi.org/10.1016/j.copbio.2019.05.010

Liu Z, Ma C, Gao C, Xu P. 2012. Efficient utilization of hemicellulose hydrolysate for propionic acid production using Propionibacterium acidipropionici. *Bioresour Technol.* 114:711–714. https://doi.org/10.1016/j.biortech.2012.02.118

Lund B, Norddahl B, Ahring B. 1992. Production of lactic acid from whey using hydrolysed whey protein as nitrogen source. *Biotechnol Lett.* 14(9):851–856. https://doi.org/10.1007/BF01029152

Luo Q, Ding N, Liu Y, Zhang H, Fang Y, Yin L. 2023. Metabolic engineering of microorganisms to produce pyruvate and derived compounds. *Molecules.* 28(3):1418. https://doi.org/10.3390/molecules28031418

Mazumdar S, Clomburg JM, Gonzalez R. 2010. Escherichia coli strains engineered for homofermentative production of D-lactic acid from glycerol. *Appl Environ Microbiol.* 76(13):4327–4336. https://doi.org/10.1128/AEM.00664-10

Mazzoli R. 2020. Metabolic engineering strategies for consolidated production of lactic acid from lignocellulosic biomass. *Biotechnol Appl Biochem.* 67(1):61–72. https://doi.org/10.1002/bab.1869

Mazzoli R. 2021. Current progress in production of building-block organic acids by consolidated bioprocessing of lignocellulose. *Fermentation.* 7(4):248. https://doi.org/10.3390/fermentation7040248

Merli G, Becci A, Amato A, Beolchini F. 2021. Acetic acid bioproduction: the technological innovation change. *Sci Total Environ.* 798:149292. https://doi.org/10.1016/j.scitotenv.2021.149292

Ning P, Yang G, Hu L, Sun J, Shi L, Zhou Y, Wang Z, Yang J. 2021. Recent advances in the valorization of plant biomass. *Biotechnol Biofuels.* 14(1):102. https://doi.org/10.1186/s13068-021-01949-3

Okano K, Zhang Q, Shinkawa S, Yoshida S, Tanaka T, Fukuda H, Kondo A. 2009. Efficient production of optically pure D-lactic acid from raw corn starch by using a genetically modified L-lactate dehydrogenase gene-deficient and α-amylase-secreting Lactobacillus plantarum strain. *Appl Environ Microbiol.* 75(2):462–467. https://doi.org/10.1128/AEM.01514-08

Okoye CO, Dong K, Wang Y, Gao L, Li X, Wu Y, Jiang J. 2022. Comparative genomics reveals the organic acid biosynthesis metabolic pathways among five lactic acid bacterial species isolated from fermented vegetables. *New Biotechnol.* 70:73–83. https://doi.org/10.1016/j.nbt.2022.05.001

Pandey AK, Sirohi R, Upadhyay S, Mishra M, Kumar V, Singh LK, Pandey A. 2021. Production and applications of polylactic acid. In: Binod P, Raveendran S, Pandey A, editors. *Biomass, Biofuels, Biochemicals.* Amsterdam, The Netherlands: Elsevier; pp. 309–357. https://doi.org/10.1016/B978-0-12-821888-4.00013-7

Parapouli M, Vasileiadi A, Afendra A-S, Hatziloukas E. 2020. Saccharomyces cerevisiae and its industrial applications. *AIMS Microbiol.* 6(1):1–32. https://doi.org/10.3934/microbiol.2020001

Passoth V, Sandgren M. 2019. Biofuel production from straw hydrolysates: current achievements and perspectives. *Appl Microbiol Biotechnol.* 103(13):5105–5116. https://doi.org/10.1007/s00253-019-09863-3

Pylak M, Oszust K, Frąc M. 2019. Review report on the role of bioproducts, biopreparations, biostimulants and microbial inoculants in organic production of fruit. *Rev Environ Sci Biotechnol.* 18(3):597–616. https://doi.org/10.1007/s11157-019-09500-5

Raj M, Devi T, Kumar V, Mishra P, Upadhyay SK, Yadav M, Sharma AK, Sehrawat N, Kumar S, Singh M. 2023. Succinic acid: applications and microbial production using organic wastes as low cost substrates. *Phys Sci Rev*: 1–17. https://doi.org/10.1515/psr-2022-0160

Ramsay JA, Aly Hassan M-C, Ramsay BA. 1998. Biological conversion of hemicellulose to propionic acid. *Enzyme Microb Technol.* 22(4):292–295. https://doi.org/10.1016/S0141-0229(97)00196-8

Rastogi M, Nandal M, Khosla B. 2020. Microbes as vital additives for solid waste composting. *Heliyon.* 6(2):e03343. https://doi.org/10.1016/j.heliyon.2020.e03343

Re A, Mazzoli R. 2023. Current progress on engineering microbial strains and consortia for production of cellulosic butanol through consolidated bioprocessing. *Microb Biotechnol.* 16(2):238–261. https://doi.org/10.1111/1751-7915.14148

Reddy ER, Babu RS. 2017. Neural network modeling and genetic algorithm optimization strategy for the production of L- asparaginase from novel Enterobacter sp. *J Pharm Sci.* 9:7.

Ren X, Liu R, Li L, Fan G, Chu Y. 2021. Health Persimmon Vinegar and Process. Thereof. CN112375650.

Roell GW, Zha J, Carr RR, Koffas MA, Fong SS, Tang YJ. 2019. Engineering microbial consortia by division of labor. *Microb Cell Factories.* 18(1):35. https://doi.org/10.1186/s12934-019-1083-3

Roy S, Majumder S, Deb A, Choudhury L. 2023. Microbial contamination of cosmetics and the pharmaceutical products, and their preservation strategies: a comprehensive review. *Nov Res Microbiol J.* 7(5):2116–2137. https://doi.org/10.21608/nrmj.2023.317346

Schroeder WL, Kuil T, Van Maris AJA, Olson DG, Lynd LR, Maranas CD. 2023. A detailed genome-scale metabolic model of Clostridium thermocellum investigates sources of pyrophosphate for driving glycolysis. *Metab Eng.* 77:306–322. https://doi.org/10.1016/j.ymben.2023.04.003

Singh A, Kumar V. 2021. Recent developments in monitoring technology for anaerobic digesters: a focus on bio-electrochemical systems. *Bioresour Technol.* 329:124937. https://doi.org/10.1016/j.biortech.2021.124937

Su H, Lin J, Tan F. 2017. Progress and perspective of biosynthetic platform for higher-order biofuels. *Renew Sustain Energy Rev.* 80:801–826. https://doi.org/10.1016/j.rser.2017.05.158

Subbalakshmi AR, Sahoo S, Biswas K, Jolly MK. 2022. A computational systems biology approach identifies SLUG as a mediator of partial epithelial-mesenchymal transition (EMT). *Cells Tissues Organs.* 211(6):689–702. https://doi.org/10.1159/000512520

Sun H, Chai L-J, Fang G-Y, Lu Z-M, Zhang X-J, Wang S-T, Shen C-H, Shi J-S, Xu Z-H. 2022. Metabolite-based mutualistic interaction between two novel clostridial species from PIT mud enhances butyrate and caproate production. *Appl Environ Microbiol.* 88(13):e0048422. https://doi.org/10.1128/aem.00484-22

Sun L, Gong M, Lv X, Huang Z, Gu Y, Li J, Du G, Liu L. 2020. Current advance in biological production of short-chain organic acid. *Appl Microbiol Biotechnol.* 104(21):9109–9124. https://doi.org/10.1007/s00253-020-10917-0

Tan J, Abdel-Rahman MA, Sonomoto K. 2017. Biorefinery-based lactic acid fermentation: microbial production of pure monomer product. In: Di Lorenzo ML, Androsch R, editors. *Synthesis, Structure and Properties of Poly(lactic acid).* Vol. 279. Cham: Springer International Publishing; pp. 27–66. https://doi.org/10.1007/12_2016_11

Tang J, Wang X, Hu Y, Zhang Y, Li Y. 2016. Lactic acid fermentation from food waste with indigenous microbiota: effects of pH, temperature and high OLR. *Waste Manag.* 52:278–285. https://doi.org/10.1016/j.wasman.2016.03.034

Wang L, Zhao B, Liu B, Yu B, Ma C, Su F, Hua D, Li Q, Ma Y, Xu P. 2010. Efficient production of l-lactic acid from corncob molasses, a waste by-product in xylitol production, by a newly isolated xylose utilizing Bacillus sp. strain. *Bioresour Technol.* 101(20):7908–7915. https://doi.org/10.1016/j.biortech.2010.05.031

Wang ZW, Saini M, Lin L-J, Chiang C-J, Chao Y-P. 2015. Systematic engineering of Escherichia coli for D-lactate production from crude glycerol. *J Agric Food Chem.* 63(43):9583–9589. https://doi.org/10.1021/acs.jafc.5b04162

Wu X, Yao H, Liu Q, Zheng Z, Cao L, Mu D, Wang H, Jiang S, Li X. 2018. Producing acetic acid of acetobacter pasteurianus by fermentation characteristics and metabolic flux analysis. *Appl Biochem Biotechnol.* 186(1):217–232. https://doi.org/10.1007/s12010-018-2732-4

Xie Y, Peng X, Li Z, Wang J, Li M, Xiao S, Zhang S. 2022. Production and fermentation characteristics of antifungal peptides by synergistic interactions with Lactobacillus paracasei and Propionibacterium freudenii in supplemented whey protein formulations. *LWT.* 164:113632. https://doi.org/10.1016/j.lwt.2022.113632

Yang H, Wang Z, Lin M, Yang S-T. 2018. Propionic acid production from soy molasses by Propionibacterium acidipropionici: Fermentation kinetics and economic analysis. *Bioresour Technol.* 250:1–9. https://doi.org/10.1016/j.biortech.2017.11.016

Yankov D. 2022. Fermentative lactic acid production from lignocellulosic feedstocks: from source to purified product. *Front Chem.* 10:823005. https://doi.org/10.3389/fchem.2022.823005

Zhao Y, Liu S, Yang Q, Han X, Zhou Z, Mao J. 2022. Saccharomyces cerevisiae strains with low-yield higher alcohols and high-yield acetate esters improve the quality, drinking comfort and safety of huangjiu. *Food Res Int.* 161:111763. https://doi.org/10.1016/j.foodres.2022.111763

Zhou C, Ma Q, Mao X, Liu B, Yin Y, Xu Y. 2014. New insights into clostridia through comparative analyses of their 40 genomes. *BioEnergy Res.* 7(4):1481–1492. https://doi.org/10.1007/s12155-014-9486-9

Zhou J, Ouyang J, Xu Q, Zheng Z. 2016. Cost-effective simultaneous saccharification and fermentation of L-lactic acid from bagasse sulfite pulp by Bacillus coagulans CC17. *Bioresour Technol.* 222:431–438. https://doi.org/10.1016/j.biortech.2016.09.119

Zhu J-Q, Zong Q-J, Li W-C, Chai M-Z, Xu T, Liu H, Fan H, Li B-Z, Yuan Y-J. 2020. Temperature profiled simultaneous saccharification and co-fermentation of corn stover increases ethanol production at high solid loading. *Energy Convers Manag.* 205:112344. https://doi.org/10.1016/j.enconman.2019.112344

Zikmanis P, Kolesovs S, Semjonovs P. 2020. Production of biodegradable microbial polymers from whey. *Bioresour Bioprocess.* 7(1):36. https://doi.org/10.1186/s40643-020-00326-6

10 Application of Enzymatic Technique for the Synthesis of Organic Acids from Different Agrobiomasses

Mukesh Singh, Ahana Bhaduri, Ranjay Kumar Thakur, and Sudip Das

10.1 INTRODUCTION

Nowadays, global environmental pollution is a significant concern, primarily due to industrialization and agricultural waste. To address these issues, it is crucial to manage these wastes in an eco-friendly and scientific manner (Singh & Kumari, 2020). Residual agricultural biomass refers to unwanted materials or by-products, originating from food and non-food portions of crops, such as crop stalks, leaves, roots, and animal waste (Ashworth & Azevedo, 2009). Sadly, much of this biomass is discarded or burnt (Tripathi et al., 2019). According to the Food and Agriculture Organization (FAO), a staggering 30% of agricultural produce, approximately 23.7 million tons per day worldwide, goes to waste during harvest, storage, and processing (Food and Agriculture Organization of the United Nations (FAO), 2017). However, this biomass contains valuable bioactive molecules and carbohydrates that can be utilized in various industries like chemicals, food, medicine, and energy (Choi et al., 2012; Singh & Kumari, 2020). By utilizing appropriate technology, biomass waste has the potential to yield around 50 billion tons of oil, serving as an eco-friendly alternative to fossil fuels and contributing to renewable energy production, which, in turn, helps reduce greenhouse gas emissions. This is achieved through the production of organic acids from agrobiomass by the application of enzymatic technologies in the realm of renewable energy (Choi et al., 2012; Tripathi et al., 2019).

The synthesis of organic acids holds immense significance across various sectors such as food, pharmaceuticals, bioplastics, cosmetics, beverages, and chemicals (Quitmann et al., 2013). Because organic acids are versatile compounds containing carboxyl groups, hydroxyls, and more, they play significant roles in metabolic pathways. Notably, these organic acids have proven effective in

DOI: 10.1201/9781003407713-10

biological control, particularly against nematodes (Jang et al., 2016; Dhanya et al., 2020). Recent market analysis predicts a substantial increase in annual demand for organic acids, reaching $12.54 billion by 2026 (Market report, Global Organic Acid Market).

Amidst global environmental challenges and a growing need for sustainable resources, the adoption of enzymatic methods for producing organic acids from diverse agrobiomass sources presents a promising solution. First and foremost, enzyme technology stands out for its environmental sustainability (Fang et al., 2020). In contrast to traditional chemical procedures that often employ harsh conditions and toxic substances, enzymes function under mild temperatures and pressures. This results in significantly reduced energy consumption and the generation of minimal waste (Katja et al., 2021; Lynd et al., 2008). Since enzymatic technology is adaptable and versatile, it can be applied to a wide range of agrobiomass materials (Gopinath & Sowmya, 2020). This adaptability enhances resource utilization and provides a means to convert various agricultural residues into valuable organic acids. The enzymes used in these processes are primarily protein-based and can be recovered and reused, enhancing economic feasibility (Jian et al., 2009). In summary, enzyme technology's many advantages make it a superior choice for the production of organic acids from agrobiomass when compared to conventional methods. These benefits contribute to more sustainable and efficient production processes, aligning with the growing demand for environmentally responsible and economically viable solutions in various industries (Jivkova et al., 2022).

10.2 BIOPROCESSING OF AGRICULTURAL BIOMASS

Agricultural biomass is recognized as a valuable resource in developing countries, with biovalorization playing a crucial role in both product development and environmental preservation (Sarkar et al., 2012; Rais & Sheoran, 2015). Crop residues, including materials like corn stover, wheat straw, and sugarcane bagasse, are considered significant sources of agricultural biomass (Sarkar et al., 2012). To manage agro-wastes effectively, they are categorized based on quality and the point of generation in the Food Supply Chain (FSC), which includes production, distribution, transportation, processing, retailing, and consumption stages (FAO, 2011). In India's organized sector, losses and wastage of agricultural products amount to 25%, 10%, and 7% during processing, distribution, and consumption, respectively (Rais & Sheoran, 2015). The process of agricultural biomass bioprocessing involves converting organic materials from plants and animals into biofuels, bioproducts, and other valuable resources. This involves a series of steps using biochemical and microbial processes to convert raw biomass into useful and sustainable materials (Chandel et al., 2018; Alvira et al., 2010). The next section will delve into agricultural biomass pretreatment, a crucial step in enhancing accessibility and digestibility for subsequent conversion processes, where various methods, including physical, microbial, chemical, and biological pretreatment, will be discussed (Sarkar et al., 2012; Chandel et al., 2018).

10.2.1 Microbial Bioprocess

Microbial bioprocessing of organic waste serves as a promising approach for environmental clean-up and the creation of valuable products. Various groups of microorganisms are utilized to transform agricultural waste into novel bioproducts. For instance, *Saccharomyces cerevisiae* is employed to convert agricultural waste into protein, primarily for animal feed production (Correia et al., 2007). *Aspergillus* species are known for producing organic acids like citric and lactic acid from agricultural residues, while *Bacillus* strains are popular for generating enzymes such as cellulase, amylase, and protease (Mussatto et al., 2012). Additionally, *Streptomyces* species are capable of producing bioactive compounds such as bafilomycin, oxytetracycline, and cephamycin from agricultural waste (Mussato et al., 2012). Different technologies are employed to extract desired products from agricultural waste, depending on its physical characteristics.

Microbial bioprocessing can be broadly categorized into two methods: solid-state fermentation (SSF) and submerged fermentation (SmF) (Ray & Ward, 2006). SSF involves the growth of microbes on solid materials derived from agricultural or horticultural residues without free liquid present, typically with a low moisture content ranging from 40% to 80% (Ali & Zulkali, 2011). SSF offers advantages over SmF as it reduces foam production and minimizes the need for controlling factors like pH, aeration, and temperature during fermentation (Couto, 2008). The selection of appropriate microorganisms for specific waste types and the optimization of physicochemical parameters are crucial in the production of value-added biological products (Panda & Ray, 2015).

10.2.2 Enzymatic Technique for the Synthesis of Organic Acids from Different Agrobiomasses

Enzymes, which are proteins, play a crucial role in accelerating chemical, biological, and biochemical reactions (Chapman & Cronan, 1999). Enzymes are typically pure proteins, although some may require a cofactor or coenzyme, in addition to their amino acid sequence, to function effectively (Nelson & Cox, 2004). Enzymes have diverse applications in various industries. Their production and stability have opened up opportunities for researchers to develop technologies for generating different products from cost-effective substrates. This discussion focuses on the utilization of microbial enzymes in the production of organic acids (Liu et al., 2019). To break down the complex structure of corn stover into fermentable sugars, a combination of cellulases and hemicellulases, produced by the lignocellulolytic fungus *Trichoderma reesei*, was employed (Lynd et al., 2002). The subsequent step involved fermenting the sugar-rich hydrolysate with a microbial culture proficient in converting sugars into organic acids, which could be bacteria like *Escherichia coli* or *Lactobacillus*. The recovery of organic acids from agrobiomass using enzymatic methods typically comprises several steps (Chen et al., 2018).

The transformation of agricultural leftovers into bioactive substances by biological means could serve as the foundation for numerous essential enterprises. Finding the right microbe for the bioconversion process might be difficult, though. Bioconversion of agricultural wastes into cellulosic biomass is a cost-effective, renewable, and untapped energy source. This process involves the degradation of lignocellulolytic materials by microorganisms like bacteria and fungi, producing extracellular cellulolytic enzymes in large quantities (Abou et al., 2010). Fungi like *Aspergillus niger* and *T. reesei* produce these enzymes, while bacteria and a few anaerobic fungus strains manufacture these enzymes in a cellulosome (Clauser et al., 2021). Higher fungi, such as Basidiomycetes, have distinct oxidative systems and ligninolytic enzymes that destroy lignocellulose. *Aspergillus* filamentous fungi are essential for the production of bio-based oxygen-associated compounds (OAs), as they are safe, non-toxic, and have increased yields, productivity, and reduced contamination risks (Ganguly et al., 2021). Over the years, efforts have been made to understand the physiological features of fungal biocatalysts' synthesis of OAs and develop *Aspergilli* genetic engineering for this purpose (Maki et al., 2009).

Here is an overview of the general process and a relevant reference:

Enzymatic hydrolysis: Agrobiomass, consisting of agricultural leftovers or plant materials, is initially subjected to enzymatic hydrolysis. Enzymes like cellulases and hemicellulases are employed to break down the complex polysaccharides within the biomass into simpler sugars. This step increases the availability of fermentable sugars for subsequent organic acid production.

Fermentation: The hydrolysate solution resulting from enzymatic hydrolysis is then subjected to fermentation. Microorganisms such as bacteria or yeast are used to convert sugars into organic acids through metabolic processes. The choice of microorganism depends on the desired type of organic acid.

Separation and recovery: Following fermentation, all organic acids must be separated and recovered from the fermentation broth. Various methods, including filtration, centrifugation, or solvent extraction, may be used to distinguish the organic acids from the biomass residue and microbial cells.

Purification: Once separated, the obtained organic acid may undergo additional purification steps to eliminate impurities and yield a high-purity product. Purification methods may include distillation, crystallization, or chromatography.

A relevant reference for the recovery of organic acids from agrobiomass using enzymatic techniques is a study conducted by Fang et al. (2020). Their research focused on the production and recovery of succinic acid from lignocellulosic biomass using enzymatic hydrolysis and fermentation. Cellulases and hemicellulases were used for hydrolysis, and *Actinobacillus succinogenes* was the fermenting microorganism. The study also explored various separation and purification techniques for the recovery and purification of succinic acid.

10.3 ORGANIC ACIDS

Organic acids are reported as the third largest category among biological products (Ali & Zulkali, 2011). Organic acid is an organic compound that is characterized by weak acidic properties and does not dissociate completely in the presence of water. Organic acids are regarded as building block chemicals which can be produced by microbial processing (Sauer et al., 2008). Citric acid, lactic acid, and acetic acid are popular among the organic acids. These organic acids are used by a broad range of industries and processing units such as food processing, nutrition and feed industry, pharmaceuticals, and oil and gas stimulation units. Microorganisms namely bacterial and fungal species are used commercially for the production of organic acids. Bacteria such as *Arthrobacter paraffinens*, *Bacillus* sp., *Lactobacillus* sp., *Streptococcus thermophilus*, and fungus like *Aspergillus* sp., *Penicillium* sp., *Yarrowia lipolytica* and related yeast species are used to produce organic acids (Shaikh & Qureshi, 2013). The bioprocessing of important organic acids from bio-wastes is discussed below.

10.3.1 Citric Acid

Among various organic acids, citric acid is the most abundant carboxylic acid having huge demand worldwide due to major applications in food and pharmaceutical industries. In England, Karls Scheels (1874) first isolated citric acid after importing lemon juice from Italy (Luciana et al., 1998). It is a natural ingredient that aids in detoxification, supporting kidney function and healthy digestion maintaining energy levels. This acid is frequently found in citrus fruits like lemons, limes, oranges, and grapefruits that are rich in citric acid. Citric acid appears in its highest amount in lemons and limes, where it can account for up to 8% of the fruit's weight, having chemical formula $C_6H_8O_7$ (2-hydroxypropane-1,2,3-tricarboxylic acid). Citric acid is a primary metabolite produced by different living organisms including microorganism biochemically (Figure 10.1).

It can moreover occur as a monohydrate or water-free (anhydrous) form. It is an important intermediary of citric acid cycle commonly operated in living systems including plant, animal, and microorganisms (Mohanty et al., 2011). Salts, esters, and the polyatomic anion, which are found in solution, are the derivatives of citrate. It has been determined to be a tribasic acid by using ^{13}C NMR spectroscopy; the pK_a of the hydroxyl group has been found to be 14.4. A citric acid solution acts as a buffer solution between about pH 2 and pH 8. At pH 7, the two species that are present are the citrate ion and mono-hydrogen citrate ion in biological systems (Goldberg et al., 2002) (Table 10.1).

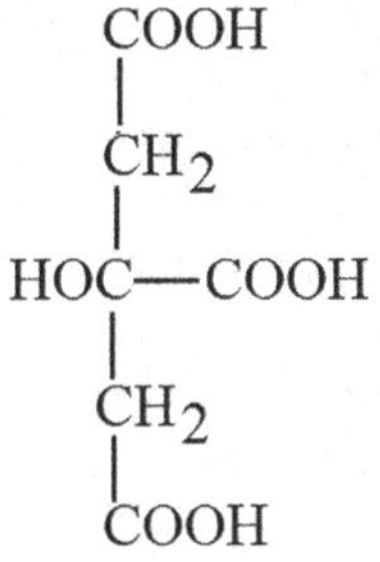

```
     COOH
     |
     CH2
     |
  HOC—COOH
     |
     CH2
     |
     COOH
```

FIGURE 10.1 Molecular structure of citric acid.

TABLE 10.1
Properties of Citric Acid

Properties	Citric Acid
Molecular Formula	$C_6H_8O_7$
Molecular Name	2-Hydroxypropane-1,2,3-tricarboxylic acid
Molecular Weight	210.14 g/mole
Occurrence	As a common metabolite in certain plants and animals
Solubility	Soluble in water (polar solvent) in pure form
Physical State	In solid state at room temperature
Production	Either by chemical (enzymatic) process or microbial fermentation process

10.3.1.1 Production of Citric Acid

Citric acid is produced via natural and synthetic methods by the process of extracting or manufacturing. It can be extracted naturally from the citrus plant, viz. orange, lemon, etc., by simple extraction techniques or through an alternative method called synthetic process which can produce citric acids either chemically through enzymatic process or biologically through fermentation using the microorganism.

Since citric acid fermentation does not produce any harmful by-products, unlike the chemical manufacture, it is regarded as a green chemistry method. The commonly used technique for citric acid production is microbial fermentation. The most popular method of synthesizing citric acid involves moist agricultural coproducts like fruit waste, starchy vegetable waste, cereal, and grains as a substrate for microbes to act. Thomas in a recent study investigated that production of citric acid by solid SSF using *A. niger* showed a promising result. The highest levels of citric acid production were obtained by solid-state fermenting fruit processing wastes (apples and grapes). *A. niger* strains that used potato or cereal grain processing coproducts as substrates produced the least amount of citric acid. Hence, it was determined that the amount of citric acid produced by *A. niger* strain depended critically on the sugar content of the substrate (West T.P, 2023).

Another study done by Dutta and coworkers concluded that banana peels, a common agricultural waste material, can be utilized as an inexpensive substitute medium for the efficient manufacture of citric acid. The research showed that under optimum condition citric acid production was maximum (0.62%) when obtained from banana peel compared to sugarcane bagasses, orange peel, and rice straw (Dutta et al., 2019). SmF technique helped to produce about 80% of the world's citric acid (Pandey & Soccol, 1998).

Numerous microorganisms are reported to produce citric acid in the presence of sugar in the medium by biotransformation process which included both bacteria and fungi. However, many of the microbes fail to produce citric acid in significant amounts. Among all microbes tested, only *A. niger* and few yeasts are employed for commercial production. The organism of choice for commercial production of citric acid is only *A. niger.*

The key features of employing this microorganism are its ease of management, potential of utilizing inexpensive raw materials, elevated product, and cost-effectiveness.

Biotransformation process: It is produced due to defective TCA or citric acid or Kreb's cycle during trophophase cell growth at high sugar content and inorganic salts (Wehmer, 1893). During this defective cycle, high amount of sugar is transported via the Glycolytic (EMP) pathway which forms Acetyl CoA which ultimately produces Citric Acid after condensation with oxaloacetate by the action of citrate synthetase enzymes. For the prevention of citric acids breakdown, it must be necessary to stop further steps of Kreb's cycle by deactivation of enzyme aconitase/isocitrate dehydrogenase. Citric acid is synthesized from acetyl-CoA and oxaloacetate in the TCA cycle via a set of enzymatic steps:

Acetyl-CoA condenses with oxaloacetate to form citrate. The enzyme citrate synthase catalyzes this reaction.

1. Isomerization occurs when citrate is isomerized to isocitrate. This conversion is catalyzed by the enzyme aconitase.
2. Isocitrate is oxidized to α-ketoglutarate, which produces NADH in the process. The enzyme isocitrate dehydrogenase catalyzes this reaction.
3. Succinyl-CoA and NADH are produced by decarboxylation of α-ketoglutarate. The enzyme α-ketoglutarate dehydrogenase catalyzes this step.
4. Substrate-level phosphorylation occurs when succinyl-CoA is converted to succinate and GDP is phosphorylated to form GTP. The enzyme succinyl-CoA synthetase catalyzes this reaction.
5. Dehydrogenation occurs when succinate is oxidized to fumarate, resulting in the production of $FADH_2$. The enzyme succinate dehydrogenase catalyzes this reaction.
6. Dehydrogenation occurs when succinate is oxidized to fumarate, resulting in the production of $FADH_2$. The enzyme succinate dehydrogenase catalyzes this reaction.
7. Malate is formed by hydrating fumarate. This conversion is catalyzed by the enzyme fumarase.
8. Dehydrogenation: Malate is oxidized to oxaloacetate, which produces NADH. The enzyme malate dehydrogenase catalyzes this reaction (Figure 10.2).

The TCA cycle represents a cyclic pathway that produces one molecule of citric acid at each turn. The primary goal of the cycle, however, is to produce energy in the form of ATP and reduce equivalents in the form of NADH and $FADH_2$. The citric acid that is produced can be extracted and utilized for a variety of purposes, including the industrial manufacturing of citric acid.

Biological method: Citric acids are produced by the involvement of microorganism in the process of fermentation. During the manufacturing of citric acid, a huge number of microorganisms, viz. fungi, yeasts, and bacteria have been employed. As the citric acid accumulation rises in appreciable amounts due to

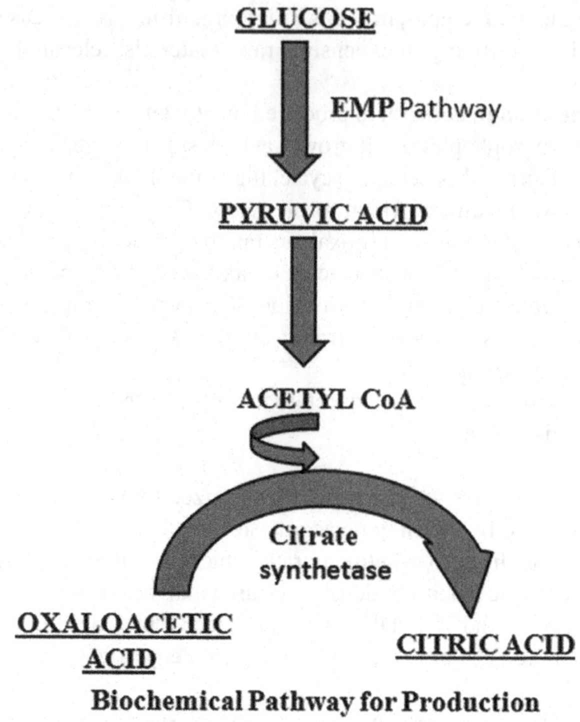

FIGURE 10.2 Biochemical (enzymatic) pathway for the production of citric acid.

drastic imbalances of energy metabolic pathway (Kreb's cycle), most of the microbes involved in production through fermentation are not able to produce commercially acceptable number of citric acid (Kubicek & Rohr, 1978). El-Masry et al. (2019) published a study on the biological synthesis of citric acid from agrobiomass. Using *A. niger*, researchers investigated the extraction of citric acid from sugarcane bagasse, a common agrobiomass. They improved the manufacturing of citric acid by optimizing fermentation conditions such as carbon source concentration, pH, and ambient temperature. The study pointed out the potential of using agrobiomass as a cost-effective and sustainable feedstock for biological citric acid production (Table 10.2).

Due to the ease of handling and high yield of citric acid from a variety of cheap raw materials, the fungus *A. niger* has remained the principal choice for commercial production of citric acid among all microorganisms (Soccol et al., 2006). Generally, submerged fermentation of molasses or sucrose is used to produce citric acid commercially, but this is costly (Soccol et al., 2006). On the other hand, a huge amount of waste has been generated during the manufacturing of different food products such as pickles, jam, and jellies by the food and agricultural industries, which creates an enormous problem concerning environmental pollution. Accordingly, the environmental problem and cost of citric acid production could be minimized by collecting these cheap wastes. These agro-wastes are well tailored to fermentation cultures due

TABLE 10.2
Citric acid (CA) Producing Microorganism (Behera et al., 2021)

Microbes	CA Producing Species	References
Bacteria	*Bacillus subtilis, Bacillus licheniformis, Arthrobacter paraffinens, Brevibacterium flavum, Corynebacterium* sp., and *Penicillium janthinellum*	Kroya Fermentation Industry (1970), Sardinas (1972), Fukuda et al. (1970)
Fungi	*Aspergillus awamori, Aspergillus wentii, Aspergillus foetidus, Aspergillus niger, Aspergillus aculeatus, Aspergillus carbonarius*, and *Penicillium janthinellum*	Karow and Waksman (1947), Roukas (1991), El Dein and Emaish (1979)
Yeasts	*Saccahromycopsis lipolytica, Candida parapsilosis, Candida tropicalis, Candida guilliermondii, Candida citroformans, Hansenula anamala, Yarrowia lipolytica*, and *Debaryomyces, Torula, Pichia, Kloekera*, and *Zygosaccharomyces*	Gutierrez et al. (1993), Kapelli et al. (1978), Oh et al. (1973)

to the manufacture of high value products, their starchy and cellulosic nature, role in solid waste management, and diminutive risk of contamination by microbes (Kumar & Jain, 2008). Besides that, various other substances could act as a substrate like apple pomace, coffee husk, wheat straw, pineapple waste, cassava bagasse, banana, sugar beet cosset, kiwi fruit peel, etc. for the production of citric acid through the biological process (Figure 10.3).

10.3.1.2 Application of Citric Acid

Citric acids are widely used in different industries, viz. food and beverage industries, pharmaceutical, cosmetics, polymer industries, etc. In food sectors, it is used as a preservative, flavouring agent, and in the processing of ice cream, candies, soft drinks, and other vital products. It imparts a pungent taste and is widely used as a flavouring agent, preservative, and acidulant in the food as well as beverage industries. Because of its versatile buildings, citric acid is also used in a variety of other industries such as pharmaceuticals, cosmetics, and cleaning products. Citric acid is used in the food and beverage sector in order to improve the flavour and acidity of products. It is present in carbonated beverages, jams, jellies, candies, and processed foods. Citric acid functions as a type of organic preservative, extending the lifespan of food by inhibiting the growth of microorganisms. It is used as cleaning and chelating agent in daily household practices. In industries, it is used for the purpose of leather tanning, electroplating, detergent manufacturing, and as a preservative for storage of buffers and blood. It has also antioxidant activities, which are essential factor to enhance shelf life of the products.

10.3.2 Acetic Acid

Acetic acid is a monocarboxylic acid commonly known as ethanoic acid having molecular formula CH_3COOH and it is the second simplest carboxylic acid (after formic acid) (Figure 10.4).

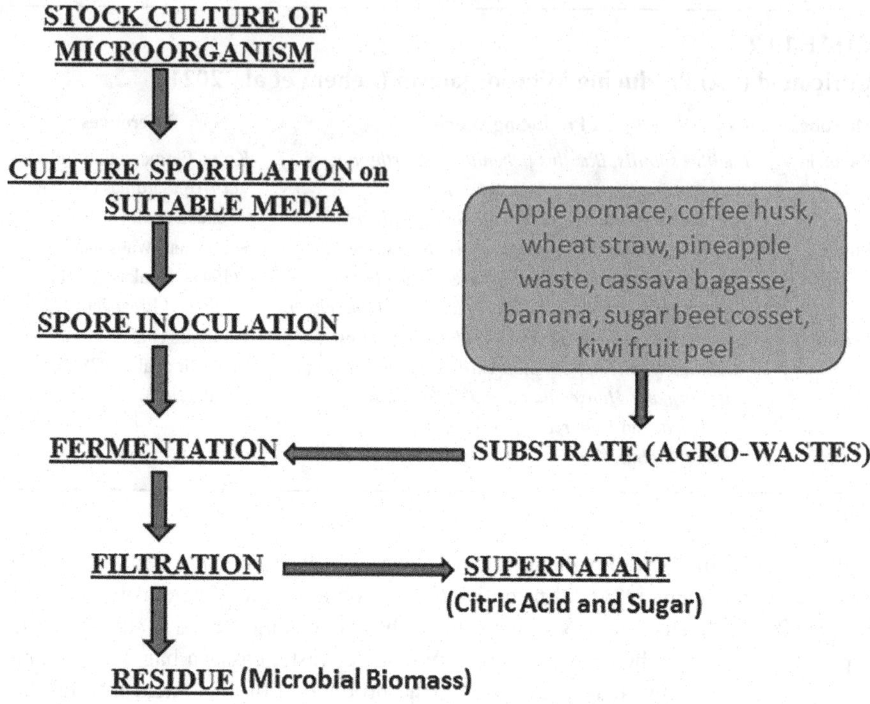

FIGURE 10.3 Biological process flow diagram for citric acid production (Soccol et al., 2006).

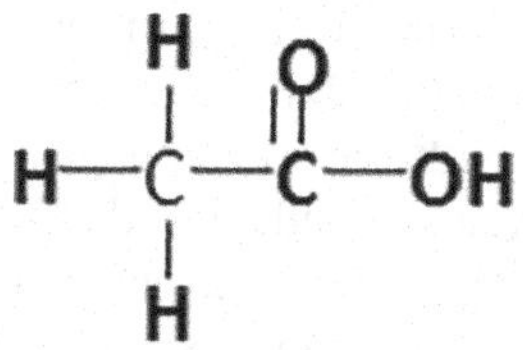

```
    H   O
    |   ||
H — C — C — OH
    |
    H
```

FIGURE 10.4 Molecular structure of acetic acid.

It is colourless liquid having pungent smell, sour taste, melts at 16.73°C, and boils at 117.9°C. Acetic acid is widely used as a food preservative and it is traditionally known as vinegar (Deshmukh & Manyar, 2020). Regarding the concentration of acetic acid in vinegar, different countries have different specifications. Vinegar is often produced by fermentation and subsequent oxidation of ethanol (Cheung et al., 2012). The production of ethanol by acetic acid bacteria is the most common natural source of acetic acid. This method is used to produce vinegar, which is essentially a dilute acetic acid solution. Various substrates, such as apples, grapes, rice, or grains, can be fermented to produce acetic acid via vinegar production. Chemical processes can also be used to produce acetic acid. The main technique is methanol carbonylation, also known as the Monsanto Corporation process. To produce acetic acid, methanol

TABLE 10.3
General Properties of Acetic acid

Properties	Acetic Acid/Ethanoic Acid/Vinegar
Molecular formula	CH_3COOH
Solubility	Soluble in water (Polar Solvent)
Pure form	Refers as Glacial acetic acid
Functional group	One carboxylic acid group (Monocarboxylic)
Taste	Sour
Odour	Pungent
Boiling point	117.9°C
Melting point	16.73°C

is combined with carbon monoxide in the presence of a catalyst, typically rhodium or iodine substance. These are the main producers of acetic acid, and the primary methods of production are biological fermentation and chemical synthesis. The source chosen is determined by factors such as availability, cost, and intended application (Table 10.3).

There are different kinds of traditional vinegar obtainable all over the world with wide application. In in vitro, vinegar inhibits the growth of various food-borne pathogens like *Bacillus cereus*, *Staphylococcus aureus*, *E. coli*, and others (Felter & Lloyd, 1898). It is used in treatment of various diseases like cholesterol, diabetes, obesity, cardiovascular disease, etc. as it has anti-inflammatory, antifungal (*A. niger*, *Cryphonectria parasitica*), and anti-bacterial (*S. aureus*, *Pseudomonas aeruginosa*, *E. coli*) properties. Some of them have higher antioxidant potential, increased fatty acid oxidation and lipolysis, and inhibited the proliferation of cancer cells (Yıldızlı et al., 2022; Baba et al., 2013). Wood vinegar acts as excellent antifungal and germicidal effect as it contains some benzene derivatives (Adfa et al., 2020). It is used as an extractant for various kinds of valuable metals like chromium, arsenic, copper, etc. (Choi et al., 2012)

10.3.2.1 Production of Acetic Acid

Acetic acid producing bacteria (*Gluconobacter*, *Acetobacter*, etc.) primarily involved in the production of vinegar, which contains more or less 4% of acetic acid (Saha & Das, 2023). The demand for vinegar is fulfilled through two manufacturing approaches: chemical and fermentative approach. Oxidation of hydrocarbon and aldehyde and methanol carbonylation is the key vinegar chemical method for manufacturing of vinegar (Deshmukh & Manyar, 2020). It can be synthesized in laboratories by artificial methods. It can also be manufactured through natural methods by passing through alcoholic beverages followed by acetic acid fermentation. Compared to other fermentation methods, the submerged process shows more productivity and high yield of products (Chozhavendan et al., 2021). The enzymatic synthesis of acetic acid is a promising method for producing it. Sen et al. (2019) published a study on the enzymatic synthesis of acetic acid. The researchers looked into the production of acetic acid using the *Gluconobacter oxydans* enzyme alcohol dehydrogenase

(ADH). The researchers optimized the reaction conditions for enzymatic acetic acid synthesis in this study, encompassing substrate concentrations, pH, temperature, and enzyme concentration. The substrate was ethanol, and the biocatalyst was ADH. The study demonstrated the feasibility of enzymatic synthesis for acetic acid production while discussing its possible applications and benefits.

Nowadays, organic wastes from diverse sources, such as dairy, food and agricultural practices, and kitchen wastes are produced on large scale. This creates serious problems on nature and responsible for environmental pollution. These wastes contain an excess amount of renewable organic components that can utilized for different purposes by the process of biotransformation. Various microorganisms are capable to convert these components into usable form.

10.3.2.2 Conventional Techniques

Most conventional methods for the production of vinegar (acetic acid) involve heterogeneous or homogeneous catalytic chemical process. It can be achieved through very simple techniques like the carbonylation of methanol (Monsanto process) or the oxidation of acetaldehyde/hydrocarbon (Pal et al., 2009) (Figure 10.5).

10.3.2.3 Fermentative Techniques

Fermentative route primarily makes use of renewable (agro-wastes) as a chief carbon source like cereals and their bran, discarded onions, pineapple peels and core, pulp, rotten and discarded fruit, fruit peels, apple pomace, whey protein from milk, beer, wine, etc. Acetic acid generated by fermentative techniques is basically tailored in the development of food products (Solieri & Giudici, 2009). Generally, this route is achieved by either submerged fermentation or surface fermentation techniques. Submerged techniques are most widely used in industries. It is a two-step process: in the first step, raw materials are fermented into alcohol by the involvement of yeasts like *S. cerevisiae*, and in the second step, these fermented products (alcohol) further fermented into acetic acid by the involvement

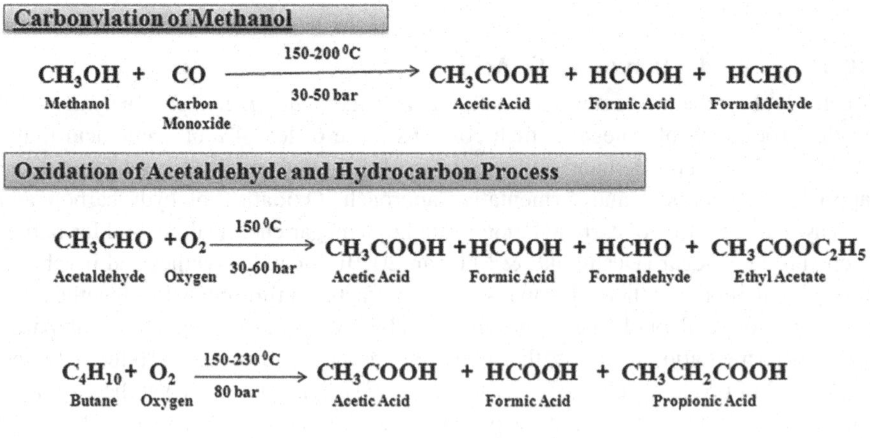

Carbonylation of Methanol

$$\underset{\text{Methanol}}{CH_3OH} + \underset{\text{Carbon Monoxide}}{CO} \xrightarrow[\text{30-50 bar}]{\text{150-200 °C}} \underset{\text{Acetic Acid}}{CH_3COOH} + \underset{\text{Formic Acid}}{HCOOH} + \underset{\text{Formaldehyde}}{HCHO}$$

Oxidation of Acetaldehyde and Hydrocarbon Process

$$\underset{\text{Acetaldehyde}}{CH_3CHO} + \underset{\text{Oxygen}}{O_2} \xrightarrow[\text{30-60 bar}]{\text{150 °C}} \underset{\text{Acetic Acid}}{CH_3COOH} + \underset{\text{Formic Acid}}{HCOOH} + \underset{\text{Formaldehyde}}{HCHO} + \underset{\text{Ethyl Acetate}}{CH_3COOC_2H_5}$$

$$\underset{\text{Butane}}{C_4H_{10}} + \underset{\text{Oxygen}}{O_2} \xrightarrow[\text{80 bar}]{\text{150-230 °C}} \underset{\text{Acetic Acid}}{CH_3COOH} + \underset{\text{Formic Acid}}{HCOOH} + \underset{\text{Propionic Acid}}{CH_3CH_2COOH}$$

FIGURE 10.5 Conventional route for the synthesis of acetic acids (Deshmukh & Manyar, 2020).

of acetic acid bacteria such as *Acetobacter aceti*, *Acetobacter pasteurianus*, *Gluconobacter*, *Gluconacetobacter*, and others.

In a recent study, it was observed that a good amount of acetic acid (19.3 g/L) was produced by simultaneous saccharification and two-step fermentation (SSTF) process compared to simultaneous saccharification and co-fermentation (SSCF) process from onion waste. Onion waste is considered as a suitable agricultural waste as it has sufficient carbohydrate content for acetic acid production using *S. cerevisiae* and *A. aceti* (Kim et al., 2019).

Another study reported that the bioconversion of apple pomace to acetic acid was done by a potential bacterial isolate *A. pasteurianus*. Apple pomace and cane molasses together yielded 14% bioethanol. Using the fermented bioethanol as a medium, 52.4 g of acetic acid/100 g of dry matter was produced (Vashisht et al., 2019) (Figure 10.6).

10.3.2.4 Applications of Acetic Acid

It is an essential industrial chemical reagent, which is used mainly in the production of cellulose acetate for photographic film, synthetic fibres and fabrics for wood glue, and polyvinyl acetate. Acetic acids in diluted form are used as descaling agents in household applications. Acetic acid is directly or indirectly applied into several chemical sectors, viz. pharmaceuticals, cosmetics, textile, polymer, food, chemical, etc. (Deshmukh & Manyar, 2020). Acetic acid is widely used as an additive and preservative in the food and beverage industry. It is commonly used to make condiments, pickles, salad dressings, sauces, and marinades. Acetic acid contributes to the sour flavour while also acting as a natural preservative by inhibiting the growth of bacteria and fungi. Acetic acid is a critical chemical intermediate in the production of a wide range of chemicals and

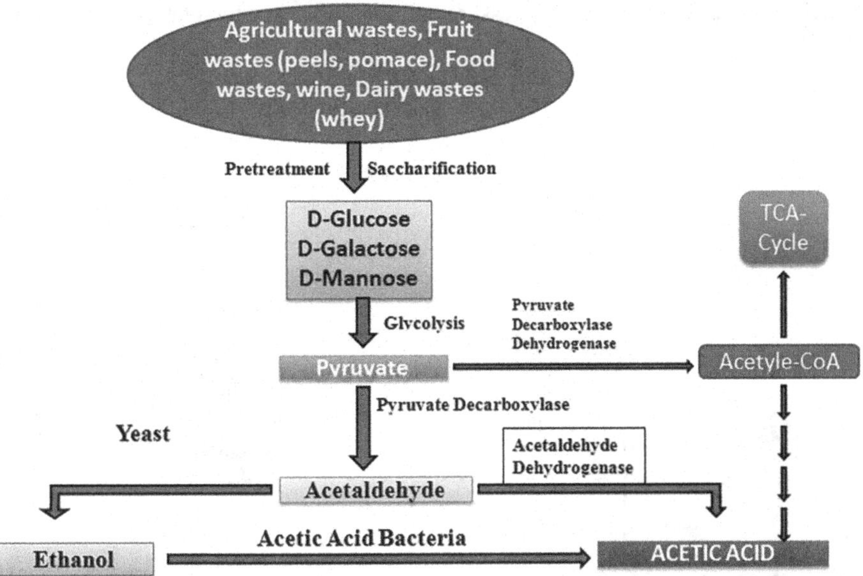

FIGURE 10.6 Fermentative route for the production of acetic acids (Saha & Das, 2023).

compounds. It is an important raw material in the production of vinyl acetate monomer (VAM), which is used to make adhesives, coatings, and polymers. Acetic acid is also used to make esters, cellulose acetate, acetic anhydride, and various solvents. Acetic acid is used in the textile industry for dyeing and finishing processes. It improves fabric colour fastness and acts as a pH regulator in textile dye baths. Acetic acid is used in the pharmaceutical industry. It's used in the production of antibiotics, vitamins, and analgesics, among other things. Acetic acid is also used in pharmaceutical formulations as a solvent. Acids containing acetic acid can be found in a variety of cleaning and household products. It is found in window cleaners, surface disinfectants, and descaling agents. Acetic acid is useful in cleaning because of its ability to dissolve mineral deposits and remove stains. The chemical acetic acid has agricultural applications, particularly as an herbicide. It is used as a non-selective weed killer in agricultural fields and non-crop areas to control unwanted plant growth. Acetic acid is commonly used in laboratories as a solvent for a variety of analytical and research applications. It is also used in the synthesis of chemical solutions and buffers. These can be considered just a few of the numerous applications of acetic acid in various industries. Its versatility and wide range of applications make it a valuable compound in many different kinds of industries (Rosenbaum et al., 2020) (Figure 10.7).

10.3.3 Malic Acid

Malic acid is regarded as an "edifice mass" for the making of biodegradable polymers. Fruit juices of apples and various fruits are good sources of malic acid. Malic acid is extensively used in food, personal care, and pharmaceutical industries. It can be used in textile finishing, water treatment, cleaning of metal, and bioavailability of calcium. The commercial production of malic acid is done by the hydration of maleic anhydride under higher pressure and temperature which produces a racemic

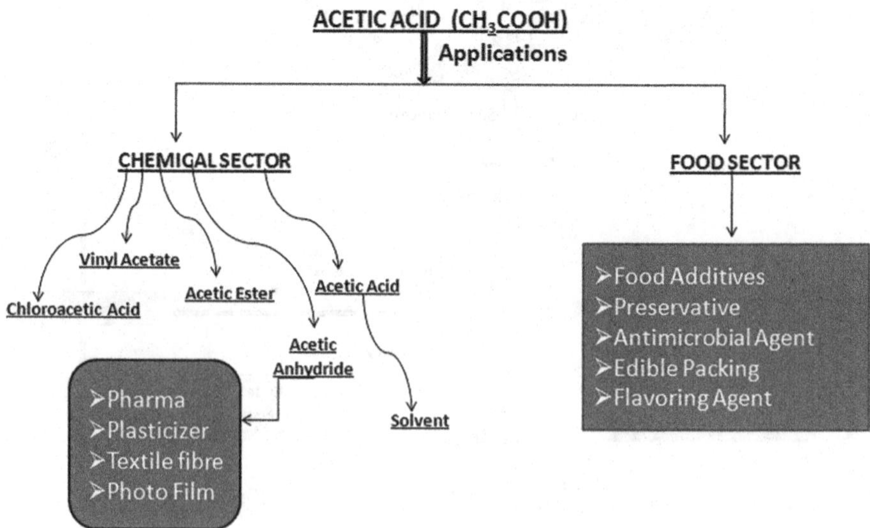

FIGURE 10.7 Applications of acetic acids (Deshmukh & Manyar, 2020).

mix of d- and l-malic acid. Alternatively, malic acid is synthesized by using the enzyme fumarate hydratase or fungus *S. cerevisiae* live cells, which catalyze the biotransformation of fumarate to malate. Malic acid is a versatile compound with a wide range of advantages and applications. It is an indispensable component due to its role in energy production, muscle pain relief, skin care, and the food industry. Understanding the potential benefits and considerations associated with malic acid can help people make informed decisions about how to incorporate it into their lifestyle and wellness routines (Attaluri et al., 2013) (Figure 10.8).

10.3.3.1 Malic Acid Production by Microorganisms

Microbial production of malic acid has added advantages over chemical synthesis (Kövilein et al., 2020). The advantages include enantiopure L-form of malic acid is only produced, and a large range of renewable substrates can be employed for microbial fermentation process. Glucose is the most favoured substrate for microbial L-malic acid production, but it is a quite expensive substrate (Song & Lee, 2006). Therefore, utilizing lignocellulose-derived bio-wastes will be possible low-cost approach. There are a number of agro-wastes that can be found easily after harvesting of crops for microbial L-malic and PMA production (Table 10.4). Recent global market of malic acid production is about 200,000 metric ton per year (Kövilein et al., 2020).

In 1962, *Aspergillus flavus* was identified as one of the potential strains for malic acid production, but later on, it was found that the strain produces aflatoxin. Later on, *Aspergillus oryzae*, whose genetic sequence similarity was same as *Aspergillus flavus*, appeared to be a promising candidate for production of malic acid (Payne et al., 2006). In addition to the improvement of cultivation by natural producers,

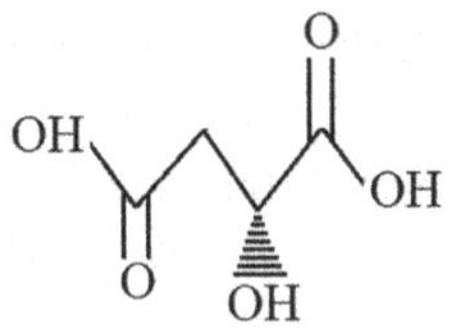

OH O OH O OH

FIGURE 10.8 Structure of malic acid.

TABLE 10.4
Malic Acid Production by Microorganisms Grown on Biofuel-Processing Coproducts or Hydrolyzed Lignocellulosic Biomass

Agro-wastes (Substrates)	Microorganisms	References
Corn straw hydrolysate	*Rhizopus delemar* HF-121	Li et al. (2014)
Wheat straw hydrolysate	*Aureobasidium pullulans* NRRL 50383	Leathers et al., (2013)
Sweet potato hydrolysate	*A. pullulans* CCTCCM2012223	Zan et al. (2013)
Sugarcane molasses	*A. pullulans* HA-4D	Xia et al. (2016)
Syngas (plant biomass)	*Clostridium ljungdahlii* DSM 13528/ *Aspergillus oryzae* DSM 1863	Oswald et al. (2016)
Crude glycerol	*Aspergillus niger* ATCC 10577	West et al. (2015)

channelling the intracellular C flux to L-malic acid synthesis through gene optimization may potentially be a promising strategy to boost microbial production capacity. The overexpression of rTCA genes, such as malate dehydrogenase mdh2, in the yeast *S. cerevisiae* led to a 3.7-fold rise in L-malic acid production. The overexpression of *A. oryzae* SpMAE1 homologue C4T318 in strain NRRL 3488 led to a more than two-fold increase in L-malate synthesis rate, further confirming the crucial significance of the L-malic acid transport (Brown et al., 2013). Besides fungal strain, genetic modification in bacterial strain also showed a promising result in malic acid production. To avoid unwanted acetate generation and to obtain a high yield of malic acid, the Moon et al. group, for example, employed a strain in which the phosphate acetyltransferase (pta) gene was removed and the phosphoenolpyruvate (PEP) carboxykinase gene from *Mannheimia succiniciproducens* was introduced (Moon et al., 2008) (Table 10.5).

10.3.3.2 Malic Acid Biosynthesis Pathway

Three possible pathways for malic acid production by microorganisms are reported. The three pathways include (i) oxidative TCA, (ii) reductive TCA (rTCA), and (iii) glyoxylate pathway. Both eukaryotes (*Aspergillus* sp., *Penicillium* sp., *Candida glabrata*, and *S. cerevisiae*) and prokaryotes (*Bacillus subtilis*), including engineered bacterial strain of *E. coli*, have these metabolic pathways. In yeast (*S. cerevisiae*), it was reported that pyruvate carboxylase and malate dehydrogenase exhibit high activities and result in enhanced malic acid production (Correia et al., 2007). *Thermobifida fusca* muC bacteria is reported to produce malic acid by conversion of phosphoenolpyruvate to oxaloacetate catalyzed by enzyme phosphoenolpyruvate carboxylase and thereafter oxaloacetate reduced to malate by enzyme malate dehydrogenase (Baba et al., 2013). The oxidative metabolic pathway of malate synthesis by microbes

TABLE 10.5
Physiochemical Properties of Malic Acid

Properties	Malic Acid
Molecular formula	$C_4H_6O_5$
Structure	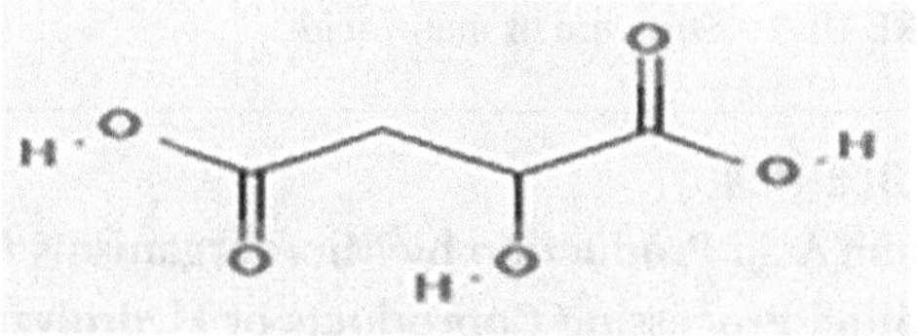
Molar mass	134.09
Solubility	Soluble in water (Polar Solvent)
Functional group	Two carboxylic acid group (Dicarboxylic)
Taste	Tart
Odour	Odourless
Boiling point	82°C
Melting point	130°C
Acidity	pK values of 3.40 and 5.11
Density	1.61 g/cm³

involves the tricarboxylic acid cycle. Acetyl-CoA undergoes condensation with oxaloacetate and enters the tricarboxylic acid cycle as citrate until malate is formed with the release of two CO_2. It is thought that *S. cerevisiae* can use this pathway to produce excess malate particularly if fumarase is overexpressed in the yeast cells (Ali et al., 2011). The second pathway of malic acid microbial synthesis involves the glyoxylate metabolic pathway in which two molecules of acetyl-CoA are used as precursor. In this pathway, isocitrate is converted into succinate and glyoxylate by isocitrate lyase enzyme. The second enzyme in the pathway is malate synthetase, which catalyzes in the presence of acetyl-CoA, glyoxylate, and water to form malate and Co-A. The primary pathway to synthesize malic acid is the reductive pathway utilized by most organisms among the three reported pathways (Figure 10.9).

10.3.3.3 Applications of Malic Acid

Malic acid has applications outside of medicine. Malic acid is well-known in the skincare industry for its exfoliating properties and ability to promote skin rejuvenation. It aids in the removal of dead skin cells, the unclogging of pores, and the improvement of skin texture, making it a popular ingredient in exfoliating and anti-ageing skincare products. Malic acid is used in the food industry as a flavour enhancer, pH regulator, and preservative. It adds a tangy flavour and acidity to food and beverages, helping to round out the flavour profile of many consumable products. Malic acid also acts as a chelating agent, assisting in the binding and stabilization of metal ions in processed foods. It is important to note that even though malic acid has many advantages, it may have some side effects and interactions. These can differ depending on the person and the type of malic acid used. To ensure safety and appropriateness, it is best to consult with a healthcare professional or a qualified practitioner before beginning any new supplement regimen. Finally, malic acid is a versatile compound with numerous advantages and applications. It is a valuable ingredient due to its role in energy production, muscle pain relief, skincare, and the food industry. Understanding the potential benefits and considerations associated with malic acid can help people make informed decisions about how to incorporate it into their lifestyle and wellness routines.

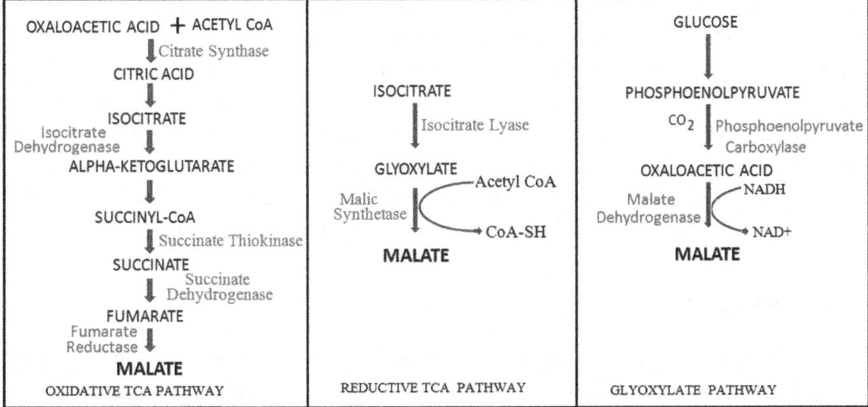

FIGURE 10.9 Oxidative, reductive, and glyoxylate pathway for malic acid production.

Food and beverage industry: Malic acid is widely used as a food ingredient and flavour enhancer. It provides a tart and sour taste, which is particularly desirable in certain food and beverage products. Malic acid is commonly found in carbonated drinks, fruit-flavoured beverages, candies, and confectionery. It also acts as an acidulant and pH regulator in processed foods, helping to maintain flavour stability and enhance microbial safety (Bastos et al., 2016).

Skincare and cosmetics: Malic acid is utilized in skincare and cosmetic formulations for its exfoliating and skin rejuvenating properties. It helps to remove dead skin cells, improve skin texture, and promote a more youthful appearance. Malic acid is often found in chemical peels, facial cleansers, toners, and anti-ageing products (Kim et al., 2009).

Pharmaceutical industry: In the pharmaceutical industry, malic acid is used as an excipient in drug formulations. It can serve as a pH-adjusting agent, enhancing the solubility and stability of certain medications. Malic acid also acts as a chelating agent, facilitating the stability of metal ions in pharmaceutical preparations (Hexsel et al., 2009).

Industrial applications: Malic acid has applications beyond the food and cosmetic industries. It is used in various industrial processes, such as metal cleaning and metal surface treatment. Malic acid's chelating properties make it effective in removing scale and rust from metal surfaces (Singh et al., 2019).

10.4 CONCLUSION

Agro-based wastes are treasure house of nutrients as well as a sustainable form for the production of bioenergy, biofuels, and industry-based products. Large-scale industrial applications of organic acids have gained global consideration for the production from low-cost substrate using biotransformation process. The present chapter highlighted the production of three important organic acids (acetic acid, citric acid, and malic acid) by microbial fermentation processes using carbohydrates as substrates for the production. The process described here is cost-effective and mainly focuses on reducing environmental pollution by biovalorization of organic waste from agriculture practices. Finally, the use of enzymatic techniques for the synthesis of organic acids from various agrobiomasses provides a sustainable and economically viable pathway. This method has the potential to meet the growing demand for organic acids while reducing environmental impact and promoting the transition to a more sustainable and circular bioeconomy. Continued research and development in this field will undoubtedly open up new avenues and advance the use of enzymatic techniques for organic acid synthesis from agrobiomass. New emerging techniques using recombinant DNA will definitely speed up the production of organic acids and bridge the gap.

ACKNOWLEDGEMENT

First and foremost, we express our sincere thanks to the authors whose works are being referred in this chapter. I would like to express my heartfelt gratitude to my esteemed co-authors for their invaluable insights, expertise, and collaboration throughout the writing process.

REFERENCES

Abou Hussein, S. D.; Sawan, O. M., The utilization of agricultural waste as one of the environmental issues in Egypt (a case study). *J. Appl. Sci. Res.*, 6, 1116–1124, 2010.

Adfa, M., Romayasa, A., Kusnanda, A. J., Avidlyandi, A., Yudha, S. S., Anon, B. C., et al., Chemical components, antitermite and antifungal activities of Cinnamomum parthenoxylon wood vinegar. *J. Korean Wood Sci. Technol.,* 48(1), 107–116, 2020.

Ali, H. K. Q., Zulkali, M. M. D., Utilization of agro-residual lignocellulosic substances by using solid state fermentation: a review. *Croat J. Food Technol. Biotechnol. Nutr.*, 6(1–2), 5–12, 2011.

Anwar, Z., Culfraz, M., Irshad, M., Agro-industrial lignocellulosic biomass a key to unlock the future bio-energy: a brief review. *J Radiat Res Appl Sci.*, 7, 163–173, 2014.

Ashworth, G., Azevedo, P., *Agricultural Wastes: Agriculture Issues and Policies Hauppauge, Nova Science Publishers.,* New York, 2009.

Attaluri, P., & Nair, K. P. (2013). The role of malic enzyme in lipogenesis and fat oxidation. In: Michael E. Symonds (Eds), *Adipose Tissue. Biology* (pp. 77–83). Springer, New York.

Baba, N., Higashi, Y., Kanekura, T., Japanese black vinegar "Izumi" inhibits the proliferation of human squamous cell carcinoma cells via necroptosis. *Nutr. Cancer*, 65(7), 1093–1097, 2013.

Bajpai, P., Bajpai, P. K., Bioconversion of lignocellulosic biomass to organic acids: advancements and challenges. *Biomass Conver. Biorefin.*, 10(1), 1–16, 2020. https://doi.org/10.1007/s13399-019-00492-6

Banerjee, A., Ghoshal, A. K. Organic acid production from potato starch waste fermentation by Aspergillus niger. *J. Chem. Technol. Biotechnol.,* 85(10), 1399–1140, 2010.

Basso, T. P., de Amorim, H. V., de Oliveira, A. J., Lopes, M. L., de Castro, A. M., Microbial production of food grade products: applications and perspectives. *Braz. J. Microbiol.*, 39(3), 369–380, 2008. https://doi.org/10.1590/S1517-83822008000300001

Bastos, D. H. M., Oliveira, T. T., Matsumoto, R. L. T., Carvalho, P. O., Ribeiro, M. L., Malic acid and citric acid as sour taste enhancers in low-calorie peach nectar. *J. Food Sci.*, 81(6), S1495–S1500, 2016.

Behera, B. C., Mishra, R., Mohapatra, S., Microbial citric acid: production, properties, application, and future perspectives. *Food Front.*, 2(1), 62–76, 2021. https://doi.org/10.1002/fft2.66.

Bhargav, S., Factors influencing the production of xylanase from A. awamori (ITCC 4857) and improvement in the xylanase production by coproduction of a xylanase inhibitory protein. *Enzyme Res.*, 2008, 1–9, 2018.

Bhargav, S., Panda, B., Ali, M., Solid-state fermentation: an overview. *Chem. Biochem. Eng.*, 22, 49–70, 2008.

Brown, S. H., Bashkirova, L., Berka, R., Chandler, T., Doty, T., McCall, K. et al., Metabolic engineering of Aspergillus oryzae NRRL 3488 for increased production of L-malic acid. *Appl. Microbiol. Biotechnol.*, 97, 8903–8912, 2013.

Bryer, S. C., Goldfarb, A. H., Effect of high dose vitamin C supplementation on muscle soreness, damage, function, and oxidative stress to eccentric exercise. *Int. J. Sport Nutr. Exerc. Metab.*, 16(3), 270–280, 2006.

Chapman, S. A., Cronan, J. E., The enzymatic biotinylation of proteins: a post-translational modification of exceptional specificity. *Trends Biochem. Sci.*, 24 (9), 359–363, 1999.

Cheung, H., Tanke, R. S., Torrence, G. P., Acetic acid. In: C. Ley (Eds), *Ullmann's Encyclopedia of Industrial Chemistry*. Wiley-VCH, Weinheim, 2012. https://doi.org/0.1002/14356007.a01_045.pub2.

Choi, C. H., Kim, K. Y., Jeong, W. S., Jeon, B. G, Jung, J. A. Z., Culfraz, M. I., Tang, J., Wang, X., Hu, Y., Zhang, Y. G., Jung, J. G., Effects of onion vinegar on the cerebral blood flow and the safety examination. *J. Physiol. & Pathol. Korean Med.*, 26(5), 657–664, 2012.

Chozhavendan, S., Aniskumar, M., Pradeepa, S., Karthika, D. S., Mathumitha, N., Converting waste agricultural, biomass into a resource, United Nations environmental programme. Review article of acetic acid production by wastes and strains. *JETIR.*, 8(12), e335–e343, 2021. http://www.jetir.org/papers/JETIR2112441.pdf

Clauser, N. M., González, G., Mendieta, C. M., Kruyeniski, J., Area, M. C., Vallejos, M. E. Biomass waste as sustainable raw material for energy and fuels. *Sustainability.*, 13, 794, 2021.

Correia, R., Magalhaes, M., Macedo, G., Protein enrichment of pineapple waste with *Saccharomyces cerevisiae* by solid state bioprocessing. *J. Sci. Ind. Res.*, 66, 259–262, 2007.

Couto, S. R., Exploitation of biological wastes for the production of value-added products under solid-state fermentation conditions. *Biotechnol J.*, 3(7), 859–870, 2008.

Desislava, J., Sathiyanarayanan, G., Harir, M., et al. Production and characterization of a novel exopolysaccharide from ramlibacter tataouinensis. *Molecules.*, 27(21), 7172. 2022. https://doi.org/10.3390/molecules27217172

Deshmukh, G., Manyar, H., Production pathways of acetic acid and its versatile applications in the food industry. In: T. P. Basso, T. O. Basso, L. C. Basso (Eds), *Biotechnological Applications of Biomass.* InTechOpen, London, 2020. https://doi.org/10.5772/intechopen.92289.

Dhanya, B. S., Mishra, A., Anuj, K. C., Verma, M. L., Development of sustainable approaches for converting the organic waste to bioenergy. *Sci. Total Environ.*, 723, 138109, 2020.

Dong, C., Huang, J., Ling, Z., Xiao, Y., Advances and prospects of organic acids production from lignocellulosic biomass. *Biores. Technol.*, 301, 122770, 2020. https://doi.org/10.1016/j.biortech.2019.122770

Du, J., Ma, C., Pei, H., Ma, Y., Xu, Y., Wang, M., Enhanced enzymatic hydrolysis of lignocellulose by optimizing enzyme complexes. *Appl. Biochem. Biotechnol.*, 179(7), 1107–1123, 2016. https://doi.org/10.1007/s12010-016-2049-x

Dutta, A., Sahoo, S., Mishra, R. R., Pradhan, B., Das, A., Behera, B. C., A comparative study of citric acid production from different agro-industrial wastes by Aspergillus niger isolated from mangrove forest soil. *Environ. Exp. Biol.*, 17, 115–122, 2019.

ElDein, S. M. N., Emaish, G. M. I., Effect of various conditions on production of citric acid from molasses in presence of potassium ferrocyanide by *A. aculeatus* and *A. carbonarius. Indian J. Exp. Biol.*, 17, 105–106, 1979.

El-Masry, H. G., Shebwy, K. I., El-Sheekh, M. M., El-Shemy, H. A., Citric acid production by Aspergillus niger using sugarcane bagasse hydrolysate. *3 Biotech.*, 9(6), 1–8, 2019. https://doi.org/10.1007/s13205-019-1710-9.

Fang, H., Enzymatic hydrolysis and fermentation for the production and recovery of succinic acid from lignocellulosic biomass. *Biores. Technol.*, 311, 123511, 2020.

Fang, X., Zuo, Z., Chen, Y., Chen, C., Zhao, Y., Xu, J., Liu, D. H., Production of succinic acid from lignocellulosic biomass: progress, challenges, and perspectives. *Biores. Technol.,* 307, 123212, 2020. https://doi.org/10.1016/j.biortech.2020.123212.

Felter, H. W., Lloyd, J. U., *Kings American Dispensatory. America.*, Available from: https://www.henriettesherbal.com/eclectic/kings/oxymel.html., 1898.

Frank, R. A., & Leeper, F. J., The enzymology of the citric acid cycle. *Crit. Rev. Biochem. Mol. Biol.*, 40(5), 327–347, 2005. https://doi.org/10.1080/10409230500302732

Fukuda, D. H., Smith, A. E., Kendall, K. L., Stout, J. R., The possible combination of sodium bicarbonate and sodium citrate as an ergogenic aid in high intensity, intermittent exercise. *Int. J. Sport Nutr. Exerc. Metab.*, 20(5), 419–434, 2010.

Fukuda, H., Susuki, T., Sumino, Y., Akiyama, S., Microbial preparation of citric acid. *German Patent.*, 2(003), 221, 1970.

Ganguly, P., Khan, A., Das, P., Bhowal, A., Cellulose from lignocellulose kitchen waste and its application for energy and environment: bioethanol production and dye removal. *Indian Chem. Eng.,* 63, 161–171, 2021.

Goldberg, R. N., Kishore, N., Lennen, R. M., Thermodynamic quantities for the ionization reactions of buffers. *J. Phys. Chem. Ref. Data.*, 31(1), 231–370, 2002. https://doi.org/10.1063/1.1416902S2CID 94614267

Gopalakrishnan, K., Prabhu, K. S., Prakash, N., *Microbial applications in bioremediation and bioconversion of xenobiotics*. Springer, New York (pp. 191–212) https://doi.org/10.1007/978((sbn))3((sbn))030((sbn))18907((sbn))9_11

Gopinath, S. M., Sowmya, K., Production of organic acids by solid-state fermentation: a review. *J Agric. Sci. Technol.*, 22, 485–492, 2020.

Gutierrez, N. A., Mckay, I. A., French, C. E., Brooks, J., Maddox, I. S., Repression of galactose utilization by glucose in the citrate-producing yeast *Candida guilliermondii. J. Ind. Microbiol.,* 11, 143–146, 1993.

Heidi, O., Line, D. H., Svein, J. H., Vincent, G. H. E., Anikó, V., Enzymatic processing of lignocellulosic biomass: principles, recent advances and perspectives. *J. Ind. Microbiol. Biotechnol.*, 47(9), 623–657, 2020. https://doi.org/10.1007/s10295-020-02301-8

Hexsel, D., Orlandi, C., Zechmeister do Prado, D., Zechmeister do Prado, D., Citric acid as an excipient in topical pharmaceutical preparations. *Clin. Cosmet. Investig. Dermatol.,* 2, 27–32, 2009. https://doi.org/10.2147/CCID.S4121

Jang, J. Y., Choi, Y. H., Shin, T. S., Kim, T. H., Shin, K. S., Park, H. W., et al., Biological control of *Meloidogyne incognita* by *Aspergillus niger* F22 producing oxalic acid. *PLoS One.*, 11, e0156230, 2016.

Jian, X., Mette, H. T., Anne, B. T., Enzymatic hydrolysis and fermentability of corn stover pretreated by lactic acid and/or acetic acid. *J Biotechnol.*,134(4), 300–305, 2009.

Jo, S. H., Kim, S. H., Lee, S., Advances in enzymatic approaches for lignocellulosic biomass utilization. *Biores. Technol.*, 291, 121890, 2019. https://doi.org/10.1016/j.biortech.2019.121890

Kapelli, O., Muller, M., Fiechter, A., Chemical and structural alterations at cell surface of *Candida tropicalis*, induced by hydrocarbon substrate. *J. Bacteriol.* 133, 952–958, 1978.

Kapoor, M., Gupta, M. N., Lipase promiscuity and its biochemical applications. *Process Biochem.*, 47(4), 555–569, 2012. https://doi.org/10.1016/j.procbio.2012.01.007

Karow, E. O., Waksman, S. A., Production of citric acid in submerged culture. *Ind. Eng. Chem.,* 39, 821–825, 1947.

Katja, V., Željko, K., Maja, L., Bioethanol production by enzymatic hydrolysis from different lignocellulosic sources. *Molecules.*, 26(3), 753, 2021. https://doi.org/10.3390/molecules26030753

Kim, B. Y., Choi, J. W., Kim, H. S., Hwang, E. S., Park, K. C., Malic acid in skincare: chemical performance and mechanism dependent on concentration. *Skin Res. Technol.*, 15(2), 204–209, 2009.

Kim, H. M., Choi, I. S., Lee, S., et al., Biorefining process of carbohydrate feedstock (agricultural onion waste) to acetic acid. *ACS Omega.*, 4(27), 22438–22444, 2019.

Kövilein, A., Christin, K., Liyin, C., Katrin, O., Malic acid production from renewables: a review. *J. Chem. Technol. Biotechnol.*, 95, 513–526, 2020.

Kroya, F. I., Citric acid prepared by fermentation. British Patent No., 1187610, 1970. https://www.scielo.br/pdf/babt/v42n3/v42n3a01.pdf

Kubicek, C. P., Rohr, M., The role of trycarboxylic acid cycle in citric acid accumulation by *Aspergillus niger. Eur. J. Appl. Microbiol. Biotechnol.*, 5, 263–271, 1978.

Kuhad, R. C., Gupta, R., Singh, A., Microbial cellulases and their industrial applications. *Enzyme Res.*, 2011, 280696, 2011. https://doi.org/10.4061/2011/280696

Kumar, A., Jain, V.K., Solid state fermentation studies of citric acid production. *Afric. J. Biotechnol.* 7, 644–650, 2008.

Lachenmeier, D. W., Nathan-Maister, D., Acetic acid: historical perspectives, myths, and potentials for the control of wine spoilage. *J. Agric. Food Chem.*, 56(13), 5577–5584, 2008. https://doi.org/10.1021/jf800465u

Leathers, T. D., Manitchotpisit, P., Production of poly (β-L-malic acid) (PMA) from agricultural biomass substrates by *Aureobasidium pullulans*. *Biotechnol Lett.*, 35, 83–89, 2013. https://doi.org/10.1007/s10529-012-1045-x

Li, X., Liu, Y., Yang, Y., Zhang, H., Wang, H., Wu, Y., et al., High levels of malic acid production by the bioconversion of corn straw hydrolyte using an isolated *Rhizopus delemar* strain. *Biotechnol. Bioprocess Eng.*, 19, 478–492, 2014. https://doi.org/10.1007/s12257-014-0047-z.

Liu, G., Zhang, J., Bao, J., Zhao, L., Enzymatic hydrolysis of corn stover for organic acid production: a review. *Biores. Bioprocess.*, 6(1), 8, 2019. https://doi.org/10.1186/s40643-019-0247-7

Liu, S., Enzymatic hydrolysis of corn stover for organic acid production. *Biotechnol. Biofuels.*, 12, 1–12, 2019.

Luciana, P. S., Vandenberghe1, L., Carlos, R. S., Pandey, A., Jean, M., Microbial production of citric acid. *Braz. Arch. Biol. Technol.*, 42 (3), 1–14, 1998.

Lynd, L., Laser, M., Bransby, D. et al. How biotech can transform biofuels. *Nat Biotechnol.*, 26, 169–172, 2008. https://doi.org/10.1038/nbt0208-169

Maki, M., Leung, K. T., Qin, W., The prospects of cellulase-producing bacteria for the bioconversion of lignocellulosic biomass. *Int. J. Biol. Sci.*, 5, 500–516, 2009.

Mihranyan, A., Strømme, M., Ek, R., Antimicrobial properties of paper impregnated with zinc oxide nanoparticles. *BioResources.*, 7(4), 5486–5496, 2012. https://doi.org/10.15376/biores.7.4.5486-5496

Mohanty, S., Behera, S., Swain, M. R., Ray, R. C., Bioethanol production from mahula (*Madhuca latifolia L.*) flowers by solid state fermentation. *Appl. Energy.*, 86, 640–644, 2009.

Moon, S. Y, Hong, S. H., Kim, T. Y., Lee, S. Y., Metabolic engineering of Escherichia coli for the production of malic acid. *Biochem. Eng. J.*, 40, 312–320, 2008.

Mussatto, S.I., Ballesteros, L.F., Martins, S., Teixeira, J.A., Use of agro-industrial wastes in solid-state fermentation processes. *Indust. Waste Intech Croatia.*, 214, 121–140, 2012.

Nagarajan, S., Jayaprakash, V., Madhavan, P., Citric acid production. In M. C. Flickinger (Eds), *Encyclopedia of Industrial Biotechnology*, John Wiley & Sons, Inc, New York, pp. 1–24, 2012.

Nanda, S., Dalai, A. K., Conversion of biomass-derived carbohydrates into organic acids: a review. *Green Chem.*, 19(21), 4917–4946, 2017. https://doi.org/10.1039/C7GC02284C

Nelson, D.L., Cox, M.M., *Principles of Biochemistry*, fourth edition. Freeman Publishers, New York, 2004.

Oh, M. J., Park, Y. J., Lee, S. K., Citric acid production by *Hansenula anamola var. anamola*. *Hanguk Sikpum Kawahakhoe Chi.*, 5, 215–223, 1973.

Oswald, F., Dorsam, S., Veith, N., Zwick, M., Neumann, A., Ochsenreither, K., Syldatk, C., Sequential mixed cultures: from syngas to malic acid. *Front. Microbiol.*, 7, 281, 2016.

Pal, P., Sikder, J., Roy, S., Giorno, L., Process intensification in lactic acid production: a review of membrane-based processes. *Chem. Eng. Process. Process Intensif.*, 48, 1549–1559, 2009. https://doi.org/10.1016/j.cep.2009.09.003.

Panda, S. K., Ray, R. C., *Microbial processing for valorization of horticultural wastes. Environmental Microbial Biotechnology*. Springer, Switzerland, pp. 203–221, 2015.

Pandey, A., Soccol, C. R., Bioconversion of biomass: a case study of lignocellulosics bioconversions in solid state fermentation. *Brazilian Arch. Biol. Technol.*, 41, 379–90, 1998.

Payne, G. A., Nierman, W. C., Wortman, J. R., Pritchard, B. L., Brown, D., Dean, R. A. et al., Whole genome comparison of Aspergillus flavus and A. oryzae. *Med Mycol.*, 44, S9–S11, 2006.

Poli, A., Di Donato, P., Abbamondi, G. R., Synthesis, properties and potential applications of biotechnologically produced microbial diacids. *Appl. Microbiol. Biotechnol.*, 90(2), 425–437, 2011. https://doi.org/10.1007/s00253-011-3161-3

Qiu, W., Wang, Q., Chen, T., Xia, X., Optimization of enzymatic saccharification of corn stover for efficient bioethanol production. *Energy Procedia.*, 152, 239–245, 2018. https://doi.org/10.1016/j.egypro.2018.09.099.

Quitmann, H., Fan, R., Czermak, P., Acidic organic compounds in beverage, food, and feed production. *Adv. Biochem. Eng. Biotechnol.*, 143, 1–141, 2014. https://doi.org/10.1007/10_2013_262

Rais, M., Sheoran, A. Scope of supply chain management in fruits and vegetables in India. *J. Food Process. Technol.*, 6, 427, 2015. https://doi.org/10.4172/2157-7110.1000427

Rani, A., Use of lignocellulolytic fungi for bioconversion of rice straw into livestock feed. *BioResources.*, 10(2), 2347–2356, 2015.

Ray, R. C., Ward, O. P., Post harvest microbial biotechnology of tropical root and tuber crops. *Microb. Biotechnol. Horticult.*, 2006, 511–552, 2006.

Rosenbaum, E., Khaw, K. S., Luo, X., A review of acetic acid as a disinfectant: efficacy, stability and safety considerations. *J. Am. Assoc. Nurse Pract.*, 32(12), 860–864, 2020. https://doi.org/10.1097/JXX.0000000000000525

Roukas, T., Production of citric acid from beet molasses by immobilized cells of *Aspergillus niger*. *J. Food Sci.*, 56, 878–880, 1991.

Saha, B. C., Cotta, M. A., Ethanol production from alkaline peroxide pretreated enzymatically saccharified wheat straw. *Biotechnology Progress.*, 24(5), 814–817, 2008. https://doi.org/10.1021/bp0704018

Saha, D., Das, P. K., Bioconversion of agricultural and food wastes to vinegar. In: A. N. Barros, J. Campos, A. Vilela (Eds), *Functional Food - Upgrading Natural and Synthetic Sources,* , 2023. https://doi.org/10.5772/intechopen.109546

Saini, R. K., Keum, Y. S., Enzymatic production of organic acids from lignocellulosic biomass: an overview. *Biores. Technol.*, 278, 146–156, 2019. https://doi.org/10.1016/j.biortech.2019.01.052

Saratale, G. D., Chen, S., Lo, Y. C., Chang, J. S., Enzymatic conversion of biomass into biofuels. *Biores. Technol.*, 304, 122994, 2020. https://doi.org/10.1016/j.biortech.2020.122994

Sardinas, J. L., Fermentative production of citric acid. *French. Patent No. 2113668*, 1972. https://www.scielo.br/pdf/babt/v42n3/v42n3a01.pdf.

Sarkar, N., Ghosh, S. K., Bannerjee, S., Aikat, K., Bioethanol production from agricultural wastes: an overview. *Renew Energy.,* 37, 19–27, 2012. https://doi.org/10.1016/j.renene.2011.06.045.

Sauer, M., Porro, D., Mattanovich, D., Branduardi, P., Microbial production of organic acids: expanding the markets. *Trends Biotechnol.*, 26(2), 100–108, 2008.

Sauer, U., Eikmanns, B. J. The PEP-pyruvate-oxaloacetate node as the switch point for carbon flux distribution in bacteria. *FEMS Microbiol. Reviews.*, 29(4), 765–794, 2005. https://doi.org/10.1016/j.femsre.2004.11.002

Sen, S. S., Ghorai, S., Nanda, A., Banerjee, U. C., Enzymatic synthesis of acetic acid using alcohol dehydrogenase from Gluconobacter oxydans. *J. Appl. Microbiol.,* 127(1), 116–125, 2019. https://doi.org/10.1111/jam.14254

Shaikh, Z., Qureshi, P., Screening and isolation of organic acid producers from samples of diverse habitats. *Int. J. Curr. Microbiol. Appl.*, 2(9), 39–44, 2013.

Sharma, N., Sahota, P. P., Singh, M. P., Organic acid production from agricultural waste. *Waste Energ. Prospects Appl.*, 2021, 415–438, 2021.

Singh, A., Kumar, P. K., Microbial production of citric acid and its applications. *Food Technol. Biotechnol.*, 52(3), 366–381, 2014.

Singh, A., Kumari, M., Sustainable production of organic acids from agricultural waste: a review. *Environ. Technol. Innov.*, 19, 100928, 2020.

Singh, D., Kumari, M., *Agricultural/Biomass Waste Management through "Green Supply Chain Way": Indian "Brickfield" Perspective.* De Gruyter, Germany, 2020. https://doi.org/10.1515/9783110628593-009.

Singh, R., Rathi, P., Dhiman, M., A comprehensive review on malic acid applications in food and non-food industries. *Int. J. Chem. Stud.*, 7(1), 527–536, 2019.

Soccol, C. R., Vandenberghe, L. P. S., Rodrigues, C. P. A., New perspectives for citric acid production and application. *Food Technol. Biotechnol.*, 44, 141–149, 2006.

Solieri, L., Giudici, P., Vinegars of the world. Springer, Verlag Milan, 1–297, 2009. https://doi.org/10.1007/978-88-470-0866-3.

Song, H., Lee, S. Y., Production of succinic acid by bacterial fermentation. *Enzyme Microb. Technol.,* 39, 352–361, 2006. https://doi.org/10.1016/j.enzmictec.2005.11.043.

Srivastava, A. K., Srivastava, S., Krebs cycle intermediates: cellular messengers influencing immunity and inflammation. *Pharmacol. Res.*, 119, 117–124, 2017. https://doi.org/10.1016/j.phrs.2017.02.005

Swain, M. R., Ray, R. C., Patra, J. K., *Citric Acid: Microbial Production and Applications in Food and Pharmaceutical Industries.* Nova Science Publisher, New York, pp. 97–118, 2011.

Tovar-Méndez, A., Miernyk, J. A., Randall, D. D., Regulation of pyruvate dehydrogenase complex activity in plant cells. *Europ. J. Biochem.*, 270(6), 1043–1049, 2003. https://doi.org/10.1046/j.1432-1033.2003.03473.x

Tripathi, N., Colin, D. H., Singh, R. S., Atkinson, C. J., Biomass waste utilisation in low-carbon products: harnessing a major potential resource. *NPJ Clim. Atmos. Sci.*, 2, 35, 2019.

Vashisht, A., Thakur, K, Baljinder, S. K., Kumar, Vinod., Yadav, S. K., Waste valorization: identification of an ethanol tolerant bacterium Acetobacter pasteurianus SKYAA25 for acetic acid production from apple pomace. *Sci. Total Environ.*, 690, 956–964, 2019.

Wang, J., He, Y., Liu, X., Yang, L., Yu, L. (2020). Acetic acid in food preservation and safety: a review. *J. Food Quality.*, 2020, 1–11, 2020. https://doi.org/10.1155/2020/6642970

Wang, Y., Wei, H., Xu, Q., Zhang, J., Xu, J., Current understanding of synergistic pretreatment of lignocellulosic biomass by hydrothermal and enzymatic hydrolysis: a review. *Biores. Technol. Rep.*, 5, 321–328, 2019. https://doi.org/10.1016/j.biteb.2019.01.008

Wehmer, C., Note surla fermentation citrique. *Bull. Soc. Chem.,* 9, 728, 1893.

West, T. P., Fungal biotransformation of crude glycerol into malic acid. *Z. Naturforsch. C.,* 70, 165–167, 2015.

West, T. P., Citric acid production by aspergillus niger using solid-state fermentation of agricultural processing coproducts. *Appl. Biosci.* 2, 1–13, 2023.

Xia, J., Xu, J., Hu, L., Liu, X., Enhanced poly 2020 (L-malic acid) production from pretreated cane molasses by *Aureobasidium pullulans* in fed-batch fermentation. *Prep. Biochem. Biotechnol.*, 46, 798–802, 2016. https://doi.org/10.1080/10826068.2015.1135464.

Xu, J., Wang, X., Zhou, Q., Liu, F., Zhao, G., Enzymatic hydrolysis of lignocellulosic biomass to sugars: engineering strategies for efficient conversion. *Biotechnol. Adv.*, 35(5), 562–576, 2017. https://doi.org/10.1016/j.biotechadv.2017.06.002

Yang, Y., Hu, C., Chen, H., Ma, R., Advances in enzymatic hydrolysis of lignocellulosic biomass for biofuel production. *Engine. Life Sci.*, 17(5), 486–498, 2017. https://doi.org/10.1002/elsc.201600217.

Yıldızlı, G., Coral, G., Ayaz, F., Anti-bacterial, anti-fungal, and anti-inflammatory activities of wood vinegar: a potential remedy for major plant diseases and inflammatory reactions. *Biomass Convers. Biorefin.,* 12, 1–10, 2022.

Zan, Z., Zou, X., Efficient production of polymalic acid from raw sweet potato hydrolysate with immobilized cells of *Aureobasidium pullulans* CCTCC M2012223 in aerobic fibrous bed bioreactor. *J. Chem. Technol. Biotechnol.,* 88, 1822–1827, 2013. https://doi.org/10.1002/jctb.4033.

Zhang, Y. H. P., Lynd, L. R., Toward an aggregated understanding of enzymatic hydrolysis of cellulose: noncomplexed cellulase systems. *Biotechnol. Bioeng.*, 88(7), 797–824, 2004. https://doi.org/10.1002/bit.20282

Zhong, C., Zhang, Z., Liu, X., Liu, S., Zhao, Z. K., Bai, F., Improving enzymatic hydrolysis of lignocellulosic biomass with ionic liquid pretreatment. *Biores. Technol.*, 292, 121959, 2019. https://doi.org/10.1016/j.biortech.2019.121959.

11 Application of Microbial Technique for the Synthesis of Biodegradable Polymers from Different Agrobiomasses

Neha Saxena and Amit Kumar

11.1 INTRODUCTION

People have gained a greater awareness of the dangers posed by pollution and the degradation of natural resources since the 1970s. As a consequence of this, scientists have been hard at work creating novel solutions, and as a consequence of this, alternative concepts have been found. The use of biodegradable polymers helps to cut down on the consumption of fossil fuels, which in turn helps to delay the rate at which carbon dioxide levels are rising in the air and surroundings (Kabasci, 2013; Abioye et al., 2018). Plastics, thanks to their cost-effectiveness and adaptability, have enabled a broad variety of uses feasible that would not have been feasible otherwise (Chae & An, 2018). In 2018, 53% of the world's bio-based polymer manufacturing came from Asia and the Pacific, making it the region with the world's largest manufacturing capability for biological polymers (Chinthapalli et al., 2019). North America accounts for 18%, followed by South America with 11% and Oceania with 1%. Europe is followed by South America. It is anticipated that Europe's market share will rise to 25% over the course of the subsequent 5 years, while the market share of the remainder of the globe is anticipated to decrease. Many different applications have found successful use for biodegradable substances. These concepts can be applied to the problem-solving processes of a number of different trades, which are not restricted to those dealing with packaging, hospitality, pharmaceuticals, cultivators, electronic gadgets, automobiles, and fabrics. In 2018, approximately 65% of the total market for bioplastics (1.2 million tons) was dedicated to the container industry (Nova-Institute, Bio-plastics Market Data, 2018). The production of standard plastic (made from petrochemicals) is anticipated to approach each year 350 million Tons by the end of this decade (European Bioplastics Report-Bioplastics Market Data, 2020).

DOI: 10.1201/9781003407713-11

The widespread application of these substances has led to the contamination of the environment. Having said that, an enormous percentage of the total amount of plastic that is manufactured is made up of plastics that can't be recycled, which accounts for approximately half of the total mass (Rhodes, 2019). Between the years 1950 and 2015, over 90% of the plastic garbage generated was not being reused, and the remaining waste was either discarded or allowed to be released into natural habitat (Geyer et al., 2017). In addition to expanding the reprocessing and repurposing of plastics that have already been manufactured, along with the use of biodegradable materials in place of various sorts of plastic goods, notably items that are only used once, and resulting from a transformation in the mindset and practices of our society as a whole, the growing trend of reprocessing and reusing plastics that have already been created is one factor that is significantly contributing to the proliferation of plastic pollution. At the same time, the supply of fossil fuels is finite, and the burning of these fuels results in the discharge of atmospheric gases. Furthermore, the UNDP advocated for an economy that is sustainable, which is defined as one in which resources from nature are used efficiently, contamination is kept to a minimum, and waste is kept to a minimum. These bio-based or biodegradable polymers are referred to as "bioplastics," and the term "bioplastics" has been in use for some time. To put it another way, even though the qualities of biodegradable polymers deteriorate and become less noticeable after being exposed to microorganisms, methane, carbon dioxide, and water, these polymers are still differentiated by the fact that they can be broken down biologically (Babu et al., 2013). Due to the limited availability of fossil resources, there is a growing need for products made from bio-based materials. These materials either help to reduce the amount of biological garbage that must be collected or cut down on the amount of pollution that is released into the environment.

Utilizing biomass from agricultural and industrial processes is seen as a viable source of raw materials for the creation of a variety of industrially important enzymes (Panesar et al., 2016; Jain et al., 2022). The use of renewable residue as substrates is beneficial since they are easy to obtain and provide copious amounts of nitrogen and carbon. Numerous cost-effective technologies are currently employed at industrial scales to harness agro-industrial residues, which promote the proliferation of microorganisms and facilitate the production of value-added products (Sadh et al., 2018). Agrobiomasses encompass a variety of materials, including bagasse, lint, seeds peel, pulp derived from oil cake (such as soy, mustard, and groundnut), rice straw, fruit pomace, wheat bran, wheat straw, sugarcane milling, corn cob, and numerous other sources (Pandit et al., 2021). The following major considerations will guide the selection of appropriate forms of agricultural waste: (i) the amount of starch present; (ii) the amount of cellulose, lignin, and hemicellulose present; (iii) the bioavailability of the substance in question, as well as its effect on food safety and agricultural supply networks; (iv) the difficulty of the synthetic approaches and the qualities of the substance that is wanted; and (v) biodegradation (Mose & Maranga, 2011; Nunes et al., 2020).

The utilization of fermentation for the production of microbial enzymes is widely regarded as a significant alternative. This process encompasses various stages, upstream and downstream, as illustrated in Figure 11.1. The proliferation of the

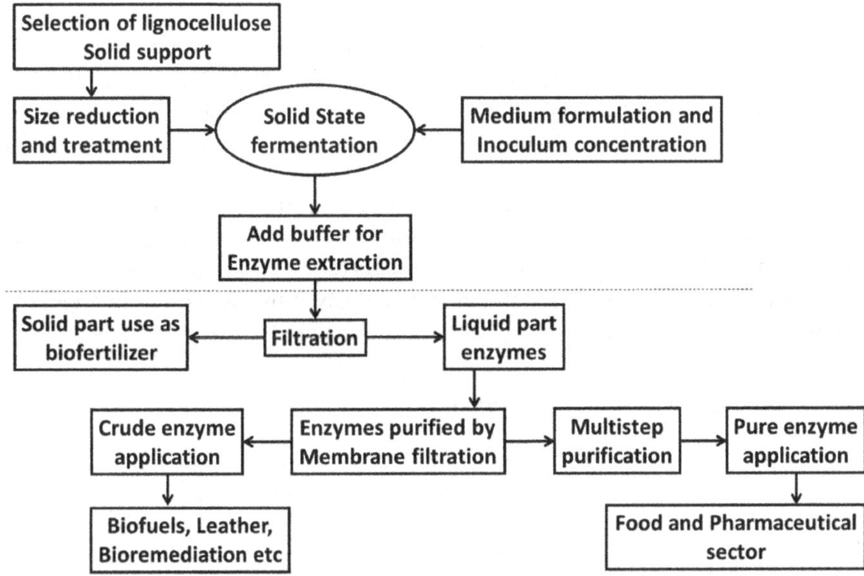

FIGURE 11.1 The image depicts a systematic representation of using lignocellulosic biomass for the production of enzymes (Raina et al., 2022).

enzyme industry occurred during the 1980s and 1990s subsequent to the discovery of microbial enzyme production. Microorganisms have the ability to generate enzymes through their own genetic coding, as well as possess the capacity to contain exogenous genes and manifest enzymes as recombined proteins (Tripathi & Shrivastava, 2019).

The categorization of enzyme utilization fundamentally delineates the manufacturing procedure. Enzymes that are employed in the production of materials, aliments, and fibers are generated in large quantities as unrefined preparations. Conversely, enzymes that are intended for application in pharmaceuticals and healthcare commodities necessitate a considerable degree of refinement. Enzymes have gotten substantial interest in the scientific community because of their potential therapeutic applications, including but not limited to fibrinolytics, thrombolytics, anticoagulants, anti-inflammatory agents, and oncolytic. Microbial enzymes have been discovered to have significant involvement in diagnostic testing, biochemical research, and ongoing treatment and illness assessment for a variety of conditions and medications, in addition to their therapeutic applications (Vachher et al., 2021). The enzyme production can be categorized into four distinct stages: the first stage is the synthesis of the enzyme; the second is the recovery of catalyst; the third is the purification of the desired enzyme, which is a crucial step in enzymology; and the final stage is the formulation of enzyme products. Previously, the methodologies for enzyme synthesis were not clearly defined. The advent of fermentation technology, specifically solid-state fermentation (SSF) and submerged fermentation (SmF), has primarily facilitated to significant production of enzymes. According to reports, the production of enzymes from filamentous organisms exhibits superior performance in the context of SSF. The production of fermented products is affected by various parameters such as pH, temperature, and fermentation type. However, the selection of appropriate microbial strains and substrate types are the most crucial factors that significantly impact their successful industrial implementation. The utilization of SSF has garnered significant interest in contemporary times as a cost-effective biotechnological approach for generating lignocellulolytic enzymes from agricultural and industrial by-products. The practice of combining wastes from agriculture and industry is commonly employed as a means to ensure that bacterial growth is supported by a sufficient array of necessary nutrients, as a single by-product may not always provide the full complement of the required nutrients.

11.2 BIODEGRADABLE POLYMERS

Polymer is a macromolecule composed of a lengthy sequence of repeating monomers. Polymers, whether they are artificially produced or occur naturally, hold significant and pervasive significance in our daily existence. This phenomenon can be attributed to the unique and distinguishable traits exhibited by each entity (Ilyas et al., 2019). These are produced through a series of reactions involving condensation and addition polymerization. Polymers can be classified into two categories, namely thermosetting and thermoplastic. The term "bio-based plastic" pertains to plastic materials that are derived from natural resources and possess the characteristic of renewability. On the other hand, the term "biodegradable plastic" pertains to the behavior exhibited by plastic materials at the end of their lifecycle. Biodegradable plastic undergoes spontaneous decomposition in the environment, primarily facilitated by microorganisms, resulting in its breakdown. Compared to conventional plastics, it exhibits greater ecological compatibility.

The production of bio-based and biodegradable polymers is an ongoing endeavor that is being made in the direction of achieving sustainability. The creation of resource-friendly bioplastics produced from bio-based or biodegradable polymers presents fresh approaches to tackling the problems of depletion of resource and plastic pollution. The current era is witnessing a major increase of the global economy, albeit one that is accompanied with irreparable damage to the natural environment. The problem of pollution caused by plastic has been identified as a crisis on a global scale that impacts all stages of the product's lifecycle, including production, disposal, and incineration (Scalenghe, 2018). There is currently work being done to produce bioplastics, which make use of components that are either naturally or chemically synthesized from renewable and oil-based resources. The goal of this work is to create materials that have complete biodegradability, high recycling value, small carbon footprint, and compostability (Babu et al., 2013). There has been a recent development in the creation of improved bioplastics that are derived from biomass, microbial/microalgal cells, and renewable waste streams. This field of research is making contemporary progress. Currently, bioplastics are employed in non-disposable carpet as well as in disposable goods including bottles, bags, straws, containers, and packaging. This advancement, in the long run, encourages the building of carbon-neutral setup for the production and management of bioplastics, with the goal of avoiding rivalry with agricultural and food reserves (Spierling et al., 2018). The estimation of worldwide bioplastics production capabilities is a difficult process due to the fluid nature of bio-based and biodegradable polymers, and the increasing investor attention in the bioplastics business. This makes the task particularly difficult. As a consequence of this, such estimations are often constructed on the basis of projections.

11.2.1 Potential Benefits and Challenges of Biodegradable Polymers

Biodegradable polymers have been employed in food packaging materials and various disposable goods for everyday use. It is anticipated that biodegradable plastics will have various applications in the future, including but not limited to their use as agricultural engineering materials, such as sandbags or mulch films, as well as materials for fisheries, such as fishing nets and fishing lines. Additionally, they may be utilized from a medical perspective as bioabsorbable materials, for instance, as scaffolds and surgical sutures (Ishii et al., 2009). Furthermore, biodegradable plastics may also serve as sanitary goods, such as paper diapers. Nonetheless, the actual practical application of these entities is impeded by numerous issues. One of the challenges is the incorporation of other polymers through blending. In instances where a biodegradable plastic lacks the requisite properties, it becomes necessary to blend it with another polymer to attain the desired characteristics. When a biodegradable polymer is combined with non-biodegradable plastic, only the biodegradable components of the plastic are capable of degrading in the atmosphere. Consequently, the dispersion of non-biodegradable plastics that undergoes fragmentation into smaller particles can lead to atmospheric pollution. In addition, the utilization of copolymers comprising both biodegradable and non-biodegradable monomers may result in heightened environmental contamination, thereby rendering them unsuitable for employment as biodegradable plastics. Regulating the pace of biodegradation and incorporating a biodegradation stimulus are two additional concerns associated with biodegradable plastics.

11.2.2 Potential Future Outlook for Biodegradable Polymers

The adverse effects of hydrocarbon-derived polymers on human health, the ecosystem, and the financial sector are increasingly being deliberated. This is especially relevant as the global petroleum prices decline, leading to reduced revenue for OPEC nations. Consequently, there is a growing impetus among researchers and industries to explore biopolymer research. Due to the unforeseen repercussions of utilizing petroleum-derived materials, academic institutions and research centers have established departments dedicated to the study of plastics. For instance, Queen Mary University in London, England is currently engaged in the development of biocomposites (Frisoni et al., 2001). As the biopolymer industry expands globally, it is imperative to address the issues of development, price, and quality, while simultaneously conducting ongoing research on better alternatives such as nanopolymer and biocomposites.

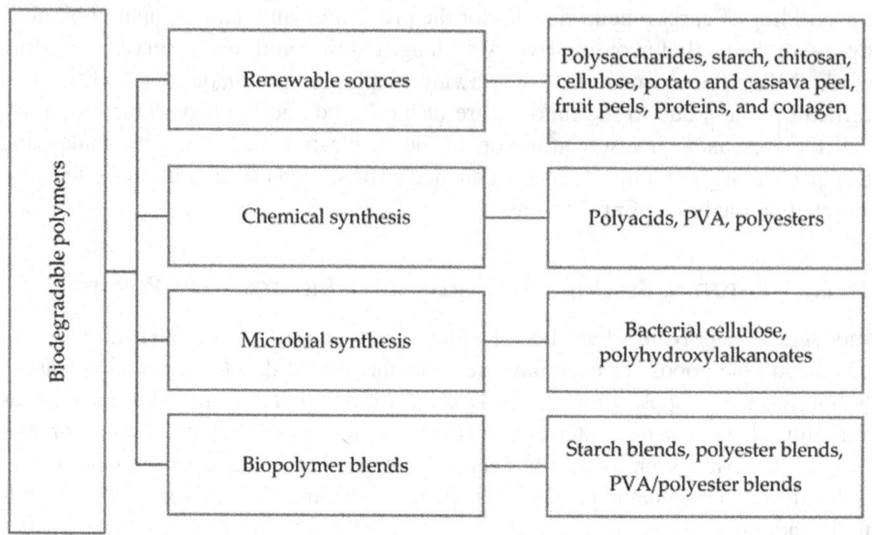

FIGURE 11.2 The image depicts the classification of production for biodegradable polymers (adapted from Satyanarayana et al., 2009).

11.3 SYNTHESIS OF BIODEGRADABLE POLYMERS

Biodegradable polymers are a special type of polymeric molecules which are environmentally friendly (biocompatible and compostable) as presented in Figure 11.2. Depending on the proposed products and the accessible materials/precursors, the manufacturing method for these biopolymers is divided into four distinct categories. The categories are biopolymer mixtures, bacterial synthesis, chemical synthesis, and sustainable sources (Satyanarayana et al., 2009). The first form of production—the creation of bio-based polymers from farm waste—is the main topic of the current debate. The parts that follow provide a short discussion of additional synthetic methods, including bacterial, chemical, and biopolymer mixtures.

11.4 MICROBIAL SYNTHESIS OF BIODEGRADABLE POLYMERS

The natural world offers a diverse array of resources, predominantly derived from plant life, owing to the efficient gathering of biomass with least input (Thakur & Voicu, 2016; Voicu & Thakur, 2021). A limited number of industries, namely those involved in the production of biofuels, paper, and wood, have utilized this extensive resource pool in contrast to contemporary naphtha chemistry (Sheldon, 2014). Notwithstanding, the diverse range of organic waste materials and the inadequacy of conversion methodologies impede the complete exploitation of natural resources, and as green sustainable method, microbial synthesis processes such as fermentation, polymerization, dimerization, etc. have been gaining importance. For example, two different methods have been used to produce succinic acid (SA): (i) a traditional chemical process using feedstock sourced from petroleum, and (ii) a biological technique using microorganisms and sustainable feed sources such as wood wastes and sugar beets (Borodina & Nielsen, 2014). Since 1546, microbial fermentation has been utilized for the production of SA, which has found applications in diverse industries such as agriculture, pharmaceuticals, and food.

11.4.1 Fermentative Production of Biodegradable Polymers

The chemical industry has demonstrated a notable degree of interdependence with the oil sector throughout the last century. The increasing expenses of petroleum and the consequential consequences to the environment have inspired many petrochemical industries to shift toward sustainable chemical synthesis approaches that depend on sustainable power plants (Jiang et al., 2017). The European Union and the United States Department of Energy have provided financial support for advanced biotechnology research, which has facilitated the exploration of novel biochemical pathways in chemical biosynthesis (Philp, 2015). Common natural resources utilized during the sugar fermentation process include refined sugars such as fructose, sucrose, and glucose, as well as starch and beetroot molasses. According to Stylianou et al. (2021, second-generation leftovers refer to the residual vegetation and biological waste materials that arise from the production processes of paper, timber, and food. These leftovers are deemed inedible.

It is required to replace pricey carbon and nitrogen resources like yeast and glucose extract with leftovers from agriculture or industry in order to accomplish sustainability manufacturing of bio-based polymers and lower the associated business expenses. This can be accomplished by using waste products from agriculture or industry. Whey from cheese, molasses made from sugarcane, maize fibers, sake lees, waste bread, bio-based cotton, crude glycerol, and other products of a similar nature may be included in this category. It's possible that these organic wastes have extra nutrients like vitamins and minerals that are essential to the body (Jiang et al., 2013; Li et al., 2021). These microorganisms can be divided into two categories: those that are not naturally occurring and those that occur naturally in capnophilic environments and are capable of reproducing (Beauprez et al., 2010). The digestive tract of ruminants is a rich source of intestinal microbes that, via anaerobic fermentation in an atmosphere of carbon dioxide, produce considerable amounts of volatile organic molecules (Gonzales et al., 2020). It has been discovered that several other species of fungi, including *Penicillium simplicissimum*, *Fusarium* sp., and *Aspergillus* sp., are also capable of producing SA. Moon et al. (2004) revealed that particular occasions have been recorded in which hybrid cultures of *Enterococcus faecalis* and *Rhizophus* sp. have generated succinic acid. These hybrid cultures were created by combining the two different bacteria. Following this step, the fumarates that were generated by the fungus were transformed to succinic acid by bacterial species through a mechanism known as species on sequence. Owing to the need of traversing two separate barriers, namely the mitochondrial and cytoplasm membranes, for excretion in the second, bacterial fermentation is a more acceptable choice for the generation of succinic acid compared to fermentation by fungus or yeast. This is because succinic acid must be excreted in the mitochondria (Jezierska & Van Bogaert, 2017).

11.4.2 Production of Biocomposites Using Solvent Extraction Method

The predominant type of agricultural waste produced by wineries is the residual fruit waste derived from the pomace of Merlot grapes. Rather than undergoing decomposition, agricultural wastes from wineries have the potential to be utilized as a feasible source of composites through the application of techniques such as pressurized liquid extraction (PLE) and solvent extraction (SE). The matrix is created through the amalgamation of extracts derived from both PLE and SE methodologies, in conjunction with polyhydroxyalkanoate of commercial-grade quality. The last stage of the production process includes the combination of the biopolymer with poly(3-hydroxybutyrat e-*co*-3-hydroxyvalerate) (PHBV), which is a co-polyester comprising hydroxyvaleric acid. This results in the creation of fully functional biocomposites. The utilization of biologically sourced substances acquired via SE leads to the decrease in tensile strength and a slight increase in the duration of the pause. The results suggest that the mechanical properties of biocomposites were influenced by the specific synthetic route or extraction method used for phenols. According to Ferri et al. (2020), SE was determined to be a more viable method in comparison to PLE.

11.4.2.1 Production of Starch-Based Polymers Using Polymerization

Polysaccharides containing starch can be found in various sources such as tubers, legumes, and agro-waste of cereals. These sources are considered to be suitable carbon antecedents for the production of bio-based polymers (Nayak, 1999). The utilization of thermoplastic starch-based polymers is a viable substitute for petroleum-based polymers, as they possess efficient reinforcement characteristics, are abundant, and have adjustable properties (Syafri et al., 2019). Starch, which is sourced from cereals, potatoes, and maize, is widely available in the biosphere (Mathiot et al., 2019). The first stage in obtaining the polymers from starch utilizing agricultural residues entails the incorporation of L-lactate and a tin-based catalyst. This leads to the production of starch-g-PPDO polymer chains that can be initiated through the introduction belonging to the category composed of poly 1,4-dioxan-2-one (PPDO)–diisocyanate (NCO), thereby inducing the polymerization process (Anjum et al., 2016). Though the process seems promising but faces certain challenges posed by material limitations and intricate synthetic pathways, the viability of starch-derived polymers on a large-scale commercial basis is subject to scrutiny due to the fact that starch sources are fundamental food staples in numerous nations. The commercial production of thermoplastics on a large scale may pose a potential threat to food security, when viewed through the lens of food security.

11.5 USES OF AGRICULTURAL WASTE-DERIVED BIOPOLYMER

The specialized uses of biopolymers in the building industry, the food packaging industry, and the automotive industry are discussed as follows.

11.5.1 Uses in Construction

The use of bio-based polymers in construction is mainly affected by reinforcing agents like nanocellulose, carbon nanofibers (CNFs), carbon nanotubes (CNTs) (Thakur & Thakur, 2015), cellulose, lignin, hemicellulose, and cellulosic micro additives (Nagarajan et al., 2020). Due to their high water permeability rates and biodegradability, polymers derived from biological resources are of the highest relevance when it comes to reinforcing. Recent advances in material science and nanotechnology have helped the building sector and spawned new uses.

11.5.2 Applications in Anti-Insect Nets and Mulching Films

Farmers employ plant-derived polymers such as polylactic acid (PLA), bio-polyethylene, polyhydroxyalkanoates (PHA), starch, and cellulose for the production of mulching films and shade nets (Briassoulis & Giannoulis, 2018; Zhang et al., 2019; Mukherjee et al., 2019). The utilization of shade netting is considered an essential component of integrated pest control due to the detrimental effects associated with the application of industrial pesticides. The reduction in pesticide usage yields favorable outcomes for both the environment and the economy, while also exhibiting superior mechanical properties compared to conventional LDPE films (Mukherjee et al., 2019). Furthermore, the implementation of nets serves to shield the plants from the detrimental impacts of ultraviolet radiation.

11.5.3 Application in Food Packaging Industry

Various polymers derived from botanical sources have been identified as viable options for utilization in the realm of food packaging. The materials mentioned in the study done by Luchese et al. (2017) encompass PLA, sugar palm nano-fibrillated cellulose (SPNFCs), blends of polybutylene adipate terephthalate (PBAT) and coffee grounds, agro-industrial waste derived from blueberries, and corn starch. The poor mechanical capabilities of PLA do not significantly affect its suitability for use as food packages, where tensile strength is not an important consideration. The capacity of PLA to serve as a substitute for non-biodegradable plastics, including polypropylene and polystyrene, in the realm of packaging is supported by its notable attributes, such as its minimal carbon footprint and various advantageous ecological effects (RameshKumar et al., 2020).

11.6 CONCLUSION

This chapter has made significant contributions to our understanding of the synthesis and applications of biopolymers, polymers that decompose, and polymers derived from sustainable crop residue resources. These sources include tomato and grape pomace, essential oils, coconut shells, green tea extracts, vegetable waste, curcumin, fruit peels, rice husks, discarded vegetables, grapefruit seed extract, municipal agro-wastes and maize and wheat starch. The major consideration of environmental impact has a substantial impact on the option for the precursor that is ultimately chosen (the type of agro-wastes). The demand for volumes of agro-waste that are viable from a business perspective is a difficulty for those working in the field of biopolymer manufacture. The absence of a reliable system for classifying and disposing of trash of this kind adds another layer of complexity to the problem. In addition, it has been discovered that the global distribution of agro-waste has an effect on the optical and mechanical properties of the synthetic polymers that have been produced.

REFERENCES

Abioye, O. P., Abioye, A. A., Afolalu, S. A., Ongbali, S. O. (2018) A review of biodegradable plastics in Nigeria. *Int. J. Mech. Eng. Technol.* 9(10), 1172–1185.

Anjum, A., Zuber, M., Zia, K. M., Noreen, A., Anjum, M. N., Tabasum, S. (2016) Microbial production of polyhydroxyalkanoates (PHAs) and its copolymers: A review of recent advancements. *Int. J. Biol. Macromol.* 89, 161–174. https://doi.org/10.1016/j.ijbiomac.2016.04.069. Epub 2016 Apr 25. PMID: 27126172.

Babu, R. P., O'connor, K., Seeram, R. (2013) Current progress on bio-based polymers and their future trends. *Prog. Biomater.* 2, 1–16.

Beauprez, J. J., De Mey, M., Soetaert, W. K. (2010) Microbial succinic acid production: Natural versus metabolic engineered producers. *Process Biochem.* 45(7), 1103–1114.

Borodina, I., Nielsen, J. (2014) Advances in metabolic engineering of yeast Saccharomyces cerevisiae for production of chemical. *Biotechnol. J.* 9(5), 609–620.

Briassoulis, D., Giannoulis, A. (2018) Evaluation of the functionality of bio-based plastic mulching films. *Polym. Test.* 67, 99–109.

Chae, Y., An, Y. J. (2018) Current research trends on plastic pollution and ecological impacts on the soil ecosystem: A review. *Environ. Pollut.* 240, 387–395.

Chinthapalli, R., Skoczinski, P., Carus, M., Baltus, W., de Guzman, D., Käb, H., Ravenstijn, J. (2019) Biobased building blocks and polymers-global capacities, production and trends, 2018–2023. *Ind. Biotechnol.* 15(4), 237–241.

European Bioplastics Report (2020) Bioplastics Market Data 2019. Global Production Capacities of Bioplastics 2019–2024. Available online: https://www.european-bioplastics.org/market/

Ferri, M., Vannini, M., Ehrnell, M., Eliasson, L., Xanthakis, E., Monari, S., Tassoni, A. (2020) From winery waste to bioactive compounds and new polymeric biocomposites: A contribution to the circular economy concept. *J. Adv. Res.* 24, 1–11.

Frisoni, G., Baiardo, M., Scandola, M., Lednická, D., Cnockaert, M. C., Mergaert, J., Swings, J. (2001) Natural cellulose fibers: Heterogeneous acetylation kinetics and biodegradation behavior. *Biomacromolecules.* 2(2), 476–482.

Geyer, R., Jambeck, J. R., Law, K. L. (2017) Production, use, and fate of all plastics ever made. *Sci. Adv.* 3(7), e1700782.

Gonzales, T. A., de Carvalho Silvello, M. A., Duarte, E. R., Santos, L. O., Alegre, R. M., Goldbeck, R. (2020) Optimization of anaerobic fermentation of Actinobacillus succinogenes for increase the succinic acid production. *Biocatal. Agric. Biotechnol.* 27, 101718.

Ilyas, R. A., Sapuan, S. M., Ibrahim, R., Abral, H., Ishak, M. R., Zainudin, E. S., Jumaidin, R. (2019) Effect of sugar palm nanofibrillated cellulose concentrations on morphological, mechanical and physical properties of biodegradable films based on agro-waste sugar palm (Arenga pinnata (Wurmb.) Merr) starch. *J. Mater. Res. Technol.* 8(5), 4819–4830.

Ishii, D., Ying, T. H., Yamaoka, T., Iwata, T. (2009) Characterization and biocompatibility of biopolyester nanofibers. *Mate.* 2(4), 1520–1546.

Jain, A., Sarsaiya, S., Awasthi, M. K., Singh, R., Rajput, R., Mishra, U. C., Shi, J. (2022) Bioenergy and bio-products from bio-waste and its associated modern circular economy: Current research trends, challenges, and future outlooks. *Fuel.* 307, 121859.

Jezierska, S., Van Bogaert, I. N. (2017) Crossing boundaries: The importance of cellular membranes in industrial biotechnology. *J. Ind. Microbiol. Biotechnol.* 44(4–5), 721–733.

Jiang, M., Xu, R., Xi, Y. L., Zhang, J. H., Dai, W. Y., Wan, Y. J., Wei, P. (2013) Succinic acid production from cellobiose by Actinobacillus succinogenes. *Bioresour. Technol.* 135, 469–474.

Jiang, M., Ma, J., Wu, M., Liu, R., Liang, L., Xin, F., Dong, W. (2017) Progress of succinic acid production from renewable resources: Metabolic and fermentative strategies. *Bioresour. Technol.* 245, 1710–1717.

Kabasci, S. (2013) Bio-based plastics: Introduction. *Bio-Based Plastics: Mater. Appl.* 2013, 1–7.

Li, C., Ong, K. L., Cui, Z., Sang, Z., Li, X., Patria, R. D., Lin, C. S. K. (2021), Promising advancement in fermentative succinic acid production by yeast hosts. *J. Hazard. Mater.* 401, 123414.

Luchese, C. L., Sperotto, N., Spada, J. C., Tessaro, I. C. (2017) Effect of blueberry agro-industrial waste addition to corn starch-based films for the production of a pH-indicator film. *Int. J. Biol. Macromol.* 104, 11–18.

Mathiot, C., Ponge, P., Gallard, B., Sassi, J. F., Delrue, F., Le Moigne, N. (2019) Microalgae starch-based bioplastics: Screening of ten strains and plasticization of unfractionated microalgae by extrusion. *Carbohydr. Polym.* 208, 142–151.

Moon, S. K., Wee, Y. J., Yun, J. S., Ryu, H. W. (2004) Production of fumaric acid using rice bran and subsequent conversion to succinic acid through a two-step process. In *Proceedings of the Twenty-Fifth Symposium on Biotechnology for Fuels and Chemicals Held May 4-7, 2003, in Breckenridge, CO* (pp. 843–855), Humana Press, Totowa, NJ.

Mose, B. R., Maranga, S. M. (2011) A review on starch based nanocomposites for bioplastic materials. *J. Mater. Sci. Eng. B.* 1(2B), 239.

Mukherjee, A., Knoch, S., Chouinard, G., Tavares, J. R., Dumont, M. J. (2019) Use of bio-based polymers in agricultural exclusion nets: A perspective. *Biosyst. Eng.* 180, 121–145.

Nagarajan, K. J., Balaji, A. N., Basha, K. S., Ramanujam, N. R., Kumar, R. A. (2020) Effect of agro waste α-cellulosic micro filler on mechanical and thermal behavior of epoxy composites. *Int. J. Biol. Macromol.* 152, 327–339.

Nayak, P. L. (1999) Biodegradable polymers: Opportunities and challenges. *J. Macromol. Sci. Rev. Macromol. Chem. Phys.* 39, 481–505.

Nova-Institute, Bio-plastics Market Data (2018) European Bioplastics: 2019. https://www.european-bioplastics.org/market/.

Nunes, L. A., Silva, M. L., Gerber, J. Z., Kalid, R. D. A. (2020) Waste green coconut shells: Diagnosis of the disposal and applications for use in other products. *J. Cleaner Prod.* 255, 120169.

Pandit, S., Savla, N., Sonawane, J. M., Sani, A. M. D., Gupta, P. K., Mathuriya, A. S., Rai, A. K., Jadhav, D. A., Jung, S. P., Prasad, R. (2021) Agricultural waste and wastewater as feedstock for bioelectricity generation using microbial fuel cells: Recent advances. *Fermentation.* 7(3), 169.

Panesar, P. S., Kaur, R., Singla, G., Sangwan, R. S. (2016) Bio-processing of agro-industrial wastes for production of food-grade enzymes: Progress and prospects. *Appl. Food Biotechnol.* 3(4), 208–227.

Philp, J. (2015) Balancing the bioeconomy: Supporting biofuels and bio-based materials in public policy. *Energy Environ. Sci.* 8(11), 3063–3068.

Raina, D., Kumar, V., Saran, S. (2022). A critical review on exploitation of agro-industrial biomass as substrates for the therapeutic microbial enzymes production and implemented protein purification techniques. *Chemosphere.* 294, 133712.

RameshKumar, S., Shaiju, P., O'Connor, K. E. (2020) Bio-based and biodegradable polymers-state-of-the-art, challenges and emerging trends. *Curr. Opin. Green Susta.* 21, 75–81.

Rhodes, C. J. (2019) Solving the plastic problem: From cradle to grave, to reincarnation. *Sci. Prog.* 102(3), 218–248.

Sadh, P. K., Duhan, S., Duhan, J. S. (2018) Agro-industrial wastes and their utilization using solid state fermentation: A review. *Bioresour. Bioprocess.* 5(1), 1–15.

Satyanarayana, K. G., Arizaga, G. G., Wypych, F. (2009) Biodegradable composites based on lignocellulosic fibers-An overview. *Prog. Polym. Sci.* 34(9), 982–1021.

Scalenghe, R. (2018). Resource or waste? A perspective of plastics degradation in soil with a focus on end-of-life options. *Heliyon.* 4(12), e00941.

Sheldon, R. A. (2014) Green and sustainable manufacture of chemicals from biomass: State of the art. *Green Chem.* 16(3), 950–963.

Spierling, S., Knüpffer, E., Behnsen, H., Mudersbach, M., Krieg, H., Springer, S., Endres, H. J. (2018) Bio-based plastics-a review of environmental, social and economic impact assessments. *J. Cleaner Prod.* 185, 476–491.

Stylianou, E., Pateraki, C., Ladakis, D., Vlysidis, A., Koutinas, A. (2021) Optimization of fermentation medium for succinic acid production using Basfia succiniciproducens. *Environ. Technol. Innov.* 24, 101914.

Syafri, E., Sudirman, Mashadi, Yulianti, E., Deswita, Asrofi, M., Abral, H., Sapuan, S.M., Ilyas, R.A., Fudholi, A. (2019) Effect of sonication time on the thermal stability, moisture absorption, and biodegradation of water hyacinth (Eichhornia crassipes) nanocellulose-filled bengkuang (Pachyrhizus erosus) starch biocomposites. *Journal of Materials Research and Technology*, 8(6), 6223–6231, ISSN 2238-7854, https://doi.org/10.1016/j.jmrt.2019.10.016.

Thakur, V. K. Thakur, M. K. (Eds). (2015) *Eco-Friendly Polymer Nanocomposites: Chemistry and Applications* (Vol. 74), Springer, New York.

Thakur, V. K., Voicu, S. I. (2016) Recent advances in cellulose and chitosan based membranes for water purification: A concise review. *Carbohydr. Polym.* 146, 148–165.

Tripathi, N. K., Shrivastava, A. (2019) Recent developments in bioprocessing of recombinant proteins: Expression hosts and process development. *Front. Bioeng. Biotechnol.* 7, 420.

Vachher, M., Sen, A., Kapila, R., Nigam, A. (2021) Microbial therapeutic enzymes: A promising area of biopharmaceuticals. *Curr. Res. Biotechnol.* 3, 195–208.

Voicu, S. I., Thakur, V. K. (2021) Aminopropyltriethoxysilane as a linker for cellulose-based functional materials: New horizons and future challenges. *Curr. Opin. Green Sustai.* 30, 100480.

Zhang, X., You, S., Tian, Y., Li, J. (2019) Comparison of plastic film, biodegradable paper and bio-based film mulching for summer tomato production: Soil properties, plant growth, fruit yield and fruit quality. *Sci. Hortic.* 249, 38–48.

12 Application of Enzymatic Technique for the Synthesis of Biodegradable Polymers from Different Agrobiomasses

Karan Kumar, Piyal Mondal, Shraddha M. Jadhav, and Vijayanand S. Moholkar*

12.1 INTRODUCTION

In the wake of the contemporary environmental crisis exacerbated by conventional polymer synthesis, enzyme-based techniques emerge as a beacon of promise, particularly in the synthesis of biodegradable polymers sourced from diverse agrarian biomass materials (Maraveas, 2020a). This transformative shift responds to the pressing need for sustainable alternatives to conventional plastics, which have long relied on finite fossil fuel resources, contributing significantly to environmental degradation (Narancic et al., 2020). The escalating demand for environmentally friendly materials underscores the urgency of adopting innovative methodologies. Globally, there has been a drastic shift in the production rate of biodegradable plastic as shown in Figure 12.1, which conveys the importance of such materials over conventional plastics for a green sustainable environment.

12.1.1 Historical Perspective: Evolution of Biopolymers

Humans, throughout history, have harnessed biopolymers to fashion essential commodities such as food, clothing, and furniture. However, the advent of the Industrial Revolution marked a pivotal shift, with fossil fuels, particularly oil, assuming a central role as the primary energy source for nearly all commercial products. Since the 1930s, the production landscape has been dominated by petroleum-derived monomers, steering diverse industries and everyday products (Aggarwal et al., 2020). The ubiquity of these materials, coupled with their adverse environmental effects, has intensified the need for a sustainable and eco-friendly alternative.

DOI: 10.1201/9781003407713-12

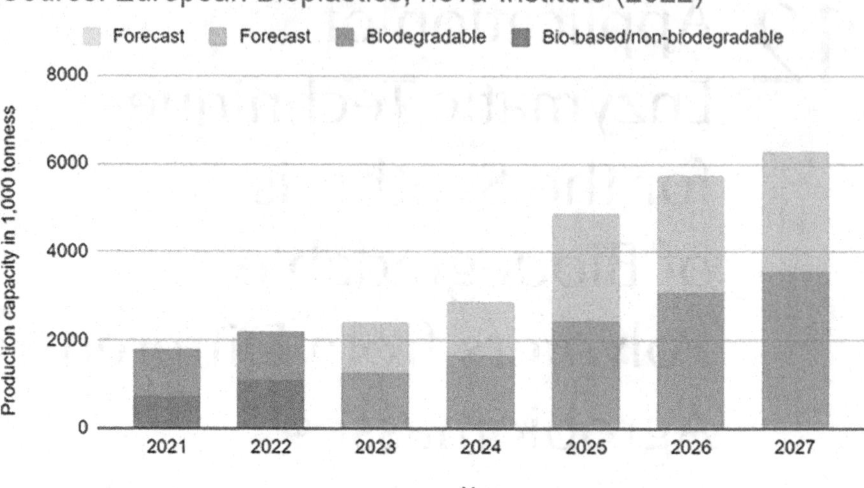

FIGURE 12.1 Bioplastics production capacity on a global scale.

The conundrum intensifies with the diminishing oil resources and the imperative to align resource usage with the planet's natural replenishment cycles. Acting sustainably necessitates a paradigm shift, and biopolymers, characterized by their renewable nature, have emerged as a compelling solution (Wu, 2012). Despite the competition with fossil-fuel-derived polymers concerning cost and functional characteristics, biopolymers prove their mettle in markets characterized by high oil prices and low feedstock costs, especially those reliant on maize and starch. The cultivation of industrial crops for non-food production underscores the potential to create biomaterials with minimal agricultural impact (Smith et al., 2016).

12.1.2 The Rise of Biopolymers: Environmental Imperatives and Advantages

Biopolymers, encompassing both synthetics produced by microbes and naturals derived from cellulose, starch, and proteins, have carved a niche in various living organisms, including microbes, animals, and plants. Despite constituting a relatively modest segment of the polymer market, biopolymers are poised to eventually replace 30%–90% of petroleum-based polymers. This prediction aligns seamlessly with the global shift toward recyclable and biodegradable products, where biopolymers increasingly find value and recognition ("Bioplastics & Biopolymers Market Worth $29.7 Billion by 2026" 2023; "Market Update 2020: Bioplastics Continue to Become Mainstream as the Global Bioplastics Market Is Set to Grow by 36 Percent over the next 5 Years – European Bioplastics e.V." 2023).

The benefits of biopolymers extend beyond their environmental friendliness. They play a pivotal role in enhancing soil quality by increasing organic matter, water retention, and nutrient absorption, while concurrently reducing chemical usage and plant diseases (Smith et al., 2016; Baranwal et al., 2022). The energy requirements for synthesizing and manufacturing biopolymers are notably lower than their synthetic counterparts, contributing to overall energy efficiency. Furthermore, their utilization of clean energy sources and the subsequent reduction in greenhouse gas emissions underscore their positive environmental impact (Tarazona et al., 2022).

12.1.3 Recent Advances and Projections: The Unfolding Biopolymer Landscape

Recent studies underscore the escalating utilization of biopolymers, particularly those derived from crops, across various industries. Notably, their unique attributes such as non-toxicity, oral protection, and wide availability position them favorably in sectors like food, medicine, and pharmaceuticals. The biopolymers sector is experiencing rapid expansion, a testament to the increasing recognition of their potential and versatility (Ibrahim et al., 2019).

Understanding the terminology associated with biopolymers, including "bioplastics," "bio-based plastics," and "biodegradable plastics," becomes paramount in navigating this evolving landscape. Projections indicate a substantial surge in global biopolymer production, reaching approximately 6.29 million tons by 2027, a marked increase from 1.79 million tons in 2021 (Aggarwal et al., 2020; Tarazona et al., 2022; "Bioplastics & Biopolymers Market Worth $29.7 Billion by 2026" 2023).

12.1.4 Bio-Based Polymers: The Bridge to a Sustainable Future

Bio-based polymers, primarily composed of polysaccharides, proteins, and fibers, represent a distinctive category within the broader realm of biopolymers. Their current material characteristics and mechanical strength differ from conventional plastics, limiting their direct applications. However, ongoing research and development endeavors strive to enhance these properties, aiming to position bio-based polymers as competitive alternatives to non-biodegradable counterparts across diverse industries (Rai et al., 2021). While biopolymers may not be entirely biodegradable, their potential for recycling marks a significant advancement. Enzymatic activity facilitates the rapid breakdown of biopolymers when released into the environment, generating natural by-products such as carbon dioxide, methane, water, biomass, and humic matter (Kumar et al., 2020). Bioplastics derived from agricultural waste gain prominence as a non-polluting and environmentally friendly substitute for non-biodegradable counterparts (Maraveas, 2020b). European Bioplastics defines bioplastics as having a biological origin, being biodegradable, or possessing both characteristics, categorizing them into biodegradable and non-biodegradable subsets. The ability of bioplastics to degrade naturally depends on the method and rate of their degradation (Narancic et al., 2020).

Biopolymers such as polyhydroxyalkanoates (PHA), polylactic acid (PLA), cellulose from starch, and their derivatives contribute to the creation of biodegradable plastics. Analogous to conventional plastics made from fossil fuels, bioplastics can be burned or recycled. Furthermore, they lend themselves to composting, showcasing a model for a zero-waste circular economy (Karan et al., 2019). In this dynamic landscape where the demands for sustainability echo loudly, enzymatic techniques stand at the forefront, guiding the synthesis of biodegradable polymers from diverse agrarian biomass sources. As the world navigates toward a future dominated by eco-friendly materials, the collaborative efforts of science, technology, and industry promise a paradigm shift where innovation and sustainability intertwine seamlessly (Maraveas, 2020a,b).

12.2 BIOPOLYMERS

Biopolymers belong to a category of polymers that can either originate from biological sources, excluding fossil fuels or be easily broken down through chemical or microbial processes in the environment (McGauran et al., 2021). This unique characteristic enables biopolymers to overcome the challenges associated with traditional plastics, which are predominantly derived from fossil fuels and are not easily disposed of in an eco-friendly manner once their usefulness has ended (Ibrahim et al., 2019; Nair et al., 2017). They are produced using natural resources and are either entirely biosynthesized by living creatures or synthesized from biomaterials (Smith et al., 2016).

12.2.1 Types of Biopolymers

Biopolymers are classified differently depending on the scale used. Natural, synthetic, and microbial biopolymers are the three types of biopolymers, as shown in Figure 12.2.

Another way to classify biopolymers is by degradability. In fact, we divide them into two families based on degradability: biodegradable and non-biodegradable, as well as bio-based and non-bio-based biopolymers. Polyesters, polysaccharides,

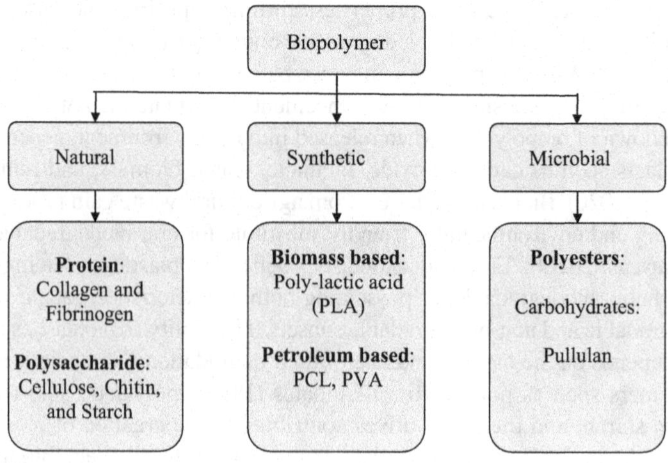

FIGURE 12.2 Classification of biopolymers based on their origin.

polycarbonates, polyamides, and vinyl polymers are additionally categorized based on the structure of their polymer backbone (Nair et al., 2017; Kumar et al., 2020). On the basis of their structure, monomers are divided into three categories: polysaccharide, protein, and nucleic acid. Based on how they react to heat, biopolymers can also be classified as elastomers, thermoplastics, and thermosets. Biopolymers are classified into several groups, each with its subgroup. Figure 12.3 depicts all of these classes (Ibrahim et al., 2019). Some of these biopolymers are listed in Table 12.1.

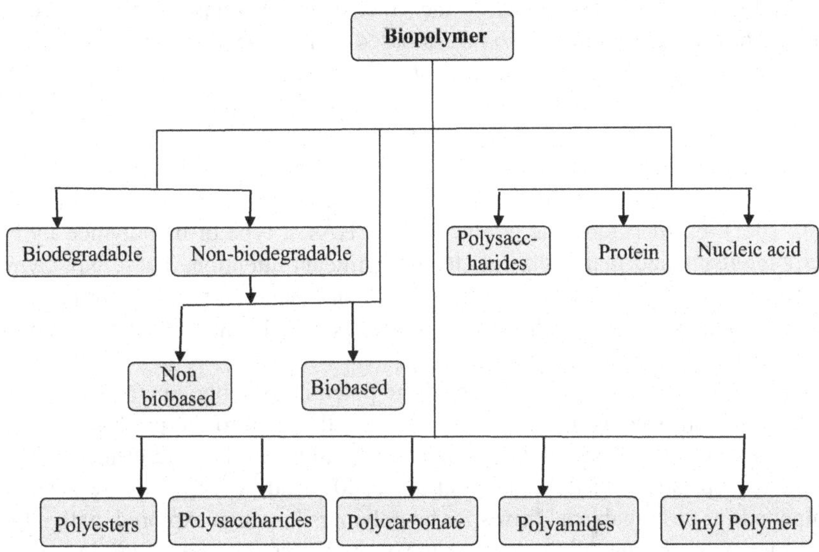

FIGURE 12.3 Biopolymer classification based on biodegradability, polymer backbone, and monomer type.

TABLE 12.1
Biopolymers, Their Substrates and Products

Biopolymer	Enzyme	Substrate	Products
Polyhydroxyalkanoates (PHA)	PHA synthase	CoA-activated monomers (e.g., 3-hydroxybutyrate)	PHA
Poly(lactic acid) (PLA)	Lipase	Lactide	PLA
Polysaccharides (e.g., cellulose, chitin)	Cellulase, chitinase	Cellulose, chitin	Glucose, N-acetylglucosamine
Polyesters (e.g., poly (3-hydroxybutyrate-co-3-hydroxy valerate))	Cutinase	Plant oils, fatty acids	Polyesters

12.2.2 Sources of Biopolymers

The diverse sources of biopolymers encompass a wide array of natural materials derived from various organisms, including livestock, plants, microbes, and agricultural waste. These sources serve as the foundation for the chemical synthesis of biopolymers, with components such as maize, wheat, potatoes, sorghum, rice, banana, yams, cassava, tapioca, corn, cotton, and barley providing the monomeric building blocks, such as oils, sugars, and amino acids (Klein and Poverenov, 2020). Notably, the most common animal sources of biopolymers are cattle, while marine sources include coral reefs, sponge-like organisms, and aquatic creatures such as lobster and shrimp. Microbiological sources, on the other hand, encompass fungi, algae, and yeasts, while carbohydrate-rich biomass-based sources include agricultural waste, paper scraps, plants, biological waste, and wood waste (Lodhi et al., 2014).

Vegetable oils, particularly those obtained from food manufacturers, have emerged as excellent natural replacements for polymer production, with examples such as sunflower, soybean, safflower, jojoba, rapeseed, castor, and meadowfoam oil containing triglycerides that are conducive to biopolymer production (Madadi et al., 2021). Furthermore, the production of PHAs, a type of biopolymer, by both microorganisms and plants, has garnered significant attention. These biopolymers are naturally made and broken down by microbial metabolisms, presenting a stark contrast to chemical and synthetic thermoplastics, which can be melted and formed (Dintcheva et al., 2020).

Recent literature has also highlighted the potential of biopolymers derived from microalgae, with studies demonstrating their contribution to the production of biopolymers such as PHA and their potential applications in the field of materials science and bioplastics (Venkateshaiah et al., 2020). Additionally, the use of natural compounds as sustainable additives for biopolymers has been explored, with a focus on naturally occurring fillers such as nano-/micro-sized layered alumino-silicate, which have shown promise in enhancing the properties of biopolymer composites (Ghosh et al., 2021).

The significance of biopolymers extends to their applications in various fields, including medicine, where bioengineering and synthetic biological methods have facilitated the production of new biopolymers with potential usage in medicines and food materials (Gunathilake et al., 2017). Furthermore, the use of biopolymers in soil treatment for environmentally friendly and sustainable geotechnical engineering has been explored, with a focus on utilizing CO_2 as an alternative source in various biopolymer production strategies (Zheng & Liu, 2019).

In the realm of nanotechnology, recent advancements have expanded the repertoire of electrospinning, with new fibers produced from emerging biopolymers such as sulfated polymers, tannin derivatives, modified collagen, and extracellular matrix extracts, showcasing the versatility and potential applications of biopolymers in tissue engineering and related fields (Chang et al., 2016).

The potential of biopolymers has also been explored in the context of renewable energy, with recent studies (Kumar et al., 2020; Tarazona et al., 2022) focusing on the characterization and analysis of biopolymers used in membrane fuel cells, highlighting their potential applications in the field of energy research. Furthermore, the production of biopolymer nanofibers from natural sources has been investigated, with a focus on utilizing biopolymers derived from both plants and animals (McGauran et al., 2021).

In summary, the sources of biopolymers are diverse and encompass a wide range of natural materials derived from various organisms, including livestock, plants, microbes, and agricultural waste. Recent studies have highlighted the potential of biopolymers derived from microalgae, natural compounds as sustainable additives for biopolymers, and their applications in various fields, including medicine, soil treatment, nanotechnology, and renewable energy. These findings underscore the versatility and potential applications of biopolymers in addressing diverse societal and environmental challenges.

12.2.3 Applications of Biopolymers

Food packaging, construction, and farming are some of the most fascinating applications for biopolymers. Applications are affected by the material's mechanical, physical, and chemical properties. For highly durable uses in agriculture and construction, biopolymers that have a substantial tensile strength/Young modulus are needed. On the other hand, flexibility/elongation at break is a crucial consideration in packaging applications. Below are detailed explanations of biopolymer applications:

Food packaging: Bio-based biopolymers that work well for food packaging applications include PLA, sugar palm nano-fibrillated cellulose (SPNFCs), coffee grounds-PBAT composites, blueberry agro-industrial waste, and corn starch. They help increase product shelf life and are safe for food (Klein & Poverenov, 2020). Additionally, these materials contribute to a reduction in the overall carbon footprint of food packaging. Delivering goods to customers in good condition is the packaging's primary objective. Applications for food packaging where tensile strength is not an important consideration are unaffected by PLA's poor mechanical properties. PLA has the potential to replace non-biodegradable plastics used in packaging such as polypropylene and polystyrene due to its low carbon footprint and other advantageous ecological effects (Maraveas, 2020a,b). Food packaging's main objectives are to safeguard and protect food from harm from the moment it is produced to the point of use, whether it be physical, chemical, or biological. Additionally, switching from oil-based packaging materials to bio-based ones gives you a competitive edge in aspects of a more sustainable and eco-friendlier image as well as better technical qualities (Baranwal et al., 2022).

Medical applications: The development of science and technology has significantly increased life expectancy. This has led to a decrease in mortality and morbidity rates thanks to a variety of inventive strategies and new technology. Biopolymers, as opposed to other polymers, are suitable for a range of biomedical applications (Gunathilake et al., 2017; Smith et al., 2016). Because biopolymers naturally break down in humans without causing any adverse side effects, they are appropriate for medical use. Because of their future applications in the biomedical industry, biopolymers have garnered much attention recently. They have a variety of uses, including those for medical devices, pharmaceutical carriers, and tissue engineering. Gelatin, for example, is a biopolymer that is often used in healthcare for dressings, as an adhesive, and so on. Furthermore, porous gelatin scaffolds and films were created with the help of gases or solvents that acted as simple porogen. Using this technique, the scaffolds can hold medication or nutrients delivered to the wound to promote healing (Gobi et al., 2021; Baranwal et al., 2022). Gums are used as nutraceutical, phenolic, antioxidant, and flavor-encapsulating compounds. Gums can improve the targeted drug delivery system's-controlled drug release functionalities because the body's digestion and absorption are limited. It is used in the pharmaceutical industry to treat hypotension brought on by bleeding or surgical shock. Additionally, it is frequently used as a tablet binder, emollient, and demulcent in cough syrups and drops (Ibrahim et al., 2019).

Agriculture: The biodegradable polymers' biodegradability is the most essential quality for this application. The most common biopolymers in this field are those based on starch. They are biodegradable and have a long enough lifespan to be helpful. In the 1930s, plastic films were first used as greenhouse covers for fumigation and mulching. Young plants must be covered because they are vulnerable to frost. The main functions of biodegradable cover films are moisture retention, soil temperature enhancement, and weed control to accelerate plant growth. At the end of the season, the film can be left in the soil, where it is biodegraded (Vroman & Tighzert, 2009; Klein & Poverenov, 2020; Chang et al., 2016).

The above section concentrates on the conventional applications of biopolymers; nowadays, biopolymers such as chitosan have emerged as water purification agents and are used for water treatment processes due to their ability to bind with contaminants and heavy metals. In the sector of cosmetics industry, biopolymers are now being extensively used as alternatives to traditional microbeads in cosmetic products, reducing environmental pollution. Furthermore, biopolymers are incorporated into automotive components as reinforcing agents in biocomposite materials, contributing to the development of lightweight and environmentally friendly vehicles.

12.3 ENZYMES FOR SYNTHESIS OF BIOPOLYMER

The synthesis of various biodegradable polymers from different agrarian biomasses using enzymatic techniques has gained significant attention due to the increasing demand for eco-friendly materials (Torres et al., 2019; Heredia-Guerrero et al., 2019; Ilyas et al., 2019). The growing concerns regarding environmental pollution caused by non-degradable synthetic polymers have led to the exploration of alternative materials that can replace them. Biopolymers offer an attractive option due to their eco-friendly nature, biodegradability, and biocompatibility (Bellenguez et al., 2022; Harmaen et al., 2015; Karan et al., 2019). However, the production of biopolymers still needs to be improved due to the high cost and complex synthetic procedures required to obtain them.

Enzymes, as biocatalysts, have several advantages over chemical catalysts, including high specificity, selectivity and efficiency, mild reaction conditions, and low environmental impact. Enzymatic techniques for synthesizing biopolymers involve using different classes of enzymes, such as lipases, proteases, and carbohydrates (Song et al., 2023; Kumar et al., 2021, 2022, 2023). The use of enzymes in the synthesis of biopolymers offers several advantages over traditional chemical synthesis methods. For example, enzymatic techniques have been found to produce biopolymers with improved properties, such as higher molecular weight, narrower molecular weight distribution, and higher thermal stability, compared to those produced by chemical methods (Hu et al., 2016; Rahmayetty et al., 2018). Additionally, enzymatic techniques have been reported to be more sustainable and cost-effective, as they can utilize renewable resources and operate under mild conditions, reducing the need for harsh chemicals and energy-intensive processes.

Lipases belong to a category of enzymes that facilitate the breakdown of ester bonds through hydrolysis. Their extensive application has been observed in the production of polyesters like PLA and PHA from renewable sources such as vegetable oils and sugars (Düşkünkorur et al., 2015; Rahmayetty et al., 2018). The utilization of lipase as a catalyst in polyester synthesis has been documented to yield high-molecular-weight polymers with a narrow distribution of molecular weights, rendering them suitable for diverse applications such as biomedical materials and packaging (Balla et al., 2021; Duchiron et al., 2015; Guzmán-Lagunes et al., 2012).

On the other hand, proteases are enzymes that catalyze the hydrolysis of peptide bonds. These enzymes have found utility in the creation of polymers based on silk fibroin. As a natural protein polymer, silk fibroin exhibits exceptional mechanical properties and compatibility with living organisms, making it an ideal material for biomedical applications. Incorporating proteases into the synthesis of silk fibroin-based polymers has been reported to produce polymers with precisely controlled molecular weights and mechanical characteristics (Nguyen et al., 2019).

Carbohydrases are a class of enzymes that catalyze the hydrolysis of carbohydrates. They have been used in the synthesis of cellulose-based polymers (Ilyas et al., 2019; Fouda-Mbanga et al., 2021; Kumar et al., 2023). Cellulose is the most abundant renewable biomass on earth and is widely used in the production of paper and textiles. The enzymatic synthesis of cellulose-based polymers offers a sustainable alternative to traditional methods that use harsh chemicals and high temperatures (Di Donato et al., 2020). The use of carbohydrate synthesis in cellulose-based polymers has been reported to produce polymers with improved thermal stability and mechanical properties (Sanchez-Vazquez et al., 2013; Sharmila et al., 2020; Vasić et al., 2021).

One of the most well-known biopolymers is PHA, which is synthesized from coenzyme A (CoA)-activated monomers such as 3-hydroxybutyrate (Arcos-Hernandez et al., 2012; Nitschke & Russell, 2013; Feder & Gross, 2010). The enzyme responsible for PHA synthesis is PHA synthase, which polymerizes the monomers into a polymer chain.

Poly(3-hydroxybutyrate) (PHB) stands out as the quintessential member within the aliphatic polyester category, akin to PHA (Soleymani Eil Bakhtiari et al., 2021; Moshood et al., 2022). It is a naturally occurring polymer formed by diverse bacterial chains, derived from inexpensive, renewable feedstock with minimal environmental impact. This polymer undergoes degradation in both anaerobic and aerobic conditions, without generating any toxic by-products (Saini et al., 2020; Tebaldi et al., 2019). The copolymer composed of hydroxybutyrate and hydroxyvalerate (HV) is referred to as poly(3-hydroxybutyrate)-hydroxyvalerate (PHBV). This polymer is characterized by a high level of crystallinity. As the proportion of HV units is elevated, there is a decrease in the melting point, glass transition temperature, crystallinity, and tensile strength, while the impact strength experiences an increase. PHBV, as a copolymer, exhibits reduced brittleness compared to PHB. Furthermore, the degradation rate of PHBV surpasses that of PHB. Notably, the synthesis of PHBV can be achieved utilizing Cutinase enzymes, which have demonstrated efficacy in catalyzing the formation of PHBV from plant oils and fatty acids (Perveen et al., 2020; Yeo et al., 2018). Different structures of major types of biopolymers are shown in Figure 12.4.

Polysaccharides such as cellulose and chitin are biopolymers that can be synthesized using specific enzymes. Cellulase enzymes are used to break down cellulose into glucose monomers, which can then be polymerized into cellulose (Lakhundi et al., 2015; Sumarno et al., 2021). Similarly, chitinase enzymes can hydrolyze chitin into N-acetylglucosamine monomers, which can be polymerized into chitin (Asghari et al., 2017).

The section indicates that the choice of enzyme and substrate for biopolymer synthesis is very important and it can affect the properties of the resulting biopolymer. For example, the use of different monomers for PHA synthesis can result in polymers with varying mechanical properties and degradation rates (Dalton et al., 2022). Additionally, the efficiency of the enzyme can affect the yield and purity of the biopolymer product. Advancements in this area can focus on enzyme engineering and metabolic engineering techniques which can specifically enhance the activity and

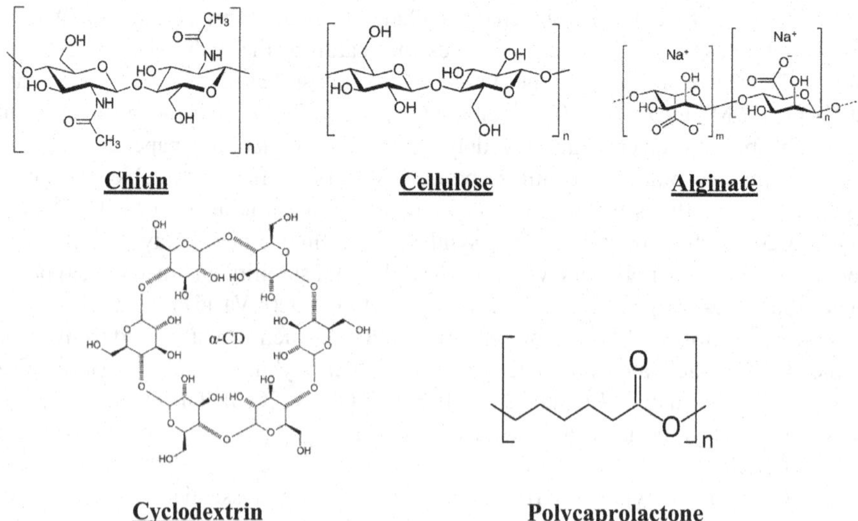

FIGURE 12.4 Chemical structures of some major biopolymers.

efficiency of enzymes used in biopolymer synthesis. With extensive research studies presently ongoing in this field of enzyme and metabolic engineering, the efficiency and effectiveness of enzyme-catalyzed biopolymer synthesis will continue to improve.

12.4 ENZYMATIC SYNTHESIS OF BIOPOLYMERS FROM AGRARIAN BIOMASS

The field of enzymatic synthesis of biopolymers from agricultural biomass is an auspicious method for the formation of sustainable and environmentally friendly polymers. Agrarian biomass contains abundant biopolymers, which are used as a renewable resource to manufacture sustainable, eco-friendly, futuristic biopolymers. Enzymes are the best choice for biopolymer synthesis because they have several benefits over chemical catalysts, including selectivity, specificity, and mild reaction conditions (Guzmán-Lagunes et al., 2012; Singh et al., 2016; Adhami et al., 2021; Engel et al., 2019).

The problem of the decomposition of biopolymers is solved by enzymes, which are primarily proteins that act as biological catalysts by accelerating chemical reactions. These enzymes catalyze the deterioration of natural polymers in the natural world which is not the case with synthetic polymers which is very hard and cost-inefficient to decompose (Rudnik & Briassoulis, 2011). Technological advancements have revealed their usefulness as biocatalysts for the functionalization and synthesis of polymers. Enzyme-based polymerization is therefore emerging as a more environmentally friendly alternative to more traditional polymer synthesis methods that use chemical catalysts (Xu et al., 2021; Kim et al., 2019; Dwi Prasetyo et al., 2020). The growth of enzyme-based polymer manufacturing is driven by the inactive search for substitutes to conventional synthesis paths that require harsh conditions for processing or metal catalysts. Extreme processing conditions lead to undesirable side reactions that severely limit the formation of a wide range of intriguing but thermally delicate vinyl, epoxy functional groups, and biological materials (Heredia-Guerrero et al., 2019; Kumar et al., 2023). Metal catalysts can have adverse environmental effects in addition to toxicity in biomedical materials (drugs, proteins, cells, etc.). The remains of the used solvents and catalysts can be very detrimental; thus, its consequence must be considered, mainly when processes for biodegradation occur, enzymatic polymerization has emerged as a critical synthetic technique for the development of biodegradable polymers alongside exceptional designs capable of meeting the requirements of cutting-edge fields such as biomedicine. The upper hand of using enzymes in polymer formation is the calm and smooth conditions of the reaction required to carefully accordance and regulate the polymerization and creation of systems, as a result of such catalysts' high enantio-, regio-, and chemoselectivity, and the low by-products produced during the synthesis (Anjum et al., 2016). Furthermore, when chemoenzymatic polymerization techniques are utilized together, the formation of macromolecules via heterogeneity of these processes can be fraught. Good action can only guarantee a portion of the effectiveness of enzymes within polyester synthesis. Competitive and sustainable enzymatic production and utilization of polyesters is feasible via correct catalyst immobilization, which allows for its recovery and reuse (Maraveas, 2020b; Rai et al., 2021; Sharma et al., 2020). Enzymatic polymer synthesis and functionalization undoubtedly provide an expanding framework for more creative and environmentally friendly polymers with unique structures to accommodate specific application requirements.

Biomass is the only renewable energy source and source of hydrocarbons for our contemporary industrialized civilization. Photosynthesis plays a significant role in biomass production in the environment, with carbohydrates accounting for 75% of the biomass. Remarkably, just 3–4% of these substances find application among humans for either nutritional or non-nutritional purposes. Currently, biomass is the most extensively utilized renewable resource, and its potential as a feedstock for upcoming green chemistry solutions is under evaluation. Using this biomass to produce fuels, chemicals, and polymers has environmental benefits and aids in reducing net CO_2 emissions, a greenhouse gas, into the atmosphere. Countries inclined to face the obstacles would benefit economically from a biomass-based energy system. Additionally, lignocellulosic biomass permits the productivity and sustainability of chemicals and fuels without impacting food supplies (referred to as second-generation fuels). In addition, lignocellulosic biomass grows more quickly and at a lower cost than food crops, and it is abundantly available, which makes it a desirable raw material for replacing fossil fuels.

12.4.1 Production of Polylactic Acid

Lactic acid, also known as 2-hydroxypropionic acid (CH_3-CHOHCOOH), is the predominant hydroxyl-carboxylic acid found in nature. It finds widespread usage in various industries including beverage production, pharmaceuticals, chemical manufacturing, and the medical sector. Due to the availability of both carboxyl and hydroxyl groups, it can be transformed into several valuable chemicals, including pyruvic acid, acrylic acid, 1,2-propanediol, and lactate esters (Wu, 2012; Takwa et al., 2011; Ojo & De Smidt, 2023). Its acclaimed usefulness for the production of PLA is exceptionally high due to the growing interest in creating biodegradable plastics. Developing eco-friendly "green" solvents (lactate esters) that can replace conventional petrochemical-based solvents is another promising lactic acid application. Growing new biological techniques and goods using lignin-based biomass biotechnology is likely to be environmentally friendly. Pretreatment, hydrolysis, fermentation, and separation are all applied to lignocellulosic substrates. A range of chemical and physical techniques have been developed and employed to enhance the breakdown of lignocellulosic compounds, which have limited bioaccessibility. These methods include acid or base treatments, as well as steam explosion, to hydrolyze the compounds into oligosaccharides prior to fermentation by microorganisms (Liu & Chen, 2015; Balla et al., 2021; Kumar & Moholkar, 2023; Kumar et al., 2022, 2023a, 2023b, 2024a, 2024b, 2024c).

Lactic acid can be generated either chemically or through microbial fermentation. The predominant method for commercial production involves bacterial fermentation of carbohydrates using homolactic organisms from the *Lactobacillus* genus, which exclusively produce lactic acid. Specifically, *Lactobacillus amylophilus*, *Lactobacillus bavaricus*, *Lactobacillus casei*, *Lactobacillus maltaromicus*, and *Lactobacillus salivarius* are the primary organisms responsible for producing the

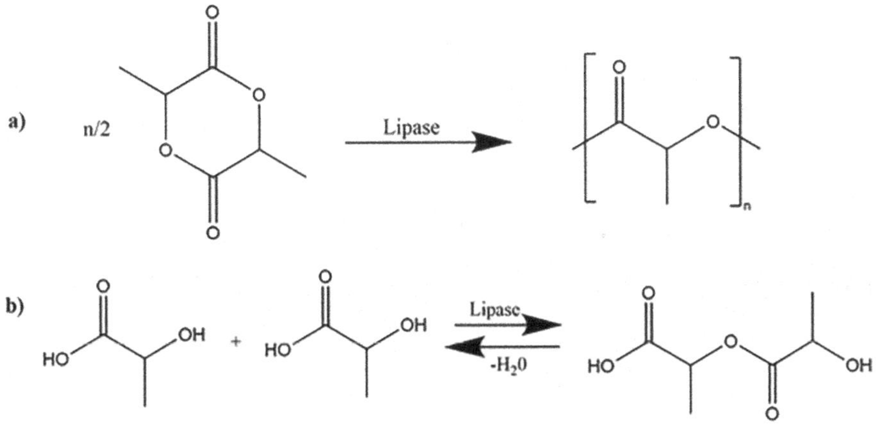

FIGURE 12.5 (a) Enzymatic ROP of lactide (eROP) and (b) polycondensation of lactic acid.

L-(+)-isomer of lactic acid (Ojo & De Smidt, 2023). On the other hand, strains like *Lactobacillus delbrueckii*, *Lactobacillus jensenii*, and *Lactobacillus acidophilus* are capable of producing the D-isomer or a mixture of both isomers. These strains exhibit effective conversion of carbon from feedstocks under typical fermentation conditions, which include a pH ranging from minimal to neutral, temperatures around 40°C, and low oxygen concentrations (Panyachanakul et al., 2020).

Enzymatic processes offer an alternative method for producing PLA, either through the ring-opening polymerization (ROP) of lactides or the direct polycondensation of lactic acid. ROP involves the propagation of cyclic monomers and is initiated through various ions. In the case of PLA production, lactide, a cyclic dimer of lactic acid, serves as the cyclic monomer (Rahmayetty et al., 2018; Takwa et al., 2011). On the other hand, lactic acid can undergo polycondensation by connecting the carboxyl and hydroxyl groups, forming water as a by-product. This process is depicted in Figure 12.5.

Lipases are the main category of enzymes used for eROPs. ROP for lipases operates as shown in the following way, according to Uyama et al. (Engel et al., 2019, Scheme 12.1): When a lactone molecule interacts with a lipase enzyme, it forms a functional acyl-enzyme intermediate (EM), which is the activated form of the monomer. The next step involves nucleophilic water molecules attacking the acyl carbon within the EM, initiating the polymerization process (Duchiron et al., 2015). In non-aqueous environments, ROP can occur because the necessary water molecules are bound to the enzyme, requiring only small catalytic amounts. As the polymer chain grows, the terminal hydroxyl group of the chain undergoes nucleophilic attack on the EM, resulting in the addition of one monomer unit to the chain. The rate-limiting step in lipase-catalyzed ROP is generally the formation of the acyl-EM (Yeniad et al., 2011).

In this section, it was envisaged that most of the current research focused on the enzymatic polymerization of PLA which favors the use of enzymatic ring-opening polymerization (eROP) as the preferred synthesis method. This preference stems from the ability to achieve a broader range of molecular weights by controlling the purity

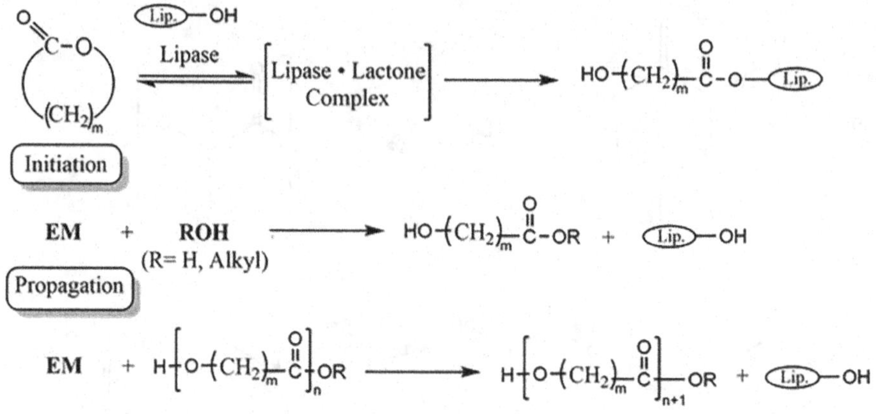

Lipase

Lipase • Lactone Complex

Initiation

EM + ROH (R= H, Alkyl) → HO-$(CH_2)_m$-C(=O)-OR + Lip.-OH

Propagation

EM + H-[O-$(CH_2)_m$-C(=O)]$_n$-OR → H-[O-$(CH_2)_m$-C(=O)]$_{n+1}$-OR + Lip.-OH

SCHEME 12.1 Lipase-catalyzed ROP mechanism.

of lactide. Conventional PLA synthesis has also followed this trend due to the advantages it offers. The primary technique for generating low-molecular-weight PLA is polycondensation, which involves straightforward tools and procedures. Considering that enzymatically synthesized polymers typically have lower molecular weights, the eROP route is considered more suitable for enzymatic production of PLA.

Advancements in the synthesis of PLA along with its industrial scale-up have been a major focus in recent studies. Scientists have been working on engineering enzymes to enhance their performance in catalyzing the polymerization of lactic acid to form PLA (Engel et al., 2019; Adhami et al., 2021). This involves modifying enzymes to increase their specificity, activity, and stability under industrial conditions. Enzymatic processes often involve microbial fermentation to produce lactic acid as a precursor to PLA. Advances in microbial strain development, fermentation processes, and optimization of culture conditions contribute to increased yields and cost-effectiveness (Kumar et al., 2022). Immobilizing enzymes on solid supports can improve their stability, reusability, and efficiency. Developing cost-effective and scalable processes is crucial for the widespread adoption of enzymatic methods in the polymer industry (Kumar et al., 2021).

12.4.2 Thermoplastic Biopolymers Based on Starch

Starch, a polysaccharide, is abundantly present in tubers, legumes, and agricultural waste from cereals. It represents a highly suitable carbon source for the production of bio-based polymers. Because of their abundant supply, effective reinforcement abilities, and adaptability, thermoplastic starch-based polymers serve as efficient substitutes for petroleum-based polymers. To produce starch-based polymers using agro-waste, the initial step involves the addition of L-lactate and a catalyst such as $Sn(Oct)_2$ (Ilyas et al., 2019; Mathiot et al., 2019; Santana et al., 2018). Another approach includes adding the poly 1,4-dioxan-2-one (PPDO)-diisocyanate (NCO) group, which initiates the polymerization process, forming starch-g-PPDO polymer chains (Santana et al., 2018). Recent research suggests that these materials are crucial for the development of sustainable food packaging due to their flexibility and lightweight characteristics. However, the properties of starch-based polymers pose challenges that vary depending on the specific starch precursor used. Sugar palms, microalgae, and jackfruit produce starch-based polymers with distinct properties (Mathiot et al., 2019; Tabasum et al., 2019). The long-term viability of starch-based polymers on a commercial scale is in doubt because of material limitations, challenging synthetic pathways, and the reality that starch sources are familiar food sources worldwide. From a security standpoint, large-scale commercial operations of thermoplastic manufacturing could harm food safety. In the section that follows, problems in the industrialization of polymers that decompose will be addressed in addition to possible solutions.

Advancements in this area involve exploring the blending of starch with other biopolymers, such as PLA or PHA, to improve the mechanical properties, processability, and overall performance of starch-based thermoplastics. The incorporation of reinforcing agents and additives, such as nanocellulose, nanoclays, or natural fibers, has been investigated to enhance the strength, stiffness, and thermal stability of starch-based thermoplastics. Advances in processing techniques, including extrusion, injection molding, and film blowing, have been made to optimize the fabrication of starch-based thermoplastics on an industrial scale.

12.4.3 Biopolymers Produced from Pineapple Peels and Tomato Pomace

A standard process is followed to extract polymers from agricultural wastes like pineapple peels and tomato pomace (Heredia-Guerrero et al., 2019; Vega-Castro et al., 2016). In this process, pineapple peels and tomato pomace are utilized. One of the initial steps is the chemical composition analysis focusing on the C/N and C/P ratios, which can predict polymer yields. Following fermentation and complete hydrolysis of the peels using H_2SO_4, along with the use of dipotassium phosphate or ammonium sulfate during fermentation, the biopolymers are extracted from the fermented peels through centrifugation at a speed of 4,000 rpm or higher. The finished product is then characterized using analytical tools such as FTIR, NMR, and GC-MS to assess its properties and composition. The only restriction is the potential harm that synthetic chemicals, including H_2SO_4, dipotassium phosphate, and ammonium sulfate, among others, may cause to the environment if used in large amounts (Yates & Barlow, 2013). Through the removal of the melting polycondensation step, the procedure for producing polymers made from bio-based materials via tomato pomace corresponds to that used to create them from pineapple peels (Maraveas, 2020a,b).

12.4.4 Biopolymer Production Problems, Polymer Availability Issues, and Precursor Availability

Different available routes for producing biodegradable polymers are characterized by an inherent incompatibility between the hydrophobic polymer matrix, which is water repellent, and the hydrophilic water fibers. This incompatibility leads to uneven dispersion and low mechanical strength in the resulting materials. Additionally, biodegradable polymers generally cannot match the mechanical properties of non-biodegradable polymers. As a result, they are typically limited to applications that do not require high mechanical strength and can accommodate their lower strength characteristics (Kibria et al., 2023; Kumar et al., 2020). To improve mechanical stability, biodegradable materials must be blended with polymers. Furthermore, biological precursors that produce high tensile strength (90 MPa), such as cellulose acetate, are not biodegradable. Recent research has attempted to address the slow rate of biodegradation by substituting bacterial cellulose for plant cellulose, which has an excellent water-holding capacity and a faster biodegradation rate.

Aside from the material shortcomings, producing plastic materials is not economically feasible compared to standard plastics. The economic sustainability of biodegradable plastics is a significant consideration, particularly in commercial applications. Cost plays a crucial role in determining the feasibility of these materials. Data indicates that plant-based polymers, such as PHA, can be up to four times more expensive than traditional polymers. Due to market dynamics where buyers prioritize the value of products over their price, bio-based polymers struggle to compete with conventional plastics in the commercial marketplace ("Bioplastics & Biopolymers Market Worth $29.7 Billion by 2026" 2023; "Market Update 2020: Bioplastics Continue to Become Mainstream as the Global Bioplastics Market Is Set to Grow by 36 Percent over the next 5 Years – European Bioplastics e.V." 2023). As a result, the economic viability of biodegradable plastics remains a challenge. Economies of scale and technological constraints are to blame for the cost problem. An estimated 335 million tons of commercial plastics, or 99.9% of all plastics, are classified as either not at all biodegradable or only partly biodegradable. Biodegradable polymers are produced on a smaller scale. The lack of scalable technologies has also impacted the cost of these materials (Maraveas, 2020b).

12.5 END-OF-LIFE TREATMENT OF BIOPOLYMERS

The durability and lifespan of bio-based polymers are influenced by factors like exposure to UV radiation, which can cause photo-oxidation, as well as heat-induced breakdown and their ability to dissolve in water. Additionally, their suitability for high-strength applications is an important consideration. When it comes to what happens to biodegradable polymers at the end of their life, there are several options available. These include natural decomposition at home, composting in industrial facilities, breaking down with the help of enzymes, disposing through specific catalytic processes, reusing the materials through chemical processes, recycling them mechanically, or subjecting them to anaerobic digestion (Aggarwal et al., 2020; Nair et al., 2017). The choice of the most appropriate end-of-life (EOF) treatment depends on the specific type of precursor used in the biodegradable polymer. Different precursors may require different approaches to ensure effective and sustainable management throughout their lifecycle. For instance, PLA is typically recycled using industrial, mechanical, or chemical composting. The choices available have aided in reducing the danger of carbon emissions and global warming. However, not every one of these methods can be reused. Previous studies (Vroman & Tighzert 2009; Abioye et al., 2019) have reported a biodegradation rate ranging from 60% to 80%. Agro-based composite materials degraded at an equivalent rate after 5 months, as did cellulose-based biopolymers after 350 days. The organisms used in the production of bacteria have an impact on the biodegradation rate as well. The biodegradation rates of *Pediococcus acidilactici* and *Enterococcus faecium* biopolymers were the lowest (50%) of all. The inclusion of reinforcing agents further slowed the process of biodegradation. The slow rates of biodegradation raise the question of whether materials are marked as "biodegradable" or just compostable (Arcos-Hernandez et al., 2012; Harmaen et al., 2015).

The main problem is that biodegradation rates differ significantly depending on the environment (marine, soil, and freshwater). Another issue is the need for knowledge about biodegradable polymers and polymers made from agricultural waste that meets these criteria. Even with fully biodegradable polymers, the environment is still risky because nanoscale substances created from broken-down polymers can contaminate water and the surrounding environment (Kibria et al., 2023; Tarazona et al., 2022).

12.5.1 Conditions for Degradation and Rates of Biopolymer Degradation

Despite being created from natural materials, not all bioplastics naturally degrade. The method and degree of deterioration determine how biodegradable they are. Biopolymers like PLA, starch, cellulose, PHA, and their derivatives are used to create biodegradable plastics. Modifying the degradation temperature, humidity, and other factors during polymer production can produce a biopolymer with regulated biodegradation potential. Pure PLA bioplastic, for example, has demonstrated biodegradability of 13%, 60%, 39%, and 63.96% in 60, 30, 28, and 90 days, respectively, under different degradation conditions such as compost, 58°C temperature, and 60% moisture. Similarly, a Cassava Mater-Bi™ bioplastic based on starch demonstrated 26.9% biodegradability in 72 days under compost, 55% moisture, aerobic, 23°C conditions (Rai et al., 2021).

12.6 METABOLIC AND ENZYME ENGINEERING TO ENHANCE THE ENZYME EFFICIENCY FOR BIOPOLYMER SYNTHESIS

Metabolic and enzyme engineering are essential for improving enzyme efficiency for biopolymer synthesis. Metabolic engineering involves modifying the metabolic pathways of microorganisms to enhance the production of substrates or enzymes, while enzyme engineering consists of the modification of enzymes themselves to improve their activity or selectivity. These techniques have been shown to be effective in improving the efficiency of enzymes used for the synthesis of a wide range of biopolymers such as PLA, PHAs, polyesters, and polysaccharides.

One approach to metabolic engineering is the manipulation of gene expression to increase the production of specific metabolites or enzymes. This can be achieved by introducing foreign genes or modifying existing genes encoding enzymes involved in the desired metabolic pathway. Directed evolution is another technique that has been used to improve the activity of enzymes used for biopolymer synthesis. It involves the random mutation of the enzyme gene, followed by screening for mutants with improved properties. This approach has been used to modify enzymes such as cutinases, which are used in the synthesis of polyesters from plant oils (Pickens et al., 2011).

Protein engineering is another approach used to improve enzyme activity. This involves the rational design of enzymes to enhance their activity or selectivity toward specific substrates. Protein engineering has been used to modify the active site of lipases, resulting in enhanced activity toward specific substrates, such as those derived from plant oils. Cofactor engineering is another approach that can be used to improve enzyme activity. This involves the manipulation of cofactor availability in the cell to enhance enzyme activity. For example, the overexpression of genes encoding for enzymes involved in the biosynthesis of cofactors like NADPH has been shown to improve the activity of some enzymes involved in biopolymer synthesis (Song et al., 2023).

Table 12.2 presents data on metabolic and enzyme engineering strategies to enhance the efficiency of enzymes involved in biopolymer synthesis. The data includes information on the biopolymer, enzyme, substrate, engineering strategy, and resulting improvements.

TABLE 12.2
Biopolymers and Their Metabolic/Enzyme Engineering Strategy

Biopolymer	Enzyme	Substrate	Metabolic/Enzyme Engineering Strategy	Result
Polyhydroxyalkanoates (PHA)	PhaC	CoA-activated fatty acid	Site-directed mutagenesis of PhaC enzyme to improve catalytic activity and substrate specificity	Increase in polymer yield and improved monomer composition
Poly(lactic acid) (PLA)	Lipase	Lactide	Optimization of enzyme expression levels and immobilization on a solid support	Increased enzyme stability and recyclability, leading to higher polymer yield
Poly(3-hydroxybutyrate-co-3-hydroxy valerate) (PHBV)	PhaC and PhaZ	CoA-activated fatty acids	Metabolic engineering of PHA biosynthetic pathway to increase precursor availability and improve monomer composition	Higher PHBV yield and improved material properties
Poly(glycolic acid) (PGA)	PGA synthase	Glycolic acid	Directed evolution of PGA synthase enzyme to improve catalytic efficiency and tolerance to substrate inhibition	Higher PGA yield and improved polymer properties
Polycaprolactone (PCL)	Lipase	ε-caprolactone	Rational design of lipase enzyme to increase substrate specificity and activity	Increased PCL yield and improved material properties

For PHA synthesis, site-directed mutagenesis of the PhaC enzyme was employed to improve catalytic activity and substrate specificity, leading to an increase in polymer yield and improved monomer composition (Karr et al., 1983). In the case of PLA synthesis, optimization of enzyme expression levels and immobilization on solid support were carried out, resulting in increased enzyme stability and recyclability, leading to higher polymer yield (Venkateswara Reddy et al., 2017).

For Poly(3-hydroxybutyrate-co-3-hydroxyvalerate) (PHBV) synthesis, metabolic engineering of the PHA biosynthetic pathway was used to increase precursor availability and improve monomer composition, leading to higher PHBV yield and enhanced material properties. For Poly(glycolic acid) (PGA) synthesis, directed evolution of the PGA synthase enzyme was employed to improve catalytic efficiency and tolerance to substrate inhibition, leading to higher PGA yield and improved polymer properties (Maia Campos et al., 2015). Finally, for Polycaprolactone (PCL) synthesis, a rational design of lipase enzyme was employed to increase substrate specificity and activity, leading to increased PCL yield and improved material properties (Wu et al., 2013).

The use of metabolic and enzyme engineering techniques has led to the development of more efficient and sustainable processes for the production of biopolymers. These techniques offer a promising approach to improving the efficiency of enzymes used for biopolymer synthesis, and their continued development and application are essential in promoting the widespread use of biopolymers and in advancing the goal of a more sustainable and environmentally friendly future.

12.7 SUMMARY, CONCLUSION, AND FUTURE PERSPECTIVES

Due to economic, environmental, and technical considerations, researchers have opted for natural and biodegradable polymers as substitutes for synthetic polymers across various applications. While the production of renewable polymers from cultivated plants poses challenges, addressing these hurdles requires interdisciplinary collaboration and scientific advancements. Enhancing our understanding of crop plants' chemical, biological, and physical behavior will yield more significant benefits from these plant sources. The success of biopolymer utilization is interconnected with progress in fields such as chemical engineering, biotechnology, genetics, and innovative experimentation. The efficient extraction of polymers and monomers from plants and their integration into biorefining facilities will be crucial. These advancements collectively contribute to the successful adoption of biopolymers, which are currently being developed for diverse fields, ranging from simple to high-demand applications like medicine.

Future prospect: The field of biopolymers is experiencing rapid growth primarily due to their inherent biodegradability, leading to their widespread use across diverse applications. Biopolymers are currently under development for a wide range of fields, encompassing both simple and high-demand sectors like medicine. The incorporation of different nanofillers into these biopolymers has been explored to enhance their mechanical and barrier properties, thereby expanding their potential applications. The utilization of biopolymers is driven by environmental concerns and the desire to replace toxic polymers, capitalizing on their excellent biocompatibility and biodegradability properties. Looking ahead, some prospects include:

- **Sustainable packaging materials:** The use of biodegradable polymers in packaging is becoming increasingly important due to the adverse environmental impacts of traditional plastic materials. Enzymatic techniques can be utilized to synthesize biodegradable polymers from various agrarian biomasses, such as cornstarch, wheat gluten, and potato starch, for the production of sustainable packaging materials.
- **Biomedical applications:** Biodegradable polymers have numerous biomedical applications, such as drug delivery systems and tissue engineering. Enzymatic techniques can synthesize biodegradable polymers from biomolecules such as chitosan, gelatin, and collagen, commonly found in agricultural biomass.
- **Agricultural and horticultural applications:** Enzymatic techniques can also be used for the synthesis of biodegradable polymers for agricultural and horticultural applications, such as controlled-release fertilizers and plant growth regulators. The use of biodegradable polymers can reduce the negative impacts of traditional chemical fertilizers and pesticides on the environment.
- **Waste management:** Enzymatic techniques can be utilized to convert agricultural and agro-industrial waste into value-added biodegradable polymers. The use of waste as a raw material for the production of biodegradable polymers can reduce the waste accumulation in landfills, leading to a more sustainable and eco-friendly approach to waste management.

  The current and next challenge of biopolymers is how to replace the full use of petrochemical-based polymers in all fields as we approach the end of the "plastic age."

In conclusion, enzymatic techniques for synthesizing biodegradable polymers from different agrarian biomasses hold great promise for various applications. With further research and development, these techniques can offer a sustainable and cost-effective alternative to traditional synthetic polymers, which have adverse environmental impacts.

ACKNOWLEDGMENTS

Dr. Karan Kumar is grateful to the Prime Minister's Research Fellowship provided by the Ministry of Education, Government of India. Ms. Shraddha Jadhav acknowledge the online WINBIOCB-2022 internship/training provided by Prof. V.S. Moholkar.

NOTE

* Corresponding author

REFERENCES

Abioye, A. A, O. O Fasanmi, D. O Rotimi, O. P Abioye, C. C Obuekwe, S. A Afolalu, and I. P Okokpujie. 2019. Review of the development of biodegradable plastic from synthetic polymers and selected synthesized nanoparticle starches. *Journal of Physics: Conference Series* 1378 (4): 042064. https://doi.org/10.1088/1742-6596/1378/4/042064.

Adhami, W., Y. Bakkour, and C. Rolando. 2021. Polylactones synthesis by enzymatic ring opening polymerization in flow. *Polymer* 230 (September): 124040. https://doi.org/10.1016/j.polymer.2021.124040.

Aggarwal, J., S. Sharma, H. Kamyab, and A. Kumar. 2020. The realm of biopolymers and their usage: an overview. *Journal of Environmental Treatment Techniques* 8 (2): 1005–1016.

Anjum, A., M. Zuber, K. M. Zia, A. Noreen, M. N. Anjum, and S. Tabasum. 2016. Microbial production of polyhydroxyalkanoates (PHAs) and its copolymers: a review of recent advancements. *International Journal of Biological Macromolecules* 89 (August): 161–174. https://doi.org/10.1016/j.ijbiomac.2016.04.069.

Arcos-Hernandez, M. V., B. Laycock, S. Pratt, B. C. Donose, M. A. L. Nikolić, P. Luckman, A. Werker, and P. A. Lant. 2012. "Biodegradation in a soil environment of activated sludge derived polyhydroxyalkanoate (PHBV)." *Polymer Degradation and Stability* 97 (11): 2301–2312. https://doi.org/10.1016/j.polymdegradstab.2012.07.035.

Asghari, F., M. Samiei, K. Adibkia, A. Akbarzadeh, and S. Davaran. 2017. Biodegradable and biocompatible polymers for tissue engineering application: a review. *Artificial Cells, Nanomedicine, and Biotechnology* 45 (2): 185–192. https://doi.org/10.3109/21691401.2016.1146731.

Balla, E., V. Daniilidis, G. Karlioti, T. Kalamas, M. Stefanidou, N. D. Bikiaris, A. Vlachopoulos, I. Koumentakou, and D. N. Bikiaris. 2021. Poly(Lactic Acid): a versatile biobased polymer for the future with multifunctional properties-from monomer synthesis, polymerization techniques and molecular weight increase to PLA applications. *Polymers* 13 (11): 1822. https://doi.org/10.3390/polym13111822.

Baranwal, J., B. Barse, A. Fais, G. L. Delogu, and A. Kumar. 2022. Biopolymer: a sustainable material for food and medical applications. *Polymers* 14 (5): 983. https://doi.org/10.3390/polym14050983.

Bellenguez, C., F. Küçükali, I. E. Jansen, L. Kleineidam, S. Moreno-Grau, N. Amin, A. C. Naj, et al., 2022. New insights into the genetic etiology of Alzheimer's disease and related dementias. *Nature Genetics* 54 (4): 412–436. https://doi.org/10.1038/s41588-022-01024-z.

"Bioplastics & Biopolymers Market Worth $29.7 Billion by 2026." 2023. Accessed January 28. https://www.marketsandmarkets.com/PressReleases/biopolymers-bioplastics.asp.

Chang, I., J. Im, and G.-C. Cho. 2016. Introduction of microbial biopolymers in soil treatment for future environmentally-friendly and sustainable geotechnical engineering. *Sustainability* 8 (3): 251. https://doi.org/10.3390/su8030251.

Dalton, B., P. Bhagabati, J. De Micco, R. B. Padamati, and K. O'Connor. 2022. A review on biological synthesis of the biodegradable polymers polyhydroxyalkanoates and the development of multiple applications. *Catalysts* 12 (3): 319. https://doi.org/10.3390/catal12030319.

Di Donato, P., V., Taurisano, A., Poli, G. G., d'Ayala, B., Nicolaus, M., Malinconinco, and G. Santagata. 2020. Vegetable wastes derived polysaccharides as natural eco-friendly plasticizers of sodium alginate. *Carbohydrate Polymers* 229 (February): 115427. https://doi.org/10.1016/j.carbpol.2019.115427.

Dintcheva, N. T., G. Infurna, M. Baiamonte, and F. D'Anna. 2020. Natural compounds as sustainable additives for biopolymers. *Polymers* 12 (4): 115427. https://doi.org/10.3390/polym12040732.

Duchiron, S. W., E. Pollet, S. Givry, and L. Avérous. 2015. Mixed systems to assist enzymatic ring opening polymerization of lactide stereoisomers. *RSC Advances* 5 (103): 84627–84635. https://doi.org/10.1039/C5RA18954C.

Düşkünkorur, H. Ö., A. Bégué, E. Pollet, V. Phalip, Y. Güvenilir, and L. Avérous. 2015. Enzymatic ring-opening (co)polymerization of lactide stereoisomers catalyzed by lipases. Toward the in situ synthesis of organic/inorganic nanohybrids. *Journal of Molecular Catalysis B: Enzymatic* 115 (May): 20–28. https://doi.org/10.1016/j.molcatb.2015.01.011.

Dwi Prasetyo, W. Z. A. P., M. R. Bilad, T. M. I. Mahlia, Y. Wibisono, N. A. H. Nordin, and M. D. H. Wirzal. 2020. Insight into the sustainable integration of bio- and petroleum refineries for the production of fuels and chemicals. *Polymers* 12 (5): 1091. https://doi.org/10.3390/polym12051091.

Engel, J., A. Cordellier, L. Huang, and S. Kara. 2019. Enzymatic ring-opening polymerization of lactones: traditional approaches and alternative strategies. *ChemCatChem* 11 (20): 4983–4997. https://doi.org/10.1002/cctc.201900976.

Feder, D., and R. A. Gross. 2010. Exploring chain length selectivity in HIC-catalyzed polycondensation reactions." *Biomacromolecules* 11 (3): 690–697. https://doi.org/10.1021/bm901272r.

Fouda-Mbanga, B. G., E. Prabakaran, and K. Pillay. 2021. Carbohydrate biopolymers, lignin based adsorbents for removal of heavy metals (Cd2+, Pb2+, Zn2+) from wastewater, regeneration and reuse for spent adsorbents including latent fingerprint detection: a review. *Biotechnology Reports* 30 (June): e00609. https://doi.org/10.1016/j.btre.2021.e00609.

Ghosh, S., D. Lahiri, M. Nag, A. Dey, T. Sarkar, S. K. Pathak, H. A. Edinur, S. Pati, and R. R. Ray. 2021. Bacterial biopolymer: its role in pathogenesis to effective biomaterials. *Polymers* 13 (8): 242. https://doi.org/10.3390/polym13081242.

Gobi, R., P. Ravichandiran, R. S. Babu, and D. J. Yoo. 2021. Biopolymer and synthetic polymer-based nanocomposites in wound dressing applications: a review. *Polymers* 13 (12): 1962. https://doi.org/10.3390/polym13121962.

Gunathilake, T. M. S. U., Y. C. Ching, K. Y. Ching, C. H. Chuah, and L. C. Abdullah. 2017. Biomedical and microbiological applications of bio-based porous materials: a review. *Polymers* 9 (12): 160. https://doi.org/10.3390/polym9050160.

Guzmán-Lagunes, F., A. López-Luna, M. Gimeno, and E. Bárzana. 2012. Enzymatic synthesis of poly-L-lactide in supercritical R134a. *The Journal of Supercritical Fluids* 72 (December): 186–190. https://doi.org/10.1016/j.supflu.2012.08.017.

Harmaen, A. S., A. Khalina, I. Azowa, M. A. Hassan, A. Tarmian, and M. Jawaid. 2015. Thermal and biodegradation properties of poly(lactic acid)/fertilizer/oil palm fibers blends biocomposites. *Polymer Composites* 36 (3): 576–583. https://doi.org/10.1002/pc.22974.

Heredia-Guerrero, J. A., G. Caputo, S. Guzman-Puyol, G.Tedeschi, A. Heredia, L. Ceseracciu, J. J. Benitez, and A.Athanassiou. 2019. Sustainable polycondensation of multifunctional fatty acids from tomato pomace agro-waste catalyzed by Tin (II) 2-ethylhexanoate. *Materials Today Sustainability* 3-4 (March): 100004. https://doi.org/10.1016/j.mtsust.2018.12.001.

Hu, Y., W. Daoud, K. Cheuk, and C. Lin. 2016. Newly developed techniques on polycondensation, ring-opening polymerization and polymer modification: focus on poly(lactic acid). *Materials* 9 (3): 133. https://doi.org/10.3390/ma9030133.

Ibrahim, S., O. Riahi, S. M. Said, M. F. M. Sabri, and S.Rozali. 2019. Biopolymers from crop plants. In *Reference Module in Materials Science and Materials Engineering*, B9780128035818115735. Elsevier. https://doi.org/10.1016/B978-0-12-803581-8.11573-5.

Ilyas, R. A., S. M. Sapuan, R. Ibrahim, H. Abral, M. R. Ishak, E. S. Zainudin, M. S. N. Atikah, et al., 2019. Effect of sugar palm nanofibrillated cellulose concentrations on morphological, mechanical and physical properties of biodegradable films based on agro-waste sugar palm (Arenga Pinnata (Wurmb.) Merr) Starch. *Journal of Materials Research and Technology* 8 (5): 4819–4830. https://doi.org/10.1016/j.jmrt.2019.08.028.

Karan, H., C. Funk, M. Grabert, M. Oey, and B. Hankamer. 2019. Green bioplastics as part of a circular bioeconomy. *Trends in Plant Science* 24 (3): 237–249. https://doi.org/10.1016/j.tplants.2018.11.010.

Karr, D. B., J. K. Waters, and D. W. Emerich. 1983. Analysis of poly-β-hydroxybutyrate in *Rhizobium Japonicum* bacteroids by ion-exclusion high-pressure liquid chromatography and UV detection. *Applied and Environmental Microbiology* 46 (6): 1339–1344. https://doi.org/10.1128/aem.46.6.1339-1344.1983.

Kibria, M. G., N. I. Masuk, R. Safayet, H. Q.Nguyen, and M. Mourshed. 2023. Plastic waste: challenges and opportunities to mitigate pollution and effective management. *International Journal of Environmental Research* 17 (1): 20. https://doi.org/10.1007/s41742-023-00507-z.

Kim, J., M. Tremaine, J. A. Grass, H. M. Purdy, R.Landick, P. J. Kiley, and J. L. Reed. 2019. Systems metabolic engineering of *escherichia coli* improves coconversion of lignocellulose-derived sugars. *Biotechnology Journal* 14 (9): 1800441. https://doi.org/10.1002/biot.201800441.

Klein, M., and E. Poverenov. 2020. Natural biopolymer-based hydrogels for use in food and agriculture. *Journal of the Science of Food and Agriculture* 100 (6): 10274. https://doi.org/10.1002/jsfa.10274.

Kumar, K., K. Roy, and V. S. Moholkar. 2021. Mechanistic investigations in sonoenzymatic synthesis of N-butyl levulinate. *Process Biochemistry* 111 (December): 147–158. https://doi.org/10.1016/j.procbio.2021.09.005.

Kumar, K., H. Shah, and V. S. Moholkar. 2022. Genetic algorithm for optimization of fermentation processes of various enzyme productions. In *Optimization of Sustainable Enzymes Production*, (pp. 121–144), edited by J. S. Eswari, and N. Suryawanshi, Chapman and Hall/CRC. https://doi.org/10.1201/9781003292333

Kumar, J. A., S. Sathish, D. Prabu, A. A. Renita, A. Saravanan, V. C. Deivayanai, M. Anish, J. Jayaprabakar, O. Baigenzhenov, and A. Hosseini-Bandegharaei. 2023. Agricultural waste biomass for sustainable bioenergy production: feedstock, characterization and pre-treatment methodologies. *Chemosphere* 331 (August): 138680. https://doi.org/10.1016/j.chemosphere.2023.138680.

Kumar, K., L. Barbora, and V. S. Moholkar. 2023a. Genomic insights into clostridia in bioenergy production: comparison of metabolic capabilities and evolutionary relationships. *Biotechnology and Bioengineering* 121 (4): 1298–1313. https://doi.org/10.1002/bit.28610.

Kumar, K., and V. S. Moholkar. 2023b. Mechanistic aspects of enhanced kinetics in sonoenzymatic processes using three simultaneous approaches. In *Sustainable Energy Generation and Storage*, edited by V. S. Moholkar, K. Mohanty, and V. V. Goud, 41–57. Singapore: Springer Nature Singapore. https://doi.org/10.1007/978-981-99-2088-4_5.

Kumar, K., P. Patro, U Raut, V Yadav, L Barbora, and VS. Moholkar. 2023c. Elucidating the molecular mechanism of ultrasound-enhanced lipase-catalyzed biodiesel synthesis: a computational study. *Biomass Conversion and Biorefinery*. https://doi.org/10.1007/s13399-023-04742-4.

Kumar, K., Anand, A., & Moholkar, V. S. (2024a). Molecular Hydrogen (H_2) Metabolism in Microbes: A Special Focus on Biohydrogen Production. In *Biohydrogen-Advances and Processes* (pp. 25–58). Cham: Springer Nature Switzerland.

Kumar, K., Jadhav, S. M., & Moholkar, V. S. (2024b). Acetone-Butanol-Ethanol (ABE) fermentation with clostridial co-cultures for enhanced biobutanol production. *Process Safety and Environmental Protection*, 185, 277–285. https://doi.org/10.1016/j.psep.2024.03.027

Kumar, K., Kumar, S., Goswami, A., & Moholkar, V. S. (2024c). Advancing sustainable biofuel production: A computational insight into microbial systems for isopropanol synthesis and beyond. *Process Safety and Environmental Protection*, 188, 1118–1132. https://doi.org/10.1016/j.psep.2024.06.024

Lakhundi, S., R. Siddiqui, and N. A. Khan. 2015. Cellulose degradation: a therapeutic strategy in the improved treatment of acanthamoeba infections. *Parasites & Vectors* 8 (1): 23. https://doi.org/10.1186/s13071-015-0642-7.

Liu, Z.-H., and H.-Z. Chen. 2015. Xylose production from corn stover biomass by steam explosion combined with enzymatic digestibility. *Bioresource Technology* 193 (October): 345–356. https://doi.org/10.1016/j.biortech.2015.06.114.

Lodhi, G., Y.-S. Kim, J.-W. Hwang, S.-K. Kim, Y.-J. Jeon, J.-Y. Je, C.-B. Ahn, S.-H. Moon, B.-T. Jeon, and P.-J. Park. 2014. Chitooligosaccharide and its derivatives: preparation and biological applications. *BioMed Research International* 2014. https://doi.org/10.1155/2014/654913.

Madadi, R., H. Maljaee, L. S. Serafim, and S. P. M. Ventura. 2021. Microalgae as contributors to produce biopolymers. *Marine Drugs* 19 (8): 466. https://doi.org/10.3390/md19080466.

Maia C., P. M. B. Gonçalves, L. R. Gaspar, G. M. S. Gonçalves, L. H. T. R. Pereira, M. Semprini, and R. A. Lopes. 2015. Comparative effects of retinoic acid or glycolic acid vehiculated in different topical formulations. *BioMed Research International* 2015: 1–6. https://doi.org/10.1155/2015/650316.

Maraveas, C. 2020a. Production of sustainable construction materials using agro-wastes. *Materials* 13 (2): 262. https://doi.org/10.3390/ma13020262.

Maraveas, C. 2020b. Production of sustainable and biodegradable polymers from agricultural waste. *Polymers* 12 (5): 1127. https://doi.org/10.3390/polym12051127.

"Market Update 2020: Bioplastics Continue to Become Mainstream as the Global Bioplastics Market Is Set to Grow by 36 Percent over the next 5 Years - European Bioplastics e.V." 2023. Accessed January 28. https://www.european-bioplastics.org/market-update-2020-bioplastics-continue-to-become-mainstream-as-the-global-bioplastics-market-is-set-to-grow-by-36-percent-over-the-next-5-years/.

Mathiot, C., P. Ponge, B. Gallard, J.-F. Sassi, F. Delrue, and N. L. Moigne. 2019. Microalgae starch-based bioplastics: screening of ten strains and plasticization of unfractionated microalgae by extrusion. *Carbohydrate Polymers* 208 (March): 142–151. https://doi.org/10.1016/j.carbpol.2018.12.057.

McGauran, T., N. Dunne, B. Smyth, E. Cunningham, and M. Harris. 2021. Poultry feather disulphide bond breakdown to enable bio-based polymer production. *Polymers from Renewable Resources* 12 (3–4), 8746. https://doi.org/10.1177/20412479211008746.

Moshood, T. D., G. Nawanir, F. Mahmud, F. Mohamad, M. H. Ahmad, and A. AbdulGhani. 2022. Sustainability of biodegradable plastics: new problem or solution to solve the global plastic pollution? *Current Research in Green and Sustainable Chemistry* 5, 100273. https://doi.org/10.1016/j.crgsc.2022.100273.

Nair, N. R., V. C. Sekhar, K. M. Nampoothiri, and A. Pandey. 2017. "Biodegradation of biopolymers." In *Current Developments in Biotechnology and Bioengineering*, edited by: A. Pandey, S. Negi and C. R. Soccol, 739–755. Elsevier. https://doi.org/10.1016/B978-0-444-63662-1.00032-4.

Narancic, T., F. Cerrone, N. Beagan, and K. E. O'Connor. 2020. Recent advances in bioplastics: application and biodegradation. *Polymers* 12 (4): 920. https://doi.org/10.3390/polym12040920.

Nguyen, T. P., Q. V. Nguyen, V.-H. Nguyen, T.-H. Le, V. Q. N. Huynh, D.-V. N. Vo, Q. T. Trinh, S. Y. Kim, and Q. Van Le. 2019. Silk fibroin-based biomaterials for biomedical applications: a review. *Polymers* 11 (12): 1933. https://doi.org/10.3390/polym11121933.

Nitschke, W., and M. J. Russell. 2013. Beating the acetyl coenzyme A-pathway to the origin of life. *Philosophical Transactions of the Royal Society B: Biological Sciences* 368 (1622), 258. https://doi.org/10.1098/rstb.2012.0258.

Ojo, A. O., and O. De Smidt. 2023. Lactic acid: a comprehensive review of production to purification. *Processes* 11 (3): 688. https://doi.org/10.3390/pr11030688.

Panyachanakul, T., T. Lomthong, W. Lorliam, J. Prajanbarn, S. Tokuyama, V. Kitpreechavanich, and S. Krajangsang. 2020. New insight into thermo-solvent tolerant lipase produced by Streptomyces Sp. A3301 for re-polymerization of poly (DL-lactic acid). *Polymer* 204, 122812. https://doi.org/10.1016/j.polymer.2020.122812.

Perveen, K., F. Masood, and A. Hameed. 2020. Preparation, characterization and evaluation of antibacterial properties of epirubicin loaded PHB and PHBV nanoparticles. *International Journal of Biological Macromolecules* 144 (February): 259–266. https://doi.org/10.1016/j.ijbiomac.2019.12.049.

Pickens, L. B., Y. Tang, and Y.-H. Chooi. 2011. Metabolic engineering for the production of natural products. *Annual Review of Chemical and Biomolecular Engineering* 2 (1): 211–236. https://doi.org/10.1146/annurev-chembioeng-061010-114209.

Rahmayetty, Y. W., Sukirno, S. F. R., E. A. Suyono, M. Yohda, and M. Gozan. 2018. Use of candida rugosa lipase as a biocatalyst for L-lactide ring-opening polymerization and polylactic acid production. *Biocatalysis and Agricultural Biotechnology* 16 (October): 683–691. https://doi.org/10.1016/j.bcab.2018.09.015.

Rai, P., S. Mehrotra, S. Priya, E. Gnansounou, and S. K. Sharma. 2021. Recent advances in the sustainable design and applications of biodegradable polymers. *Bioresource Technology* 325 (April): 124739. https://doi.org/10.1016/j.biortech.2021.124739.

RameshKumar, S., P. Shaiju, K. E. O'Connor, and R. B. P. 2020. Bio-based and biodegradable polymers - state-of-the-art, challenges and emerging trends. *Current Opinion in Green and Sustainable Chemistry* 21 (February): 75–81. https://doi.org/10.1016/j.cogsc.2019.12.005.

Rudnik, E., and D. Briassoulis. 2011. Comparative biodegradation in soil behaviour of two biodegradable polymers based on renewable resources. *Journal of Polymers and the Environment* 19 (1): 18–39. https://doi.org/10.1007/s10924-010-0243-7.

Saini, D. K., H. Chakdar, S. Pabbi, and P. Shukla. 2020. Enhancing production of microalgal biopigments through metabolic and genetic engineering. *Critical Reviews in Food Science and Nutrition* 60 (3): 391–405. https://doi.org/10.1080/10408398.2018.1533518.

Sanchez-Vazquez, S. A., H. C. Hailes, and J. R. G. Evans. 2013. Hydrophobic polymers from food waste: resources and synthesis. *Polymer Reviews* 53 (4): 627–694. https://doi.org/10.1080/15583724.2013.834933.

Santana, R. F., R. C. F. Bonomo, O. R. R. Gandolfi, L. B. Rodrigues, L. S. Santos, A. C. D. S. Pires, C. P. de Oliveira, R. da Costa Ilhéu Fontan, and C. M. Veloso. 2018. Characterization of starch-based bioplastics from jackfruit seed plasticized with glycerol. *Journal of Food Science and Technology* 55 (1): 278–286. https://doi.org/10.1007/s13197-017-2936-6.

Sharma, P., V. K. Gaur, S.-H. Kim, and A. Pandey. 2020. Microbial strategies for bio-transforming food waste into resources. *Bioresource Technology* 299 (March): 122580. https://doi.org/10.1016/j.biortech.2019.122580.

Sharmila, G., C. Muthukumaran, N. M. Kumar, V.M. Sivakumar, and M. Thirumarimurugan. 2020. Food waste valorization for biopolymer production. In *Current Developments in Biotechnology and Bioengineering*, edited by S. Varjani, A. Pandey, E. Gnansounou, S. K. Khanal, S. Raveendran, 233–249. Elsevier. https://doi.org/10.1016/B978-0-444-64321-6.00012-4.

Singh, R., M. Kumar, A. Mittal, and P. K. Mehta. 2016. Microbial enzymes: industrial progress in 21st century. *3 Biotech* 6 (2): 174. https://doi.org/10.1007/s13205-016-0485-8.

Smith, A. M., S. Moxon, and G. A. Morris. 2016. Biopolymers as wound healing materials. In *Wound Healing Biomaterials*, 261–287. Elsevier. https://doi.org/10.1016/B978-1-78242-456-7.00013-1.

Soleymani, E. B., S. Saeed Karbasi, and E. B.Toloue. 2021. Modified poly (3-hydroxybutyrate)-based scaffolds in tissue engineering applications: a review. *International Journal of Biological Macromolecules* 166 (January): 986–998. https://doi.org/10.1016/j.ijbiomac.2020.10.255.

Song, Z., Q. Zhang, W. Wu, Z. Pu, and H.Yu. 2023. Rational design of enzyme activity and enantioselectivity. *Frontiers in Bioengineering and Biotechnology* 11 (January): 1129149. https://doi.org/10.3389/fbioe.2023.1129149.

Sumarno, P. N. T., B. Airlangga, N. E. Mayangsari, and A. Haryono. 2021. The degradation of cellulose in ionic mixture solutions under the high pressure of carbon dioxide. *RSC Advances* 11 (6): 3484–3494. https://doi.org/10.1039/D0RA07154D.

Tabasum, S., M. Younas, M. A. Zaeem, I. Majeed, M. Majeed, A. Noreen, M. N. Iqbal, and K. M. Zia. 2019. A review on blending of corn starch with natural and synthetic polymers, and inorganic nanoparticles with mathematical modeling. *International Journal of Biological Macromolecules* 122 (February): 969–996. https://doi.org/10.1016/j.ijbiomac.2018.10.092.

Takwa, M., M. W. Larsen, K. Hult, and M. Martinelle. 2011. Rational redesign of candida antarctica lipase B for the ring opening polymerization of D,D-Lactide. *Chemical Communications* 47 (26): 7392. https://doi.org/10.1039/c1cc10865d.

Tarazona, N. A., R. Machatschek, J. Balcucho, J. L. Castro-Mayorga, J. F. Saldarriaga, and A. Lendlein. 2022. Opportunities and challenges for integrating the development of sustainable polymer materials within an international circular (bio)economy concept. *MRS Energy & Sustainability* 9 (1): 28–34. https://doi.org/10.1557/s43581-021-00015-7.

Tebaldi, M. L., A. L. C. Maia, F. Poletto, F. V. de Andrade, and D. C. F. Soares. 2019. Poly(-3-hydroxybutyrate-co-3-hydroxyvalerate) (PHBV): current advances in synthesis methodologies, antitumor applications and biocompatibility. *Journal of Drug Delivery Science and Technology* 51 (June): 115–126. https://doi.org/10.1016/j.jddst.2019.02.007.

Torres, F. G., S. Rodriguez, and A. C. Saavedra. 2019. Green composite materials from biopolymers reinforced with agroforestry waste." *Journal of Polymers and the Environment* 27 (12): 2651–2673. https://doi.org/10.1007/s10924-019-01561-5.

Vasić, K., Ž. Knez, and M. Leitgeb. 2021. Bioethanol production by enzymatic hydrolysis from different lignocellulosic sources. *Molecules* 26 (3): 753. https://doi.org/10.3390/molecules26030753.

Vega-Castro, O., J. Contreras-Calderon, E. León, A. Segura, M. Arias, L. Pérez, and P. J. A. Sobral. 2016. Characterization of a polyhydroxyalkanoate obtained from pineapple peel waste using ralsthonia eutropha. *Journal of Biotechnology* 231 (August): 232–238. https://doi.org/10.1016/j.jbiotec.2016.06.018.

Venkateshaiah, A., V. V. T. Padil, M. Nagalakshmaiah, S. Wacławek, M. Černík, and R. S. Varma. 2020. Microscopic techniques for the analysis of micro and nanostructures of biopolymers and their derivatives." *Polymers* 12 (3), 512. https://doi.org/10.3390/polym12030512.

Venkateswar Reddy, M., Y. Mawatari, R. Onodera, Y. Nakamura, Y. Yajima, and Y.-C. Chang. 2017. Polyhydroxyalkanoates (PHA) production from synthetic waste using pseudomonas pseudoflava : PHA synthase enzyme activity analysis from P. pseudoflava and P. palleronii. *Bioresource Technology* 234 (June): 99–105. https://doi.org/10.1016/j.biortech.2017.03.008.

Vroman, I., and L. Tighzert. 2009. Biodegradable polymers. *Materials* 2 (2): 307–344. https://doi.org/10.3390/ma2020307.

Wu, C., Z. Zhang, F. He, and R. Zhuo. 2013. Enzymatic synthesis of poly(ε-Caprolactone) in monocationic and dicationic ionic liquids. *Biotechnology Letters* 35 (6): 879–885. https://doi.org/10.1007/s10529-013-1160-3.

Wu, C.-S. 2012. Preparation, characterization, and biodegradability of renewable resource-based composites from recycled polylactide bioplastic and sisal fibers. *Journal of Applied Polymer Science* 123 (1): 347–355. https://doi.org/10.1002/app.34223.

Xu, Z., C. Pan, X. Li, N. Hao, T. Zhang, M. J. Gaffrey, Y. Pu, et al., 2021. Enhancement of polyhydroxyalkanoate production by co-feeding lignin derivatives with glycerol in pseudomonas putida KT2440. *Biotechnology for Biofuels* 14 (1): 11. https://doi.org/10.1186/s13068-020-01861-2.

Yates, M. R., and C. Y. Barlow. 2013. Life cycle assessments of biodegradable, commercial biopolymers-a critical review. *Resources, Conservation & Recycling* 78: 54–66.

Yeniad, B., H. Naik, and A. Heise. 2011. Lipases in polymer chemistry. *Advances in Biochemical Engineering/Biotechnology* 125: 69–95. https://doi.org/10.1007/10_2010_90.

Yeo, J. C. C., J. K. Muiruri, W. Thitsartarn, Z. Li, and C. He. 2018. Recent advances in the development of biodegradable PHB-based toughening materials: approaches, advantages and applications. *Materials Science & Engineering. C, Materials for Biological Applications* 92 (November): 1092–1116. https://doi.org/10.1016/j.msec.2017.11.006.

Zheng, H., and G. Liu. 2019. Sugar-based biopolymers as novel imaging agents for molecular magnetic resonance imaging. *WIREs Nanomedicine and Nanobiotechnology* 11 (4). https://doi.org/10.1002/wnan.1551.

13 Application of Microbial Technique for Synthesis of Food-Grade Agents from Different Agrobiomasses

An Overview

Indrani Paul and Priyanka Sarkar

13.1 INTRODUCTION

The unchecked global population increase in the past few decades has led to serious challenges that are being faced to achieve Sustainable Development Goals specially in the areas of the environment and food security (Leong et al., 2021). Agrobiomass derived from agricultural sources is an abundant source of organic material that can have multiple applications and can become the fulcrum to sustainable development in the upcoming century (Sidana et al., 2022; Hanumante et al., 2022). An estimated 30% of the world's total food production is lost in the agro-food industry (Awasthi et al., 2022). Demand for new innovative strategies for appropriate utilization of the waste biomass is need of the hour. Agrobiomass includes crop residues (such as leaves, stalks, and husks), animal manure, dedicated energy crops (like miscanthus and switchgrass), dairy products, vegetables, fruits, meat, poultry, and many other organic materials produced through farming practices (Ramady et al., 2022; Iram et al., 2023).

The general composition of agrobiomass consists of three main polymers – cellulose (40%–50%), hemicellulose (20%–30%), and lignin (20%–35%) (Selo et al., 2021). These can be utilized by different microbial strains as their source of carbon and energy under specific conditions, depending on the nature and enzymatic machinery. Lignin is most abundant in agro-wastes since cellulose and hemicelluloses are used as raw materials for mainstream industrial products such as paper and ethanol. Besides, lignin being the most recalcitrant component of agrobiomass is most tough to deconstruct and is a major challenge for microbial players in utilizing lignocellulosic biomasses (Li et al., 2022). Breaking down of the components of the agrobiomass is dependent on the presence of valuable enzymes that include laccases, cellulases, hemicellulases, lignin peroxidases, and manganese peroxidases (Nargotra et al., 2022).

 DOI: 10.1201/9781003407713-13

Multiple options of varied agrobiomasses and their easy availability have made them a preferable choice for fermentation-based production of different food additives (Bala et al., 2023). Different types of low-cost agrobiomasses depending on their availability and chemical composition are selected for the purpose (Ueda et al., 2022; Diaz & Blandino, 2022). Bagasse, vegetable and fruit peel offs, pulp remains, pomaces, molasses, oil cakes, etc. have been reported to serve as economically suitable substrates for production of wide variety of food additives. Selection of the microbial strains meant for the bioconversion of these agro-wastes into valorizable food additives depends on their metabolic capability to utilize these wastes that in turn depends on the enzymatic machinery of the strain.

It's crucial for food-grade agents to stick to the strict regulatory guidelines before they are approved for use in food products. Regulatory bodies like the U.S. Food and Drug Administration (FDA) and the European Food Safety Authority (EFSA) establish and monitor the safety standards for food-grade agents to ensure consumer safety. Derived from a biological source ensures the safety of such bio-based products in comparison with their chemical counterparts (Atta et al., 2022; Dey et al., 2022). The variety of components that come under the category of food-grade agents may include food additives (for enhancing quality and appearance), preservatives, sweeteners, colouring and flavouring agents, thickeners, pH regulators, emulsifiers, and antioxidants (Sarkar et al., 2020; Sadiq et al., 2019; Sen et al., 2019). Agrobiomass being rich in nutrition for the thriving of microorganisms that are potent producers of these class of components can be an excellent sustainable substrate for meeting the global demand for food-grade agent production.

This chapter focuses on different classes of agrobiomasses that have been reported to serve a promising substrate for food-grade agent production with the help of specifically abled microorganisms via microbial fermentation. Further, the emphasis has been laid upon the enzymatic requirements to break down the recalcitrant lignin component of the lignocellulosic biomass to make the cellulose and hemicellulose fractions available for bioconversion into value-added food-grade agents. The various categories of food-grade agents and their microbial sources have also been elaborated. The various factors affecting microbial fermentation process, bioreactor designs, optimization parameters, and purification methods have been highlighted. Besides, the importance of genetic engineering for enhancing yield and ensuring better production is a point of great interest, and a lot of research is going on for the betterment of industrialization and for increasing the overall output of the process. The wide implications of different food-grade agents have also been detailed in the upcoming sections.

13.2 TYPES AND APPLICATIONS OF FOOD ADDITIVES AND THEIR PRODUCTION FROM DIFFERENT AGROBIOMASSES

According to the Codex Alimentarius Standard CAC/GL 36-1989, food-grade agents can be classified into certain categories, depending on their utility. Agrobiomasses from different sources have been exploited and utilized as nutritional supplements for selected metabolically capable microbial strains to produce value-added food-grade

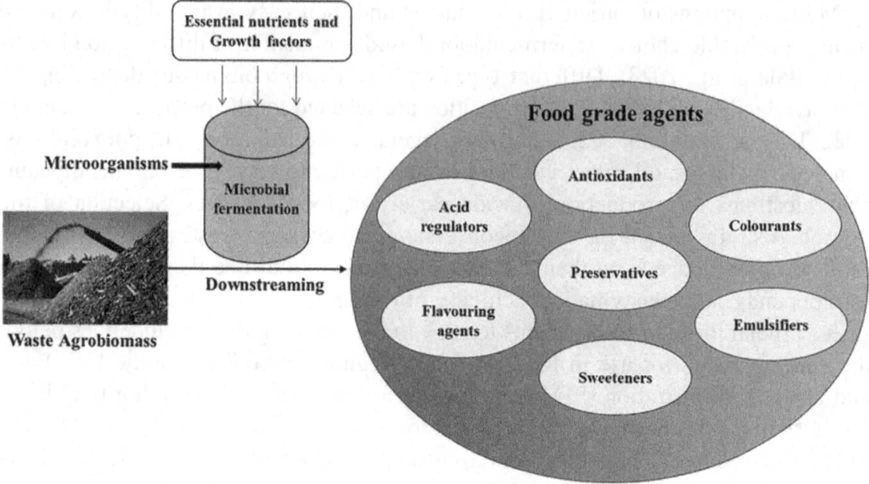

FIGURE 13.1 Production of different food-grade agents from waste agrobiomass via action of microorganisms.

agents (Atta et al., 2022; Sarkar et al., 2020; Figure 13.1). The following section discusses the commonly known food-grade agents, their production from cheap and easily available agrobiomass, and their possible applications and health benefits.

13.2.1 Types of Different Food-Grade Agents

13.2.1.1 Acid Regulators

Acidifiers generally find a place in food additives due to their ability to alter rheological character of food along with its taste, consistency, and texture. These are generally organic acids which are produced from biomass through various biochemical and chemical processes. Very common examples of acid regulators are citric acid, lactic acid, tartaric acid, calcium lactate, acetic acid, malic acid, fumaric acid, etc. A group of homofermentative (*Streptococcus* sp., *Lactococcus* sp., *Enterococcus* sp., etc.) and heterofermentative bacteria (*Oenococcus* sp., *Leuconostoc* sp., *Lactobacillus* sp., etc.) have been reported to mediate biotransformation of multiple agricultural wastes such as cotton textile wastes, coffee waste, cassava bagasse, and sugarcane bagasse into valuable lactic acid, which can also be further processed to obtain calcium lactate (Singh et al., 2023; El-Sheshtawy et al., 2022; John et al., 2006; Abedi & Hashemi, 2020). Apple pomace, peanut shells, banana peel, pineapple pomace and sugar sweetened beverage wastes, etc. serve as excellent sources for Citric acid production (Mores et al., 2023; Ali et al., 2016; Karthikeyan & Sivakumar, 2010; Kareem et al., 2010). Bao et al. (2022) showed how it is possible to obtain multiple organic acids from one source (lactic acid and acetic acid) via the use of specific bacterial consortium. Although fermentative production of fumaric and malic acids from lignocellulosic biomasses has been successfully obtained via fungal strains such as *Aspergillus* sp., *Pseudozyma* sp., and *Ustilago* sp. (West, 2023; Mondala, 2015), several bacterial strains have also been exploited for the purpose. Production of organic acids from wastes has a wide scope for large-scale commercialization in future.

13.2.1.2 Antioxidants

Antioxidants are most commonly used in foods that are oil/fat based (also called fatty foods) in order to prevent oxidative deterioration resulting in defect in colour and aroma. Antioxidant properties have been reported from different agrobiomasses derived components in recent times. The antimicrobial and antioxidant properties of agrobiomass-derived bioactive components have gained popularity in the present scenario. One of the recent examples is carotenoids, derived from vegetable waste having excellent antioxidant properties along with anticancer applications (Sinha et al., 2023). In another study, the fermented acai fruit pulp (*Euterpe oleracea*) was found to have multiple beneficial roles by the organic acids derived from the biomass along with potential antioxidant properties (Liu et al., 2023). Among the most frequently used antioxidants, ascorbic acid ranks high as far as microbial fermentation technology-based production is concerned. *Aspergillus flavus* and *Aspergillus tamarii* have produced ascorbic acid from brewery spent grain, while other strains such as *Blastoschizomyces* sp. and *Kluyveromyces marxianus* have given up to a yield of 6.5 g/L from corn straw waste using submerged fermentation technology (Banjo et al., 2018; Kareem et al., 2019). There is still a considerable space for exploration of production of antioxidants from agro-waste using microbial fermentation technology, and research are going on to discover new suitable substrates for this purpose.

13.2.1.3 Colourants

The use of food colours for enhancing appearance is a very well-known practice and this has increased the demand for green bio-based colouring agents such as biopigments in the food industry. The most commonly produced bacterial pigments are carotenoids, prodigiosin, tambjamines, melanins, quinones, and violacein. Agrobiomass has been an evident source of biopigment production (Jameel et al., 2023). Microbial strains that are observable pigment producers may be bacterial (*Gordonia jacobaea*, *Serratia marcescens* – red, *Chromobacterium* sp. – purple, *Erwinia chrysanthemi*, *Vogesella indigofera* – blue, *Micrococcus* – yellow, *Kocuria* sp. – red-yellow, and *Pseudomonas* sp. – green) and fungal (*Talaromyces* sp., *Monascus* sp., *Neurospora intermedia*, *Streptomyces* sp., etc.) having robust metabolism to utilize a broad range of substrates which may include nutrient-enriched agro-industrial wastes (Grewal et al., 2022; Hafez et al., 2023). They are reported to have antioxidant, antimalarial, and anticarcinogenic properties, which make them suitable for food applications.

13.2.1.4 Emulsifiers

Biomolecules that have a combination of both hydrophilic (protein, peptide, and carbohydrate) and hydrophobic (lipid chains and fatty acids) moieties, i.e., they are amphiphilic in nature, have the potential to reduce surface tensions at oil-water interface (Sharon et al., 2023; Paul et al., 2022). This property is exploited and they serve as great emulsifying agents in food products such as bread cheese, chewing gums, and chocolates. These biomolecules that may be lipopeptides or glycolipids in nature are produced by different bacterial and fungal strains using agro-industrial wastes, especially carbon source-rich biomasses like bagasses, vinasses, consisting of lignocellulosic components (Das et al., 2021). Solid-state or submerged fermentation techniques have been adapted by many research groups depending on the requirement of the microbial strain. Strains such as *Pseudomonas* sp., *Bacillus* sp., and *Lactobacillus* sp. have been found to produce good quality biosurfactant/emulsifiers using sugarcane bagasse as the sole source of carbon (Nagtode et al., 2023; Barbosa et al., 2023). Wastes composed of oil or fatty acids, oil cakes, and olive waste can serve as excellent source for biosurfactant or emulsifier production. There is room for wide research scope in this area, since meeting the growing demands is still a challenge when it comes to cheap and eco-friendly biosurfactant production from low-cost suitable sources like agro-waste residues.

13.2.1.5 Preservatives

The use of preservatives as food additives had been an essential as well as controversial practice for years. Spoilage prevention is mandatory for providing a good shelf life to the food for its large-scale marketization, but at the same time, preservatives have been associated with setting negative health impacts on consumers. In order to overcome this dilemma, the use of organic and green preservatives of biological origin has gained attention from food technologists and researchers and has become an emerging area of active research. Preservatives belonging to this category may include organic acids such as acetic, benzoic, propionic, and sorbic acids along with bacteriocins like nisin and enzymes that disrupt cell walls/cell membranes like lysozyme (Mores et al., 2023; Singh et al., 2023; West, 2023; Blekas, 2016). These have been derived microbially by the help of fermentation technology using varied agro-wastes such as fruit pomaces, pulps, peel offs, lignocellulosic biomasses, brewery wastes, whey, cull potatoes, soyabean meal waste, and molasses. The involvement of bacterial strains from different genus, such as *Bacillus* sp., *Lactobacillus* sp., *Leuconostoc mesenteroides*, and *Enterococcus* sp., have been reported in this regard by food scientists and is now one of the pioneer measures towards healthier, safer, and long-lasting food stocks on commercial level (Bali et al., 2014).

13.2.1.6 Sweeteners

Energy-reduced edible products marked with "No added Sugar" require sweeteners to maintain the taste and are widely used in the food industry. This category of food additives includes mainly sugar alcohols like xylitol, mannitol, erythritol, lactitol, maltitol, etc. (Iram et al., 2023; Cuadrado-Osorio, 2022). Hor and research group showed how the vinasse part of sugarcane bagasse after ethanol extraction can be further utilized for Xylitol production by the help of strain *Candida guilliermondii* (Hor et al., 2022; Kumar et al., 2022). Martinez Miranda et al. (2022) have reported the production of high-quality mannitol by heterofermentative lactic acid bacteria from cheap agro-wastes, which can be utilized as commercial sweeteners in food industry. High yield of erythritol has been found using oil crop waste using a genetically modified strain of *Yarrowia lipolytica* via the one-step solid-state fermentation technology (Liu et al., 2019). In many instances, the production of sugar alcohols is not a single-step process and may need a catalytic conversion that involves the use of enzymes, while on other cases, it may be accompanied along with bioethanol production (Park et al., 2016). The production of sugar alcohols from agro-industrial wastes is still in its infancy and needs further exploration for low-cost production and downstream processing that may be achieved by exploiting new novel isolates that metabolically fit into the requirements of the bioprocess.

13.2.1.7 Flavouring Agents

Food flavouring agents are components used to enhance or impart flavour to foods and beverages. They are essential in the culinary arts to create a wide range of tastes and aromas. Flavouring agents can be natural or synthetic, and they come in various forms, including liquids, powders, extracts, and essential oils. Common flavouring agents like vanilla essence, derived from vanillin, have been shown to be derived from a variety of agricultural biomass that can have wide commercial applications in food industry (Nayak et al., 2022; Gomathi et al., 2023). Compounds that act as precursors of vanillin are also of much importance, which may include ferulic acid and hydroxycinnamic acid (Subramani et al., 2023), might be synthesized via microorganisms that have the metabolic ability for bioconversion of these precursors to vanillin. These precursors can be biosynthesized from different agro-biomasses, thus bringing down the capital investment for mass production of such flavouring agents. Despite all the given facts, this area still needs to be explored since the yield and economic suitability need optimization for meeting up with the required demands (Table 13.1).

13.2.1.8 Nutraceutical Prebiotics

The implementation of novel food products to obtain health benefits has gained worldwide attention. Applications of food products in preventative medicine in the form of nutraceutical prebiotics are an area of utmost importance in the current global scenario where malnutrition and contamination in food quality are serious issues to overcome (Vishvakarma et al., 2023). Nutraceuticals are food or food products that provide health benefits beyond basic nutrition (Damian et al., 2022). These products are typically derived from natural sources and contain biologically active compounds that have potential therapeutic or health-promoting properties. Nutraceuticals are often taken as supplements or included in functional foods and beverages (Chandra et al., 2022). The residual biomass that is obtained after industrial processing of agrobiomass contains manno-oligosaccharides (MOS), which can be successfully utilized in the production of nutraceutical prebiotics. The main backbone of mannans is composed of mannose residues that are linked via β-1,4 linkages. Different by-products from MOS are formed as a result of cleaving off the mannan backbone via enzymes called Endo β-1,4 mannanases (Hlalukana et al., 2021). Similarly, xylo-oligosaccharides have also been found to be obtained from lignocellulosic biomasses (Dong et al., 2023; Ravichandra et al. 2022).

An array of MOS by-products aids in stimulating the microbial population that dwell within the gastrointestinal tract. As a result of stimulation, the microbes tend to produce beneficial fermentation products that help in improving host health. Some oligosaccharides are broken down to yield nutraceutical components (Martins et al., 2023). Mannans are one of the most abundant hemicellulosic components in

TABLE 13.1
Different Agrobiomasses that have been Reported to Serve as Potential Substrates for Different Food Additives

Sl no.	Agro-Waste	Microbe	Food Additive	Application	Yield	Reference
1.	Olive mill waste	*Aspergillus niger*	Citric acid	pH regulator	3.6 g/L/day	Massadeh et al. (2022)
2.	Oil crop waste	*Yarrowia lipolytica*	Erythritol	Sweetener	185.4 mg/gds	Liu et al. (2019)
3.	Sugarcane vinasse	*Candida guilliermondii*	Xylitol	Sweetener	10.2 ± 1.12 g/L	Hor et al. (2022)
4.	Corn straw waste	*Blastoschizomyces* spp. and *Kluyveromyces marxianus*	Ascorbic acid	Antioxidant	5.4 and 4.2 g/L	Kareem et al. (2019)
5.	Sugarcane bagasse	*Pseudomonas aeruginosa*	Biosurfactant	Emulsifier	0.92 ± 0.06 g/L	Das and Kumar (2019)
6.	Coconut coir	*Bacillus aryabhattai*	Vanillin	Flavouring agent	0.533 ± 0.03 g/100 g	Paul et al. (2023)
7.	Soybean meal	*Rhodopseudomonas faecalis*	Carotenoids	Antioxidant Biopigment	1.17 ± 0.01 g/L	Patthawaro et al. (2020)
8.	Wheat straw and pinewood sawdust	*Paracoccus aminophilus CRT1* and *Paracoccus kondratievae CRT2*	Carotenoids	Biopigment	631.33 and 758.82 µg/g	Pyter et al. (2022)

agro-wastes such as palm kernel press cake and spent coffee grounds. The prebiotics selectively stimulate probiotic microbial strains that include *Lactobacilli, Bifidobacterium* sp., and *Saccharomyces boulardii*, preventing the colonization of toxin-producing microbes such as *Escherichia coli, Clostridium, Streptococcus pneumonia*, etc. Prebiotics are primarily carbohydrates or sugar alcohols (Saini et al., 2022). Various groups have individually established the roles played by mannanases derived from *Talaromyces trachyspermus, Aspergillus quadrilineatus*, and *Streptomyces* sp. to have successfully produced efficient mannanases from different agro-wastes such as coffee wastes, copra meal, and reducing sugars (mannotetrose, mannopentose, mannobiose, and other oligosaccharides) (Narisetti et al., 2022).

13.2.2 Applications and Health Benefits of Common Food-Grade Agents

The fundamental purpose of using food additives is to increase the quality of food in terms of nutritional value as well as physical appearance. Food-grade agents have been extensively used for getting health benefits as well. The major industrial applications of food additives are in the food and pharmaceutical industry. Due to the richness of sources within agro-industrial wastes, they can be considered as suitable substrates to be turned into essential bioactive compounds. Valorization is achieved via various physical, chemical, biological, and other pretreatment methods. Microbial biotechnology and nanotechnology have played pioneer roles in the conversion of waste agro-biomass into cost-effective value-added products in pharmaceutical sectors, organic fertilizers, vermicompost, nutraceutical prebiotics, and biofuels (Bala et al., 2023).

Bioactive compounds play essential role in enhancing nutritional value of food supplements as well as have medicinal properties and thus increase the overall quality of conventional foods (Garcia et al., 2022). Often they contribute to increase shelf life of dietary supplement, thus increasing economic suitability and convenience to consumers. The pharmacological relevance of these products makes them commercially suited for manufacturing and marketing. Examples of commonly used food additives of pharmaceutical importance include antimicrobials, antioxidants, and antibiotic grade products. While nutraceuticals, bio-nanoparticles, biodiesels, and phytoactive compounds can be obtained after extraction processes, pretreatment and hydrolysis are required for the generation of bioenzymes and bioethanol. On the other hand, biogas and biofertilizers can be obtained only after anaerobic digestion. Commonly targeted industries are:

- Fruits and vegetable industries
- Oleaginous industries
- Cereal industry
- Winery industry

  Peels and pomaces, seeds, pulps, molasses, fibres, and bagasses are generally obtained as waste by-products of these industrial units (Bala et al., 2021). A variety of food additives and bioactive components can be obtained from these wastes. Bioactive compounds include carotenoids, phytosterols, flavonoids, essential oils, tocopherols, organosulfur compounds, and tannins. While food additives like prebiotics, fortifying compounds, dairy supplements, bakery products, vegetable oils, etc. can be obtained from these wastes.

Food additives of pharmaceutical importance have been extracted from different sources and they find a wide range of applications via various sustainable technological approaches adapted by food technologists worldwide. These may include antibiotics, anticancer, and antioxidants.

Antibiotics aid in killing germs at significantly low doses. Agricultural by-products laden with nutritional components are rich sources that can serve as substrates for the production of antibiotics. Strains such as *Streptomyces speibonae* OXS1 produced oxytetracycline from cocoyam peels (domestic food waste) as reported by Ezejiofor et al. (2012). Other examples may include Rifamycin B and tetracycline that have been obtained from fermentation of low-cost carbon wastes such as oil pressed cake and peanut shells, respectively (Vastrad et al., 2011; Ezejiofor et al., 2012). Applications of the desired fermentation techniques for individual waste type and microbes involved in the biconversion can facilitate the large-scale production of valuable antibiotic components. Solid-state fermentation approaches have emerged to become one of the best adapted approaches for the production and optimization of the fermentation bioprocess, which further ensure economic suitability of the process. The variety of sugars, organic and inorganic compounds, and phenolics derived from agro-waste residues have special immunomodulatory, anti-inflammatory, and antimicrobial properties. Example of one such waste is lemon and orange peels and pomaces, which are commonly derived from the orange and lemon juice industries; these are rich sources of flavonoids, hydroxycinnamic acids, flavanone glycosides, flavanone, and flavone aglycons (Fermoso et al., 2018). Chlorogenic acid is one of the major by-products from potato waste with pharmaceutical properties (Joly et al., 2021).

The antioxidant properties of fruit seeds and peels have widely been explored till date. The presence of a high amount of polyphenolic compounds contributes to the antioxidant potential of the sources. Examples may include compounds like punicalin, punicalagin, and ellagitannins (Mourtzinos et al., 2019). They are also reported to serve as therapeutically active components that help in the management of cardiovascular diseases, inflammatory and degenerative disorders (Jimenez et al., 2020). Vitamin C (ascorbic acid) is another food additive that has antioxidant effects and has been reported to have been produced by different microbial strains (mentioned in Section 13.2.2). Besides, biopigments like carotenoids, prodiogiosins, lycopenes, etc. have been reported to have anticancer properties as well (Majumdar et al., 2020). These not only help in enhancing the outer appearance of food by imparting colour, but they also provide health benefits and save consumers from the adverse effects of chemical colouring agents.

13.3 MICROBES INVOLVED IN BIOTRANSFORMATION OF AGROBIOMASS INTO FERMENTED FOOD-GRADE AGENTS

Microbial strains that include divergent genus of fungus and bacteria that engage in converting agrobiomasses into valorizable components via fermentation can be key tools in transformation of waste to wealth. Bioconversion of agro-wastes to form new metabolites that can be beneficial to the food industry has also achieved a lot of success in past few years, given the safety and sustainability of the bioproducts and the bioprocesses. Large-scale implementations of these fermentative technologies for the industrial production of valorizable compounds can pioneer and contribute to the modern implications of microbial fermentation technology in applied bioprocessing and industrialization of the fermented bioproducts for benefit of the population as a whole. The emerging utilities of fermenter technology can comprehend bacteria, fungi, and consortium of multiple microbes. Depending on their specifications, parameters are optimized, nutritional sources are standardized, and keeping economic suitability in mind, downstream processing is done.

13.3.1 Bacterial Monocultures

Given the robust nature of bacterial strains to tolerate wide range of physiological conditions like pH, temperature, salinity, etc., lesser incubation time, flexibility to grow in varied nutritive media, and easier downstreaming, these are preferable over fungal alternatives. A variety of different food additives can be produced by fermentation technology using such robust strains isolated from different sources and can be implemented for large-scale production (Nain et al., 2022). One of the commonest examples of food additive is cellulose polymer/nanocellulose used as gelling, setting, food fixing agent or fat replacer, and bacterial cellulose is an excellent alternative (Oliveira et al., 2021). Production of vanillin as a flavouring agent has been reported by different bacterial strains like *Bacillus* sp. by Paul et al. (2023). Das and Kumar et al. (2019) have shown the successful production of biosurfactant using sugarcane bagasse as substrate mediated by *Pseudomonas aeruginosa.* Acid-producing strains have gained attention in this regard due to the broad implications of organic acids as pH stabilizers which play important roles in maintaining texture and shelf life of the food. Besides, biopigments can be utilized as food colours that enhance the appearance of food (Ueda et al., 2022). Carotenoid production has been reported from soyabean meal by *Rhodopseudomonas faecalis*, while prodigiosin pigment has been produced from lignocellulosic biomass wastes by *Serratia marcescens* (Patthawaro et al., 2020; Majumder et al., 2020).

13.3.2 Fungus

Fungal strains possess the advantage of higher yield per unit of biomass as compared to bacterial production. Many strains belonging to diverse genus like *Aspergillus niger* producing antioxidants, *Monascus purpureus* producing pigments, and *Yarrowia lipolytica* producing lactic acid, propionic acids, and sweeteners like erythritol have been utilized for large-scale production of these essential food additives

(Torres- Leon et al., 2019; Jirasatid et al., 2019; Wang et al., 2019; Liu et al., 2019). The robust nature of fungal strains has made them preferable option for propagation. *Candida guilliermondii* is one of the well-reported fungal strain known for its ability to produce xylitol which is widely used as sweetener (Hor et al., 2022). *Monascus ruber* is another fungal variety that is reportedly involved in the biopigment production using seven different agro-industrial wastes (Sayed et al., 2022). Economically important fungal strains have been isolated from waste materials itself due to their metabolic capacity to utilize the wastes as their sole source of nutrients and this characteristic is exploited in their fermentation applications for production of these industrially marketable products (Troiano et al., 2022).

13.3.3 Consortium Cultures

The use of pure monocultures has been in practice all these years, but for better outcomes, yield, and faster fermentation, consortium cultures are gaining popularity (Schwalm et al., 2019). A consortium culture containing multiple strains from *Lactobacillus* sp. showed successful production of lactic acid from food waste (Chenebault et al., 2022). A bacterial consortium co-culture of two bacterial strains namely *Alternaria alternata* and *Alternaria ornatus* yielded 0.45 mg/mL of citric acid using agro-waste materials by solid-state fermentation technology. Similar studies have been reported by Schwalm et al. in 2019 who reportedly developed a consortium culture comprised of *Clostridium beijerinckii* and *Yokenella regensburgei* for large-scale metabolite production from food waste that included food additive supplements like lactate and butyrate via anaerobic fermentation technology. In a report by Pyter et al. (2022), *Paracoccus aminophilus* CRT1 and *Paracoccus kondratievae* CRT2 were shown to produce valuable biopigment, carotenoids, from cheap agricultural wastes, such as wheat straw and pie sawdust. The use of bacterial consortium for generation of food additives from agro-wastes still remains in its infancy and paves path for researchers to explore more in this area. When it comes to consortium cultures, fungal strains too can form co-metabolic partners to utilize common substrate and produce valuable substances. A unique combination of *Blastoschizomyces* spp. and *Kluyveromyces marxianus* have been shown to produce valuable antioxidant ascorbic acid using corn straw waste (Kareem et al., 2019). The diversity in metabolic infrastructure of microbes present within a community can be utilized in favour of driving a bioprocess with increased efficiency due to the participation of multiple enzymes and the interplay among their systems. Standardization of microbial consortium cultures is a challenging task to achieve, but this strategy can lead to maximal results in short time span.

13.4 CHEMICAL COMPOSITION OF AGRO-WASTE BIOMASS

Agro-waste biomass mainly consists of three polysaccharides – cellulose, hemicellulose, lignin, and some soluble sugars. The percentage of each component varies between different biomasses and their pretreatments and downstreaming that have undergone during industrial processing (Nazar et al., 2022). Where wastes such as rye straw (19.0%–30.8%), soyabean stalks (19.8%), and sunflower seed hulls (29.4%) are high in lignin content as compared to corn cobs (6.1%), rice straw (8.3%–9.9%), and barley straw (9.6%–13.8%), which are much lesser in lignin. Similarly, some are rich in cellulose content (e.g., apple pomace, sugarcane bagasse, and olive mill waste) compared to others that have lesser cellulosic component (e.g., grape pomace, pumpkin oil cake, wheat bran, and flax oil cake). The more the content of lignin, it is more recalcitrant to deconstruction (Hanumante et al., 2022). Besides these, the waste biomass also contains other components that are much richer in nutrients and may include lipids, proteins, pectins, and polyphenols. Depending upon the chemical composition, the microbial strains are selected in order to ensure maximum utilization of the biomass as substrate, and this in turn depends on different classes of enzymes that are being expressed in those microbes at the time of fermentation (Selo et al., 2021).

13.5 ADVANCED FERMENTATION TECHNOLOGY FOR BIOCONVERSION OF AGROBIOMASS WASTES INTO VALORIZABLE FOOD-GRADE AGENTS

The complex enzymatic pool of microorganisms is involved in bioconversion/biotransformation of the agro-waste lignocellulosic biomass into valorizable food additives that are fermentable sugars, sugar acids/phenolic derivatives, etc. in terms of chemical nature (Selo et al., 2021; Ueda et al., 2022; Saroj et al., 2018). These mainly include the lignocellulolytic, cellulolytic, and hemicellulolytic enzymes (Figure 13.2).

The structure of lignin is a complex one, with an aromatic branched skeleton. It is an amorphous hydrophobic substance and optical in nature. This component is mainly responsible for the structural rigidity in plants and their percentage varies depending on the species, space localization, and maturity of the plant. It is impermeable, recalcitrant by nature, and found to be tolerant to microbial attacks. Microbes need a very strong enzymatic system to utilize lignin as its carbon source which may consist mainly of peroxidases and laccases (Ahmad et al., 2022). Laccases are multi-copper

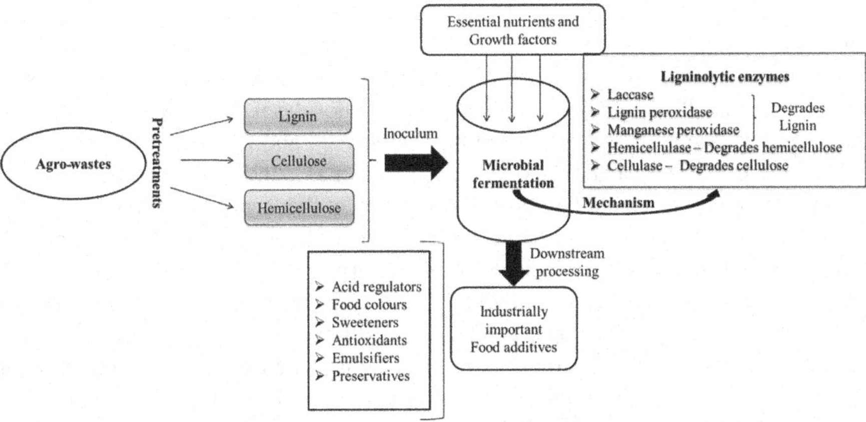

FIGURE 13.2 Overview of microbial fermentation technology for production of food-grade agents from agro-waste.

glycoproteins in nature (Selo et al., 2021). They generally utilize molecular oxygen and mechanistically drive their reactions via a radical-catalyzed reaction. Laccases find their possible applications in multiple industries including pulp and paper (delignification), textile (bleaching of dyes), and food and beverage (colour modification). Many microorganisms such as some fungal species (*Trametes versicolor, Trametes pubescens, Pleurotus eryngii*, etc.) are well-known laccase producers.

Lignin peroxidases (LiP) use H_2O_2 for catalysis reaction and are water soluble glycosylated in nature. Some fungal species like *Ionotus obiquus* are versatile producers of all categories of ligninolytic enzymes when subjected to solid-state fermentation (Contreras et al., 2023; Xu et al., 2018). Different lignocellulosic biomasses like sugarcane bagasse, peanut shells, cassava peel offs, wheat bran, etc. have been shown to be bioconverted to valuable food by-products by microbial fermentation technology. This mainly exploited the presence of ligninolytic enzyme produced by these strains under optimum physiological conditions and required nutritional demands.

Fungal strains such as *Ascomycetes* and *Basidiomycetes* produce manganese peroxidases. These are glycosylated heme enzyme and are extracellular in nature. They work by chelation of manganese ions which in turn use H_2O_2.

Bacterial strains such as *Pseudomonas fluorescens, Pseudomonas aeruginosa, Planococcus* sp., and *Serratia marscecesns* are well-known producers of ligninolytic enzymes and can be exploited to produce different food-grade additives, such as aromatic compounds and vanilla essence vanillin, as well as biopigments and emulsifiers from lignocellulosic food/agro-waste substrates like sugarcane bagasse, peel offs, pomaces, etc. (Majumdar et al., 2020).

Cellulases work by breaking the linkages and converting long chain cellulosic structures into simple sugar units. This makes it easier for further fermentation by the help of same microbes that produce cellulases or different sets of microbes. Cellulolytic enzymes can be categorized as exoglucanases/cellobiohydrolases, endoglucanases, and β-glucosidases/cellobioses. One of the very popular genus that produce cellulolytic enzymes is *Trichoderma* sp. Others like *Trichoderma viridae, Ganoderma lucidum*, and *Streptomyces* sp. have been found to have utilized cheap substrate corn stover waste to produce valuable cellulases (Shreshtha et al., 2022; Shahzadi et al., 2014; Ilic et al., 2023).

Hemicellulases are less prevalent for industrial utilization except in some cases. These include β-mannanases, xylanases, β-xylosidases, arabinofuranosidases, etc. Different bacterial species including *Bacillus* sp. and *Bacillus halodurans* are reported to produce xylanases and laccases (Mou et al., 2023; Gupta et al., 2015).

Special emphasis should be given to scaling up of the fermentation process, since application on a large scale is possible when bulk production is done. Designing proper fermenter, optimizing physiological parameters (pH, temperature, aeration speed, etc.), and choosing the correct microbial strain and substrate are important criteria for the scale-up strategies (Verma et al., 2022). Although the wild strains are capable of producing the particular products, genetic engineering can produce high yield of up to 10.2 ± 1.12 g/L (Xylitol by *Candida guilliermondii* using sugarcane vinasse), as demonstrated by Hor et al. (2022). Other additives like citric acid (pH regulators) have been produced by *Aspergillus niger* with a yield capacity of 3.6 g/L/day using olive mill waste (Massadeh et al., 2022). High production capacity of biopigment carotenoids of up to 1.17±0.01 g/L has been shown by *Rhodopseudomonas faecalis*, utilizing soyabean meal as a substrate (Patthawaro et al., 2020). Despite all the efforts, the mass production of food additives is still in its infancy and has an area of huge research scope.

With the advancement of genetic engineering, the manipulation of genes to get better yield and quality of value-added components is an area of research that has become much popular. Food-grade agents have been produced in increased volumes by altering levels of gene expression of related genes or untargeted gene mutations within the target microbes along with compelling them to utilize cheap substrates like agro-waste materials. Altered gene expression or genetic manipulation of lignocellulolytic genes can lead to cheap agro-waste utilization (Nargotra et al., 2022). Genes responsible for producing enzymes involved in sugar alcohol production, which have been in practice over the last decade now (xylitol, mannitol, erythritol, etc.), can be engineered for enhancing their yield from respective substrates, as shown by Park et al. back in 2016. Raman et al. (2022) have reported genetically modified lactic acid-producing bacteria (via CRISPR CAS-based genome editing technology) to produce higher yields of polylactate, which can make them promising agents for soil bioremediation as well as active ingredients of biofertilizers. This can be further implemented for food-grade lactic acid. Genetically engineered *E. coli* has been used in the de novo synthesis of vanillic acid through the shikimate cascade. The synthesized vanillic acid is then converted to vanillin through a reduction reaction. The next-generation genetic and fermentation engineering has also been used for the production of colouring (lycopene, indigoidine, violacein carminic acid, and leghemoglobin) and flavouring (butyl butyrate, limonene, vanillin, humulone, and taurine) agents as shown by Seo and Jin (2022).

13.6 CHALLENGES FACED DURING THE PRODUCTION PROCESS

The major focus in the utilization of wastes and by-products for any purpose is the bioconversion into valorizable products to meet the current market needs and making it economically friendly for the industries as well as consumers (Selo et al., 2021; Ueda et al., 2022). Besides, there is a high demand to meet the cleaner, greener sustainable goals in due course of the whole process. There is a requirement for toxicity analysis of the produce after completion of downstreaming to ensure the safety and reliability for consumption.

The key challenges faced during the valorization of wastes include the following:

13.6.1 Choosing the Correct Substrate

It is important to understand the possibility of getting beneficial products from a particular substrate, depending on the nutritional richness of the source. Not all wastes contain enough components to serve the purpose. Besides, there are several factors that determine the potential of a substrate that includes climate, amount of rainfall, geographical location, type of soil, and growth conditions. The same type of substrate may vary from place to place depending upon all these factors. Hence, choosing the correct substrate depending on the purpose is extremely important to drive the feasibility and applicability of the process. Substrates like lemon and orange peels and pomaces can be excellent substrates for the production of ascorbic acid (Vitamin C) that has remarkable antioxidant potential (Banjo et al., 2018; Kareem et al., 2019). Lignocellulosic biomasses such as sugarcane bagasse and potato peel offs can be targeted for the production of vanillin (vanilla essence) which has great applicability in the bakery and beverage industry and is a well marketed product. Substrates rich in carbohydrates and sugar alcohols can serve as carbon-rich source for the production of many other food supplements such as biopigments, biosurfactants, humidifiers, and sweeteners (Priyadarshinee et al., 2019; Majumdar et al., 2020). Depending upon the purpose and type of product desired, the substrate needs to be chosen very carefully, which is one of the key factors that determine the possible outcome of the fermentation process.

13.6.2 Choice of Microbial Strain

Different microbes have different metabolic potentials, which are mainly driven by the type of enzymatic machinery that is actively expressed within its cellular pool. Two different microbial strains can utilize the same substrate in different ways to produce different fermented products. There is versatility among many microbes that can utilize multiple components of a substrate. Lignocellulosic biomasses which are composed of celluloses, hemicelluloses, and lignin are utilized by microbes that have the ability to produce cellulases, hemicellulases, and ligninolytic enzymes (lignin peroxidase, manganese peroxidase, laccase, etc.). Some microbes can produce multiple enzymes and are more promising in terms of metabolite production. Common food additives derived from lignocellulosic biomasses include biopigments (prodigiosin, carotenoids, lycopene, etc.), biosurfactants, bioemulsifiers, vanillin, and phenolic compounds. Bacterial strains like *Pseudomonas* sp., *Serratia* sp., *Planococcus* sp., *Bacillus* sp., etc. and fungal strains such as *Aspergillus*, *Pleurotus*, and *Phanerochaete*, are well-known utilizers of several lignocellulosic biomasses (Majumdra et al., 2020; Paul et al., 2022). The basic mechanism of these fermentation processes (solid-state fermentation or submerged fermentation) is the fact that the microbial strain chosen must utilize the substrate as its carbon source and participate in the bioconversion of the substrate into valorizable components.

13.6.3 Process Optimization

Any bioprocess or fermentation technology must be designed in a way that brings out maximum benefit with minimum investment of time, energy, and resources. Different physiological parameters that determine the practical applicability of the fermentation process include temperature, pH, aeration speed, inoculum volume, and addition of nutrients and growth factors. There are experimental approaches via different optimization strategies (One variable at a time – OVAT, Response surface methodology – RSM, Plackett Burman method, etc.) that help in choosing the correct physiological and nutritional parameters for getting the maximum yield out of the process with minimum investment (Javed et al., 2022; Raina et al., 2022). This helps in the maximum utilization of resources with minimum investments. Proper optimization of process parameters is tedious and is an ultimate challenge for fermentation and food technologists globally. New developments are taking place every day and new technologies are getting invented to ensure health benefits for consumers along with considerable economic benefit for the industries.

13.6.4 Toxicity Assessment

When it comes to edible products, the first parameter that needs to be satisfied is its safety and non-toxicity. Since the process of microbial fermentation involves the use to microbial strains and several other chemical components (solvents, reagents, minerals, etc.) in the bioprocess, ensuring its safety is a matter of great concern. The bioproduct needs to undergo safety assessments both in plant and animal systems to confirm that it has no adverse impact on the environment, animal, and human health (Paul et al., 2022). Maintaining the non-toxic nature of the product is a big challenge for biotechnologists without which regularization, marketing, and industrial production will not be approved.

13.6.5 Economic Suitability

The biggest challenge concerning large-scale implementation of any production technology is to make it economically suitable. Production optimization and ensuring maximum yield are keys to achieve cost-effective outcomes (Diaz & Blandino, 2022). There is an added advantage of working with waste substrates since the cost of initial raw material is nil in those cases. Choosing wastes as substrates is in itself a fundamental strategy towards green sustainable cost-friendly production of valorizable components. Choosing the proper strain that consumes less time and nutrients adds to the minimum exploitation of resources, energy, and manpower to reach adequate supply in a cost-effective manner. It is a big challenge for biotechnologists worldwide to be particular in keeping all parameters at par with cost involved in the process to achieve maximum benefit.

13.7 FUTURE PROSPECTS

Agro-wastes, generated from varied industrial sources, are increasing in quantity with growth in industrialization and increasing global food requirements. The bulk amount of nutritionally enriched dumping material can be utilized to produce valuable components for the mass. This will serve two primary purposes. Firstly, it will aid in environmental clean-up by reducing the waste burden of colossal industrial units that would otherwise cause pollution and acquire space, and the never-ending collection of wastes will not provide anything beneficial to the population as a whole. Secondly, waste treatment and valorization of the waste material can open new insights to form a parallel industrial segment that will help in increasing the economic turnover of the industries and the nation as a whole. The cost of waste material is zero/minimum, which makes the bioprocess a cost-effective approach to achieve sustainable goals. Biotechnologists around the globe will find a key solution to manage waste and a parallel increase in the overall economic output. The biologically active components that can be generated from the bioconversion process can have utility in multiple fields, with food industry being one of the most relevant sectors. Different food additives derived from agro-industrial wastes will help in increasing food quality and thus will pave a new pathway in the conversion of waste into wealth. New microbial strains that can contribute in the bioconversion process can be isolated, explored, and genetically engineered for better outcomes, and the optimization of process parameters can be done to get the maximum benefits out of minimum investments. Designing fermenters according to the requirements of the microbe as well as the nature of waste is a key challenge faced by researchers. Standardization of such strategic approaches can help individual industries to utilize their own waste materials for the improvement of their own products or generate new products, thus unleashing the potential of biological agents to get industrial and environmental benefits in the future.

13.8 CONCLUSION

The utilization of advanced microbial fermentation technology for the production of food-grade agents from inexpensive agricultural biomass represents a promising approach to address the growing demands of the food industry. As a result, such wastes need to be recycled, reprocessed, and reused via suitable and appropriate techniques to produce value-added products that will be viable, economic, and environmentally friendly. These strategies not only contribute to reducing the adverse impacts associated with conventional chemical processes, but also provide economic benefits by efficiently converting readily available biomass resources into valuable food additives. This chapter has explored various aspects of this technology, including the selection of suitable microorganisms, optimization of fermentation conditions, and downstream processing techniques. The versatility and adaptability of microbial fermentation in transforming diverse agricultural feedstocks into a wide range of food-grade agents such as enzymes, organic acids, and flavour compounds have been highlighted. Moreover, the use of agricultural biomass as a feedstock for microbial fermentation not only mitigates waste and boosts sustainable agricultural practices but also increases food security by diversifying the sources of food additives. Novel and customized product development to meet specific market demands ensuring the availability of high-quality food ingredients for consumers is the key aspect of this work. It is imperative to continue advancing our understanding of microbial physiology, genetic engineering techniques, and process optimization to further improve the efficiency, yield, and quality of food-grade agents produced through microbial fermentation. By harnessing the power of microorganisms to convert cheap agricultural biomass into valuable food-grade agents, a greener and more resilient food supply chain along with meeting the evolving needs of consumers is ensured.

REFERENCES

Abedi, E., Seyed Mohammad Bagher, H. (2020). Lactic acid production - producing microorganisms and substrates sources-state of art. *Heliyon*, 6, e04974.

Ahmad, N., Aslam, S., Hussain, N., et al. (2022). Transforming lignin biomass to value: Interplay between ligninolytic enzymes and lignocellulose depolymerization. *BioEnergy Research*, 16, 1–18.

Ali, S. R., Anwar, Z., Irshad, M., Mukhtar, S., Warraich, N. T. (2016). Bio-synthesis of citric acid from single and co-culture-based fermentation technology using agro-wastes. *Journal of Radiation Research and Applied Sciences*, 9, 57–62.

Atta, O. M., Manan, S., Shahzad, A., Ul-Islam, M., Ullah, M. W., Yang, G. (2022). Biobased materials for active food packaging: A review. *Food Hydrocolloids*, 125, 107419.

Awasthi, M. K., Sindhu, R., Sirohi, R., et al., (2022) Agricultural waste biorefinery development towards circular bioeconomy. *Renewable and Sustainable Energy Reviews*, 158, 112122.

Bala, S., Garg, D., Sridhar, K., Inbaraj, B. S., Singh, R., Kamma, S., Tripathi, M., Sharma, M. (2023). Transformation of agro-waste into value-added bioproducts and bioactive compounds: Micro/nano formulations and application in the agri-food-pharma sector. *Bioengineering (Basel)*, 10(2), 152. https://doi.org/10.3390/bioengineering10020152.

Bala, A., Raugei, M., Teixeira, C. A., Fernández, A., Pan-Montojo, F., Fullana-i-Palmer, P. (2021). Assessing the environmental performance of municipal solid waste collection: A new predictive LCA model. *Sustainability*, 13(11), 5810. https://doi.org/10.3390/su13115810

Bali, V., Panesar, P. S., Bera, M. B. (2014). Trends in utilization of agro-industrial byproducts for production of bacteriocins and their biopreservative applications. *Critical Reviews in Biotechnology*, 36, 1–11.

Banjo, T., Kareem, S., Popoola, T., Akinloye, O. (2016). Microbial production of ascorbic acid from brewery spent grain (BSG) by *Aspergillus flavus* and *Aspergillus tamarii*. *Food and Applied Bioscience Journal*, 6(2), 93–105.

Chandra, S., Saklani, S., Kumar, P., Kim, B., Coutinho, H. D. (2022). *Nutraceuticals: Pharmacologically active potent dietary supplements. BioMed Research International*, 2022, 2051017.

Chenebault, C., Roman, M., Eric, T., Renaud, E., Benjamin, P., (2022). Lactic acid production from food waste using microbial consortium: Focus on key parameters for process upscaling and fermentation residues valorization. *Bioresource Technology*, 354, 127230.

Contreras, E., Flores, R., Gutiérrez, A., et al. (2023). Agro-industrial wastes revalorization as feedstock: Production of lignin-modifying enzymes extracts by solid-state fermentation using white rot fungi. *Preparative Biochemistry & Biotechnology*, 53(5), 488–499.

Cuadrado-Osorio, P. D., Ramírez-Mejía, J. M., Mejía-Avellaneda, L. F., et al., (2022). Agro-industrial residues for microbial bioproducts: A key booster for bioeconomy. *Bioresource Technology Reports*, 20, 101232.

Damián, M. R., Cortes-Perez, N. G., Quintana, E. T., et al., (2022). Functional foods, nutraceuticals and probiotics: A focus on human health. *Microorganisms*, 10(5), 1065.

Das, N., P. K. Jena, D. Padhi, M. Kumar Mohanty, G. Sahoo (2021). A comprehensive review of characterization, pretreatment and its applications on different lignocellulosic biomass for bioethanol production. *Biomass Conversion and Biorefinery*, https://doi.org/10.1007/s13399- 021-01294-3.

Dey, S., Nagababu, B. H. (2022). Applications of food color and bio-preservatives in the food and its effect on the human health. *Food Chemistry Advances*, 1, 100019.

Diaz, A. B., Blandino, A. (2022). Value-added products from agro-food residues. *Foods*. 11(5), 766. https://doi.org/10.3390/foods11050766

Dong, Z., Bai, X., Xu, D., Li, W. (2023). Machine learning prediction of pyrolytic products of lignocellulosic biomass based on physicochemical characteristics and pyrolysis conditions, *Bioresource Technology*, 367, 128182, ISSN 0960-8524, https://doi.org/10.1016/j.biortech.2022.128182.

El-Sheshtawy, H. S., Fahim, I., Hosny, M., El-Badry, M. A. (2022). Optimization of lactic acid production from agro-industrial wastes produced by *Kosakonia cowanii. Current Research in Green and Sustainable Chemistry*, 5, 100228.

Ezejiofor, T. I. N., Duru, C. I., Asagbra, A. E., Ezejiofor, A. N., Orisakwe, O. E., Afonne, J. O., Obi, E. (2012). Waste to wealth: Production of oxytetracycline using Streptomyces species from household kitchen wastes of agricultural produce. *African Journal of Biotechnology*, 11, 10115–10124.

Fermoso, F. G., Serrano, A., Alonso-Fariñas, B., Fernandez-Bolaños, J., Borja, R., Rodríguez-Gutiérrez, G. (2018). Valuable compound extraction, anaerobic digestion, and composting: A leading biorefinery approach for agricultural wastes. *Journal of Agricultural and Food Chemistry*, 66, 8451–8468.

Gomathi, S., Rameshpathy, M. (2023). Valorization of agro-waste residues into bio-vanillin a comprehensive review. *Industrial Crops and Products*, 205, 117522.

Grewal, J., Mikołaj Woła̧, C., Weronika, P., Namrata, J., Lukasz, D., Kumar, P. (2022). Colourful treasure from agro- industrial wastes: A sustainable chassis for microbial pigment production. *Frontiers in Microbiology*, 13, 832918.

Guía-García, J. L., Charles-Rodríguez, A. V., Reyes-Valdés, M. H., Ramírez-Godina, F., Robledo-Olivo, A., García-Osuna, H. T., Cerqueira, M. A., Flores-López, M. L. (2022). Micro and nanoencapsulation of bioactive compounds for agrifood applications: A review. *Industrial Crops and Products*, 186, 115198.

Gupta, V., Garg, S., Capalash, N., Gupta, N., Sharma, P. (2015). Production of thermo-alkali-stable laccase and xylanase by co- culturing of Bacillus Sp. and B. Halodurans for biobleaching of kraft pulp and deinking of waste paper. *Bioprocess and Biosystems Engineering*, 38, 947–956.

Hafez, A. I., Ali, H. M., Sabry, R. M., et al., (2023). Generation of novel, hygienic, inhibitive, and cost-effective nanostructured Core-shell pigments. *Progress in Organic Coatings*, 175, 107325.

Hanumante, N., Maitre, N. (2023). Can agro-biomass valorization be the fulcrum for sustainable development? *Frontiers in Agronomy*, 5, 1203543.

Hlalukana, N., Magengelele, M., Malgas, S., Pletschke, B. I. (2021). Enzymatic conversion of mannan-rich plant waste biomass into prebiotic mannooligosaccharides. *Foods*, 10(9), 2010. https://doi.org/10.3390/foods10092010.

Hor, S., Kongkeitkajor, M. B., Reungsang, A. (2022). Sugarcane bagasse-based ethanol production and utilization of its vinasse for xylitol production as an approach in integrated biorefinery. *Fermentation*, 8, 340

Ilić, N., Milić, M., Beluhan, S., Dimitrijević-Branković, S. (2023). Cellulases: From lignocellulosic biomass to improved production. *Energies*, 16(8), 3598.

Iram, A., Ozcan, A., Turhan, I., Demirci, A. (2023). Production of value-added products as food ingredients via microbial fermentation. *Processes*, 11(6), 1715.

Jameel, M., Umar, K., Parveen, T., Ismail, I. M. I., et al., (2023). Extraction of natural dyes from agro-industrial waste. In: Bhawani, S. A., Khan, A., Ahmad, F. B. (Eds), *Extraction of Natural Products from Agro-Industrial Wastes*, pp. 197–216. Elsevier, Amsterdam, The Netherlands.

Javed, S., Azeem, M., Mahmood, S., Al-Anazi, K. M., Farah, M. A., Ali, S., Ali, B. (2022). Biotransformation of agricultural wastes into lovastatin and optimization of a fermentation process using response surface methodology (RSM). *Agronomy*, 12(11), 2848.

Jimenez-Lopez, C., Fraga-Corral, M., Carpena, M., García-Oliveira, P., Echave, J., Pereira, A. G., Lourenço-Lopes, C., Prieto, M. A., Simal-Gandara, J. (2020). Agriculture waste valorisation as a source of antioxidant phenolic compounds within a circular and sustainable bioeconomy. *Food & Function*, 11, 4853–4877.

Jirasatid, S., Limroongreungrat, K., Nopharatana, M. (2019). Monacolin K, pigments and citrinin of rice pasta byproducts fermented by Monascus purpureus. *International Food Research Journal*, 26, 1279–1284.

John, R. P., Nampoothiri, K. M., Pandey, A. (2006). Solid-state fermentation for L-lactic acid production from agro wastes using *Lactobacillus delbrueckii*. *Process Biochemistry*, 41, 759–763.

Joly, N., Souidi, K., Depraetere, D., Wils, D., Martin, P. (2021) Potato byproducts as a source of natural chlorogenic acids and phenolic compounds: Extraction, characterization, and antioxidant capacity. *Molecules*, 26, 177.

Kareem, S. O., Akpan, I., Alebiowu, O. O. (2010). Production of citric acid by Aspergillus niger using pineapple waste. *Malaysian Journal of Microbiology*, 6(2), 161–165.

Kareem, S. O., Banjo, T. T., Omisore, A. E. (2019). Production of ascorbic acid by *Blastoschizomyces spp* and *Kluyveromyces marxianus* from corn straw waste using submerged fermentation. *Crawford Journal of Applied and Natural Science*, 1, 29–37.

Karthikeyan, A., Sivakumar, N. (2010). Citric acid production by Koji fermentation using banana peel as a novel substrate. *Bioresour Technol.*, 101(14), 5552–5556. https://doi.org/10.1016/j.biortech.2010.02.063. Epub 2010 Mar 9. PMID: 20219361.

Kumar, K., Singh, E., Shrivastava, S. (2022). Microbial xylitol production. *Applied Microbiology and Biotechnology*, 106(3), 971–979.

Leong, H. Y., Chang, C.-K., Khoo, K. S., et al., (2021). Waste biorefinery towards a sustainable circular bioeconomy: A solution to global issues. *Biotechnology for Biofuels*, 14, 87.

Li, N., An, X., Xiao, X., An, W., Zhang, Q. (2022). Recent advances in the treatment of lignin in papermaking wastewater. *World Journal of Microbiology and Biotechnology*, 38(7), 116.

Liu, X., Yan, Y., Zhao, P., Song, J., Yu, X., Wang, Z., Xia, J., Wang, X. (2019). Oil crop wastes as substrate candidates for enhancing erythritol production by modified Yarrowia lipolytica via one-step solid state fermentation. *Bioresource Technology*, 294, 122194.

Liu, W. Y., Wang, X., Ren, J., et al., (2023). Preparation, characterization, identification, and antioxidant properties of fermented acaí (Euterpe oleracea). *Food Science & Nutrition*, 11(6), 2925–2941.

Majumdar, S., Paul, I., Dey, S., Dutta, S., Mandal, T., Mandal, D. D. (2020). Biotransformation of paper mill sludge by Serratia marcescens NITDPER1 for prodigiosin and cellulose nanocrystals: A strategic valorization approach. *Biochemical Engineering Journal*, 164, 107766. https://doi.org/10.1016/j.bej.2020.107766

Martínez-Miranda, J. G., Chairez, I., Durán-Páramo, E. (2022). Mathematical modeling characterization of mannitol production by three heterofermentative lactic acid bacteria, *Food and Bioproducts Processing*, 135, 11–32, ISSN 0960-3085, https://doi.org/10.1016/j.fbp.2022.06.003.

Martins, M., Ávila, P. F., Poletto, P., Goldbeck, R. (2023). Polysaccharide degradation for oligosaccharide production with nutraceutical potential for the food industry. In: Goldbeck, R., Poletto, P. (Eds), *Polysaccharide Degrading Biocatalysts*, pp. 335–363. Academic Press, London.

Massadeh, M. I., Fandi, K., Al-Abeid, H., et al., (2022). Production of citric acid by aspergillus niger cultivated in olive mill wastewater using a two-stage packed column bioreactor. *Fermentation*, 8(4), 153. https://doi.org/10.3390/fermentation8040153

Mondala, A. H. (2015). Direct fungal fermentation of lignocellulosic biomass into itaconic, fumaric, and malic acids: Current and future prospects. *Journal of Industrial Microbiology and Biotechnology*, 42, 487–506.

Mores, S., de Souza Vandenberghe, L. P., Martinez-Burgos, W. J. et al., (2023). Simultaneous reuse and treatment of sugar-sweetened beverage wastes for citric acid production. *Journal of Food Science and Technology*, 60, 1–7.

Mou, L., Pan, R., Liu, Y., Jiang, W., et al., (2023). Isolation of a newly Trichoderma asperellum LYS1 with abundant cellulase-hemicellulase enzyme cocktail for lignocellulosic biomass degradation. *Enzyme and Microbial Technology*, 171, 110318.

Mourtzinos, I., Goula, A. (2019). Polyphenols in agricultural byproducts and food waste. In Watson, R. R., Ed. *Polyphenols in Plants: Isolation, Purification and Extract Preparation*. Academic Press: London, UK.

Nagtode, V. S., Cardoza, C., Yasin, H. K. A., et al., (2023). Green surfactants (Biosurfactants): A petroleum-free substitute for sustainability— comparison, applications, market, and future prospects. *ACS Omega*, 8(13), 11674–11699.

Nain, N., Katoch, G. K., Kaur, S., et al., (2022). Production of biopigments from agro-industrial waste. In: Anal, A. K., Panesar, P. S. (Eds), *Valorization of Agro-Industrial Byproducts*, pp. 143–160. CRC Press, Boca Raton, FL.

Nargotra, P., Sharma, V., Lee, Y. C., et al., (2022). Microbial lignocellulolytic enzymes for the effective valorization of lignocellulosic biomass: A review. *Catalysts*, 13(1), 83. https://doi.org/10.3390/catal13010083

Narisetty, V., Parhi, P., Mohan, B., Hazeena, S. H., et al., (2022). Valorization of renewable resources to functional oligosaccharides: Recent trends and future prospective. *Bioresource Technology*, 346, 126590.

Nayak, J., Basu, A., Dey, P., Kumar, R., et al., (2022). Transformation of agro-biomass into vanillin through novel membrane integrated value-addition process: A state-of-art review. *Biomass Conversion and Biorefinery*, 13, 1–24.

Oliveira, Z., Zhou, Y., Yue, W., Qin, W., Dong, H., Vasanthan, T. (2021). Nanostructures of protein-polysaccharide complexes or conjugates for encapsulation of bioactive compounds. *Trends in Food Science and Technology*, 109, 169–196.

Park, Y.-C., Oh, E. J., Jo, J.-H., Jin, Y.-S., Seo, J.-H. (2016). Recent advances in biological production of sugar alcohols. *Current Opinion in Biotechnology*, 37, 105–113.

Patthawaro, S., Lomthaisong, K., Saejung, C. (2020) Bioconversion of agro-industrial waste to value-added product lycopene by photosynthetic bacterium *Rhodopseudomonas faecalis* and its carotenoid composition. *Waste and Biomass Valorization*, 11, 2375–2386.

Paul, I., Mandal, T., Mandal, D. D. (2022). Assessment of bacterial biosurfactant production and application in Enhanced Oil Recovery (EOR)-A green approach. *Environmental Technology & Innovation*, 1(28), 102733.

Paul, V., Rai, N., Agarwal, A., Gautam, V., Tripathi, A. D. (2023). Valorization of lignocellulosic waste (coconut coir) for bio-vanillin production having antioxidant and anticancer activity against human breast cancer cells (MCF-7). *Industrial Crops and Products*, 205, 117502.

Pyter, W., Grewal, J., Bartosik, D., Drewniak, L., Pranaw, K. (2022). Pigment production by paracoccus spp. strains through submerged fermentation of valorized lignocellulosic wastes. *Fermentation*, 8(9), 440. https://doi.org/10.3390/fermentation8090440

Raina, D., Kumar, V., Saran, S. (2022). A critical review on exploitation of agro-industrial biomass as substrates for the therapeutic microbial enzymes production and implemented protein purification techniques. *Chemosphere*, 294, 133712.

Ramady, H., Brevik, E. C., Bayoumi, Y., et al., (2022). An overview of agro-waste management in light of the water-energy-waste nexus. *Sustainability*, 14(23), 15717. https://doi.org/10.3390/su142315717

Raman, J., Kim, J-S., Choi, K. R., Eun, H., Yang, D., Ko, Y-J., Kim S-J. (2022). Application of Lactic Acid Bacteria (LAB) in Sustainable Agriculture: Advantages and Limitations. *International Journal of Molecular Sciences*, 23(14), 7784. https://doi.org/10.3390/ijms23147784

Ravichandra, K., Balaji, R., Devarapalli, K., Batchu, U. R., et al., (2022). Enzymatic production of prebiotic xylooligosaccharides from sorghum (Sorghum bicolor (L.) xylan: Value addition to sorghum bagasse. *Biomass Conversion and Biorefinery*, 13, 1–9.

Sadiq, F. A., Yan, B., Tian, F., Zhao, J., Zhang, H., Chen, W. (2019). Lactic acid bacteria as antifungal and anti-mycotoxigenic agents: A comprehensive review. *Comprehensive Reviews in Food Science and Food Safety*, 18(5), 1403–1436.

Saini, J. K., Himanshu, Hemansi, Kaur, A., Mathur, A. (2022). Strategies to enhance enzymatic hydrolysis of lignocellulosic biomass for biorefinery applications: A review, *Bioresource Technology*, 360, 127517, ISSN 0960-8524, https://doi.org/10.1016/j.biortech.2022.127517.

Sarkar, A., Dickinson, E. (2020). Sustainable food-grade Pickering emulsions stabilized by plant-based particles. *Current Opinion in Colloid & Interface Science*, 49, 69–81.

Saroj, P., Manasa, P., Narasimhulu, K. (2018). Characterization of thermophilic fungi producing extracellular lignocellulolytic enzymes for lignocellulosic hydrolysis under solid-state fermentation. *Bioresources and Bioprocessing*, 5, 31.

Sayed, E. S. R., Gach, J., Olejniczak, T., Boratyński, F. (2022). A new endophyte Monascus ruber SRZ112 as an efficient production platform of natural pigments using agro-industrial wastes. *Scientific Reports*, 12(1), 12611.

Schwalm N. D., Mojadedi W., Gerlach E. S. et al. (2019) Developing a microbial consortium for enhanced metabolite production from simulated food waste. *Fermentation*. 5, 98. https://doi.org/10.3390/fermentation5040098

Šelo G., Planinić M., Tišma M., Tomas S., Koceva Komlenić D., Bucić-Kojić A. (2021) A comprehensive review on valorization of agro-food industrial residues by solid-state fermentation. *Foods*, 10(5), 927. https://doi.org/10.3390/foods10050927

Sen, T., Barrow, C. J., Deshmukh, S. K. (2019). Microbial pigments in the food industry-challenges and the way forward. *Frontiers in Nutrition*, 6, 7.

Seo S. O., Jin Y. S. (2022). Next-generation genetic and fermentation technologies for safe and sustainable production of food ingredients: Colors and flavorings. *Annual Review of Food Science and Technology*, 13, 463–488.

Shahzadi, T., Anwar, Z., Iqbal, Z., Anjum, A., Aqil, T., Bakhtawar Afzal, A., Kamran, M., Mehmood, S., Irshad, M. (2014). Induced production of exoglucanase, and Glucosidase from fungal co-culture of T. Viride and G. Lucidum. *Advances in Bioscience and Biote chnology*, 5, 426–433.

Sharon, A. B., Ahuekwe, E. F., Nzubechi, E. G., et al. (2023). Statistical optimization strategies on waste substrates for solving high-cost challenges in biosurfactants production: a review. In *IOP Conference Series: Earth and Environmental Science* (Vol. 1197, No. 1, p. 012004). IOP Publishing, Bristol.

Shrestha, S., Chio, C., Khatiwada, J. R., et al., (2022). Formulation of the agro-waste mixture for multi-enzyme (pectinase, xylanase, and cellulase) production by mixture design method exploiting Streptomyces sp. *Bioresource Technology Reports*, 19, 101142.

Sidana, A., & Yadav, S. K. (2022). Recent developments in lignocellulosic biomass pretreatment with a focus on eco-friendly, non-conventional methods. *Journal of Cleaner Production*, 335, 130286.

Singh, J., Sharma, A., Sharma, P., Tomar, G. S., et al., (2023). Production of ethanol, lipid and lactic acid from mixed agrowastes hydrolysate. *Natural Product Research*, 37(15), 2575–2582.

Sinha, S., Das, S., Saha, B., Paul, D., Basu, B. (2023). Anti-microbial, anti-oxidant, and anti-breast cancer properties unraveled in yeast carotenoids produced via cost-effective fermentation technique utilizing waste hydrolysate. *Frontiers in Microbiology*, 13, 1088477.

Subramani, G., Manian, R. (2023). Production of ferulic acid from dried bamboo shoots for the biotransformation into vanillin using a novel microbe Enterobacter aerogenes. *Biomass Conversion and Biorefinery*, 13, 1–15.

Torres-León, C., Ramírez-Guzmán, N., Ascacio-Valdés, J., Serna-Cock, L., dos Santos Correia, M. T., Contreras-Esquivel, J. C., et al. (2019). Solid-state fermentation with *Aspergillus niger* to enhance the phenolic contents and antioxidative activity of Mexican mango seed: A promising source of natural antioxidants. *LWT*, 112, 108236. https://doi.org/10.1016/j.lwt.2019.06.003

Troiano, D., Orsat, V., Dumont, M. J. (2022). Solid-state co-culture fermentation of simulated food waste with filamentous fungi for production of bio-pigments. *Applied Microbiology and Biotechnology*, 106(11), 4029–4039.

Ueda, J. M., Pedrosa, M. C., Heleno, S. A., Carocho, M., Ferreira, I. C. F. R., Barros, L. (2022). Food additives from fruit and vegetable byproducts and bio-residues: A comprehensive review focused on sustainability. *Sustainability*, 14: 5212.

Vastrad, B. M., Neelagund, S. E. (2011). Optimization and production of neomycin from different agro industrial wastes in solid state fermentation. *International Journal of Pharmaceutical Sciences and Drug Research*, 3, 104–111.

Verma, D. K., Thakur, M., Singh, S., Tripathy, S., et al., (2022). Bacteriocins as antimicrobial and preservative agents in food: Biosynthesis, separation and application. *Food Bioscience*, 46, 101594.

Vishvakarma, P., Mandal, S., Verma, A. (2023). A review on current aspects of nutraceuticals and dietary supplements. *International Journal of Pharma Professional's Research (IJPPR)*, 14(1), 78–91.

Wang, Z. P., Wang, Q. Q., Liu, S., Liu, X. F., Yu, X. J., Jiang, Y. L. (2019). Efficient conversion of cane molasses towards high- purity isomaltulose and cellular lipid using an engineered Yarrowia lipolytica strain in fed-batch fermentation. *Molecules* 24, 1228. https://doi.org/10.3390/molecules24071228

West, T. P. (2023). Citric acid production by Aspergillus niger using solid-state fermentation of agricultural processing coproducts. *Applied Biosciences*, 2(1), 1–13. https://doi.org/10.3390/applbiosci2010001

Xu, X., Lin, M., Zang, Q., Shi, S. (2018). Solid state bioconversion of lignocellulosic residues by inonotus obliquus for production of cellulolytic enzymes and saccharification. *Bioresource Technology*, 247, 88–95.

14 Strategic Approaches for Agricultural Biomass Conversion to Bioalcohol Using Membrane-Based Microbial Technology

Prashant Shukla, Indrani Paul, Apurba Sarkar, Sayantan Ghosh, and Priyanka Sarkar

14.1 INTRODUCTION

Bioalcohol production from agrobiomass has gained significant attention as a renewable solution to address the global problems on energy and environment (Gupta & Verma, 2016). Agrobiomass, comprising agricultural and forestry residues, offers an abundant and readily available feedstock for bioalcohol synthesis. The exponentially increasing global requirements for energy, along with the finite nature of fossil fuel reserves, have heightened the demands for alternative and renewable energy sources (Kour et al., 2019). Conventional fossil fuels are associated with environmental concerns such as greenhouse gas emissions, air pollution, and climate change, hence bioalcohols, including bioethanol, biobutanol, and biomethanol, derived from agrobiomass, offer a promising and sustainable energy option (Gautam et al., 2020; Zhen et al., 2020; Kohse-Höinghaus et al., 2010).

Agricultural residues like corn stover, wheat straw, rice husks, and sugarcane bagasse and forestry residues like wood chips and sawdust are considered as abundant waste materials. Utilizing these agro-wastes for bioalcohol production promotes waste reclamation and satiates waste management challenges. Another beneficial aspect of bioalcohol production from agrobiomass is its prospective significance in reducing greenhouse gas discharges (Heinimö et al., 2011). Consequently, bioalcohol-based fuels have a reduced carbon footprint compared to traditional fossil fuels, contributing to climate change.

Membrane-based microbial systems play an important role in the production of pure bioalcohol from different agrobiomass resources (Figure 14.1). This technique encompasses the use of specific membranes to separate and concentrate microbial-produced bioalcohols, such as ethanol or butanol, from fermentation brews. The fermentation takes place in a controlled environment, utilizing a sugar solution obtained by pretreatment of agro-based biomass and subsequent enzymatic

DOI: 10.1201/9781003407713-14

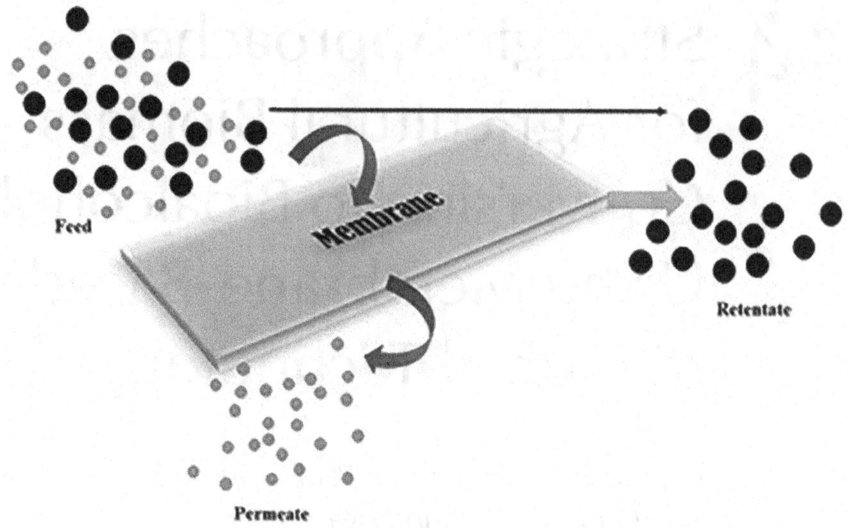

FIGURE 14.1 Principle of membrane-based separation technique.

hydrolysis with the selected microorganisms. The microbes utilize the sugars as a source of energy to finally break them into the preferred end products.

The process of fermentation takes place in specifically designed bioreactors, where the fermentation of agrobiomass takes place (Marchenko et al., 2017). They offer a precise environment for culturing microorganisms and the translation of fermentable sugars into ethanol or other alcohol-based biofuels. Physiological conditions such as temperature, pH, and agitation speed can be monitored in a bioreactor during the process of fermentation and can be modified as per requirement (Priya et al., 2016). This approach offers numerous benefits, including greater product purity, reduced energy consumption, and enhanced process efficacy.

Corn stover, Sugarcane bagasse, Wheat straw, Switchgrass, and Cassava are examples of common agrobiomass feedstocks utilized for bioalcohol synthesis. Important steps involved in the production process include pretreatment, fermentation followed by membrane separation. Many third-world countries would become self-reliant in energy production with the development of technologies that can reduce cost in the collection of agro-wastes and the production of biofuels. The quest for bioalcohol production from agrobiomass has already led to noteworthy developments in biorefinery technology. Biorefineries can proficiently transform agro-waste into numerous value-added products, including bio-based chemicals, biofuels, and biomaterials.

Exploiting possibilities of utilizing agrobiomass for bioalcohol production can embrace sustainable practices in agricultural sectors. Crop remains left in the field after harvest can be collected and changed into bioalcohol, promoting the notion of circular economy and sustainable agriculture (Bochtis et al., 2020). Utilization of bioalcohol production from agrobiomass has the prospects to encourage rural development by generating new job opportunities in biomass gathering, pretreatment, and biorefinery procedures. In this context, the progress can lead to enhanced financial settings in rural communities. This chapter explores the background and significance of bioalcohol production from agrobiomass, discussing the driving factors, gains, and environmental benefits of membrane-based fermentation as a promising technology.

14.2 AGROBIOMASS AS A RENEWABLE FEEDSTOCK FOR PRODUCTION OF BIOALCOHOL

Multi-purpose agrobiomass fuels are available in huge quantities for several benefits that include gasification (branches, twigs, wood shavings, roots, bark, sawdust, etc.). A wide array of agricultural residues (maize cobs, coconut shells, cereal straws, rice husks, cotton ginning, olive kernels, coconut husks, etc.) can be included in this category. These fuels vary in their physical, morphological, and chemical properties, thus leading to demands on the method of gasification, hence different reactor designs and gasification technologies are required (Nemtsov & Zabaniotou, 2008).

Based on the source of biomass, biofuels are generally classified as follows: (i) First-generation ethanol: Ethanol produced from edible sources. (ii) Second-generation bioethanol: Ethanol produced from non-edible feedstocks. (iii) Third-generation Algae: Derived from algae and marine organisms. (iv) Fourth-generation Algae: Bioethanol produced from genetically enhanced algae. First-generation bioethanol is the most abundant globally produced biofuel (Aghaei et al., 2022). Sugarcane and corn are among the most common sources of first-generation ethanol. Agro-wastes like sugarcane bagasse are rich in cellulose, hemicellulose, and lignin, which are firmly bound together to form the rich biomass. Another abundantly available agrobiomass is corn stover, one of its by-products, which is one of the top three agricultural wastes in the world.

Producing bioalcohol from agrobiomass involves the conversion of various feedstocks into bioethanol or other alcohol types through processes like fermentation. The composition of agrobiomass can vary based on the specific feedstock used. Table 14.1 shows the composition of different agrobiomass feedstocks commonly used for bioalcohol production.

14.3 PRINCIPLE OF MEMBRANE-BASED MICROBIAL TECHNIQUE

Membrane separation technologies play a significant role in different stages of bioethanol production, such as fermentation, purification, and concentration. After breaking down the agro-waste biomass to release sugars, yeast or bacteria are introduced to convert the sugars into ethanol. Membrane technologies like microfiltration (MF) and ultrafiltration (UF) can be used to clarify the fermentation broth and separate biomass solids, cells, and suspended particles from the liquid, allowing for a clearer and more efficient fermentation process (Espamer et al., 2006).

In permeate purification, after the fermentation, the mixture contains ethanol along with water, by-products, and impurities. Nanofiltration (NF) and reverse osmosis (RO) can be used to purify the permeate by separating water and ethanol from smaller impurities (Figure 14.1). The membranes allow smaller molecules like water and ethanol to pass through while rejecting larger molecules and contaminants (Roca et al., 2010). The ethanol obtained after fermentation may need to be concentrated

TABLE 14.1
Different Agrobiomass Feedstocks Commonly Used for Bioalcohol Production

S. no.	Agrobiomass Feedstock	Cellulose (%)	Hemicellulose (%)	Lignin (%)	Protein (%)	Lipids (%)	Water (%)	References
1.	Corn Starch	35–45	25–35	25	2–4	6	14	Chinnaswamy and Hanna (1990)
2.	Sugarcane	40–45	25–30	20–25	2	1	40–45	Pandey et al. (2000)
3.	Wheat	32–41	22–28	20–22	10–15	2.4–3.8	12	Ali et al. (2022)
4.	Switchgrass	30–50	20–40	10–20	5	1–2	15–30	Sanderson et al. (1997)
5.	Rice straw	35–40	20–30	15–20	2–5	>2	Variable	Sharma et al. (2017)
6.	Wheat Straw	35–45	25–30	25–35	15–20	>1	Variable	Kosowska-Golachowska et al. (2018)
7.	Algae	<2	5–10	10	50–70	50%	70% to over 90%,	Frey et al. (2007)

to achieve the desired ethanol content. Technologies like pervaporation and vapor permeation can be employed. Pervaporation uses a membrane to selectively remove ethanol as vapor from a liquid mixture, while vapor permeation involves transporting ethanol vapor through a membrane by maintaining a vapor pressure difference (Farahi et al., 2018).

For producing fuel-grade ethanol, further dehydration is often necessary to reduce the water content. Azeotropic distillation, which combines distillation with a membrane separation step, is commonly used. Membrane-based dehydration processes, such as dehydration using pervaporation or vapor permeation, can also help remove the last traces of water from the ethanol (Anderson & Rahman, 2018). In some cases, membrane bioreactors can be integrated into the fermentation process. MBRs combine biological treatment with membrane separation, resulting in high-quality permeate while retaining biomass and microorganisms within the reactor. This approach enhances fermentation efficiency and simplifies downstream processing (Anderson & Rahman, 2018).

The specific choice of membrane separation technology depends on several factors including the composition of the agro-waste feedstock, the desired quality of the final ethanol product, energy efficiency, and economic considerations. Integrating membrane processes into bioethanol production from agro-waste can enhance the overall efficiency of the process, reduce energy consumption, and contribute to the sustainable production of biofuels.

14.4 INTEGRATION OF MICROBIAL FERMENTATION WITH MEMBRANE SEPARATION FOR ENHANCED BIOALCOHOL PRODUCTION

The integration of microbial fermentation with membrane separation is a promising approach for enhancing bioalcohol production from agro-waste. Agro-waste, such as agricultural residues, food processing by-products, or organic waste, contains complex carbohydrates that can be broken down by microorganisms through fermentation. Microbes can convert these carbohydrates into bioalcohols like ethanol or butanol. Fermentation processes often have limitations such as low product yield, incomplete substrate utilization, and the presence of inhibitory compounds that can affect microbial growth and product formation. Additionally, the fermentation broth might contain impurities, microbial cells, and other particulates that need to be separated from the desired bioalcohol product (Hossain et al., 2017).

Membrane separation involves using semipermeable membranes to selectively separate components in a liquid mixture based on their size, charge, or other properties. In the context of bioalcohol production, membranes can be used to separate the fermentation broth into a permeate (desired product) and a retentate (unwanted components) (Imamoglu & Sukan, 2014). Membrane separation helps maintain a higher concentration of bioalcohol in the fermentation broth, which can lead to improved yields. The integration of this technique can increase process efficiency and reduce energy consumption compared to traditional distillation-based separation methods (Dutta & Suresh Kumar, 2023).

14.5 ESSENTIAL PREREQUISITES FOR PROCESSING

Reaching the desired efficacy of the bioprocess is a key challenge and requires specific initial requirements. Choosing the correct microorganisms is the key to efficient fermentation (Sivaramakrishnan et al., 2021). The selection depends on several factors, including the type of biomass, the composition of the hydrolysate, the desired end product, and the process conditions. Microorganisms have specific competencies to exploit biomass; some may be suitable for lignocellulosic biomass, while others may outshine with starchy or sugary biomass.

14.5.1 Selection of Agrobiomass

The choice of suitable agrobiomass is fundamental as it has direct bearings on the type and yield of bioalcohols that can be produced. The carbohydrate content of feedstock should be high, mostly composed of sugars that are fermentable like glucose, xylose, or other saccharides (Ho et al., 2013). This is because these carbohydrates are the substrates for microbial fermentation, which on action of microbes are converted into bioalcohols. The feedstock should be readily obtainable in adequate quantities to sustain the production of bioalcohol on a large scale. Justifiable feedstock choices, such as crop residues or dedicated energy crops, are favored to reduce the influence on food production and land use (Allen et al., 2014). It is possible that some feedstocks may be available only in a particular season, and their availability may affect the overall production schedule of bioalcohol. The cost of the feedstock plays an important role in the overall economics of bioalcohol production (Kasmuri et al., 2017). Economically feasible feedstock choices are vital for the commercial realization of the process.

14.5.2 Types of Microorganisms Used

Selecting the proper microorganisms is crucial for efficient fermentation to take place (Sivaramakrishnan et al., 2021). The choice of organisms depends on several factors, including the type of biomass, the composition of the hydrolysate, the desired end product, and the process conditions. For biofuel production, such as ethanol or butanol, yeast or specific bacteria strains (e.g., *Saccharomyces cerevisiae* for ethanol) are commonly used, while for biogas production, anaerobic microorganisms like methanogenic archaea are employed (Klocke et al., 2008). Strains with higher yields and faster fermentation rates are generally preferred.

Genetic engineering can be utilized to transform microorganisms to improve their performance or to introduce particular pathways that can result in desired product synthesis. Yeasts such as *S. cerevisiae* are commonly used for ethanol production from sugars and starchy biomass (Azhar et al., 2017). *Zymomonas mobilis* has been reported to have efficacy in producing high volume ethanol by fermenting glucose and sucrose (Panesar et al., 2006). *Escherichia coli* is considered a model bacterium which can be engineered for the production of numerous products, including organic acids and biofuels (Atsumi & Liao, 2008). In the race of fermenting complex biomass, the role of eukaryotic microorganisms cannot be ignored. Filamentous

Fungi (e.g., *Aspergillus* and *Trichoderma*) can produce enzymes, such as cellulases and hemicellulases, which assist in the hydrolysis of complex biomass (Kumar & Verma, 2020). Other important microbes may include *Clostridium* species that can be used for the production of organic acids and solvents from lignocellulosic biomass (Dolejš et al., 2014), which can be later converted to alcohol using a separate set of microorganisms.

14.5.3 Selection of Membrane Type

Specialized membranes are utilized to separate desired products from the fermentation process in a membrane-based microbial system. This results in enhancing the purity of the products and simply the downstream processing. When the process is coupled with the use of membranes, it ensures the purity of alcohol which is renewable and environment-friendly fuel. Membranes with appropriate pore size, material composition, and the desired selectivity are chosen depending on the final product. Since membranes are typically semipermeable, they will allow some components to pass through while retaining the others. The objective is to have a membrane that allows the desired bioalcohol to permeate through while refusing water, sugars, and microorganisms.

The separation unit has the selected membrane present in between the feed solution and the permeate. As bioalcohol is small molecules with very specific characteristics, the membrane allows them to pass through the pores or through other selective mechanisms which are then collected on the permeate side (Mao et al., 2019). Further, based on the membrane separation process, the permeate may be subjected to purification steps, such as distillation or molecular sieves for higher purity in the final product. It is possible that some portion of the unpermeated feed solution, which does have residual sugars, may be recycled back into the fermentation vessel to improve the overall yield of bioalcohol production. This helps in reducing waste and maximizing the efficiency of the overall process.

14.5.4 Enzymes Used in Enzymatic Hydrolysis

Enzymes play a critical role in the hydrolysis of cellulose and hemicellulose for the production of fermentable sugars for biofuel production. These enzymes are typically isolated from microorganisms (fungi and bacteria), which naturally produce lignocellulolytic enzymes proficient of breaking down complex polysaccharides. The main classes of enzymes used in cellulose and hemicellulose hydrolysis are cellulases and hemicellulases, respectively.

Cellulases like endoglucanases are enzymes that cleave internal β-1,4-glycosidic bonds inside cellulose chains, arbitrarily breaking down cellulose into shorter fragments; cellobiohydrolases like exoglucanases act on the cellulose chain from the ends, releasing cellobiose (a dimer of glucose) or other cello-oligosaccharides as products, while β-glucosidases hydrolyze cellobiose and other cello-oligosaccharides into glucose molecules, which are the chief fermentable sugars for biofuel production (Nitsos et al., 2018).

Hemicellulases like xylanases target hemicellulose, particularly xylan, and break the β-1,4-glycosidic bonds, releasing xylose as the primary product; arabinofuranosidases act on arabinose side chains in hemicellulose, releasing arabinose units; mannanases hydrolyze mannan, another type of hemicellulose, into mannose and other oligosaccharides; and galactanases break down galactan into galactose and shorter galactooligosaccharides (Polprasert et al., 2021). For competent hydrolysis of cellulose and hemicellulose, enzyme cocktails comprising a combination of these enzymes are often used. This is because cellulose and hemicellulose have complex and interconnected structures, and the synergistic role of different enzymes is needed for complete hydrolysis. During the hydrolysis process, these enzymes act on the lignocellulosic biomass, breaking down cellulose and hemicellulose into soluble sugars, such as glucose, xylose, mannose, arabinose, and galactose. These fermentable sugars can then be fermented by microorganisms, such as yeasts or bacteria, to produce biofuels like ethanol or other value-added products.

14.6 PRETREATMENT OF AGROBIOMASS

If any step is crucial in utilizing agrobiomass for the production of bioalcohols, it is pretreatment as without it microorganisms will never be able to convert the biomass into useable products (Ramos et al., 2021). Especially lignocellulosic biomass needs pretreatment so as to make the complex structure available for enzymatic hydrolysis thereby increasing the yield of fermentable sugars. The objective of pretreatment methods is to break down these obstacles and improve the efficiency of subsequent enzymatic hydrolysis. Pretreatment of agrobiomass can be physical, chemical, or biological in nature.

14.6.1 Physical Pretreatment

Physical pretreatments include washing, drying, grinding/milling followed by steam explosion, where the initial processes are meant to bring down the particle size for easy accessibility and increased surface area and for effectual steam penetration and treatment. The severity of the treatment (i.e., temperature, time, and pressure) must be optimized for each precise biomass type to evade unnecessary degradation and loss of fermentable sugars (Negro et al., 2003). Steam explosion has been reported by many as their choice of pretreatment for lignocellulosic biomass (El-Zawawy et al., 2011). By partially removing lignin, steam explosion makes cellulose and hemicellulose more vulnerable to enzymes thereby increasing the yield of fermentable sugars. Apart from making cellulose more accessible, steam explosion also makes it vulnerable to enzymatic hydrolysis by decreasing its natural crystalline nature.

14.6.2 Chemical Pretreatment

Chemical pretreatment methods are opted for further processing of the physically pretreated agrobiomass that enhances the efficacy of the biomass to be accessible for enzymatic hydrolysis and microbial fermentation processes. Chemical pretreatments include the use of acids, alkalis, and ozone, and they also implicate specific methods like organosolv and ammonia fiber expansion processes. Dilute acid hydrolysis is a commonly used pretreatment method for agrobiomass, especially lignocellulosic materials (Kehsav et al., 2016). This pretreatment involves the use of a dilute acid solution at elevated temperatures to break down the complex structure of the biomass. However, dilute acid hydrolysis also has some disadvantages, such as the potential formation of inhibitory compounds during the process, which can affect the fermentation step (Behera et al., 2014).

Ammonia Fiber Expansion (AFEX) is another chemical pretreatment process used in the production of cellulosic biochemicals and biofuels from lignocellulosic biomass including forestry waste, agricultural residues, and energy crops (Lau et al., 2010). The primary objective of AFEX is to break down the complex lignocellulosic structure of biomass and make it more available for enzymatic hydrolysis, thereby increasing the efficiency of converting the biomass into fermentable sugars. Alkaline pretreatment is a chemical process used in many industries, predominantly in the production of biofuels to maximize the conversion of lignocellulosic biomass into fermentable sugars which involves treating lignocellulosic feedstocks, such as agricultural residues and energy crops, with alkaline solutions to disintegrate the complex structure of the biomass and make it more accessible to enzymatic hydrolysis (Harun et al., 2011). The sugars are tightly bound within the lignin matrix, which acts as a defensive barrier, making them difficult to access and extract (Hoşgün et al., 2017). Alkaline pretreatment unsettles the lignin structure and transforms the lignocellulosic constituents through different mechanisms.

Besides acid- and alkali-based methods, Organosolv pretreatment, a chemical process used to break down lignocellulosic biomass into its individual components, namely cellulose, hemicellulose, and lignin, has also been explored. This process involves the use of organic solvents, usually a combination of water and organic alcohols or organic acids, to selectively dissolve and remove lignin from the biomass (Amiri & Karimi, 2015). Organosolv pretreatment is mostly useful for producing high-quality lignin and facilitating the conversion of cellulose and hemicellulose into biofuels, or other value-added products.

Ozonolysis pretreatment can effectively break down the components of the biomass, thus making the cellulose more accessible to enzymatic hydrolysis. The agrobiomass is first subjected to size reduction like some other methods described before to increase the surface area and thus enhance the efficiency of the ozonolysis reaction (Sulfahri et al., 2020). The pretreated biomass is then exposed to ozone gas, which freely reacts with the unsaturated bonds in lignin and hemicellulose. The reaction breaks down these complex polymers, leading to the formation of smaller fragments and increasing chances of enzymatic hydrolysis of cellulose. The specific conditions for ozonolysis, such as ozone concentration, reaction time, and temperature, can be adjusted based on the characteristics of the agrobiomass and the desired outcomes of the pretreatment (Rashid et al., 2021).

14.6.3 Biological Pretreatment

Biological pretreatment of agrobiomass is a process that involves the use of microorganisms to disrupt and alter the lignocellulosic structure of the biomass before subsequent conversion into biofuels, biochemicals, or other value-added products (Gupta & Verma, 2015). This approach utilizes the enzymatic capabilities of microorganisms to break down complex polymers like lignin and make cellulose more available to further processing steps. The first and foremost step is to select suitable microorganisms that possess the enzymatic machinery to degrade lignocellulosic materials efficiently. Fungi (brown rot, white rot, etc.) are commonly used due to their ability to produce a wide range of lignocellulolytic enzymes (Ma et al., 2010). The selected microorganisms are inoculated onto the agrobiomass, providing them with an environment appropriate for their growth and enzyme production. The inoculated biomass is allowed to undergo a pre-determined incubation time during which the microorganisms grow and produce lignocellulolytic enzymes. The lignocellulolytic enzymes produced by the microorganisms, including cellulases, hemicellulases, and ligninases, act on the lignocellulosic components of the biomass, breaking down lignin and hemicellulose and making cellulose more accessible. Selectivity, mild conditions, and environmental sustainability are some of the advantages associated with the process. Some microorganisms can produce valuable coproducts during the pretreatment process, such as certain bioactive compounds or enzymes (Gupta & Verma, 2015). Some pretreatments may necessitate costly equipment or harsh chemicals, which may affect the overall viability of the process. It's important to identify the benefits and limitations of each pretreatment method and select the one that best suits the particular agro-waste and the subsequent steps in the biofuel production process. Proper pretreatment can improve the enzymatic accessibility of agro-waste, and thus, the overall efficiency of biofuel production is boosted.

14.7 MICROBIAL FERMENTATION OF AGROBIOMASS

The bioprocess technologies involving microbial fermentation permit efficient and cost-effective industrial production (Manikandan et al., 2022) of valorizable products. The purity and reliability of the final alcohol product are maintained by the use of bioreactors as they can be sterilized (Mahboubi et al., 2018). Nutrient optimizations can be achieved in bioreactors ensuring that the microorganisms have access to the required nutrients for optimum growth and ethanol production (González-Figueredo et al., 2018). The use of bioreactors has been a success as they allow a broad range of agro-based biomass feedstocks to be processed, thus making it a flexible technology that can employ different biomass sources for alcohol production (Boro et al., 2022). The regulated environment in bioreactors permits for greater cell densities and amplified fermentation rates. This causes improved overall effectiveness and higher production of alcohol from the agro-based biomass (Sakhtah et al., 2019). For scaling up of the process, bioreactors are necessary as they allow for larger volumes of agro-based biomass, making them appropriate for commercial-scale production of alcohol fuels.

Monitoring and optimizing the fermentation process for bioalcohol production is critical to ensure maximum yield, product quality, and cost-effectiveness. Bioalcohol, such as ethanol, is usually produced through microbial fermentation of sugars derived from biomass (Priya et al., 2016). As the microorganisms grow at specific temperature range, enzymes too work at a very narrow optimum range. Therefore, for maximum yield, monitoring and maintaining a consistent temperature within the optimal range is vital for the process (Brethauer & Wyman, 2010). The concentration of the biomass and substrate (sugars) has to be monitored regularly during fermentation (George et al., 1993). Monitoring the reduction of sugars and the build-up of alcohol will help determine the fermentation progress.

TABLE 14.2
Microorganisms with Specific Enzymes Acting on Substrates Differently

S. No.	Organism	Enzyme	Substrate	Reference
1.	*Trichoderma harzianum* LMLBP07 and *Trichoderma longibrachiatum* LMLSAUL	Cellulase	Banana Pseudostem	Legodi et al. (2021)
2.	*Trichoderma reesei* ZU-02 and *Aspergillus niger* ZU-07	Cellulase Cellobiase	Corncob	Chen et al. (2007)
3.	*Trichoderma reesei* ATCC 26921	Cellulase	*Chlorococcum* sp.	Hariun and Danquah (2011)
4.	*Chrysosporium lucknowense* UV18-25	Cellobiohydrolase, endoglucanases II and V; β-glucosidase, xylanase II	Avicel, cotton, and pretreated Douglas fir wood	Gusakov et al. (2007)
5.	*Trichoderma reesei* ZU-02 and *Aspergillus niger* ZU-07	Cellulase Cellobiase	Maize straw	Polprasert et al. (2021)
6.	White-rot fungi	Cellulase	Various	Tian et al. (2012)
7.	*Aspergillus aculeatus*	Pectinase	*Chlorella vulgaris*	Kim et al. (2014)

Fermentation strategies for bioalcohol production from a variety of agrobiomass can differ due to dissimilarities in the composition and structure of the biomass. Each type of agrobiomass may require specific pretreatment, enzyme hydrolysis, and fermentation conditions for the optimization of bioalcohol yields (Table 14.2).

14.8 MEMBRANE SEPARATION TECHNOLOGIES IN BIOALCOHOL PRODUCTION

Membrane separation technologies play a central role in bioalcohol production processes by facilitating the separation and purification of bioalcohols from fermentation broths or other feed streams. These technologies offer several advantages, such as energy efficacy, simplicity of scale-up, and environmental sustainability. There is an array of different membrane separation methods for achieving the desired purity and yield (Table 14.3). These include microfiltration, ultrafiltration, nanofiltration, RO and pervaporation techniques.

14.8.1 Microfiltration for Biomass Separation from Fermentation Mixture

Pervaporation and distillation can be used for the separation of bioethanol from the fermentation broth, but the solid materials (debris) and yeast cause various problems. Suspended solid materials and yeast cells present in the fermentation broth can accumulate, and this leads to reduction of the bioethanol production efficiency. Studies

TABLE 14.3
Parameters of Different Membranes

Membrane	Microfiltration	Ultrafiltration	Nanofiltration	Reverse Osmosis	Pervaporation
Pore Size	0.1–10 mm	1–100 nm	1–100 Å	1 nm	No specific pore size
Separation mechanism	Sieving and size exclusion	Size exclusion and charge-based interactions	Size exclusion, charge-based interactions, and some degree of solute diffusion	Solvent diffusion through a semipermeable membrane	Selective permeation and evaporation
Selectivity	Low selectivity; primarily used for solids separation	Moderate selectivity; effective for separating larger molecules	Moderate to high selectivity; suitable for removing salts and small organic molecules	High selectivity; effective for desalination and solvent purification	High selectivity; efficient for separating volatile components
Energy consumption	Low energy consumption due to larger pore size	Moderate energy consumption	Moderate energy consumption	Consumption of high energy due to the need for high pressure	Moderate energy consumption
Applications	Clarification, sterilization, and removal of suspended solids	Protein concentration, enzyme purification, and removal of macromolecules	Softening water, color removal, and partial desalination	Desalination, water purification, and solvent recovery	Bioalcohol recovery, volatile component separation, and aroma recovery
Cost	Relatively low cost	Moderate cost	Moderate cost	Higher cost due to energy requirements	Moderate to higher cost depending on the specific application
References	Kartawiria et al. (2015)	Meixner et al. (2015)	Weng et al. (2009)	Gautam and Menkhaus (2014)	Khalid et al. (2019)

have shown that the yeast content in the feed for pervaporation can result in biofouling (Kusworo et al., 2023; Faria et al., 2017). Microfiltration is an effective membrane separation technology used for biomass separation from fermentation mixtures. In the context of biofuel production, this process is mostly useful for clarifying the fermentation broth and separating the biomass (microorganisms or cells) from the liquid phase containing the desired biofuel product.

The microfiltration process involves the use of membranes with relatively large pore sizes (in the range of 0.1–10 µm) that can effectively remove suspended solids, microbial cells, and large particles from the fermentation mixture. The separation is based on size exclusion, a process in which particles larger than the membrane pore size are retained on the membrane surface, while smaller molecules pass through the membrane.

After fermentation, broth contains the liquid phase with the biofuel product, as well as microbial cells or biomass and other solid particles. This mixture is often turbid due to the presence of suspended solids. The fermentation broth is then subjected to microfiltration by passing it through microfiltration membranes. The large pores in the membranes capture and retain the biomass (microorganisms and larger particles), leaving behind a clarified liquid containing the desired biofuel product. The permeate, which is the clarified liquid passing through the microfiltration membranes, is collected for further processing, such as downstream purification or distillation to concentrate the biofuel product. The biomass retained on the microfiltration membranes can be processed or recycled, depending on the specific requirements of the fermentation process. For example, the biomass can be concentrated and recycled back into the fermentation tank to improve productivity.

There are several advantages of microfiltration-based methods that include selective removal of biomass and suspended solids while retaining the desired biofuel product in the liquid phase. It requires moderately low pressures and temperatures, making it an energy-efficient separation process compared to other separation techniques like centrifugation or sedimentation. This is integrated into continuous fermentation processes, permitting continuous production and minimizing downtime. Microfiltration is simply scalable, making it appropriate for both laboratory-scale research and large-scale industrial applications. Properly designed and optimized microfiltration systems can reduce membrane fouling, improving the longevity and efficiency of the process. Besides this, different membranes have been fabricated for the microfiltration process which can help in purification of fermentation broth.

14.8.2 Ultrafiltration for Protein and Microbial Cell Removal from Fermentation Mixture

Ultrafiltration is a membrane separation technique usually used for protein and microbial cell removal from permeates. This process employs membranes with pore size of 0.001–0.1 µm, making it effective in separating macromolecules like proteins and microbial cells from the liquid phase. Ultrafiltration offers advantages such as high selectivity, gentle operating conditions, and the potential for continuous processing.

The fermentation broth is fed into an ultrafiltration system that contains the ultrafiltration membrane. The membrane acts as a physical barrier, allowing only molecules smaller than the membrane pore size to pass through while retaining larger molecules, including proteins and microbial cells. The fermentation broth flows through the ultrafiltration membrane forming two streams. The concentrated stream containing the retained proteins and microbial cells is called the "retentate," and the purified liquid stream is called the "permeate." Ultrafiltration effectively separates proteins and microbial cells from the fermentation broth, producing permeate with a reduced protein content and a clarified liquid phase. If desired, the retentate, which contains the concentrated proteins and microbial cells, can be further processed to concentrate the product for downstream applications (Garcia et al., 2011).

Ultrafiltration selectively separates macromolecules like proteins and microbial cells while allowing smaller molecules, such as the desired biofuel or biochemical, to pass through. Ultrafiltration operates at relatively low pressures and temperatures, minimizing the risk of product denaturation or degradation.

Ultrafiltration has been employed for the removal of cells and other debris from fermentation broth for the production of biobutanol (Behara et al., 2023; Garcia et. al., 2011).

14.8.3 Nanofiltration for Concentration and Impurity Removal

Nanofiltration is a membrane filtration process that functions at the nanometer scale (Kumar et al., 2017). It is extensively used for concentration and impurity exclusion from various liquid mixtures, including fermentation mixtures. During fermentation, several valuable biomolecules, such as enzymes, organic acids, bioactive compounds, and proteins, are produced. Nanofiltration can be engaged to concentrate these biomolecules from the fermentation broth, leading to higher product yields and reduction of the overall volume of the final product (Dey et al., 2020). Fermentation mixtures often comprise undesirable by-products, salts, small molecules, and other impurities. Nanofiltration membranes have selective permeability, allowing smaller impurities to pass through while retaining larger molecules. This selective removal helps purify the fermentation mixture and improves the quality of the final product. In some cases, fermentation processes result in the formation of an aqueous solution with valuable products. Nanofiltration can be used to recover water from such solutions while leaving behind the concentrated valuable biomolecules. This water recovery can lead to reduced waste and potentially save on resources. Nanofiltration can also be used for fractionation purposes, where diverse components of the fermentation mixture are separated based on their molecular size and charge. This is mainly useful when multiple valuable components need to be separated and purified. Concentration and impurity removal using nanofiltration can help to minimize the complexity of downstream processing steps, resulting in a more efficient and cost-effective overall process. Several reports have been published where nanofiltration has been used either singly or in combination of other methods for purification of bioalcohols (Kumar et al., 2017; Dey et al., 2020).

14.8.4 RO for Bioalcohol Purification

RO is another membrane filtration process that can be utilized for the purification of bioalcohol from fermentation mixtures (Lipnizki, 2010). RO functions by applying pressure to a solution to force water molecules through a semipermeable membrane, while larger molecules and impurities are rejected and remain on the feed side. RO can effectively eliminate impurities, including salts, sugars, organic acids, and other by-products, from the fermentation mixture. This results in a highly purified bioalcohol product. The permeate obtained from RO contains mostly purified water, which can be reused after recovery in the fermentation process. This water recovery can contribute to the overall process efficacy and reduced water consumption. RO can concentrate the bioalcohol present in the fermentation mixture by removing water and other unwanted components. Concentrating the bioalcohol can lead to a more efficient downstream purification process. The process is generally considered more energy-efficient in comparison to conventional distillation processes for alcohol purification. This is particularly relevant for bioalcohol production, where sustainability and energy conservation are crucial considerations. It can be operated in a continuous mode, allowing for a steady and continuous production of purified bioalcohol without the need for frequent batch operations. RO can be applied with microfiltration and ultrafiltration for pretreatment. One can find different membrane processes used in the bioalcohol industry in the article of Lipnizki (2010).

14.8.5 Pervaporation for Bioalcohol Recovery

Pervaporation is a membrane-based separation technology which can be used for the recovery of bioalcohol from fermentation mixtures. This method is specific as it uses nonporous membranes. The separation depends on the membrane properties and the chemical characteristics of the mixture like diffusivity into the membrane and the capacity to form hydrogen bonds. In this method, the concentration gradient formed by vacuum pressure is the main driving force for the separation to take place (Zentou et al., 2019). Pervaporation operates on the principle of selective permeation of a liquid mixture through a membrane, which is followed by the evaporation of the permeate on the other side of the membrane.

This process is particularly valuable for separating volatile components, such as bioalcohols, from non-volatile components present in the fermentation mixture. The fermentation mixture is brought into contact with a pervaporation membrane that exhibits selectivity toward the desired bioalcohol. The membrane allows the bioalcohol to pass through while selectively rejecting water, organic acids, sugars, and other impurities present in the fermentation mixture. Once the bioalcohol permeates through the membrane, it is collected on the other side of the membrane, typically in a vapor form. The collected vapor can then be condensed to attain the purified bioalcohol product. Pervaporation method has been employed by various workers to achieve high purification of bioalcohols. Dadi et al. (2018) have successfully purified ethanol from fermented coffee by-products using alcohol-selective pervaporation membranes.

The pervaporation setup includes feed tank, heater, feed pump, membrane module, vacuum pump, and condenser cold trap. The type of membrane material is largely chosen as per the type of the selective component. The hydrophobic membrane generally selects organic compounds which are recovered in the permeate. In cases when the membrane is hydrophilic, the mixture liquid in the feed will be dehydrated and the recovery of water in the permeate will be possible (Zentou et al., 2019).

14.9 ADVANTAGES AND BENEFITS OF MEMBRANE-BASED MICROBIAL TECHNIQUE

Depending on the reports and findings, it is evident that membrane-based microbial techniques have promising benefits over other conventional methods. The versatile range of raw substrates that can be utilized in this approach can help overcome the scarcity of resources when it comes to bioethanol production. These substrates are often waste biomass like agro-industrial waste materials which can be valorized as per the availability of compatible microorganisms that have the caliber to degrade or utilize the components within the substrate (El-Zawawy et al., 2011). Depending on the composition and characteristics of substrates, a wide array of microorganisms can be selected as per availability, thus making the process more acceptable. Besides being highly specific, these processes also ensure an energy-saving strategic approach for the purpose (Laval, 2010).

One of the major drawbacks of the conventional fermentation processes is product inhibition which results in the retarded growth of microorganisms (Gryta et al., 2000). The best methodology to overcome this problem is the implication of a multistage bioreactor system (integrated) for the continuous production of bioethanol, resulting in a complete replacement of the conventional system. One membrane-based system can replace multiple conventional steps that include centrifugation, adsorption, neutralization, ion exchange, acidification, and distillation (Pal et al., 2018).

Proper selection of membranes based on molecular weight cut-off, compatibility with the fermentation process, and fouling resistance is crucial. Fouling of membranes due to microbial growth or deposition of solids can impact separation efficiency and require regular cleaning or replacement. Process optimization is essential to balance fermentation and separation conditions for maximum bioalcohol production (Khalid et al., 2019).

In conclusion, integrating microbial fermentation with membrane separation offers a way to enhance bioalcohol production from agro-waste by improving yields, product purity, and overall process efficiency. However, successful implementation requires careful design, selection of appropriate membranes, and continuous optimization to achieve desired outcomes.

Conventional approaches require huge space for maintenance and operation which is drastically brought down when membrane-based technologies are used. Yield of product can be varied depending upon the availability of raw materials and market demand in the case of membrane-based fermentation technology. Another positive implication of the membrane-based technology is the possibility to reuse and recycle residual raw substrate that includes biomass, microbial cells, small ionic components, and water. This makes the process more sustainable, eco-friendly, and safe for use (Koschikowski et al., 2003). As a result of all these factors, there is a huge drip in cost consumption leading to shooting up of the economic benefit factor.

14.10 CHALLENGES AND LIMITATIONS OF USING MEMBRANE-BASED FERMENTATION METHODS

Despite several advantages that make membrane-based fermentation methods a promising approach for the production of bioalcohol from agrobiomass, there can be some limitations pertaining to the operation and management of these processes. Overcoming these challenges using different strategies without compromising the yield and quality of the product is another vast area of research interest.

One of the primary drawbacks of the process lies in the frequent clogging of membranes due to interference from coarser materials that can cause hindrance to the process as a whole, which is also termed as membrane fouling. Unwanted materials often get accumulated on the membrane surface impeding its performance and there is a reduction in the flux (Padhan et al., 2023). Clogging can also be caused as a result of microbial growth (biological fouling) or deposition of inorganic salts (scaling) or organic compounds (organic fouling), forming a layer on the membrane surface thus making it impermeable to fermented broth; hence, methods of cleaning need to be developed (Ahmed et al., 2023).

In order to design an efficient membrane-based fermenter system, it is essential to maintain the proper pore size or molecular weight cut-off size of the membrane. There have been instances when the overall efficiency of the system has been seen to get compromised due to leakage of important molecules such as the enzymes itself. Leakage of enzymes is a serious problem faced, which must be overcome in order to avoid enzyme refilling and to ensure that the enzymes can be recycled and reused in the next batch of fermentation. This will reduce not only the complexity but also the overall cost consumption of the process (Fan et al., 2020).

Optimization of physiological parameters that play a crucial role in the rate of performance and feasibility of the process should be properly maintained throughout the process, failing which proper yield coefficient cannot be reached. This can bring down the production volume and prominently shoot up the production cost (Padhan et al., 2023). The physiological parameters that are most vital to the proper functioning of the membrane-based fermentation methods may include pH, temperature, agitation speed, inoculum volume, and pretreatment process of the biomass. Appropriate parameters must be set in order to carry on the process efficiently. Heating up of the system may occur during the process and must be checked from time to time. The release of extracellular metabolites often alters the pH of the media and can be extremely derogatory to the fermentation outcome (Su et al., 2021). Overcoming these challenges is key to designing, executing, and maintaining a proper membrane-based fermenter system for achieving expected outcomes.

14.11 FUTURE PROSPECTS AND RESEARCH DIRECTIONS

The implications of membrane-based fermentation technology methods for bioethanol production from agrobiomass can have a huge impact on the global concept of bioenergy production using cheap waste materials. Agro-waste must be reused for the betterment of society on a larger scale without causing any harmful adversities to the environment. Sustainable alternatives to conventional methods are the key to achieve a green future. Cheap agro-wastes that are rich sources of valorizable biomolecules can be implicated for the production of economically important bioalcohol that has several applications.

Pretreatment being a vital step in the production process has many aspects to be considered in order to improve the efficiency of the overall process. Selection of the proper microbial strain with compatible enzymatic machinery that can utilize the agrobiomass to the maximum possible extent is an essential step and needs to be explored further. Many pretreatment enzymes that include delignifying enzymes like laccases, manganese peroxidases, lignin peroxidases, etc. can be used in crude or purified form in the process. Identifying such microbes is very important to initiate the pretreatment process of any agrobiomass. Besides, reducing the use of any chemical component is also a task to be considered, which will further increase the sustainability of such processes.

Despite numerous researches that have already taken place, many areas have remained unexplored and are prospective areas for future research. Exploring new agrobiomass materials from different sectors is highly encouraged. Isolation of new microbial strains from several sources, including respective waste sites, solid wastes, or waste effluents, might be considered since the possibility for the occurrence of such microbes is higher in those areas. Besides, overcoming the limitations of the existing methods is also an area of concern. Designing of the proper bioreactors, membrane immobilization, and optimization of physiological process parameters must be studied to enhance the future prospects of this promising technique.

14.12 CONCLUSION

The current global scenario with respect to fast replenishing rates of conventional fuel sources and demand for new technologies that project promising strategies for alternative energy generation is the need of the hour. The use of advanced fermentation technologies in combination with membrane-based technologies can play pioneer roles in addressing this crisis. The proper selection of substrate, microorganism, and specific membrane forms the basis for achieving success in this technology. Optimized process parameters and appropriate designing of the fermenters are extensive areas of research to increase the chances of product yield in less time and minimum investment. The use of waste agrobiomass material as the cheap low-cost substrate for bioethanol production using membrane-based fermentation technology can be a milestone toward achieving sustainable energy alternatives following the concept of "waste to wealth." These advanced technologies not only set new insights into the goal toward a sustainable future but also bring new opportunities to boost the national economy.

REFERENCES

Aghaei, S., Alavijeh, M. K., Shafiei, M., & Karimi, K. (2022). A comprehensive review on bioethanol production from corn stover: Worldwide potential, environmental importance, and perspectives. *Biomass and Bioenergy*, 161, 106447.

Ahmed, M. A., Amin, S., & Mohamed, A. A. (2023). Fouling in reverse osmosis membranes: Monitoring, characterization, mitigation strategies and future directions. *Heliyon*, 9, e14908.

Ali, M., Alabdulkarem, A., Nuhait, A., et al. (2022). Characteristics of agro waste fibers as new thermal insulation and sound absorbing materials: Hybrid of date palm tree leaves and wheat straw fibers. *Journal of Natural Fibers*, 19(13), 6576–6594.

Allen, B., Kretschmer, B., Baldock, D., et al. (2014). Space for energy crops-assessing the potential contribution to Europe's energy future. *Report produced for BirdLife Europe, European Environmental Bureau and Transport & Environment*. IEEP, London, 61.

Amiri, H., & Karimi, K. (2015). Improvement of acetone, butanol, and ethanol production from woody biomass using organosolv pretreatment. *Bioprocess and Biosystems Engineering*, 38, 1959–1972.

Anderson, S., & Rahman, P. K. (2018). Bioprocessing requirements for bioethanol: sugarcane vs. sugarcane bagasse. In Pathak, V. M., Navneet, N. (Eds), *Handbook of Research on Microbial Tools for Environmental Waste Management*. IGI Global, Hershey, PA, pp. 48–56.

Atsumi, S., & Liao, J. C. (2008). Metabolic engineering for advanced biofuels production from *Escherichia coli*. *Current Opinion in Biotechnology*, 19(5), 414–419.

Azhar, S. H. M., Abdulla, R., Jambo, S. A., et al. (2017). Yeasts in sustainable bioethanol production: A review. *Biochemistry and Biophysics Reports*, 10, 52–61.

Behera, S., Arora, R., Nandhagopal, N., & Kumar, S. (2014). Importance of chemical pretreatment for bioconversion of lignocellulosic biomass. *Renewable and Sustainable Energy Reviews*, 36, 91–106.

Bochtis, D., Achillas, C., Banias, G., & Lampridi, M. (Eds.). (2020). *Bio-economy and Agri-production: Concepts and Evidence*. Academic Press, London.

Boro, M., Verma, A. K., Chettri, D., et al. (2022). Strategies involved in biofuel production from agro-based lignocellulose biomass. *Environmental Technology & Innovation*, 28, 102679.

Brethauer, S., & Wyman, C. E. (2010). Continuous hydrolysis and fermentation for cellulosic ethanol production. *Bioresource Technology*, 101(13), 4862–4874.

Chen, M., Xia, L., & Xue, P. (2007). Enzymatic hydrolysis of corncob and ethanol production from cellulosic hydrolysate. *International Biodeterioration & Biodegradation*, 59(2), 85–89.

Chinnaswamy, R., & Hanna, M. (1990). Macromolecular and functional properties of native and extrusion-cooked corn starch. *Cereal Chemistry*, 67(5), 490–499.

Dadi, D., Beyene, A., Simoens, K., & et al. (2018). Valorization of coffee byproducts for bioethanol production using lignocellulosic yeast fermentation and pervaporation. *International Journal of Environmental Science and Technology*, 15, 821–832.

Dey, P., Pal, P., Kevin, J., & Das, D. (2020). Lignocellulosic bioethanol production: prospects of emerging membrane technologies to improve the process: A critical review. *Reviews in Chemical Engineering*, 36(3), 333–367. https://doi.org/10.1515/revce-2018-0014

Dolejš, I., Rebroš, M., & Rosenberg, M. (2014). Immobilisation of Clostridium spp. for production of solvents and organic acids. *Chemical Papers*, 68, 1–14.

Dutta, S., & Suresh Kumar, M. (2023). Potential use of thermophilic bacteria for second-generation bioethanol production using lignocellulosic feedstocks: a review. *Biofuels*, 2023, 1–14.

Espamer, L., Pagliero, C., Ochoa, A., & Marchese, J. (2006). Clarification of lemon juice using membrane process. *Desalination*, 200(1–3), 565–567.

El-Zawawy, W. K., Ibrahim, M. M., Abdel-Fattah, Y. R., et al. (2011). Acid and enzyme hydrolysis to convert pretreated lignocellulosic materials into glucose for ethanol production. *Carbohydrate Polymers*, 84(3), 865–871.

Fan, R., Burghardt, J. P., Prell, F., Zorn, H., & Czermak, P. (2020) Production and purification of fructo-oligosaccharides using an enzyme membrane bioreactor and subsequent fermentation with probiotic Bacillus coagulans. *Separation and Purification Technology*, 251, 117291.

Farahi, A., Najafpour, G., & Ghoreyshi, A. (2018). Composite multi wall carbon nano tube polydimethylsiloxane membrane bioreactor for enhanced bioethanol production from broomcorn seeds. *International Journal of Engineering*, 31(4), 516–523.

Faria, A. F., Liu, C. H., Xie, M., et al. (2017). Thin-film composite forward osmosis membranes functionalized with graphene oxide-silver nanocomposites for biofouling control. *Journal of Membrane Science*, 525, 146–156.

Frey, J. W., Caskey, B. J., & Lowe, B. S. (2007). *Relations of Principal Components Analysis Site Scores to Algal-Biomass, Habitat, Basin-Characteristics, Nutrient, and Biological-Community Data in the West Fork White River Basin*, Indiana, 2001: US Geological Survey.

Hossain, N., Zaini, J. H., & Mahlia, T. (2017). A review of bioethanol production from plant-based waste biomass by yeast fermentation. *International Journal of Technology*, 8(1), 291–319.

García, V., Päkkilä, J., Ojamo, H., et al. (2011) Challenges in biobutanol production: How to improve the efficiency? *Renewable and Sustainable Energy Reviews*, 15(2), 964–980. https://doi.org/10.1016/j.rser.2010.11.008.

Gautam, A., & Menkhaus, T. J. (2014). Performance evaluation and fouling analysis for reverse osmosis and nanofiltration membranes during processing of lignocellulosic biomass hydrolysate. *Journal of Membrane Science*, 451, 252–265.

Gautam, P., Upadhyay, S. N., & Dubey, S. K. (2020). Bio-methanol as a renewable fuel from waste biomass: Current trends and future perspective. *Fuel*, 273, 117783.

George, S., Larsson, G., & Enfors, S. O. (1993). A scale-down two-compartment reactor with controlled substrate oscillations: Metabolic response of Saccharomyces cerevisiae. *Bioprocess Engineering*, 9, 249–257.

González-Figueredo, C., Flores-Estrella, R. A., & Rojas-Rejón, O. A. (2018). Fermentation: Metabolism, kinetic models, and bioprocessing. *Current Topics in Biochemical Engineering*, 2018, 1–17.

Gryta, M., Morawski, A. W., & Tomaszewska, M. (2000). Ethanol production in membrane distillation bioreactor. *Catalysis Today*, 56, 159–165.

Harun, R., Jason, W. S. Y., Cherrington, T., & Danquah, M. K. (2011). Exploring alkaline pre-treatment of microalgal biomass for bioethanol production. *Applied Energy*, 88(10), 3464–3467.

Heinimö, J., Ranta, T., Malinen, H., & Faaij, A. (2011). *Forest Biomass Resources and Technological Prospects for the Production of Second-generation Biofuels in Finland by 2020*. Research Report/Lappeenranta University of Technology. Faculty of Technology. LUT Energy.

Ho, S. H., Huang, S. W., Chen, C. Y., Hasunuma, T., Kondo, A., & Chang, J. S. (2013). Bioethanol production using carbohydrate-rich microalgae biomass as feedstock. *Bioresource Technology*, 135, 191–198.

Hoşgün, E. Z., Berikten, D., Kıvanç, M., & Bozan, B. (2017). Ethanol production from hazelnut shells through enzymatic saccharification and fermentation by low-temperature alkali pretreatment. *Fuel*, 196, 280–287.

Imamoglu, E., & Sukan, F. V. (2014). The effects of single and combined cellulosic agrowaste substrates on bioethanol production. *Fuel*, 134, 477–484.

Kartawiria, I. S., Syamsu, K., Noor, E., & Legowo, E. H. (2015). Microbial contamination reduction of sorghum (Sorghum bicolor L.) stalk juice as bioethanol raw material through microfiltration process. *Iconiet Proceeding*, 1(2), 170–175.

Kasmuri, N. H., Kamarudin, S. K., Abdullah, S. R. S., et al. (2017). Process system engineering aspect of bio-alcohol fuel production from biomass via pyrolysis: An overview. *Renewable and Sustainable Energy Reviews*, 79, 914–923.

Khalid, A., Aslam, M., Qyyum, M. A., et al. (2019). Membrane separation processes for dehydration of bioethanol from fermentation broths: Recent developments, challenges, and prospects. *Renewable and Sustainable Energy Reviews*, 105, 427–443.

Kim, K. H., Choi, I. S., Kim, et al. (2014). Bioethanol production from the nutrient stress-induced microalga Chlorella vulgaris by enzymatic hydrolysis and immobilized yeast fermentation. *Bioresource Technology*, 153, 47–54.

Klocke, M., Nettmann, E., Bergmann, I., Mundt, K., Souidi, K., Mumme, J., & Linke, B. (2008). Characterization of the methanogenic Archaea within two-phase biogas reactor systems operated with plant biomass. *Systematic and Applied Microbiology*, 31(3), 190–205.

Kohse-Höinghaus, K., Oßwald, P., Cool, T. A., et al. (2010). Biofuel combustion chemistry: from ethanol to biodiesel. *Angewandte Chemie International Edition*, 49(21), 3572–3597.

Koschikowski J, Wieghaus M, & Rommel M. (2003). Solar thermal-driven desalination plants based on membrane distillation. Desalination 156295-304.

Kosowska-Golachowska, M., Otwinowski, H., Wolski, K., et al. (2018). Oxy-fuel combustion of wheat straw pellets in a lab-scale fluidized bed combustor. Paper presented at the Renewable Energy Sources: Engineering, Technology, Innovation: ICORES 2017.

Kour, D., Rana, K. L., Yadav, N., et al. (2019). Technologies for biofuel production: cCurrent development, challenges, and future prospects. *Prospects of renewable bioprocessing in future energy systems*, 2019, 1–50.

Kruyt, B., Van Vuuren, D. P., de Vries, H. J., & Groenenberg, H. (2009). Indicators for energy security. *Energy Policy*, 37(6), 2166–2181.

Kumar, B., & Verma, P. (2020). Application of hydrolytic enzymes in biorefinery and its future prospects. *Microbial Strategies for Techno-Economic Biofuel Production*, 2020, 59–83.

Kumar, R., Ghosh, A. K., & Pal, P. (2017) Fermentative energy conversion: Renewable carbon source to biofuels (ethanol) using Saccharomyces cerevisiae and downstream purification through solar driven membrane distillation and nanofiltration. *Energy Conversion and Management*, 150, 545–557. https://doi.org/10.1016/j.enconman.2017.08.054.

Kusworo, T. D., Yulfarida, M., Kumoro, A. C., & Utomo, D. P. (2023) Purification of bioethanol fermentation broth using hydrophilic PVA crosslinked PVDF-GO/TiO2 membrane. *Chinese Journal of Chemical Engineering*, 55(3), 123–136.

Lau, M. J., Lau, M. W., Gunawan, C., & Dale, B. E. (2010). Ammonia fiber expansion (AFEX) pretreatment, enzymatic hydrolysis, and fermentation on empty palm fruit bunch fiber (EPFBF) for cellulosic ethanol production. *Applied Biochemistry and Biotechnology*, 162, 1847–1857.

Laval, A. (2010). Membrane process opportunities and challenges in the bioethanol industry. *Desalination*, 250, 1067–1069.

Legodi, L. M., LaGrange, D. C., Jansen van Rensburg, E. L., & Ncube, I. (2021). Enzymatic hydrolysis and fermentation of banana pseudostem hydrolysate to produce bioethanol. *International Journal of Microbiology*, 2021, 1–14.

Lipnizki, F (2010) Membrane process opportunities and challenges in the bioethanol industry. *Desalination*, 250(3), 1067–1069. https://doi.org/10.1016/j.desal.2009.09.109.

Ma, F., Yang, N., Xu, C., Yu, H., Wu, J., & Zhang, X. (2010). Combination of biological pretreatment with mild acid pretreatment for enzymatic hydrolysis and ethanol production from water hyacinth. *Bioresource Technology*, 101(24), 9600–9604.

Mahboubi, A., Cayli, B., Bulkan, G., Doyen, W., De Wever, H., & Taherzadeh, M. J. (2018). Removal of bacterial contamination from bioethanol fermentation system using membrane bioreactor. *Fermentation*, 4(4), 88.

Manikandan, A., Muthukumaran, P., Poorni, S., Priya, M., Rajeswari, R., Kamaraj, M., & Aravind, J. (2022). Microbial approaches for bioconversion of agro-industrial wastes: A review. In: Aravind, J., Kamaraj, M., Karthikeyan, S. (eds) *Strategies and Tools for Pollutant Mitigation: Research Trends in Developing Nations*, Springer, Cham, 151–180.

Mao, H., Zhen, H. G., Ahmad, A., et al. (2019). Highly selective and robust PDMS mixed matrix membranes by embedding two-dimensional ZIF-L for alcohol permselective pervaporation. *Journal of Membrane Science*, 582, 307–321.

Marchenko, V., Sorokin, A., Sidelnikov, D., & Panasenko, A. (2017). Investigation in process of fermentation medium mixing in bioreactor. Engineering for Rural Development. https://doi.org/10.22616/ERDev2017.16.N174

Meixner, K., Fuchs, W., Valkova, T., et al. (2015). Effect of precipitating agents on centrifugation and ultrafiltration performance of thin stillage digestate. *Separation and Purification Technology*, 145, 154–160.

Negro, M. J., Manzanares, P., Oliva, J. M., et al. (2003). Changes in various physical/chemical parameters of Pinus pinaster wood after steam explosion pretreatment. *Biomass and Bioenergy*, 25(3), 301–308.

Nemtsov, D., & Zabaniotou, A. (2008). Mathematical modelling and simulation approaches of agricultural residues air gasification in a bubbling fluidized bed reactor. *Chemical Engineering Journal*, 143(1–3), 10–31.

Nitsos, C., Matsakas, L., Triantafyllidis, K., Rova, U., & Christakopoulos, P. (2018). Investigation of different pretreatment methods of Mediterranean-type ecosystem agricultural residues: Characterisation of pretreatment products, high-solids enzymatic hydrolysis and bioethanol production. *Biofuels*, 9(5), 545–558.

Padhan, B., Ray, M., Patel, M., & Patel, R. (2023). Production and bioconversion efficiency of enzyme membrane bioreactors in the synthesis of valuable products. *Membranes*, 13, 673.

Pal, P., Kumar, R., Ghosh, A. K. (2018). Analysis of process intensification and performance assessment for fermentative continuous production of bio-ethanol in a multi-staged membrane-integrated bioreactor system. *Energy Conversion and Management*, 171, 371–383.

Pandey, A., Soccol, C. R., Nigam, P., & Soccol, V. T. (2000). Biotechnological potential of agro-industrial residues. I: sugarcane bagasse. *Bioresource Technology,* 74(1), 69–80.

Panesar, P. S., Marwaha, S. S., & Kennedy, J. F. (2006). Zymomonas mobilis: an alternative ethanol producer. *Journal of Chemical Technology & Biotechnology: International Research in Process, Environmental & Clean Technology*, 81(4), 623–635.

Polprasert, S., Choopakar, O., & Elefsiniotis, P. (2021). Bioethanol production from pretreated palm empty fruit bunch (PEFB) using sequential enzymatic hydrolysis and yeast fermentation. *Biomass and Bioenergy*, 149, 106088.

Priya, A., Dureja, P., Talukdar, P., Rathi, R., Lal, B., & Sarma, P. M. (2016). Microbial production of 2, 3-butanediol through a two-stage pH and agitation strategy in 150 l bioreactor. *Biochemical Engineering Journal*, 105, 159–167.

Ramos, A., Monteiro, E., & Rouboa, A. (2022). Biomass pre-treatment techniques for the production of biofuels using thermal conversion methods-A review. *Energy Conversion and Management*, 270, 116271.

Sakhtah, H., Behler, J., Ali-Reynolds, A., Causey, T. B., Vainauskas, S., & Taron, C. H. (2019). A novel regulated hybrid promoter that permits autoinduction of heterologous protein expression in Kluyveromyces lactis. *Applied and Environmental Microbiology*, 85(14), e0054219.

Sanderson, M. A., Egg, R. P., & Wiselogel, A. E. (1997). Biomass losses during harvest and storage of switchgrass. *Biomass and Bioenergy*, 12(2), 107–114.

Sharma, A., Giri, S. K., Kartha, K. R., & Sangwan, R. S. (2017). Value-additive utilization of agro-biomass: Preparation of cellulose triacetate directly from rice straw as well as other cellulosic materials. *RSC Advances*, 7(21), 12745–12752.

Sivaramakrishnan, R., Shanmugam, S., Sekar, M., et al. (2021). Insights on biological hydrogen production routes and potential microorganisms for high hydrogen yield. *Fuel*, 291, 120136.

Su, H., Zhang, Q., Yu, K., et al. (2021) A novel neutral and mesophilic β-glucosidase from coral microorganisms for efficient preparation of gentiooligosaccharides. *Foods*, 10, 2985.

Sulfahri, M. S., Langford, A., & Tassakka, A. C. M. A. (2020). Ozonolysis as an effective pretreatment strategy for bioethanol production from marine algae. *BioEnergy Research*, 13, 1269–1279.

Tian, X. F., Fang, Z., & Guo, F. (2012). Impact and prospective of fungal pre-treatment of lignocellulosic biomass for enzymatic hydrolysis. *Biofuels, Bioproducts and Biorefining*, 6(3), 335–350.

Weng, Y. H., Wei, H. J., Tsai, T. Y., et al. (2009). Separation of acetic acid from xylose by nanofiltration. *Separation and Purification Technology*, 67(1), 95–102.

Zentou, H., Abidin, Z. Z., & Yunus, R., et al. (2019). Overview of alternative ethanol removal techniques for enhancing bioethanol recovery from fermentation broth. *Processes*, 7(7), 458.

Zhen, X., Wang, Y., & Liu, D. (2020). Bio-butanol as a new generation of clean alternative fuel for SI (spark ignition) and CI (compression ignition) engines. *Renewable Energy*, 147, 2494–2521.

15 Membrane-Based Enzymatic Technique for the Synthesis and Purification of Bioalcohols from Agrobiomass

A Review

Sujoy Bose, Ajay K. Shakya, and Chandan Das

15.1 INTRODUCTION

As global climate change and environmental degradation become a significant concern in the world due to the continuous diminution of fossil energy reserves and the emission of greenhouse gases and various environmental pollutants (lead, carbon monoxide, nitrogen oxides, particulate matters, volatile organic compounds, and sulphur, etc.), an alternate source of energy, i.e., renewable energy, plays a vital role in the present and future period to overcome the issue of global warming and to stabilize the climate by managing carbon dioxide emissions (Chandrasekhar et al., 2021; Singh et al., 2021).

Based on the findings of the Global Energy and CO_2 Status Report (2017) published by the International Energy Agency (IEA), it is projected that global primary energy consumption will significantly increase by 55% from 2005 to 2030, with an average annual growth rate of 1.8%. Additionally, worldwide renewable energy sources are expected to have a 15% rise by 2040, as stated by the U.S. Energy Information Administration in 2013. India also aims to meet 50% of their energy requirements from renewables (500 GW) by 2030 (Report on Renewable Energy in India, 2022). Renewable energy sources such as wind (40.03 GW), solar (48.55 GW), hydropower (51.34 GW), biopower (10.62 GW), and nuclear power (6.78 GW) can generate electrical energy and replace fossil fuels. Consequently, a lot of research has been done recently on finding green alternatives to support renewable fuels and energy.

Bioalcohols such as biomethanol, bioethanol, biopropanol, and biobutanol are renewable and sustainable products derived from the green agrobiomass, which may

 DOI: 10.1201/9781003407713-15

lead to less carbon footprint. Bioethanol is the most important and commonly used biofuel among all bioalcohols. It has received significant attention among researchers, scientists, and world policymakers due to its renewability, biodegradability, low toxicity, and better quality of exhaust gas emission. The green agrobiomass utilized for bioalcohol production is grouped into (i) sugar and starchy biomass, (ii) lignocellulosic biomass (LCB), (iii) triglycerides, and (iv) hydrocarbon. Sweet crops, such as sugarcane and its derivatives, palm and fruit juice, sugar beet, and starchy crops, including wheat, barley, rice, potato, corn, and root crops, are categorized as sugar-sweetened beverages. Sugar-sweetened beverages are the principal sources of first-generation (1G) bioethanol. Second-generation (2G) bioethanol primarily originates from LCB, predominantly sourced from forest wood waste, sawdust, agro-industrial leftovers, aquatic plants, and grasses. Triglycerides or lipids are derived from several sources, such as microalgal oils, animal fat waste, edible vegetables, and non-edible oils. However, their use in bioalcohol synthesis is infrequent. Hydrocarbon biofuels are frequently generated by several technological processes such as hydrotreating, gasification, pyrolysis, and other biochemical and thermochemical methods (Goswami et al., 2022).

The primary components of lignocellulosic bio-waste consist of cellulose (35%–50%), hemicellulose (20%–35%), and lignin (5%–30%). Cellulose is an inherent linear polymer characterized as a polysaccharide complex consisting primarily of glucose units. Hemicelluloses represent a diverse collection of biopolymers. The composition of this substance includes a diverse array of monosaccharide subunits, such as arabinose, mannose, galactose, glucose, and xylose. Lignin is the second most prevalent organic polymer, following cellulose, within the natural environment. The substance is an intricate, non-crystalline polymer composed of phenylpropanoid constituents, specifically p-coumaryl, coniferyl, and sinapyl alcohol. According to Deshavath et al. (2022), the bacterium metabolizes the C5 and C6 carbon compounds found in cellulose and hemicellulose, producing diverse bioproducts. There has been significant encouragement for developing second-generation biofuels using lignocellulosic bio-waste in recent years. This bio-waste is obtained from many sources, such as agricultural leftovers and wastes, forest residue, wood-processing waste, municipal garbage (organic), and energy crops grown explicitly for this purpose. However, a significant obstacle in manufacturing bioethanol from lignocellulosic bio-waste lies in pretreating the feedstock to enhance the accessibility of carbohydrates inside the lignocellulose for conversion purposes. Bioethanol production encounters significant hurdles, including substantial technological risk, elevated production costs, and political and policy concerns that offer limited potential rewards. These challenges necessitate careful consideration and resolution.

In addition to the various problems, it has been observed that LCB stands out as a highly favoured option in terms of economic viability, abundance, renewability, and cost-effectiveness compared to fossil-based resources. Consequently, it has become a potentially viable replacement for fossil fuels based on petroleum. LCB is an economically viable and easily accessible source of plant material, frequently obtained as a by-product or waste from agro-industrial processes. Furthermore, it has been observed that various components of LCB can be economically harnessed to produce a wide range of value-added products, as highlighted by Velvizhi et al. (2022).

Bioethanol is a commonly employed advanced liquid biofuel that is a viable alternative to conventional fossil fuels. The primary means of production include the hydrolysis of cellulose derived from LCB and the fermentation of sugars obtained from various lignocellulosic sources. Acids or enzymes are employed in biomass pretreatment to facilitate biomass conversion into sugars. The primary techniques used to extract sugars from agricultural biomass include concentrated acid hydrolysis, dilute acid hydrolysis, and enzymatic hydrolysis. Enzymatic hydrolysis is a crucial unit operation that has garnered much attention in recent times owing to its substantial impact on the economic aspects of the process. Enzymatic hydrolysis encounters various challenges, including but not limited to enzyme inhibition through feedback mechanisms, presence of oligosaccharides, inhibitors derived from biomass, lack of synergistic effects, limitations in mass transfer, water constraints, and non-productive interactions between enzymes and lignin (Huang et al., 2022). The primary advantages of bioethanol over other biofuels lie in its biodegradability and decreased toxicity. Bioethanol primarily functions as a fuel additive to blend with petrol, either on its own for light cars or in combination with petrol in a predetermined ratio. This blending process enhances the octane number, improving the fuel's quality.

Furthermore, bioethanol exhibits several advantages when employed as a substitute for fossil fuels. These include a notably elevated combustion rate, resulting in comparatively cleaner exhaust emissions. Additionally, bioethanol demonstrates minimal pollution levels and emits fewer greenhouse gases. Moreover, it possesses the advantage of being easily dilutable and less poisonous. Bioethanol as a fuel in colder temperatures presents difficulties due to its low vapour pressure, resulting in vehicle cold start issues. The extent of CO_2 emissions during production remains a subject of debate.

Bioethanol production has received greater attention than other bioalcohol studies for several reasons:

- Ethanol is a high-octane fuel that can be used as a gasoline additive and significantly reduce carbon emissions when blended with gasoline. This has made it a popular choice for countries that aim to meet their renewable fuel standards and reduce their reliance on fossil fuels.
- Ethanol is produced from renewable biomass feedstock such as com sugarcane and cellulosic 107 materials, which are abundant and easily accessible. This makes it a promising alternative to traditional fossil fuels, which are finite and associated with several environmental concerns.
- Ethanol is a well-established industry that has been in existence for many years. The knowledge, infrastructure, and regulatory framework for ethanol production are already in place, which makes it easier to scale up production and bring it to market.
- The by-products of ethanol production, such as distillers' grains, can be used as animal feed, adding value to the process and making it more economically feasible.
- Finally, the environmental benefits, the abundance of feedstock, the established industry, and the economic feasibility of ethanol production have made it a more attractive option than other Bio-alcohol studies.

65% of all biofuels are made up of bioethanol, which guarantees the country's energy and financial security. By 2026, the yearly global demand for biofuels is anticipated to increase by 28% to 186 billion litres. Advanced biofuels reduce greenhouse gases by 39%–46%, accounting for 5% of the world's energy use. They also continue to enhance environmental benefits. In terms of volume produced, the United States leads the world in bioethanol production. Due to the epidemic, Asia is still expected to surpass the United States and Europe. By 2026, Asia will surpass Europe and the United States in the production of biofuels, accounting for about 30% of new bioethanol produced. Most of the growth in Asia is attributable to India's recently revised National Policy on Biofuel (NPB, 2018), which suggested a recommended blending target of 20% ethanol in petrol. India's demand for bioethanol has tripled over the last five years, and by 2026, it is expected to be the third-largest bioethanol consumer in the world (IEA, 2021; Renewables 2021).

Pretreating biomass is typically the first step in turning LCB into chemicals or fuels. This process lowers cellulose crystallinity and removes hemicellulose from cellulose, making cellulose easier to hydrolyze. Although a more expensive procedure, it can boost productivity, encourage cellulose accessibility, and enhance overall process efficiency, facilitating biomass conversion into value-added products in subsequent steps. Pretreatment techniques can generally be divided into four groups: physical, chemical, physicochemical, and biological (Velvizhi et al., 2022). Physical procedures include pyrolysis, milling, microwaves, ultrasonography, and mechanical extrusion. It is more straightforward, but has the disadvantage of using a lot of energy. The chemical approaches for the pretreatment of LCB include ammonia fibre explosion, steam explosion, Organosolv, liquid hot water, and ionic liquid. A few examples of physicochemical pretreatment are CO_2 explosions, ammonia fibre explosions, and steam explosions. The biomass is split into these methods by applying mechanical forces and chemical reactions. Whole-cell and enzymatic pretreatments, which are biological, are energy-efficient and environmentally friendly methods. Several other pretreatment techniques have been established, including nitrobenzene (Wang et al., 2018), sulphite pulping (Fache et al., 2016), enzymatic hydrolysis (Guo et al., 2018), alkaline treatment (Sun et al., 2018), acid hydrolysis (García et al., 2017), chemical crosslinking (Sathawong et al., 2018), and fungal hydrolysis (Xu et al., 2018). Enzymes can be employed in bioconversion in two different forms: immobilized or freely distributed. There are benefits and drawbacks to both tactics by nature. The enzyme's immobilization helps regulate its separation from the reaction mixture by preserving pH and heat stability. However, 10% regained activity is a rare result of immobilized enzymes, which could be detrimental (Kamrat et al., 2007).

Enzyme-catalyzed bioconversion can typically be carried out continuously or in batches. Continuous enzyme-catalyzed bioconversion has an advantage over batch-wise bioconversion since it does not need to stop the catalytic reaction to recover the products. Several reactor types, including packed bed reactors (PBR), continuously stirred tank reactors (CSTR), and microchannels (MC), can be used for continuous enzyme-catalyzed bioconversion. For CSTR, PBR, and MC processes, recovering the enzyme at the end of the process and immobilizing it might be expensive.

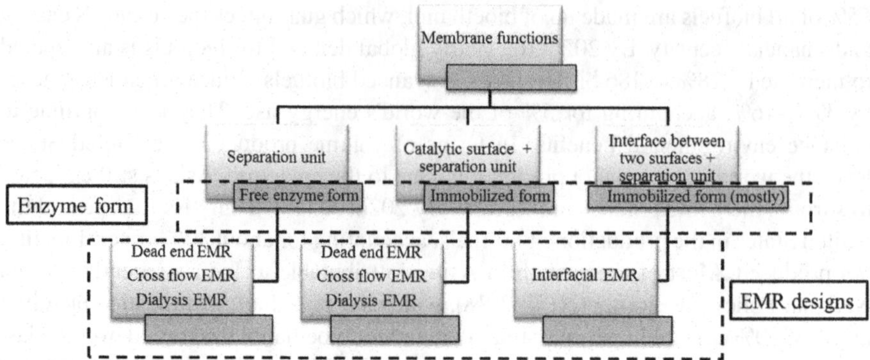

FIGURE 15.1 Configurations of EMR based on membrane functions. (Reproduced with copyright permission from Sitanggang et al., 2022).

Compared to conventional technologies, the membrane as a separator and reactor—that is, a membrane reactor—may be the greatest option for producing bioethanol. Alternatively, the enzymatic membrane reactor (EMR) is considered a unique and economical reactor design that can manage product separation and catalytic reaction simultaneously, especially for the hydrolysis of giant molecules. EMR can be categorized depending on how enzymes and substrates make contact. These mechanisms fall into three categories: interfacial contact, contact by diffusion, and direct contact. Additionally, according to the hydrodynamic and filtration properties of the membrane functions (as shown in Figure 15.1), EMR can be divided into the following groups (Sitangganga et al., 2022). These groups include:

i. Dead-end EMR
ii. Cross-flow EMR
iii. Dialysis EMR
iv. Interfacial/contactor EMR (see Figure 15.2a–d)

In an EMR, the membrane functions with the following benefits:

- Selectively separate the products from the enzymatic reaction mixture.
- Control the addition of substrate to the enzymatic reaction mixture.
- Strengthen the contact between substrate and enzyme.

The size exclusion EMR was primarily researched in energy, medicines, and agro-food. These days, enzymatic biofuel production has greatly investigated the use of EMR. The extent to which freely distributed enzyme molecules are retained is contingent upon selecting the molecular weight cut-off of the membrane. During the hydrolysis of macromolecules, it is possible for the substrates to undergo rejection and for the products to enter.

The technological and commercial challenges of producing conventional bioalcohol are compiled in this article. On the other hand, recent developments in membrane-based enzymatic methods for the more environmentally friendly synthesis and

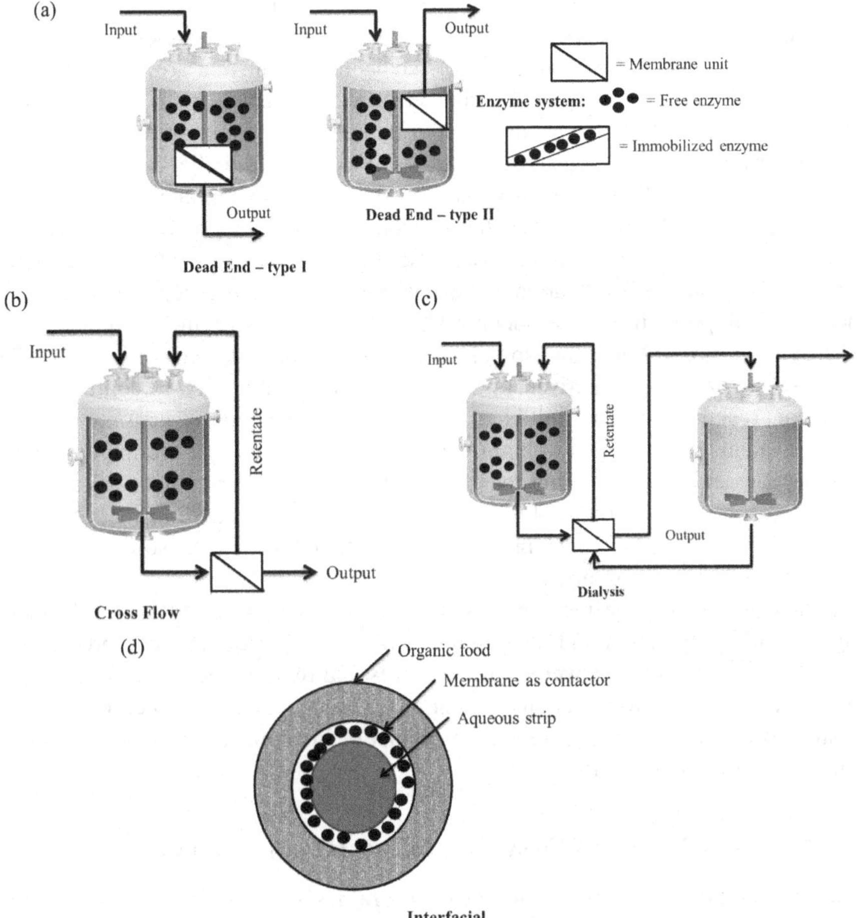

FIGURE 15.2 Schematic diagrams of existing enzymatic membrane reactor system where the enzyme is freely dispersed (a–c) and immobilized (d). (Reproduced with copyright permission from Sitanggang et al., 2022).

purification of bioalcohols from agro-based LCB were highlighted. To concentrate on the demands for future research and perspectives on membrane-based technologies in the synthesis and purification of bioalcohols, technological obstacles have also been addressed collectively.

15.2 APPLICATION OF MEMBRANES FOR THE PRODUCTION OF BIOALCOHOLS

Membrane technology is beneficial in synthesizing bioalcohols to purify and separate fermentation products. It is a common practice to obtain bioalcohols like ethanol, butanol, and propanol by fermentation of various feedstocks, including corn, sugarcane, and LCB. The required bioalcohol, water, and other contaminants are removed from the fermentation broth using membrane-based techniques. The most commonly used membrane processes for bioalcohol production are microfiltration (MF), ultrafiltration (UF), nanofiltration (NF), and pervaporation (PV). MF separates suspended solids, bacteria, and yeast from the fermentation broth. UF is used to separate proteins and other macromolecules. NF separates smaller molecules, such as salts and sugars. PV is a process that uses a selective membrane to separate the desired bioalcohol from the fermentation broth by vaporization. Using membrane technology in bioalcohol production offers several advantages over traditional separation methods such as distillation and evaporation. These advantages include lower energy consumption, higher product purity, and reduced environmental impact. The production of bioalcohols can also benefit from using membrane technology, which can raise the process's effectiveness, economy, and sustainability. In commercially significant reactions, product-inhibited and equilibrium reaction systems are particularly prevalent. Reactions can obtain very high conversions that are impossible in traditional reactors by removing reaction products as they form. Both organic and inorganic materials can be used to make the membrane. Because of their superior thermal stability, inorganic membranes are better suited for organic solvents and can withstand high reaction temperatures.

Comparing membrane reactors to conventional reactors, the following benefits might be noted: Reaction and Separation should be combined into one process for the following reasons: (i) lower separation costs and recycle needs; (ii) higher conversions per pass due to the augmentation of reactions that are product- or thermodynamically constrained; (iii) regulated contact of incompatible reactants; and (iv) removal of unwanted side reactions.

15.2.1 Issues Related to Conventional Bioalcohol Production

Conventional bioalcohol production faces several issues related to feedstock availability and competition, energy-intensive production, environmental impacts, low energy efficiency, and limited scalability. The primary feedstocks used for bioalcohol production, such as corn, sugarcane, and wheat, compete with food and animal feed production. This competition can lead to higher prices and supply constraints. The conventional bioalcohol production process involves a substantial amount of energy, particularly in the form of fossil fuels, to power the various stages of the process, such as harvesting, transportation, and fermentation. Conventional bioalcohol production can also have significant environmental impacts, including soil erosion, water pollution, and greenhouse gas emissions. This limitation makes it challenging to produce bioalcohol on a large scale to meet the increasing demand for renewable fuels. The problems depicted in Figure 15.3a are thoroughly explained below.

15.2.1.1 Feedstock Availability

Traditional bioalcohol production depends on a small proportion of feedstocks, which might rival these resources and restrict the industry's capacity to scale up. This rivalry may raise food costs and increase food insecurity. A few examples of lignocellulose include switch grass, maize stalks, wood, herbaceous crops, waste

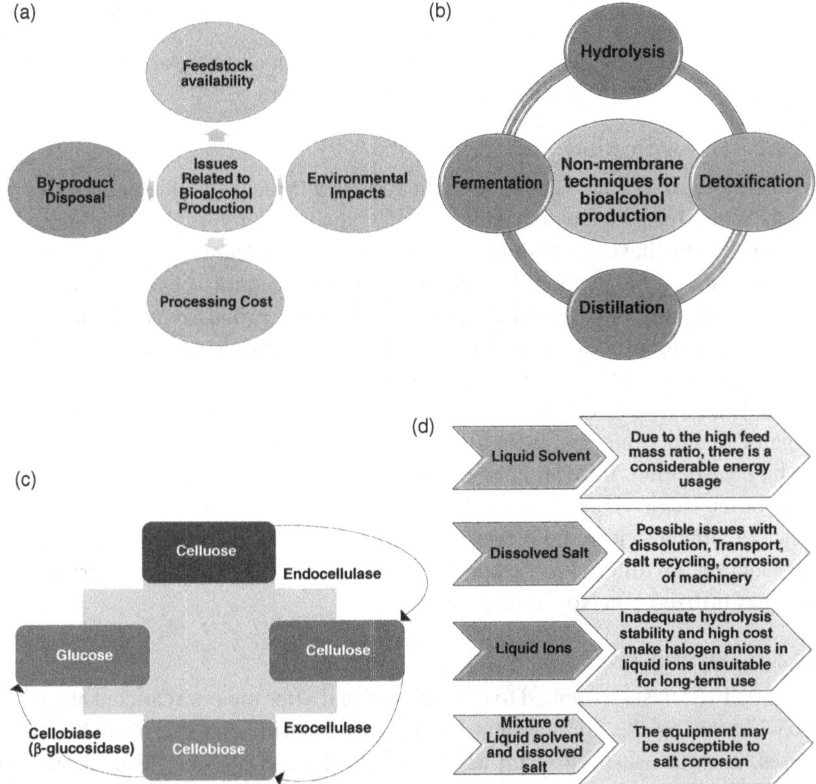

FIGURE 15.3 (a) Concerns with the production of bioethanol, (b) non-membrane techniques available for bioalcohol production, (c) mode of action of components of cellulose, and (d) challenges in the separation of bioalcohol during extractive distillation.

paper and paper products, agricultural and forestry leftovers, pulp and paper mill trash, municipal solid waste, and food sector waste. LCB comprises cellulose, hemicellulose, lignin, protein, ash, and trace amounts of extractives (Toor et al., 2020).

15.2.1.2 Environmental Impacts

Fossil fuel exhaustion and environmental change brought on by greenhouse gas emissions from fossil fuel burning. The carbon cycle naturally regenerates fossil fuels at a rate that is much slower than the rate at which they are currently consumed. Most fossil fuel reserves are concentrated in a limited number of nations, which further contributes to the unsustainable nature of their production. Moreover, increased greenhouse gas emissions result from burning fossil fuels and changes in land use brought on by human activity, exacerbating global warming.

15.2.1.3 Processing Cost

In a recent study, LCB is considered feedstock because it is readily available and inexpensive to acquire to lower manufacturing costs. Regrettably, the procedure is still economically undesirable because of the high processing costs (Sindhu et al., 2019).

15.2.2 Types of Non-Membrane Techniques for Bioalcohol Production

The non-membrane approaches for producing bioalcohol are illustrated in Figure 15.3b. The methods are described in detail below.

15.2.2.1 Detoxification

When using LCB as a feedstock, detoxification is a critical step in synthesizing bioalcohol. Several substances included in LCB, including lignin, hemicellulose, and other extractives, have been shown to hinder the fermentation process and lower the production of bioalcohol. The detoxification procedure seeks to eliminate or minimize these inhibitory substances to increase the productivity of the bioalcohol manufacturing process. The objective of detoxification is to completely remove harmful substances from pretreated biomass or hydrolysates, including chemicals that could impede fermentation and diminish the efficacy of enzymes. At present, a range of in situ strategies, including solvent extraction, ion exchange, adsorption, microbial adaptation, utilization of microbial consortium or tailored microbes, and various other methods, are commonly utilized to mitigate inhibitors in biomass and enhance bioethanol yield and productivity. The methodologies employed in bioalcohol production are tailored to the specific pretreatment, hydrolysis, and fermentation techniques utilized (Da Nogueira et al., 2021).

15.2.2.2 Hydrolysis

Enzymatic hydrolysis is applied to the raw material after the pretreatment stage. This approach converts the pretreatment LCB polysaccharides into fermentable sugars, pentoses, and hexoses. Although acids and alkalis can also be utilized, enzymes are typically used to catalyze the hydrolysis of cellulose and hemicellulose (Tsai et al., 2014). Cellulases are enzymes that can hydrolyze cellulose into glucose monomers. The various cellulase component action methods are shown in Figure 15.3c. Exo-1,4-β-D-glucanase randomly cleaves intermolecular links in cellulose, whereas endo-1,4-β-D-glucanase removes monomers and dimers off the end of the glucose chain. β-glucosidase hydrolyzes cellobiose, glucose dimers, and other short cellulose oligomers into glucose monomers. All three components must work together for a native cellulose polymer to hydrolyze completely into glucose monomers. Numerous bacteria and fungi produce cellulases (Srivastava et al., 2018).

15.2.2.3 Fermentation

The pretreatment and hydrolysis processes are crucial for a fermentation process to be improved. Even if the process of fermentation is a natural one, the conversion of fermentable sugars into alcohol still requires the presence of microbes. Several industrial yeasts, including *S. cerevisiae*, were employed for these purposes, primarily in the wine, brewery, and other alcoholic beverage production industries. The sugar-based biofuel industry also utilizes the primary fermentative strain, *S. cerevisiae*. When the cellulose-based material is processed, it is transformed into fermentable sugars and made accessible for either enzymatic or acidic hydrolysis. Such a cellulose-based material turns into broth when combined with water. *S. cerevisiae* is typically used in batch fermentation to convert hexose carbohydrates, such as glucose, into ethanol under anaerobic circumstances and at a predetermined temperature (Naghshbandi et al., 2019). Depending on the lignocellulosic hydrolysate and the microorganisms' kinetics, fermentation can occur continuously, in batches, or in fed-batch mode. In a continuous process, a reactor containing active microorganisms is continually pumped with a substrate that has feed and culture media. The batch process is a straightforward procedure seen as a closed culture system with a finite beginning supply of nutrients. These cultures contain fermentation-producing microorganisms as inoculants. Because fed-batch reactors have numerous benefits from continuous and batch processes, they are commonly utilized in industrial applications. The fed-batch procedure can raise the amount of maximum viable cells in suspension, prolong the life of the culture, and accumulate more concentrated products (Gutiérrez-Antonio et al., 2017). Combining the fermentation and hydrolysis steps in a single reactor is possible using various methods and techniques (Szambelan et al., 2018). Enzymatic hydrolysis and fermentation can each be going through under ideal circumstances using a process known as separate hydrolysis and fermentation. When lignocellulosic materials are used in pretreatment methods that mix diluted acid with high temperatures, simultaneous saccharification and fermentation (SSF) is a successful method to generate bioethanol. The SSF process increases the hydrolysis rate while reducing enzyme loading, leading to higher bioethanol yields with little to no danger of contamination. The key benefits of the SSF process include faster hydrolysis rates due to the transformation of sugars, low enzyme concentration with greater product yields, smaller reactor volumes, and shorter process times. Moreover, the microbes instantly devour sugars, causing fermenters to have low sugar content (Luiza et al., 2020). The formation of cellulose, its hydrolysis, and the fermentation of glucose are all included in the direct microbial conversion process. Such a procedure streamlines the manufacturing process and decreases the number of reactors, lowering expenses (Prasad et al., 2019).

15.2.2.4 Distillation

The transformation of LCB into fuel-grade bioalcohol requires distillation and dehydration. Distillation is a method that efficiently separates a component chemical (like ethanol) from a fermentation broth by a series of selective evaporation and condensation processes based on differences in their volatilities (Kang et al., 2019). Usually comprising more than 80% of the dry weight, the post-fermentation mixture has an extremely high-water content. Consequently, an enormous amount of energy is required to concentrate ethanol to a 96% level, leading to exorbitant costs (Haghighi Mood et al., 2013). The outcome, comprising around 37% bioethanol, undergoes a concentration process within a rectification column to achieve a concentration level of approximately 95%. This is achieved when the bioethanol is subjected to dehydration, producing a high-quality dry product containing a minimum volume of 99.5% ethanol (Hamelinck et al., 2005). To extract bioethanol from a fermentation broth, several techniques have been developed, including vacuum distillation, chemical dehydration, azeotropic distillation (AD), diffusion distillation, extractive distillation (ED), PV, and adsorption distillation. The strategies, effectiveness, and operational expenses of these approaches differ.

15.2.3 Challenges in Synthesizing and Purification of Bioalcohol Using Traditional Techniques

Bioalcohols, such as ethanol and butanol, can be synthesized through the fermentation of biomass using microorganisms, such as yeast and bacteria. However, synthesizing and purifying bioalcohols using traditional techniques can present several challenges.

15.2.3.1 Separation from Water

Bioalcohols are highly soluble in water, which can make it challenging to separate the alcohol from the aqueous solution during the purification process. This can require energy-intensive distillation or dehydration methods.

15.2.3.2 Azeotropic Distillation

In order to produce a new, heterogeneous, lower-boiling azeotrope, AD breaks azeotropes to which another volatile material, such as an entrainer, solvent, or mass-separating agent, is added. When selecting entrainers, health and safety factors, including the potential for benzene toxicity and the thermal characteristics of cyclohexane, must be considered. Due to its drawbacks—expensive capital expenditures, high energy requirements, and potential health and safety risks—the aforementioned AD system is used less frequently for ethanol separation and purification (Huang et al., 2010).

15.2.3.3 Extractive Distillation

A miscible, extremely flammable, or generally non-volatile solvent is usually used in the vapour–liquid separation process, or ED, to increase the relative volatility of the components that need to be separated. The chemical industry often uses this method to separate mixtures with close boiling points or azeotropes. The added component as a separating agent can be an ionic liquid, a hyper-branched polymer, dissolved salt, a standard liquid solvent combined with dissolved salt, or any combination. The challenges of bioalcohol separation during ED using different separating agents are shown in Figure 15.3d.

15.2.3.4 Extractive Fermentation

The extractive fermentation method employs in situ extraction to eliminate ethanol and additional inhibitory chemicals from the product. Removing inhibitory effects induced by ethanol and other inhibitors enhances the ethanol output. Various biocompatible solvents have been examined for in situ ethanol extraction from beer spirits. These solvents include oleyl alcohol (Weilnhammer & Blass, 1994), n-dodecanol (Boluda et al., 2005), isoamyl acetate, and iso-octyl alcohol (Koullas et al., 1999). Isoamyl acetate and nonanoic acid are two other potential solvents that exhibit biocompatibility. To mitigate the inhibitory effects of ethanol production, the application of oleyl alcohol was employed for in situ ethanol extraction inside a continuous fermentation process for ethanol synthesis. This process used the thermophilic and anaerobic bacteria known as *Clostridium thermohydrosulfuricum*. According to Weilnhammer and Blass (1994), the ethanol yield obtained via extractive fermentation is twice as high as conventional fermentations that do not incorporate in situ extraction.

15.2.3.5 Contamination

There is a chance that other microorganisms could contaminate the fermentation process and outcompete the targeted strains, resulting in lower yields and contaminants in the final product. It is well known that pretreating lignocelluloses with diluted acid hydrolysis and steam explosion releases many poisonous chemicals to microorganisms and appears to lower fermentation productivity and yield. Other compounds released include fermentable pentose and hexose sugars. To produce a high-quality bioalcohol product, different purifying procedures may be required.

Detoxification is a crucial stage, and choosing the proper detoxification procedure is vital because the process can be costly and account for a significant amount of the overall cost of producing ethanol. For instance, a study revealed that detoxification accounted for 22% of the cost of producing ethanol using willow as the feedstock (Von Sivers et al., 1994). In general, there are three main categories of inhibitors: phenolic compounds, furan derivatives, and aliphatic acids. Moreover, different inhibitory compounds and variable concentrations of inhibitors result from using various biomasses as feedstock, pretreatment techniques, and fermentation organisms. In addition to optimizing the pretreatment and hydrolysis process to reduce the formation of inhibitors, it is required to detoxify hydrolysates before fermentation or perform in situ detoxification to increase hydrolysate efficiency fermentation (Palmqvist & Hahn-Hagerdal, 2000).

15.2.3.6 Energy Consumption

Traditional synthesizing and purifying bioalcohol techniques can be energy-intensive, increasing production's overall cost and environmental impact. Ordinary distillation is straightforward to operate, but it requires a lot of trays and uses a lot of energy. There are fewer trays required for AD than for regular distillation. However, because many azeotropic agents are present, they must be vaporized in the column, which consumes much energy.

15.2.3.7 Low Yield

Traditional fermentation techniques may produce bioalcohols with relatively low yields, which raises the cost of production. This may be because the metabolic pathways of the utilized microorganisms are constrained or because there are problems with the supply and quality of the biomass feedstock. Three types of inhibitors are released when lignocellulose is processed with acid or alkali at high temperatures: Inhibitors of microbe development and ethanol yield during fermentation include (i) furan derivatives, (ii) phenolic compounds, and (iii) weak acids (He et al., 2012).

15.2.3.8 Chemical Impurities

Depending on the fermentation methods employed, chemical contaminants, including aldehydes, esters, and organic acids, may be present in bioalcohols. The performance and quality of the bioalcohol product may be adversely affected by these contaminants. For example, compared to AD, ED cannot produce products with high purity because the solvent exiting the bottom of the solvent recovery column generally retains certain impurities that affect the separation effect. In order to address these issues, scientists are looking into novel approaches to manufacturing bioalcohol, such as genetically modifying microbes, creating more effective separation and purification procedures, and using renewable energy sources.

15.2.4 Enzymatic Membrane Reactor

The process of enzyme-catalyzed biotransformation can be carried out either batch-wise or continuously. In batch processing, pausing the catalytic reaction to retrieve the products is necessary before commencing a new cycle. Consequently, the complex setup and shutdown procedures may result in more unproductive operating time and significant labour expenditures (Prazeres & Cabral, 1994). Moreover, product quality fluctuations from batch to batch could happen. With various reactor designs, an ongoing enzyme-promoted bioconversion can be carried out. Higher expenses to extract and immobilize the enzyme at the end of the process are typical constraints for MC, CSTR, and PBR operations (Sitanggang et al., 2022).

Chemical processes occur in the EMR, a bioreactor combining membrane filtration and enzyme catalysis. In EMRs, enzymes accelerate processes in a continuous-flow system and are immobilized on a membrane surface. The membrane is a barrier that permits the deliberate separation of products from reactants and shields the enzymes from inactivation. The size of the molecules contributing to the reaction and the necessary separation level will determine which membrane is used. In particular, EMR is regarded as a novel reactor design for the hydrolysis of big molecules that can simultaneously promote catalytic reaction and product separation. As a result, a more economical method can be achieved. The membrane in the EMR served as both a separation unit and a catalytically active surface. This expands the potential uses for EMR, particularly in multi-phase processes (Rios et al., 2004).

15.2.5 Prospects and Challenges of Producing Bioalcohol Using an EMR

Producing bioalcohol using an EMR has several potential advantages over traditional fermentation processes.

15.2.5.1 Future Prospects

Enzymes can be used in many industries to perform efficient and reasonably priced bioconversions because of their advantageous catalytic qualities (Aehle et al., 2008). Several hydrolytic and isomerization enzymes, such as lipases, proteases, carbohydrates, isomerases, and transferases, are presently employed in industry, according to Hari Krishna et al. (2002). The main applications for these enzymes are in manufacturing medications, cleansers, food and feed products, and, to a lesser degree, agricultural chemicals. Separating the enzymes from the product allows for continuous reaction, which is one of the key benefits of EMR. The continuous operation made possible by the EMR requires less setup time for the start-up and shutdown processes than batch-wise operation (Rios et al., 2004)—a more economical process results from cutting down on idle time. Moreover, continuous operation results in higher productivity and economic viability (Prazeres & Cabral, 1994). When product or substrate inhibition, subsequent hydrolysis by the relevant enzyme, or any side reaction leading to product degradation occurs, it is advantageous to continuously separate the product from the reacting mixture (Sitanggang et al., 2014). Also, the product withdrawal may advantageously cause the reaction equilibrium to change in favour of the product (Padt et al., 1991). By recycling unreacted substrates, enzyme membrane reactors can boost the output of bioalcohol synthesis, improve conversion rates, and lessen the need for expensive downstream purification procedures.

15.2.5.2 Challenges

15.2.5.2.1 Membrane Fouling

Fouling is a phenomenon encompassing cake formation, adsorption, gel formation, and pore-clogging caused by chemicals on the surface or within the pores. The occurrence of particle adsorption within pore channels can lead to pore blockage, resulting in an instantaneous decrease in flux (Prazeres & Cabral, 1994). Reversible and irreversible fouling can reduce flux under constant pressure conditions or, if the flux remains constant, an increase in pressure drop. As a result, catalytic productivity may be severely impacted, and the energy needed to operate and clean the membrane may also increase.

The reaction mixture's content, the membrane's characteristics, and the hydrodynamic circumstances affect the type of membrane fouling (Sitanggang et al., 2022). Macromolecules like polysaccharides, lipids, and proteins tend to build a tiny layer on the membrane surface when an EMR is operating, particularly in a dead-end filtration mode. A gel layer is produced when one of these macromolecules forms a significant three-dimensional crosslinked structure.

15.2.5.2.2 Enzyme Stability

Enzymes can be sensitive to changes in temperature, pH, and other environmental factors, which can affect their activity and stability over time. This can lead to a decrease in reactor performance and overall bioalcohol yield.

15.2.5.2.3 Economics

The initial cost of setting up an EMR can be high due to the need for specialized equipment, enzymes, and membranes. This can make it difficult to justify the investment in smaller-scale production facilities. Consequently, it can be difficult to allocate funds for the setup of the reactor, making it a less viable option for smaller projects.

15.2.5.2.4 Regulatory Hurdles

The production of bioalcohol using enzymatic membrane reactors may be subject to regulatory restrictions, such as safety concerns, regarding the use of enzymes or the need for specific permits or certifications.

15.3 MEMBRANE-BASED BIOALCOHOL PURIFICATION TECHNOLOGY

Bioalcohols can be extracted and purified from fermentation broth or other sources using membrane-based bioalcohol purification technology. Using this method, contaminants and water are selectively removed from the bioalcohol mixture, yielding a more concentrated and purer bioalcohol output. Microalgae harvesting for third-generation bioethanol production is the first possible membrane use. Pretreatment of biomass is necessary to render the carbohydrates present in the biomass accessible for the subsequent conversion into second- and third-generation bioethanol. The second possible use of membrane technology is in the purification and concentration of pre-hydrolysates subsequent to pretreatment but before the fermentation process.

The sugar solution can be concentrated, and fermentation inhibitors can be eliminated using different types of membrane filters. An NF process in combination with UF has been considered for enzyme recovery and other value-added products. Low-concentration bioethanol is transported for pre-concentration and PV after fermentation. To execute continuous fermentation, PV and fermentation have been merged. Utilizing UF and NF techniques throughout the hybrid process holds promise for potentially eliminating yeast and fermentation inhibitors (Wei et al., 2014).

- ***Membrane used in microalgae harvesting*:**

  Microalgae are gaining increasing recognition as a viable feedstock for bioethanol production due to their composition, including proteins and carbohydrates, which can serve as carbon substrates during fermentation. Microalgae cells possess a comparatively brief harvesting cycle, ranging from 1 to 10 days, enabling abundant resources to meet ethanol production requirements (Avagyan, 2008).
- ***Membrane used in fermentation*:**

  The saccharification and fermentation processes are essential for bioethanol generation in every instance. Nevertheless, the straightforward sugar content in pre-hydrolysates for second- and third-generation bioethanol derived from biomass is sometimes limited due to the diverse pretreatment techniques and hydrolysis efficiency (Bura et al., 2002). In addition, it has been observed that subjecting biomass to steam explosion or weak acid pretreatment results in the production of fermentation inhibitors, which subsequently impede fermentation progress (Yan et al., 2009). Low levels of ethanol can be attributed to the presence of low quantities of fermentable sugars in the pre-hydrolysates, as well as the presence of fermentation inhibitors. These factors increase operational costs and energy consumption during subsequent purifying processes. Therefore, it is necessary to detoxify the pre-hydrolysate to remove inhibitors and focus on improving the sugar content before its utilization for ethanol fermentation. This approach aims to enhance the efficiency of pre-hydrolysate fermentation. Conventional sugar concentration or detoxification procedures include evaporation, solvent extraction, over-liming, activated charcoal adsorption, and ion exchange. However, it should be noted that most of these methodologies have certain limitations that must be considered. These limitations include the potential increase in processing expenses, the added complexity in the production of lignocellulose-to-ethanol, the generation of additional waste by-products, the requirement for extended processing durations, and the loss of sugars (Palmqvist & Hahn-Hägerdal, 2000).

The most crucial element is that sugar cannot be concentrated while inhibitors are removed using traditional procedures. These two problems can be resolved by using membranes as a separation technology. Moreover, studies have shown that the membrane process's initial investment and operational costs were lower than those associated with evaporator installation (Murthy et al., 2005). Reverse osmosis, UF, NF, and membrane distillation (MD) are current membrane configurations used in these processes. A microporous hydrophobic membrane serves as a physical substrate for separating a heated solution, consisting of either a liquid or a gas mixture, in the thermally induced process commonly referred to as MD.

- ***Membrane used in bioalcohol recovery*:**

  For instance, the final fermenter in conventional corn-to-ethanol processes contains more than 10% ethanol by weight. The maximum bioethanol concentration that the fermentation microorganisms can tolerate is 10 wt.%. As a result, removing the ethanol is necessary for further production. However, compared to bioethanol made from corn, bioethanol from cellulosic biomass probably offers lower product concentrations (5 wt.%). The ethanol should be concentrated before the refining procedure to conserve energy and money (Nomura et al., 2002).

  It is a good idea to allow for the development of other membrane-based technologies that can reduce energy use and costs while still producing ethanol (Sukitpaneenit & Chung, 2011). Compared to conventional distillation-based technologies, membrane-based bioalcohol purification technology has several advantages, including lower energy usage, prices, and enhanced product purity. Furthermore, the method is easily scaled up or down to fit various production volumes, making it a flexible and effective bioalcohol purification technology.

15.3.1 Conventional Purification Technology and Its Challenges

Using various techniques, the separation of bioalcohol from fermentation broths is broadly categorized based on equilibrium, and particle affinity has been extensively studied in recent years.

15.3.1.1 Absorption

Separating desired gas components or particles from a gas mixture into a liquid solvent phase is frequently accomplished through absorption. Biorefineries extensively use absorption to remove acid gases like H_2S and CO_2 from syngas before converting syngas into products like methanol and fuel. Physical absorption and chemical absorption are the two main types of absorption. Commercially, physical absorption takes acid gases like CO_2 and H_2S out of syngas to produce hydrogen, ammonia, and methanol. The Selexol method uses dimethyl ethers of polyethylene glycol at relatively high pressure (2.07–13.8 MPa), and the Rectisol process uses cold methanol at −40°C and 2.76–6.89 MPa to separate H_2S and CO_2, which are among the most prevalent physical absorption approaches (Kohl & Nielsen, 1997).

15.3.1.2 Distillation

A typical separation technique in the chemical and biological industries is distillation. There are several distillation techniques for separating liquid mixtures: ordinary distillation, AD, and ED. For instance, Distillation is the standard technique for removing and concentrating ethanol. Balat et al. (2008) noted that another distillation-related component lacks usability. The demerits of ordinary distillation techniques are as follows:

- **Energy requirements:** Energy requirements are higher at low alcohol feed concentrations compared to higher concentrations.
- **Azeotrope:** Simple distillation is ineffective in producing higher amounts of ethanol. At 95.6% ethanol by weight, ethanol and water form an azeotrope at atmospheric pressure.
- **High temperature:** High temperature hinders the reuse of salts and microorganisms and deactivates proteins and enzymes.

15.3.1.3 Liquid–Liquid Extraction

Liquid–liquid extraction is a widely employed method of separation that entails the transfer of desired constituents from the initial liquid phase to the solvent phase by utilizing one or more solvents in combination. The process of liquid–liquid extraction is employed to effectively remove biofuels and chemicals from liquid mixtures that are in a diluted state. For example, it has been demonstrated that chromatography can separate bioalcohols and carboxylic acids (Bressler & Braun, 1999) from the fermentation broths. These substances are known to be toxic to the microorganisms involved in the fermentation process of biomass hydrolysates (Grzenia et al., 2011). Additionally, chromatography can effectively remove impurities from biodiesel derived from used cooking oils (Berrios et al., 2011).

15.3.1.4 Supercritical Fluid Extraction

Supercritical fluid extraction involves extracting necessary dissolved chemicals from either a solid matrix or a liquid solution using a supercritical fluid in its supercritical state. Supercritical CO_2 is widely utilized as a supercritical fluid in the food, pharmaceutical, and chemical industries because of its non-polar or hydrophobic nature. According to Huang et al. (2012), supercritical carbon dioxide effectively extracts hydrophobic constituents from biomass.

15.3.1.5 Membrane Separation

It is possible to remove inhibitors such as acetic acid with a membrane. Cellulose and hemicellulose are typically hydrolyzed to mono sugars and then fermented to generate the necessary products during the bioconversion of LCB. Acetic acid is released from the acetate in biomass during hemicellulose hydrolysis. Acetic acid must be eliminated from the hydrolysate before fermentation because it hinders the ensuing fermentation. According to Wickramasinghe and Grzenia (2008), anion exchange membrane outperformed anion exchange resin regarding separation performance, throughput, and product loss when removing acetic acid from biomass hydrolysates.

15.3.1.6 Pervaporation

Recently, a significant amount of academic research has been conducted on extracting bioalcohol from fermentation broths through membrane-based PV technology. PV is a membrane-based process technology that separates liquid molecules at the molecular level. This technique integrates both membrane permeability and evaporation to achieve efficient separation. The aforementioned attributes of the system include a high level of selectivity, affordability, energy efficiency, safety, and environmental sustainability. This technology presents a feasible option for separating liquid mixtures in many industries, such as biorefinery, petrochemical, pharmaceutical, and other relevant sectors. During the PV process, liquid feed mixtures are exposed to either a single side of a non-porous polymeric membrane or a molecularly porous inorganic membrane. Due to chemical changes induced by vacuum or gas purge, it is observed that components with higher permeability are absorbed into the membrane, undergo diffusion across it, and then evaporate upon penetration. The process of separating different components is accomplished by using the differences in sorption and diffusion properties shown by the various constituents of the feed. The separation operations are predominantly affected by the partial vaporization of the molecules present in the feed and the occurrence of a phase change in the penetrates (Huang et al., 2013).

15.3.1.7 Challenges

Distillation, low-temperature crystallization, adsorption, extraction, and chromatography are conventional separation methods for liquid mixtures; among them, the primary refinery process continues to be distillation. However, the methods mentioned above are typically neither inexpensive nor practicable for the complete bioalcohol separation process due to their energy-intensive nature, detrimental environmental impact, and convoluted operating procedure. For example, a standard distillation method is insufficient for small-scale operations, especially when azeotropic mixtures are present. The development of biorefinery concepts and hybrid technologies is required to increase the effectiveness of the biofuel separation process (Balat et al., 2008).

15.3.2 Bioalcohol Separation through Different Membrane Processes

The process of separating alcohols generated from biological feedstocks, such as ethanol, from the fermentation of sugar, water, and other contaminants is known as bioalcohol separation. PV, vapour permeation, and MD are a few membrane methods that can separate bioalcohols. The utilization of microalgae for bioethanol production in the third generation represents a significant advancement in membrane applications. Using membrane filtration techniques such as MF and UF is a viable method for the extraction of microalgae. Pretreatment of biomass is necessary to enhance the accessibility of carbohydrates for the subsequent conversion into second- and third-generation bioethanol. The second prospective application of the membrane involves the purification and concentration of pre-hydrolysate after pretreatment and fermentation processes.

15.3.2.1 Membrane Distillation

MD is a thermal separation process wherein a solution, consisting of a liquid or a gas mixture, is subjected to physical separation utilizing a microporous hydrophobic membrane. Different configurations of MD processes, including direct contact, air gap, and vacuum, have been employed to achieve the concentration of sugar solutions and the elimination of inhibitors. The act of rejecting sugar is typically highly effective. The flux, feed flow rate, feed temperature, and beginning sugar concentration are contingent upon the type of membrane utilized (Al-Asheh et al., 2006). Furthermore, the MD method can use low-grade heat or alternative energy sources, including geothermal, solar, and waste hot water and steam (Wang et al., 2009). Due to ethanol's comparatively higher partial pressure concerning water, the permeation of ethanol vapour through membrane holes is facilitated to a greater extent.

Meanwhile, it can separate ethanol and water using MD (Wei et al., 2014). The possibility of using a direct contact tubular MD module to separate ethanol from an ethanol–water mixture was first confirmed by Franken et al. (2019). Popular configurations for ethanol generation include direct and air gap MD in conjunction with fermentation. Udriot et al. (1989) combined direct MD and fermentation to separate ethanol from the culture medium. The analysis revealed an 87% improvement in ethanol productivity from 0.99 to 1.85 g/L h^{-1}. Tomaszewska et al. (2013) reported that the production rate was between 2.5 and 4 g $dm^{-3}h^{-1}$ in the case of fermentation integrated with the MD, compared to a lower rate of 0.8–2 g $dm^{-3}h^{-1}$ in the case of batch fermentation with saccharose. According to Lewandowicz et al. (2011), the continuous process was facilitated, sugars were fermented more thoroughly, the osmotic pressure in the fermentation broth was reduced, glycerol synthesis was reduced, and the number and viability of yeast cells increased, leading to a 15.5% increase in ethanol production.

15.3.2.2 Pervaporation

Unlike MD, commercial uses of PV systems for extracting water from concentrated alcohol solutions are available (Chapman et al., 2008). PV is a technique to separate liquid mixtures by partially vaporizing them via a non-porous membrane. The solution-diffusion model describes the separation mechanism. Ethanol has a wider kinetic diameter (0.57 nm) than water (0.32 nm), so the control method is the best option for recovery. Thus, picking the suitable membrane material for the separation system is essential. In their study, He et al. (2012) examined the performance of PV systems within a biorefinery context. The generation of bioethanol and the subsequent dehydration process are considered key aspects of the PV industry and are widely recognized as crucial components in the whole bioethanol manufacturing process. The study conducted by Gaykawad et al. (2013) examined the utilization of PV technology in ethanol production from LCB. Hence, it can be posited that PV represents a viable approach capable of enhancing the performance of the bioethanol unit. In the PV approach, a supplementary phase transition occurs when the fluid feed is transferred from the liquid phase to the vapour phase at the permeate side of the membrane. Despite the utilization of a vacuum pump within this system, it has been observed that the mass transfer increases. In contrast, the vacuum pressure on the permeate side decreases (Abels et al., 2013).

15.3.2.3 Hybrid Processes

Fermentation and hybrid PV technology are used to increase ethanol production. Continuous fermentation is more desirable than the batch process because of increased productivity, better process control, and higher yields. Yet, the intrinsic restrictions of cell washout and product inhibition in traditional continuous fermentation significantly negatively impact the process' productivity. Continuous in situ elimination of ethanol can help such product-inhibited fermentations become more productive. PV is easy to use, highly selective, and non-toxic to fermentative microbes compared to conventional methods like gas stripping and adsorption. Moreover, it could be less expensive than distillation. However, hybrid membrane PV with fermentation is feasible (Mori et al., 1990). After discussing the benefits and drawbacks of different configurations, Lipnizki et al. (2000) recommended utilizing an external PV unit because of its high efficiency and simple maintenance. A hybrid fermentation–PV approach has improved ethanol productivity over batch fermentation.

The hybrid fermentation–PV process for producing bioethanol has been the subject of various studies that have examined its economic evaluation in terms of investment and output costs. Most research has found hybrid PV–bioreactor technologies for elevated ethanol productivity and low energy utility have good economic potential. Due to the comparatively high cost of the PV unit, these reports have shown that the hybrid fermentation–PV process's overall cost of producing ethanol was slightly surpassing that of continuous fermentation (Vankelecom et al., 1997).

15.3.2.4 Membrane Bioreactors for Hydrolysis

Applying membrane reactors for the hydrolysis of cellulose via enzymatic hydrolysis has attracted interest in many studies and research to work on different aspects of the process. Membrane reactors provide a unique advantage of simultaneous enzyme recovery and glucose removal, which is lacking in conventional batch and continuous reactors like CSTR and PFR. Integrating a reactor and a membrane separation unit offers many advantages for effective enzymatic hydrolysis of cellulose, which is heavily limited under the conventional batch or continuous reactors. The significant benefits of product removal via membrane reactors during enzymatic hydrolysis are summarized as follows (Andric et al., 2010):

- Membrane separation can allow cellulases to be utilized more extended by retaining the enzymes in the reactor. Cellulases are enzymes that break down cellulose, a complex carbohydrate found in plant cell walls, into simpler sugars that can be used as a source of energy by microorganisms or for industrial processes such as biofuel production. However, cellulases are often expensive, and their activity can be inhibited by pH, temperature, and product inhibition. Using membrane separation, the cellulases can be physically separated from the reaction mixture, allowing for continuous enzyme reuse and reducing the risk of deactivation or loss. The membrane separation process involves using a semipermeable membrane to allow the passage of specific molecules while blocking others selectively. In the case of enzyme retention, the membrane would be designed to allow the passage of smaller molecules, such as sugars and other reaction products, while retaining larger enzyme molecules in the reactor. This can be achieved through various membrane technologies such as UF, NF, and RO. Furthermore, membrane separation can potentially improve the efficiency and cost-effectiveness of cellulase-based processes by allowing for continuous enzyme reuse and reducing enzyme deactivation or loss.

- Cellulose is a complex carbohydrate made up of repeating units of glucose. The accumulation of glucose in the reaction mixture can inhibit its conversion into simpler sugars and other products. Product inhibition occurs when the end product of a reaction accumulates in the reaction mixture, causing a decrease in the activity of the enzymes involved in the reaction. In the case of cellulose conversion, glucose is the end product that can accumulate and inhibit the activity of cellulase enzymes. One approach to minimize product inhibition and improve conversion is to remove glucose from the reaction mixture. This can be achieved through various separation technologies, such as membrane separation or adsorption onto a solid support. Researchers combined membrane separation and enzymatic hydrolysis to convert cellulose into glucose and xylose. They found that removing glucose from the reaction mixture improved xylose yield. Another approach is to use different enzymes or a combination of enzymes that can convert glucose into other products. For example, glucose oxidase can convert glucose into gluconic acid, a precursor for various chemical products. Minimizing product inhibition by removing glucose from the reaction can improve cellulose conversion into other products. It can be achieved through multiple separation technologies or using different enzymes or enzyme cocktails.
- Membrane separation can selectively allow the passage of specific molecules while blocking others. It can separate glucose and other products from the reaction mixture while retaining unconverted cellulose and enzymes in the reactor. Membrane separation and enzymatic hydrolysis combination can achieve a high-purity glucose product. The cellulose can be hydrolyzed using cellulase enzymes to produce glucose and other sugars. The resulting mixture can then be passed through a membrane that selectively allows the passage of smaller molecules, such as glucose and other sugars, while retaining larger molecules, such as unconverted cellulose and enzymes, in the reactor. The permeate collected at the other end of the membrane will contain a high concentration of glucose product free from impurities such as unconverted cellulose and enzymes. This stream can be further purified using other separation technologies such as chromatography or crystallization. By obtaining a high-purity glucose product, subsequent fermentation processes can be more efficient and effective, as the presence of impurities such as unconverted cellulose and enzymes can negatively affect the fermentation process. The high-purity glucose stream can be used as a feedstock for various industrial processes such as biofuel production or chemical synthesis.

15.3.2.4.1 Modes of Operation and Types of Membranes

Researchers have thoroughly investigated the variety of modes of functioning and types of membranes organized in different circumstances. The issue of inhibiting glucose build-up within the batch reactor can be efficiently addressed using membrane reactors for hydrolysis. These reactors facilitate the removal of glucose in two distinct modes: continuous and non-continuous. The primary emphasis of scholarly investigation lies in examining hydrolysis within membrane reactors featuring continuous product extraction to establish a comparative analysis with hydrolysis conducted in batch reactors (Andric et al., 2010). The cellulose conversion achieved by hydrolysis in membrane reactors exhibits a notably greater efficiency than in batch reactors.

Nevertheless, the various methods of product extraction employed in membrane reactors, whether continuous or intermittent, exhibit a negligible disparity in cellulose conversion. The article by He et al. (2012) explains the three types of membranes, namely UF, NF, and MF, as depicted in Table 15.1. The membrane utilized in the reactor plays a crucial role in determining the efficiency of a membrane bioreactor system in both enzyme retention and product permeation. In enzymatic hydrolysis applications, Pino et al. (2018) typically refer to two distinct membrane types: UF and NF membranes.

UF with a molecular weight cut-off of 10 kDa is a commonly employed method in hydrolysis for the concurrent recovery of enzymes and removal of products. This approach is favoured due to the complete exclusion of enzyme cellulases and the unrestricted passage of glucose (Mores et al., 2001). Additionally, NF is often employed to enhance the glucose concentration before the subsequent fermentation process. Malmali et al. (2015) discovered a phenomenon of enzyme loss occurring on the membrane's permeate side during MF use. This loss is attributed to the MF membrane's more significant molecular weight cut-off, which allows enzyme molecules to pass through.

- **Microfiltration**

  The effectiveness of MF membranes in biorefineries was suggested by He et al. (2012). Furthermore, the recovery of lignin and hemicelluloses, the retrieval of enzymes, the manufacture of biogas and biodiesel, and the generation of acetic acid are all interconnected with the bioethanol production process.
- **Ultrafiltration**

  In a biorefinery, UF membranes can be used for the recovery of lignin and hemicelluloses, enzyme recovery, the generation of biogas and biodiesel, the harvesting of algae, and the generation of acetic acid. Other performances that can be considered in bioethanol synthesis include the recovery of lignin and hemicelluloses, enzyme recovery, and acetic acid generation. Developing renewable and eco-friendly biofuels is made possible by the biological conversion of LCB into fuel-grade ethanol. Three key processes comprise the biorefinery process: pretreatment, hydrolysis, and fermentation. One of the most extensive production costs is the enzymes (cellulases) necessary to hydrolyze cellulose into fermentable sugars; they

TABLE 15.1
Types of Membranes

Type of Membranes	Separation Principle	Structure	Materials	Thickness (μm)	Pore size (nm)	Configuration	Driving Force
MF	Sieving mechanism	Symmetric porous	Polymeric and ceramic	~10–150	~50–10,000	Flat sheet/plate and frame, tubular, hollow fibre	Pressure
UF	Sieving mechanism	Asymmetric porous	Polymeric and ceramic	~150	1–100	Flat sheet/plate and frame, tubular, capillary, spiral wound, hollow fibre	Pressure
NF	Solution diffusion	Composite	Polymeric	1/150	< 2	Flat sheet/plate and frame, tubular, spiral wound	Pressure

account for about 50% of the hydrolysis process' overall cost and 20% of the overall cost of ethanol production (Knutsen et al., 2002). Cellulases may be recovered and recycled from the hydrolyzed solution via membrane separation, making recovering enzymes an essential step in bioethanol synthesis.

- **Nanofiltration**

  NF membranes exhibit a range of notable capabilities, including the recovery of lignin and hemicellulose, the elimination of fermentation inhibitors, and the facilitation of biodiesel generation in a biorefinery. Biorefinery technology encompasses numerous applications, with NF membranes finding multiple utilizations. This approach finds applications in recovering lignin and hemicelluloses, eliminating fermentation inhibitors, and manufacturing biodiesel. Researchers have worked on separating acetic acid from xylose by the NF method. They observed that acid hydrolyzes lignocellulose to release sugars and several derivatives (Weng et al., 2009). The hydrolysate's sugars are subsequently fermented to produce ethanol. Before fermentation, acetic acid should be removed from the hydrolysates because it is thought to be an inhibitor that limits ethanol production. Using a synthetic acetic acid–xylose solution as the model, acetic acid and xylose were separated using a Desal-5 DK NF membrane. It was discovered that the separation performance was influenced by both the solution pH and the applied pressure. Depending on the pH of the solution and the applied pressure, the measured retention of xylose and acetic acid ranged from 28% to 81% and 6.8% to 90%, respectively. Moreover, acetic acid retention was negatively retained only when xylose was present.

The findings indicated that the separation of xylose and acetic acid depends on intermolecular interactions (Weng et al., 2009). The NF membrane was used in experiments to separate an ethanol/water combination. Moreover, Gautam et al. (2014) observed the efficacy of NF membranes; they were assessed for their capacity to remove phenolic chemicals, organic and mineral acids, and furans from sugar.

15.3.3 Bioalcohol Purity Using Membrane Refining

The study considers various membrane processes and feed source types in producing bioethanol. For upcoming biorefineries, bioethanol, in particular, is a noteworthy product. The sharp rise in the price of crude oil and the intensity of the world's petroleum demand has led to a growing interest in ethanol produced from biomass (Ghasemzadeh et al., 2014). Using diverse designs and stages of the bioethanol production process, the membrane separation technique can be effectively used for separation or purification. Refineries that produce bioethanol can increase their efficiency by utilizing many benefits of separation procedures. Additionally, these techniques are environmentally sustainable because they produce less wastewater.

Moreover, one significant limitation that membrane separation techniques can meet is the high purity of the finished product, particularly bioethanol. Therefore, it is made clear by contrasting the many process alternatives that are accessible that using membrane systems in conjunction with a biochemical plant—such as a bioethanol production plant—can result in considerable advantages. Thus, using various membrane technology configurations per the previously indicated benefits can demonstrate greater potential in multiple applications, particularly for manufacturing bioethanol (Ghasemzadeh et al., 2014).

EMR's primary benefit is its capacity to promote continuous reaction by separating enzymes from the final product. The continuous operation enabled in the EMR needs less setup time for the start-up and shutdown processes than batch-wise operation (Rios et al., 2004). A more economical process results from cutting down on idle time—moreover, continuous operation results in higher productivity and economic viability (Prazeres & Cabral, 1994). When product or substrate inhibition, subsequent hydrolysis by the relevant enzyme, or any side reaction leading to product degradation occurs, it is advantageous to continuously separate the product from the reacting mixture (Sitangganga et al., 2014). Furthermore, the product removal may favourably tip the balance of the reaction towards the product (Padt et al., 1991).

Membrane reactors equipped with glucose product elimination enable the concurrent removal of supplementary lower-molecular compounds that hinder the process, such as those arising from substrate pretreatment, which could ultimately impact the biocatalyst's performance (Andric et al., 2010).

15.4 CONCLUSIONS AND OUTLOOK

The synthesis and purification of bioalcohols from agrobiomass involve many significant strategies to achieve high yield, reduce production costs, and ensure environmental sustainability. However, it was shown that membrane-based enzymatic methods were the most appropriate for synthesizing and purifying bioalcohols compared to other methods. Nevertheless, EMRs are subject to limitations that can hinder their utilization in many sectors. These limitations include diminished enzyme activity resulting from inactivation and washout, as well as the occurrence of fouling events.

From this review, it can further be concluded that

- The utilization of agricultural LCB exhibits potential as a viable feedstock for the synthesis of bioalcohol, provided appropriate pretreatment techniques are employed.
- Bioalcohols derived from LCB, a sustainable feedstock, have the potential to generate a virtually limitless supply of energy. Moreover, their utilization may offer significant environmental and societal benefits compared to conventional fossil fuels.
- Using membrane-based enzymatic methodologies holds significant potential for producing bioalcohol that is environmentally sustainable, intelligent, and cost-effective.
- Further examination of technological advancements in pretreatment, enzymatic hydrolysis, fermentation, and distillation is necessary to enhance the economic and environmental efficiency of bioalcohol production from LCB.
- However, further examination is necessary to explore current advancements in bioalcohol synthesis and purification techniques, aiming to maximize bioalcohol production and purification for cost-effectiveness and environmentally sustainability. This is crucial to promote the widespread adoption of biofuels globally.

NOMENCLATURE

1G	First-generation
2G	Second-generation
AD	Azeotropic distillation
CSTR	Continuous stirred tank reactor
ED	Extractive distillation
EMR	Enzymatic membrane reactor
IEA	International Energy Agency
LCB	Lignocellulosic biomass
MC	Microchannel
MD	Membrane distillation
MF	Microfiltration
NF	Nanofiltration
PBR	Packed bed reactor
PV	Pervaporation
SSF	Simultaneous saccharification and fermentation
UF	Ultrafiltration

REFERENCES

Abels, C., Carstensen, F., Wessling, M. (2013) Membrane processes in biorefinery applications, *J. Membr. Sci.* 444 285–317.

Aehle, W., Misset, O. (2008) *Enzymes for Industrial Applications*. Biotechnology, Wiley-VCH Verlag GmbH, New York, pp. 189–216.

Al-Asheh, S., Banat, F., Qtaishat, M., Al-Khateeb, M. (2006) Concentration of sucrose solutions via vacuum membrane distillation, *Desalination* 195 60–68.

Andric, P., Meyer, A.S., Jensen, P.A., Dam-Johansen, K. (2010) Reactor design for minimizing product inhibition during enzymatic lignocellulose hydrolysis: II. Quantification of inhibition and suitability of membrane reactors, *Biotechnol. Adv.* 28 (3) 407–425.

Avagyan, A. (2008) A contribution to global sustainable development: inclusion of microalgae and their biomass in production and bio cycles, *Clean Techn. Environ. Policy* 10 313–317.

Balat, M., Balat, H., Öz, C. (2008) Progress in bioethanol processing, *Progress Energy Combust. Sci.* 34 551–573.

Berrios, M., Martín, M.A., Chica, A.F., Martín, A. (2011) Purification of biodiesel from used cooking oils, *Appl. Energy* 88 (11) 3625–3631.

Boluda, N., Gomis, V., Ruiz, F., Bailador, H. (2005) The influence of temperature on the liquid-liquid-solid equilibria of the ternary system water + ethanol + 1-dodecanol, *Fluid Phase Equilibria.* 235 99–103.

Bressler, E., Braun, S. (1999) Separation mechanisms of citric and itaconic acids by water-immiscible amines, *J. Chem. Technol. Biotechnol.* 74 (9) 891–896.

Bura, R., Mansfield, S., Saddler, J., Bothast, R. (2002) SO_2-catalyzed steam explosion of corn fiber for ethanol production, *Appl. Biochem. Biotechnol.* 98–100 59–72.

Chandrasekhar, K., Naresh Kumar, A., Kumar, G., Kim, D.-H., Song, Y.-C., Kim, S.-H. (2021) Electro-fermentation for biofuels and biochemicals production: current status and future directions, *Bioresour. Technol.* 323 124598.

Chapman, P.D., Oliveira, T., Livingston, A.G., Li, K. (2008) Membranes for the dehydration of solvents by pervaporation, *J. Membr. Sci.* 318 (1–2) 5–37.

Da Nogueira, C., de Araújo Padilha, C.E., de Medeiros Dantas, J.M., de Medeiros, F.G.M., de Araújo Guilherme, A., de Santana Souza, D.F., dos Santos, E.S. (2021) In-situ detoxification strategies to boost bioalcohol production from lignocellulosic biomass, *Renew. Energ.* 180 914–936.

Deshavath, N.N., Mogili, N.V., Dutta, M., Goswami, L., Kushwaha, A., Veeranki, V.D., Goud, V.V. (2022) Role of lignocellulosic bioethanol in the transportation sector: limitations and advancements in bioethanol production from lignocellulosic biomass, In: *Waste-to-Energy Approaches towards Zero Waste*, Edited by: C. M. Hussain, S. Singh, L. Goswami. Elsevier, Amsterdam, The Netherlands, pp. 57–85.

Fache, M., Boutevin, B., Caillol, S. (2016) Vanillin production from lignin and its use as a renewable chemical, *ACS Sustain. Chem. Eng.* 4 (1) 35–46.

Franken, A. C. M., Nolten, J. A. M., Mulder, M. H. V., Smolders, C. A. (2019) Ethanol-water separation by membrane distillation: effect of temperature polarization, In: *Synthetic Polymeric Membranes*, De Gruyter, Berlin, Germany, pp. 531–540.

García, A., Spigno, G., Labidi, J. (2017) Antioxidant and biocide behaviour of lignin fractions from apple tree pruning residues, *Ind. Crops. Prod.* 104 242–252.

Gautam, A.K., Menkhaus, T.J. (2014) Performance evaluation and fouling analysis for reverse osmosis and nanofiltration membranes during processing of lignocellulosic biomass hydrolysate, *J. Membr. Sci.* 451 252–265.

Gaykawad, S.S., Zha, Y., Punt, P.J., van Groenestijn, J.W., van der Wielen, L.A.M., Straathof, A.J.J. (2013) Pervaporation of ethanol from lignocellulosic fermentation broth, *Bioresour. Technol.* 129 469–476.

Ghasemzadeh, K., Jafarharasi, N., Vousoughi, P. (2014) An overview on the bioethanol production using membrane technologies, *Int. J. Membrane Sci. Techno.* 1 (1) 9–22.

Goswami, L., Kayalvizhi, R., Dikshit, P.K., Sherpa, K.C., Roy, S., Kushwaha, A., Kim, B.S., Banerjee, R., Jacob, S., Rajak, R.C. (2022) A critical review on prospects of bio-refinery products from second and third generation biomasses, *Chem. Eng. J.* 448 137677.

Grzenia, D. L., Qian, X., da Silva, S. S., Wang, X., Wickramasinghe, S. R. (2011) Chapter 10- membrane extraction for biofuel production, In: *Membrane Science and Technology*, Edited by: S. Ted Oyama, S. M. Stagg-Williams. Elsevier, Amsterdam, The Netherlands, pp. 213–233.

Guo, M., Jin, T., Nghiem, N.P., Fan, X., Qi, P.X., Jang, C.H., Shao, L., Wu, C. (2018) Assessment of antioxidant and antimicrobial properties of lignin from corn stover residue pretreated with low-moisture anhydrous ammonia and enzymatic hydrolysis process, *Appl. Biochem. Biotechnol.* 184 (1) 350–365.

Gutiérrez-Antonio, C., Gómez-Castro, F.I., de Lira-Flores, J.A., Hernández, S. (2017) A review on the production processes of renewable jet fuel, *Renew. Sustain. Energy Rev.* 79 709–729.

Haghighi Mood, S., Hossein Golfeshan, A., Tabatabaei, M., Salehi Jouzani, G., Najafi, G.H., Gholami, M., Ardjmand, M. (2013) Lignocellulosic biomass to bioethanol, a comprehensive review with a focus on pretreatment, *Renew. Sustain. Energy Rev.* 27 77–93.

Hamelinck, C.N., Van Hooijdonk, G., Faaij, A.P. (2005) Ethanol from lignocellulosic biomass: techno-economic performance in short-, middle-and long-term, *Biomass Bioenergy* 28 384–410.

Hari Krishna, S. (2002) Developments and trends in enzyme catalysis in nonconventional media, *Biotechnol. Adv.* 20 239–267.

He, Y., Bagley, D.M., Leung, K.T., Liss, S.N., Liao, B.-Q. (2012) Recent advances in membrane technologies for biorefining and bioenergy production, *Biotechnol. Adv.* 30 (4) 817–858.

He, Y., David, M.B., Kam, T.L., Steven, N.L., Bao-Qiang, L. (2012) Recent advances in membrane technologies for biorefining and bioenergy production, *Biotechnol. Adv.* 30 (4) 817–858.

Huang, H-J., Ramaswamy, S., Tschirner, U.W., Ramarao, B.V. (2010) Chapter 10- Separation and purification processes for lignocellulose-to-bioalcohol production, In: *Bioalcohol Production, Woodhead Publishing Series in Energy*, Edited by: K. W. Waldron, pp. 246–277.

Huang, H. J., Ramaswamy, S. (2012) Chapter 3- Separation and purification of phytochemicals as co-products in biorefineries, In: *Biorefinery Co-Products: Phytochemicals, Primary Metabolites and Value-Added Biomass Processing*, Edited by: C. V. Stevens. John Wiley and Sons, New York, pp. 37–53.

Huang, H.-J., Bandaru, V.R., Ramaswamy, S. (2013) Separation and Purification Technologies in Biorefineries. John Wiley & Sons, New York.

Huang, C., Jiang, X., Shen, X., Hu, J., Tang, W., Wu, X., Yong, Q. (2022) Lignin-enzyme interaction: a roadblock for efficient enzymatic hydrolysis of lignocellulosics, *Renew. Sustain. Energy Rev.* 154 111822.

IEA. (2021) Renewables 2021, IEA, Paris, License: CC BY 4.0. URL: https://www.iea.org/reports/renewables-2021.

Kamrat, T., Nidetzky, B. (2007) Entrapment in *E. Coli* improves the operational stability of recombinant β-glycosidase CelB from pyrococcus furiosus and facilitates biocatalyst recovery, *J. Biotechnol.* 129 69–76.

Kang, K.E., Jeong, J.-S., Kim, Y., Min, J., Moon, S.-K. (2019) Development and economic analysis of bioethanol production facilities using lignocellulosic biomass, *J. Biosci. Bioeng.* 128 475–479.

Knutsen, J.S., Davis, R.H. (2002) Combined sedimentation and filtration process for cellulase recovery during hydrolysis of lignocellulosic biomass, *Appl. Biochem. Biotechnol.* 98 1161–1172.

Kohl, A. L., Nielsen, R. B. (1997) Chapter 11-absorption of water vapor by dehydrating solutions, In: *Gas Purification*, 5th ed., Edited by: A. L. Kohl, R. B. Nielsen. Houston: Gulf Publishing, pp. 946–1021.

Koullas, D.P., Umealu, O.S., Koukios E.G. (1999) Solvent selection for the extraction of ethanol from aqueous solutions, *Sep. Sci. Technol.* 34(11) 2153–2163.

Lewandowicz, G., Białas, W., Marczewski, B., Szymanowska, D. (2011) Application of membrane distillation for ethanol recovery during fuel ethanol production. *J. Membr. Sci.* 375 (1–2) 212–219.

Lipnizki, F., Hausmanns, S., Laufenberg, G., Field, R., Kunz, B. (2000) Use of pervaporation-bioreactor hybrid processes in biotechnology, *Chem. Eng. Technol.* 23 (7) 569–577.

Luiza, A., Rempel, A., Cavanhi, V.A.F., Alves, M., Deamici, K.M., Colla, L.M., Costa, J.A.V. (2020) Simultaneous saccharification and fermentation of spirulina sp. and corn starch for the production of bioethanol and obtaining biopeptides with high antioxidant activity, *Bioresour. Technol.* 301 122698.

Malmali, M., Stickel, J., Wickramasinghe, S.R. (2015) Investigation of a submerged membrane reactor for continuous biomass hydrolysis, *Food Bioprod. Process.* 96 189–197.

Mores, W.D., Knutsen, J.S., Davis, R.H. (2001) Cellulase recovery via membrane filtration, *Appl. Biochem. Biotechnol.* 91 297–309.

Mori, Y., Inaba, T. (1990) Ethanol production from starch in a pervaporation membrane bioreactor using Clostridium thermohydrosulfuricum, *Biotechnol. Bioeng.* 36 (8) 849–853.

Murthy, G.S., Sridhar, S., Shyam Sunder, M., Shankaraiah, B., Ramakrishna, M. (2005) Concentration of xylose reaction liquor by nanofiltration for the production of xylitol sugar alcohol, *Sep. Purif. Technol.* 44 221–228.

Naghshbandi, M.P., Tabatabaei, M., Aghbashlo, M., Gupta, V.K., Sulaiman, A., Karimi, K., Moghimi, H., Maleki, M. (2019) Progress toward improving ethanol production through decreased glycerol generation in saccharomyces cerevisiae by metabolic and genetic engineering approaches, *Renew. Sustain. Energy Rev.* 115 109353.

Nomura, M., Bin, T., Nakao, S. (2002) Selective ethanol extraction from fermentation broth using a silicalite membrane, *Sep. Purif. Technol.* 27 59–66.

Padt, A., Riet, K., (1991) Membrane bioreactors, In: *Chromatographic and Membrane Processes in Biotechnology*, Edited by: C. A. Costa, J. S. Cabral. Springer, Netherlands, Dordrecht, pp. 443–448.

Palmqvist, E., Hahn-Hagerdal, B. (2000) Fermentation of lignocellulosic hydrolyzates I: inhibition and detoxification, *Bioresour. Technol.* 7417–7424.

Pino, M.S., Rodríguez-Jasso, R.M., Michelin, M., Flores-Gallegos, A.C., Morales-Rodriguez, R., Teixeira, J.A., Ruiz, H.A. (2018) Bioreactor design for enzymatic hydrolysis of biomass under the biorefinery concept, *Chem. Eng. J.* 347 119–136.

Prasad, R.K., Chatterjee, S., Mazumder, P.B., Gupta, S.K., Sharma. S., Vairale, M.G., Datta, S., Dwivedi, S.K., Gupta D.K. (2019) Bioethanol production from waste lignocelluloses: a review on microbial degradation potential, *Chemosphere.* 231 588–606.

Prazeres, D.M.F., Cabral, J.M.S. (1994) Enzymatic membrane bioreactors and their applications, *Enzym. Microb. Technol.* 16 738–750.

Report on Renewable Energy in India. (2022) Press Information Bureau, Ministry of New and Renewable Energy, Government of India. URL: Press Information Bureau (pib.gov.in).

Rios, G.M., Belleville, M.P., Paolucci, D., Sanchez, J. (2004) Progress in enzymatic membrane reactors- a review, *J. Membr. Sci.* 242 189–196.

Sathawong, S., Sridach, W., Techato, K. (2018) Lignin: isolation and preparing the lignin-based hydrogel, *J. Environ. Chem. Eng.* 6 5879–5888.

Sindhu, R., Binod, P., Pandey, A., Ankaram, S., Duan, Y., Awasthi, M.K. (2019) Biofuel production from biomass: toward sustainable development, In: *Current Developments in Biotechnology and Bioengineering: Waste Treatment Processes for Energy Generation*, Edited by: S. Kumar, R. Kumar, and A. Pandey. Elsevier, Amsterdam, The Netherlands, pp. 79–92.

Singh, A., Srivastava, S., Rathore, D., Pant, D. (Eds.) (2021) *Environmental Microbiology and Biotechnology: Volume 2: Bioenergy and Environmental Health.* Springer Nature, Singapore.

Sitanggang, A.B., Drews, A., Kraume, M. (2014) Continuous synthesis of lactulose in an enzymatic membrane reactor reduces lactulose secondary hydrolysis, *Bioresour. Technol.* 167 108–115.

Sitanggang, A.B., Drews, A., Kraume, M. (2022) Enzymatic membrane reactors: Designs, applications, limitations and outlook, Chem. Eng. Process. 180 108729.

Srivastava, N., Srivastava, M., Mishra, P.K., Gupta, V.K., Molina, G., Rodriguez-Couto, S., Manikanta, A., Ramteke, P.W. (2018) Applications of fungal cellulases in biofuel production: advances and limitations, *Renew. Sustain. Energy Rev.* 82 2379–2386.

Sukitpaneenit, P., Chung, T.-S. (2011) PVDF/nanosilica dual-layer hollow fibers with enhanced selectivity and flux as novel membranes for ethanol recovery, *Ind. Eng. Chem. Res.* 51 978–993.

Sun, S., Liu, F., Zhang, L., Fan, X. (2018) One-step process based on the order of hydrothermal and alkaline treatment for producing lignin with high yield and antioxidant activity, *Ind. Crops. Prod.* 119 260–266.

Szambelan, K., Nowak, J., Szwengiel, A., Jeleń, H., Łukaszewski, G. (2018) Separate hydrolysis and fermentation and simultaneous saccharification and fermentation methods in bioethanol production and formation of volatile by-products from selected corn cultivars, *Ind. Crop. Prod.* 118 355–361.

Tomaszewska, M., Białończyk, L. (2013) Production of ethanol from lactose in a bioreactor integrated with membrane distillation, *Desalination* 323 114–119.

Toor, M., Smita, S.K., Sandeep, K.M., Narsi, R.B., Mathimani, T., Rajendran, K., Pugazhendhi, A. (2020) An overview on bioethanol production from lignocellulosic feedstocks, *Chemosphere*. 242 125080.

Tsai, C.-T., Meyer, A.S. (2014) Enzymatic cellulose hydrolysis: enzyme reusability and visualization of β-glucosidase immobilized in calcium alginate, *Molecules* 19 19390–19406.

Udriot, H., Ampuero, S., Marison, I.W., von Stockar U. (1989) Extractive fermentation of ethanol using membrane distillation, *Biotechnol. Lett.* 11 509–514.

Vankelecom, I.F.J., Kinderen, J.D., Dewitte, B.M., Uytterhoeven, J.B. (1997) Incorporation of hydrophobic porous fillers in PDMS membranes for use in pervaporation, *The J. Phys. Chem. B*. 101 (26) 5182–5185.

Velvizhi, G., Balakumar, K., Shetti, N.P., Ahmad, E., Pant, K.K., Aminabhavi, T.M. (2022) Integrated biorefinery processes for conversion of lignocellulosic biomass to value added materials: Paving a path towards circular economy, *Bioresour. Technol.* 343 126151.

Von Sivers, M., Zacchi, G., Olsson, L., Hahn-Hagerdal, B. (1994) Cost analysis of ethanol production from willow using recombinant Escherichia coli, *Biotechnol. Prog*. 10 555–560.

Wang, X., Zhang, L., Yang, H., Chen, H. (2009) Feasibility research of potable water production via solar-heated hollow fiber membrane distillation system, *Desalination* 247 403–411.

Wang, Y., Sun, S., Li, F., Cao, X., Sun, R. (2018) Production of vanillin from lignin: the relationship between β-O-4 linkages and vanillin yield, *Ind. Crops Prod.* 116 116–121.

Wei, P., Cheng, L.-H., Zhang, L., Xu, X.-H., Chen, H.-L., Gao, C. (2014) A review of membrane technology for bioethanol production, *Renew. Sustain. Energy. Rev*. 30 388–400.

Weilnhammer, C., Blass, E. (1994) Continuous fermentation with product recovery by in-situ extraction, *Chem. Eng. Technol.* 17 365–373.

Weng, Y.-H., Wei, H.-J., Tsai, T.-Y., Chen, W.-H., Wei, T.-Y., Hwang, W.-S., Wang, C.-P., Huang, C.-P. (2009) Separation of acetic acid from xylose by nanofiltration. *Sep. Purif. Technol.* 67(1) 95–102.

Wickramasinghe, S.R., Grzenia, D.L. (2008) Adsorptive membranes and resins for acetic acid removal from biomass hydrolysates, *Desalination* 234 (1–3) 144–151.

Xu, X., Lin, M., Zang, Q., Shi, S. (2018) Solid state bioconversion of lignocellulosic residues by Inonotus obliquus for production of cellulolytic enzymes and saccharification, *Bioresour. Technol.* 247 88–95.

Yan, L., Zhang, H., Chen, J., Lin, Z., Jin, Q., Jia, H. (2009) Dilute sulfuric acid cycle spray flow-through pretreatment of corn stover for enhancement of sugar recovery, *Bioresour. Technol.* 100 1803–1808.

16 Application of Membrane-Based Technique for the Synthesis of Acetic Acid from Different Agrobiomasses

Santoshi Mohanta, Anuradha Upadhyaya, Jayato Nayak, and Sankha Chakrabortty

16.1 INTRODUCTION

In the last century, huge growth in the petroleum industry resulted in rapid depletion of petroleum reservoirs which nowadays raises concern over global energy demand. The rapidly rising population and their increased energy demand have led to extensive research on the conversion of renewable and carbon-neutral biomass resources into energy. Moreover, the limited availability of renewable sources needs proper utilization of waste generated from the agriculture sector, industry, and domestic sources [1]. Organic wastes are opulent in both organic as well as inorganic nutrients, possess high moisture, and are highly accessible. This can be considered as a potential source of bioenergy and biochemical production by anaerobic digestion [2]. Four independent metabolic reactions are performed by a mixed microbial community in the anaerobic digestion process such as hydrolysis, acidogenesis, acetogenesis, and methanogenesis [3]. The mixed microorganisms decay the organic waste into low-molecular-weight intermediates like hydrogen and volatile fatty acids in the acidogenesis stage and methane in the methanogenesis stage. The intermediate products such as volatile fatty acids are gaining plenty of interest as they possess more applications compared to the end product methane [4]. The market assessment of volatile fatty acid for 1 ton (USD 150) is way higher compared to methane (USD 31) [5]. The volatile fatty acids such as acetic (ethanoic), propionic (propanoic), butyric (butanoic), valeric (pentanoic), and caproic (hexanoic) acids are precious short- and medium-chain carboxylates that find common application in food and beverage as well as chemical and fuel sector.

 DOI: 10.1201/9781003407713-16

Acetic acid is one of the significant volatile fatty acids and is a crucial component that goes to the top 50 commodities of the chemical industry. Furthermore, several industrial reports predicted a rise in the acetic acid market at a rate of 5% per year, which may lead to a demand of 16 million in 2020 in various parts of the world [6,7]. A leading publication house (https://www.statista.com/statistics/1245203/acetic-acid-market-volume-worldwide/) reported that the demand for acetic acid was swelled to 17.48 million during 2022 and further forecasted to grow around 23.6 million by the year 2030.

It is a transparent, colorless carboxylic acid having corrosive properties, a sour taste, and a pungent smell. The name acetic acid was taken from the Latin term "Acetum", which means vinegar. It is also commonly called vinegar, typically used for food preservation, a great solvent, and a feedstock of different commercially available chemicals. Acetic acid is a very flexible and extensively used compound in numerous sectors [8]. It is used to produce polyvinyl acetate, which is widely applied in the synthesis of paint and glue. It is also used as the primary component in resins and emulsion polymers, as well as a major compound in coatings, fabrics, cables, etc. [9,10].

As a replacement for depleting petroleum resources, the biotechnological production of acetic acid from rich and sustainable cellulosic biomass by acidogenic fermentation is more and more important [11]. Various types of biomass wastes are used to produce ethanoic acid. The main source of ethanoic acid includes different types of fruits and fruit peels, agricultural wastes, i.e., crop stems, bagasse, leaves, roots, seeds, and nut shells, which are mainly thrown, burned, or left in fields. The main components of these materials are cellulose, hemicellulose, and lignin. The synthesis process mainly required three types of sugar, mainly extracted from sugarcane, sugar beet, corn, etc. Other sources of biomass waste are rotten and abandoned vegetables, leaves, skin, pulp, banana leaf, pith, grapes, pineapple shells and core, giant reed, and apple pomace [12]. The fermentation process employs a mixed biomass culture that is simple to handle under sterile conditions; however, it is necessary to be attentive to several operation parameters to maximize acetic acid production. The parameters such as temperature, pH, hydraulic retention time, optimal organic loading rate (OLR), and substrate concentration need to be maintained [13]. For instance, the OLR has to be maintained between 2 and 15 g-volatile solid (VS)/L·day [13]. Additionally, the acidogenic fermentation of carbohydrate-rich substrates (e.g., food waste (FW) and agricultural residues) must be performed in acidic and neutral pH ranges [14]. The operational parameters of AF include temperature, pH, hydraulic retention time, and substrate concentration.

Numerous studies have reported the synthesis process and key operational parameters of acetic acid production; however, these cannot be simply scaled up because of the limitations in energy requirement during the recovery step [15]. The main limitations are (i) processing cost which is nearly 30%–40% of the production cost, (ii) complication of composition of the fermentation broth mixture, (iii) low concentration in fermented stream, (iv) presence of suspended solids, etc. Various conventional separation and recovery techniques are utilized such as distillation, extraction, ion exchange, and membrane-based techniques. Among these, membrane-based techniques are found to be promising because of their low energy requirement, high selectivity, and environment-friendly characteristics [15].

This chapter provides a detailed review of the importance of acetic acid and its various utilities, its production from agricultural biomass, factors affecting acidogenic fermentation, and highlighting the membrane-based recovery techniques. Finally, the prospects and challenges involved in membrane techniques are emphasized.

16.2 IMPORTANCE AND APPLICATION OF ACETIC ACID

Acetic acid is one of the most significant volatile fatty acids which is consumed worldwide. It is used in different fields as the source material. It finds potential application in the food industry as a solvent and in the preparation of some food products [16]. Furthermore, it is used in the medicinal industry, chemical synthesis, and agricultural purposes. It is the base material for the synthesis of different types of value-added chemicals such as acetanilide and acetic anhydride, which is applied in the production of dyes, explosives, perfumes, etc. This can also be utilized in the production of acetyl chloride, polyvinyl acetate, butyl acetate, chloroacetic acid, etc. It is also a main component in the synthesis of poly(vinyl alcohols), poly(vinyl butyral), and poly(vinyl formal) [17]. Acetic acid is further used in the production of purified terephthalic acid (PTA) which is the primary material for manufacturing of polyester fibers [18]. All the major application areas are elaborated in Figure 16.1.

16.3 SYNTHESIS ROUTES OF ACETIC ACID

Production of ethanoic acid includes chemical synthesis and fermentation process. Different types of chemical synthesis include methanol carbonylation, acetaldehyde oxidation, and partial oxidation of ethane. In the methanol carbonylation process,

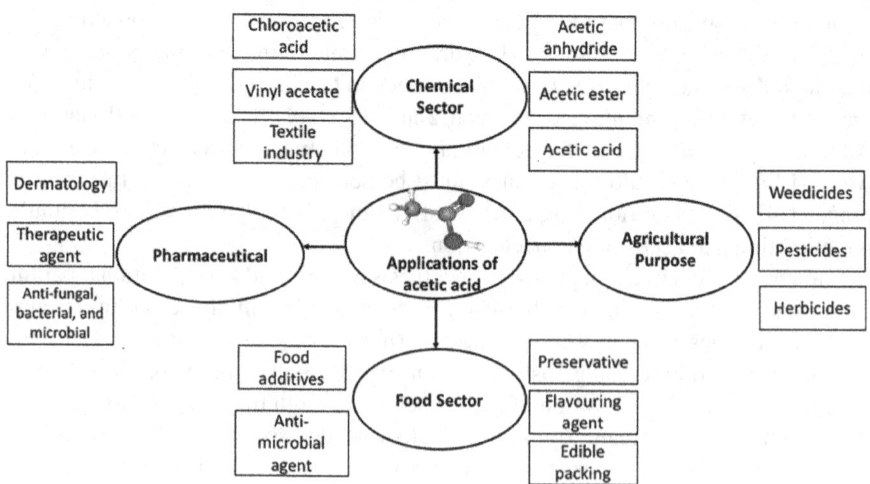

FIGURE 16.1 Various applications of acetic acid [19].

carbon monoxide and methanol react in the presence of a carbonyl catalyst made of rhodium to yield acetic acid as a result. In the acetaldehyde oxidation method, the acetaldehyde gets oxidized in the presence of a cobalt- or chromium-based metal catalyst to produce acetic acid [20]. In the partial oxidation of ethanol, molybdenum-vanadium-based catalyst is used for the reaction to get acetic acid as the final product. These chemical processes include products with some by-products, i.e., water, acetic acid, isobutyl acetate, butane, and propionic acid, which are harmful to the environment and cause pollution [20].

16.3.1 Fermentation Method

The fermentation process is a biological process that includes acetous and alcoholic fermentation. During alcoholic fermentation, sugars are converted into acetic acid under anaerobic conditions in the presence of homoacetogenic bacteria. In acetous fermentation, alcohol oxidizes into acetic acid in the presence of acetic acid bacteria, performed under an aerobic environment. Considering the technological aspects, fermentation is of two types: slow process and quick process [19]. The slow process involves the Orleans process and surface fermentation process, while the quick process involves a submerged process and the immobilized cell fermentation method as shown in Figure 16.2.

16.3.1.1 Orleans Method

The name of the method is extracted from the French word "Orlean" and it is one of the oldest traditional methods used for the synthesis of ethanoic acid from biomass. The technique is a very slow and dilatory process. This method is mainly used in different parts of the globe to synthesize acetic acid by taking different sources. The culture initiates with a higher quality acetic acid to which wine is introduced on a weekly basis. The process is done regularly from the initial weeks till 4 weeks. After 4 weeks, the product can be removed. Depending on the parameters, including the

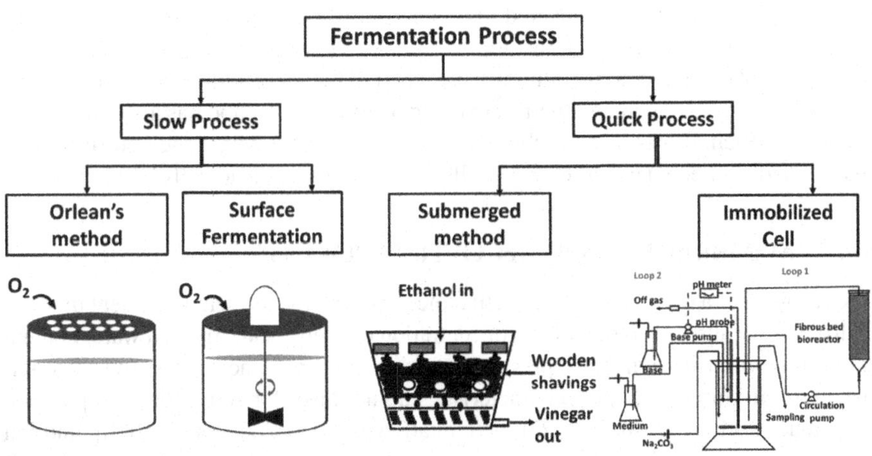

FIGURE 16.2 Synthesis of acetic acid by fermentation process [17,19].

temperature, chemical composition, microorganism used, and amount of air provided, the fermentation takes 8–14 weeks [21]. The major challenge encountered during this process is the introduction of additional liquid to the container without disturbing the buoyant culture. However, this problem can be addressed by utilizing a glass tube that connects to the bottom of the cylinder.

16.3.1.2 Generator Method/Surface Fermentation Method

The generator method is also known as the trickling method, which is an example of the surface fermentation method. This method is established to correct the disadvantages of the Orleans process. This method improved the interaction between acetic acid synthesis microbes and the material. To get the target amount of product, substrate was continuously added over the fermentation, and plenty amount of air circulated throughout the device to control the reaction's temperature. The generator, which is an upright cylinder filled with raw materials and fitted with devices that allow the alcoholic solution to drop by passing the shavings, is an essential component of this process [8].

16.3.1.3 Submerged Method

This fermentation method is cost-effective, modern, and one of the most popular techniques for ethanoic acid production. It has a high production with a high oxidation rate. This method produces acetic acid more effectively and 30 times quicker than the Orleans method. It consumes less space and results in a higher percentage of production with a reaction rate of 98%. This method includes a 10,000–40,000 L tank, air delivery apparatus, cooling apparatus, foam apparatus, and loading unloading valves. The three stages of this method include a loading of the basic material and introducing it to the medium, fermentation, and then unloading the fermented medium [17].

16.3.1.4 Immobilized Cell Fermentation

This method has a working capacity between 0.4 and 1 L. A fibrous-bed bioreactor is mostly used to produce ethanoic acid from fixed cells or immobilized cells. The equipment is made up of a jacketed glass column filled with a cotton towel or terry fabric. There is a recirculation reactor of capacity 5 L whose main function is maintaining the pH and a basal medium was attached to the bioreactor. The high output is due to the suspended cell and their high cell density. This method has a self-renewing property, which allows it to reach a dynamic steady state where the equilibrium is maintained between developing new cells and eliminating dead cells [19].

16.4 MEMBRANE-BASED RECOVERY PROCESS

The recovery of a relevant amount of ethanoic acid from the fermented effluent mixture is difficult because of: (i) the lower concentration of acetic acid, (ii) high water content in the mixture, (iii) significant presence of other carboxylic acids, and (iv) the complexity of mixtures [22]. Such characteristics do not favor for conventional separation techniques such as distillation [23], precipitation, adsorption, extraction [24], and ion exchange due to the high energy consumption and CO_2 emission [3]. However, today's current researchers are paying more interest in applying membrane-based separation and purification techniques to recover a significant quantity of acetic acid. This technique eliminates several intermediate steps resulting in reducing the processing time and operating cost [25,26]. Additionally, the membranes offer improved selectivity, better surface-to-volume ratio, and good process control [10]. The membrane processes can be broadly classified as follows: (i) Pressure-driven processes such as Microfiltration (MF), Ultrafiltration (UF), Nanofiltration (NF), and Reverse osmosis (RO) and (ii) Non-Pressure-driven processes such as forward osmosis (FO), Pervaporation (PV), and Membrane Distillation (MD) process. In this section, we will discuss the application of all these techniques for the recovery and purification of acetic acid.

16.4.1 Pressure-Driven Membrane Processes

Pressure-driven membrane processes are utilized to separate acetic acid and purify it from the fermented effluents and are found to be quite cost-effective. The effluents are in general complex mixture of various volatile fatty acids and other impurities like residual sugar, colloids, proteins, etc. These impurities tend to cause fouling in membranes, which may retard the process efficiency and performance. This drives the requirement of pretreatment of fermented effluents before being treated. The UF and MF membranes are found to alleviate the fouling agents present in the fermented effluents before being processed further [27,28]. Thus, these membranes can be used as a clarifier in the pretreatment step of recovery and purification.

Initially, the UF membranes were applied in a system to separate low-molecular-weight acid compounds from the fermented sewage sludge. In the end, it was found that the process successfully clarifies the sludge, and the permeate stream contains concentrated volatile fatty acids and very low suspended solids. However, the separation performance was reduced with time due to the sedimentation of insoluble suspended solids on the UF membrane surface [29]. Recently, Pervez et al. [30,31] developed a MF- and UF-based membrane bioreactor which enables to separate larger particles from the anaerobically digested effluents. The reactor setup was operated 98 days and can be used for the long-term formation of volatile fatty acids. Moreover, the UF membranes (50 kDa) can successfully retrieve a very high amount of minerals and volatile fatty acids. Among them, ethanoic acid was the primary component. In another study, Zacharof et al. [32] treated the effluents from agricultural residue for recovery and purification of acetic acid and butyric acid using MF and NF membranes. They applied a crossflow MF unit to prevent it from fouling and were successfully able to produce ethanoic acid and butanoic acid of the following concentration: 21.08 and 15.81 mM. These were further processed using a variety of NF membranes.

Out of all the pressure-driven membrane techniques, NF and RO are most prominent membrane separation techniques which can efficiently purify and separate various carboxylic acids like acetic and butyric acids in aqueous solution. The efficiency of separation of NF and RO membranes is high because of the negative charge of carboxylic acids and their low molecular weight. However, for dense RO membranes, the separation is mainly dominated by the size effect and is less affected by the physicochemical characteristics of the feed. The separation of acetic acid from monosaccharides using NF/RO was reported by Zhou et al. [33]. The influence of operating parameters such as pH and temperature on the separation factor was compared for both types of membranes. The findings confirm that, for NF membranes, the electrostatic interaction between the acid and membrane surface, as well as the pH value, mostly regulated the transport of acetic acid. However, RO membranes are much less affected by the pH of feed solution [34,35]. In a recent work by Domingos et al. [36], the application of polyamide-based NF/RO membranes for concentrating volatile fatty acid was reported. They utilized pilot-scale spiral wound modules, and the performance of these membranes was assessed at different operating conditions. From the results, it was confirmed that RO membranes are likely to be unaffected by the pH level of acidogenic effluents; however, for NF membranes, pH is a critical parameter and concentration polarization might be resulted across these membranes. While pH has less effect in RO membranes, other parameters such as applied pressure or operating temperature might affect the process. Previous studies have shown that with increase in applied pressure, acetic acid retention increases, and the reverse happens with an increase in temperature [37].

16.4.2 Non-Pressure-Driven Membrane Processes

FO process has proven to be another promising method of volatile fatty acid recovery and purification due to its exceptional mass transfer properties, low hydraulic pressure requirement, and the presence of reverse draw solution diffusion [38]. Additionally, it also shows a lesser fouling tendency and better fouling reversibility compared to other processes. This technique is frequently utilized in desalination processes, and very few literatures are reported with carboxylic acid or acetic acid separation. In the FO process, water flows through the FO membrane from the feed side to the draw solution and dilutes the draw solution. This process may operate in two ways depending on the orientation of the selective layer of FO membrane. It is known as FO mode when the selective layer of membrane faces the feed solution, and if it faces the draw solution, it is known as pressure retarded osmosis (PRO) mode. When a feed solution having organic compounds is allowed to pass through the FO process, the organic compounds are rejected by the membrane, which stay in the feed side only. Surge in the concentration of organic compounds on the feed side declines the water flux which eventually reduces the overall FO driving force. Membrane fouling influences the deposition of organic compounds on the membrane surface; therefore, changing the mode of operation has no appreciable effect on FO water flux.

The results reported by Blandin et al. suggest that volatile fatty acids with pH 7.5 recovery in the FO system show the highest rejection of 90% with wastewater and acetic acid showed lesser rejection (33%) at pH of 4 [37]. The rejection was mainly governed by steric forces. When pH increases (alkaline pH), ionization of acids becomes significant which ultimately results in higher rejection. Concentration of volatile fatty acid in consecutive batch operations is observed to follow the following trend: acetic acid > propionic acid > butyric acid and so on. In other studies, it can also be observed that the pH value has a significant effect on volatile fatty acid rejection rate, reverse salt flux, and has less effect on FO water flux, similar to NF membranes, as stated earlier [39].

MD is another important non-pressure-driven membrane technique utilized for the separation of various components from liquid mixture. It is mainly a temperature-driven process where the molecules are separated by the MD membrane based on their volatilities. The membrane prevents the passage of liquids through it and allows the gas and vapors to pass through the membrane pores. It has gained lots of interest because of its lower electricity demand and low sensitivity to concentration polarization [40]. However, if the feed contains a mixture of volatile components, then separation in MD becomes challenging compared to desalination by MD, where the volatile water is separated from salts. This is why very few studies are reported on the separation and purification of volatile fatty acids from the fermentation broth. A detailed study reported by Khiter et al. [41] showed the separation efficiency of MD toward various carboxylic acids (formic acid, acetic acid, and succinic acid). Various operating parameters such as temperature have greater influence on permeate flux compared to acid type, concentration, and pH of the feed solution. Another study by Gryta [42] showed the potential application of polypropylene membrane-based MD for the separation of acetic acid from the glycerol fermentation broth. This eventually helps in an increase in the conversion of glycerol without any adjustment of pH. After about 2 years of operation, the PP membranes show a partial wetting of the membrane wall, which results in a decrease in productivity. Accordingly, utilizing superhydrophobic membranes in MD would mitigate the membrane fouling and result in improved organic acid recovery.

PV is considered to be an energy-saving membrane technique that allows the separation of some close-boiling point mixtures and temperature-sensitive compounds present in very low concentration such as acetic acid/water and ethanol/water, which are difficult to separate by conventional techniques. Several works have been reported on the industrial application of PV technology. This technique is nowadays hugely applied in the dehydration of acetic acid, which can be seen from the increasing number of publications for this purpose. The development of microporous inorganic membranes (sol–gel-derived ceramic and zeolite membranes) helped to separate components from a harsh mixture. Additionally, these membranes are proved to have more potential for dehydration of acetic acid compared to polymeric membranes. PV technique shows no unfavorable effect on the microorganisms present in the feed solution during anaerobic fermentation process. To improve the performance of dehydration of acetic acid, Su et al. [43] developed a mixed matrix membrane by incorporating metal-organic frameworks (MOFs). The results show that the increase in MOF loading results in improved water permeability and selectivity. Additionally, the effect of acid concentration and temperature was also investigated. With an increase in temperature, the vapor pressure difference between feed and permeate side also increases which results in an increase in water flux.

16.5 FUTURE PROSPECTIVE AND CHALLENGES

The membrane-based techniques discussed show promising results to recover, purify, and concentrate the acidogenic fermented products especially acetic acid. Additionally, the selectivity of acetic acid can be improved. However, further research needs to be done to attain a scalable and economic recovery method. More careful control and handling of complex fermented stream is necessary to avoid membrane scaling and fouling. Fouling is the major concern of membrane-based separation techniques, which significantly affects the process efficiency and performance. Pretreatment of the fermented effluent is often becoming an essential step to avoid membrane fouling [44]. In the case of pressure-driven membrane processes such as MF and UF systems, the feed stream is pretreated and reduces the fouling, as discussed earlier. Also, the dense hydrophobic also helps in reducing the fouling factor of membrane systems. The major advantage of these membrane-based techniques is the flexibility of being integrated into any existing conventional recovery technique. Hybrid membrane processes are also possible to enhance the efficiency such as FO-MD. In FO-MD hybrid process, FO acts as a pretreatment for the membrane process and may help in reducing fouling. And MD will be applicable as a downstream separation process for resulting high-purity products.

Although some of the recovery techniques result in high purity, recovery of acetic acid such as MD, RO, and PV is quite expensive. Accordingly, a cost–benefit analysis for each method is necessary before application.

16.6 CONCLUSION

Organic wastes, especially biomass, are plentily available across the globe and are rich in organic components. This can be considered as a potential source of bioenergy and biochemical production by anaerobic digestion. The intermediates such as volatile fatty acids are being explored presently because of their wide range of applications. Among all volatile fatty acids, Acetic acid is the most demanded one. The production of acetic acid by fermentation is mainly favored in the acidic-neutral pH range. Additionally, temperature and optimal OLR are considered to be crucial for microbial growth and hence should be maintained at optimum levels. Membrane-based technology may act as a promising recovery technology because it enhances the productivity and selectivity of acetic acid. Among the operating conditions, the pH of the fermented effluent may affect the recovery of acetic acid. The membrane techniques such as FO, MD, and PV may advance the recovery of acetic acid due to their unique advantages such as less fouling and more selectivity. Similarly, RO, UF, and NF membranes can act as pretreatment steps to separate the suspended particle loading in the fermented effluents. Furthermore, the integration of various membranes can substantially improve the recovery efficiency.

REFERENCES

1. Agler MT, Wrenn BA, Zinder SH, Angenent LT (2011) Waste to bioproduct conversion with undefined mixed cultures: The carboxylate platform. *Trends Biotechnol* 29:70–78.
2. Sekoai PT, Ghimire A, Ezeokoli OT, et al (2021) Valorization of volatile fatty acids from the dark fermentation waste streams: A promising pathway for a biorefinery concept. *Renew Sustain Energy Rev* 143:110971.

3. Sukphun P, Sittijunda S, Reungsang A (2021) Volatile fatty acid production from organic waste with the emphasis on membrane-based recovery. *Fermentation* 7:159.
4. Tampio EA, Blasco L, Vainio MM, et al (2019) Volatile fatty acids (VFAs) and methane from food waste and cow slurry: comparison of biogas and VFA fermentation processes. *GCB Bioenergy* 11:72–84.
5. Parchami M, Wainaina S, Mahboubi A, et al (2020) MBR-assisted VFAs production from excess sewage sludge and food waste slurry for sustainable wastewater treatment. *Appl Sci* 10:2921.
6. Pal P, Nayak J (2017) Acetic acid production and purification: Critical review towards process intensification. *Sep Purif Rev* 46:44–61. https://doi.org/10.1080/15422119.2016.1185017
7. Karekar SC, Srinivas K, Ahring BK (2022) Batch screening of weak base ion exchange resins for optimized extraction of acetic acid under fermentation conditions. *Chem Eng J Adv* 11:100337. https://doi.org/10.1016/j.ceja.2022.100337
8. Merli G, Becci A, Amato A, Beolchini F (2021) Acetic acid bioproduction: The technological innovation change. *Sci Total Environ* 798:149292.
9. Li H, Cheng JQ, Ng TY, et al (2004) A meshless hermite-cloud method for nonlinear fluid-structure analysis of near-bed submarine pipelines under current. *Eng Struct* 26:531–542. https://doi.org/10.1016/j.engstruct.2003.12.005
10. Sarchami T, Batta N, Berruti F (2021) Production and separation of acetic acid from pyrolysis oil of lignocellulosic biomass: A review. *Biofuels Bioprod Biorefining* 15:1912–1937.
11. Tammali R, Seenayya G, Reddy G (2003) Fermentation of cellulose to acetic acid by Clostridium lentocellum SG6: Induction of sporulation and effect of buffering agent on acetic acid production. *Lett Appl Microbiol* 37:304–308. https://doi.org/10.1046/j.1472-765X.2003.01397.x
12. Saha S, Kurade MB, El-Dalatony MM, et al (2016) Improving bioavailability of fruit wastes using organic acid: An exploratory study of biomass pretreatment for fermentation. *Energy Convers Manag* 127:256–264. https://doi.org/10.1016/j.enconman.2016.09.016
13. Ramos-Suarez M, Zhang Y, Outram V (2021) Current perspectives on acidogenic fermentation to produce volatile fatty acids from waste. *Rev Environ Sci Bio/Technol* 20:439–478.
14. Menzel T, Neubauer P, Junne S (2020) Role of microbial hydrolysis in anaerobic digestion. *Energies* 13:5555.
15. Atasoy M, Owusu-Agyeman I, Plaza E, Cetecioglu Z (2018) Bio-based volatile fatty acid production and recovery from waste streams: Current status and future challenges. *Bioresour Technol* 268:773–786.
16. Sengun IY, Karabiyikli S (2011) Importance of acetic acid bacteria in food industry. *Food Control* 22:647–656.
17. Deshmukh G, Manyar H (2020) Production pathways of acetic acid and versatile applications in food industry. *Biotechnol Appl Biomass*. https://doi.org/10.5772/intechopen.92289
18. Aghapour Aktij S, Zirehpour A, Mollahosseini A, et al (2020) Feasibility of membrane processes for the recovery and purification of bio-based volatile fatty acids: A comprehensive review. *J Ind Eng Chem* 81:24–40.
19. Vidra A, Németh Á (2018) Bio-produced acetic acid: A review. *Period Polytech Chem Eng* 62:245–256.
20. Pal P, Nayak J (2017) Acetic acid production and purification: Critical review towards process intensification. *Sep Purif Rev* 46:44–61.
21. Grewal HS, Tewari HK, Kalra KL (1988) Vinegar production from substandard fruits. *Biol Wastes* 26:9–14. https://doi.org/10.1016/0269-7483(88)90145-0

22. de Oliveira Carneiro L, Matos RPDS, Ramos WB, et al (2022) Sustainable design and optimization of the recovery process of acetic acid from vinasse: Operational, economic, and environmental analysis. *Chem Eng Process - Process Intensif* 181:109176. https://doi.org/https://doi.org/10.1016/j.cep.2022.109176
23. Lei Z, Li C, Li Y, Chen B (2004) Separation of acetic acid and water by complex extractive distillation. *Sep Purif Technol* 36:131–138.
24. Almhofer L, Paulik C, Bammer D, et al (2023) Contaminations impairing an acetic acid biorefinery: Liquid-liquid extraction of lipophilic wood extractives with fully recyclable extractants. *Sep Purif Technol* 308:122869. https://doi.org/https://doi.org/10.1016/j.seppur.2022.122869
25. Mahboubi A, Ylitervo P, Doyen W, et al (2016) Reverse membrane bioreactor: Introduction to a new technology for biofuel production. *Biotechnol Adv* 34:954–975.
26. Abels C, Carstensen F, Wessling M (2013) Membrane processes in biorefinery applications. *J Memb Sci* 444:285–317.
27. Kanagaraj P, Nagendran A, Rana D, Matsuura T (2016) Separation of macromolecular proteins and removal of humic acid by cellulose acetate modified UF membranes. *Int J Biol Macromol* 89:81–88.
28. Chen G-E, Sun W-G, Kong Y-F, et al (2018) Hydrophilic modification of PVDF microfiltration membrane with poly (ethylene glycol) dimethacrylate through surface polymerization. *Polym Plast Technol Eng* 57:108–117.
29. Longo S, Katsou E, Malamis S, et al (2015) Recovery of volatile fatty acids from fermentation of sewage sludge in municipal wastewater treatment plants. *Bioresour Technol* 175:436–444.
30. Pervez MN, Mahboubi A, Hasan SW, et al (2022) Microfiltration and ultrafiltration as efficient, sustainable pretreatment technologies for resource recovery. In: Naddeo V, Choo KH, Ksibi M (eds). *Water-Energy-Nexus in the Ecological Transition: Natural-Based Solutions, Advanced Technologies and Best Practices for Environmental Sustai.* Springer International Publishing, Cham, pp 279–281.
31. Pervez MN, Mahboubi A, Uwineza C, et al (2022) Feasibility of nanofiltration process for high efficient recovery and concentrations of food waste-derived volatile fatty acids. *J Water Process Eng* 48:102933. https://doi.org/https://doi.org/10.1016/j.jwpe.2022.102933
32. Zacharof MP, Lovitt RW (2014) Recovery of volatile fatty acids (VFA) from complex waste effluents using membranes. *Water Sci Technol* 69:495–503. https://doi.org/10.2166/wst.2013.717
33. Zhou F, Wang C, Wei J (2013) Separation of acetic acid from monosaccharides by NF and RO membranes: Performance comparison. *J Memb Sci* 429:243–251. https://doi.org/10.1016/j.memsci.2012.11.043
34. Zacharof M-P, Mandale SJ, Williams PM, Lovitt RW (2016) Nanofiltration of treated digested agricultural wastewater for recovery of carboxylic acids. *J Clean Prod* 112:4749–4761. https://doi.org/https://doi.org/10.1016/j.jclepro.2015.07.004
35. Bellona C, Drewes JE (2005) The role of membrane surface charge and solute physico-chemical properties in the rejection of organic acids by NF membranes. *J Memb Sci* 249:227–234.
36. Domingos JMB, Martinez GA, Morselli E, et al (2022) Reverse osmosis and nanofiltration opportunities to concentrate multicomponent mixtures of volatile fatty acids. *Sep Purif Technol* 290:120840. https://doi.org/10.1016/j.seppur.2022.120840
37. Zhou F, Wang C, Wei J (2013) Simultaneous acetic acid separation and monosaccharide concentration by reverse osmosis. *Bioresour Technol* 131:349–356. https://doi.org/https://doi.org/10.1016/j.biortech.2012.12.145
38. Yang Y, Gao X, Li Z, et al (2018) Porous membranes in pressure-assisted forward osmosis: Flux behavior and potential applications. *J Ind Eng Chem* 60:160–168.

39. Jung K, Lee D, Seo C, et al (2015) Permeation characteristics of volatile fatty acids solution by forward osmosis. *Process Biochem* 50:669–677.
40. Diaby AT, Byrne P, Loulergue P, et al (2017) Design study of the coupling of an air gap membrane distillation unit to an air conditioner. *Desalination* 420:308–317. https://doi.org/https://doi.org/10.1016/j.desal.2017.08.001
41. Khiter A, Balannec B, Szymczyk A, et al (2020) Behavior of volatile compounds in membrane distillation: The case of carboxylic acids. *J Memb Sci* 612:118453. https://doi.org/10.1016/j.memsci.2020.118453
42. Gryta M (2016) The application of membrane distillation for broth separation in membrane bioreactors. *J Membr Sci Res* 2:193–200.
43. Su Z, Chen JH, Sun X, et al (2015) Amine-functionalized metal organic framework (NH 2-MIL-125 (Ti)) incorporated sodium alginate mixed matrix membranes for dehydration of acetic acid by pervaporation. *RSC Adv* 5:99008–99017.
44. Masse L, Massé DI, Pellerin Y (2008) The effect of pH on the separation of manure nutrients with reverse osmosis membranes. *J Memb Sci* 325:914–919.

17 Application of Various Thermal Treatment Methodologies for Conversion of Agrobiomass into Value-Added Products

Rakesh Upadhyay, Ekta Chaturvedi, and Kulbhushan Samal*

17.1 INTRODUCTION

Increasing population, industrialization, and depleting fossil fuel reserves have generated an urgent need of finding sustainable and economic alternatives to fulfil future demands of energy and fuels. The use of biomass to produce different biofuels and value-added products has become a leading pathway recently due to its wide abundance and negligible costs (Verma et al., 2012). Biomass is a renewable organic material which is mainly derived from plants and animals. It can be categorized as agricultural, forestry/wood-based, biogenic derivatives of municipal solid waste, and human and animal residues. As per EIA, this bioenergy consumption is expected to double in different sectors by 2030. It also involves the replacement of conventional cooking methods by modern technologies including bioethanol and biogas digesters, which meant providing clean and sustainable cooking to approximate 350 million households by 2030. Also, biomass-derived biofuels are expected to provide 27% of total demand for transportation fuels by 2050 which is currently 4% of total renewable energy provided by liquid biofuels only. Likewise, biomass-based power generation is projected to reduce the CO_2 emissions by 1.3 GT annually by 2050 (Hodgson et al., 2022). These are diverse application scenarios of bioenergy, which are widely distributed in different sectors as a clean and sustainable alternative.

Biomass can be converted into different products using different chemical, biological, and thermochemical processes. Direct combustion is the commonly employed technique for biomass processing, and it can be applied to produce energy in the form of heat for industrial processes and to supply power for steam turbines (Zhang & Zhang, 2019). Chemical methods like esterification and transesterification are chiefly

DOI: 10.1201/9781003407713-17

used for biodiesel production using different vegetable oils, waste cooking oils, fats, and greases by converting them into esters (Hariprasath et al., 2016; Singh et al., 2021). Likewise, bioethanol is being produced from biomass using fermentation, which is an effective biological method that has been used in various distilleries and refineries since the 1990s (Lee, 1997). Also, biogas and biomethane are generated in several waste treatment plants and other processes by anaerobic digestion which is one of the key biological techniques for conversion of biomass into value-added products (Bundhoo, 2018; Bundhoo & Mohee, 2018). Overall, biological methods can produce both liquid and gaseous products, whereas liquid products are dominant when using chemical conversion methods (Matthews et al., 2016). There are certain limitations associated with the application of biological methods that can be overcome by thermochemical methods, so thermochemical techniques are preferred over biological methods, especially in the case of large-scale productions (Yu et al., 2021). The motivation behind prioritizing thermochemical processes over other methods lies in its versatile multiple characteristics towards sustainable energy production. For example, various types of biomass feedstocks can be processed through thermochemical processes, whereas biological methods are restricted to some specific types of biomasses. These processes often provide higher energy efficiency than other methods and give products of high energy density, which is one of the key challenges associated with the adoption of bioenergy at the global level. Also, thermochemical processes have low input requirements and offer flexibility in producing a wide range of end products (Moura et al., 2022). Overall, each biomass conversion method has its specific advantages, but the combination of versatility, flexibility, high efficiency, waste management capabilities with less detrimental environmental impacts makes thermochemical conversion processes a captivating preferred choice for advancements in bioenergy towards sustainable energy future (Saidi & Faraji, 2024). Various advantages of thermochemical methods over biological are discussed in Table 17.1.

Thermochemical methods include pyrolysis, liquefaction, and gasification which are nothing but thermal decomposition processes of biomass materials and they occur in closed reactors at different operating conditions (Zhang & Zhang, 2019). Gasification is a high-temperature conversion process that can be utilized to mainly produce gaseous fuels, whereas pyrolysis produces all three solid, liquid, and gaseous products whose percentage in products varies as per the operating conditions. Products of pyrolysis can be tuned as per the requirements by controlling operating conditions, whereas liquefaction mainly produces liquid fuels. The need for these thermochemical processes mainly arises from the growing demand for sustainable renewable energy alternatives which could be a feasible solution of future energy security and environmental issues (Carregosa et al., 2023). These methods allow for the effective utilization of a wide range of unutilized vast biomass waste materials and can reduce our dependence on fossil fuels. Thermochemical conversion also enhances the energy density of biomass like bio-oil produced from pyrolysis having high energy content than feedstock biomass (Zhou et al., 2023). Similarly, syngas, a product of gasification, also has tremendous potential to produce high-energy biofuels. Thermochemical conversion processes play the most crucial role in biomass waste disposal and help in mitigating the harmful hazardous impacts of landfilling

TABLE 17.1
Comparative Analysis of Thermochemical and Biological Conversion Techniques of Biomass

S.No.	Characteristics	Thermochemical Methods	Biological Methods
1.	Type of feedstock	Any type of biomass can be processed	Different enzymes and microbes are used for the conversion thus certain specified biomass can be treated
2.	Reaction Time	It can be completed in few minutes and can extend up to few hours	Time-consuming as compared to thermochemical methods
3.	Reaction products	It gives a broad range of valuable products due to chemical reaction that occurs	It is limited to few specific products
4.	Productivity	High productivity	Limited productivity due to biological nature
5.	Environmental constraints	Not dependent on different environmental conditions like weather and temperature as performed in different reactors by maintaining certain temperature and process conditions	Dependent on sunlight and natural conditions due to the involvement of enzymes and microbes

and open-field burning of these wastes. It also diminishes greenhouse gas (GHG) emissions and contributes in reducing environmental pollutions and other fatal impacts. The integration of thermochemical approaches into utilization and conversion of biomass boosts energy security and can lead to more robust and self-sufficient energy infrastructures in the future (Okolie et al., 2022). This chapter focuses on various thermochemical methods of biomass processing and discusses all parameters in detail. It also addresses the challenges and limitations associated with both processes and possible solutions, which can be found to make these methods feasible for commercial-level production.

17.2 THERMOCHEMICAL TREATMENT METHODOLOGIES

Thermochemical techniques are those that employ heat for the chemical conversion of biomass into different value-added products. These processes mainly include combustion, gasification, liquefaction, and pyrolysis (Verma et al., 2012). Torrefaction is also a thermochemical conversion and the slow form of pyrolysis which produces carbon-rich solid fuels. All these different techniques produce a certain phase of products at specific temperature ranges. The main techniques of thermochemical conversion, their characteristics, and dominant products are discussed in Figure 17.1. A comparative table discussing the basic parameters of all thermochemical techniques is shown in Table 17.2.

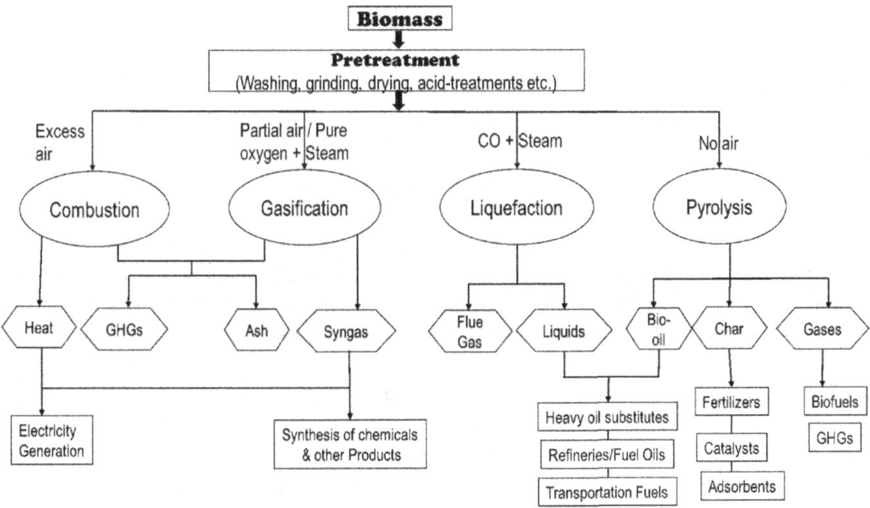

FIGURE 17.1 General layout of different thermochemical techniques, their products, and applications in different areas.

TABLE 17.2
Comparative Analysis of All Thermochemical Conversion Processes

Parameter	Pyrolysis	Gasification	Combustion	Liquefaction
Principle	Thermal decomposition in absence of oxygen	Partial oxidation in limited oxygen environment	Rapid oxidation in excess oxygen	Conversion in the presence of a solvent or catalyst
Operating Temperature	400°C–600°C for slow pyrolysis, up to 1,000°C for fast	700°C–1,000°C	Above 800°C	250°C–500°C.
Products	Biochar, bio-oil, gas	Syngas ($CO + H_2$), char	CO_2, water, ash, heat	Bio-oil, biogas, solid residues
Yield Distribution	Varies with temperature	Syngas (20%–40% CO, 20%–30% H_2), tar, char	CO_2 major fraction	Liquid biofuels, gas, solid residues
Carbon-to-Gas Ratio	High char content, lower gas yield	Higher gas yield, lower char	Complete combustion; high CO_2 emissions	Lower char, liquid and gas products
Energy Efficiency	Moderate to high, influenced by process design	Moderate, affected by gasification efficiency	High, but less efficient for electricity	Moderate, affected by conversion efficiency
Environmental Impact	Potential for bio-oil contaminants; lower GHG	Syngas contaminants, tar formation; lower GHG	High CO_2 emissions; incomplete combustion	Moderate, depends on solvent and conditions

(Continued)

TABLE 17.2 (*Continued*)
Comparative Analysis of All Thermochemical Conversion Processes

Parameter	Pyrolysis	Gasification	Combustion	Liquefaction
Feedstock Compatibility	Broad range, suitable for various biomass types	Versatile, suitable for diverse biomass feeds	Limited by combustion properties; woody fuels	Versatile, adaptable to various feedstocks
Catalyst Utilization	Can be used in some cases to influence product	Catalysts may be used for tar reduction	Not applicable in traditional combustion	Catalysts may enhance liquefaction efficiency
Applications	Biochar as soil fertilizer, bio-oil as fuel	Syngas for power generation, biochar for soil	Heat and power generation	Liquid biofuels for transportation

17.2.1 Combustion

It is the oldest thermochemical conversion technology, and it includes the burning of biomass materials in the oxygen-rich environment at high temperature ranges. Combustion is exothermic in nature, producing CO_2 and steam (H_2O vapor) as the main products due to the reaction occurring between the hydrocarbons and oxygen present in the biomass samples (Demirbas, 2007). Here, the produced heat during the reaction can provide up to 90% of total energy required for biomass conversion. This released heat can also be used for cooking and several other purposes in rural areas. A high temperature range of 800°C–1,000°C is required for complete combustion of the biomass. Except heat generation, electricity production is another significant application of combustion of biomass with the help of steam turbine, which employs the use of steam which is produced from biomass combustion process (Nussbaumer, 2003).

Factors that contribute to the overall efficiency of any combustion process can be categorized as follows:

- Physical characteristics of biomass (like density, surface area, particle, pore size, etc.)
- Chemical characteristics of biomass (like composition, calorific value, etc.)
- Thermal properties of biomass (like conductivity, moisture content, specific heat, etc.)
- Pretreatment processes (like drying, acid leaching, shredding, etc.)

These factors combinedly affect the overall cost and efficiency of the biomass combustion process. This method is mostly employed in power production units and steam boilers, but setting up combustion-based plants is neither commercially viable nor technically efficient as it is very specific in nature and even a type of biomass used can affect its performance in a bigger way (Dhahak et al., 2019; Morris et al., 2018). So overall, this process should be done in a controlled environment by taking care of all the surrounding conditions to make this a technically feasible and cost-effective technology with the significant potential to reduce our energy dependence on burning of fossil fuels.

17.2.2 Liquefaction

Liquefaction is a thermochemical technique that converts biomass straight into liquid products with the help of a solvent. It is also called as hydrothermal liquefaction and usually occurs at high pressure and temperature conditions (Behrendt et al., 2008). Hydrothermal liquefaction is usually performed at temperature and pressure ranges of 150°C–400°C and 4–22 MPa in the presence of water as solvent. This technique is also employed at a commercial scale to produce different chemicals like acids and furans from carbohydrate-rich biomass materials, whereas lignin-based biomass produces aromatic-rich organic fractions dominantly. Unlike carbohydrate-rich biomass, commercial application of lignin-rich biomass is restricted due to its low solubility in water (Ding et al., 2023). Liquefaction is a complex process that contains different stages like depolymerization, dehydration, isomerization, hydrolysis, etc. Selection of proper solvent for the liquefaction depends on process conditions and operating parameters (Castello et al., 2018). Water is commonly the preferred liquefaction solvent due to its wide abundance and low cost. It also can convert high moisture-containing biomass materials into liquids. Production of highly viscous liquids is the only limitation faced while using water as solvent, which can be overcome in the presence of organic solvent, resulting in the low viscosity of produced bio-oils (Gollakota et al., 2018). However, the use of an organic solvent to replace water is the current trending area of research which should mandatorily have a strong hydrogen donating tendency. Recently, the use of organic solvents like methanol, ethanol, etc. is being promoted, which aid in lignin decomposition and produce cyclic aromatics and phenols in the presence of H_2/N_2. Hydrothermal liquefaction is a less explored area and sometimes produces 40%–80% of solid products. The main factors that are responsible for the performance are:

- Temperature
- Pressure
- Solvent medium

Except these, heating rate and residence time are other parameters that also have a significant impact on the efficiency of liquefaction process (de Caprariis et al., 2017). The use of catalysts to accelerate the reaction mechanism of hydrothermal liquefaction is also being done to effectively increase the yield of bio-oil. K_2CO_3 is one of the commonly used liquefaction catalysts, which increases yield and HHV of produced bio-oils. It showed better performance with a high percentage of lignin and produced products that have similar properties as petro-diesel (Naeem et al., 2023). Production of bio-oil is mainly dependent on the temperature of the process and is maximum around 300°C (Akhtar & Amin, 2011). This is the only thermochemical process by which the produced heat can be recycled due to the solvent media present during the reaction. Initial drying of biomass is not required in hydrothermal liquefaction which is the key advantage of this process over pyrolysis. Recently, this method has its wide applicability in super- and subcritical-level water technologies due to the operating conditions of high temperature and high pressure. Even if this process has several advantages over other thermochemical conversion techniques, including its environment-friendly nature, there are still various challenges like optimization of operating conditions, selection of proper solvent, and high viscosity of bio-oil products that must be addressed properly before using this technology for commercial-level applications (Yang et al., 2019).

Solvothermal liquefaction is the other liquefaction technique that is widely applied to produce superior liquid products. This process has a wide range of solvents and catalysts which can be used to increase the yield and upgrading the quality of liquid oils. Oxalic acid, salicylic acid, sodium hydroxide, lime, etc. are common catalysts used in the solvothermal liquefaction process (Ding et al., 2023). Bifunctional catalysts like propylamine are also explored in a recent study which not only aid in the breaking of hemicellulose and lignin linkages but also contributed in breaking the inter- and intramolecular bonding of celluloses. Although it is an efficient technology for the production of liquid products, challenges like volatility, flammability, catalyst, and product recovery make it energy-intensive and high-cost process.

17.2.3 Gasification

Gasification is a high-temperature thermochemical process of converting biomass into gaseous fuel mixtures of different gases. It is one of the attractive techniques to produce hydrogen-rich products from biomass. Gasification occurs in the presence of a gasifying agent or an oxidizing solvent like air, oxygen, and steam, which initially breaks down the biomass molecules into smaller molecules and then into the product gaseous mixtures of CO, CO_2, CH_4, and H_2 (Kumar et al., 2009).

Gasification is a partial combustion of biomass in the presence of air or oxygen. Gasification of biomass materials converts it into synthesis gas, which can be further used to generate heat/electricity or to produce gasoline/diesel range hydrocarbons in the presence of Fe- and Co-based catalyst by Fisher–Tropsch synthesis reactions (Qureshi et al., 2024). It can also participate in methanol formation in the presence of Cu catalysts. The addition of small amounts of O_2 initiates the endothermic steam-reforming reactions as shown in Table 17.3. Gasification usually occurs in the temperature range of 600°C–900°C, which is above the pyrolysis range but below the combustion temperature range. Lower heating value range gaseous products are obtained in the presence of air, whereas the mid-range of heating values is obtained in the oxygen environment (Schmieder et al., 2000). Different reactions that occur during gasification are shown in Table 17.3. Gasification results in 80%–85% yield of gaseous and volatile products, whereas char and non-volatile fractions are recovered in the remaining percentages. Yield of char or unburned coke increases in the product if there is an incomplete burning of biomass due to the presence of high amount of moisture into the feedstock (Kirubakaran et al., 2009). A suitable gasification catalyst is necessary to avoid excessive generation of tars, which can be converted into synthesis gas in the presence of catalysts. Catalysts should be regenerative and have the capability of removing undesired tars. They should be cost-effective and have a low deactivation tendency (Li et al., 2023). Dolomites are commonly used gasification catalysts that are low cost and have strong catalytic activity due to the presence of iron oxides and alumina. It also helps in reducing tar concentration inside the reactors by converting

TABLE 17.3
Different Reactions Occurring During the Gasification Process

S.No.	Reaction Category	Reaction
1.	Overall gasification process	Biomass + Steam/Oxygen + Heat = $CO + H_2 + CH_4 + CO_2$ + other side products
2.	Oxidization	$C + \frac{1}{2}O_2 \rightarrow CO$ $C + O_2 \rightarrow CO_2$ $C + H_2O \leftrightarrow CO + H_2$
3.	Water gas shift	$CO + H_2O \leftrightarrow CO_2 + H_2$
4.	Methanation	$CO + 3H_2 \leftrightarrow CH_4 + H_2O$ $CO_2 + 4H_2 \rightarrow CH_4 + 2H_2O$
5.	Steam Reforming	$CH_4 + H_2O \leftrightarrow CO + 3H_2$ $CH_4 + \frac{1}{2}O_2 \rightarrow CO + 2H_2$

them into gases. Olivine, Metals, and alkali metals are a few other good gasification catalysts, but deactivation affects their catalytic performance. Nickel-based supported catalysts are also used during water gas shift reactions, which not only alter the chemical composition of produced gas but also discard the tar and methane from the reactors; however, coke deposition and sintering caused frequent deactivation of Nickel. Overall, the factors that affect the gasification and its products are as follows:

- Types of biomasses and their properties (size, density, composition, moisture content, etc.).
- Gasification operating conditions and types of gasifiers used (temperature, pressure, etc.).
- Types of oxidizing agents used and their flow rates into the reaction.
- Types of gasification catalysts used.

Gasification process is made up of three steps including upstream processing, gasification, and downstream processing steps, respectively. Here, upstream includes all pretreatment processes that are required to prepare the feedstock as per the required operating conditions. It includes drying of biomass, densification, and size reduction. After feedstock preparation, gasification is the next step at specified operating conditions of temperature and pressure. The final step of downstream processing involves the purification and separation of obtained gaseous fuels. It also includes the further processing of obtained gases through steam-reforming and other reactions as per the end uses. Finally, the obtained gases are utilized in different industrial applications like gas turbines, burners, fuel cell, etc. Most of the gasification reactions are endothermic in nature, and thus, the energy is required which is supplied by the oxidation of biomass materials in the presence of air or oxygen. There are different types of gasifiers being used for gasification like updraft, fluidized bed, downdraft, entrained flow type, etc. depending upon the operating conditions and types of feedstocks.

Gasification has several advantages over direct combustion which also includes high efficiency as compared to the combustion process. Even low-value biomass feedstock can be converted into electricity as well as transportation fuels. GHG and NO_x emissions are less in gasification than combustion and can easily be separated to improve the quality of obtained products. Gasification can produce high-energy density products by deoxygenating the biomass feedstocks. It also increases the ratio of hydrogen to carbon of the obtained fuels (H/C). Overall, gasification has significant potential than other conversion technologies due to its wide applicability and range of obtained products (Arpia et al., 2022; Yakaboylu et al., 2015). Purification and conversion of the obtained product gases into valuable biofuels are the challenges associated with the commercial-level applications of gasification process and need to be addressed to make it a technically efficient and economical process (Wang et al., 2008).

17.2.4 Pyrolysis

Pyrolysis is a thermal decomposition of biomass materials in the absence of oxygen and the most preferred method over others nowadays to obtain gaseous, liquid, and solid fuels (Yogalakshmi et al., 2022). The yield of these pyrolysis products depends on the operating conditions and can be tuned as per the requirements (Roy & Dias, 2017). The type of biomass used also affects the type of product obtained and the yield of that dominant product. Pyrolysis can be categorized into the following three categories based on the operating conditions:

- Slow pyrolysis: large residence time from few hours to few days with low heating rate of 5°C–10°C/min.
- Fast pyrolysis: residence time up to 10 seconds to high heating rate of 50°C–150°C/min.
- Flash pyrolysis: very high heating rate to lowest residence time.

Here, slow pyrolysis results in high yield of solid products and usually occurs at 200°C–300°C, whereas fast pyrolysis results in high yield of liquid products in the temperature range of 300°C–600°C, and temperatures above 600°C result in the highest yield of gaseous products. Characteristics of different types of pyrolysis and their dominant products are shown in Table 17.4. That is why this method is mostly used because its products can easily be tuned as per the requirements by simply controlling the operating conditions of the pyrolysis (Fahmy et al., 2020).

TABLE 17.4
Product Distribution and Operating Conditions of Different Types of Pyrolysis

S.No.	Pyrolysis Type	Temperature Range (°C)	Heating Rate (°C/sec)	Residence Time	Char (%)	Liquid (%)	Gas (%)
1.	Slow	250°C–400°C	0.1	Few hours to few days	35	30	35
2.	Fast	400°C–700°C	10	1–10 seconds	12	75	13
3.	Flash	700°C–900°C	100	Less than 1 seconds	10	20	70

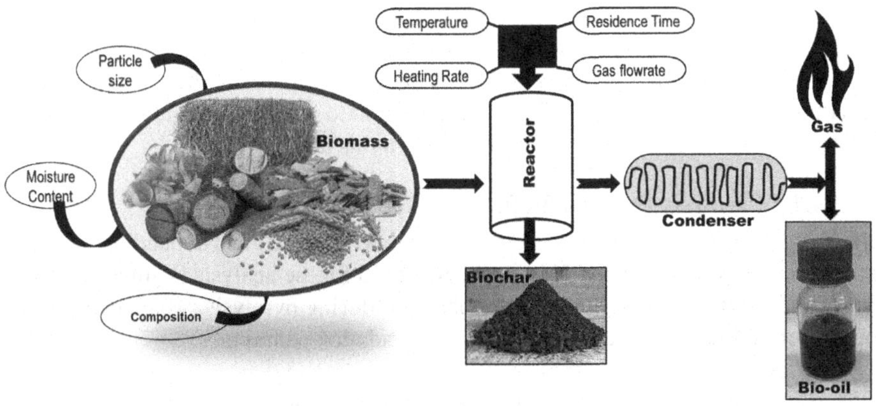

FIGURE 17.2 A schematic diagram of pyrolytic process and important parameters affecting the yield of products.

The main advantage of the pyrolysis process is that it can be considered as zero waste process because all the products obtained from the pyrolysis have their wide applications in different sectors (Hoang et al., 2021). Pyrolysis process and main factors that affect the product yield and characteristics of the pyrolysis products are shown in Figure 17.2 (Chen et al., 2021; Kumar et al., 2020; Zhou et al., 2021).

17.2.4.1 Co-pyrolysis

Co-pyrolysis is a technique in which biomasses are combined with different plastic polymers in varying ratios for pyrolysis. With the increasing population and industrialization, the demand and generation of plastic wastes have increased tremendously around the globe and proper disposal and recycling are another big issue. This co-pyrolysis provides a pathway for the utilization of both waste plastics and biomass and their conversion into different valuable products (Ojha & Vinu, 2018; Salvilla et al., 2020).

There are various advantages of co-pyrolysis over pyrolysis technique including product enhancement and low energy consumption as follows:

- Catalytic co-pyrolysis is one of the promising technologies to produce aromatics directly from different biomass wastes.
- This technique critically improves the yield of aromatics, which is very low during pyrolysis.
- Co-pyrolysis improves both the quality and quantity of liquid products obtained during biomass pyrolysis.
- Water content is found to be lower in the case of co-pyrolysis oil, which is very high (around 70%–80%) in the case of biomass-derived bio-oils.
- Co-pyrolysis decreases the percentage of undesired coke in the products and enhances the fraction of petroleum derivatives.
- Catalytic co-pyrolysis shows synergistic behaviour of plastics and may decrease their decomposition temperature in the presence of biomass.

Although co-pyrolysis has several advantages over convention pyrolysis, in-depth studies are still required for scaling and commercialization of the process (Wang et al., 2021). Modelling and recent AI/ML techniques can also be incorporated to level up this process in future. Detailed insights into the mechanism of the process may also contribute in leading advancements in this area.

17.2.4.2 Pyrolysis-Derived Biofuels

Biofuels derived from pyrolysis vary in their properties and yield depending upon the process conditions and type of biomass being used for the analysis (Vamvuka, 2011). Three different phases of products are obtained during pyrolysis, which can further be converted into valuable products using upgradation techniques:

- **Biochar:** It is the solid form product of pyrolysis and a carbon-rich material. It dominates in the pyrolysis products in the case of slow pyrolysis (Chaiwong et al., 2013; Lee et al., 2013). Also, low-temperature pyrolysis of biomass with high residence time also results in higher yield of biochar. Initially, biochar was considered as the least preferred side product of pyrolysis, but recently, it has emerged as a potential substituent as fertilizer in agriculture. It has multiple other applications in agriculture and environment (Chen et al., 2017; Husain et al., 2022; Kazimierski et al., 2022). Here, the type of feed and operating process conditions are the two crucial factors, which determine the characteristics of the biochar and thus play an important role in identifying its application in different areas (Tripathi et al., 2016). It can increase the soil productivity and can also reduce the level of contaminants. It can be used in the treatment of wastewater and in many other areas. Wide applications of biochar are mainly due to its specific characteristics of high surface area, good adsorption efficiency, porosity, and ion exchange tendency.
- **Bio-oil:** Liquid product obtained from the pyrolysis of biomass is called bio-oil. It is brownish to black in colour and have strong pungent smell. It is a complex mixture of different organic compounds and contains a very high percentage of water, which makes it unsuitable for different applications. High oxygenated compounds are another factor that limits its application and conversion into different products. Alcohols, ethers, acids, esters, phenols, ketones, aldehydes, and aromatics are the dominant components of bio-oil. Till now, more than 300 different compounds have been identified in bio-oils through characterization. The exact composition of bio-oils cannot be determined due to the presence of a large number of compounds. Thus, the molecular weight of bio-oils typically ranges between 18 and 5,000, but it may also be on higher side depending upon the components present and type of biomass feed. Bio-oils are not stable in structure and may start degrading upon storage; so, they cannot be directly used after recovery. Thus, upgrading of bio-oils is a mandatory step for their application in different areas (Alvarez-Chavez et al., 2019; Fermoso et al., 2017). They can also be converted into biodiesels, kerosene, and other fuels through esterification and other chemical reactions. Bio-oils can also be blended with diesels in different ratios in vehicles (Hansen et al., 2020; Mei et al., 2020). It has significant

potential as transportation fuel and can also be employed as flavour enhancer in food industries. It can also be used as fuel oil for combustion in different boilers and furnaces of refineries and industries (Machado et al., 2022; Zhang et al., 2013). Even though it is highly recommended for transportation, its application requires upgrading before any use, both qualitatively and quantitatively. Thus, possibilities of conversion of bio-oils into biofuels at the commercial level still need to be explored comprehensively.

- **Gases:** Gases that cannot be condensed during pyrolysis are recovered as gaseous products of the process. Pyrolytic gases mainly contain CO, CO_2, CH_4, NO_x, SO_x, H_2, H_2S, etc. depending upon the type of feed material being pyrolyzed (Honus et al., 2014). Gases are the dominant product during biomass pyrolysis and they contain GHGs in higher percentages. Thus, exposure to these gases for long hours may create health issues. These gases are highly undesirable and can be the major pollutant in the environment. These gases are mostly used for pre-heating purposes of biomass and can be used for energy recovery of the pyrolysis process (Honus et al., 2018; Saravana Sathiya Prabhahar et al., 2020). Hydrogen is also found to be present in minor fractions in these gases, but can be produced in higher percentage up to 70%–80% of total gases evolved by using suitable catalyst and favourable operating conditions (Honus et al., 2016; Nnabuife et al., 2022; Onarheim et al., 2020). Production of clean hydrogen fuel from biomass pyrolysis may lead the path for the effective utilization and conversion of undesirable pyrolytic gaseous fractions.

17.3 ROLE OF CATALYSTS IN THERMOCHEMICAL CONVERSION PROCESSES

Recently, pyrolysis has emerged as a promising thermochemical technique for the conversion of biomass into different fuels. However, the products obtained from pyrolysis have certain limitations due to their composition and specific properties. Application of these products in different areas requires a mandatory step of upgrading before their conversion into value-added products. Catalytic pyrolysis is one of those techniques, which improves the quality and quantity of bio-oils during pyrolysis. The use of suitable catalyst has the potential to convert the uncondensed undesirable gases of pyrolysis into clean hydrogen-rich fuels (Norinaga et al., 2012). Thus, catalyst plays a very significant role in converting biomass into biofuels and can be proved as a leading pathway in the commercialization of thermochemical techniques. There are various parameters associated with these catalysts that determine the characteristics of obtained products and efficiency of these catalysts during the process (Bhoi et al., 2020). Catalyst to biomass ratio, catalysts activation temperatures, biomass type, and catalyst mixing methods are few of the parameters that critically affect the performance of catalysts. There are two ways in which catalytic pyrolysis can be performed as follows:

- **In situ mode:** In this mode of mixing, catalysts and biomass are mixed in single reactor and pyrolysis is performed. This method is cost-effective due to the requirement of single reactor, but improper contact between catalyst and material results in poor heat transfer, which ultimately affects the products. Catalyst poisoning is also one disadvantage of this mode, which limits the applicability of this mode for high-quality products.

- **Ex situ mode:** This mode is preferred to obtain high-value products from catalytic pyrolysis. This method requires a separate reactor for catalyst contact with products. This method provides better control on operating conditions and heat requirements of the process. In this mixing mode, initially pyrolysis is performed at a suitable temperature range, followed by a catalytic reaction or cracking is performed usually at high temperatures to obtain high-quality products. High capital cost requirement is the only limitation associated with this mixing mode of catalytic pyrolysis.

Catalysts play the very important role in optimizing products yields, enhancing product quality and increasing the conversion efficiency of thermochemical conversion processes. They also contribute towards the development of sustainable bioenergy systems by addressing several challenges like formation of undesired tar and production of high-value biofuels (Li et al., 2023). The addition of catalysts in pyrolysis results in deoxygenation reaction, which reduces the oxygen content of bio-oil, improving its quality and stability for storage purposes. Similarly, in gasification, catalysts influence the composition and quality of syngas, which further determines the conversion of syngas into different biofuels. During liquefaction, catalysts promote hydrogenation reactions, which upgrade the quality of liquid fuels. Regenerative-type catalysts with large stability and low poisoning tendency also reduce operational costs and minimize waste generation during the processes. Catalysts also contribute to cleaner thermochemical conversion processes by causing reduction in harmful emissions.

There are various types of catalysts which have been explored by different researchers previously (Garba et al., 2018; Norouzi et al., 2021; Tawalbeh et al., 2021). Zeolite is the most adopted catalyst for the pyrolysis, which can also be impregnated with different metals for better catalytic efficiency (Rahman et al., 2018). Silica-based catalyst has also been employed for upgrading of fuels. A recent work highlighted different types of multifunctional catalysts that can be used for biomass conversion techniques (Naeem et al., 2023). Biochar obtained from pyrolysis has also been proven to be an effective catalyst to produce biofuels. Catalyst deactivation, selectivity, and cost of the catalysts are the main challenges that need to be addressed to utilize the potential benefits of catalysis in thermochemical conversion methodologies. Overall, low-cost, environment-friendly, and regenerative-type catalysts are the need of the hour in the current scenario, which can be utilized for upscaling the process (Cheng et al., 2016).

17.4 CONCLUSIONS

This chapter has covered all the thermochemical techniques like pyrolysis, gasification, liquefaction, and combustion, which are generally used for the conversion of biomass into different biofuels. These techniques produce different types of products depending upon the process operating conditions and feedstock material. Gasification is employed for heat and power generation, whereas liquefaction is preferred where liquid products are dominant. Pyrolysis is a technique that gives all three phases: solid, liquid, and gaseous type products, which further require upgrading to obtain valuable products. The versatility of thermochemical processes like pyrolysis, gasification, and liquefaction enables the conversion of a diverse range of feedstocks into valuable high-energy density products. High energy efficiency, low GHG emissions, and the ability to address biomass waste disposal with less harmful environmental impacts place them on top of sustainable resilient energy solutions. Thermochemical processes not only offer solution for the effective biomass utilization but also contribute to economic developments and technological advancements in bioenergy sectors. Thermochemical methods can be the game changer option for transition into bio-based economy, and their strategic implementation is crucial for maximizing bioenergy benefits and promoting its adoption globally.

NOTE

* Corresponding author

REFERENCES

Akhtar, J., & Amin, N. A. S. (2011). A review on process conditions for optimum bio-oil yield in hydrothermal liquefaction of biomass. *Renewable and Sustainable Energy Reviews*, 15(3), 1615–1624. https://doi.org/10.1016/j.rser.2010.11.054

Alvarez-Chavez, B. J., Godbout, S., Palacios-Rios, J. H., Le Roux, É., & Raghavan, V. (2019). Physical, chemical, thermal and biological pre-treatment technologies in fast pyrolysis to maximize bio-oil quality: A critical review. *Biomass and Bioenergy*, 128, 105333. https://doi.org/10.1016/j.biombioe.2019.105333

Arpia, A. A., Nguyen, T. B., Chen, W. H., Dong, C. Di, & Ok, Y. S. (2022). Microwave-assisted gasification of biomass for sustainable and energy-efficient biohydrogen and biosyngas production: A state-of-the-art review. *Chemosphere*, 287, 132014. https://doi.org/10.1016/j.chemosphere.2021.132014

Behrendt, F., Neubauer, Y., Oevermann, M., Wilmes, B., & Zobel, N. (2008). Direct liquefaction of biomass. *Chemical Engineering and Technology*, 31(5), 667–677. https://doi.org/10.1002/ceat.200800077

Bhoi, P. R., Ouedraogo, A. S., Soloiu, V., & Quirino, R. (2020). Recent advances on catalysts for improving hydrocarbon compounds in bio-oil of biomass catalytic pyrolysis. *Renewable and Sustainable Energy Reviews*, 121, 109676. https://doi.org/10.1016/j.rser.2019.109676

Bundhoo, Z. M. A. (2018). Microwave-assisted conversion of biomass and waste materials to biofuels. *Renewable and Sustainable Energy Reviews*, 82, 1149–1177. https://doi.org/10.1016/j.rser.2017.09.066

Bundhoo, Z. M. A., & Mohee, R. (2018). Ultrasound-assisted biological conversion of biomass and waste materials to biofuels: A review. *Ultrasonics Sonochemistry*, 40, 298–313. https://doi.org/10.1016/j.ultsonch.2017.07.025

Carregosa, I. S. C., Carregosa, J. de C., Silva, W. R., Santos, T. M., & Wisniewski, A. (2023). Thermochemical conversion of aquatic weed biomass in a rotary kiln reactor for production of bio-based derivatives. *Journal of Analytical and Applied Pyrolysis*, 173, 106048. https://doi.org/10.1016/j.jaap.2023.106048

Castello, D., Pedersen, T. H., & Rosendahl, L. A. (2018). Continuous hydrothermal liquefaction of biomass: A critical review. *Energies*, 11(11), 3165. https://doi.org/10.3390/en11113165

Chaiwong, K., Kiatsiriroat, T., Vorayos, N., & Thararax, C. (2013). Study of bio-oil and bio-char production from algae by slow pyrolysis. *Biomass and Bioenergy*, 56, 600–606. https://doi.org/10.1016/j.biombioe.2013.05.035

Chen, Y., Zhang, X., Chen, W., Yang, H., & Chen, H. (2017). The structure evolution of biochar from biomass pyrolysis and its correlation with gas pollutant adsorption performance. *Bioresource Technology*, 246, 101–109. https://doi.org/10.1016/j.biortech.2017.08.138

Chen, D., Gao, D., Huang, S., Capareda, S. C., Liu, X., Wang, Y., Zhang, T., Liu, Y., & Niu, W. (2021). Influence of acid-washed pretreatment on the pyrolysis of corn straw: A study on characteristics, kinetics and bio-oil composition. *Journal of Analytical and Applied Pyrolysis*, 155, 105027. https://doi.org/10.1016/j.jaap.2021.105027

Cheng, S., Wei, L., Zhao, X., & Julson, J. (2016). Application, deactivation, and regeneration of heterogeneous catalysts in bio-oil upgrading. *Catalysts*, 6(12), 6120195. https://doi.org/10.3390/catal6120195

de Caprariis, B., De Filippis, P., Petrullo, A., & Scarsella, M. (2017). Hydrothermal liquefaction of biomass: Influence of temperature and biomass composition on the bio-oil production. *Fuel*, 208, 618–625. https://doi.org/10.1016/j.fuel.2017.07.054

Demirbas, A. (2007). Combustion of biomass. *Energy Sources, Part A: Recovery, Utilization and Environmental Effects*, 29(6), 549–561. https://doi.org/10.1080/009083190957694

Dhahak, A., Bounaceur, R., Le Dreff-Lorimier, C., Schmidt, G., Trouve, G., & Battin-Leclerc, F. (2019). Development of a detailed kinetic model for the combustion of biomass. *Fuel*, 242, 756–774. https://doi.org/10.1016/j.fuel.2019.01.093

Ding, Z., Kumar Awasthi, S., Kumar, M., Kumar, V., Mikhailovich Dregulo, A., Yadav, V., Sindhu, R., Binod, P., Sarsaiya, S., Pandey, A., Taherzadeh, M. J., Rathour, R., Singh, L., Zhang, Z., Lian, Z., & Kumar Awasthi, M. (2023). A thermo-chemical and biotechnological approaches for bamboo waste recycling and conversion to value added product: Towards a zero-waste biorefinery and circular bioeconomy. *Fuel*, 333, 126469. https://doi.org/10.1016/j.fuel.2022.126469

Fahmy, T. Y. A., Fahmy, Y., Mobarak, F., El-Sakhawy, M., & Abou-Zeid, R. E. (2020). Biomass pyrolysis: Past, present, and future. *Environment, Development and Sustainability*, 22(1), 17–32. https://doi.org/10.1007/s10668-018-0200-5

Fermoso, J., Pizarro, P., Coronado, J. M., & Serrano, D. P. (2017). Advanced biofuels production by upgrading of pyrolysis bio-oil. *Wiley Interdisciplinary Reviews: Energy and Environment*, 6(4), 245. https://doi.org/10.1002/wene.245

Garba, M. U., Musa, U., Olugbenga, A. G., Mohammad, Y. S., Yahaya, M., & Ibrahim, A. A. (2018). Catalytic upgrading of bio-oil from bagasse: Thermogravimetric analysis and fixed bed pyrolysis. *Beni-Suef University Journal of Basic and Applied Sciences*, 7(4), 776–781. https://doi.org/10.1016/j.bjbas.2018.11.004

Gollakota, A. R. K., Kishore, N., & Gu, S. (2018). A review on hydrothermal liquefaction of biomass. *Renewable and Sustainable Energy Reviews*, 81, 1378–1392. https://doi.org/10.1016/j.rser.2017.05.178

Hansen, S., Mirkouei, A., & Diaz, L. A. (2020). A comprehensive state-of-technology review for upgrading bio-oil to renewable or blended hydrocarbon fuels. *Renewable and Sustainable Energy Reviews*, 118, 109548. https://doi.org/10.1016/j.rser.2019.109548

Hariprasath, P., Selvamani, S. T., Vigneshwar, M., Palanikumar, K., & Jayaperumal, D. (2016). Comparative analysis of cashew and canola oil biodiesel with homogeneous catalyst by transesterification method. *Materials Today: Proceedings*, 16, 1357–1362. www.sciencedirect.com

Hoang, A. T., Ong, H. C., Fattah, I. M. R., Chong, C. T., Cheng, C. K., Sakthivel, R., & Ok, Y. S. (2021). Progress on the lignocellulosic biomass pyrolysis for biofuel production toward environmental sustainability. *Fuel Processing Technology*, 223, 106997. https://doi.org/10.1016/j.fuproc.2021.106997

Hodgson, D., Bains, P., & Moorhouse, J. (2022). *Bioenergy - Analysis - IEA*. https://www.iea.org/reports/bioenergy

Honus, S., Juchelkova, D., Campen, A., & Wiltowski, T. (2014). Gaseous components from pyrolysis - characteristics, production and potential for energy utilization. *Journal of Analytical and Applied Pyrolysis*, 106, 1–8. https://doi.org/10.1016/j.jaap.2013.11.023

Honus, S., Kumagai, S., Němček, O., & Yoshioka, T. (2016). Replacing conventional fuels in USA, Europe, and UK with plastic pyrolysis gases - Part I: Experiments and graphical interchangeability methods. *Energy Conversion and Management*, 126, 1118–1127. https://doi.org/10.1016/j.enconman.2016.08.055

Honus, S., Kumagai, S., Fedorko, G., Molnár, V., & Yoshioka, T. (2018). Pyrolysis gases produced from individual and mixed PE, PP, PS, PVC, and PET-Part I: Production and physical properties. *Fuel*, 221, 346–360. https://doi.org/10.1016/j.fuel.2018.02.074

Husain, Z., Shakeelur Raheman, A. R., Ansari, K. B., Pandit, A. B., Khan, M. S., Qyyum, M. A., & Lam, S. S. (2022). Nano-sized mesoporous biochar derived from biomass pyrolysis as electrochemical energy storage supercapacitor. *Materials Science for Energy Technologies*, 5, 99–109. https://doi.org/10.1016/j.mset.2021.12.003

Kazimierski, P., Januszewicz, K., Godlewski, W., Fijuk, A., Suchocki, T., Chaja, P., Barczak, B., & Kardaś, D. (2022). The course and the effects of agricultural biomass pyrolysis in the production of high-calorific biochar. *Materials*, 15(3), 15031038. https://doi.org/10.3390/ma15031038

Kirubakaran, V., Sivaramakrishnan, V., Nalini, R., Sekar, T., Premalatha, M., & Subramanian, P. (2009). A review on gasification of biomass. *Renewable and Sustainable Energy Reviews*, 13(1), 179–186. https://doi.org/10.1016/j.rser.2007.07.001

Kumar, A., Jones, D. D., & Hanna, M. A. (2009). Thermochemical biomass gasification: A review of the current status of the technology. *Energies*, 2(3), 556–581. https://doi.org/10.3390/en20300556

Kumar, R., Strezov, V., Weldekidan, H., He, J., Singh, S., Kan, T., & Dastjerdi, B. (2020). Lignocellulose biomass pyrolysis for bio-oil production: A review of biomass pre-treatment methods for production of drop-in fuels. *Renewable and Sustainable Energy Reviews*, 123, 109763. https://doi.org/10.1016/j.rser.2020.109763

Lee, J. (1997). Biological conversion of lignocellulosic biomass to ethanol. *Journal of Biotechnology*, 56, 1-24.

Lee, Y., Park, J., Ryu, C., Gang, K. S., Yang, W., Park, Y. K., Jung, J., & Hyun, S. (2013). Comparison of biochar properties from biomass residues produced by slow pyrolysis at 500°C. *Bioresource Technology*, 148, 196–201. https://doi.org/10.1016/j.biortech.2013.08.135

Li, S., Wu, Y., Dao, M. U., Dragoi, E. N., & Xia, C. (2023). Spotlighting of the role of catalysis for biomass conversion to green fuels towards a sustainable environment: Latest innovation avenues, insights, challenges, and future perspectives. Chemosphere, 318, 137954. https://doi.org/10.1016/j.chemosphere.2023.137954

Machado, H., Cristino, A. F., Orišková, S., & Galhano dos Santos, R. (2022). Bio-oil: The next-generation source of chemicals. *Reactions*, 3(1), 118–137. https://doi.org/10.3390/reactions3010009

Matthews, L. R., Niziolek, A. M., Onel, O., Pinnaduwage, N., & Floudas, C. A. (2016). Biomass to liquid transportation fuels via biological and thermochemical conversion: Process synthesis and global optimization strategies. *Industrial and Engineering Chemistry Research*, 55(12), 3205–3225. https://doi.org/10.1021/acs.iecr.5b03319

Mei, D., Guo, D., Wang, C., Dai, P., Du, J., & Wang, J. (2020). Evaluation of esterified pyrolysis bio-oil as a diesel alternative. *Journal of the Energy Institute*, 93(4), 1382–1389. https://doi.org/10.1016/j.joei.2019.12.008

Morris, J. D., Daood, S. S., Chilton, S., & Nimmo, W. (2018). Mechanisms and mitigation of agglomeration during fluidized bed combustion of biomass: A review. *Fuel*, 230, 452–473. https://doi.org/10.1016/j.fuel.2018.04.098

Moura, P., Henriques, J., Alexandre, J., Oliveira, A. C., Abreu, M., Gírio, F., & Catarino, J. (2022). Sustainable value methodology to compare the performance of conversion technologies for the production of electricity and heat, energy vectors and biofuels from waste biomass. *Cleaner Waste Systems*, 3, 100029. https://doi.org/10.1016/j.clwas.2022.100029

Naeem, M., Imran, M., Latif, S., Ashraf, A., Hussain, N., Boczkaj, G., Smułek, W., Jesionowski, T., & Bilal, M. (2023). Multifunctional catalyst-assisted sustainable reformation of lignocellulosic biomass into environmentally friendly biofuel and value-added chemicals. *Chemosphere*, 330, 138633. https://doi.org/10.1016/j.chemosphere.2023.138633

Nnabuife, S. G., Ugbeh-Johnson, J., Okeke, N. E., & Ogbonnaya, C. (2022). Present and projected developments in hydrogen production: A technological review. *Carbon Capture Science and Technology*, 3, 100042. https://doi.org/10.1016/j.ccst.2022.100042

Norinaga, K., Kudo, S., & Hayashi, J.-I. (2012). Applications of catalysis in the selective conversion of lignocellulosic biomass by pyrolysis. *Kyushu University Global COE Program Journal of Novel Carbon Resource Sciences*, 6, 1–8. https://www.researchgate.net/publication/260384794

Norouzi, O., Taghavi, S., Arku, P., Jafarian, S., Signoretto, M., & Dutta, A. (2021). What is the best catalyst for biomass pyrolysis? *Journal of Analytical and Applied Pyrolysis*, 158, 105280. https://doi.org/10.1016/j.jaap.2021.105280

Nussbaumer, T. (2003). Combustion and co-combustion of biomass: fundamentals, technologies, and primary measures for emission reduction. *Energy and Fuels*, 17(6), 1510–1521. https://doi.org/10.1021/ef030031q

Ojha, D. K., & Vinu, R. (2018). Copyrolysis of lignocellulosic biomass with waste plastics for resource recovery. *Waste Biorefinery: Potential and Perspectives*, 17, 349–391. https://doi.org/10.1016/B978-0-444-63992-9.00012-4

Okolie, J. A., Epelle, E. I., Tabat, M. E., Orivri, U., Amenaghawon, A. N., Okoye, P. U., & Gunes, B. (2022). Waste biomass valorization for the production of biofuels and value-added products: A comprehensive review of thermochemical, biological and integrated processes. *Process Safety and Environmental Protection*, 159, 323–344. https://doi.org/10.1016/j.psep.2021.12.049

Onarheim, K., Hannula, I., & Solantausta, Y. (2020). Hydrogen enhanced biofuels for transport via fast pyrolysis of biomass: A conceptual assessment. *Energy*, 199, 117337. https://doi.org/10.1016/j.energy.2020.117337

Qureshi, T., Farooq, M., Munir, M. A., Imran, S., Javed, M. A., Sohoo, I., Sultan, M., Rehman, A. U., Farhan, M., Asim, M., & Andresen, J. M. (2024). Structural and thermal investigation of lignocellulosic biomass conversion for enhancing sustainable imperative in progressive organic refinery paradigm for waste-to-energy applications. *Environmental Research*, 246, 118129. https://doi.org/10.1016/j.envres.2024.118129

Rahman, M. M., Liu, R., & Cai, J. (2018). Catalytic fast pyrolysis of biomass over zeolites for high quality bio-oil - A review. *Fuel Processing Technology*, 180, 32–46. https://doi.org/10.1016/j.fuproc.2018.08.002

Roy, P., & Dias, G. (2017). Prospects for pyrolysis technologies in the bioenergy sector: A review. *Renewable and Sustainable Energy Reviews*, 77, 59–69. https://doi.org/10.1016/j.rser.2017.03.136

Saidi, M., & Faraji, M. (2024). Thermochemical conversion of neem seed biomass to sustainable hydrogen and biofuels: Experimental and theoretical evaluation. *Renewable Energy*, 221, 119694. https://doi.org/10.1016/j.renene.2023.119694

Salvilla, J. N. V., Ofrasio, B. I. G., Rollon, A. P., Manegdeg, F. G., Abarca, R. R. M., & de Luna, M. D. G. (2020). Synergistic co-pyrolysıs of polyolefin plastics with wood and agricultural wastes for biofuel production. *Applied Energy*, 279, 115668. https://doi.org/10.1016/j.apenergy.2020.115668

Saravana Sathiya Prabhahar, R., Nagaraj, P., & Jeyasubramanian, K. (2020). Promotion of bio oil, H2 gas from the pyrolysis of rice husk assisted with nano silver catalyst and utilization of bio oil blend in CI engine. *International Journal of Hydrogen Energy*, 45(33), 16355–16371. https://doi.org/10.1016/j.ijhydene.2020.04.123

Schmieder, H., Abeln, J., Boukis, N., Dinjus, E., Kruse, A., Kluth, M., Petrich, G., Sadri, E., & Schacht, M. (2000). Hydrothermal gasification of biomass and organic wastes. *Journal of Supercritical Fluids*, 17, 145–153. www.elsevier.com/locate/supflu

Singh, D., Sharma, D., Soni, S. L., Inda, C. S., Sharma, S., Sharma, P. K., & Jhalani, A. (2021). A comprehensive review of biodiesel production from waste cooking oil and its use as fuel in compression ignition engines: 3rd generation cleaner feedstock. *Journal of Cleaner Production*, 307, 127299. https://doi.org/10.1016/j.jclepro.2021.127299

Tawalbeh, M., Al-Othman, A., Salamah, T., Alkasrawi, M., Martis, R., & El-Rub, Z. A. (2021). A critical review on metal-based catalysts used in the pyrolysis of lignocellulosic biomass materials. *Journal of Environmental Management*, 299, 113597. https://doi.org/10.1016/j.jenvman.2021.113597

Tripathi, M., Sahu, J. N., & Ganesan, P. (2016). Effect of process parameters on production of biochar from biomass waste through pyrolysis: A review. *Renewable and Sustainable Energy Reviews*, 55, 467–481. https://doi.org/10.1016/j.rser.2015.10.122

Vamvuka, D. (2011). Bio-oil, solid and gaseous biofuels from biomass pyrolysis processes-An overview. *International Journal of Energy Research*, 35(10), 835–862. https://doi.org/10.1002/er.1804

Verma, M., Godbout, S., Brar, S. K., Solomatnikova, O., Lemay, S. P., & Larouche, J. P. (2012). Biofuels production from biomass by thermochemical conversion technologies. *International Journal of Chemical Engineering*, 2012, 542426. https://doi.org/10.1155/2012/542426

Wang, L., Weller, C. L., Jones, D. D., & Hanna, M. A. (2008). Contemporary issues in thermal gasification of biomass and its application to electricity and fuel production. *Biomass and Bioenergy*, 32(7), 573–581. https://doi.org/10.1016/j.biombioe.2007.12.007

Wang, Z., Burra, K. G., Lei, T., & Gupta, A. K. (2021). Co-pyrolysis of waste plastic and solid biomass for synergistic production of biofuels and chemicals-A review. *Progress in Energy and Combustion Science*, 84, 100899. https://doi.org/10.1016/j.pecs.2020.100899

Yakaboylu, O., Harinck, J., Smit, K. G., & de Jong, W. (2015). Supercritical water gasification of biomass: A literature and technology overview. *Energies*, 8(2), 859–894). https://doi.org/10.3390/en8020859

Yang, J., (Sophia) He, Q., & Yang, L. (2019). A review on hydrothermal co-liquefaction of biomass. *Applied Energy*, 250, 926–945. https://doi.org/10.1016/j.apenergy.2019.05.033

Yogalakshmi, K. N., Poornima Devi, T., Sivashanmugam, P., Kavitha, S., Yukesh Kannah, R., Sunita, V., AdishKumar, S., Gopalakrishnan, K., & Rajesh Banu, J. (2022). Lignocellulosic biomass-based pyrolysis: A comprehensive review. *Chemosphere*, 286, 131824. https://doi.org/10.1016/j.chemosphere.2021.131824

Yu, I. K. M., Chen, H., Abeln, F., Auta, H., Fan, J., Budarin, V. L., Clark, J. H., Parsons, S., Chuck, C. J., Zhang, S., Luo, G., & Tsang, D. C. W. (2021). Chemicals from lignocellulosic biomass: A critical comparison between biochemical, microwave and thermochemical conversion methods. *Critical Reviews in Environmental Science and Technology*, 51(14), 1479–1532. https://doi.org/10.1080/10643389.2020.1753632

Zhang, J., & Zhang, X. (2019). The thermochemical conversion of biomass into biofuels. *Biomass, Biopolymer-Based Materials, and Bioenergy: Construction, Biomedical, and other Industrial Applications*, 2019, 327–368. https://doi.org/10.1016/B978-0-08-102426-3.00015-1

Zhang, X. S., Yang, G. X., Jiang, H., Liu, W. J., & Ding, H. S. (2013). Mass production of chemicals from biomass-derived oil by directly atmospheric distillation coupled with co-pyrolysis. *Scientific Reports*, 3, 1120. https://doi.org/10.1038/srep01120

Zhou, X., Moghaddam, T. B., Chen, M., Wu, S., Zhang, Y., Zhang, X., Adhikari, S., & Zhang, X. (2021). Effects of pyrolysis parameters on physicochemical properties of biochar and bio-oil and application in asphalt. *Science of the Total Environment*, 780, 146448. https://doi.org/10.1016/j.scitotenv.2021.146448

Zhou, Y., Remón, J., Pang, X., Jiang, Z., Liu, H., & Ding, W. (2023). Hydrothermal conversion of biomass to fuels, chemicals and materials: A review holistically connecting product properties and marketable applications. *Science of the Total Environment*, 886, 163920. https://doi.org/10.1016/j.scitotenv.2023.163920

18 Biochar Revisited

A Systematic and Critical Review on Coupling of Biochar with Anaerobic Digestion and Its Role in Bioenergy Production

Sanjana Pal, Koustav Saha, and Ritesh Pattnaik

GRAPHICAL ABSTRACT

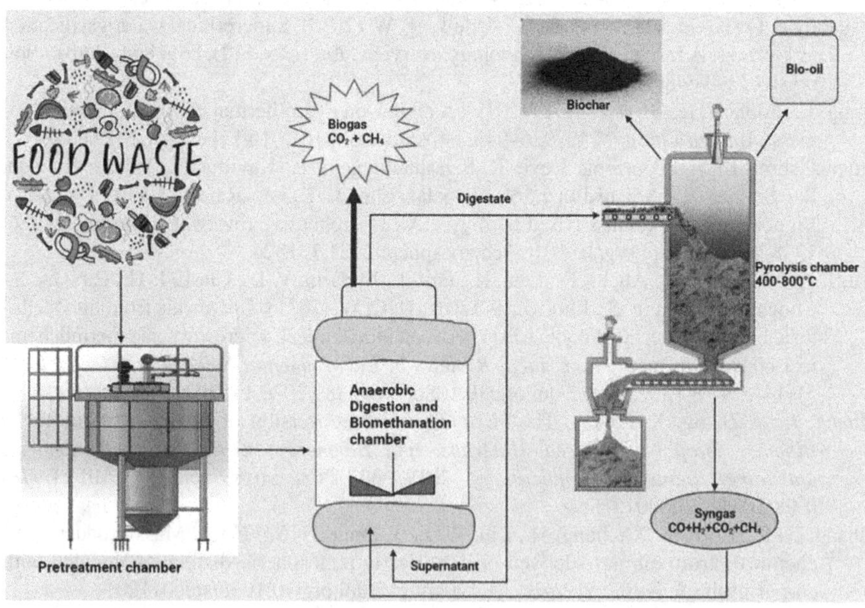

18.1 INTRODUCTION

Waste generation is mainly associated with the transformation of raw materials into useful products. The accumulation of wastes in the natural environments has increased significantly owing to fast-growing economy and increase in population (Khalid et al., 2011;

 DOI: 10.1201/9781003407713-18

Elsayed et al., 2020; Fodah et al., 2021). Waste has been produced by four main industries, including: (i) municipal solid wastes produced by homes, offices, schools, shops and institutions; (ii) industrial solid wastes, which primarily include paper, packaging materials, waste from food processing, oils, solvents, resins, paints and sludges; (iii) agricultural wastes, which include animal waste, paddy husk, straw and coir fibre; and (iv) hazardous wastes, which primarily include by-products of broad spectrum of industrial, agricultural and manufacturing processes, nuclear establishments, hospitals and healthcare facilities (Maina et al., 2017). According to the Food And Agriculture Organization of the United Nations, almost 33% of the food (1.3 billion tonnes per year) is wasted all over the world (Chang et al., 2006; Kumar et al., 2020). Global energy requirements due to increase in population have paved the way for the production of new eco-friendly nature, cost-effective, easy to prepare carbon-rich product derived from organic biomass called biochar, produced mainly through pyrolysis, a thermochemical process (Moreno-Riascos & Ghneim-Herrera, 2020).

Earlier, the indigenous people of the Amazon lived on tropical infertile soil called oxisol, but then they added biochar with oxisol to form terra preta or black soil which made the soil fertile, and hence, it illustrates the origin of biochar (Shareef & Zhao, 2016; de Oliveira et al., 2020; Moreno-Riascos & Ghneim-Herrera, 2020). But then it was discovered that biochar can be utilized all over the world. The concept and motivation for studying biochar came from its potential to address many of the problems exhibited in today's world, including waste management, soil deterioration, renewable energy and climatic changes (Gu et al., 2020).

Since its inception, biochar has been used as a soil amendment to improve soil quality by increasing carbon sequestration, aggregate stability and cation exchange capacity (CEC), as well as reducing soil toxicity by modifying the pH (Fodah et al., 2021). Biochar considering its eco-friendly features has shown promising applications in other environmental aspects such as improving the composting process and quality of compost, increasing carbon content, reducing toxicity level of the compost and serving as a platform to produce carbon-based materials that were vastly used for energy storage and conversion (Qian et al., 2015).

Several studies suggest that biochar could be potentially used as an additive for enhancing the performance of anaerobic digestion (AD) which is by far the most successful technologies that can convert organic materials into biogas and digested slurry (Huiru et al., 2019; Lansing et al., 2019). The AD process can be mainly divided into four phases namely hydrolysis, acidogenesis, acetogenesis and methanogenesis (Ghodrat et al., 2018) having impact on biochar showcasing different mechanisms for each phase: the process is mainly dependent on the interactions between different microorganisms that are able to carry out the four above-mentioned stages (Wang et al., 2021). With eco-compatible biochar serving the purpose of an additive in AD process, the lag phase of methanogenesis shortened, the methane production rate increased and the chemical oxygen demand removal capacity also increased (Wang, Liu et al., 2018; Wang, Li et al., 2018; Wang, Fang et al., 2018).

This review mainly demonstrates thorough information about the numerous technologies, process conditions and feedstock materials utilized for biochar production, as well as their influence on the physicochemical properties and practical applications of biochar. Furthermore, this paper gives an outline of the impacts of biochar on different phases of AD and its uses on AD process (Tables 18.1 and 18.2).

TABLE 18.1
Impact of Biochar on Different Phases of AD

Phases of Anaerobic Digestion	Impacts of Biochar	Mechanism	References
Hydrolysis	Positive impact on the degraded organic matters. Improving hydrolysis efficiency of protein, polysaccharide and lipid Activation and enrichment of hydrolytic bacteria and enzymes Enhance process efficiency	Activation of hydrolases such as proteases, α-glucosidase, dextranase, cellulose and amylase Efficient disruption of the cell walls of insoluble matter	Duan et al. (2019), Lou et al. (2019)
Acidogenesis and acetogenesis	Accumulation of VFAs in excess concentration leading to decrease in pH and disturbed microbial activity Acting as an acidogenesis–acetogenesis regulator for proper organic acid generation Promoting the production of easily utilized acetate	Anaerobic microorganisms for acetic production	Wang et al. (2013)
Methanogenesis	Enhancing methane content in biogas plants and methane production rate Reduction in lag phase and greater methane production Stable methane production at high loading rate Enhance enzymatic activity and VFA yield	Supporting for microbial colonization Colonization of methanogens	Li et al. (2019)

18.2 RECYCLING OF ORGANIC WASTE

Organic wastes mostly comprise municipal, food and cattle wastes. They have high moisture content, which increases the volume of the waste and eventually becomes a problem with the increase in population. The urge for recycling the organic wastes was indisputable. The main goal was to extract the biodegradable part and convert it into manure or fertilizer. There are several ways where organic wastes can be recycled – the food wastes can be reused as animal feed, composting to decompose the organic waste and recycling of macronutrients into the soil humus; AD also helps convert organic biomass into biogas and nutrient-rich residue (Adekunle et al., 2011; Xu et al., 2018).

The organic wastes are collected and dumped on the dumpsite followed by decontamination where the non-biodegradable wastes are separated to avoid its harmful effects. They are then shredded or stacked and stabilized prior to the recycling following which they are screened and graded for their different applications

TABLE 18.2
Application of Biochar in AD Process

Food Waste Types/AD Substrate	Biochar Feedstock	Biochar Method	Biochar Synthesis Conditions	AD Process Parameters	Conclusion/Impacts on Addition of Biochar	References
Fish processing wastes	Bamboo hydrochar	Hydrothermal carbonization	Hydrothermal carbonization at 220°C, 2.2MPa	Batch AD at 37±1°C for 36 days	Biomethanation performance and yield depends upon the content of HTC. Reduced risk of spreading of antibiotic resistance genes	Choe et al. (2019)
Chicken manure (CM) and kitchen waste (KW)	Vermicompost	Pyrolysis	Temperature at 500°C	Batch AD at 35±1°C	Reduction in acidification during KW digestion Improvement in process stability during CM Improved buffering capacity	Wang et al. (2017), Pan et al. (2019)
Lipid-rich, protein-rich and carbohydrate- rich substrates	Sewage sludge-derived biochar	Pyrolysis	Temperature > 600°C	Batch AD incubated at 35°C for 45 days	Shortening lag phase Improving methane production rate	Xu et al. (2020)
Brewer's spent grain (BSG)	Sawdust, dried BSG	Pyrolysis	Temperature at 300°C	Batch AD at 37°C	Improved biogas production rate Increase methane production by the addition of biochar VFAs degradation	Wang et al. (2018), Dudek et al. (2019)

(Continued)

TABLE 18.2 (*Continued*)
Application of Biochar in AD Process

Food Waste Types/AD Substrate	Biochar Feedstock	Biochar Method	Biochar Synthesis Conditions	AD Process Parameters	Conclusion/Impacts on Addition of Biochar	References
Food waste comprising mango, tomato, papaya and banana	Sewage sludge digestate	Pyrolysis	Temperature at 350°C or 550°C	Batch AD at 37°C, for 50 hours at 100 rpm	Inoculum–substrate ratio and pyrolysis temperature mainly influence the efficiency of biochar	Ambaye et al. (2020)
Food Waste	Fruitwoods	Pyrolysis	Temperature at 800°C–900°C	Batch AD at 35°C	Shorten the lag phase. Improve the methane production rate Enhancing the degradation rate of dissolved organics and VFAs	Cai et al. (2016)
Liquid Food waste	Rice husk biochar (RHB) and palm tree-based biochar (PTB)	Pyrolysis	Temperatures at 450°C and 550°C	Batch AD	Improve biochar stability, favourable for syntrophic volatile fatty acid (VFA) oxidation and thereby increase the cumulative methane yield	Ovi et al. (2022)
Wastewater treatment plant sludge	Corn stover	Gasification	Temperature at 850°C	Batch AD	Increase in methane production	Shen et al. (2015)

(Sharma et al., 2019). This recycling helps to produce biogas; composts can be used as biofertilizers, increasing soil fertility as it provides various nutrients to the soil, which in turn enhances plant growth. Thus, less garbage on the dumpyard decreases harmful release of greenhouse gases.

Too much composting produces odours, which cause air pollution, whereas an excess of compost on soil can lead to the accumulation of metal elements like lead and cadmium, which can affect plant as well as human health and growth. Bioaerosols can affect the plant workers and the residents of the area.

So, we aim to modify this recycling process through the production of biochar, keeping the disadvantages in mind and approaching for a betterment of the environment as well as the society.

18.3.1 Food Wastes and Its Correlation with Biochar

There has been a huge generation of municipal solid wastes (MSW) due to rapid growth in population, urbanization and modern living standards. World Hunger is on the rise; nevertheless, an estimated 1/3 of all food produced globally is lost or goes to waste. One- third of food produced for human consumption is lost or wasted globally, which amounts to around 1.3 billion tonnes per year according to Food and Agriculture Organization of the United Nations (Loss & Waste, 2019). Fruit and Vegetable losses have been projected to be approximately 12 and 21 million tonnes, respectively, by Ministry of Food Processing Industries (MFPI) costing to be over 4.4 billion USD, with a total food value loss and waste produce of nearly 11 billion USD (Kumar et al., 2020).

Food wastes have great potential to be utilized as raw materials for the production of various value-added products like biochar, biofuels and chemicals. The pyrolysis of food wastes with the aim of producing biochar has been the key point of several studies. This review focuses on the production of value-added biochar products with the help of pyrolysis. Along with it, the study also explores different available pyrolysis processes for the production of biochar from food waste through effective pyrolysis and meeting the specifications required.

Biochar is a carbon-rich stable solid which is created through pyrolysis as a result of decomposition of organic feedstock materials or raw materials like forest, animal compost, food and vegetable wastes, and plant leftovers at high temperatures under an oxygen-free conditions (Lehmann & Joseph, 2015). Different studies have shown biochar production from different types of food and animal wastes.

Pyrolysis of banana and orange peels was accomplished within temperature range of 400°C–500°C that was then investigated as adsorbent in Palm Oil Mill Effluent treatment (Zhou et al., 2017). The POME treatment exhibited considerable decreased concentration of BOD, chemical oxygen demand (COD), TSS and oil and grease compared to the fresh POME (Panwar et al., 2012; Mamimin et al., 2015; Lam et al., 2018).

Fresh and dehydrated banana peels were used as biomass feedstock materials to manufacture highly effective sorbent biochar by a one-step hydrothermal carbonization technique at 230°C for 2 hours with easy process, mild environment and saving cost. The dried banana peel biochar and fresh banana peel biochar both showed high lead adsorption capacities of 359 mg/g and 193 mg/g, respectively (Zhou et al., 2017).

Pineapple, pomelo and sweet-lime derived biochar were also developed for adsorption of hexavalent chromium (Cr(VI)) (Wang et al., 2016; Wu et al., 2017).

The activated biochar produced from pomelo peel waste (PPAB) and litchi peel waste was prepared utilizing pomelo peel (PP) wastes and litchi peel waste as the raw materials for the removal of methyl orange colouring and malachite green in wastewater (Zhang et al., 2019; Wu et al., 2020).

18.3.2 Agricultural Influences of Biochar

Biochar has gained considerable interest due to its many benefits and potential uses in soils. Its particles resemble those found in "Terra Preta de Indio" (Shareef & Zhao, 2016), a particularly fertile anthropogenic soil discovered near the ruins of a pre-Columbian civilization located near the Amazon basin (Glaser, 2007). Given its high adsorption capacity and promise to reduce greenhouse gas emissions, biochar has also demonstrated the ability to improve the soil's ability to store water and nutrients when used in agriculture. Researchers nowadays are attempting to mimic "Terra Preta" by applying biochar to agricultural soils. The addition of biochar to soil not only paved the way for constructing the soil's organic fraction but also used as means for carbon sequestration. Numerous studies have demonstrated that biochar's high surface area and other physical and chemical characteristics have resulted in strong adsorption power, which has improved the availability of nutrients in soil.

pH is considered to be one of the most important soil parameters because of its influence on physical, chemical, biological and geological processes, as well as its relation with soil fertility (Yin et al., 2019). pH fluctuations induce a number of changes in the soil environment, altering the availability of nutrients for plant growth and antimicrobial activity, which is why biochar has come to the attention of many researchers since it improves soil quality by reducing soil acidity. Additionally, biochar can enhance soil structure by creating micro-aggregates, which increases water retention, enhances solubilization, improves nutrient retention and transport, and decreases nutrient leaching, all of which have an impact on crop performance. Biochar increases the structure, porosity (Du et al., 2018), CEC (Tan et al., 2017) and aggregation of the soil facilitating tillage. In addition to the aforementioned parameters, numerous other factors such as feedstock type, pyrolysis temperature and addition rate influence the nutritional status in biochar additions. Biochar can also directly provide higher nutrient content such as phosphorus, calcium, potassium and magnesium, as well as affect soil microbial activity due to changes in soil properties. The physicochemical, biological and geological aspects of soil are all interrelated and are affected by biochar.

18.3.3 Environmental Benefits and Impacts

Biochar had been long recognized as an efficient tool to mitigate climatic changes (Ippolito et al., 2012). Application of biochar to soils not only benefits as an organic fertilizer but also used in climatic change. Organic wastes in most of the areas are left to decompose naturally, which leads to carbon release in the atmosphere in the form of CO_2 hence causing greenhouse effect. But owing to lack of oxygen in the pyrolysis process which is required for biochar formation, the carbon from the organic wastes is used up in the biochar, and hence, it decreases the greenhouse gas emission into the atmosphere. Moreover, it also reduces the emission of nitrous oxide and carbon dioxide from the soil thereby reducing the overall GHG in the atmosphere. The overall/net GHG impact on biochar is influenced by the variations in primary crop productivity, increase in humification, and methane and nitrous oxide emission. Furthermore, the GHG emissions arising from the production of biochar, transport and application of soils also have an impact on biochar.

18.4 METHODS OF BIOCHAR PRODUCTION

The need of biochar in various applications nowadays has led to the conversion of more biomass into biochar. The preparation and production of biochar can be done by various types of processes (as shown in Figure 18.1) which are described below:

18.4.1 Thermochemical Conversion Method

This is one of the most common techniques for production of biochar. It is a very efficient method to convert biomass into biofuels. Thermochemical conversion of biomass can be performed via three primary pathways namely pyrolysis, gasification

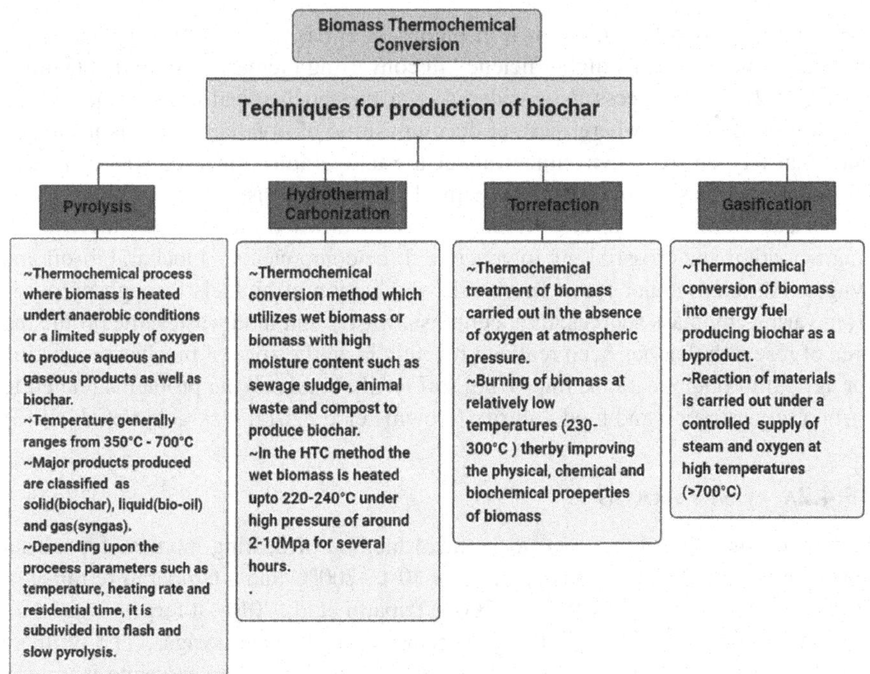

FIGURE 18.1 Process development of biochar production.

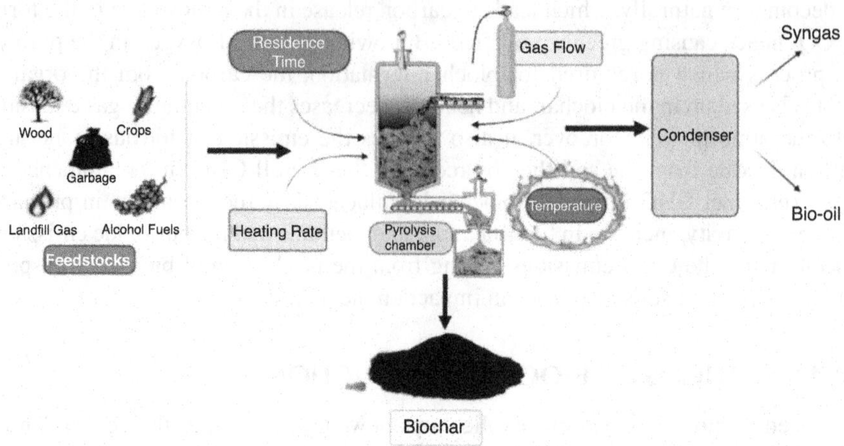

FIGURE 18.2 Schematic of a typical pyrolysis process.

and direct combustion based on their temperature, pressure, duration of their heating and vapour residence time. This method mainly involves structure degradation of biomass with oxygenic or anoxygenic atmosphere at high temperature.

18.4.2 Pyrolysis

The process of pyrolysis (as shown in Figure 18.2) has received immense attention in recent times due to its high efficiency in converting biomass into biofuel (Jahirul et al., 2012). This process is considered as a thermochemical conversion method (Panwar et al., 2012), where thermal decomposition of organic materials is carried out in an oxygen-free environment under the temperature range of 300°C–700°C. During this process of pyrolysis, the natural polymeric constituents, i.e., lignin, cellulose and hemicellulose, are thermally broken down through depolymerization, fragmentation and crosslinking to generate three components – biochar, bio-oil and syngas. Although major focus was on the production of biofuels through pyrolysis from various biomass sources, lignocellulosic wastes and food wastes, the promising area of research that has been realized recently is on the zone of biochar production for agricultural applications, improving soil fertility and solving problems related to carbon sequestration and food security (Downie et al., 2012; Azargohar et al., 2013)

18.4.2a Fast Pyrolysis

Fast pyrolysis is the direct thermochemical method of heating biomass for a brief length of time of 1–10 seconds at a rate of 10°C–200°C and a temperature range of 800°C–1,300°C (Demirbas & Arin, 2002; Tripathi et al., 2016). It targets at producing maximum yield of bio-oil in comparison to biochar and syngas. The ultimate objective of this technique is to rapidly decompose biomass by exposing it to high temperatures, which shortens the residence time and favours the production of biochar (Mohan et al., 2006).

18.4.2B Slow Pyrolysis

Slow pyrolysis has been considered to be one of the best and feasible methods for the production of high-quality biochar deliberated for agricultural use with authentic and congruous product quality. This process of slow pyrolysis is carried out at low heating rate of 0.1–1°C/s with pyrolysis temperature of around 400°C–500°C for an approximate time span of 5–30 minutes. The longer residence time in this process makes it a perfect candidate for greater and better yield of biochar production in comparison to other pyrolysis and carbonization processes (Demirbas & Arin, 2002; Tripathi et al., 2016).

18.4.3 Hydrothermal Carbonization

Hydrothermal carbonization (as shown in Figure 18.3), a thermochemical conversion process carried out at a lower temperature in the range of 180°C–350°C, is viewed as a cost-effective process for the wastes, forest and agricultural residues, algal biomass, animal and human wastes, etc. The product formed at the end of this process is called hydrochar which is a slurry comprised of two-phase mixture (solid and liquid) that creates a demarcation between the product formed from dry processes like pyrolysis and gasification. During hydrothermal carbonization, biomass decomposition is characterized by some reactions, including hydrolysis and dehydration.

18.4.4 Gasification

Gasification is a thermochemical method (as shown in Figure 18.4) of decomposition which converts the carbonaceous materials into syngas, i.e., mixture of CO, H_2, CO_2, CH_4 and traces of hydrocarbons in the presence of oxygen-deficient conditions at a high temperature of more than 700°C (You et al., 2017).

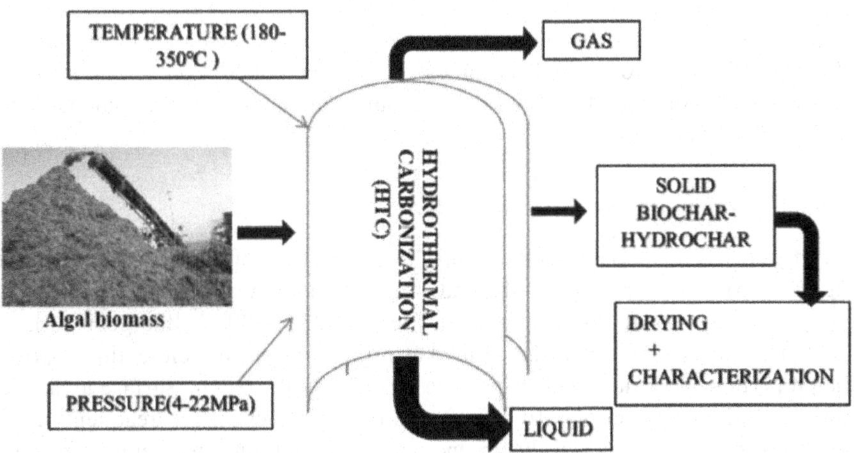

FIGURE 18.3 Process of hydrothermal carbonization (HTC).

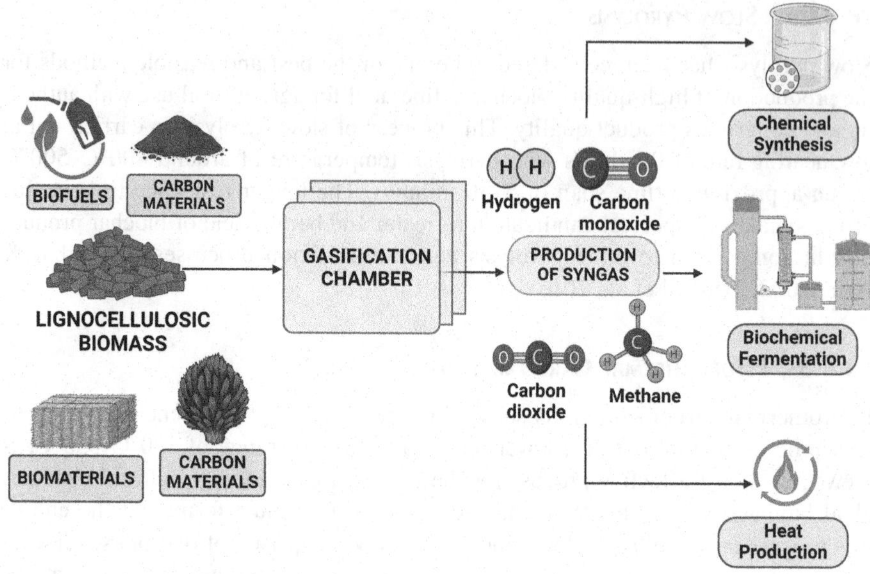

FIGURE 18.4 Gasification process in biochar production.

$$\text{GASIFICATION} = [\text{DRYING} + \text{PYROLYSIS} + \text{PARTIAL OXIDATION} + \text{REDUCTION}]$$

The production of biochar from the gasification process is quite less in comparison to the other thermochemical processes mentioned above, which can be understood by the conversion of carbon into carbon monoxide (CO) because of its partial oxidation conditions. The formation of biochar as a by-product of the gasification process is limited to maximize the energy recovery. Gasification produces much more amount of syngas and lower level of emissions than any other conventional methods, i.e., pyrolysis, hydrothermal carbonization and combustion. Although varied temperatures are employed in this biomass decomposition process, the total temperature is in the range of 500°C–700°C and is characterized by a number of processes, including hydrolysis, dehydration, decarboxylation, aromatization and recondensation (Turner et al., 2008; Guan et al., 2016).

18.4.5 Torrefaction

Torrefaction is a modern method for the production of biochar (as shown in Figure 18.5) which surfaced at the end of the 20th century. The process employs burning of biomass at low temperature of around 230 –300°C (Bourgeois & Doat, 1984; Yu et al., 2017), hence also termed as mild pyrolysis; therefore, this improves the properties of biomass such as particle size, moisture content, surface area, heating rate and energy density. Torrefaction is the thermochemical treatment of biomass which is carried out in the absence of oxygen within a temperature range of 230°C–300°C, residence time of less than 30 minutes and heating rate of less than 50°C/min (Pentananunt et al., 1990).

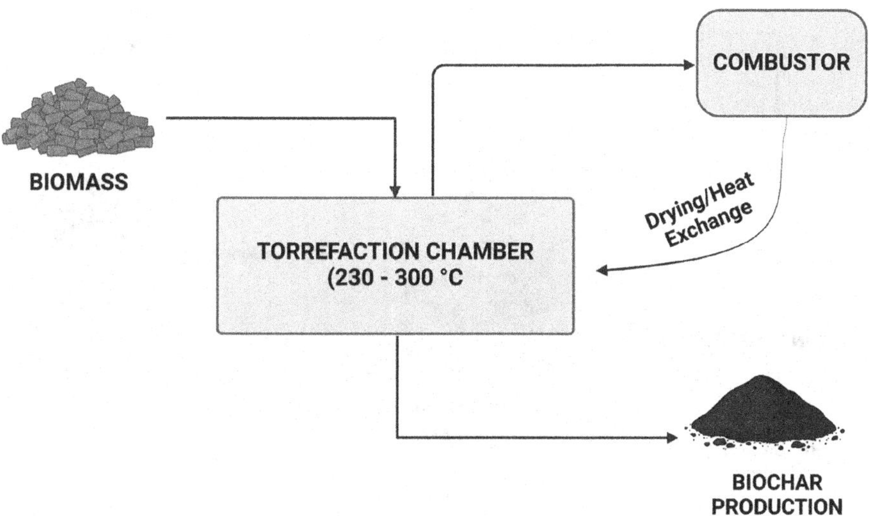

FIGURE 18.5 Torrefaction process for the production of biochar.

18.4.6 Flash Pyrolysis

Flash pyrolysis is an altered and enriched version of quick pyrolysis. This process is employed with biomass decomposition at temperatures ranging from 900°C to 1,200°C within a very short time period mostly less than a minute at a considerable heating rate of more than 1,000°C/sec difference; the overall temperature lies in the range of 500°C–700°C characterized by some reactions which include hydrolysis, dehydration, decarboxylation, aromatization and recondensation (Demirbas & Arin, 2002; Li et al., 2013). Though all these conditions lead to a high bio-oil yield, it reduces considerable amount of yield in the production of biochar.

18.4.7 Microwave Pyrolysis

Production of biochar though microwave pyrolysis (as shown in Figure 18.6) has come up as an advanced and more promising technique in recent times. It is more rapid and has quite a lot of advantages over conventional technique (Gabhane et al., 2020) of pyrolysis considering the fact that microwave technology has provided some great thermal characteristics due to its rapid, selective and uniform heat generation, thus decreasing the temperature which therefore improved the steam gasification. Temperature requirement in the case of biochar production through microwave pyrolysis has been found out to be 200°C less than conventional pyrolysis, and hence, it is better and more advantageous method than conventional pyrolysis producing similar results. Currently researchers are focusing more on this method for biochar production. A recent experiment found by a research team (Hossain et al., 2017) showed that 450°C temperature for maximum biochar production and 700°C for maximum hydrogen production, 400W microwave power and residence time of 4–6 minutes are the best and suitable conditions.

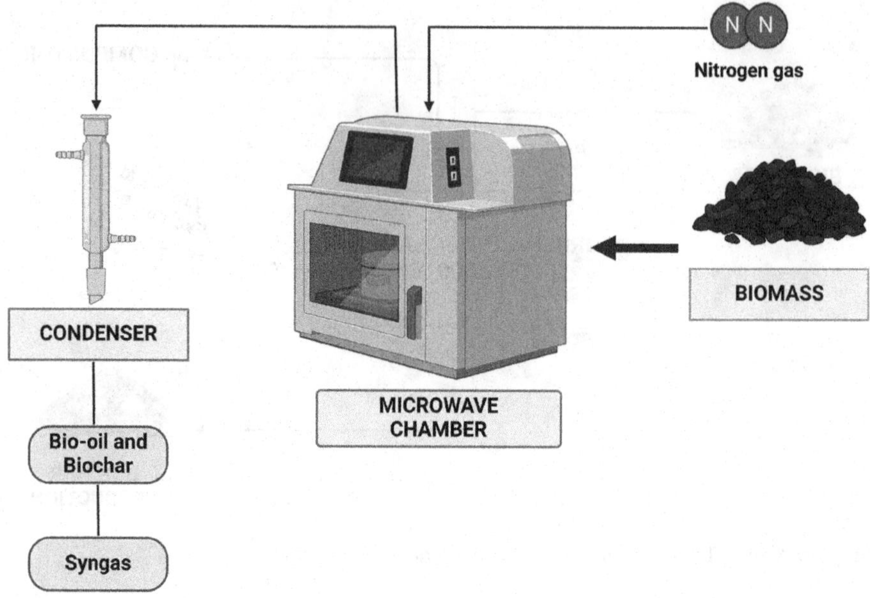

FIGURE 18.6 Microwave-assisted pyrolysis of biochar.

18.5 CUSTOMIZING BIOCHAR FOR AD

The recent emerging application of biochar as an additive in AD of agricultural waste and sludge has been investigated by many researchers in the last few years. The addition of biochar to AD processes not only escalates the removal of COD but also diminishes the latency phase of methanogenesis, thereby leading to the higher production of methane. Apart from this, increasing the elemental composition of solid residue of AD has been another application of biochar in AD process which can then be further processed and used as fertilizer in agricultural field. In general, AD is a very complex process to accomplish that is reliant on many factors including feedstock properties, temperature, heating rate and many other process parameters. Designing the physicochemical properties of biochar requires a full understanding of the physiochemical characteristics of biochar.

18.5.1 Feedstock Characteristics

A suitable balanced composition of micro- and macronutrients is required for the growth and survival of microorganisms in the AD process. The C/N ratio in the range of 16–25/1 is recommended for AD process although reports imply that out-of-the-range values of C/N/P/S ratio of 1,000/20/5/3 would be adequate for AD to proceed. Considering the methanogenic route, different types of metalloenzymes are engaged in the AD process, and consequently, different micronutrients are necessary. The addition of micronutrients at an appropriate combination is more suitable for an AD process because a high concentration of micronutrients can inhibit the

AD process. Pan et al. (2019) observed that higher lignin and fixed C content in feedstock have led to the synthesis of biochar having better solid surface area and higher surface functional groups. Liu et al. (2019) and Wang et al. (2020) reported that hydrochar synthesized from cow manure feedstock exhibited lower C content and higher N content than corn stalk.

18.5.2 Temperature

The AD process is generally performed at thermophilic (55°C–70°C) or mesophilic (32°C–45°C) temperature configurations. A higher oxygen content is exhibited for biochar synthesized at lower temperature of around 300°C–500°C, and elevated temperature of around 500°C–1,000°C reduces the contents of H and N (Igalavithana et al., 2017). Temperature plays an essential impact in determining the biochar quality. Temperature of around 450°C–600°C is suitable for the production of biochar considering the nature and type of feedstocks, and it is also quite practicable. Cantrell et al. (2012) reported that the yield of biochar would be uniform if the temperature > 400°C, but there is a decline in yield of biochar if temperature is < 400°C which is due to the loss of volatile matter and non-condensable gases like methane and carbon dioxide.

18.5.3 Heating Rate, Resident Time and Carrier Gas

Wang et al. (2018) in his research had illustrated the fact that a marginal rise in temperature from 2°C to 20°C can elevate the H:C and O:C ratio in biochar at 250°C and 350°C, which further creates an alteration in the functional groups. The importance of carrier gas in the pyrolysis process of production of biochar along with its physicochemical properties was demonstrated (Lee et al., 2017). The synthesis of biochar under the influence of CO_2 showed higher solid surface area compared to that of N_2.

18.6 CONTROLS ON BIOCHAR FOR AD PROCESSES

The parameters that influence biochar on AD process have been comprehensively recorded in this literature review. These factors include porosity, specific surface area (SSA), CEC, pH, redox properties and surface functional groups which play a crucial role during the AD process and are discussed below (Rafiq et al., 2016; Qiu et al., 2019; Chiappero et al., 2020; Frias Flores, 2020; Tomczyk et al., 2020).

18.6.1 Porosity or Pore Size

Biochar porosity or pore size is recognized as one of the essential parameters in relation with bacteria participating in the AD process. The physicochemical properties of biochar have been impacted by the pyrolysis temperature, which further offers a direct impact on the porosity of the biochar. Proliferation of aerobic and anaerobic microbes of sizes 0.30–13 μm for bacteria, 2–80 μm for fungus and 7–30 μm for protozoa is also attributable to the pore size of biochar, which helps the microhabitats to do so (Faruqui et al., 2010; Luz, Cordiner et al., 2018). Porous biochar facilitates biofilm formation thus acting as a shield for the selective enrichment of efficient microorganisms involved in AD process under acid-stress conditions.

18.6.2 Specific Surface Area

SSA of biochar is also considered as one of the most important aspects which influence biochar for the AD process (Ding et al., 2014; Tomczyk et al., 2020). A higher surface area and porosity of biochar promotes microbial attachment, and therefore, it has been reported to have a potential impact on the direct interspecies electron transfer (DIET). For the contact of biochar with microorganisms and AD substrate (Qin et al., 2020), a higher SSA is beneficial and SSA of biochar is hence a very important determinant in the extent of soil microbial attachment.

18.6.3 Cation Exchange Capacity

Biochar has a strong CEC, and reports have suggested that due to this property biochar can efficiently diminish the inhibition effect of ammonia and thus accelerate the biomethanization performance under ammonium and acid-stress conditions. Nevertheless, ammonia inhibition was restricted during the AD of agricultural digestates on addition of biochar (Mumme et al., 2014).

18.6.4 pH

pH has a major role to play in biochar conductivity and microbial association in the AD process (Kumar et al., 2021). With increase in pyrolysis temperature, biochar ash content in biochar increases which thereby increases the pH and values fall due to ash content and ebullition of acidic functional groups (Yin et al., 2019; Ren et al., 2020).

Li et al. (2017) suggested that biochar could elevate the alkaline behaviour of AD (minimum $pH \geq 6$). It has been observed that biochar could be a potent accelerator in the case of methanogenesis which could hence improve the operating capacity with higher OLRs. Cantrell et al. (2012) in their study had proved that a higher pH with a high pyrolysis temperature is favourable for biochar generated from dry manure, which has also been confirmed in many studies. With the recent studies on biochar, it has been decrypted that biochar from food wastes had lowest pH when pyrolysis was done at temperature of 500°C and the highest pH was found in the case of digestate biochar.

18.6.5 Redox Properties

Biochar consists of carbon in both amorphous and graphite form, along with labile organic molecules and mineral phases that can act as an electron acceptor or donor (Chacón et al., 2020). After contacting the soil, soil organic matter and minerals, roots and the root hairs, redox reactions may take place, which results in producing biochar rich in organic minerals and thus enhancing the nutrient cycling process especially nitrogen and phosphorus along with the formation of methane and nitrous oxide. The redox properties of biochar have also been considered important parameters in the AD process, and these are governed mainly by their surface functional groups and presence of free radicals, metals and metal oxides (Joseph et al., 2015). Free radicals affect the redox behaviour of biochar, and redox active metals namely Fe and Mn oxides are present in feedstocks showing its existence in many oxidation states and thus serving as both electron donors and acceptors (Dieguez-Alonso et al., 2019).

18.6.6 Surface Functional Groups

Biochar surface composition contains different functional groups such as −OH, −COOH, C=O and −NHx, which determines the adsorption rate of biochar and also helps in nutrient retention as well as removal of contaminants (Kumar et al., 2020). Hydrophobic behaviour of biochar can be determined by the presence and addition of O- and N-containing functional groups (Fagbohungbe et al., 2017). Qambrani et al. (2017) mentioned that due to pyrolytic conditions, there has been an alteration in the functional groups (−OH, =CH2, C=O, C=C and −CH3), thereby enhancing the biochar hydrophobic interactions.

18.7 CHALLENGES AND FUTURE PERSPECTIVES

Various studies have been carried out on the applications and impacts of biochar in the AD process, but still there are many voids to be filled in many cases due to lack of information. Current studies mainly focus on diverse methods to investigate the relationship between the biochar properties, its impact and the enhancement effect on microorganisms, but there is no mention about the mathematical quantification in the above-mentioned relationship to estimate the economic viability of addition of biochar in AD process, products formed and biogas production rate. New and developed processes should be implemented to inhibit the excessive application of biochar in the AD process, and desirable biochar production conditions should be specified to make the biochar-amended AD process much more productive and cost-effective.

18.8 CONCLUSION

Based on the preliminary study, it is reviewed and revised that conventional waste management methods are not able to cope up with the enormous quantity of food waste and municipal waste generation. So, recycling of valuable food waste or biomass and conversion into biochar, biogas, biofuel is found effective not only for the environment but also for the human community.

Biochar production unveils quite a broad range of biomass utilized as the feedstocks and pyrolyzed by various procedures, which are influenced by significant properties namely feedstock, pH, CEC, temperature, pyrolysis conditions, etc. These properties in turn aid in the growth of microbiota, which enhances the methanogenesis process, producing methane biogas through the process of AD. Nevertheless, the main focus of development of biochar in future is solely dependent on refining and sharpening the various properties of biochar, which also makes it a promising solution for pollutant removal. In this review, we have also touched on the varied impacts of biochar addition for anaerobic digestion of food waste. Biochar addition to AD process supplies essential trace elements, shortens the lag phase of AD, enhances the production of methane, buffering capacity and also reduces the dissolved concentration of toxic compounds produced.

ABBREVIATIONS

AD Anaerobic digestion
BOD Biochemical oxygen demand
BSG Brewer's spent grain
CEC Cation exchange capacity
CM Chicken manure
CO Carbon monoxide
COD Chemical oxygen demand
DIET Direct interspecies electron transfer
GHGs Greenhouse gases
HTC Hydrothermal carbonization
KW Kitchen waste
MFPI Ministry of food processing industries
OLR Organic loading rate
POME Palm Oil Mill Effluent
PP Pomelo peel
SSA Specific surface area
TSS Total suspended solids
VFA volatile fatty acids

REFERENCES

Adekunle, I. M., et al. (2011). "Recycling of organic wastes through composting for land applications: A Nigerian experience." *Waste Management & Research* 29(6): 582–593.

Ambaye, T. G., et al. (2020). "Anaerobic digestion of fruit waste mixed with sewage sludge digestate biochar: Influence on biomethane production." *Frontiers in Energy Research* 8: 31.

Azargohar, R., et al. (2013). "Evaluation of properties of fast pyrolysis products obtained, from Canadian waste biomass." *Journal of Analytical and Applied Pyrolysis* 104: 330–340.

Bourgeois, J. and J. Doat (1984). Torrefied wood from temperate and tropical species. Advantages and prospects. Bioenergy 84. Proceedings of Conference 15–21 June 1984, Goteborg, Sweden. Volume III. Biomass Conversion, Elsevier Applied Science Publishers, Amsterdam, The Netherlands.

Cai, J., et al. (2016). "Effects and optimization of the use of biochar in anaerobic digestion of food wastes." *Waste Management & Research* 34(5): 409–416.

Cantrell, K. B., et al. (2012). "Impact of pyrolysis temperature and manure source on physicochemical characteristics of biochar." *Bioresource Technology* 107: 419–428.

Chacón, F. J., et al. (2020). "Enhancing biochar redox properties through feedstock selection, metal preloading and post-pyrolysis treatments." *Chemical Engineering Journal* 395: 125100.

Chang, J. I., et al. (2006). "Composting of vegetable waste." *Waste Management & Research* 24(4): 354–362.

Chiappero, M., et al. (2020). "Review of biochar role as additive in anaerobic digestion processes." *Renewable and Sustainable Energy Reviews* 131: 110037.

Choe, U., et al. (2019). "Effect of bamboo hydrochar on anaerobic digestion of fish processing waste for biogas production." *Bioresource Technology* 283: 340–349.

de Oliveira, E. A., et al. (2020). "Legacy of amazonian dark earth soils on forest structure and species composition." *Global Ecology and Biogeography* 29(9): 1458–1473.

Demirbas, A. and G. Arin (2002). "An overview of biomass pyrolysis." *Energy Sources* 24(5): 471–482.

Dieguez-Alonso, A., et al. (2019). "Designing biochar properties through the blending of biomass feedstock with metals: Impact on oxyanions adsorption behavior." *Chemosphere* 214: 743–753.

Ding, W., et al. (2014). "Pyrolytic temperatures impact lead sorption mechanisms by bagasse biochars." *Chemosphere* 105: 68–74.

Downie, A., et al. (2012). "Biochar as a geoengineering climate solution: Hazard identification and risk management." *Critical Reviews in Environmental Science and Technology* 42(3): 225–250.

Du, Z., et al. (2018). "Peanut-shell biochar and biogas slurry improve soil properties in the North China Plain: A four-year field study." *Scientific Reports* 8(1): 1–9.

Duan, X., et al. (2019). "New method for algae comprehensive utilization: Algae-derived biochar enhances algae anaerobic fermentation for short-chain fatty acids production." *Bioresource Technology* 289: 121637.

Dudek, M., et al. (2019). "The effect of biochar addition on the biogas production kinetics from the anaerobic digestion of brewers' spent grain." *Energies* 12(8): 1518.

Elsayed, M., et al. (2020). "Innovative integrated approach of biofuel production from agricultural wastes by anaerobic digestion and black soldier fly larvae." *Journal of Cleaner Production* 263: 121495.

Fagbohungbe, M. O., et al. (2017). "The challenges of anaerobic digestion and the role of biochar in optimizing anaerobic digestion." *Waste Management* 61: 236–249.

Faruqui, A., et al. (2010). "The impact of informational feedback on energy consumption-A survey of the experimental evidence." *Energy* 35(4): 1598–1608.

Fodah, A. E. M., et al. (2021). "Bio-oil and biochar from microwave-assisted catalytic pyrolysis of corn stover using sodium carbonate catalyst." *Journal of the Energy Institute* 94: 242–251.

Frias Flores, C. B. (2020). "Application of biochar as an additive to enhance biomethane potential in anaerobic digestion." Thesis. Rochester Institute of Technology, https://repository.rit.edu/cgi/viewcontent.cgi?article=11540&context=theses

Gabhane, J. W., et al. (2020). "Recent trends in biochar production methods and its application as a soil health conditioner: A review." *SN Applied Sciences* 2(7): 1–21.

Ghodrat, A. G., et al. (2018). Waste management strategies; the state of the art. *Biogas*, Springer, New York: 1–33.

Glaser, B. (2007). "Prehistorically modified soils of central Amazonia: A model for sustainable agriculture in the twenty-first century." *Philosophical Transactions of the Royal Society B: Biological Sciences* 362(1478): 187–196.

Gu, J., et al. (2020). "Anaerobic co-digestion of food waste and sewage sludge under mesophilic and thermophilic conditions: Focusing on synergistic effects on methane production." *Bioresource Technology* 301: 122765.

Guan, G., et al. (2016). "Catalytic steam reforming of biomass tar: Prospects and challenges." *Renewable and Sustainable Energy Reviews* 58: 450–461.

Hossain, M. A., et al. (2017). "Optimization of process parameters for microwave pyrolysis of oil palm fiber (OPF) for hydrogen and biochar production." *Energy Conversion and Management* 133: 349–362.

Huiru, Z., et al. (2019). "Technical and economic feasibility analysis of an anaerobic digestion plant fed with canteen food waste." *Energy Conversion and Management* 180: 938–948.

Igalavithana, A. D., et al. (2017). "Advances and future directions of biochar characterization methods and applications." *Critical Reviews in Environmental Science and Technology* 47(23): 2275–2330.

Ippolito, J. A., et al. (2012). "Environmental benefits of biochar." *Journal of Environmental Quality* 41(4): 967–972.

Jahirul, M. I., et al. (2012). "Biofuels production through biomass pyrolysis-a technological review." *Energies* 5(12): 4952–5001.

Joseph, S., et al. (2015). "The electrochemical properties of biochars and how they affect soil redox properties and processes." *Agronomy* 5(3): 322–340.

Khalid, A., et al. (2011). "The anaerobic digestion of solid organic waste." *Waste Management* 31(8): 1737–1744.

Kumar, H., et al. (2020). "Fruit and vegetable peels: Utilization of high value horticultural waste in novel industrial applications." *Molecules* 25(12): 2812.

Kumar, M., et al. (2020). "Critical review on biochar-supported catalysts for pollutant degradation and sustainable biorefinery." *Advanced Sustainable Systems* 4(10): 1900149.

Kumar, M., et al. (2020). "Ball milling as a mechanochemical technology for fabrication of novel biochar nanomaterials." *Bioresource Technology* 312: 123613.

Kumar, M., et al. (2021). "A critical review on biochar for enhancing biogas production from anaerobic digestion of food waste and sludge." *Journal of Cleaner Production* 305: 127143.

Lam, S. S., et al. (2018). "Pyrolysis production of fruit peel biochar for potential use in treatment of palm oil mill effluent." *Journal of Environmental Management* 213: 400–408.

Lansing, S., et al. (2019). "Food waste co-digestion in Germany and the United States: From lab to full-scale systems." *Resources, Conservation and Recycling* 148: 104–113.

Lee, J., et al. (2017). "Pyrolysis process of agricultural waste using CO2 for waste management, energy recovery, and biochar fabrication." *Applied Energy* 185: 214–222.

Lee, J., et al. (2018). "Hydrothermal carbonization of lipid extracted algae for hydrochar production and feasibility of using hydrochar as a solid fuel." *Energy* 153: 913–920.

Lehmann, J. and S. Joseph (2015). *Biochar for Environmental Management: Science, Technology and Implementation*. Routledge, London.

Li, L., et al. (2013). *An Introduction to Pyrolysis and Catalytic Pyrolysis: Versatile Techniques for Biomass Conversion*. Elsevier, Amsterdam, The Netherlands.

Li, H., et al. (2017). "Mechanisms of metal sorption by biochars: biochar characteristics and modifications." *Chemosphere* 178: 466–478.

Li, Y., et al. (2019). "Enhancement of methane production in anaerobic digestion process: A review." *Applied Energy* 240: 120–137.

Liu, Z., et al. (2019). "Comparative production of biochars from corn stalk and cow manure." *Bioresource Technology* 291: 121855.

Loss, F. and F. Waste (2019). "Food and Agriculture Organization of the United Nations." URL: https://www.fao.org/food-loss-and-food-waste/flw-data.

Lou, L., et al. (2019). "Adsorption and degradation in the removal of nonylphenol from water by cells immobilized on biochar." *Chemosphere* 228: 676–684.

Luz, F. C., et al. (2018). "Biochar characteristics and early applications in anaerobic digestion-a review." *Journal of Environmental Chemical Engineering* 6(2): 2892–2909.

Luz, F. C., et al. (2018). "Ampelodesmos mauritanicus pyrolysis biochar in anaerobic digestion process: Evaluation of the biogas yield." *Energy* 161: 663–669.

Maina, S., et al. (2017). "From waste to bio-based products: A roadmap towards a circular and sustainable bioeconomy." *Current Opinion in Green and Sustainable Chemistry* 8: 18–23.

Mamimin, C., et al. (2015). "Effect of operating parameters on process stability of continuous biohydrogen production from palm oil mill effluent under thermophilic condition." *Energy Procedia* 79: 815–821.

Mohan, D., et al. (2006). "Pyrolysis of wood/biomass for bio-oil: A critical review." *Energy & Fuels* 20(3): 848–889.

Moreno-Riascos, S. and T. Ghneim-Herrera (2020). "Impact of biochar use on agricultural production and climate change. A review." *Agronomía Colombiana* 38(3): 367–381.

Mumme, J., et al. (2014). "Use of biochars in anaerobic digestion." *Bioresource Technology* 164: 189–197.

Ovi, D., et al. (2022). "Effect of rice husk and palm tree-based biochar addition on the anaerobic digestion of food waste/sludge." *Fuel* 315: 123188.

Pan, J., et al. (2019). "Effects of different types of biochar on the anaerobic digestion of chicken manure." *Bioresource Technology* 275: 258–265.

Pan, J., et al. (2019). "Achievements of biochar application for enhanced anaerobic digestion: a review." *Bioresource Technology* 292: 122058.

Panwar, N., et al. (2012). "Thermo chemical conversion of biomass-eco friendly energy routes." *Renewable and Sustainable Energy Reviews* 16(4): 1801–1816.

Pentananunt, R., et al. (1990). "Upgrading of biomass by means of torrefaction." *Energy* 15(12): 1175–1179.

Qambrani, N. A., et al. (2017). "Biochar properties and eco-friendly applications for climate change mitigation, waste management, and wastewater treatment: A review." *Renewable and Sustainable Energy Reviews* 79: 255–273.

Qian, K., et al. (2015). "Recent advances in utilization of biochar." *Renewable and Sustainable Energy Reviews* 42: 1055–1064.

Qin, Y., et al. (2020). "Specific surface area and electron donating capacity determine biochar's role in methane production during anaerobic digestion." *Bioresource Technology* 303: 122919.

Qiu, L., et al. (2019). "A review on biochar-mediated anaerobic digestion with enhanced methane recovery." *Renewable and Sustainable Energy Reviews* 115: 109373.

Rafiq, M. K., et al. (2016). "Influence of pyrolysis temperature on physico-chemical properties of corn stover (Zea mays L.) biochar and feasibility for carbon capture and energy balance." *PLoS One* 11(6): e0156894.

Ren, S., et al. (2020). "Hydrochar-facilitated anaerobic digestion: Evidence for direct interspecies electron transfer mediated through surface oxygen-containing functional groups." *Environmental Science & Technology* 54(9): 5755–5766.

Shareef, T. M. E. and B. Zhao (2016). "The fundamentals of biochar as a soil amendment tool and management in agriculture scope: An overview for farmers and gardeners." *Journal of Agricultural Chemistry and Environment* 6(1): 38–61.

Sharma, B., et al. (2019). "Recycling of organic wastes in agriculture: An environmental perspective." *International Journal of Environmental Research* 13(2): 409–429.

Shen, Y., et al. (2015). "Producing pipeline-quality biomethane via anaerobic digestion of sludge amended with corn stover biochar with in-situ CO2 removal." *Applied Energy* 158: 300–309.

Tan, Z., et al. (2017). "Returning biochar to fields: A review." *Applied Soil Ecology* 116: 1–11.

Titirici, M.-M., et al. (2007). "Back in the black: Hydrothermal carbonization of plant material as an efficient chemical process to treat the CO2 problem?" *New Journal of Chemistry* 31(6): 787–789.

Tomczyk, A., et al. (2020). "Biochar physicochemical properties: Pyrolysis temperature and feedstock kind effects." *Reviews in Environmental Science and Bio/Technology* 19(1): 191–215.

Tripathi, M., et al. (2016). "Effect of process parameters on production of biochar from biomass waste through pyrolysis: A review." *Renewable and Sustainable Energy Reviews* 55: 467–481.

Turner, J., et al. (2008). "Renewable hydrogen production." *International Journal of Energy Research* 32(5): 379–407.

Wang, Q., et al. (2013). "The effect of a buffer function on the semi-continuous anaerobic digestion." *Bioresource Technology* 139: 43–49.

Wang, C., et al. (2016). "Sorption behavior of Cr (VI) on pineapple-peel-derived biochar and the influence of coexisting pyrene." *International Biodeterioration & Biodegradation* 111: 78–84.

Wang, D., et al. (2017). "Improving anaerobic digestion of easy-acidification substrates by promoting buffering capacity using biochar derived from vermicompost." *Bioresource Technology* 227: 286–296.

Wang, C., et al. (2018). "Role of biochar in the granulation of anaerobic sludge and improvement of electron transfer characteristics." *Bioresource Technology* 268: 28–35.

Wang, G., et al. (2018). "Synergetic promotion of syntrophic methane production from anaerobic digestion of complex organic wastes by biochar: Performance and associated mechanisms." *Bioresource Technology* 250: 812–820.

Wang, J., et al. (2018). "Effect of zero-valent iron and trivalent iron on UASB rapid start-up." *Environmental Science and Pollution Research* 25(1): 749–757.

Wang, P., et al. (2020). "Enhancement of biogas production from wastewater sludge via anaerobic digestion assisted with biochar amendment." *Bioresource Technology* 309: 123368.

Wang, Z., et al. (2021). "Effect of biochar addition on the microbial community and methane production in the rapid degradation process of corn straw." *Energies* 14(8): 2223.

Wu, Y., et al. (2017). "Activated biochar prepared by pomelo peel using H 3 PO 4 for the adsorption of hexavalent chromium: Performance and mechanism." *Water, Air, & Soil Pollution* 228(10): 1–13.

Wu, J., et al. (2020). "High-efficiency removal of dyes from wastewater by fully recycling litchi peel biochar." *Chemosphere* 246: 125734.

Xu, F., et al. (2018). "Anaerobic digestion of food waste-challenges and opportunities." *Bioresource Technology* 247: 1047–1058.

Xu, Q., et al. (2020). "Effects of biochar addition on the anaerobic digestion of carbohydrate-rich, protein-rich, and lipid-rich substrates." *Journal of the Air & Waste Management Association* 70(4): 455–467.

Yin, C., et al. (2019). "Sludge-based biochar-assisted thermophilic anaerobic digestion of waste-activated sludge in microbial electrolysis cell for methane production." *Bioresource Technology* 284: 315–324.

You, S., et al. (2017). "A critical review on sustainable biochar system through gasification: Energy and environmental applications." *Bioresource Technology* 246: 242–253.

Yu, K. L., et al. (2017). "Recent developments on algal biochar production and characterization." *Bioresource Technology* 246: 2–11.

Zhang, B., et al. (2019). "Removal of methyl orange dye using activated biochar derived from pomelo peel wastes: performance, isotherm, and kinetic studies." *Journal of Dispersion Science and Technology* 41: 125–136.

Zhou, N., et al. (2017). "Biochars with excellent Pb (II) adsorption property produced from fresh and dehydrated banana peels via hydrothermal carbonization." *Bioresource Technology* 232: 204–210.

19 Valorization of Agrobiomass to Ferulic Acid

Sourab Paul, Jayato Nayak, and Sankha Chakrabortty

19.1 INTRODUCTION

Ferulic acid is a phenolic compound that belongs to the family of hydroxycinnamic acid. It is found in families of *Ranunculaceae* and *Gramineae*, which includes species such as reed root, *Ligusticum chuanxiong* (Zhang, 2009), *Angelica* (Wang, 2012), *Cimicifuga* (Pan et al., 2000), and *Rhizoma sparganii* (Zhang et al., 2011). It was first isolated from *Ferula foetida* (Hlasiwetz, 1866) and underwent chemical synthesis in 1925, which revealed its structure in both trans- and cis-isomeric forms (Dutt, 1925; Nethaji, 1988). The molecule known as 4-hydroxy-3-methoxycinnamic acid ($C_{10}H_{10}O_4$) has an unsaturated side chain that is transformed into a resonance-stabilized phenoxy radical through cis–trans isomerization (Rosazza et al., 1995). Ferulic acid is a phenolic derivative phytochemical that serves as a covalent side chain in plant cell wall components, particularly lignocelluloses, where it forms crosslinks with lignin and polysaccharides to increase cell wall rigidity (Strack, 1990).

Ferulic acid being a potent antioxidant and anti-inflammatory agent plays a significant role in medicine. Being a naturally occurring antioxidant, it is essential for scavenging free radicals from the body (Kikuzaki, 2002; Natalia, 2013; Ou & Kwok, 2004). These radicals are connected to oxidative stress and several chronic disorders, including cancer and heart issues. Studies suggest that ferulic acid may possess neuroprotective qualities that mitigate the effects of neurodegenerative illnesses such as Alzheimer's. Because of its anti-inflammatory properties, it is also a suitable choice for treating inflammatory illnesses like arthritis and inflammatory bowel syndrome. It is frequently used in skin care products due to its photoreactive properties, which help against the UV radiation damage of the skin. Additionally, it has antimicrobial activity, making it a preferred choice for treating infections (Akihisa et al., 2000; Graf, 1992; Ou & Kwok, 2004; Rukkumani et al., 2004).

Traditionally, in the chemical synthetic pathway, ferulic acid was produced by condensation of vanillin and malonic acid using the catalytic activity of piperidine (Adams & Bockstahler, 1952), involving about 21 days. Agricultural biomass is a great renewable source for phytochemicals with bioactive compounds as it is mostly found in the cell wall of the plants (Antonopoulou et al., 2002). It is created by the plant during the biosynthesis of lignin. In fact, ferulic acid used to be present in

DOI: 10.1201/9781003407713-19

plant tissues in both free and conjugate forms. Common foods high in ferulic acid include whole grains, beetroot, eggplant cabbage bananas, coffee, spinach, orange juice, citrus fruits, bamboo shoots, and broccoli. Furthermore, coffee, spices, and plant extracts are also rich source of ferulic acid. Grain bran, fruit peels, and vegetable peels and roots have a particularly high concentration of ferulic acid. The detailed availability of ferulic acid at various concentrations found in the agricultural by-products is mentioned in Table 19.1. Ferulic acid may be linked to mono/disaccharides, hydroxyl acids, or pectic and hemicellulose polysaccharides, depending on the source. When it comes to vegetables, such as burdock and aubergine, free ferulic acid is more common than in cereals (Zhao & Moghadasian, 2008). Studies show that the alkaline treatment and enzymatic hydrolysis are efficient methods for extracting ferulic acid from agricultural biomass. The aim of this review is to provide a detailed outline of various routes for the extraction of ferulic acids from the agrobiomass.

TABLE 19.1
Concentrations of Ferulic Acid in Different Agro-Products and Commercial Foods

Agrobiomass	Ferulic Acid Contents (mg/100 g)	References
	Grains	
Refined corn bran	2,610–3,300	Saulnier et al. (1995), Zhao et al. (2005)
Barley extract	1,358–2,293	Madhujith and Shahidi (2006)
Soft and hard wheat bran	1,351–1,456	Liyana-Pathirana and Shahidi (2006)
Rice endosperm cell wall	910	Shibuya (1984)
Fine wheat bran agrobiomass	530–540	Kroon et al. (1997), Andreasen et al. (2001)
Rye bran	280	Mattila et al. (2005), Andreasen et al. (2001)
Corn, dehulled kernels	174	Adom and Liu (2002)
Whole wheat kernels	64–127	Nishizawa et al. (1998), Adom and Liu (2002)
Whole-wheat flour	89	Mattila et al. (2005)
Whole grain rye flour	86	Mattila et al. (2005)
Whole brown rice	42	Nishizawa et al. (1998), Adom and Liu (2002)
Corn flour	38	Mattila et al. (2005)
Whole oats	25–35	Adom and Liu (2002), Mattila et al. (2005)
Whole grain barley flour	25–34	Nishizawa et al. (1998), Mattila et al. (2005)
Oat bran	33	Mattila et al. (2005)
	Fruits	
Grapefruit	10.7–11.6	Mattila et al. (2006)
Orange	9.2–9.9	Mattila et al. (2006)
Banana	5.4	Mattila et al. (2006)
Berries	0.25–2.7	Mattila et al. (2006)
Rhubarb	2	Mattila et al. (2006)
Plum, dark	1.47	Mattila et al. (2006)
Apples	0.27–0.85	Mattila et al. (2006)

(Continued)

TABLE 19.1 (*Continued*)
Concentrations of Ferulic Acid in Different Agro-Products and Commercial Foods

Agrobiomass	Ferulic Acid Contents (mg/100 g)	References
	Vegetables	
Bamboo shoots	243.6	Nishizawa et al. (1998)
Water dropwort	7.3–34	Sakakibara et al. (2003)
Eggplant	7.3–35	Sakakibara et al. (2003)
Red beet	25	Mattila and Hellstrom (2007)
Burdock	7.3–19	Sakakibara et al. (2003)
Soyabean	12	Mattila and Hellstrom (2007)
Peanut	8.7	Mattila and Hellstrom (2007)
Spinach/frozen	7.4	Mattila and Hellstrom (2007)
Red cabbages	6.3–6.5	Mattila and Hellstrom (2007)
Tomato	0.29–6	Bourne and Rice-Evans (1997), Mattila and Hellstrom (2007)
Radish	4.6	Mattila and Hellstrom (2007)
Broccoli	4.1	Mattila and Hellstrom (2007)
Carrot	1.2–2.8	Mattila and Hellstrom (2007)
Parsnip	2.2	Mattila and Hellstrom (2007)
Mizuna	1.4–1.8	Sakakibara et al. (2003)
Pot grown basil	1.5	Mattila and Hellstrom (2007)
Chinese cabbage	1.4	Mattila and Hellstrom (2007)
Pot grown lettuces	0.19–1.4	Mattila and Hellstrom (2007)
Green bean/fresh	1.2	Mattila and Hellstrom (2007)
Avocado	1.1	Mattila and Hellstrom (2007)
	Commercial foods and beverages	
Sugar beet pulp	800	Micard et al. (1997)
Popcorn	313	Nishizawa et al. (1998)
Whole grain rye bread	54	Mattila et al. (2005)
Whole grain oat flakes	25–52	Mattila et al. (2005), Nishizawa et al. (1998)
Sweet corn	42	Nishizawa et al. (1998)
Pickled red beet	39	Mattila and Hellstrom (2007)
Rice, brown, long grain parboiled	24	Mattila et al. (2005)
Coffee	9.1–14.3	Mattila et al. (2006), Nardini et al. (2002)
Boiled spaghetti	13.6	Nishizawa et al. (1998)
Pasta	12	Mattila et al. (2005)
White wheat bread	8.2	Mattila et al. (2005)

Reproduced with copyright from Zhao & Moghadasian (2008).

19.2 GLOBAL MARKET OF FERULIC ACID

With a projected value of US$ 67.8 million in 2022 and a Compound Annual Growth Rate (CAGR) of 6.74%, the global ferulic acid market is predicted to reach roughly US$ 130.1 million by 2032. North America holds 13.5% of the market share, while Europe is leading by having 24.2% share of the global ferulic acid market as new healthcare treatments are driving significant development in Europe, the Asia Pacific region—headed by China, India, and Japan—is expected to surpass US$ 32 million by 2032 as a result of increased packaged food consumption. Applications for ferulic acid include food, cosmetics, and medicines; the cosmetics industry is growing at an exponential rate because of ferulic acid's anti-aging properties (Future market insights, 2022).

19.3 CLASSICAL ROUTES TOWARDS THE SYNTHESIS OF FERULIC ACID

Previously, the chemical synthesis of ferulic acid was performed by mixing vanillin and malonic acid in a condensation reaction that was assisted by piperidine and heated to 70°C for 3.5 weeks (Adams & Bockstahler, 1952). This method was utilized to produce ferulic acid. It is important to note that this process produces ferulic acid isomers, both trans and cis. This process of generating ferulic acid is considered to be a lengthy and laborious one (it can take up to 21 days), and it necessitates the usage of a substantial amount of hazardous chemical reactants (malonic acid), solvents (piperidine), and catalysts (benzylamines), which in turn necessitates further purification stages (Ou & Kwok, 2004). Despite the ferulic acid that the reaction may take up to 3 weeks to complete, the yield is quite high. The enhancements that Da and Xu (1997) made to this method included the addition of benzylamine and methylbenzene, as well as a temperature range of 85°C–95°C for the reaction. Additionally, the upgraded method resulted in an increase in yield, in addition to reducing the response time to 2 hours.

A significant amount of phenolic compound which is found in plant cell wall is ferulic acid. It ranges from 5 g/kg in wheat bran, 9 g/kg in sugar beet pulp, 15–28 g/kg of rice bran oil, and 25 g/kg in corn kernel (Kroon et al., 1997; Bunzel et al., 2005; Zhang & Xu, 1997; Buranov & Mazza, 2009).

The commercial production of ferulic acid is currently from rice bran pitch, the waste stream left over after making rice bran oil. It is extracted from wheat bran and maize bran using concentrated alkali solution (Di Silvestro et al., 2022). The alkali solution is generally comprised of NaOH or KOH solutions of concentration 0.2–1 M. It is fundamentally a process of liquid–liquid extraction, where during further processing of extract, ferulic acid comes out as the final product. There are some of the major demerits of the conventional processes for such biochemical production.

It turns out that the extraction of ferulic acid from plant resources can be challenging while employing liquid–liquid extraction technology. Because of the large range of possible quantities of ferulic acid in plants and the components of those plants, it can be difficult to locate the optimal extraction method that would result in the highest possible yields. In addition, it may be challenging to locate and acquire suitable plant sources that have a high concentration of ferulic acid on their bodies. From a financial and environmental point of view, it is possible that it would not be possible to cultivate the majority of these plants on a vast scale. Ferulic acid is available from plant sources; however, it is still mostly unavailable. There is a correlation between the changing of the seasons and the concentration of ferulic acid in plants, which in turn has an effect on the overall output. With proper timing, one may extract the maximum amount of ferulic acid from the harvest. Some biotechnological challenges are presented by the use of microbial or enzymatic processes for the production of ferulic acid. These challenges include problems regarding the efficiency, cost-effectiveness, and scalability of the process. To provide further clarification, extremely high levels of purity are necessary for the vast majority of the applications of ferulic acid in the fields of medicine and cosmetics. When contaminants or impurities are present in the finished product, it is possible for its effectiveness and safety to be impaired. Last but not least, worries regarding the prices of such products may have an impact on their economic sustainability. For instance, the costs associated with the production of ferulic acid, which include the costs of extraction, purification, and processing, might occasionally have a considerable impact. Finding methods that are efficient in terms of cost is really necessary in order to achieve widespread acceptance.

As a result of the increasing number of industrial applications for the chemical, the bioprocessing and milling industries are searching for environmentally friendly alternatives to the conventional sources of ferulic acid resources. Because of such previously discussed issues, it would be beneficial to discover more sources of ferulic acid and to devise more environmentally friendly methods for releasing it. The hydrolysates that are produced as a result of the enzymatic hydrolysis of by-products have been mentioned by certain individuals as they have the potential to be refined into ferulic acid if the conditions are favorable. Consequently, this might be utilized in the production of natural and cost-effective high-performance liquid chromatography (HPLC) standards (a combination of ferulic acid isomers or pure trans-ferulic acid), inhibitors of lipid oxidation, agents with antibacterial properties intended for use in pharmaceuticals, and components for use in food (Ahuja, 2019; Magnani et al., 2014; Pei et al., 2016).

19.4 EMERGING ROUTES FOR FERULIC ACID SYNTHESIS

19.4.1 Chemical Extraction

Due to the drawbacks of the conventional production of ferulic acid, continuous research on eco-friendly, less time-consuming, high-yield, low-cost production is an aim of modern research. Since natural sources like plants have an abundant source of ferulic acid, various routes have been developed to extract the ferulic acid through enzymatic, alkaline, and acidic extraction (Tilay et al., 2008). Study by Tilay et al. (2008) focused on to maximize the extraction of esterified ferulic acid (EFA) from a variety of agricultural wastes, such as maize, rice, wheat bran and straw, sugarcane bagasse, pineapple, orange, and pomegranate peels. The challenges that lie in the extraction are due to their complex nature and susceptibility to oxidation and hydrolysis. This study optimized the alkaline extraction of EFA along with response surface methodology (RSM), which resulted in a 1.3-fold increase in extraction efficiency. The purity was achieved with the help of polymeric adsorbents, especially Amberlite XAD-16, and preparative high-performance thin-layer chromatography (HPTLC) was performed, which significantly improved the purity of ferulic acid.

The study by Sibhatu et al. (2021) focused on the extraction of ferulic acid from the agro-industrial waste, specifically brewer's spent grains (BSG), which constitute around 85% of the brewery by-products. They have employed a novel approach to effectively remove hemicellulose and solubilize lignin by pretreating with diluted sulfuric acid and then treating with sodium hydroxide, which improves the ferulic acid extraction. The Box–Behnken design, which is a statistical experimental design and a type of Response surface methodology, was used in order to maximize the yield concentration by maximizing the optimal conditions like reaction time, temperature, and hydrolyzing acid concentrations. The statistical analytical tools make it easier to identify the ideal circumstances required to maximize the yield and ferulic acid concentrations. The sugars identified in the BSG hydrolysate composition study include galactose, glucose, fructose, maltose, sucrose, xylose, mannose, and arabinose. HPLC analysis reveals that some polyphenolic acids are absent following treatment, whereas ferulic acid, p-coumaric acid (pCA), and sinapic acid are found to be abundant. 46.17 mg/100g BSG is the maximum ferulic acid content that can be produced using the optimized extraction process.

Another study by Cantillo-Pérez et al. (2017) focused on converting the agro-waste generated during the barley malting process into a valuable product. They aimed to extract ferulic acid by alkaline hydrolysis process with varying concentrations of sodium hydroxide and temperatures of the hydrolysis process to optimize the yield of the ferulic acid extraction. They have achieved the higher concentration of yield (0.074%) at lower temperature and diluted sodium hydroxide concentrations. A mathematical model was developed containing variable factors like temperature and sodium hydroxide concentrations to predict the concentration of yield to evaluate the extraction process's scalability and economic viability; simulation and economic analysis were also carried out. A number of steps were included in the simulation, such as washing, drying, sulfuric acid pretreatment, alkaline hydrolysis, and ferulic acid separation from contaminants. The cost of raw materials, operating expenses, and possible sales revenue from ferulic acid were all considered in the economic evaluation. However, the large investment expenditures and the low selling price needed to cover these costs meant that the current extraction procedure was not commercially sustainable, according to the economic analysis. Even though the extraction yield was higher than in the earlier research, the procedure required a higher selling price for ferulic acid in order to be profitable. Truong et al. (2017) focused on the synthesis of ferulic acid from γ-oryzanol, which is a component found in the rice bran oil (RBO) soap stock. They used hydrolysis technique to break down the γ-oryzanol into ferulic acid. To produce the maximum yield, various parameters like concentration of γ-oryzanol, the ratio of potassium hydroxide (KOH) to γ-oryzanol, and the ratio of ethanol (EtOH) to ethyl acetate (EtOAc) as solvents were taken into consideration. In their study, they found that 80% ferulic acid yield was obtained with higher starting γ-oryzanol concentrations and a

KOH/γ-oryzanol ratio of 10/1. Additionally, it was discovered that the maximum ferulic acid content was obtained with a 5/1 EtOH/EtOAc solvent mixture. They studied the effect of the temperature and ultrasonic irradiation, and found that at higher temperature (75°C), the yield of ferulic acid was significantly higher and the use of ultrasonic irradiation further increased the yield, implying that the ultrasound and heating worked well together to speed up the hydrolysis process, which can be used for large-scale industrial manufacturing. Furthermore, the researchers looked into the hydrolysis of extracts from RBO soap stock that included γ-oryzanol and compared with the genuine γ-oryzanol, where they have achieved slightly lower yield of 74%. Result shows that the technique is found pretty effective in manufacturing of ferulic acid from RBO soap stock extracts.

19.4.2 Enzymatic Extraction

There have been several studies reported on the release of ferulic acid from plant cell walls via the usage of feruloyl esterases synthesized by microorganisms. A subclass of carboxylesterase known as feruloyl esterases is capable of liberating ferulic acid from a variety of esterified materials, such as polysaccharides, methyl ferulate, and feruloylated oligosaccharides (Kroon et al., 1999). *Bacillus* species, *Aspergillus niger*, *Pycnoporus cinnabarinus*, *Clostridium thermocellum*, *Lactobacilli*, *Streptomyces avermitilis*, *Pseudomonas fluorescens*, and *Brettanomyces anomalus* are among the bacterial, fungal, and yeast microorganisms that secrete these enzymes. Despite the lack of economic utility, feruloyl esterases have been thoroughly studied in order to prepare ferulic acids. Microorganisms capable of secreting feruloyl esterases first undergo screening. Second, the structure, optimal pH and temperature, stability-affecting variables, and the effect of the substrate and polysaccharide-degrading enzymes on the release of ferulic acid are among the attributes of the enzymes that were identified (Prates et al., 2001; Williamson & Vallejo, 1997; Saulnier et al., 2001). After that, the impact of several elements, including ferulic acid, carbon and nitrogen resources, on the synthesis of feruloyl esterases was investigated (Faulds & Williamson, 1997). Finally, a procedure for extracting ferulic acid from fermentation solutions was developed (Edlin et al., 1998). Future research will concentrate on enhancing fermentation technology and genetic engineering to produce more feruloyl esterases.

The studies by Juhnevica-Radenkova et al. (2021) were focused on the production of ferulic acid through enzymatic hydrolysis from wheat and rye bran, which is a by-product of flour production. About 85% of the cultivated grain is used for flour, and a by-product bran that is produced is underutilized as it is composed of complex structure of cellulose, hemicellulose, and lignin. To overcome this complexity, biotechnological method-based approach was taken by using a multi-enzyme complex, Viscozyme® L, which is produced by *Aspergillus aculeatus*, serving as a key enzyme to partially break down water-insoluble dietary fiber in bran. Beyond its uses in technology, enzymatic hydrolysis promotes the formation of healthy bacteria in the human digestive system, which has positive health effects. The study reports that after 24 hours of enzymatic hydrolysis, 11.3 g/kg of ferulic acid from wheat bran and 8.6 g/kg from rye bran were extracted which surpass the yield obtained from alkali-assisted hydrolysis or micro-saponification. This study not only contributes to the efficient production of ferulic acid, but also the broader field of sustainable biorefining of agro-industrial by-products.

The study by Faulds and Williamson (1993) aims to extract ferulic acid from plant polysaccharides by ferulic acid esterase derived from *Streptomyces olivochromogenes.* They aim to see the effectiveness of ferulic acid esterase against O-[5-O-(trans-feruloyl)-α-L-arabinofuranosyl]-(1–3)-O-β-xylopyranosyl-(1–4)-D-xylopyranose (FAXX), one of the pectins found in sugar beetroot and wheat bran. The experimental data shows that the free ferulic acid with a K_m of 0.24 mM and a V_{max} of 0.134 U/mg was released from the FAXX which is found in wheat bran by ferulic acid esterase. Moreover, the enzyme did not require pectinase to release 0.25% of covalently bound ferulic acid from sugar beetroot pectin. The findings demonstrate that ferulic acid esterase efficiently releases free ferulic acid from compounds produced from wheat bran and that it prefers a number of substrates, including O-[5-O-(trans-feruloyl)-α-L-arabinofuranosyl]. It also releases part of the covalently bound ferulic acid from the sugar beetroot pectin. The study examines the role of ferulic acid esterase in breaking down the polysaccharides containing feruloyl groups, which are found in sugar beet pectin and wheat bran.

Another study by Xiros et al. (2009) focused on extracting ferulic acid from BSG using a crude enzyme extract obtained from *Fusarium oxysporum.* The aim of the study was to understand the effect of combining different enzymes on the extraction of ferulic acid from BSG. As compared to using individual enzymes, the results show that more than twice as much ferulic acid was released when the crude enzyme extract was utilized. They extracted 94.5 ± 1.5 μmol/g of ferulic acid from BSG using the crude extract. This indicates that using a mixture of enzymes is the most effective method for extracting ferulic acid from complex substrates like BSG. They also found that the release of ferulic acid from BSG was significantly influenced by the proteases present in the crude extract. Ferulic acid release was greatly increased when proteases such as papain and alcalase 2.5 L were applied to BSG. Furthermore, they discovered a novel finding that ferulic acid might actually limit the activity of one of the enzymes involved in its own release.

In a recent study, scientists focused on combining two techniques, hydrothermal pretreatment with enzymatic hydrolysis for extracting ferulic acid from reed straws; these techniques efficiently break down the lignocellulose, hemicellulose, and lignin, along with the disruption of cell walls or cellulose fibers to release the phenolic compounds. The highest yield obtained for this technique is 0.69 mg/g of dry reed straws. The extracted phenolic compounds demonstrated excellent scavenging activity of hydroxyl radical and DPPH radicals (Qian et al., 2023). The study shows that a combination of enzymatic hydrolysis by cellulase and xylanase (CXEH) and fed-batch enzymatic hydrolysis (FBEH) is an effective way of extracting ferulic acid from the corn straw. The enzymes cellulase and xylanase hydrolyze cellulose and hemicellulose in plant biomass, respectively. CXEH helps in the effective breakdown of corn straw's complex structure into simpler sugars and other components. In the FBEH operation, they introduced 150 mL of enzymatic hydrolysis solution, comprising a 1.5% enzyme concentration and a 5:4 (v/v) ratio of cellulase to xylanase, into the existing 250 mL batch enzymatic hydrolysis solution. Additionally, the substrate loading was maintained at a 2.0% level. It is found that through FBEH, the maximum yield that can be extracted is 2,178.58 and 2,710.17 mg/L of ferulic acid (Qian et al., 2022).

Feruloyl esterase, secreted by the inoculum isolated from the mixed culture of soil, was able to react with the Feruloyl polysaccharide (polymers) of banana stem waste (BSW) to break into cellulose, hemicellulose, and lignin. Optimization approach was taken forward by using 2^5 full factorial design in Design expert software on fermentation temperature, agitation, water-to-banana stem waste proportion, substrate-to-microbes proportion, and time. The highest ferulic acid outcome was observed as 1.2187 mg ferulic acid per gram of BSW at the optimum operating temperature of 26°C and shaking speed of 150 rpm. By providing optimized equal proportion of water-to-banana stem waste and equal substrate-to-inoculum ratio, and 24 hours of time, the best aforementioned output was obtained (Mohd Sharif et al., 2021).

19.4.3 Synthesis through Gene Recombinant Approach

The recombinant technology by genetically modifying the microorganisms can enhance the production of thermostable esterase enzyme called acetylxylan esterase (AXE), which is required for ferulic acid extraction from lignocellulosic agricultural biomass (Huang et.al., 2011). First, the AXE esterase gene from *Thermobifida fusca* NTU22 was cloned into a *Yarrowia lipolytica* P01g host strain, which was then cultivated to assess extracellular esterase production levels. To check the efficiency of the enzyme for extracting ferulic acid, they have performed hydrolysis test on various agricultural biomass substrates like rice bran, wheat bran, bagasse, xylan, and corncob. To maximize the release of ferulic acid from the biomass, *T. fusca* xylanase (Tfx) and AXE esterase were sequentially cultivated, either alone or in combination. In their study, they found that the maximum yield was obtained from corncob which is 10.2 μM for *T. fusca* xylanase (Tfx) and 328.0 μM for AXE esterase, and in combination of both the enzyme however leads to a lower yield production around 265.0 μM. This work shows that using genetically modified yeast to produce enzymes may effectively extract ferulic acid from agricultural waste, raising the prospect of environmentally friendly biorefinery operations. The integration of enzymatic treatment with the cold-adapted carboxylesterase catalysis is an effective way to breakdown the lignocellulosic component of the wheat bran to produce ferulic acid (Cao et al., 2023). In *Arthrobacter soli* Em07, the gene responsible for cold-adapted carboxylesterase is identified and cloned in *E. coli*. The optimum condition for the enzyme to operate is as low as 10° C temperature and a pH of 7. The destarched wheat bran (DSWB) was pretreated with the xylanase enzyme and then further treated with the cold-adapted carboxylesterase to synthesize ferulic acid. In this study, it was found that from 20 mg pretreated DSWB, 235.8 μg of ferulic acid can be produced, which proves the effectiveness of the technique.

In another attempt of genetic engineering, the bifunctional endoglucanase/carboxylesterase enzyme which is found in *Sphingobacterium soilsilvae* Em02 is identified and expressed in *E. coli*. They found that in the temperature range of 40°C and in the pH range of 6.0–7.0, the bifunctional enzyme shows better enzymatic activity. These enzymes can degrade the DSWB. First, the endoglucanase can break down the glycosidic bonds in DSWB cellulose structure to a smaller component of oligosaccharides. Further, the carboxylesterase enzyme hydrolyzes the ester bond and releases ferulic acid, which were previously bound to xylan polymers. Through this technique, 49.4 µg of ferulic acid was produced from 20 mg of DSWB. It signifies the effectiveness of the bifunctional enzyme (Fang et al., 2023). The gene coding for feruloyl esterases was identified in the *Lactobacillus crispatus* S524, and through genetic modification, it was heterogeneously expressed in *E. coli* BL21(DE3). The enzyme shows maximum activity at a high temperature of 65°C and a pH of 7.0. 199 µg of ferulic acid was extracted from 0.2 g of DSWB using the synthesized feruloyl esterases. It was observed that the genetically modified *E. coli* produced 1.86 mg of ferulic acid when supplemented with a medium containing 2 g of DSWB. It shows that the ability to synthesize ferulic acid is present in both newly synthesized feruloyl esterases and genetically modified *E. coli* strain.

In the metagenome assembled genome of soil sample, a putative gene was discovered in 33 bins, coding for the feruloyl esterases; it was then expressed in *E. coli* after codon optimization and single peptide truncation. The study attempted to extract ferulic acid from the distilled spent grain with husk (DSGH). The optimal condition for the enzyme is to be found at higher temperature of 50°C and slightly acidic pH of 6.0 with the presence of 4-nitrophenyl trans-ferulate as the substrate. Through this technique, 89.25 µg of yield was generated from 1 g of distilled spent grain with husk, which is comparatively lower than other agricultural waste, but it shows a promising technique for the valorization of brewery waste (Zhang et al., 2024).

19.4.4 Synthesis through Fermentation

Lv, (2021) genetically engineered *E. coli* for the fermentative production of ferulic acid. The objective of the research was to introduce new genes and modify the activity of existing genes in the ferulic acid-producing bacterium. This gives the bacteria a novel pathway for producing ferulic acid. They chose particular genes because they have been shown to be crucial for the production of ferulic acid. These genes instruct the microbes on how to produce enzymes such as tyrosine ammonia-lyase (TAL), p-coumarate 3-hydroxylase (SAM5), and caffeic acid O-methyltransferase (COMT). The production of ferulic acid by different strains of *E. coli* was studied. It was seen that the production of ferulic acid by JM109(DE) strain was higher than BL21(DE3) strain. The use of weaker promoter such as T5 was more helpful in the fermentation to produce ferulic acid. They have studied the impact of cofactors on the production of ferulic acid; in this process, they found that they could enhance the production by upregulating the activity of genes involved in the synthesis of Nicotinamide adenine dinucleotide phosphate (NADHP) and introducing a gene from another organism that produces S-Adenosyl-L-methionine. The addition of the metK gene and the co-expression of the pntAB genes with the ferulic acid pathway increased the yield, which leads to the conclusion that the increase of cofactor availability by optimizing pathway enzyme expression levels can increase the production of ferulic acid by *E. coli*. In simple terms, in *Escherichia coli*, tyrosine ammonia-lyase converts

L-tyrosine to pCA, which is then hydroxylated to caffeic acid and methylated to create ferulic acid with the help of some cofactors. The engineered *E. coli* strain produced 212 mg/L of ferulic acid with 11.8 mg/L of caffeic acid residual, according to the flask fermentation data. This is the highest ferulic acid yield that *E. coli* strains have ever achieved.

Another study by Effendi et al. (2023) aimed to genetically modify and construct artificial pathway in *E. coli* through modular design for the synthesis of ferulic acid. They focused on both the upstream and downstream processes. The objective of module (M1), the upstream step, was to create pCA, a precursor of FA, from tyrosine. The researchers focused on using hydroxylation and methylation processes to transform pCA into ferulic acid in downstream module (M2). This step required the optimization of heterologous gene expression for the enzymes involved in these reactions. Codon optimization and mRNA secondary structure analysis were employed to enhance gene expression and enzyme activity. The further crucial component of the process was the optimization of cofactor supply. S-adenosylmethionine (SAM) and flavin adenine dinucleotide ($FADH_2$) are the types of cofactors that are necessary for ferulic acid production. In 48 hours, a significant ferulic acid yield of 972.6 mg/L with 89.4% conversion was attained when combined with simultaneous optimization of culture conditions. They even explored the cost-effective way for pCA source: spent coffee ground was utilized to extract pCA, which resulted in yields of 74.4–75.3 mg/L ferulic acid, with a conversion rate of 17.8%–18.0%. Though this conversion rate is less, it provides a cheaper and alternate way for pCA production. In another study, (Zhou et al., 2022) by using Luria-Bertani (LB) medium and M9Y medium, in a fed batch conditions, *E. coli* produced 5.09 g/L of ferulic acid due to the presence of cofactors like SAM and $FADH_2$, which is about 20 times higher than that produced without the use of cofactors.

19.5 FUTURE PERSPECTIVE FOR FERULIC ACID SYNTHESIS

Continuous research has been going on to advance the synthesis of ferulic acid through biotechnological advancement like genetically modified microorganisms to produce enzymes that can be employed for ferulic acid extraction from the complex molecular structure or genetically modified microorganisms like *E. coli* to synthesize ferulic acid through a series of biochemical process. However, more advancements need to be made in the case of enzyme production like feruloyl esterases for the efficient ferulic acid release. There can be further optimization of variable factors like pH, reaction time, and temperature to uplift the yield of final product. A few literatures available on the use of statistical approach like Response Surface Methodology, Box–Behnken design, and full factorial design, which can help to identify the optimal conditions for the extraction of ferulic acid. However, there is still extreme lack in scale-up confidence which is significantly observed. Detailed economic assessment of the methodologies followed are pretty scanty. However, there is acute dearth of published papers that exhibit mathematical modeling and simulation to predict a pathway toward industrial automation. These are the research gaps that were found through the exhaustive literature review on the topic, which will be addressed in future approaches.

19.6 CONCLUSION

In conclusion, due to the significant importance of ferulic acid in global health, nutrition, and environmental sustainability, there has been tremendous research that led to the development in the synthesis of ferulic acid from the agricultural by-products. It is seen that the chemical-based pathway is not acceptable in ferulic acid synthesis. Solvent extraction of ferulic acid uses harsh chemicals, which may contaminate the final product. In cutting-edge technologies, enzymatic extraction of ferulic acid, recombinant enzyme development and extraction of ferulic acid from agrobiomass, and recombinant fungal fermentation for direct production of ferulic acid are being emphasized. Through the enzymatic extraction method, using feruloyl esterase in extracting ferulic acid from brewery spent grains, wheat bran, etc. led to the utilization of agricultural waste. Among all, regarding the consumption of agrobiomass, recombinant enzyme development from the gene-modified yeast species is in the forefront of final importance. However, as the new age is progressing toward genetically modified species development, the fermentative production of ferulic acid using recombinant species is currently of utmost desire. Employing this methodology, the producers have no need to depend on the specific plant-based resources, but get empowered to produce ferulic acid from more diversified agrobiomass. Despite all these advancements in scalability, cost-effectiveness has remained a great challenge that requires further research.

REFERENCES

Adams, R., & Bockstahler, T. E. (1952). Preparation and reaction of hydroxycinnamic acids and esters. *Journal of the American Chemical Society*, 74(21), 5346–5348.

Adom, K. K., & Liu, R. H. (2002). Antioxidant activity of grains. *Journal of Agricultural and Food Chemistry*, 50, 6182–6187.

Ahuja, K., & Mamtani, K. (2019). Global Market Insights. Retrieved from https://www.gminsights.com/industry-analysis/naturalferulic-acid-market

Akihisa, T., Yasukawa, K., Yamaura, M., Ukiya, M., Kimura, Y., Shimizu, N., & Arai, K. (2000). Triterpene alcohol and sterol ferulates from rice bran and their anti-inflammatory effects. *Journal of Agricultural and Food Chemistry*, 48, 2313–2322.

Andreasen, M. F., Kroon, P. A., Williamson, G., & Garcia-Conesa, M. T. (2001). Esterase activity able to hydrolyze dietary antioxidant hydroxycinnamates is distributed along the intestine of mammals. *Journal of Agricultural and Food Chemistry*, 49, 5679–5684.

Antonopoulou, I., Sapountzaki, E., Rova, U., & Christakopoulos, P. (2022). Ferulic acid from plant biomass: A phytochemical with promising antiviral properties. *Frontiers in Nutrition*, 8, 777576.

Bourne, L. C., & Rice-Evans, C. A. (1997). The effect of the phenolic antioxidant ferulic acid on the oxidation of low density lipoprotein depends on the pro-oxidant used. *Free Radical Research*, 27, 337–344.

Bunzel, M., Ralph, J., Funk, C., & Steinhart, H. (2005). Structural elucidation of new ferulic acid-containing phenolic dimers and trimers isolated from maize bran. *Tetrahedron Letters*, 46(35), 5845–5850.

Buranov, A. U., & Mazza, G. (2009). Extraction and purification of ferulic acid from flax shives, wheat and corn bran by alkaline hydrolysis and pressurised solvents. *Food Chemistry*, 115(4), 1542–1548.

Cantillo-Pérez, N., Villarraga-Palencia, F., González Delgado, A., Ojeda Delgado, K., Sánchez Tuiran, E., & Paz Astudillo, I. (2017). Ferulic acid production from agroindustrial waste of barley malting process. *International Journal of ChemTech Research*, 10(4), 226–232.

Cao, L., Xue, D., Liu, X., Wang, C., Fang, D., Zhang, J., & Gong, C. (2023). Ferulic acid production from wheat bran by integration of enzymatic pretreatment and a cold-adapted carboxylesterase catalysis. *Bioresource Technology*, 385, 129435

Da, Y. F., & Xu, Y. P. (1997). Synthesis of trans-ferulic acid. *Chinese Journal of Pharmacy*, 28, 188–189.

Di Silvestro, R., Kosik, O., Chatzigragkou, A., Lovegrove, A., Shewry, P., & Charalampopoulos, D. (2022). Project Report No. 641, Production of ferulic acid from wheat bran. Retrieved from https://projectblue.blob.core.windows.net/media/Deferulic acidult/Research%20Papers/Cereals%20and%20Oilseed/2022/PR641%20final%20project%20report.pdf

Dutt, S. (1925). General synthesis of α-unsaturated acids from malonic acid. *Quarterly Journal of the Chemical Society*, 1, 297–301.

Edlin, D. A. N., Narbad, A., Gasson, M. J., Dickinson, J. R., & Lloyd, D. (1998). Purification and characterization of hydroxycinnamate decarboxylase from *Brettanomyces anomalus*. *Enzyme and Microbial Technology*, 22, 232–239.

Effendi, S. S. W., & Ng, I. S. (2023). High value ferulic acid biosynthesis using modular design and spent coffee ground in engineered *Escherichia coli* chassis. *Bioresource Technology*, 384, 129262.

Fang, D., Xue, D., Liu, X., Cao, L., Zhang, J., & Gong, C. (2023). Concurrent production of ferulic acid and glucose from wheat bran by catalysis of a putative bifunctional enzyme. *Bioresource Technology*, 369, 128393.

Faulds, C. B., & Williamson, G. (1993). Release of ferulic acid from plant polysaccharides by ferulic acid esterase from *Streptomyces olivochromogenes*. *Carbohydrate Polymers*, 21(2–3), 153–155.

Faulds, C. B., & Williamson, G. (1997). Influence of ferulic acid on the production of feruloyl esterases by *Aspergillus niger*. *FEMS Microbiology Letters*, 157, 239–244.

Future Market Insights. (2022). Retrieved from https://www.futuremarketinsights.com/reports/ferulic-acid-market.

Graf, E. (1992). Antioxidant potential of ferulic acid. *Free Radical Biology & Medicine*, 3, 435–513.

Hlasiwetz, H., & Barth, L. (1866). Mittheilungen aus dem chemischen Laboratorium in Innsbruck I, Ueber einige Harze (Zersetzungsproducte derselben durch schmelzendes Kali). *Liebig's Annalen der Chemie*, 138, 61–76.

Huang, Y. C., Chen, Y. F., Chen, C. Y., et al. (2011). Production of ferulic acid from lignocellulolytic agricultural biomass by Thermobifida fusca thermostable esterase produced in Yarrowia lipolytica transformant. *Bioresource Technology*, 102(17), 8117–8122.

Juhnevica-Radenkova, K., Kviesis, J., Moreno, D. A., Seglina, D., Vallejo, F., Valdovska, A., & Radenkovs, V. (2021). Highly-efficient release of ferulic acid from agro-industrial by-products via enzymatic hydrolysis with cellulose-degrading enzymes: part i-the superiority of hydrolytic enzymes versus conventional hydrolysis. *Foods*, 10(4), 782.

Kikuzaki, H., Hisamoto, M., Hirose, K., Akiyama, K., & Taniguchi, H. (2002). Antioxidant properties of ferulic acid and its related compounds. *Journal of Agricultural and Food Chemistry*, 50, 2161–2169.

Kroon, P. A., Ferulic Acidulds, C. B., Ryden, P., Robertson, J. A., & Williamson, G. (1997). Release of covalently bound ferulic acid from fiber in the human colon. *Journal of Agricultural and Food Chemistry*, 45, 661–667.

Kroon, P. A., Garcia-Conesa, M. T., Fillingham, I. J., Hazlewood, G. P., & Williamson, G. (1999). Release of ferulic acid dehydrodimers from plant cell walls by feruloyl esterases. *Journal of the Science of Food and Agriculture*, 79, 428–434.

Liyana-Pathirana, C. M., & Shahidi, F. (2006). Importance of insoluble bound phenolics to antioxidant properties of wheat. *Journal of Agricultural and Food Chemistry*, 54, 1256–1264.

Lv, H., Zhang, Y., Shao, J., Liu, H., & Wang, Y. (2021). Ferulic acid production by metabolically engineered Escherichia coli. *Bioresources and Bioprocessing*, 8(70). https://doi.org/10.1186/s40643-021-00423-0

Madhujith, T., & Shahidi, F. (2006). Optimization of the extraction of antioxidative constituents of six barley cultivars and their antioxidant properties. *Journal of Agricultural and Food Chemistry*, 54, 8048–8057.

Magnani, C., Isaac, V. L. B., Correa, M. A., & Salgado, H. R. N. (2014). Caffeic acid: A review of its potential use in medications and cosmetics. *Analytical Methods*, 6, 3203–3210.

Mattila, P., Pihlava, J. M., & Hellstrom, J. (2005). Contents of phenolic acids, alkyl- and alkenylresorcinols, and avenanthramides in commercial grain products. *Journal of Agricultural and Food Chemistry*, 53, 8290–8295.

Mattila, P., Hellstrom, J., & Torronen, R. (2006). Phenolic acids in berries, fruits, and beverages. *Journal of Agricultural and Food Chemistry*, 54, 7193–7199.

Mattila, P., & Hellstrom, J. (2007). Phenolic acids in potatoes, vegetables, and some of their products. *Journal of Food Composition and Analysis*, 20, 152–160.

Micard, V., Grabber, J. H., Ralph, J., Renard, C. M. G. C., & Thibault, J. F. (1997). Dehydrodiferulic acids from sugar-beet pulp. *Phytochemistry*, 44, 1365–1368.

Mohd Sharif, N. S. A., Jamaluddin, M. F., & Zaino, N. (2021). Factorial analysis of ferulic acid production from biowaste. *Materials Today: Proceedings*, 46, 1763–1769.

Nardini, M., Cirillo, E., Natella, F., & Scaccini, C. (2002). Absorption of phenolic acids in humans after coffee consumption. *Journal of Agricultural and Food Chemistry*, 50, 5735–5741.

Natalia, N. R., Claire, D., Lullien, P. V., & Valerie, M. (2013). Exposure or release of ferulic acid from wheat aleurone: Impact on its antioxidant capacity. *Food Chemistry*, 141, 2355–2362.

Nethaji, M., Pattabhi, V., & Desiraju, G. R. (1988). Structure of 3-(4-hydroxy-3-methoxyphenyl)-2-propenoic acid (ferulic acid). *Acta Crystallographica Section C*, 44, 275–277.

Nishizawa, C., Ohta, T., Egashira, Y., & Sanada, H. (1998). Ferulic acid contents in typical cereals (in Japanese with abstract in English). *Nippon Shokuhin Kaga*, 45, 499–503.

Ou, S., & Kwok, K. C. (2004). Ferulic acid: Pharmaceutical functions, preparation and applications in foods. *Journal of Science, Food and Agriculture*, 84, 1261–1269.

Pan. R. L., Chen, D. H., Shen, L. G., et al. (2000). Determination of ferulic acid and isoferulic acid in rhizoma cimicifugae by HPLC. *Chinese Journal of Pharmaceutical Analysis*, 47, 396–398.

Pei, K., Ou, J., Huang, J., & Ou, S. (2016). p-Coumaric acid and its conjugates: Dietary sources, pharmacokinetic properties and biological activities. *Journal of Science, Food and Agriculture*, 96, 2952–2962.

Prates, J. A. M., Tarbouriech, N., Charnock, S. J., Fontes, C. M. G. A., Ferreira, L. M. A., & Davies, G. J. (2001). The structure of the feruloyl esterase module of xylanase 10B from Clostridium thermocellum provides insights into substrate recognition. *Structure*, 9(12), 1183–1190.

Qian, S., Gao, S., Li, J., Liu, S., Diao, E., Chang, W., Liang, X., Xie, P., & Jin, C. (2022). Effects of combined enzymatic hydrolysis and fed-batch operation on efficient improvement of ferulic acid and p-coumaric acid production from pretreated corn straws. *Bioresource Technology*, 366, 128176. https://doi.org/10.1016/j.biortech.2022.128176.

Qian, S., Guan, T., Li, L., Diao, E., Fan, J., Chen, S., & Liang, X. (2023). Efficient production of ferulic acid and p-coumaric acid from reed straws via combined enzymatic hydrolysis and hydrothermal pretreatment. *Food and Bioproducts Processing*, 140, 122–131

Rosazza, J. P. N., Huang, Z., Dostal, L., Volm, T., & Rousseau, B. (1995). Biocatalytic transformations of ferulic acid: An abundant aromatic natural product. *Journal of Industrial Microbiology*, 15, 457–471.

Rukkumani, R., Aruna, K., Suresh, V. P., & Padmanabhan, M. V. (2004). Hepatoprotective role of ferulic acid: A dose-dependent study. *Journal of Medicinal Food*, 7, 456–461.

Sakakibara, H., Honda, Y., Nakagawa, S., Ashida, H., & Kanazawa, K. (2003). Simultaneous determination of all polyphenols in vegetables, fruits, and teas. *Journal of Agricultural and Food Chemistry*, 51, 571–581.

Saulnier, L., Marot, C., Chanliaud, E., & Thibault, J. F. (1995). Cell-wall polysaccharide interactions in maize bran. *Carbohydrate Polymers*, 26, 279–287.

Saulnier, L., Marot, C., Elgorriaga, M., Bonnin, E., & Thibault, J. F. (2001). Thermal and enzymatic treatments for the release of free ferulic acid from maize bran. *Carbohydrate Polymers*, 45, 269–275.

Shibuya, N. (1984). Phenolic-acids and their carbohydrate esters in rice endosperm cell-walls. *Phytochemistry*, 23, 2233–2237.

Sibhatu, H. K., Jabasingh, S. A., & Yimam, A. (2021). Ferulic acid production from brewery spent grains, an agro-industrial waste. *LWT - Food Science and Technology*, 135, 110009.

Strack, D. (1990). Metabolism of hydroxycinnamic acid conjugates. *Bulletin Liaison Groupe Polyphenols*, 15, 55–64.

Tilay, A., Bule, M., Kishenkumar, J., & Annapure, U. (2008). Preparation of ferulic acid from agricultural wastes: Its improved extraction and purification. *Journal of Agricultural and Food Chemistry*, 56(17), 7644–7648.

Truong, H. T., Do Van, M., Duc Huynh, L., Thi Nguyen, L., Do Tuan, A., Le Xuan Thanh, T., Duong Phuoc, H., Takenaka, N., Imamura, K., & Maeda, Y. (2017). A method for ferulic acid production from rice bran oil soapstock using a homogenous system. *Applied Sciences*, 7(8), 796.

Wang, Z. H. (2012). Determination of ferulic acid in Angelica and research of quality standards. *Hebei Journal of Traditional Chinese Medicine*, 34, 1058–1063.

Williamson, G., & Vallejo, J. (1997). Chemical and thermal stability of ferulic acid esterase-III from *Aspergillus niger*. *International Journal of Biological Macromolecules*, 21, 163–167.

Xiros, C., Moukouli, M., Topakas, E., & Christakopoulos, P. (2009). ferulic acidctors affecting ferulic acid release from Brewer's spent grain by Fusarium oxysporum enzymatic system. *Bioresource Technology*, 100(23), 5917–5921.

Zhang, X. N., & Xu, P. Y. (1997). An improved technology to extract ϒ-oryzanol from rice bran oil. *Cereal Feeding Industry*. 9, 36–37.

Zhang, X. L. (2009). Determination of ferulic acid in rhizoma chuanxiong by HPLC. *Chinese Pharmaceutical Afferulic acidirs*, 23, 469–471.

Zhang, H. B., Gao, Y., Liang, Q. L., et al. (2011). HPLC determination of free and total ferulic acid in sparganium stoloniferum. *Journal of Nanjing University of Traditional Chinese Medicine*, 27, 169–171.

Zhang, J., Tang, H., Yu, X., Xue, D., Li, M., Xing, X., Chen, H., Chen, J., Wang, C., & Gong, C. (2024). Co-production of ferulic acid and p-coumaric acid from distiller grain by a putative feruloyl esterases discovered in metagenome assembled genomes. *Journal of Cleaner Production*, 439, 140814

Zhao, Z., & Moghadasian, M. H. (2008). Chemistry, natural sources, dietary intake and pharmacokinetic properties of ferulic acid: A review. *Food Chemistry*, 109(4), 691–702.

Zhou, Z., Zhang, X., Wu, J., Li, X., Li, W., Sun, X., Wang, J., Yan, Y., Shen, X., & Yuan, Q. (2022). Targeting cofactors regeneration in methylation and hydroxylation for high level production of Ferulic acid. *Metabolic Engineering*, 73, 247–255. https://doi.org/10.1016/j.ymben.2022.08.007

20 Agrobiomass Valorization through a Sustainability Prism

Neeraj Hanumante and Neeta Maitre

20.1 INTRODUCTION

Agriculture plays a pivotal role in civilizational growth and stability. It contributes to about 20.2% of the Gross Value Added (GVA) of India for the year 2020–2021. Globally, agriculture consumes more than 90% of the total freshwater (Hoeskstra et al., 2012). Studies have shown that water needed for agricultural production would lead to increased water stress (Pastor et al., 2019). Like other industrial goods, agricultural produce also needs transportation (carbon footprint of transportation reference). Production processes of fertilizers and pesticides, used in agriculture, are energy intensive (Jaiswal & Agrawal, 2020). Thus, present-day agriculture has a non-trivial carbon footprint. Moreover, there are several other impacts of excessive use of fertilizers such as water pollution, eutrophication, loss of biodiversity, and productivity reduction. Considering all these factors, it is prudent to maximize the utility of agricultural production. Agrobiomass valorization is the process of retaining the value of residual biomass from the agricultural processes which otherwise would have been disposed of. In this chapter, we will look at the process of agrobiomass valorization from a sustainability perspective. First, we look at different ways to classify agrobiomass valorization processes based on the products and the feed. Then, we will describe concepts associated with sustainability and quantification routes. Lastly, we discuss how agrobiomass valorization influences sustainability and how we can quantify it.

Sustainability can be defined as the ability to maintain or support a process over time. The attitude that encourages "satisfying today's needs without compromising to future generations". There are three pillars of sustainability: Social, Economical, and Environmental. Agriculture being an important part of the ecosystem contributes to all these pillars. This contribution requires articulation of the process that can gear up the valorization of agricultural by-products. Figure 20.1 depicts valorization with respect to sustainability. This figure shows the Venn diagram, which mentions two key terms "circular economy" and "circular bioeconomy".

Conventional economic model typically consists of four steps. Extraction of material from nature, for example, metals, wood, and non-metallic minerals. Making products from these materials, for example, a car, a chair, and plastic carry bags. These products

DOI: 10.1201/9781003407713-20

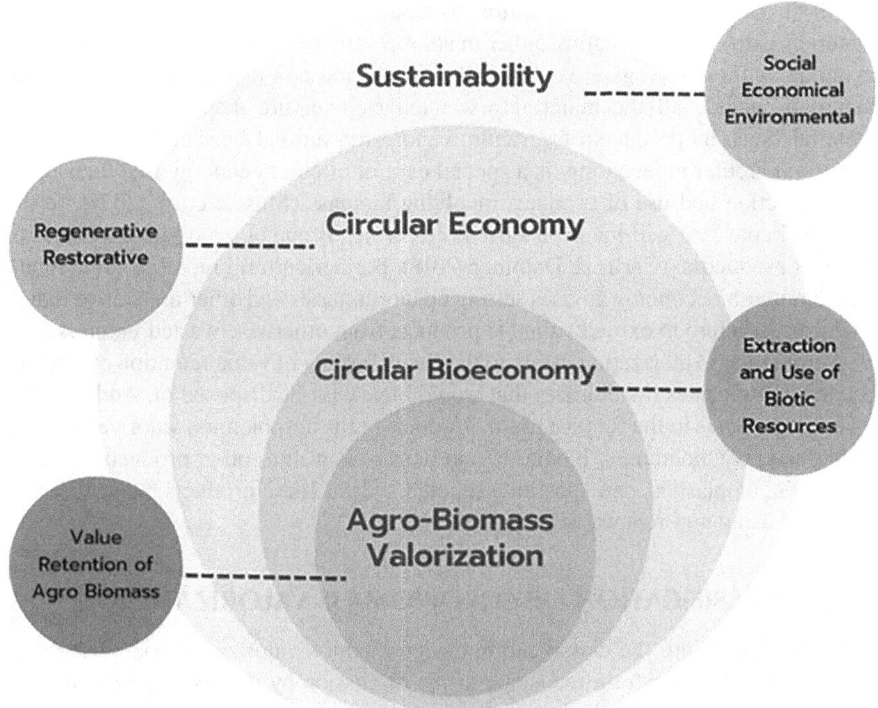

FIGURE 20.1 A Bird's-eye view for agrobiomass valorization.

are used by the consumers. At the end of their lifetime, the products are removed from the economy. Typically, the discarded products are disposed of at the landfill.

In contrast to the extract-make-use-dispose linear economy, the circular economy is regenerative and restorative in nature (Ellen MacArthur Foundation, 2012). It relies on three Rs of circular economy: Reduce, Reuse, and Recycle (Kirchherr et al., 2017; Rizos, 2017; Manickam & Duraisamy, 2019). The first two Rs advice reduction of the demand for the product itself. Consider examples of plastic carry bags. Reduce principle of circular economy would suggest the usage of cotton carry bags instead of plastics ones. Here, the overall demand for plastic carry bags is reduced because of substitution by cotton carry bags. Reuse principle, on the other hand, suggests reusing the plastic carry bags upon acquisition. Thus, the reuse principle does not eliminate the demand for plastic carry bags completely. However, because the functional life of existing bags is extended, future demand for the same would be reduced. The last principle is Recycle. At the end of life of the plastic bags, they can be collected and recycled to extract the plastic resins. Thus, the need for the production of new resins can be avoided.

To summarize, the Reduce principle suggests present and future demand elimination through product substitution, Reuse leads to reduction in the future demand for the products, and lastly, Recycle reduces virgin raw material used to produce the products.

All the products that bring comfort to modern lifestyle can trace their origin to materials extracted from nature either in abiotic or biotic form. Abiotic form refers to materials without biological contents like metals and non-metallic minerals. Biotic materials include all the materials extracted from nature that contain biological materials such as products of agriculture, forestry, animal husbandry, and fishing industry. Circular bioeconomy is a special case of circular economy that focuses on the production and use of products involving biomass (Muscat et al., 2021). In the present chapter, we will focus on agriculture, which is one of the dominant routes of biomass production (Carus & Dammer, 2018). For agricultural biomass value retention, circular bioeconomy advises setting up biorefineries and other innovative industrial infrastructure to extract valuable products from otherwise wasted biomass.

Agrobiomass valorization caters to the Recycle route of value retention of circular economy. It recycles the biomass that would otherwise be disposed of. Additionally, it also contributes to the Reduce route. Products of the agrobiomass valorization such as biofuels like bioethanol, biodiesel, and biogas, as well as other products like biochemicals, bioplastics, can substitute the demand for these products conventionally produced using non-renewable resources.

20.2 CLASSIFICATION OF AGROBIOMASS VALORIZATION

Before we jump into the classification of agrobiomass valorization routes and processes, we would constrain the scope of the discussion by understanding how agriculture functions and at which points there is an opportunity to extract value from residual biomass. This relationship between agricultural products and agrobiomass valorization is depicted in Figure 20.2.

Agricultural plants can typically be classified as annuals and perennials. Annuals are the plants that complete their lifecycle during the season, that is, the sowing, growth, and harvesting are done in a single season. Consider the example of rice. The conventional usage and lifecycle of the rice plant ends with harvesting. It cannot produce anymore rice once the existing crop is harvested. On the other hand, consider the example of apples, which are perennial. After harvesting, the lifecycle of an apple tree would continue, and it can produce apples in the next season as well.

Significant quantities of resources are utilized in growth (like water and fertilizers) and protection (like pesticides) of agricultural plants. When annual crops like rice, wheat, or cotton are harvested, the remaining plant cannot continue production and needs to be disposed of to prepare for the next season. Thus, a significant quantity of biomass could be discarded. This residual biomass is produced using the resources as mentioned earlier. It is prudent to extract the maximum value from this residual biomass. However, for perennial plants like apple and guava, the scope of biomass valorization would be limited at the point of harvesting as the lifecycle of the plant has not yet been completed. However, when these products are processed in the factories to extract pulps and juices, the residual biomass is discarded. There is a significant opportunity of biomass valorization at this juncture.

To summarize, for the purpose of the current discussion, we consider agrobiomass production to include two components. First, the agricultural residue, which is waste biomass following the harvest of the annual crops. After harvesting, the crops, fruits,

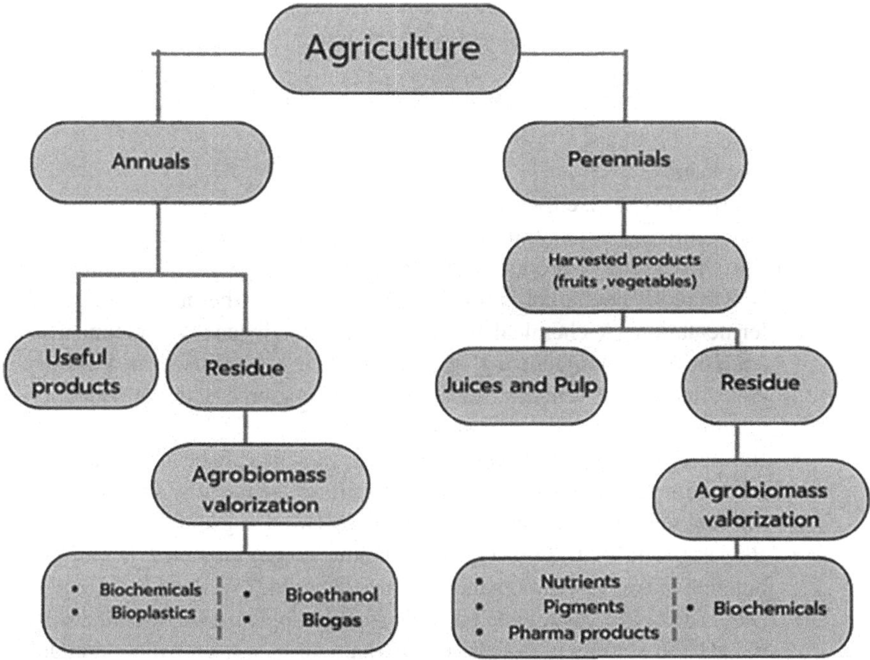

FIGURE 20.2 Agricultural crops and agrobiomass valorization.

and vegetables are processed in agro-processing industries. These processes also produce waste biomass. This is the second component of the agrobiomass used as feed. We can use the type of feed taken in to classify agrobiomass valorization routes and processes as follows:

- Post-harvest residual biomass (crops)
- Post-processing residual biomass (fruits and vegetables)

Taghizadeh-Alisaraei et al. (2022) have summarized the biomass classes used as feed for fuel production. In addition to the post-harvest residues, they also include animal waste. However, here we do not include animal waste as a feed as the current scope is limited to agriculture and does not include animal husbandry.

Like any other manufacturing process, agrobiomass valorization needs an input, that is a feed, and produces an output, that is a product. We can also classify the routes of biomass valorization based on what products are obtained at the end. One of the important and significant products of agrobiomass valorization is biofuel, either in the form of bioethanol, biodiesel, or biogas. Additionally, certain biochemicals and bioplastics can also be produced. Lastly, nutrients, pigments, and pharma products can be obtained from the post-processing residual biomass. Thus, based on the products produced, we can classify agrobiomass valorization routes and processes as follows:

- Energy products:
  - Bioethanol
  - Biodiesel
  - Biogas
- Non-energy products:
  - Bioplastics
  - Pharma products: enzymes and nutrients
  - Biochemicals: dyes, pigments, and specialty chemicals
- Let's go through these products one by one:
  - Bioethanol: Bioethanol is the fuel produced from the process of sugar fermentation. A chemical reaction between ethylene and steam also leads to the manufacturing of bioethanol. It can be used as a substitute of petrol for vehicles. It is a renewable resource with the benefit of greenhouse gas (GHG) emission (Murali et al., 2022).
  - Biodiesel: Biodiesel is a biodegradable fuel. It can be manufactured from animal fat, vegetable oil, and restaurant grease. It is a clean-burning fuel that produces less toxic pollutants (Ma Fangrui et al., 1999).
  - Biogas: On decomposition of organic matter, biogas is formed. It is a combination of two gases, methane and carbon dioxide (Weiland Peter, 2010).
  - Bioplastics: Bioplastics can be made from biological materials like softwoods and starchy materials. Plants that can be used for the manufacture of bioplastics are potato, corn, wheat, and sugarcane.
  - Biochemicals: Biochemicals include dyes, pigments, and specialty chemicals. They can conventionally be considered as the chemical processes occurring in living beings.

The division depicted in Figure 20.3 clearly shows the useful or directly consumable products. These products are considered as "main products". For example, wheat, rice, and apples can be considered as the main products of agriculture. These main

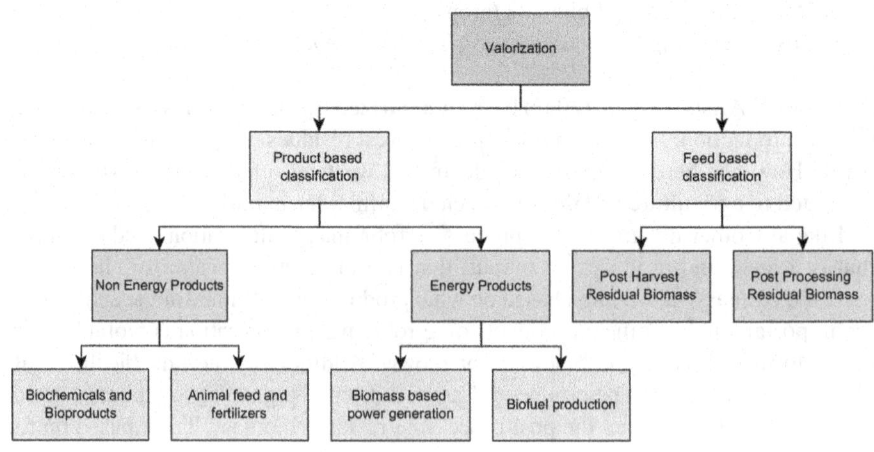

FIGURE 20.3 Classification of agrobiomass valorization.

products can be classified as crops and fruits. The crops can be processed in order to get the consumable product and the fruits can be consumed in the form of juice or pulp. The noticeable point here is that the fruits result in the by-product in the form of residue after the generation of juice or pulp. The by-products can be considered as the discarded materials of agriculture. For example, organic materials are produced from agronomic activities.

The classification highlights the term "Agrobiomass valorization" in both streams of the products generated from agriculture. This leads to the act of valorization of the by-products generated; "Valorization" is the process of stating or thinking that something is valuable. This thought is important to understand the need of agrobiomass valorization to satisfy the needs of today and tomorrow.

20.3 THE "SUSTAINABILITY" PRISM

The significance of agrobiomass valorization is more when it is visualized through the sustainability prism.

The World Commission on Environment and Development introduced the concept of sustainable development (World Commission on Environment and Development, 1987) as development that meets the needs of the present generation without compromising the ability of future generations to meet their own needs. This definition focuses on intergenerational equity, with an implicit assumption that the present generation of humanity is carrying out development in the best possible way. However, intragenerational equity of resources is equally important. In the context of climate change, the intragenerational and intergenerational equity were beautifully woven together in the form of Common But Differentiated Responsibilities, which were adopted under the aegis of the United Nations Framework Convention on Climate Change (UNFCCC) in the Rio Declaration of 1992. Later on, following multiple deliberations and scientific studies, sustainable development goals were adopted by the countries in the world under Agenda 2030. These are 17 distinct but interrelated goals with actionable objectives. Progress toward sustainable development can be tracked using these goals.

Sustainable development is supported by three pillars: economic development, social progress, and environmental protection. It may be noted that these three pillars correspond to different disciplines of scientific studies. Thus, it is an interdisciplinary concept. Sustainability is a concept that is used interchangeably vis-a-vis sustainable development. However, it has also been used to explicitly refer to balancing of progress in any one dimension of sustainability, for example, economic sustainability and environmental sustainability (Sartori et al., 2014). Sustainability has developed as a science in itself (Prugh & Assadourian, 2003; Kuhlman & Farrington, 2010). Agrobiomass valorization is thus the process of valuing agricultural by-products in order to provide sustainable development to the future generation.

Here, we are examining the sustainability of a process. Sustainability can be considered as a local phenomenon that can be applied at the global level through SDGs (Sustainable Development Goals). We examine how the system changes in the context of sustainability when agrobiomass valorization is introduced. Again, refer to the sustainability prism shown in Figure 20.4. "Sustainability" is a prism that scatters several hidden colors of agrobiomass valorization in the form of various benefits. These benefits can be classified into three categories:

- Social
- Environmental
- Economical
- Let's see these three categories, along with the SDGs to which the category contributes:
  1. **Social:** The agrobiomass valorization helps to generate new opportunities for farmers as well as local people. It will open up with the new market and industry which will be able to contribute to the transport and infrastructure field of businesses. Integration of agrobiomass valorization with the refineries will give rise to the "biorefineries". Biorefineries provide facilities to convert the biomass into value-added products majorly in the form of biofuels and bioenergy. This will ultimately contribute to SDG-9, which is basically related to industry, innovation, and infrastructure. Improvement in this SDG will surely help to improve our social well-being.
  2. **Environmental:** The agrobiomass valorization will help in reducing the carbon footprint. The already grown plants that otherwise would have been disposed of can be utilized in a regenerative manner. This will help in a positive impact on climate change and thus will contribute to SDG-13.
  3. **Economical:** This perspective has a deep-rooted impact as it directly contributes to economic growth. Agrobiomass valorization majorly can be seen as a "waste to wealth" solution through sustainability, and it will be added to SDGs 7 and 8 of sustainable development. SDG-7

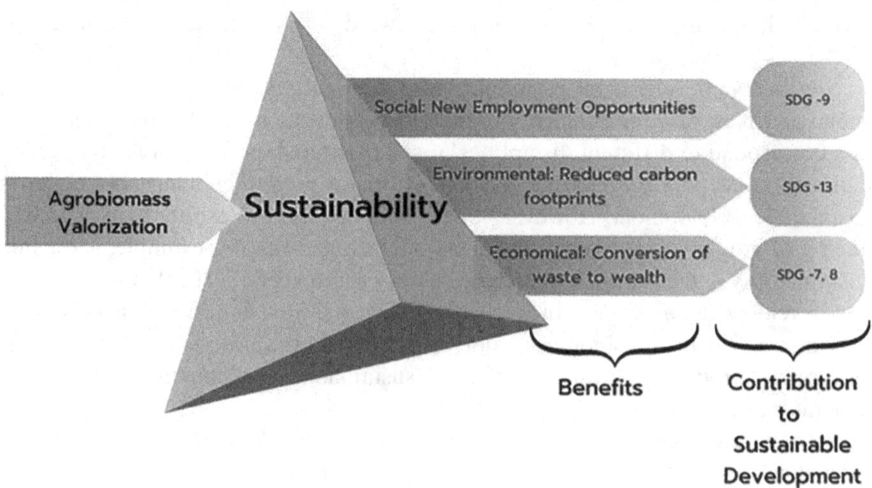

FIGURE 20.4 The "sustainability" prism.

specifically mentions affordable and clean energy, while SDG-8 refers to economic growth. The waste out material from agriculture, if utilized in generating biofuels, is an economic booster and provides the path for the economic growth.

Thus "sustainability" prism gives you a way to look at agrobiomass from these three different perspectives which will depend upon how you turn the prism for one's good.

Till now, we have seen sustainability as a more or less abstract concept. However, there are tools to quantify the sustainability of a system (Singh et al., 2012). Indicators and indices are generally used to disburse sustainability-related information to masses. Indicators refer to a single piece of information regarding a physical quantity, whereas an index is a combination of multiple indicators. For example, in the context of environmental sustainability, carbon footprint indicates the quantity of GHG emitted during the production of a product, in an activity or a process, by an organization, industry, or region. On the other hand, the Environmental Performance Index (EPI) (Hsu & Zomer, 2016) incorporates multiple indicators. Examples include carbon footprint, air pollution (e.g., particulates emission), water (e.g., water access), and agriculture (e.g., pesticide use).

20.4 CONCLUSION

Agrobiomass valorization plays a key role in biotic resource management and use. Here, we refer to sustainability as a concept applied to systems of different disciplines (environmental, economic, social) and scales (products and processes, industries, nations, global) with sustainable development as a special case of global scale involving all three dimensions.

Though, at present, it mainly produces biofuels from residues, other products like biochemicals and bioplastics are gaining traction. Contributions of agrobiomass valorization to sustainable development are significant and are expected to grow. Specifically, the fuel production from agrobiomass valorization serves as a means for sustaining the ever-increasing global demand for energy using renewable resources. To conclude, agrobiomass valorization is an important driving force to accelerate sustainable development.

REFERENCES

Carus, M., and L. Dammer. "The circular bioeconomy-concepts, opportunities, and limitations." *Industrial Biotechnology* 14, no. 2 (2018): 83–91.

Ellen MacArthur Foundation. *Towards the Circular Economy*. Ellen MacArthur Foundation, Cowes, England, 2012.

Hsu, A., and A. Zomer. "Environmental performance index." Wiley StatsRef: Statistics Reference Online (2016): 1–5. https://doi.org/10.1002/9781118445112.stat03789.pub2

Jaiswal, B., and M. Agrawal. "Carbon footprints of the agriculture sector." In *Carbon Footprints: Environmental Footprints and Eco-design of Products and Processes*, pp. 81–99. Springer, Singapore, 2020. https://doi.org/10.1007/978-981-13-7916-1_4

Kirchherr, J., D. Reike, and M. Hekkert. "Conceptualizing the circular economy: An analysis of 114 definitions." *Resources, Conservation and Recycling* 127 (2017): 221–232.

Ma, F., and M. A. Hanna. "Biodiesel production: a review." *Bioresource Technology* 70, no. 1 (1999): 1–15.

Manickam, P., and G. Duraisamy. "3Rs and circular economy." In: Muthu, S. S. (ed), *Circular Economy in Textiles and Apparel*, pp. 77–93. Woodhead Publishing, Cambridge, 2019. https://doi.org/10.1016/B978-0-08-102630-4.00004-2

Murali, G., and Y. Shastri. "Life-cycle assessment-based comparison of different lignocellulosic ethanol production routes." *Biofuels* 13, no. 2 (2022): 237–247.

Muscat, A., E. M. de Olde, R. Ripoll-Bosch, H. H. E. Van Zanten, T. A. P. Metze, C. J. A. M. Termeer, M. K. van Ittersum, and I. J. M. de Boer. "Principles, drivers and opportunities of a circular bioeconomy." *Nature Food* 2, no. 8 (2021): 561–566.

Pastor, A. V., A. Palazzo, P. Havlik, H. Biemans, Y. Wada, M. Obersteiner, P. Kabat, and F. Ludwig. "The global nexus of food-trade-water sustaining environmental flows by 2050." *Nature Sustainability* 2, no. 6 (2019): 499–507.

Singh, R. K., H. R. Murty, S. K. Gupta, and A. K. Dikshit. "An overview of sustainability assessment methodologies." *Ecological Indicators* 15, no. 1 (2012): 281–299.

Taghizadeh-Alisaraei, A., A. Tatari, M. Khanali, and M. Keshavarzi. "Potential of biofuels production from wheat straw biomass, current achievements and perspectives: a review." *Biofuels* 14, no. 1 (2022): 1–14.

Weiland, P. "Biogas production: current state and perspectives." *Applied Microbiology and Biotechnology* 85 (2010): 849–860.

World Commission on Environment and Development. Our Common Future. Oxford University Press, Oxford, 1987.

Index

ABE 25, 66, 97, 114, 118, 192, 196, 234, 295
absence of oxygen 32, 36, 113, 391, 396, 416
absorption 272, 275, 280, 361, 373, 440
abundance 3, 62, 125, 140, 156, 181, 347, 348, 388, 393
accessibility 2, 13, 74, 75, 77, 78, 81–82, 137, 151, 166–167, 180–183, 238, 330, 332, 347, 349, 363
accessible surface area 180
acetaldehyde oxidation 379
acetate 147, 203–204, 213, 227, 236, 249–250, 252, 288, 356, 362, 377–379, 386, 408, 432
acetic acid 59, 120, 149, 168, 170, 180–181, 202–205, 207, 209, 213, 217, 218, 221, 225–226, 231, 234, 236, 241, 245, 246–250, 254, 255, 256–257, 259–260, 302, 345, 362, 367, 369, 375, 377, 378–387
Acetobacter 203, 205, 209, 217, 221, 236, 247, 249, 260
acetogenesis 407–408
acetoin 141, 144, 159, 208, 218
acetone 25, 42, 49, 63, 66, 97, 118, 169, 192, 194, 196, 221, 234, 295, 341
acetylxylan esterase 131, 435
acid
  extraction 432, 435, 437
  hydrolysis 150, 164, 168, 217, 331, 348, 349
  leaching 392
  pretreatment 22, 30, 41, 76, 77, 83, 87, 88, 91, 137, 343, 360, 432
  regulators 302
  transport 252
acidic 4, 30, 38, 87, 129, 137, 144, 168, 210, 212, 240, 259, 354, 377, 384, 420, 431, 436
acidification 338, 426
acidity 202, 204, 245, 252–253, 412
acidogenesis 23, 376, 407, 408
acidogenic fermentation 377
acidulant 213, 245
Acinetobacter 51
acrylic acid 284
Actinobacteria 51
Actinomycosis 51
activated
  biochar 412, 426
  charcoal adsorption 360
adaptability 13, 205, 209, 238, 261, 286, 316
adaptive evolution 219
additive 25, 59, 141, 206, 213, 249, 308, 309, 344, 348, 407, 418, 422, 423
adsorbent 27, 46, 411
adsorption 40, 42, 44, 46, 48, 49, 63, 68, 134, 137, 172, 180, 181, 189, 338, 354, 355, 359, 360, 363, 365, 380, 398, 402, 412, 421, 423–424, 426
  efficiency 398
  rate 421
advanced
  biofuels 116, 341, 349, 402
  fermentation 65, 340
aeration 84, 110, 221, 225, 239, 312, 314
aerobic 1, 51, 62, 81, 110, 131, 174, 176, 179, 208, 211, 221, 260, 282, 289, 379, 419
agglomeration 3, 4, 403
aggregate stability 407
agitation 78, 110, 153, 182, 191, 225, 324, 344, 435
agrarian biomass 276, 283
agricultural
  biomass 2, 4, 6, 8, 10, 12, 14, 16, 18, 20, 22, 26, 28, 30, 32, 34, 36, 38, 40, 42, 46, 48, 50, 52, 54, 56, 58, 60, 62, 64, 66, 68, 70, 74, 76, 78, 80, 82, 84, 86, 88, 90, 94, 96, 98, 100, 102, 104, 106, 108, 110, 112, 114, 116, 118, 120, 124, 126, 128, 130, 132, 134, 136, 138, 142, 144, 146, 148, 150, 152, 154, 156, 158, 160, 164, 166, 168, 170, 172, 174, 176, 178, 180, 182, 184, 186, 188, 190, 192, 194, 196, 198, 200, 204, 206, 208, 210, 212, 214, 216, 218, 220, 222, 224, 226, 228, 230, 232, 234, 236, 237, 238, 240, 242, 244, 246, 248, 250, 252, 254, 256, 258, 260, 262, 264, 266, 268, 270, 272, 274, 276, 278, 280, 282, 283, 284, 286, 288, 290, 292, 294, 296, 298, 302, 304–306, 308, 310, 312, 314, 316, 318, 320, 322–345, 348, 350, 352, 354, 356, 358, 360, 362, 364, 366, 368, 370, 372, 374, 378, 380, 382, 384, 386, 390, 392, 394, 396, 398, 400, 402–404, 408, 410, 412, 414, 416, 418, 420, 422, 424, 426, 427, 428, 430, 432, 434–436, 438–440, 444, 446, 448, 450
  biostimulant 141, 158
  coproducts 242
  feedstocks 316
  residue 381, 444

agricultural (*cont.*)
  waste 4, 13, 54, 58, 65, 68, 72, 78, 84, 86, 93, 96, 97, 138, 169, 200, 203, 207, 211, 237, 239, 242, 249, 255, 259, 262, 268, 271, 275, 278, 279, 289, 317, 418, 424, 436, 438
agrobiomass 44–45, 50–62, 72–75, 82–84, 93–94, 96–99, 102, 109, 113, 165–170, 179, 180, 182, 188, 190–191, 203, 211–213, 220–221, 225, 237–240, 244, 254, 300–303, 305, 309, 311, 323–325, 328, 330–333, 339–340, 346–347, 370, 388, 427–429, 431, 433, 435, 437, 438–439, 441–449
air pollution 323, 411, 449
alcalase 434
alcohol 19, 25–28, 36, 38, 49, 97, 101–102, 112, 114, 118, 132, 133–134, 159, 163–164, 169, 189–190, 193, 196, 205, 217–218, 247–248, 259, 313, 324–325, 329, 332, 337, 342–343, 347–348, 354, 356, 362, 364, 373, 379, 414, 438
aldehyde 102, 114, 118, 193, 196, 247
algae 6, 18, 83, 93, 163, 167, 173, 278, 325, 345, 367, 401, 423, 424
alkali 3, 20, 23, 29, 41, 52, 54, 70, 76, 82, 147, 149, 151, 152, 169, 179, 318, 331, 342, 357, 395, 430, 434
alkaline
  hydrolysis 169, 432, 438
  pretreatment 45, 76, 87, 89, 117, 138, 193, 331
allelopathic 108
α-1,4 linkages 83
alternative
  energy 40, 113, 116, 123, 171, 364
  fuels 107
Alzheimer's 293, 427
Amazon basin 412
amensalism 109
amino acids 14, 20, 60, 99, 153, 184–185, 202, 204, 210, 227, 278
ammonia 30, 76, 138, 151, 167, 219, 330, 331, 343, 349, 361, 372, 420, 436
ammonia fiber expansion (AFEX) 30, 331, 343
ammonium sulfate 49, 64, 83, 287
amorphous 15–17, 19, 38, 65, 76, 81, 165, 170, 185, 186, 311, 420
  cellulose 76, 185–186
  hydrophobic substance 311
amylase 82–86, 88, 148, 150, 214, 234, 239, 408
amylopectin 83, 150
amylose 83, 150
anaerobic 1, 23, 34, 36, 51, 52, 59, 62, 63, 69, 85, 101, 107, 123, 131, 137, 166, 176, 200, 210, 215–216, 218–219, 221, 235, 240, 267, 270, 282, 288, 307, 310, 318, 328, 355–356, 376, 379, 383–385, 389, 406–408, 419, 421–426
  digestion 1, 36, 59, 63, 69, 85, 288, 307, 318, 376, 384, 389, 407, 421–426
  fermentation technology 310
analgesics 250
analysis 3, 21, 38–39, 50, 69–70, 105–107, 112, 116–118, 120, 132, 134–135, 137, 138–139, 141, 158, 183, 187, 193, 195, 207, 219, 225, 229, 230, 234, 236, 238, 278, 287, 295, 298, 313, 342, 344, 364, 367, 372, 373, 375, 384, 385, 386, 390, 391, 392, 398, 402, 423, 432, 437–438, 440, 449
anhydrous ethanol 123, 125
animal 1, 25, 69, 93, 152, 206, 211–212, 218, 237, 239, 241, 278, 300, 315, 347, 348, 352, 388, 407, 408, 411, 415, 444–446
  feed 93, 218, 239, 348, 408
  husbandry 206, 444–445
  manure 300
  residues 388
  waste 93, 237, 407, 411, 445
anion exchange 46, 47, 362
anthropogenic 122, 412; soil 412
antibiotics 197, 206, 207, 212, 250, 308
anticancer 303, 308, 320
anticoagulants 263
antifreeze 140–142
antifungal 236, 247, 255, 321
antimalarial 303
antimicrobial 193, 231, 258, 303, 308, 372, 427
antioxidant 90, 142, 166, 205, 245, 247, 301, 303, 307–309, 308, 310, 314, 319–320, 322, 372–374, 427–440
apple 78, 160, 231, 245, 248, 249, 260, 302, 311, 372, 377, 444
  pomace 245, 248, 249, 260, 302, 311, 377
APR 26, 35, 270
aquatic 6, 93, 167, 176, 278, 347, 401
aqueous 26, 29, 41–43, 64, 112, 114, 120, 159, 169, 285, 336, 356, 373, 381
arabinofuranosidases 174–175, 312, 329
arabinose 17–18, 50, 55, 74, 98, 132, 149, 151, 159, 165, 168, 174, 179, 205, 228, 330, 347, 432
AraR 54
archaea 22, 51, 62, 328, 343
archaeon 112, 115
arginine 208
aroma 123, 209, 303
aromatic 31, 33, 393, 397, 398
  branched skeleton 311
  compounds 35, 151, 312
  hydroxycinnamyl alcohol 19
  polymer 19
aromatization 416, 417
arthritis 427
Arthrobacter soli Em07 435
artificial pathway 437

Ascomycota 52
ascorbic acid 19, 22, 303, 308, 310, 314, 317, 319
ash 3, 4, 15, 22–23, 166, 353, 391, 420
aspergillosis 212
Aspergillus 52, 55, 66, 68, 81, 83–86, 88, 115, 131–132, 177, 178, 203, 205, 211–212, 221, 228, 230, 234, 239–241, 245, 251–252, 255–260, 267, 302–303, 307, 309, 312, 314, 317, 319, 322, 329, 333, 433, 439, 441
Aspergillus niger 55, 81, 84, 206, 212, 218, 228, 242, 244, 247
assays 50–51, 183, 187, 208
atmospheric 4, 37, 86, 262, 265, 362, 405
ATPase 219
Aureobasidium melanogenum 58
AXE esterase 435
azeotrope 356, 362
azeotropic 35–357, 134, 189, 327, 371
  distillation 327, 355–356, 371

Bacillus 51, 52, 54, 64, 68, 71, 82, 119, 121, 131, 132, 144, 156–159, 161, 177–178, 233, 236, 239, 241, 245, 247, 252, 304, 309, 312, 314, 318, 341, 433
bacteria 18, 22, 36, 38, 45, 50–52, 56, 59–60, 62, 64, 67, 73, 77, 81–82, 94, 97–98, 109, 113, 119, 129, 131, 142, 145, 149, 151, 158, 167, 168, 173–176, 179, 194, 202–207, 209–211, 213–216, 218–219, 221, 225–226, 228, 230, 231–232, 239–247, 249, 252, 258–259, 267, 289, 302, 304, 309, 313, 319–321, 325, 328–330, 341, 352, 354, 356, 379, 385, 408, 419, 433, 436
  cellulose 70, 209, 288, 309
  monocultures 309
  pigments 303
  strains 215, 303–304, 309, 310, 312, 314
bagasse 33, 40, 43–44, 61, 75, 82, 84, 85, 87, 88, 89, 94, 115–117, 120, 128, 151, 156, 159–160, 166, 203, 213, 231, 236, 238, 244–245, 256, 262, 301–302, 304, 309, 311–312, 314, 318, 321, 323, 324, 325, 341, 344, 377, 402, 423, 431, 435
bamboo
  biomass 149
  shoots 322, 428, 429
banana peel 302, 319, 412
barley malting process 432, 439
barley straw 16, 21, 87, 120, 311
basal medium 380
Basidiomycetes 68, 86, 240, 312
Basidiomycota 52
batch
  enzymatic hydrolysis 434
  fermentation 150, 153, 231, 260, 322, 354, 364, 365
  operation 231, 440
  process 355, 365
  reactor 367
beetroot 267, 428, 434
benzene toxicity 356
Bescii 55, 64, 66
β-d glucose 83
β-glucosidases 83
beverage industry 203–204, 212–213, 249, 254, 314
Bifidobacterium 177, 178, 307
bifunctional
  catalysts 41
  enzyme 436, 439
bifurcation analysis 229
binders 13
binding domain 185
bioabsorbable 265
bioactive components 303, 307
bioaerosols 411
bioalcohol 92–97, 99–103, 105–107, 109, 111–113, 115, 117, 119, 121, 126, 162–166, 172–179, 182–183, 188–191, 195, 323–329, 331–333, 335, 337–339, 341, 343, 345, 347–349, 351–361, 363, 365, 367, 369–373, 375
bio-based 140, 141, 143, 145, 147, 149, 151, 153, 155, 157, 159, 161
biobutanol 25, 66, 118, 195–197, 200, 234, 295, 323, 336, 342, 346
biocatalyst 160, 248, 297, 370, 373
biochar 13, 14, 21, 34, 123, 391, 392, 398, 400, 402–409, 411–426
biochemical 14, 16–17, 19, 32, 36, 38, 39, 42, 62, 86, 88–89, 108, 127, 129, 130, 137, 187, 190, 201, 208, 209, 229, 238, 239, 244, 257, 267, 299, 302, 319, 336, 342, 344, 347, 369, 376, 384, 405, 422, 430, 437
biocompatible 126, 266, 293
biocomposites 68, 265, 268, 270, 272, 294
bioconversion 23–24, 41, 55–56, 65, 87, 95–96, 98, 107, 110, 117, 157, 166, 176, 195, 199, 200, 231, 240, 249, 255, 257, 258–259, 301, 305, 309, 311, 313, 314, 316, 320, 322, 341, 343, 344, 349, 358, 362, 375
biodegradability 37, 60, 137, 264, 269, 277, 280, 289, 291, 292, 298, 347, 348
biodegradable 1, 28, 59, 133, 213, 217–218, 236, 250, 261–267, 269–277, 279–285, 287–289, 291–299, 408, 446
  fuel 446
  plastics 213, 217, 265, 269, 275, 279, 288, 289, 296
  polymers 59, 261–262, 264–267, 271, 273, 276, 280, 283, 287–289, 291–293, 296–298

biodegradable (*cont.*)
potential 289
rate 288–289
solvent 28
biodelignifying 55
biodiesel 13, 26, 59, 64, 95, 118–120, 137, 146, 152–153, 158, 192, 193, 196–197, 199, 231–232, 295, 343, 362, 367, 369, 371, 389, 402, 405, 444–446, 450
biodiversity 32, 442
bioeconomy 72, 85, 193, 232, 254, 271, 294, 317, 319, 402, 424, 442, 444, 449–450
bioemulsifiers 314
bioenergy 1, 4, 5, 20, 22, 23, 66, 70, 84, 87, 88, 93, 117–118, 138, 158, 210, 232, 233, 236, 254, 256, 270, 295, 317, 339–340, 344, 345, 372–373, 374, 376, 384–385, 388–389, 400, 401, 402, 404, 405, 406, 422, 448
bioenzymes 307
bioethanol 39, 42, 64–65, 67–68, 71–78, 83, 86–90, 96–97, 101, 115–125, 127, 129–139, 150, 159–160, 190, 192–193, 195–200, 249, 256–259, 298, 305, 307, 317, 323, 325, 327, 333, 338, 340–349, 353–355, 360–361, 363–365, 367, 369–375, 388–389, 444–446
biofertilizers 307, 313, 411
biofilm formation 420
biofouling 342
biofuel 2, 13, 16, 18, 21–23, 25–26, 40–41, 50, 53, 59, 63, 70, 72–73, 78, 81, 83, 85, 87–88, 90, 92, 94–95, 107, 111, 113–114, 117, 120, 122, 125–129, 135–139, 164, 192, 195, 197, 231–232, 234–235, 251, 260, 325, 328–329, 332, 335–336, 341, 343, 347–350, 354, 363, 365–366, 372, 374, 386, 402, 404, 414, 421, 423, 445
biogas 5, 34, 36, 45, 59, 64, 70, 76, 78, 137–138, 166, 307, 328, 343, 367, 385, 388–389, 391, 407–408, 411, 421, 423–424, 426, 444–446, 450
biogenic derivatives 388
biological 1, 4, 13, 19–20, 22, 36–38, 52, 67, 69, 73–74, 77, 85–86, 89–90, 95, 117, 119–120, 127, 129, 136, 140–145, 147, 149, 151–152, 156, 159, 165, 167–168, 183, 193–195, 197, 202, 204, 229, 231–232, 235–240, 241, 243–247, 256–257, 261–262, 266–267, 269, 278–279, 283–284, 288, 291, 293, 296–299, 301, 307, 316, 320, 327, 330–331, 339, 342–344, 349, 362–363, 367, 379, 389–390, 401, 403–404, 412, 423, 440–441, 444, 446
fouling 339
methods 38, 52, 89, 168, 183, 278, 389–390
pretreatment 19, 73, 77, 90, 149, 165, 238, 331
process 38, 245–246, 379
products 36, 240
route 140–141, 144–145, 156
synthesis 144, 244, 293
treatment 136, 152, 167, 327
biology 64, 67, 88, 92–93, 105, 107–108, 111–112, 116, 119, 186, 193, 195–196, 198, 205, 226, 229–231, 235, 255, 439
biomass 1–2, 4–6, 8, 10, 12–14, 15–18, 19–46, 48, 50–52, 54–56, 58–66, 68, 70–78, 80–98, 100–102, 104, 106–108, 110, 112, 114, 116–122, 124–132, 134, 136–139, 141–142, 144–152, 154, 156–160, 163–176, 178–182, 184–186, 188–190, 192–201, 203–204, 206–208, 210, 212–214, 216–218, 220–222, 224, 226–228, 230–232, 234–238, 240, 242, 244, 246, 248–252, 254–260, 262, 264–266, 268, 270, 272, 274–276, 278–284, 286, 288–290, 292, 294–296, 298–312, 314–318, 319–324, 325–345, 347–350, 352, 354, 356–364, 366, 368–376, 377–380, 382, 384–386, 388–390, 392–405, 407–408, 410–418, 420–428, 430, 432, 434–436, 438–440, 442, 444–446, 448, 450
composition 146, 402
conversion 21–22, 37, 68, 84, 86, 89–91, 95, 118, 169, 171, 196, 232, 295, 317, 320, 321, 325, 327, 329, 331, 333, 335, 337, 339, 341, 343, 345, 348–349, 389, 392, 400, 403–404, 422, 424
decomposition 416, 417
digestibility 141, 148
energy potential 4
feedstock 38, 89, 227, 348, 357, 396, 411
gathering 324
materials 29, 32, 78, 83, 389, 392–396
molecules 394
pretreatment 30, 77, 139, 147, 157, 198, 385
processing 373, 388, 390
production 59, 227, 284, 444
recalcitrance 171, 198, 201
resources 4, 6, 39, 107, 163, 230, 316, 342, 376
separation 333, 335
solids 325
sources 24, 39, 163–164, 167, 213, 276, 332, 414
type 127, 330, 399
utilization 55, 82, 119, 199, 401
valorization 63, 120, 121, 318, 320, 404, 444, 445

waste 13, 38, 237, 256, 259, 260, 377, 389, 401, 405
biomaterials 24, 166, 276, 294, 296–297, 324
biomedical applications 279, 281, 292
biomethane 36, 197, 389, 422–423, 425
biomethanol 323, 346
biomolecules 49, 173, 292, 303, 336
bio-oil 414
biopigment 303, 310, 312
bioplastic 271, 289, 298
biopolymer 74, 150, 265–266, 268–269, 275–279, 280, 282–283, 289–291, 293–295, 297, 405
biopower 346
bioprocess 68, 70, 87, 121, 138, 145, 158–160, 231, 234, 236, 239, 258, 271, 305, 308–310, 314–316, 318, 328, 332, 341–342
bioproduct 140, 315, 384
biopropanol 346
bioreactor 67, 84, 111–112, 130, 138, 149, 150, 159, 172, 214, 217, 221, 232, 260, 301, 319, 324, 338, 341–344, 358, 365, 373–374, 380–381, 386
biorefinery 21–22, 39–40, 62, 67–68, 78, 84–87, 89–91, 93, 95, 118–120, 135, 137–141, 146–147, 153, 156, 160, 196, 198, 230, 232, 236, 295, 317–324, 343, 363–364, 367, 369, 371, 373–375, 384, 386, 402, 404, 424, 435
bioremediation 63, 68, 70, 108, 210, 257, 313
biosensors 67, 209
biosorption 131
biosurfactant 304, 309, 320
biosynthesis 115, 165, 235, 252, 267, 290, 322, 427, 439
biosynthetic 108, 109, 115, 123, 167, 193, 226, 235, 290–291
biotechnology 21, 22, 63–71, 85–90, 114–120, 138, 139, 160, 186, 192–200, 201, 206, 211, 212, 225, 231, 233, 258, 259, 270, 271, 284, 291, 293–299, 317–322, 341, 343, 344, 371, 373, 374, 403, 449, 450
biotic resource 449
biotransformation 55, 63, 73, 242–243, 248, 251, 254, 260, 302, 309, 318–319, 322, 358
biovalorization 238, 254
Bispora sp. 82, 88
black soil 407
blending 125, 218, 265, 287, 298, 348, 349, 423
blockage 25, 359
boilers 392, 399
bottlenecks 105–107, 109, 164, 179
bovis 52, 207
Brettanomyces anomalus 433
brewery waste 436
brittleness 282
broccoli 428–429
broth 47–49, 62, 64, 133–137, 145, 156, 159, 217, 240, 325, 327, 333, 335–336, 339, 343, 345, 352, 354–355, 359, 364, 372, 374, 377, 383, 387
brown rot 19, 332
buffer exchange 46
building blocks 59, 203, 270
buoyant culture 380
butadiene 28, 141–142
2,3-butanediol 140
butanol 25, 41–42, 59, 66, 70, 84, 92, 94, 95, 97, 99, 107, 113–118, 141, 167, 175, 190, 192–194, 195–196, 221, 234–235, 295, 323, 327–328, 341, 345, 351, 356
butene oxides 142
butyl
acetate 378
butyrate 313
butyric acid 59, 381–382

cadmium 411
caffeic acid 436–437, 440
cake formation 359
calcium 47, 76, 86, 129, 133, 139, 175, 204, 250, 302, 375, 412
carbonate 47, 76
Caldicellulosiruptor 55, 64, 66, 132
calorific value 2, 26, 392
capnophilic 267
caproic acid 233
carbohydrases 281
carbohydrate 6, 19, 21, 25, 27, 29, 32, 33, 35–37, 39–40, 45, 53, 56, 70, 74, 82, 83, 87, 89, 113, 127, 138, 145–146, 157, 159, 161, 169, 194, 198–199, 206, 208, 210, 212, 214, 220, 225, 249, 254, 257–258, 276, 278, 281, 284, 293–294, 296, 303, 307, 314, 327–328, 341–342, 347, 354, 360, 363, 365–366, 377, 393, 426, 439, 441
carbon 1, 3, 4, 6, 13, 14, 15, 21, 23–27, 30–31, 33–37, 39–43, 47, 59, 62, 69, 74, 76–77, 83, 94–95, 97, 99, 101–102, 109, 116, 119, 121–124, 130–131, 133, 135–136, 138, 140–142, 144–147, 150, 153, 166, 169, 176, 195, 202–203, 205, 211–219, 228, 231, 244–248, 254, 256–257, 259, 261–262, 264–265, 267–269, 275, 277, 279, 284–286, 288, 298, 304, 308, 311, 314, 323, 342–343, 346–348, 353, 360, 362, 376, 378–379, 390–392, 394, 396, 398, 401–402, 404, 407, 411–416, 419–426, 433, 442, 446, 448–449

carbon (*cont.*)
cycle 353
dioxide 1, 26, 30, 33, 36, 47, 59, 95, 97, 122–124, 218, 261–262, 267, 275, 298, 346, 362, 413, 419, 446
emissions 26, 288, 348
footprint 94, 264, 269, 279, 323, 347, 442, 448, 449
incineration 4
monoxide (CO) 122, 416
sequestration 407, 412, 414
source 83, 97, 99, 109, 130, 141, 145–147, 150, 153, 176, 203, 205, 211, 217, 228, 231, 244, 248, 286, 304, 311, 314
substrates 360
carbonaceous 26, 33, 415
carbonated 245, 254
carbonization processes 415
carbonylation 246–248, 378
carbonyl catalyst 379
carboxylate 206, 376, 384
carboxylesterase 433, 435–436, 439
carboxylic 28, 29, 30, 202, 204–205, 213, 219, 228, 241–242, 245, 247, 252–253, 257, 284, 362, 377, 380–383, 386, 387
carbs 214
cardiovascular diseases 308
carminic acid 313
carotenoid production 309
carotenoids 303, 307–308, 310, 313, 314, 321
carrier gas 419
cas9 54–55, 62, 184, 193, 195, 226, 228, 233
cassava 78, 83, 89, 119, 150, 156, 158–159, 245, 278, 289, 302, 312, 324
catalysis 32, 37–39, 41, 65, 67, 76, 89, 138, 185, 192–193, 198, 232, 294, 297, 312, 342, 358, 372, 400, 403–404, 435, 439
catalyst 21, 24, 26–27, 32–33, 35–36, 38–39, 41–43, 63–64, 68, 76–77, 90, 117, 160, 170, 189, 194, 196, 199, 204, 240, 247–248, 268, 281, 283, 286, 293, 297, 319–320, 370, 373, 379, 391–395, 399–400, 402, 404–405, 423–424, 430
catalytic conversion 22, 28, 32, 41, 43, 304
pyrolysis 21, 399–401, 424
cattle waste 408
cell 2, 4, 12, 14–15, 16, 17, 18–20, 21–22, 23–25, 26–27, 28–30, 31–32, 33–34, 35–36, 38–40, 41–44, 45–46, 48–49, 51–56, 58–59, 61–65, 66–70, 71–74, 75–78, 81–85, 86–90, 91–92, 93, 95–97, 98, 105, 107–108, 109, 111–117, 118–121, 122, 124–127, 128–133, 136–140, 144–149, 151–153, 156–158, 160, 163–171, 172–177, 178–183, 184–186, 190–195, 196–200, 201, 203–206, 208–214, 217, 219–221, 226, 229–233, 234–236, 239–240, 243, 245, 247, 249–250, 251–254, 255–260, 262–264, 268–270, 271–274, 276–277, 278–279, 281–284, 287–290, 292–295, 298, 300–305, 309–314, 317–322, 325, 327–331, 332–333, 335–336, 338–339, 341–345, 347–349, 352–355, 357, 360–366, 367–374, 375, 377, 379–380, 385–386, 394–395, 402–405, 408, 414, 424, 426–427, 428, 430, 432–436, 439–441, 450
growth 144, 243
wall anatomy 181
wall rigidity 427
washout 365
cellobiohydrolase 58, 73, 81, 87, 174, 194, 312, 329, 333
cellobiose 15, 58, 66, 81, 169, 174, 270, 312, 329, 354
cells 18, 26, 28, 46, 59, 61, 64, 70, 93, 108, 132, 133, 160, 204, 229, 235, 240, 247, 251, 253–255, 259–260, 264, 271, 278, 283, 293, 320, 325, 327, 333, 335–338, 355, 360, 364, 380, 424
cellular metabolism 204
cellulase 16, 51, 53, 55–58, 65–69, 71, 73, 78, 81–87, 89–90, 95, 97–98, 107, 113, 116, 129, 131, 137, 139, 149, 151–152, 166, 172, 174–181, 184–185, 192, 194, 198, 201, 206, 209–210, 212, 214, 217, 239–240, 257–260, 277, 282, 300, 312, 314, 318, 320–321, 329, 332–333, 354, 365–367, 369, 373–374, 434
cellulolytic 51, 52, 54–55, 62–64, 66, 68–69, 71, 83, 86, 97–98, 107, 117, 120, 149, 152, 171, 174, 177–180, 194, 239–240, 259, 264, 311–313, 320–322, 329, 332, 375

Cellulomonas 51, 131, 177
Cellulomonas fimi 51, 177
Cellulomonas flavigena 51
cellulose 2, 12, 14, 15, 16, 17, 18–20, 21, 22–23, 25–26, 29–30, 31–33, 35–36, 38–39, 40–45, 51–53, 55–56, 58, 61–62, 64–66, 68, 70–73, 74–76, 77–78, 81–83, 86–88, 90, 92, 95, 97–98, 108–109, 111–112, 114–116, 119–120, 126–133, 137–139, 149, 151, 158, 163–170, 171–176, 179–186, 191–195, 197–200, 201, 209, 210, 214, 217, 219–220, 232, 234–235, 240, 249–251, 256, 260, 262, 268–270, 272, 274, 276–277, 279–282, 287–289, 294–295, 298–301, 309–311, 314, 317, 319, 325, 329–332, 341, 344, 347–349, 352–355, 357, 360, 362, 365–366, 367, 369, 371, 373–375, 377, 385, 386, 394, 403, 408, 414, 427–439
  acetate 249–250, 386
cellulosic 4, 12, 15–17, 19–23, 25–27, 29, 31–35, 38–45, 51, 54–56, 62–65, 67–68, 70–78, 82–87, 88–93, 95–96, 98, 107, 108, 113–116, 118–119, 121–122, 124–125, 127, 130, 136–140, 145–148, 151–152, 156–158, 160, 163–167, 169, 173–176, 179, 182, 184–185, 190–197, 198–201, 203, 205, 213, 217, 221, 230–231, 234–236, 240, 245, 251, 255–260, 263, 268, 271, 284, 298, 300–305, 309, 311–312, 314, 317–322, 328–332, 341–345, 347–348, 355, 361, 371–375, 377, 385, 402–405, 414, 435
Cellulosilytica 52
cellulosome 71, 176, 186, 194–195, 197, 199, 240
central liming tank 47
centrifugal force 46–47
centrifugation 44, 46–50, 62, 214, 240, 287, 335, 338
cereal industry 307
chelating agent 253
chelation of manganese ions 312
chemical
  conversion 3, 32–33, 95–96, 137, 157, 389–392, 394, 400–401, 404–405, 413, 414, 425
  energy 403
  extraction 431
  fertilizers 60, 292
  heterogeneity 165
  impurities 357
  industry 95, 267, 356, 377
  pathway 229, 267
  reduction 35
  route 40, 140, 142, 145
  synthesis 141, 158, 203, 205, 247, 251, 266, 267, 278, 281, 366, 378, 427, 430
chimeric 55
chiral 142, 218
chitinase enzymes 282
chitosan 23, 133, 272, 280, 292
Chlamydomonas reinhardtii 78
Chlorella vulgaris 333, 343
chloroacetic acid 378
chlorogenic acid 319
chlorohydrination 142
CHNS/O analyzer 15
chromatography 44, 46–50, 69–70, 190, 217, 240, 295, 362, 432
chromogenic 208
chronic disorders 427
Cimicifuga 427, 440
circular economy 65, 68, 94, 140, 146, 156, 270, 276, 324, 375, 442–444, 449, 450
citrate 204, 228, 241, 243, 253, 256, 257
citric acid 202–205, 212, 221, 225, 228, 241–246, 254–259, 260, 302, 312, 319–321, 322
citrus fruits 204, 212, 241, 428
clarification 82, 112, 341, 431
clarified liquid 335, 336
clarifier 381
clean energy 140, 275, 449
climate change 1, 113, 146, 323, 346, 407, 412, 425, 448
clogging 44, 253, 339, 359
cloning 54, 201
closed reactors 389
Clostridium 42, 45, 51–52, 58, 97–98, 108, 114–117, 120, 131–132, 138, 161, 177–178, 193, 195–197, 199, 209, 221, 232, 235, 251, 307, 310, 329, 341, 356, 373, 385, 433, 440
C/N ratio 418
CO_2 70, 424, 425
Codex Alimentarius Standard 301
codon 184, 185
  optimization 184, 200, 436, 437
cofactor 175, 239, 290, 436–437, 441
  engineering 290
coffee 17, 100, 116, 245, 269, 279, 302, 307, 337, 341, 428–429, 437, 439–440
coke deposition 395
cold start issues 348
Coli 55, 68, 115–119, 157, 160, 195, 199, 202–205, 221, 228, 231, 234, 236, 239, 247, 252, 258, 260, 290–296, 303, 307, 313, 319, 328, 341, 373, 375, 428–429, 435–437, 439–440
collagen 276, 278, 292
colloid 42, 47, 317, 321, 381
colour 123, 246, 250, 301, 303, 308–309, 312–313, 318, 398, 412

colourants 303
column chromatography 46
combustion 3, 5, 14, 22–23, 26, 32–34, 68, 88, 123, 124, 162, 343, 348, 388, 391–392, 394, 396, 399–400, 402–405, 414, 416
commensalism 109
commercial 4–5, 22, 33, 37, 39, 47, 58, 76–77, 81–83, 95, 133, 135, 150–152, 168, 190–193, 204–206, 208–209, 218–220, 230, 233, 241–244, 250, 268, 273, 284, 287–288, 298, 303–305, 307, 328, 332, 350, 352, 361, 364, 377, 390, 392–394, 396, 398–399, 428–430, 432, 440
  sweeteners 304
  viability 47, 190
complex
  biomass 101, 329
  carbohydrate 56, 194, 212, 214, 225, 365, 366
  fermented stream 384
  mixture 48–49, 96, 165, 220, 381, 398
  polymers 331
  process 225, 393, 418
  structure 51, 53, 95, 98, 126, 165, 239, 330–331, 433–434
compost 1, 166, 235, 264, 266, 275, 288–289, 307, 318, 407, 408, 411, 422, 426
computational
  modeling 106, 183
  models 229
concentrated
  acid hydrolysis 348
  volatile fatty acids 381
concentration polarization 382
condensation 35, 130, 133–134, 138, 169, 243, 253, 264, 285–287, 294, 416–417, 427, 430
condenser cold trap 337
conductivity 14, 392, 420
configuration 4, 15, 27, 37, 189, 350, 361, 364–365, 370, 419
coniferyl alcohol 19
conjugate forms 428
consistency 34, 52, 84, 181, 302
consortia 53, 77, 108–112, 115–116, 119, 120, 171, 235
constituent sugars 98, 165, 171, 172
consumable products 253
consumers 134–135, 304, 308, 313, 315–316, 443
contaminant 44, 46–50, 60, 131, 325, 352, 357, 363, 391, 398, 421, 431–432
continuous fermentation 335, 360, 365, 375
  mode 337
  operation 358, 370
  process 134, 168, 335, 355, 364
  reaction 370
continuously stirred tank reactors (CSTR) 349
control 36, 53, 68, 70, 75–78, 101, 110, 111, 121, 124, 135, 142, 144, 153, 157, 173, 186, 189, 202, 204, 217, 221, 227–229, 234, 238–239, 250, 257, 269, 280, 285, 292, 323, 342, 350, 364–365, 380–381, 384–385, 389, 392, 400, 419
convenience 83, 96, 307
conventional
  chemical processes 316
  methods 170, 171, 238, 248, 321, 338, 339, 365, 416
  purification 361
  pyrolysis 417
  reactors 352
  recovery 384
  separation 363, 380
  techniques 54, 164, 168, 248, 383
conversion 3, 14, 21–23, 24, 26, 28, 32–33, 35–38, 40–43, 53, 55–56, 59, 65, 67–68, 70, 73–77, 78, 83–90, 91, 93, 95–96, 98, 107, 110, 113, 115, 117–119, 127, 131, 137–138, 141, 144, 151, 157–159, 161–162, 166–171, 176, 179, 189, 194–197, 199–200, 210, 220–221, 231–232, 235, 238, 240, 243, 249, 252, 255, 257–260, 266, 271, 285, 295, 301, 304–305, 307–309, 311, 313–318, 320–323, 325, 327, 329, 331, 333, 335–337, 339, 341, 343–345, 347–349, 352, 354–355, 358–360, 362–363, 366–367, 375–376, 383–384, 388–392, 394, 396–399, 400–406, 407, 413–414, 416, 421–425, 437
cooking 13, 65, 362, 371, 388–389, 392, 405
cooling 47, 189, 380
copolymer 23, 29, 40, 270, 282, 293
copra meal 307
coproducts 242, 251, 260, 322, 332
corn 16, 19, 23–25, 44, 54, 55, 65, 71, 75–78, 82, 84, 87–88, 91, 94, 100, 120, 124–125, 128, 135, 137–138, 165–168, 197, 203, 212–217, 232–234, 236, 238–239, 251, 257–258, 262, 269, 271, 278–279, 292, 296, 298, 303, 310–312, 319, 323–325, 333, 340–342, 347, 351–352, 361, 371–375, 377, 402, 419, 423–426, 428–430, 434–435, 438, 440, 446
corrosion 3, 14, 76, 96, 170
cosmetics 60, 140–141, 202, 209, 213, 235, 237, 245, 249, 254, 280, 430–431, 440

cost 1, 19, 21, 26, 33, 38–40, 44–47, 52–54, 58, 60, 62, 64, 67, 72, 74, 76–78, 82–86, 88, 91–92, 94, 96, 107, 111, 112, 115, 122–123, 125, 127, 132–136, 140–143, 145–147, 149, 156, 159, 162–163, 165–167, 169–171, 173, 180–184, 188–191, 193, 194, 197, 203, 205, 207, 218, 220–221, 230, 235–240, 243–244, 247, 251, 254, 261–264, 274, 280–284, 286, 288, 292, 297, 301, 304–308, 315–318, 321, 324, 328, 332, 336, 338–340, 347, 352–355, 357–361, 365–367, 369–370, 373–375, 377, 380–384, 388, 392–394, 400, 407, 411–412, 415, 421, 431–432, 437–438
cotton
  carry bags 443
  straw 44, 94
  textile wastes 302
coumaryl 19, 165, 347
covalent 45, 74, 92, 126, 127, 131, 133, 172, 427, 434, 439
covalently bound ferulic acid 434, 439
CRISPR 54–55, 62, 69, 184, 193, 195, 197, 226, 228, 233, 313
crop 1, 5–6, 14, 21, 23, 30, 40, 56, 59, 74, 78, 89, 93–94, 97, 107, 113, 115, 120, 125, 136–137, 141, 153, 159, 163, 166–169, 183, 195, 197, 203, 212–213, 232, 237–238, 250–251, 255, 259, 269, 274–275, 284, 291, 294, 300, 304, 318–320, 324, 328, 331, 341, 347, 352, 361, 364, 372, 374–375, 377, 383, 412–414, 444–445, 447
crosslinking 151, 172, 349, 414
crude
  enzyme extract 434
  glycerol 120, 153, 156, 158–160, 231–233, 236, 251, 260, 267
  oil 140, 142, 369
  product 46
crystallinity 15, 75–76, 95, 126, 129, 150, 169, 179, 183, 195, 282, 349
crystallization 44, 47–48, 50, 68–70, 170, 189, 214, 240, 363, 366
culture
  conditions 144, 286, 437
  media 355
cutinase enzymes 282
CXEH 434
cycle stability 14
cyclic
  aromatics 393
  dimer 285
  monomers 285
cyclisation 35
cytoplasm 267

dairy
  products 18, 140, 213, 300
  supplements 308
debris 46, 48, 333, 336
decanter 47
decarboxylation 144, 243, 416, 417
decomposition 30, 34, 41–42, 76, 131, 166, 264, 283, 288, 389, 391, 393, 396–397, 411, 414–417, 446
deconstruction 29, 40, 68, 113, 311
decontamination 408
dedicated energy crops 300, 328
deformation 4
degenerative disorders 308
degradation 19, 21–23, 34, 36, 42, 51, 52, 55, 66, 69, 72, 74–77, 81, 82, 95, 98, 111, 112, 116, 126, 127, 152, 167, 169, 170, 174, 176, 181, 182, 189, 192, 194–196, 198–199, 209, 240, 261–265, 270–275, 282–283, 288–289, 293–297, 298, 319, 320, 330, 336, 341, 346, 359, 370, 374, 414, 424, 426
degree of polymerization 15, 50, 137, 169
dehydration 25, 27, 33, 133–134, 141, 199, 327, 355–356, 364, 372, 383, 393, 415–417
dehydrogenation 126, 141, 243
delignification 30, 119, 129, 192, 197
delignifying enzymes 340
de novo synthesis 313
denser 43, 48, 337, 414
densification 395
deoxygenation 400
dependence 59, 72, 95, 113, 122, 140, 199, 203, 267, 389, 393
depletion 92, 264, 376
deposition 338–339, 382, 395
derivatives 4, 29, 141, 145, 241, 247, 275, 278, 289, 296, 298, 311, 347, 357, 369, 388, 397, 401
desalination 341, 343, 371, 374–375, 382, 383, 387
descaling 249
desiccation 134
destarched wheat bran (DSWB) 435
detection units 46
detectors 46
detoxification 41, 76, 130, 151–152, 197, 241, 354, 357, 360, 374
dextranase 408
DFF 27–28
DHMF 27–28
diafiltration 46, 62

dialysis EMR 350
dicarboxylic acid 28, 204–205, 213, 219, 228
diffusion 134, 181, 350, 355, 363–364, 382
diffusivity 337
digestibility 22, 30, 41, 77, 95, 138, 141, 148, 167, 197, 200–201, 238, 296
digestion 1, 14, 23, 32, 36, 59, 63, 69, 85, 169, 241, 280, 288, 307, 318, 376, 384, 389, 406–408, 421–426
digestive system 433
dilatory process 379
dilute 19, 21–23, 25, 30, 41, 76, 87, 88, 123, 128, 137, 149, 168–171, 198, 246, 249, 330–331, 348, 355, 357, 362, 375, 382, 432
dilute acid hydrolysis 168, 331, 348
dioxide 1, 26, 30, 33, 36, 47, 59, 95, 97, 122–124, 218, 261–262, 267, 275, 298, 346, 362, 413, 419, 446
direct
  combustion 388, 396, 414
  contact 350, 364
  microbial conversion 355
disaccharides 145, 149, 171, 181
discarded wood 25
disposal 68, 146, 156, 218, 264, 271, 389, 397, 401
disruption 76, 108, 408, 434
dissolved
  impurities 47
  salt 356
distillation 25, 94, 123, 133–138, 163, 183, 189, 214, 240, 327, 329, 333, 335, 337–338, 342–343, 352–357, 362–365, 370–375, 377, 380–381, 386–387, 405
distilled spent grain with husk (DSGH) 436
distilleries 389
diversity 1, 32, 66, 93, 97, 110, 172, 200, 203, 208, 310, 442
DMF 27, 28
DMTHF 27, 28
dominant products 390, 396
downstream 44, 47–49, 62, 67, 69, 112–113, 120, 138, 142, 144–145, 156, 158, 161, 170, 188–190, 205, 219, 262, 305, 309–310, 313, 316, 327, 329, 335–337, 343, 359, 384, 395, 437
DPPH radicals 434
driving force 183, 337, 382, 449
drug delivery systems 292
dry
  ash content 15
  biomass 15, 17, 34
drying 44, 47, 49, 147, 151–152, 330, 392–393, 395, 416, 432
duckweed 93
dumpsite 408
durability 32, 288
dynamic 35, 108–111, 160, 192, 229, 257, 276, 288, 350, 352, 380

E. coli 55, 205, 228, 247, 252, 313, 373, 435–437
ecology 23, 28, 108, 116, 200, 423, 425
economic 22, 44–45, 49, 53, 56, 58–60, 62, 72, 76, 78, 94, 107–108, 111–112, 119, 123, 127, 133–135, 137–140, 145–146, 158, 162–163, 166, 168–170, 173, 182, 188–189, 200, 219–220, 230, 232–233, 236, 238, 254, 272, 284, 288, 291, 301, 305, 307, 310, 313, 315–316, 327–328, 338, 340, 343, 347–348, 350, 353, 358–359, 365, 370, 372–373, 384, 388, 396, 401, 421, 423, 431–433, 437, 442, 447–449
economy 26, 62–65, 68–69, 72, 77, 85, 94, 115, 140, 146, 153, 156, 192, 193, 211, 231, 232, 254, 262, 269–271, 276, 294, 298, 317, 319, 324, 340–341, 352, 375, 401–402, 406, 424, 442–444, 449, 450
ecosystem 125, 135, 173, 270, 344, 442
edges 48
edible products 123, 304, 315
effective separation 358
efficacy 85, 124, 130, 157, 185, 227, 234, 259, 282, 324, 328, 330, 333, 337, 354, 369
efficiency 1, 3, 21, 23, 26, 31, 38, 46, 47, 48, 53, 58, 62, 70, 75, 77, 82, 85, 87, 93, 95, 97, 99–102, 105, 107–110, 112, 113, 144, 147, 151, 162, 165, 171, 176, 179–184, 185, 188–191, 203, 205, 220, 221, 229, 275, 281–286, 289–291, 310, 316, 327, 329–333, 335, 338–340, 342, 344, 349, 352, 357, 360, 363–367, 369–370, 381, 383–384, 389, 391–393, 396, 398–401, 408, 414, 426, 431–432
efficient technology 394
effluent treatment 135, 411
EH techniques 165, 171
elastomers 277
electrical energy 26
electricity demand 383
electrode materials 13
electron acceptor 420
electron donor 233, 421
electroplating 245
electrospinning 278

elemental composition 14, 15, 418
elemental constituents 4
elimination of inhibitors 364
ellagitannins 308
emerging routes 431
EMF 27
emission 3, 13, 14, 22, 25, 26, 34, 60, 69, 88, 92, 94, 95, 116, 122, 124, 136, 142, 145, 146, 191, 203, 237, 275, 284, 288, 323, 346–348, 352–353, 380, 388, 390–391, 400–401, 404, 412–413, 416, 446, 449
emulsifiers 301, 303, 304, 314
emulsifying agents 303
enantiopure 157, 158, 251
endocellulases 81
endoglucanase 51, 55, 58, 64, 73, 81, 90, 174, 177, 201, 312, 329, 333, 435, 436
endospores 210
end product 21, 47, 48, 50, 52, 59, 62, 174, 190, 209, 324, 328, 366, 376, 389
energy 1–2, 4–6, 13–15, 19–26, 32–33, 37–44, 48, 51, 53, 59, 62, 64, 66, 68, 70–72, 74–77, 84–85, 87–90, 92–97, 99, 101, 107, 109, 113, 116–120, 122–125, 127–130, 134–135, 136–147, 149, 151, 156, 158–159, 162–164, 166, 170–171, 176, 179, 189–193, 196–200, 203–205, 210, 212, 220–221, 231–233, 235–238, 240–241, 243–244, 251, 253–259, 267, 270–271, 273, 275, 278–279, 281, 284, 295, 298, 300, 304, 314–315, 317, 320, 323–324, 327–328, 331, 333, 335, 337–347, 349–350, 352, 355–365, 370–377, 378, 380, 383–386, 388–397, 399–407, 416, 422–426, 442, 446, 448–449
engineer 20, 21, 23, 41, 45, 54–56, 58, 59, 61–63, 65–71, 85, 87–91, 93, 96–99, 100–102, 105–108, 110–121, 132, 136, 138, 139, 157, 158, 160, 163, 164, 173, 183–189, 192–201, 205, 206, 218–221, 225–228, 230–236, 240, 252, 255, 258, 260, 265, 270, 278, 280, 282, 283, 286, 289–297, 299, 301, 312, 313, 316–319, 328, 341–344, 374, 401, 403, 405, 422–424, 433, 435–437, 439–441
enhanced bioalcohol production 101, 105, 107, 327
enhancements 430
Enterococcus faecalis 232, 267
Enterococcus faecium 289
enthalpy 124
entrainer 356
environment 1, 13–15, 19–22, 24–25, 29, 32, 45, 47, 50, 53, 55, 59–60, 62–64, 66, 68–69, 71, 74–75, 77–78, 84–85, 87–89, 92–94, 99, 101, 107, 110–112, 115, 118–119, 122, 126, 133–134, 136, 140–143, 146, 149, 156, 166, 170–171, 173, 176, 181, 184, 189, 193–194, 196–197, 199–200, 203, 206–212, 216, 220, 225–228, 230, 234, 237–238, 244, 248, 254, 256, 258, 262, 264–265, 267, 269, 272–276, 278–280, 282–285, 287, 289, 291–293, 295, 297–298, 300, 315–317, 320–321, 323–325, 329, 332–333, 339–341, 344, 346–350, 352–353, 357, 359, 369–370, 374, 378–379, 386, 389–392, 394, 399–407, 411–412, 414, 421, 423–426, 431, 435, 438, 442, 447–450
enzymatic 21–22, 30–31, 38–41, 51–52, 62, 65, 72–77, 78–79, 81–85, 87–92, 97, 100, 115–120, 126, 129, 131–133, 137, 139, 141, 144–145, 147, 149–152, 157–158, 160, 162–165, 167, 169, 171, 173, 175–183, 185–187, 189, 191–200, 201, 210, 217, 225, 227, 233, 237–244, 245, 247–258, 259–260, 273, 275–277, 279–281, 283, 285–287, 289, 291–301, 311, 314, 318, 321, 323, 329–332, 340–351, 353–355, 357–359, 361, 363, 365–367, 369–375, 408, 428, 431, 433–436, 438–441
  membrane reactor (EMR) 350
enzyme 14, 16, 18, 22, 24, 30–32, 36–38, 42, 49, 51–56, 58, 60–63, 65–69, 71, 73–75, 77, 78, 81–90, 92, 95–98, 100–102, 109, 111, 113, 117, 118, 120, 126, 129, 131–133, 138, 139, 144, 147–152, 163–166, 171–180, 181–199, 200–201, 206, 209, 210, 212, 214, 217, 218, 225–228, 233, 235, 238–240, 243, 247, 248, 251–257, 260, 262–264, 271–273, 277, 280–283, 285, 286, 288–290, 291, 295–298, 300, 304, 305, 307, 310–314, 316, 317, 320–322, 329, 330, 332, 333, 336, 339–341, 343, 344, 348–351, 354, 355, 358–360, 362, 365–367, 369–373, 375, 390, 408, 418, 433–439, 446
equilibrium 42, 72, 199, 352, 359, 361, 380
erosion 352
Escherichia 55, 68, 115–117, 119, 157, 160, 195, 199, 203, 221, 231, 234, 236, 239, 258, 295, 307, 328, 341, 375, 436, 439–440

essential
component 14, 26, 99
oils 269, 305, 308
ester 21, 23, 26, 28, 39–42, 51, 59, 64, 120, 129, 131, 141, 152, 153, 172, 193–195, 199, 201, 232, 236, 241, 247, 250, 258, 268, 270, 276–277, 281, 283–284, 289, 357, 378, 388–389, 398–399, 402–403, 423, 431, 433–441
ethanoic acid 245, 247, 377–381
ethanol 13, 19, 21, 24–26, 28, 30, 37, 39–40, 42–43, 49, 59, 61–63, 64–68, 70–72, 73–78, 83–84, 86–90, 92, 94–97, 99–102, 107–108, 113, 115–125, 127–130, 131–139, 141, 144, 147–148, 150, 156, 159–160, 162, 164, 169, 172–173, 175, 190–199, 200, 206, 208–209, 213, 218, 221, 226, 234, 236, 246–249, 256–260, 295, 298, 300, 304–305, 307, 317–318, 321, 323–325, 327–328, 330, 332–333, 337–338, 340–349, 353–357, 360–365, 367, 369–375, 378–379, 383, 388–389, 393–394, 403, 432, 444–446, 450
ethers 142, 361, 398
ethyl acetate (EtOAc) 432
eukaryotic microorganisms 328
eutrophication 173, 442
evaporation 124, 133, 169, 337, 352, 355, 360
excipient 254, 257
execution 129
exfoliant 202–213
exfoliating 253–254
exhaust
emissions 348
gas emission 347
exoamylase 83
exocellulases 81
exogenous 263
exoglucanases 51, 174, 177, 312, 329
expression 55, 64–67, 98, 102, 105–107, 115, 119, 160, 184, 186, 197, 200, 227–228, 232, 252, 272, 289–291, 313, 344, 436–437
ex situ mode 400
exterior 48
external PV unit 365
extracellular
hydrolase 82
metabolites 339
vesicles 49
extraction 25, 44, 47, 49, 64, 67, 89, 112, 129, 148–149, 159, 190, 220, 240, 242, 244, 268, 291, 304, 307, 318–319, 354, 356, 360, 362–363, 367, 372–375, 377, 380, 385–386, 428, 430–433, 435, 437–442
extractive fermentation 356, 375
extractives 15, 166, 353, 354, 386
extracts 160, 268–269, 278, 305, 317, 362, 428, 433
extrusion 75, 85, 91, 271, 296, 341, 349

fabric production 28
factories 67, 115, 120, 226, 232, 234–235, 444
failure 3, 109, 110, 203
farming practices 300
fast pyrolysis 34, 43, 396, 401, 404, 414, 422
fat replacer 309
fats 145, 146, 152, 166, 389
fatty acids 23, 145, 149–153, 207, 211, 277, 282, 290, 294, 303, 304, 376–384, 385–387, 422, 423
FAXX 434
Fe and Mn oxides 421
fed batch conditions 437
feed 3, 4, 13, 16, 17, 22, 25, 28–31, 34–35, 38, 41–42, 47, 53, 56, 59, 62, 66, 67, 72, 78, 85, 87, 89–90, 92–95, 97–100, 102, 107, 113, 119, 120, 123–125, 129–130, 135–136, 138, 140–141, 143, 145–147, 149, 151–153, 155–159, 161, 165–166, 168, 173, 183, 184, 188, 190–192, 195, 197, 205, 206, 208, 211, 213, 218–221, 227, 236, 239, 241, 244, 257, 259, 266–267, 271, 274, 282, 284–285, 295, 298, 316–317, 323–325, 327–329, 331–333, 335, 337–338, 341–342, 347–355, 357–358, 360, 362–364, 366, 369–370, 375, 377, 381–384, 389–390, 392, 394–396, 398–401, 405, 407–408, 411–412, 414, 418–423, 425, 441–442, 445
feedstock 4, 13, 16, 17, 22, 25, 28–31, 34–35, 38, 56, 59, 66–67, 72, 78, 85, 87, 89–90, 92–95, 97–100, 102, 107, 113, 119–120, 123–125, 130, 135–136, 138, 140–141, 143, 145–147, 149, 151–153, 155–159, 161, 165–166, 168, 173, 183, 184, 188, 190–192, 195, 197, 205, 219–221, 227, 236, 244, 257, 266, 271, 274, 282, 284–285, 295, 316–317, 323–325, 327–328, 331–332, 341–342, 347–348, 351–352, 353–354, 357, 360, 363, 366, 370, 375, 377, 389–390, 392, 394–396, 400–401, 405, 407, 411–412, 414, 418–423, 425
fermentable 24, 41, 44–45, 47, 49–51, 52–54, 55–57, 58–63, 65, 67, 69, 71–73, 75, 77, 79, 81, 83, 85, 87, 89, 91, 96–98, 123, 126, 130, 131, 136, 140, 146, 150–151, 194, 239–240, 311, 324, 328–331, 354, 357, 367

fermentable sugars 41, 44–45, 47, 49–50, 51–53, 54–57, 58–60, 61–63, 65, 67, 69, 71, 73, 75, 77, 79, 81, 83, 85, 87, 89, 91, 96–98, 123, 126, 130–131, 146, 150–151, 239, 311, 329–331, 354, 367
fermentation 14–15, 22–23, 25, 29–32, 36–37, 41–42, 44–47, 48–49, 51–57, 59–64, 65, 67, 69–71, 73, 78, 83–87, 88–90, 92–97, 99–103, 105–107, 109, 111, 113–115, 117–123, 130–137, 139–141, 144, 145, 147, 149–150, 152–153, 156–161, 163, 165, 167, 170, 172–173, 182, 189, 192–197, 199, 203, 205–211, 213–219, 220–221, 225–234, 236, 239–249, 251, 254–258, 260, 262–264, 267, 270–272, 284–287, 301, 303–305, 308–313, 314–319, 320–325, 327–332, 333–344, 345, 348, 351–352, 354–358, 359–362, 363–366, 367, 369–374, 375, 377–380, 383–386, 389, 423, 433, 435–438, 446
fermentative 92, 156, 159, 170, 207, 218–219, 231, 234–236, 247–249, 259, 267, 270–271, 302, 304, 309, 319, 343–344, 354, 365, 436, 438
fermented
  acai fruit pulp 303
  bioproducts 309
  effluent 380, 381, 384
  products 264, 314, 383
fermenter 78, 309, 312, 316, 339–340, 355, 361
fertilizer 14, 60, 67, 93, 230, 292, 294, 307, 313, 392, 398, 408, 411, 412, 418, 442
Ferula foetida 427
ferulic acid 131, 305, 322, 427–441
feruloylated oligosaccharides 433
feruloyl esterase 201, 433, 435–441
feruloyl groups 434
feruloyl polysaccharide 435
fiber 2, 19, 43, 54, 55, 65, 71, 75–76, 147, 151, 166–167, 186, 198, 214, 263, 267–268, 270–271, 275, 278–279, 287, 294, 298, 330–331, 341, 343, 371, 374–375, 378, 423, 433–434, 439
fibrinolytics 263
fibrous 19, 85, 126, 221, 232, 260, 380
filamentous fungi 52, 83, 84, 97, 98, 199, 211, 240, 322
filter 27, 44, 46, 48, 70, 195, 214, 217, 360
  clogging 44
filtration 44, 46–50, 62, 64, 70, 182, 217, 240, 325, 333, 335–337, 342–345, 350, 352, 359, 363, 367, 369, 371–373, 375, 381, 386
financial considerations 5
finite fossil fuel resources 273
Fischer–Tropsch 34
flammability 394
flammable 3, 356
flash chromatography 46
flash pyrolysis 396, 417
flavanone 308
flavone aglycons 308
flavonoids 308
flavour
  compounds 316
  enhancer 253, 254, 399
flavouring agent 245, 301, 305, 309
flax oil cake 311
flexibility 126, 279, 287, 309, 384, 389
flour production 433
flow 4, 395
fluctuations 110, 140, 358, 412
fluid feed 364
fluorogenic 208
flux 100–102, 105–106, 107, 117, 120, 144, 218, 236, 252, 259, 339, 359, 364, 374, 382, 383, 386
flying 3
food 1, 6, 16–17, 22, 23, 27, 29, 42, 56, 59–60, 63, 65–67, 69, 71–72, 78, 86–87, 89–90, 93–94, 108, 116, 119, 123–125, 140–141, 145–146, 149–150, 153, 157, 159–160, 161, 163–165, 166, 173, 175, 186, 195, 197, 201–207, 208–209, 211–213, 217–218, 228, 230–232, 236–238, 241, 244–251, 253–257, 259–260, 262, 265, 267–269, 271, 273–275, 278–279, 284, 287, 293, 295, 297, 300–305, 307–321, 322, 327–328, 344, 350, 352–353, 358, 362, 373, 376–378, 385–386, 399, 407, 408, 411, 414, 420–425, 426, 428–431, 438–441, 450
FO process 382
forest residues 1, 207
forestry 5, 6, 56, 72, 146, 166–167, 203, 298, 323, 331, 353, 388, 444
forward osmosis (FO) 381
fossil 1, 24–26, 34, 45, 58–61, 93–96, 107, 113, 122–125, 134, 136, 142, 156, 162–163, 167, 190, 211, 237, 261–262, 273–276, 284, 323, 346–348, 352–353, 370, 388–389, 393
fouling 3, 20, 46, 48, 189, 335, 338, 339, 340, 342, 359, 370, 372, 381, 382, 383, 384
FO water flux 382
fractionation 23, 76–78, 90, 198, 336
fraction collectors 46
fragmentation 30, 127, 265, 414
fragments 48–49, 81, 174, 175, 176, 331

free
  enzymes 172
  fatty acids 152, 153
  ferulic acid 428, 434, 441
  radicals 126, 421, 427
fructose 83, 97, 132, 148–150, 153, 169, 267, 432
fruit 6, 204, 212, 213, 217, 241, 250, 259, 300, 307, 377, 385, 428, 440, 441, 444, 445, 447
  peels 248, 269, 377, 428
  pomaces 304
  seeds 308
fuel 1–3, 5, 13, 14, 16, 17, 18, 20–21, 22–27, 29, 33–35, 36–39, 40–45, 50, 52–53, 56, 58–60, 61–64, 66–68, 70–74, 76, 78, 81–87, 88–90, 92–97, 107, 108, 111, 113–117, 118–120, 122–127, 128–129, 133–141, 146, 150, 156, 158–161, 162–165, 167, 182, 190, 192–197, 198–200, 201, 206, 211, 213, 216–217, 231, 232, 234–235, 237–238, 251, 254, 256, 258, 260–262, 266, 270–271, 273–276, 278, 284, 294–295, 298, 307, 319, 323–325, 327–332, 335–336, 340–349, 350, 352–355, 361–367, 370–374, 376, 385–386, 388–396, 398–405, 411, 413–424, 425, 444–446, 448–450
full factorial design 435, 437
fumarase 228, 243, 253
fumarate 206, 243, 251, 267
fumaric acid 28, 30, 202, 204, 213, 232, 271, 302
function 22, 29, 38, 41, 48, 50, 60, 70–71, 81, 85, 86, 106, 108, 111–112, 130, 133, 134, 140, 175, 183–188, 193–195, 197, 199, 201–202, 205, 208, 213, 229, 234, 238–239, 241, 245, 247, 252, 255, 259, 268, 270, 272, 274, 280, 283–285, 293–294, 305, 317, 319–320, 336–342, 348, 350, 367, 380, 387, 394, 400, 404, 419–421, 425–426, 435–436, 439–440, 443–444
fungal
  fermentation 319, 438
  species 129, 131, 241, 312
fungus 88, 97, 139, 208, 211, 239, 240–241, 244, 251, 267, 309, 419
furan 26–28, 31, 41, 129, 131, 172, 174–175, 178, 200, 220, 312, 329, 357, 369, 393, 434
furfural 26–28, 30, 33, 39–41, 61, 77, 120, 151–152, 169, 170, 181
furnaces 399
Fusarium 52, 132, 136, 267, 434, 441
Fusca 52, 177, 252, 435, 439
fusion 3, 4, 22, 134, 181, 185, 350, 355, 363–364, 382
galactans 18, 20, 175
galactonate 206
galactose 17–18, 50, 74, 98, 132, 149, 165, 174–175, 330, 347, 432
gas 1, 2, 3, 5, 13, 14, 20, 21, 23, 25, 26, 27, 32, 33, 34, 35, 36, 38, 40, 41, 42, 43, 44, 45, 46, 49, 50, 59, 60, 61, 64, 69, 70, 75, 76, 78, 82, 84, 85, 87, 88, 89, 92, 94, 95, 97, 114, 115, 116, 117, 120, 122, 123, 124, 125, 128, 134, 137, 138, 141, 142, 145, 151, 156, 158, 159, 160, 162, 166, 191, 199, 203, 204, 210, 213, 223, 231, 232, 233, 236, 237, 238, 241, 242, 244, 245, 251, 255, 256, 258, 262, 275, 280, 284, 296, 301, 302, 304, 305, 307, 309, 311, 312, 314, 318, 321, 323, 324, 325, 328, 331, 341, 343, 344, 346, 347, 348, 349, 352, 353, 361, 363, 364, 365, 367, 373, 377, 383, 385, 388, 389–392, 394–396, 399–405, 400–405, 407, 408, 411, 412, 413, 414, 415–416, 417, 419, 421, 422, 423, 424, 426, 431, 435, 439, 440, 444, 445, 446, 450
gastrointestinal tract 232, 305
gel 17, 18, 21, 48–50, 64, 70, 113, 132–133, 150, 152, 160, 170–172, 280, 283, 285–286, 292, 294–295, 309, 318, 337, 344, 374, 383, 427, 441
  formation 359
  layer 359
gelatin 150, 172, 280, 292
gelling agent 18
gene 6, 13, 15, 20–21, 23, 25–27, 28–30, 31–32, 33–34, 35–36, 37–39, 41–42, 45–46, 50–52, 54–55, 56, 58–59, 61–62, 65–70, 72–73, 75–77, 78, 81–83, 84–85, 87, 92, 94–97, 98–100, 101–102, 105–107, 108, 110–113, 114–118, 120–123, 125–126, 128, 131–136, 142, 144, 146–147, 149, 151, 153, 156–158, 160, 162–170, 172–173, 176, 180–185, 186–189, 192, 193, 195–200, 203–206, 208–211, 214–219, 221, 225–229, 232–235, 238–240, 244, 247–252, 255, 262–264, 267, 270–271, 275, 282–287, 289–295, 297, 300–302, 304, 307–308, 310, 312–316, 318, 321–325, 328, 337–342, 345–347, 349, 353, 355–358, 360, 362–364, 367, 369–372, 374, 376, 380–381, 384, 388–389, 391–392, 394, 397–403, 405–407, 408, 411, 414, 417–421, 427, 430–439, 441–443, 447–449
  engineered *E. coli* 313, 436

genome 54–55, 58, 62, 93, 102, 105–106, 115, 118, 173, 183–185, 193, 195–196, 199, 208, 210, 218–219, 230, 233–236, 258, 313, 436, 441
genomics 92, 229, 235
gentle operating conditions 335
geographical location 313
geological aspects 412
geotechnical 278, 293
geothermal 13, 364
geothermal energy 13
GH43 54, 71
GHG 122, 390, 391, 396, 399, 401, 413, 422, 446, 449
GHG emissions 401, 413
giant molecules 350
glass transition temperature 282
glass tube 380
global climate change 346
global demand 301, 349, 449
global energy 346, 376, 407
global food requirements 315
global scale 264, 274, 449
global warming 288, 346, 353
Gluconacetobacter 249
gluconic acid 65, 366
Gluconobacter 217, 247, 249, 259
glucose 15–18, 36, 45, 50, 55, 56, 58, 62, 65, 72, 74, 75, 77, 81, 83, 92, 95, 96, 98, 99, 116, 126, 131, 132, 146, 148, 149, 150, 151, 152, 153, 158, 160, 165, 168, 169, 174, 179, 197, 201, 207, 218, 221, 226, 227, 228, 251, 257, 267, 277, 282, 328, 329, 341, 347, 354, 355, 365, 366, 367, 370, 432, 439
glycerine 26, 39
glycerol
  biorefineries 146
  synthesis 364
glycidic feedstocks 145, 149, 153
glycogen 83
glycols 33
glycolysis 144, 204, 218, 235
glycoproteins 29, 312
glycoside linkages 15
glycosidic bonds 51, 72, 83, 149, 165, 174–175, 329, 436
glyoxylate 252–253
grain 6, 16, 18, 25, 74, 84, 88, 150, 152, 156, 158, 163, 166, 195, 212, 242, 246, 303, 317, 348, 422–423, 428–429, 432–433, 436, 438, 440, 441
gramineae 427
grape pomace 311
graphite 420
GRAS microorganisms 144
greases 389
green 2, 4, 24–26, 37–38, 40–41, 43, 49, 60, 63–65, 68–69, 85, 87, 92, 94, 122, 135, 139, 141–142, 145, 147, 149, 156–157, 159, 191, 194–196, 198, 203, 231–233, 237, 242, 258–259, 266, 269, 271–273, 275, 280, 284, 294, 296–298, 303–304, 313, 315–316, 318, 320, 323, 340, 346–349, 352–353, 390, 403, 411–413, 422, 424, 429, 446
greenhouse 2, 25–26, 60, 69, 92, 94, 122, 142, 145, 191, 203, 237, 275, 284, 323, 346, 348–349, 352–353, 390, 411–413, 422, 446
grinding 19, 75, 151
Gross Value Added (GVA) 442
growth 3, 5, 14, 28, 45, 48, 53, 55, 60–62, 66, 70, 73, 78, 90, 99–101, 107–111, 117, 140, 144, 163–164, 166–167, 202–204, 206, 215, 220–221, 226, 232, 239, 243, 245, 247, 249–250, 264, 272, 280, 283, 291–292, 313–315, 327, 332, 338–339, 346, 349, 376, 384, 411–412, 418, 421, 430, 442, 444, 448–449
guaiacyl 19, 127
guava 444
gums 280, 303

halogenation 33
harmful emissions 400
harsh chemicals 281, 438
harvest 3, 6, 67, 76, 94, 216, 220, 237, 251, 259, 324, 344, 352, 360, 367, 431, 444–445
harvesting 6, 67, 76, 94, 220, 251, 352, 360, 367, 444
  of algae 367
  cycle 360
hazardous 1, 21, 29, 37, 96, 204, 207, 389, 407, 430
hazardous impacts 389
hazardous waste 207, 407
health and safety 356
health benefits 86, 302, 305, 307, 315
heat and power 5, 392, 401
heater 337
heat generation 392
heating rate 34, 393, 396, 414–419
heating value 6, 13, 15, 141, 394
heat integration 143
heat stability 349
hemicellulase 51, 53, 69, 78, 83–85, 90, 98, 116, 137, 149, 151–152, 172, 174–179, 194, 206, 209–210, 217, 239–240, 300, 312, 314, 320, 329, 332

hemicellulose 12, 14–20, 22–23, 25, 30–33, 41, 43–45, 51–53, 56, 61, 66, 72–78, 82, 92, 95, 97–98, 108–109, 116, 119, 126–132, 149, 151, 158, 163–169, 171–175, 179–180, 182–183, 191, 198, 217, 219–220, 234–235, 262, 268, 300, 314, 325, 329–332, 347, 349, 354, 362, 367, 369, 377, 394, 414, 428, 432–435
herbaceous plants 2
herbicide 108, 204, 250
heterofermentative lactic acid bacteria 304, 319
heterogeneous 76, 126, 163, 165, 270, 356, 402, 436
heterologous genes 102, 228
heteropolysaccharide 17
hexane 49, 356
hexanoic acid 211
hexavalent chromium (Cr(VI)) 412
hexose 17, 56, 78, 107, 126, 131–132, 144, 149, 165, 354, 357
holocellulose 19
homoacetogenic bacteria 379
homolactic organisms 284
homologue 252
homopolysaccharide 15
horizontal gene transfer (HGT) 110
hot water treatment 148
human
  health 124, 207, 226, 265, 315, 317, 411
  waste 13, 415
Humicola insolens 82, 177
humidity 289
humification 413
humulone 313
husks 13, 93, 100, 116, 166, 269, 300, 323, 325
hyacinth 93, 272, 343
hybrid 32, 41–42, 120, 190, 267, 294, 341, 344, 360, 363, 365, 373, 384
  membrane processes 384
  process 190, 360, 365, 373, 384
  PV technology 365
hydratase 251
hydrocarbon 14, 21, 25–26, 34–35, 37, 40–41, 142, 247–248, 257, 265, 284, 347, 392, 394, 401–402, 415
hydrochar 415, 419, 423–425
hydrogen 13–15, 22–24, 26, 30, 33, 35–36, 40–42, 59, 63–64, 66, 78, 100, 102, 114–115, 118, 120, 126, 128, 131, 137, 141, 144, 151, 156, 158, 166, 170, 174, 192–193, 196, 202, 210, 218, 227–228, 231, 233–234, 241, 243, 247, 252, 259–260, 295, 337, 344, 361, 376, 393–394, 396, 399–401, 404, 417, 423–425
hydrolases 58, 81, 174, 312, 329, 408
hydrolysate 22, 30–31, 33, 43, 56, 78, 114, 118, 140, 145–146, 151–152, 156–160, 192, 196, 205, 231, 234–235, 239–240, 251, 256, 260, 321, 328, 341–343, 354–355, 357, 360, 362–363, 369, 372, 375, 431–432
hydrolysis 19, 21–22, 29–35, 38–39, 41–42, 45, 51–52, 55–56, 58, 62, 65, 71–73, 76–78, 81–83, 85, 87, 89–91, 96, 100, 107, 113–116, 119–121, 126, 130–131, 133, 137–139, 142, 147, 149–153, 158, 162–164, 167–173, 175, 177–182, 184–186, 188–189, 191–192, 195–201, 208, 214, 216–217, 225, 227, 240, 256–258, 260, 281, 284, 287, 298, 307, 321, 324, 329–331, 333, 341, 343–345, 348–350, 354–355, 357–358, 360, 362, 365–367, 369–376, 385, 393, 407–408, 415–417, 428, 431–435, 438–440
hydrolytic enzyme 51, 71, 74, 77, 82, 126, 151, 184–185, 199, 343, 439
hydrolyze 19, 58, 74, 98, 131–133, 147–152, 165, 167–171, 174–176, 214, 251, 282, 284, 329–330, 349, 354, 362, 366–367, 369, 434, 436, 438
hydrophilic group 15
hydrophobic
  group 15
  interactions 421
  membrane 338, 361, 383
  nature 362
hydropower 346
hydrothermal 21–22, 35, 87, 120, 137, 149, 169, 171, 198, 260, 374, 393, 401–405, 411, 415–416, 422, 424–426, 440
  carbonization 415–416, 422, 424–426
  liquefaction 21–22, 35, 137, 393, 401–402
  pretreatment 120, 169, 171, 440
hydroxybutyrate 150, 160, 277, 281–282, 290–291, 295, 297–298
hydroxycinnamic acid 305, 308, 438, 441
hydroxyl
  acids 428
  group 18, 126, 241, 284, 285
  radical 169, 434
hydroxylation 437, 441
hydroxymethyl 27–28, 30, 33, 41, 61, 152, 169, 181
hydroxyvalerate 268, 282, 291, 298
hyperthermic 55, 69
hyphae 211

immobilization 172, 182, 195, 198, 231, 283, 290–291, 340, 349
immobilized
  cell fermentation 379, 380
  enzymes 63, 172, 349

impacts of biochar 407–408, 421
improper contact 400
impurities 44, 46–50, 62, 153, 171, 190, 240, 325, 327, 336–337, 357, 362, 366, 381, 431
incomplete
  burning 394
  hydrolysis 81, 170
incubation time 309, 332
indicators 4, 343, 449, 450
indices 4, 449
indigoidine 313
inducible promoters 186
inducing 55, 142, 268
industrial 15, 21, 23, 26–29, 33, 35, 38–40, 47, 52, 54, 55, 60–61, 66–67, 69, 72, 77, 81, 83–84, 87–90, 101, 105, 106, 108, 112, 115–116, 120, 123, 128, 132–134, 139–143, 145–147, 153, 156–159, 170–173, 175, 182–184, 187, 191–200, 202, 206–212, 215–217, 221, 231–237, 243, 249, 254–258, 262, 264, 269–271, 273–274, 279, 284, 286–288, 292, 297, 300–305, 307, 309–310, 312, 315–320, 321–322, 332, 335, 338, 343, 344, 347, 354–355, 365–366, 371, 377, 383, 388, 395, 403, 405, 407, 424, 431–434, 437, 439, 441–442, 449
  application 28, 54, 69, 77, 89, 90, 116, 141, 175, 200, 208, 217, 232, 235, 254, 257, 307, 335, 371, 383, 395, 405, 424, 431
  crops 21, 23, 89, 115, 120, 195, 197, 274, 318, 320
  waste 89, 140, 147, 158, 193, 233, 256–258, 269, 271, 279, 292, 303–304, 307, 310, 316–318, 320–322, 338, 343, 432, 439, 441
  yeast 354
industry 1, 25, 27, 29, 38, 50, 59, 66, 69, 82–83, 95, 108, 125, 140, 146, 156, 162–165, 172, 175–178, 182, 202–209, 212–213, 219, 231, 245, 249–254, 256, 261, 263–265, 267–269, 276, 279–280, 286, 300, 303–305, 307, 309, 314, 316, 319, 321, 337, 343, 348, 352, 354, 356, 358, 364, 376–378, 385, 430, 438, 441, 444, 448–449
infections 142, 206, 210–212, 295, 427
inflammatory bowel syndrome 427
infrastructure 94, 133, 135–136, 310, 348, 390, 444, 448
inhibition 53, 81, 118, 131, 133, 183, 189, 196–197, 200–201, 210, 215, 290–291, 338, 348, 358, 365–366, 370–371, 420
inhibitors 31, 41, 45, 61, 63, 65, 76, 109, 132, 151, 159, 163, 165, 181–182, 192, 198, 348, 354, 356, 357, 360–362, 364, 369, 431
inhibitory 30–31, 45, 53, 76, 99, 101–102, 105, 109, 113, 147, 151, 170, 180–181, 220–221, 231, 255, 327, 331, 354, 356–357
innovation 63, 68, 73, 114, 122, 192, 230, 234, 276, 320, 341, 343, 385, 448
inoculation 94
inoculum volume 314, 339
input 5, 77, 135, 191, 266, 389, 445
insect pests 142
in situ extraction 183, 356, 400
instability 45, 61, 106, 110, 123, 162
integration 37, 48, 56, 58, 65, 94, 106–107, 120, 135, 137, 139, 143, 146, 168, 172, 192–193, 227, 291, 294, 327, 384, 390, 435, 439, 448
interfacial contact 350
interference 339
intergenerational equity 447
intermediate 27–28, 33, 37, 62, 77, 95, 108, 141, 204, 213, 228, 249, 260, 285, 376, 381
International Energy Agency (IEA) 346
intracellular 144, 151, 153, 219, 229, 252
intracellular redox balance 151
intragenerational equity 447
intramolecular bonding 394
inversely 48
investment 94, 173, 191, 305, 314–315, 316, 340, 361, 432
ion exchange 46–47, 69, 131, 214, 338, 354, 360, 362, 377, 385, 398, 407, 420, 422
ionic liquid 19, 23, 43, 74, 76, 91, 147, 151, 156, 260, 298, 349, 356
ionization 257, 382
isoamyl acetate 356
isocitrate 243, 253
isocitrate lyase 253
isomerization 243, 358, 393, 427
isomers 143–144, 285, 293–294, 430–431

jacketed glass column 380
JM109(DE) strain 436

kenaf 25, 31, 41–42
keratin 14
kerosene 162, 398
ketones 76, 398
key enzymes 98, 102, 144
kinetics 21–22, 40, 42, 66, 118, 138, 171, 196, 232, 236, 270, 295, 355, 402, 423
kitchen waste valorization 150
Kluyveromyces marxianus 58, 303, 310, 319

laccase 55, 85, 119, 129, 175–176, 178, 197, 300, 311–312, 314, 318, 340
lactate dehydrogenase (LDH) 218, 227

lactate esters 284
lactic acid 28, 45, 59–60, 64–67, 119, 202–204, 206–207, 213–215, 216, 218–219, 221, 225–227, 230–236, 239, 241, 257–258, 269, 275–277, 284–285, 290, 293–294, 296–297, 302, 304, 309–310, 313, 317, 319–321
lactide 277, 285–286, 290, 293–294, 297–298
Lactobacillus 206–207, 241, 302, 304, 307, 310, 433
  acidophilus 285
  bavaricus 284
  casei 68, 284
  crispatus S524 68, 284, 436
  delbrueckii 285, 319
  jensenii 285
  maltaromicus 284
  salivarius 284
lead 1, 3, 16, 25–26, 30, 32, 35–36, 45, 52–55, 76, 82, 95, 102, 105–110, 124, 132, 142, 144, 147, 169–170, 173, 179–182, 185, 189, 191, 202, 205, 212, 218, 221, 265, 268, 283, 287, 290–292, 310, 313, 318, 324–325, 327, 331, 333, 336–338, 346–347, 349, 352, 355, 359, 364, 370, 377, 388, 390, 398–399, 408, 411–413, 417–418, 423, 430, 435–436, 442–443, 446–447
leghemoglobin 313
lentinus 52
less toxic pollutants 446
levulinic 27, 170, 181, 213
lifecycle 264, 288, 444
lifetime 3, 443
lignin 2, 6, 12, 14–17, 19–23, 25, 27, 29–32, 35, 41, 44–45, 51–53, 55, 61, 65, 67–68, 72–75, 76–78, 86, 88, 90, 92, 95, 97–98, 109, 117, 119, 126–129, 137, 147, 149, 151, 164–169, 171, 173–176, 178–181, 192, 194–197, 201, 217, 220, 238, 240, 262, 268, 284, 294, 298–300, 310–312, 314, 317, 319, 325, 330–332, 340, 347, 348, 353–354, 367, 369, 372–375, 377, 393–394, 414, 419, 427, 432–435
ligninases 51, 53, 149, 332
ligninolytic enzyme 68, 174–176, 196, 312, 314, 317
lignocellulolytic 51–54, 62–63, 69, 83, 98, 120, 149, 239–240, 259, 264, 311, 320–321, 329–332
  bacteria 51, 149
  enzymes 51–52, 62, 69, 83, 120, 264, 320–321, 329, 332
lignocellulose 15, 29, 35, 38, 41, 43, 52, 62, 64, 68, 70, 78, 86, 95, 98, 111–112, 114, 116, 120, 138–139, 158, 164–166, 176, 179, 181, 192, 194–195, 197–199, 201, 232, 234, 240, 251, 256, 317, 341, 347, 352, 357, 360, 369, 371, 373–374, 403, 427, 434
lignocellulosic 4, 15, 17, 19–22, 25–27, 29, 31, 33, 35, 38–45, 51, 54, 56, 62–65, 67–68, 70–78, 82–91, 92–93, 95–96, 98, 108, 113–116, 118–122, 124–125, 127, 130, 136–140, 145–148, 151–152, 156–158, 163–167, 169, 173, 174, 176, 179, 184–185, 190, 192–198, 200, 203, 213, 217, 221, 231, 234, 236, 240, 251, 255–260, 263, 271, 284, 298, 300–305, 309, 311–312, 314, 317–322, 328–332, 341, 342, 347–348, 355, 371–375, 385, 402–405, 414, 435
Ligusticum chuanxiong 427
limonene 63, 313
linear economy 443
linkages 15, 18, 74, 83, 127, 168, 174–175, 305, 312, 375, 394
lipases 149, 152, 198, 281, 285, 290, 294, 299, 358
lipid 6, 43, 49, 120, 145–146, 175, 204, 210, 212, 303, 311, 321–322, 347, 359, 408, 424, 426, 431
lipolytica 56, 58, 67, 241, 245, 304, 309, 322, 435, 439
liquefaction 21–22, 32–33, 35, 40, 89, 116, 137, 148, 150, 389–394, 400–402, 405
liquid–liquid partition 46
lycopene 308, 313–314, 320
lytic polysaccharide monooxygenases 82

macromolecules 14, 22, 67, 69, 85, 120, 193, 270, 283, 293–294, 297–298, 335–336, 350, 352, 359, 441
macronutrients 408, 418
magnesium 412
maize 25, 33, 61, 164, 267–269, 274, 278, 325, 333, 352, 430–431, 438, 441
malachite green 412
malate 243, 251–253
maleic acid 30
malnutrition 305
malonic acid 427, 430, 439
maltose 149, 432
manganese 55, 86, 129, 175, 178–179, 300, 312, 314, 340
  peroxidase 129, 178–179, 300, 314, 340
mannanases 17, 131, 174–175, 177, 194, 305, 307, 312, 330

mannose 17, 18, 50, 55, 74, 98, 99, 132, 165, 174, 179, 305, 330, 347, 432
manufacturing 24–26, 28, 32–35, 37–38, 45, 50, 60, 62, 84, 94, 107, 112, 121, 205–212, 217, 227, 234, 242–245, 247, 261, 266–267, 275, 283, 284, 287, 307, 347, 353–357, 364, 369–370, 378, 407, 433, 445–446
manure 6, 13, 23, 93, 300, 387, 408, 419–420, 422, 424, 425
marine organisms 325
mass 1–28, 29–46, 48, 50–56, 58–68, 70–78, 80–87, 89–96, 97–102, 104, 106–114, 116–122, 124–139, 141–142, 144–154, 156–165, 166–176, 178–182, 184–186, 188–195, 196–204, 206–208, 210–214, 216–218, 220–222, 224–231, 232, 234–240, 242, 244, 246, 248, 250–252, 254–259, 260–264, 266, 268, 270–275, 276, 278, 280–284, 286, 288, 290, 292, 294–296, 298, 300–305, 306–319, 320–331, 332–341, 343–350, 348, 352, 354, 356–358, 360–364, 366, 368–374, 376–380, 382, 384–385, 386–390, 392–396, 397–405, 407–408, 410–418, 420–425, 426–450; transfer 188–189
mathematical model 298, 319, 344, 432, 437
maximizing bioenergy benefits 401
maximum yield 152, 314–315, 332, 414, 432, 435
MCAX 54
meat 49, 138, 149–150, 157, 300, 325, 327, 329, 335–338, 363, 364, 366–367, 381, 383, 387
media 26–29, 32–34, 37, 39, 41, 46, 48, 60, 62–63, 68, 70, 77, 85, 93, 95, 99–101, 108–109, 126, 141, 175, 197, 204, 209–210, 213, 228, 232, 235, 241, 249, 257, 260, 285, 302, 303, 309, 313, 339, 355, 372, 376, 381, 393, 425, 439
medicines 123, 206, 278, 350, 430
membrane 27, 44, 46, 48, 50, 67, 69–70, 131, 134, 138, 161, 172, 182–183, 189, 199, 209, 258, 267, 272, 278, 304, 320, 323–325, 327, 329, 333, 335–344, 346, 350–362, 363–367, 369–376, 377–387
mesophilic 131, 344, 419
mesoporosity 27
mesoporous 27, 42, 403
metabolic 45, 54, 56, 58–59, 61–68, 93, 96–99, 100–102, 105–110, 111, 113–118, 120, 132, 136, 138–141, 158, 160, 195–200, 202–205, 210, 216, 219–220, 225–236, 240, 244, 252–255, 258, 270, 282–283, 289–291, 295, 297, 301, 305, 310, 314, 341–342, 357, 374, 376, 441
  engineering 45, 54, 56, 58–59, 62–63, 66, 67, 93, 97, 99–102, 105, 113–118, 120, 136, 138–139, 160, 195, 205, 219–220, 225–228, 230, 233–234, 258, 270, 282–283, 289–290, 295, 297, 341, 441
  fluxes 100–102, 105, 106–107
metabolism 54, 61, 66, 99, 100, 102, 105, 107, 118, 127, 144, 158, 196, 204, 207, 210, 219–220, 227–229, 233, 278, 295, 303, 342, 441
metabolite 84, 109–110, 121, 144, 199, 209, 236, 241–242, 289, 309–310, 314, 321, 339, 373
  production 121, 310, 314, 321
metagenome 436, 441
  metagenomics 92
metal catalyst 283, 379
metalloenzymes 418
metals 1, 3, 247, 280, 294, 395, 400, 421, 423, 442, 444
metamorphosis 28
methane 13, 23, 36, 59, 84, 122, 197, 203, 262, 275, 376, 385, 389, 395, 407–408, 413, 418–426, 446
methanogenesis 376, 407, 408, 418, 420
methanogenic archaea 328, 343
methanogens 408
methanol 25–28, 30, 40, 42–43, 138, 169, 246–248, 323, 342, 346, 361, 378–379, 393–394
  carbonylation 246–247, 378
  formation 394
methoxylation 19
methyl
  ethyl ketone 158–159
  ferulate 433
  orange 412, 426
methylation 153, 437, 441
metK gene 436
microaerobic conditions 144
microalgae 67, 90, 107, 137, 192, 195, 232, 271, 278–279, 287, 296, 342, 360, 363
  harvesting 67, 360
microbeads 280
microbes 13, 19, 36, 38, 45, 51, 65–66, 93–94, 96–97, 108–109, 118, 123, 132–133, 149, 163, 173–174, 194, 196, 203, 205, 211, 218, 221, 225–228, 233, 235, 239, 242, 244–245, 252, 267, 274, 278–279, 295, 305, 307–314, 324, 327–329, 340, 354–355, 358, 365, 380, 390, 419, 435–436

microbial 43–47, 49, 51–53, 51–57, 55, 57, 59, 61–71, 63, 65, 66, 67, 69, 71, 73–74, 77–78, 82–83, 87–88, 90, 92–95, 93–95, 97–103, 99–101, 103, 105–117, 107, 109, 111, 113–115, 117, 118–121, 119, 121, 122, 129–133, 130, 135, 136, 138, 140, 144, 147, 151, 158–159, 173, 176, 192–194, 196, 199–200, 202–203, 205, 207–209, 211, 213, 215, 217, 219–221, 223, 225–229, 228, 231–236, 238–242, 239, 242, 251–255, 254, 256–261, 262–267, 263, 267, 269–272, 271, 276, 278, 284, 286, 293, 295, 297, 300–305, 301, 303, 307–313, 309, 311–312, 314–321, 315–316, 318, 320
  enzymes 61
  fermentation 45, 323, 325, 327–328, 327–330, 332, 335–344, 338–339, 338
  growth 53, 354–355, 372, 374, 376, 384–385, 408, 412, 420, 426–427, 431, 439
microbiome 108, 200
microbiota 210, 236, 421
microchannels (MC) 349
microfiltration 48–49, 333, 335, 337, 342, 352, 367, 371, 381, 386
microhabitats 419
micronutrients 418
microorganism 1, 6, 29, 31, 36–37, 45, 51–53, 56, 58–60, 62–63, 69–70, 73–74, 77, 81–85, 90–94, 96–98, 100–102, 105, 109–111, 113, 123, 131–132, 144–145, 147, 150, 170, 172, 176, 181–182, 184, 195, 200, 203, 205, 211, 213, 217, 219–221, 225, 226, 228, 230, 233, 239–245, 248, 251–252, 262–264, 267, 278, 284, 289, 301, 302, 305, 311–312, 316–317, 324, 327–330, 332–333, 335, 338, 340, 344, 355–357, 361–362, 365, 376, 380, 383, 407–408, 418, 420–421, 433, 435, 437
microwave 31, 74, 77, 88, 128, 137, 149, 151–152, 183, 349, 401, 405, 417–418, 423, 423; pyrolysis 417
mild pyrolysis 416
milling industries 431
mineral phases 420
miscanthus 16, 21, 25, 93, 166, 300
mitochondrial 208, 267
mixture 4, 6, 20, 21, 25–26, 31, 34, 47–49, 67, 82, 96, 115, 123, 128, 134, 142, 147, 152, 165, 172, 180, 189–190, 201, 214, 217, 220, 266, 285, 298, 321, 325, 327, 333, 335–338, 349–350, 355–356, 359–366, 370, 377, 380–381, 383, 386, 394, 398, 415, 433–434
ML techniques 398
mobile phase 46, 49
modelling 135, 137, 160, 344, 398
mode of operation 382
modification 30, 52, 54, 100–107, 129, 169, 183, 186, 193, 219, 227, 229, 255, 289, 294, 312, 386, 424, 436
modular design 437, 439
moisture 33, 36, 73, 75, 84, 95, 188, 239, 272, 280, 289, 372, 376, 392–394, 408, 416
molasses 25, 117, 121, 125, 140, 147, 149–150, 153, 160–161, 195, 203, 218, 236, 244, 249, 251, 256, 259–260, 267, 301, 304, 307, 322
molecular
  mechanisms 229
  oxygen 312
  sieves 329
  size 189, 336
  weight 14, 19, 26, 29, 35, 74, 242, 281, 285–286, 293, 338–339, 350, 367, 381, 398
molecular mass 15, 124
molecules 1, 14–15, 18–19, 22, 27, 32, 46, 48–50, 63–64, 67, 69, 72, 81, 85, 90, 119–120, 128, 133, 137, 145, 150–151, 161, 169, 173, 175, 179–180, 187, 189, 193, 198, 200, 208, 210, 226, 229, 233–234, 237, 253, 256–257, 266–267, 270, 283, 285, 292–294, 297–298, 303, 319, 322, 325, 329, 335–337, 339–340, 350, 352, 358–359, 363, 365–366, 375, 383, 394, 420, 424, 441
monitoring 108, 111, 226, 235, 332, 340
monocarboxylic 245, 247
monomer 17–18, 51, 72, 78, 98, 126–127, 129, 131, 133, 167–168, 181, 236, 250, 264–265, 277–278, 281–282, 285, 290–293, 354
monosaccharide 17, 29, 31, 78, 149, 174, 194, 347, 386
mono sugars 362
morphology 76
mRNA secondary structure 437
multifunctional catalysts 400
multiple enzymes 191, 310, 314
municipal solid waste 1, 166, 199, 219, 317, 353, 388, 407, 411
mutagenesis 54, 187, 290, 291
mutation 61, 106, 110, 173, 185, 187–188, 193, 219, 289, 313
mutualism 109

NADH 144, 151, 243, 436
nanocellulose 42, 268, 272, 287, 309

nanofibers 268, 270, 279
nanofiltration 49, 325, 333, 336, 342–343, 345, 352, 369, 371–373, 375, 381, 386
nanopolymer 265
nanoporous structure 13
nanoscale substances 289
nanotechnology 278, 279, 307
NaOH 19, 39, 76, 152, 169, 233, 430
naphtha cracking 142
national economy 340
national ethanol fuel program 125
natural decomposition 288
natural resource utilization 141
negative 1, 3, 14, 19, 31, 47, 170, 173, 207, 209, 220, 292, 304, 365–366, 369, 381
  charge 47, 381
nematodes 238
Neosartorya fischeri 77, 178
neurodegenerative illnesses 427
Neurospora crassa 97
neutralism 109
neutralization 76, 96, 169
Niger 55, 66, 81, 84–85, 88, 131, 177–178, 192, 203, 206, 212, 218, 228, 234, 240, 242, 244–247, 251, 255–260, 270, 309, 312, 319, 322, 333, 422, 433, 439, 441
nitrogen 13, 14, 83, 99, 100–101, 124, 212, 215, 234, 262, 267, 346, 420, 433
nitrous oxide 122, 413, 420
nonanoic acid 356
novel isolates 305
NOx 26, 85, 97, 122, 124, 247, 255, 346, 379, 414, 416, 422, 427
nuclear power 346
nucleases 54
nucleic acid 46, 277
nucleophilic attack 285
nutraceutical 280, 305, 307, 317, 319
  prebiotics 305, 307
nutrient 14, 45, 61, 73, 99, 100–101, 109–110, 114, 119, 145, 161, 166, 211, 215, 221, 227, 254, 264, 267, 275, 280, 303, 310–311, 314, 332, 342–343, 355, 387, 408, 411–412, 418, 420–421, 445–446
nutrition 13, 69, 90, 206, 217, 241, 284, 297, 301, 305, 307–309, 312–315, 319, 321, 412, 438

obligate anaerobe 210
observable pigment producers 303
octane 25, 123–124, 138, 141, 348
octane number 123–124, 141, 348
–OH 15, 169, 421
oil 2, 13–15, 19, 21–23, 26, 34–35, 41, 43, 60, 64–67, 84–85, 93, 96, 108, 111, 116, 119–126, 134, 138–140, 142, 145, 152–153, 157, 160, 162, 166, 176, 189–190, 197, 200, 204, 206, 208, 211–212, 230, 237, 241, 246–247, 252, 256–257, 262, 264, 267, 269–272, 273–274, 277–282, 289–290, 293–294, 297, 301, 303–305, 308, 311, 313, 319, 320, 347, 352, 356, 362, 369, 371, 383, 385, 389–393, 394, 397–405, 407–408, 411–414, 417, 420, 422–426, 429, 430, 432, 435, 436, 439, 441, 446
oleaginous feedstock 145, 152–153, 156
oleyl alcohol 356
oligomers 30, 51, 78, 171, 200, 354
oligosaccharides 18, 29, 86, 174, 181, 199, 284, 305, 307, 318, 320–321, 329, 330, 341, 344, 348, 433
olive waste 304
omics 92, 105, 107, 115, 117, 137, 145, 159, 229–230, 234–235, 328, 359
oncolytic 263
optical purity 142, 144–145, 219
optimal condition 183, 432, 436
optimal OLR 384
optimization 44–45, 53, 61, 66, 70, 84–85, 93, 97, 99–102, 105–107, 111, 113, 115, 118, 120, 137, 156, 158, 161, 181, 184–185, 190–191, 196–200, 203, 220–227, 229, 232–235, 239, 258, 270, 272, 286, 290–291, 295, 301, 305, 308, 314–316, 318, 321–322, 332–340, 386, 394, 403, 422–423, 435–437, 440
optimizing fermentation 51, 227, 244
organic
  acids 30, 37, 45, 50, 59, 61, 76, 84, 97, 139, 147, 169, 202, 231, 232, 234, 237–241, 254–259, 302–304, 309, 316, 328–329, 331, 336, 337, 341, 357, 386
  compounds 28, 130, 202, 259, 338–339, 346, 382, 398
  fouling 339
  precursors 26
  solvent 67, 76, 77, 90, 123, 151, 331, 352, 393
  waste 5, 6, 36, 239, 248, 254, 256, 266–267, 327, 376, 384–385, 404, 408, 412, 422, 424–426

organism 1, 6, 29, 31, 36–37, 45, 51–53, 56, 58–63, 69–70, 73–74, 77, 81–85, 90–94, 96–98, 100–102, 105–111, 113, 123, 131–132, 144–147, 150, 170, 172, 176, 181–182, 184, 186, 195, 200, 202–205, 208–209, 211–213, 217, 219–221, 225–230, 233, 239–245, 248, 251–253, 262–264, 267, 274, 278–279, 281, 284, 288–289, 301–302, 305, 311–312, 316–317, 324–325, 327–333, 335, 338, 340, 344, 355–357, 361–362, 365, 376, 380, 383, 407–408, 418, 420–421, 433, 435–437
organosolv 77, 86, 138, 169, 192, 194, 330–331, 341, 349
 pretreatment 138, 331, 341
Orleans method 379, 380
outlook 157, 265, 270, 370, 374
output 14, 19, 25, 46–47, 93, 124, 215, 301, 316, 356, 359–360, 365, 380, 431, 435, 445
overall
 cost 58, 171, 339, 357, 365, 369, 392
 process efficiency 110, 113, 338
overexpressed 55, 253
overexpression 55, 102, 115, 228, 252, 290
oxaloacetate 243, 252, 259
oxidases 51, 174–176, 300, 311–312, 340
oxidation 19, 30, 33, 39, 52, 70, 144–145, 153, 174–175, 209, 213, 218, 246–248, 255, 288, 378–380, 391, 395, 416, 421, 431, 438
oxidative
 delignification 30
 deterioration 303
 stress 204, 255, 427
oxides 3, 29, 38, 124, 142, 346, 394, 421
oxidizing agents 76, 395
 solvent 394
oxisol 407
oxygen 2, 4, 13–15, 32, 34–36, 77, 82, 101, 110–113, 121, 124, 144, 159, 174, 176, 202, 204, 206, 209–212, 215, 217, 227, 240, 285, 312, 391–398, 400, 407, 411, 413–416, 419, 422, 425
ozone 30, 85, 330–331
ozonolysis 19, 30, 77, 90, 331, 345

propanediol 149, 158, 284
pyruvic acid 284

quality 1, 14, 18, 20, 33, 44, 47, 50, 77, 136, 166, 168, 170–171, 182–183, 188–190, 212, 216, 220, 227, 236, 238, 260, 265, 268, 272, 274, 280, 301, 304–305, 307, 313, 316, 327, 331–332, 336, 339, 347, 348, 355, 357–358, 379, 394, 396–400, 401, 404, 407, 412, 415, 419, 424–425, 441
quantification 50, 137, 371, 421, 442
quick process 379
quick pyrolysis 417
quorum 112

racemic 250
rainfall 313
Ranunculaceae 427
rapeseed 74–75, 93, 119, 128, 278
rate 14, 45, 48, 50, 61, 78, 83, 135, 140, 152, 179–182, 185, 203, 215, 221, 252, 261, 273, 275, 282, 285, 288–289, 339, 346, 348, 351, 353, 355, 364, 377, 380, 382, 393, 396, 407–408, 412, 414–419, 421–422, 430, 437
rational design 107, 111, 183, 186–188, 290–291, 297
raw beet juice 47
raw material 26, 128, 130, 211, 217, 230, 250, 256, 292, 315, 342, 354, 443
raw substrates 338
RBO soap stock extracts 433
reaction 25, 29–30, 32, 34–39, 65, 78, 81–82, 142, 144, 168–170, 180–181, 183, 199, 217, 243, 248, 281, 283, 312–313, 331, 349–352, 358–359, 365–366, 370, 373, 379–380, 390, 392–395, 400, 430, 432, 437–438, 446
reactor 76, 87, 110–112, 189, 232, 325, 327, 342, 344, 349–355, 350, 358–359, 365–367, 371, 373–374, 380–381, 400–401; design 111–112
recalcitrance 71, 74, 91–92, 95, 121, 171, 198, 201
recalcitrant 42, 62, 130, 151, 165, 300–301, 311
recombinant 54, 58, 64, 67, 116–117, 119, 120, 193, 195, 226, 272, 373, 375, 435, 438
recombined 263
recondensation 416, 417
recovery 19, 25, 47–49, 67, 77, 88, 120, 143, 159, 168, 170, 189–192, 197, 199, 203, 229, 239–240, 256, 263, 320, 336–338, 345, 357, 360–365, 367, 373–375, 377–384, 386, 394, 398, 402, 404, 416, 424–425
rectification column 355
rectisol process 361
recyclability 68, 147, 170, 172, 290, 291
recycle 338, 352, 443, 444
redox 419–420, 422, 424
reduce 26, 59, 62, 95, 102, 110, 124, 125, 129, 147, 163, 165, 171–174, 180, 182–184, 187, 190–191, 203, 213, 229, 237, 243, 262, 292, 303, 324, 327–328, 335, 339, 348–349, 357, 359, 361, 370, 388–389, 393, 398, 400, 412, 443–444
 sugars 40, 90, 152, 167, 307
reed root 427
reed straws 434, 440

Reesei 55–56, 58, 65, 71, 81, 97–98, 117, 131, 163, 177–178, 201, 239, 240, 333
refinement 47, 61, 113, 263
refineries 93, 294, 369, 389, 399, 448
reforestation 5
reforming 26, 35, 40, 41–42, 395, 423
regeneration 144, 189, 192, 294, 402, 441
regenerative 394, 400, 443, 448
regularization 315
regulatory
 bodies 301
 framework 348
 guidelines 301
 hurdles 359
reinforcing agents 280, 287, 289
rejection 350, 382, 386
rejuvenating 254
renewability 264, 347
renewable 4–5, 13, 21–26, 32, 37, 39–44, 51, 59, 64, 66, 68, 70–72, 74, 78, 86, 89–90, 92, 94–96, 98, 101, 107, 113–116, 119, 125, 138–143, 145, 147, 149, 151, 153, 155, 156–157, 159, 161, 163, 166, 190–195, 198–199, 203, 205, 211, 217, 219, 221, 231, 237, 240, 248, 251, 262, 264, 270, 274, 278–284, 291, 296–298, 317, 323, 325, 329, 341–346, 348, 352, 358, 367, 372, 374, 376, 388–389, 401–405, 422–425, 427, 444, 446, 449
reservoir 4, 95, 122, 136, 209
residence time 30, 393, 396, 398, 414, 415, 416, 417
residual 6, 47, 93, 141, 145, 153, 156, 237, 255, 267, 268, 305, 329, 338, 381, 437, 442, 444, 445
residue 6, 118, 123, 126, 136, 240, 262, 269, 347, 381, 408, 418, 444, 447
resilience 19, 51, 81, 185, 203, 230
resilient food supply chain 316
resin 47, 49, 362
resource 2, 4, 6, 13, 21, 25, 26, 39, 52, 56, 59, 68, 72, 74, 107, 108, 109, 113, 119, 122, 129, 133, 138, 140, 145, 150, 156, 163, 165, 167, 170, 171, 172, 191, 203, 219, 230, 238, 261, 262, 264, 266, 267, 269, 270, 273, 274, 276, 281, 296, 297, 298, 314, 315, 316, 320, 323, 336, 338, 342, 347, 352, 360, 376, 377, 424, 430, 431, 433, 438, 444, 447, 449
respiration and fermentation 144
respiratory process 144
restorative 443
retarded growth of microorganisms 338
retentate 327, 336
retention 32, 49, 68, 166, 275, 280, 365, 367, 369, 377, 382, 412, 421, 444
reusability 139, 172, 286, 375
reuse 283, 294, 320, 338, 362, 365–366, 443
reuse and recycle 338
reverse osmosis 67, 340, 342, 361, 381, 386–387
reverse salt flux 382
Rhizoma sparganii 427
Rhodococcus 52, 231
Rhodopseudomonas faecalis 320
rice 16, 44, 66, 70, 75–78, 82, 86, 89, 91, 94, 119, 121, 145, 166, 242, 246, 259, 262, 269, 271, 278, 311, 319, 323, 325, 344, 347, 404, 425, 428–432, 435, 438, 441, 444, 446
rye bran 428, 433

saccharides 82, 151, 168, 328
saccharification 31, 65, 71, 75, 77, 83, 85, 87, 89–90, 95–96, 101, 117, 120, 133, 148, 150, 152–153, 156, 168, 193–194, 197, 233, 236, 249, 258, 322, 342, 355, 360, 371, 373–375
saccharine biomass 149
saccharolytic 96–98, 181
Saccharomyces 56, 58, 65, 83, 96, 116–117, 120, 132, 158, 160, 195, 197, 205, 208, 221, 233–239, 256, 270, 307, 328, 342, 343, 374
 boulardii 307
 cerevisiae 56, 58, 83, 96, 116–117, 120, 132, 158, 197, 208, 233–239, 256, 328, 342, 343, 374
 pastorianus 58, 65
Saccharomyces cerevisiae 56, 58, 61, 83, 96, 98, 100, 206, 208–209, 228, 248–249, 251–253, 328, 354
safety 21, 65–66, 118, 170, 189, 196, 234, 236, 253–255, 259–262, 287, 295, 301, 309, 313, 315, 321, 356, 359, 363, 404, 431
salinity 51, 309
salts 3, 30, 46, 99, 241, 243, 336–337, 339, 352, 362, 383
sawdust 13, 77, 166, 310, 323, 325
scaffold 280, 297
 scaffolding protein 176, 186
scalability 85, 101, 112–113, 190, 205, 352, 431–432, 438
scaling 48, 61, 94, 111–112, 135, 173, 188–189, 312, 332, 339, 384, 398
scraps 33, 93, 278
screening 43, 66, 69, 101, 105, 107, 183, 187, 199–200, 227, 259, 271, 289, 296, 385, 433
sedimentation 335, 373, 381
seeds 6, 13, 17–18, 22, 42, 262, 307, 308, 342, 377
selection 32, 48, 52, 56, 61, 78, 99, 100, 105–107, 180, 185, 203, 205, 220–221, 226, 229, 239, 262, 264, 301, 316, 328–329, 338, 340, 373, 393–394, 422

selective removal 335, 336
selectivity 48, 163, 183, 185, 189, 281, 283, 289–290, 294, 329, 332, 335, 337, 363, 374, 378, 381, 383–384, 400
Selexol method 361
semipermeable membrane 337, 365
sensing 38
sensors 27, 63
separate 46–50, 88, 96, 133, 172, 189–190, 217, 267, 323, 325, 327, 329, 350, 352, 355–356, 359, 361–364, 366, 369–370, 374, 381, 383–384, 400
separation
  efficiency 48, 383
  factor 381
  performance 362, 369, 381
  technique 324, 335, 362, 369
  from water 356
sequencing 92, 184
setup time 358, 370
sewage sludge 381, 385–386, 422–423
shaking speed 435
shelf life 204, 245, 279, 304, 307, 309
shredding 392
side product 166, 398
sieves 46, 329
signal transduction 229
silk fibroin 281, 296
simulation 40, 135, 137, 344, 432, 437
sinapic acid 432
sinapyl 19, 165, 347
single peptide truncation 436
single reactor 355, 400
sintering 395
size 45, 48–50, 61, 75–78, 128–130, 147, 169, 180, 182, 189, 197, 327, 329–331, 335–336, 339, 350, 358, 381, 392, 395, 416, 419
skin care products 427
slag 3, 23
slow pyrolysis 34, 391, 396, 398, 401, 415
sludge 137, 293, 319, 381, 385–386, 418, 422–426
soap stock 432, 433
soil
  amendment 93, 166, 407, 425
  bioremediation 313
  deterioration 407
  fertility 13, 411–412, 414
  humus 408
  microbial activity 412
  quality 166, 190, 274, 412
solar 4, 13, 32, 38–39, 343, 346, 364, 375
soluble sugars 174, 310, 330
solvent 28, 46–47, 49, 67, 77, 85, 90, 108, 123, 131, 141, 148, 151, 189–190, 240, 242, 247, 250, 252, 268, 296, 354, 356–357, 360, 362, 373, 377–378, 391, 393–394, 433, 438
  extraction 47, 240, 268, 354, 360, 438
  media 393
  mixture 67, 433
solvothermal liquefaction 394
sonication 31, 129, 272
sorbent biochar 411
sorghum 16, 25, 43, 75, 78, 88, 160, 166, 231, 278, 321, 342
sorption 363, 423–424, 426
soybean 59, 93
soy whey wastewater 49
species 4, 32, 51–52, 86, 88, 95–97, 100, 107, 109, 115, 129, 131–132, 136, 142, 176, 203–204, 207–208, 211–212, 215–216, 220–221, 235–236, 239, 241, 245, 267, 311–312, 318, 329, 422–423, 433, 438
specific
  bacterial consortium 302
  heat 124
specificity 32, 85, 171, 173, 182–186, 205, 255, 281, 283, 286, 290–291
spectroscopy 241
speed 46, 48, 75, 78, 124, 132, 138, 159, 254, 287, 312, 314, 324, 339, 433, 435
spent coffee ground 307, 437, 439
Sphingomonas 52
stability 14, 25, 54, 58, 85–86, 92, 102, 105–107, 110–111, 123, 171–173, 181, 183–189, 194, 203, 239, 254, 272, 281, 286–288, 290–291, 293, 349, 352, 359, 373, 400, 407, 424, 441–442
stabilization 18, 253
starch 6, 20, 38, 45, 66, 72, 78, 82–83, 87–89, 123, 145–146, 148, 150, 156, 214, 221, 234, 255, 262, 267–272, 274–276, 279–280, 286–287, 289, 292, 294, 296–298, 341, 373
starchy
  biomass 150, 328, 347
  materials 147, 150, 446
stationary phase 46, 49, 190
statistical experimental design 432
steam 19, 21, 31, 33–36, 42, 52, 74, 115, 128, 151, 164, 167, 169, 179, 181, 183, 197, 217, 284, 296, 330, 344, 349, 357, 360, 364, 371, 388, 392, 394–395, 417, 423, 446
steam explosion 19, 21, 31, 42, 52, 74, 115, 128, 151, 164, 167, 169, 179, 181, 183, 197, 217, 284, 330, 344, 349, 357, 360, 371
stems 2, 6, 42, 285, 377
stereospecificity 144
steric forces 382
sterilization 150, 220
storage 3, 14, 23, 27, 42, 66, 94–95, 118, 166–167, 196, 204, 237, 245, 295, 344, 398, 400, 403, 407

strain 45, 53–56, 58, 61, 64, 67, 85, 99–102, 105–107, 109, 115, 121, 133, 139, 153, 156–157, 159–160, 197, 199, 203, 215–216, 219–221, 226–227, 229–230, 232, 234, 236, 242, 251–252, 258, 286, 301, 304, 310, 312, 314–315, 322, 340, 354, 435–437
development 219, 286
optimization 99, 101–102, 105–107, 115
selection 61, 203, 220–221, 226, 229
stability 102, 105–107
straw 3, 6, 13, 16, 19, 21–22, 25, 40–41, 44, 70, 74–78, 82–87, 89–91, 93–94, 100, 119–120, 128, 138, 151, 159, 166, 197, 203, 235, 238, 242, 245, 251, 258–259, 262, 303, 310–311, 319, 323–324, 333, 341, 343–344, 402, 407, 426, 431, 434, 450
streams 37, 56, 60, 62–63, 78, 85, 264, 333, 336, 384–385, 447
Streptococci 207, 208
Streptomyces 52, 131–132, 176–178, 239, 296, 303, 307–308, 312, 318, 321, 433, 434, 439
olivochromogenes 439
stress 45, 61, 108, 109, 115, 120, 197, 204, 255, 343, 420, 427, 442
structural rigidity 19, 92, 311
submerged 67, 73, 84, 86, 88, 225, 234, 239, 244, 247–248, 257, 303–304, 314, 320, 373, 379, 380
fermentation 67, 73, 84, 86, 225, 239, 244, 248, 303, 304, 314, 320
substituent 398
substitute 27, 30, 35, 59, 122, 242, 268–269, 275, 320, 348, 444
substrate 40, 42, 56, 64, 73, 81, 88, 98, 100–102, 105–106, 109, 111–113, 117, 129, 133, 141, 146, 150, 159, 168, 171–172, 174, 176, 179–183, 185, 188, 194, 200–201, 205, 215, 217–219, 225, 227, 242–243, 245, 248, 251, 254, 257, 277, 282, 290–291, 301, 309–314, 319, 327, 332–333, 338, 340, 342, 350, 355, 358, 361, 370, 377, 380, 420, 433, 435–436, 440
succinate 243, 253
succinic acid 28–29, 59, 202–203, 205–206, 213, 219–220, 228, 235, 240, 256, 260, 266, 267, 270–272, 383
sucrose 50, 70, 132, 148–150, 153, 244, 267, 328, 371, 432
sugar 17, 20, 24, 26–28, 30, 32–33, 35–36, 39, 45, 47, 50–51, 54–56, 59, 62–63, 65, 67, 69–70, 72, 74–76, 78, 83, 87, 89, 115–116, 123, 125, 129–133, 136, 139, 145–146, 150–152, 159–160, 168–169, 175, 179, 194, 203, 205–208, 212–218, 221, 239, 242–245, 267, 269–270, 279, 287, 294, 299, 302–305, 307, 311–314, 320, 323, 341, 347–348, 354–355, 360–361, 363–364, 369, 373, 375, 377, 381, 429–430, 434, 440, 446
sugarcane 21, 25, 33, 40, 44, 50, 61, 67, 75, 78, 82–85, 88–89, 94, 116–117, 120–121, 124–125, 128, 150, 153, 156, 159, 160, 165, 166, 193, 203, 213, 238, 242, 244, 251, 256, 262, 267, 302, 304, 309, 311–312, 314, 318, 323, 324–325, 341, 344, 347, 352, 377, 431, 446
sulfolobus 51
sulfonic 202
sulphur 124
sulphuric acid 30, 128–129
sunflower 13, 93
supercritical 35, 294, 362, 404–405
fluid extraction 362
superhydrophobic membranes 383
superior liquid products 394
supernatant 46, 120
supply 1–2, 4, 60, 72, 110, 122, 125, 144, 162, 215, 238, 259, 262, 286, 315–316, 352, 355, 357, 370, 388, 437
surface 3, 13, 31, 49, 56, 58, 71, 74–75, 95, 99, 114, 117, 126, 129–130, 147, 151, 169, 172, 180, 182–183, 193, 201, 248, 250, 254, 257, 303, 314, 318, 330–331, 335, 339, 358–359, 379–382, 386, 392, 398, 412, 416, 419–422, 425, 432, 437
area 13, 74–75, 95, 126, 129–130, 147, 151, 169, 180, 182–183, 330–331, 392, 398, 412, 416, 419, 420, 422, 425
fermentation 379–380
functional groups 419, 421
surfactant 47
surgical 265, 280
suspended
cell 380
particles 325
solids 335, 352, 377, 381, 422
sustainability 29, 45, 60, 62, 65, 68, 88, 90, 107, 114, 171, 192, 238, 256, 264, 267, 276, 284, 288, 293–296, 298, 309, 317, 320–322, 332–333, 340, 352, 363, 370, 402, 431, 438, 442–445, 447–450

sustainable 2, 5, 13, 21–25, 40, 42, 44–45, 53, 58–64, 66–68, 70–71, 74, 85, 87, 89–90, 92–96, 101, 106, 107, 113, 115–118, 121–122, 134–136, 139–141, 143, 145–149, 151, 153, 155–157, 159–161, 164, 166–167, 173, 189, 191–200, 201, 203, 205, 207, 211, 213, 219–221, 227, 230–234, 238, 244, 254, 256, 259, 262, 266–267, 269, 272–275, 278–279, 281, 283, 287–288, 291–297, 300, 301, 308, 315–321, 323–324, 327, 338–342, 346, 369–371, 374, 377, 385, 386, 388–389, 400–405, 422–426, 432, 434, 447–449
  agriculture 320, 324
  development 59, 74, 140, 160, 173, 300, 371, 374, 447, 448, 449
  development goals 140, 160, 300, 447
  economy 63, 115, 192, 231
  energy 2, 21–24, 64, 66, 70–71, 85, 90, 92–93, 113, 116, 118, 192–193, 196, 295, 317, 323, 340–342, 389, 401–404, 422, 423, 425
sutures 265
sweeteners 45, 304, 309, 314
swelling 169, 204
switch grass 16, 352
synergistic activity 186
syngas 14, 21, 25–26, 34, 40, 42–43, 114–115, 199, 251, 258, 361, 389, 391–392, 400, 414–416
syngas production 199
synthesis 2, 4, 6, 8, 10, 12, 14, 16, 18, 20, 22, 24, 25–33, 34–37, 38–45, 46–49, 50–57, 58–65, 66–73, 74–81, 82–83, 84–87, 88–91, 92–97, 98–105, 106–113, 114–121, 124, 126, 128, 130, 132–136, 138, 141, 142, 144, 146, 148, 150, 152, 154, 156, 158, 160, 162–165, 166–171, 172–179, 180–185, 186–191, 192–197, 198–205, 206–213, 214–219, 220–225, 226–231, 232–237, 238–245, 246–254, 255–260, 261–265, 266–269, 270–275, 276–284, 285–290, 291–296, 297–300, 301–305, 306–309, 310–319, 320–323, 324, 326, 328, 330, 332, 334, 336, 338, 340, 342, 344, 346–347, 348–355, 356–358, 359–362, 363–365, 366–369, 370–373, 374–378, 379–385, 386–387, 390, 392, 394, 396, 398, 400, 402, 403, 404, 408, 410, 412, 414, 416, 418–420, 422, 424, 426, 427, 428, 430, 431–440, 444, 446, 448, 450
  gas 26, 41, 43, 394
  and purification 347, 349, 351, 353, 355, 357, 359, 361, 363, 365, 367, 369–371, 373, 375
synthetic 1, 68, 92, 105, 107–108, 111–112, 111–117, 116, 120, 140–141, 173, 186, 193, 196, 200, 205–206, 226, 230, 230; polymers 269, 242, 249, 259, 262, 266, 268–269, 275–276, 278, 280, 283, 287, 291–292; rubber 140–141, 291–294, 298, 305, 369, 372, 427; biology 92
syringyl 19, 127
systemic resistance 142
systems biology 93, 105, 107, 119, 229, 231, 235

tannins 308
tar concentration 394
taste 123, 218, 245–247, 252, 254–255, 302, 304, 377
  and aromas 305
taurine 313
TCA cycle 228–229, 243
TCP 3, 4
technically feasible 393
technological
  advancements 283, 370, 401
  aspects 379
  risk 347
temperature 3–4, 22, 30–31, 33–37, 38, 61, 75–76, 78, 81, 84, 86, 101, 108, 110–111, 122, 124, 128, 133, 135, 141, 144, 148, 151, 169–170, 173, 181–185, 189, 210, 215–218, 221, 225, 227, 236, 239, 242, 244, 248, 250, 264, 280, 282, 289, 309, 312, 314, 324, 330–332, 339, 342, 359, 362–364, 371–372, 377, 380–384, 389–398, 400, 402, 411–422, 425, 430, 432–437
tensile
  industry 202, 250
  strength 126, 269, 279, 288
texture 18, 126, 206–207, 253–254, 302, 309
therapies 49, 67
thermal 13, 21, 26, 34, 38, 40, 42, 74–77, 128, 134, 138, 194, 198, 271–272, 281, 287, 294, 341, 343–344, 352, 356, 364, 388–389, 391–397, 399, 401, 403–405, 414, 417, 441
Thermoactinomyces thalophilus 87
Thermobifida 52, 177, 252, 435, 439
Thermocellum 52, 58, 98, 108, 116–117, 133, 138, 177–178, 199, 209–210, 232, 235, 433, 440

thermochemical 3, 25, 32–33, 35–36, 39, 41–42, 95–96, 388–394, 399–401, 403–405, 413–416
Thermofilum 51
Thermogladius 51
thermophilic 85, 88, 107, 112, 115, 117, 131, 157, 159, 161, 210, 233, 321, 341, 356, 419, 423–424, 426
thermoplastic 264, 268, 286–287
  biopolymers 286
thermosets 277
thermostability 90, 92, 198
thermostable esterase enzyme 435
throughput 46, 48, 62, 69, 92, 105, 107, 183, 187, 197, 199–200, 229, 362
tissue engineering 293, 297
tobacco waste 152, 157
tocopherols 308
tolerance 45, 54, 61, 97, 99, 101–102, 105, 110, 115, 142, 205, 208, 218–219, 226, 230, 233, 290–291
torrefaction 390, 416, 417, 425
toxic 1, 29, 31, 46, 61, 75, 77, 109, 110, 124, 131, 133, 141, 147, 163, 169, 197, 209, 238, 240, 282, 292, 315, 362, 365, 422, 446
  assessment 315
  toxicity 60, 109, 114, 147, 192, 205, 275, 283, 313, 315, 347, 348, 356, 407
traditional methods 54, 281, 379
Trametes 52, 178, 312
transaldolase 228
transcription 54, 55, 66
  analysis 230
  engineering 193, 230
transesterification 26, 59, 120, 152–153, 199, 388, 402
transformation 24–25, 29, 32–33, 35–40, 43, 47, 62, 73, 106, 170–171, 239, 262, 309, 311, 317, 320, 355, 406
transketolase 228
transportation 13, 41, 59, 94, 107, 134, 136, 238, 352, 372, 388, 392, 396, 399, 403, 442
trays 357
treatment 21, 27, 30–31, 37, 40, 42, 52, 66, 73, 78, 82, 84–85, 88–89, 100, 119, 128–130, 135–136, 141, 145, 147–150, 152, 167, 169–170, 181, 183, 199, 208, 214, 217, 225, 227, 232, 247, 250, 254, 263, 278–280, 288, 293–295, 315, 319–320, 327, 330, 342–345, 348–349, 374, 385–391, 393, 395–399, 401, 403, 405, 411, 416, 425, 428, 432, 435
tribasic acid 241
tricarboxylic 228, 241–242, 253
Trichoderma 52, 55, 65, 68, 71, 81–82, 88, 97, 107, 117, 131–132, 163, 177–178, 201, 239, 312, 320, 329, 333
  reesei 55, 65, 71, 81, 97, 117, 131, 163, 177–178, 201, 333
  viride 82
trickling method 380
triglycerides 149, 152–153, 278, 347
types
  of biomass 1, 15, 377, 389
  of catalysts 400
  of gasifiers 395
  of membranes 367, 382
  of microorganisms 328

ultrafiltration membrane 336
ultrasonic irradiation 433
ultrasound 31, 40, 76, 77, 88, 91, 118–120, 151, 196–197, 199, 295, 401, 433
ultraviolet radiation 269
undesired coke 397
undesired tar 400
United States 1, 25, 107, 124, 163–164, 349, 424
upgrading 156, 160, 198, 259, 394, 398–402, 425
used cooking oils 362
U.S. Energy Information Administration 164
utilization 13, 46, 55–56, 59, 63, 67, 70, 71, 78, 82–84, 88, 92–99, 100–102, 105, 107, 109, 113, 115, 119–120, 136–138, 141, 147, 170, 192, 199, 211, 219, 225, 231, 234, 238–239, 255, 257, 262–265, 268–269, 271, 275, 283, 291–292, 300, 311–318, 324, 327, 344, 354, 360, 363–364, 370, 376, 389–390, 392, 397, 401–402, 423–425
UV radiation 288, 427

vacuum 49, 134, 138, 194, 337, 355, 363–364, 371
  distillation 355
  pressure 337, 364
valorizable 309, 314
valorization 63, 67, 71, 86, 88, 94, 115, 120–121, 150, 160, 200, 234, 258, 260, 297, 307, 313, 315–321, 341, 384, 404, 427, 429, 431, 433, 435–437, 439, 441–445, 446–449
valuable biopigment 310
vanilla essence 305, 312, 314
vanillic acid 181, 313
vanillin 305, 309, 312–314, 318, 320, 322, 372, 375, 427
vapor
  form 337
  liquid separation 356

vapor (*cont.*)
permeation 327, 363
phase 364
pressure difference 327, 383
vectors 112, 120, 186, 226, 295, 403
vegetable
oils 123, 278, 308, 389
peels 424
waste 150, 232, 242, 269, 303, 422
vermicompost 307, 426
versatility 97, 113, 250, 275, 278, 279, 314, 316, 389, 401
vessel 48, 133, 168, 170, 329
VFA yield 408
vinasse 304, 312, 318, 386
violacein 303, 313
virgin raw material 443
viscosity 17–18, 124, 150, 180, 393–394
viscozyme® L 433
vitamin C 255
volatile 3, 23, 28, 34, 134, 208, 267, 337, 346, 356, 374, 376–378, 381–387, 394, 419, 422
fatty acids 23, 376–378, 381–387, 422
volatilities 134, 355, 383
volumetric productivity 219

waste 1, 4–6, 13, 15–16, 19, 21–23, 26, 38, 40, 42–43, 54, 56, 58–60, 63–66, 68, 70, 72, 77–78, 83–86, 87–90, 92–94, 96–97, 116–117, 120–121, 123, 125, 127–128, 137–141, 145, 147, 149–153, 156–161, 166, 169–170, 195, 196, 199–200, 203, 207, 211–213, 218–219, 230–231, 232, 235–239, 242, 244–245, 249, 254–259, 260, 262, 264, 266–271, 275, 276, 278–279, 286, 289, 292, 294–298, 300, 302–304, 307–308, 309, 310–312, 313, 315–319, 320–325, 327, 329, 331–332, 336, 338–342, 347, 352–353, 360, 364, 372, 374, 376–377, 384–386, 388–389, 397, 400–404, 405–408, 411–412, 418, 421–426, 430, 432, 435–436, 438–439, 441, 444, 445, 448–449
valorization 86, 150, 260, 297
water 13, 27, 49, 71, 119, 234, 271, 294, 319, 369, 382, 385–386, 398, 412, 425–426
water 1, 14, 18, 20, 23, 25–27, 30–35, 37, 40, 42, 49, 59–60, 73, 78, 84, 86, 93–94, 107, 119, 126, 134, 142, 148–152, 158, 166, 169–170, 173, 198, 200, 202, 204, 217, 225, 240–242, 247, 250, 252–253, 262, 269, 272, 274–275, 280, 285, 287–289, 303, 312, 320, 325, 327, 329, 331, 336–338, 343, 348–349, 352, 354–356, 359, 362–364, 369, 371–373, 375, 379–383, 386, 391, 393–395, 397–398, 405, 412, 424, 426, 429, 433, 435, 442, 444, 449–450
waxes 49, 166
wet disk milling 77
wet oxidation 30
wheat 16, 19, 21, 25, 41, 44, 74, 75, 77, 82–85, 88, 90, 94, 100, 125, 128, 138, 145, 151, 159, 197, 203, 213, 233, 238, 245, 251, 259, 262, 269, 278, 292, 310–312, 323–324, 341, 343, 347, 352, 428–440, 444, 446, 450
bran 82, 85, 203, 262, 311, 312, 428–431, 433–435, 438–439
gluten 292
straw 16, 19, 25, 41, 44, 74, 75, 83–85, 90, 94, 100, 128, 138, 151, 159, 197, 203, 238, 245, 251, 259, 262, 310, 323–324, 341, 343, 450
whey 49, 71, 145, 149–150, 157, 231–234, 236, 248, 267, 304
white rot 317, 332
willow 93, 357, 375
wood 2–6, 13, 14, 17, 19, 22, 25, 33, 77, 85, 93, 128, 165–167, 166–167, 194, 198, 247, 249, 255, 260, 266–267, 278, 323, 325, 333, 344, 347, 352, 386, 388, 404, 414, 422, 425, 442
chips 13

xylan 22, 33, 54–56, 69, 78, 82, 88, 117, 126, 131, 172, 174, 321, 329, 435–436
xylanase 18, 21–22, 54–56, 69, 73, 82–84, 87–88, 113, 151, 181, 206, 255, 318, 321, 333, 434–435, 440
xylanimonas 52
xylitol 27–28, 33, 36, 41–42, 68, 77, 199, 234, 236, 304, 310, 312–313, 318–319, 373
xylooligosaccharides 18, 321

xylose 17–18, 31, 33, 41, 48, 50, 54–56, 66, 71, 74–75, 82–83, 89, 96, 98–99, 116, 120, 132–133, 149, 151, 156, 158–159, 165, 168, 174, 179, 197, 205–206, 216–217, 228, 234, 236, 296, 328–330, 345, 347, 366, 369, 373, 375, 432

Yarrowia 56, 58, 67, 241, 245, 304, 309, 319, 322, 435, 439
Yarrowia lipolytica 56, 58, 67, 241, 304, 309, 322, 435, 439
yeast 42, 56, 59, 62, 68, 69, 71, 83, 94, 96, 120, 133, 139, 150, 158, 204, 206, 208–209, 217–218, 225–226, 240–241, 252–253, 257, 267, 270–271, 321, 325, 328, 333, 335, 341, 342–344, 352, 356, 360, 364, 433, 435, 438
yield 21, 25, 29, 34, 44, 46–48, 58, 60–62, 73, 75, 77–78, 83–84, 90, 94–95, 100, 102, 108–109, 116–117, 131–132, 141, 144, 150–153, 158–159, 165–168, 170, 173–174, 179–180, 182–183, 190, 203, 206, 214–220, 228, 230, 236–240, 244, 247, 252, 272, 281–282, 290–291, 301, 303–305, 309, 310, 312–316, 327–333, 338–340, 344, 354, 356–357, 359, 366, 370, 374–375, 379, 391, 393–398, 401, 408, 414–417, 419, 424, 430–437

zeolite 137, 383, 400
Zymomonas mobilis 97, 120, 132, 328, 344

Printed in the United States
by Baker & Taylor Publisher Services